HUMAN BIOLOGY

BOOKS IN THE WADSWORTH BIOLOGY SERIES

HUMAN BIOLOGY

CECIE STARR

Belmont, California

BEVERLY McMILLAN

Gloucester, Virginia

Wadsworth Publishing Company

An International Thomson Publishing Company

Belmont • Albany • Bonn • Boston
Cincinnati • Detroit • London • Madrid
Melbourne • Mexico City • New York
Paris • San Francisco • Singapore
Tokyo • Toronto • Washington

For S.K.A. and J.A.M.—B.C.M.

BIOLOGY PUBLISHER: Jack Carey
DEVELOPMENT EDITOR: Mary Arbogast
EDITORIAL ASSISTANT: Kristin Milotich
PRODUCTION EDITOR: Jerilyn Emori
MANAGING DESIGNER: Cloyce Wall
INTERIOR DESIGNER: Gary Head
PRINT BUYER: Karen Hunt
ART EDITORS: Kevin Berry, Kelly Murphy
PERMISSIONS EDITOR: Marion Hansen
COPY EDITOR: Alan Titche
ARTISTS: Artemis (Betsy Palay), Lewis Calver, Raychel Ciemma, Robert Demarest, Hans & Cassady, Inc.

(Hans Neuhart), Darwen Hennings, Vally Hennings, Carlyn Iverson, John W. Karapelou, Leonard Morgan, Palay/Beaubois, Precision Graphics (Jan Flessner, Jim Gallagher), Nadine Sokol, Kevin Somerville, Lloyd Townsend
PRODUCTION ARTISTS: Craig Hanson, Natalie Hill, Carole Lawson
PHOTO RESEARCHER: Stuart Kenter
COVER DESIGNER: Cloyce Wall
COVER PHOTO: Carolina Biological Supply Company/Phototake
COMPOSITOR: York Graphic Services, Inc.
PRINTER: The Banta Company

Library of Congress Cataloging-in-Publication Data
Starr, Cecie.
    Human biology/Cecie Starr, Beverly McMillan.
      p.   cm.
    Includes index.
    ISBN 0-534-20208-X
    1. Human biology. I. McMillan, Beverly. II. Title.
QP34.5.S727   1995
612—dc20
                              94-10833

For more information, contact Wadsworth Publishing Company:

Wadsworth Publishing Company
10 Davis Drive, Belmont, California 94002, USA

International Thomson Publishing Europe
Berkshire House 168-173, High Holborn
London, WC1V 7AA, England

Thomas Nelson Australia
102 Dodds Street
South Melbourne 3205, Victoria, Australia

Nelson Canada
1120 Birchmount Road
Scarborough, Ontario, Canada M1K 5G4

International Thomson Editores
Campos Eliseos 385, Piso 7
Col. Polanco, 11560 México D.F. México

International Thomson Publishing GmbH
Königswinterer Strasse 418, 53227 Bonn, Germany

International Thomson Publishing Asia
221 Henderson Road, #05-10 Henderson Building
Singapore 0315

International Thomson Publishing Japan
Hirakawacho Kyowa Building, 3F
2-2-1 Hirakawacho, Chiyoda-ku, Tokyo 102, Japan

BRIEF CONTENTS

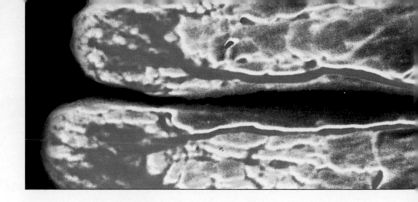

DETAILED CONTENTS

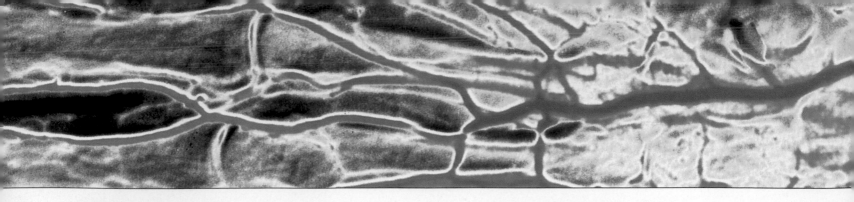

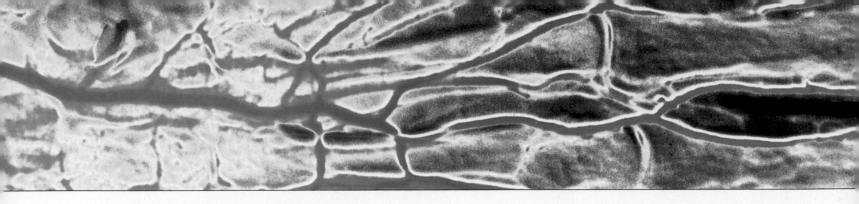

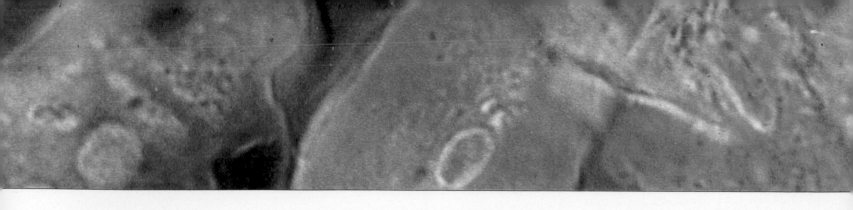

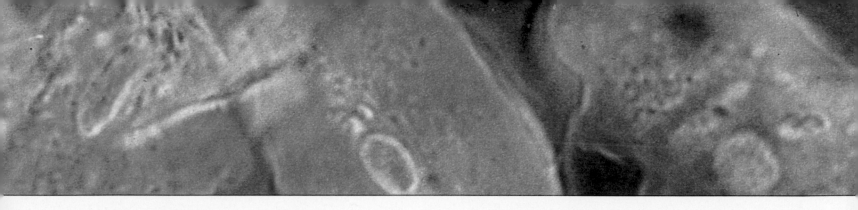

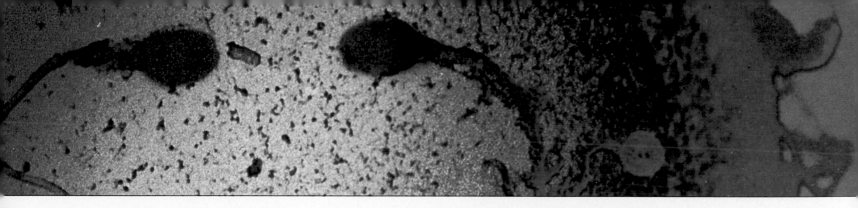

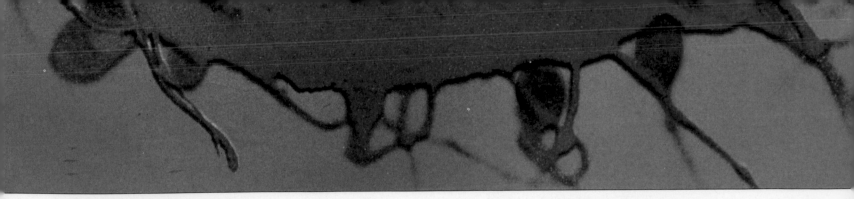

PREFACE

Humans are unique among all other living things on earth in that we actively, insatiably, ask questions about things. Of the questions that relate to biology, many are practical—how our bodies work, how life-style choices, genes, disease organisms, and environmental events affect our health and well-being. Other concerns are more philosophical—where we fit in the larger world, what our individual and collective responsibilities are, what challenges we face as a diverse society in a rapidly unfolding future.

When we sat down several years ago to chart the course of a new text in human biology, we knew that both kinds of questions must shape our work, because the people who will use this book are more than just students of biological science as it applies to humans. Our readers are consumers, and voters. They are citizens of a larger world in which scientists and nonscientists alike are now confronting staggeringly complex biological issues—genetic engineering of humans and other life forms, gene therapies as potential treatments and even cures for human diseases, how to prevent and treat AIDS and cancers and how to combat antibiotic-resistant microbes, how to cope with global warming and other ecological impacts of human populations—to name only a few. Society will not be able to solve these problems and create a satisfactory future for all humankind until more of us have a firm grasp of core biological concepts and can think critically about how those concepts apply to human life.

One of us, Cecie Starr, has spent twenty years researching, writing and rewriting general biology texts. Her recipe for a superior biological science text has been straightforward and effective, and it has been tested by more than 2 million students around the world. The recipe goes like this: Begin with accurate, up-to-date science, identifying the key concepts and topics that reflect current research in all major fields pertaining to human biology. Add engaging writing that draws students into the topic at hand. Clearly explain key concepts and the links between them, so that students can connect facts and ideas into a coherent picture of interrelated biological processes and events. Develop illustrations that are as easy to follow as the text, accurate and closely integrated with key concepts. Include plenty of relevant examples of how biological concepts and research discoveries apply to health, medical, and environmental issues. Finally, do all

of this within a framework of scientific inquiry, giving enough examples of problem solving and experiments to familiarize students with a scientific approach to interpreting the world. Students can use the same powerful approach to pick their way logically through today's environmental, medical, and social minefields.

Human Biology bears all these Starr trademarks, with the emphasis on anatomy and physiology that a course in human biology demands. For this enterprise Cecie Starr has joined forces with Beverly McMillan. Bev brings with her the potent combination of profound respect for and love of her subject, and fifteen years' involvement in the writing and development of college biology texts.

TEXT OBJECTIVES AND HIGHLIGHTS

Today, no biology textbook can meet the needs of its users unless the writing is clear, engaging, logical, and respectful of the intelligence of students. For each major field of inquiry, key concepts must be identified, and the topics selected must give students a good picture of the current state of knowledge as well as research trends and gaps in understanding. The structure and function of tissues, organs, and organ systems—the anatomy and physiology of the human body—must be explained in enough detail that students can build a working vocabulary about the parts and processes of life. Moreover, the theme that structure follows function must be reinforced, chapter after chapter.

Highlights

Beyond these basic goals, we set and met four additional objectives.

Chapter Opener Vignettes First, at the start of each chapter we wrote a lively or sobering vignette—a short "story" that introduces students to key concepts in a nonthreatening way and blends science with applications to human life. For example, the Chapter 7 vignette, "Russian Roulette, Immunological Style," describes how early experiments in providing immunity to smallpox were carried out by ancient Chinese physicians and an intrepid ambassador's wife. Chapter 13's vignette, "Boys and

Girls," introduces the fascinating topic of how human sex is determined. In Chapter 20, which deals with cancer as a case study of genes and human disease, "The Body Betrayed" sketches the biological basis of cancer and outlines cancer's variable impacts on adults and children, men and women.

Focus Essays Second, wherever possible without breaking the logical flow of text discussions, we include examples of applications, emphasizing human health concerns. In addition, each chapter contains one or more concise *Focus* essays that highlight links between human biology and a broad spectrum of student concerns. For example, the dedicated instructors who responded to our questionnaires reported something we already suspected: Their students are keenly aware of environmental issues and concerned about our collective impact on the biosphere. So, with the help of environmental science author Tyler Miller, Jr., we developed *Focus on Environment* essays that present balanced discussions of environmental concerns such as pollution of groundwater supplies (page 218), noise pollution (page 263), bioremediation of toxic spills by microbes (page 451), and a host of other topics.

Focus on Wellness boxes explore health concerns such as weight control and eating disorders (page 126), risk factors for heart disease (page 161), and practical strategies for avoiding sexually transmitted diseases (page 347).

Focus on Science essays describe scientific methods, or the process of scientific inquiry—microscopy, use of gel electrophoresis in genetics research, how mtDNA research can shed light on human origins. We include these essays to underscore that science is more than test tubes and white-coated technicians—it is a way of knowing about the world.

Choices: Biology and Society boxes delve into current questions relating to personal choice and bioethics, such as fetal tissue transplants, screening of embryos for genetic traits, and the decision to smoke cigarettes.

Human Impact Features Third, after consulting dozens of instructors dedicated to both science *and* teaching, we included *Human Impact* features in major text sections. These five illustrated "short subjects" address key health and social issues that warrant fuller treatment than is possible in the context of a well-integrated text chapter. They cover the global public health crisis related to the spread of infectious disease and multidrug resistance in microbes; global undernutrition and malnutrition; the promise and issues related to human gene therapy; drug effects and drug abuse; and physiological effects of aging.

System Summary Diagrams Finally, we developed a new type of illustration for human biology texts—system summary diagrams in the core chapters on human anatomy and physiology. These visual summaries integrate structure and function, so students don't have to jump back and forth from text, to tables, to an illustration in order to get an overview of how an organ system is put together and what its assorted parts do. Even the individual descriptions are arranged to reflect each system's structural and functional organization.

Illustrations

Often we are asked how we come up with such beautiful, pedagogically sound illustrations for our biology books. All it takes, really, is a twenty-year obsession with writing and creating illustrations at the same time, as inseparable units. Researching, developing, and integrating illustrations with the text takes almost as much time as writing the text itself.

Visual Summaries Illustrations having step-by-step written descriptions that are positioned within the art itself are one of the hallmarks of our books. When we break down information into a series of illustrated steps, students find this far less threatening than a complex, "wordless" diagram. *In this book, all major concepts have their own visual summaries.*

A few examples are the visual summaries for mitosis (page 362), meiosis (page 366), protein synthesis (page 418), antibody-mediated immunity (page 179), and neural function (page 230).

Zoom-Sequence Illustrations Many illustrations progress from macroscopic to microscopic views. See, for example, Figures 4.15 through 4.17, which integrate the text discussion of skeletal muscle components and their organization and function, and Figures 9.4 and 9.5, which link the gross structure of the kidney with the functioning of nephrons and the glomerulus.

Color-Coding Careful use of color helps students track information on hard-to-visualize molecular topics. For example, wherever proteins appear in illustrations they are color-coded green, carbohydrates pink, lipids yellow and gold, DNA blue, RNA orange.

Icons Icons, or small pictorial representations, relate the topic of an illustration to the bigger picture. For example, in Chapter 2 small but detailed representations of a cell or organelle remind students that a particular event proceeds in the cytoplasm, the nucleus, or in mitochondria.

Our Approach to Writing

In writing this book we were committed to doing everything in our power to ensure that the science would be accurate, up-to-date, and representative of a broad consensus of authoritative opinion. To this end, each chapter's text and illustrations were exhaustively reviewed by content specialists and instructors of human biology courses.

Accuracy and scientific rigor are essential, but they are not enough in a modern text. Today, students often pick up any kind of science textbook with apprehension. If it's science, many students fear, it's going to be hard, pedantic, and *boring*. Hence, we worked to make the book more than a dry expanded outline that piled fact upon fact. In short, we wanted it to be *interesting* to read. What a concept! The chapter opening vignettes are the first volley in the fight to instruct *and* intrigue students. Classroom testing shows that our strategy is paying off—the vignettes pique student interest while serving as effective tools for introducing key concepts and placing each chapter in a meaningful context.

In the main body of each chapter, creating an interesting book meant tossing words out of the manuscript whenever they got in the way of clarity. It meant being vigilant for confusing jargon and dedicating ourselves to the simple declarative sentence—while making sure that needed terms and definitions were there. It meant including memorable analogies *where they are appropriate* and balancing examples of applications with the unimpeded flow of concepts. We have tried especially hard to avoid distracting digressions—in the words of one appreciative biology student, to "put in the stuff and leave out the fluff."

STUDY AIDS

We are strong believers in helping students learn, not simply telling them they must learn (somehow). Hence, a list of *key concepts* follows the introductory vignette for each chapter. Boldfaced *summary statements* of concepts within the text itself help keep readers on track. Several end-of-chapter study aids reinforce the key concepts. Each chapter has a *summary* in list form, conceptual *review questions, critical thinking questions* that emphasize practical applications of chapter concepts, objective *self-quiz questions*, a listing of *key terms*, and *readings*. Page numbers tie review questions and key terms to relevant text pages.

Genetics problems in the unit on inheritance help students grasp essential principles; we used human examples as much as possible. The *glossary* includes bold-faced terms as well as some other basic vocabulary; whenever it seemed potentially useful, a glossary definition includes a pronunciation guide and a notation of the term's origins. When students need to look up a particular subject, the comprehensive *index*, with its fine division of topics, will be an invaluable tool.

Appendix I is a periodic table of the elements. We included it at the behest of instructors who report that the table facilitates their teaching of basic concepts in chemistry. Appendix II provides answers to the genetics problems in Chapters 17 and 18, and Appendix III provides answers to self-quiz questions. Appendix IV is a key containing suggestions for students who work the critical thinking questions, and here we have departed from texts that include full answers to such questions in the book itself. Instead of being temptingly available at the back of the book, suggested answers to these questions are provided to instructors. The in-text key suggests an approach to thinking through a plausible answer for each question. We hope it will stimulate students to analyze and pull together different facets of what they have learned, and so make their grasp of concepts that much stronger.

A COMMUNITY EFFORT

Authors can write accurately and often well about their field of interest, but it takes more than this to deal with the full breadth of biological science as it relates to the human body. For us, it has taken a far-flung network of thoughtful reviewers, researchers, photographers, and illustrators. We salute their commitment to quality in science education.

Charles Booth, Doug Burks, Joe Connell, Vic Chow, Lisa Danko, Gail Hall, Michael Henry, Karen McMahon, Rodney Mowbray, Sally Sapp Olson, and Robert Sullivan gave large chunks of time over many months helping us better understand what students need to know. Ed Alcamo, Fred Delcomyn, Kate Denniston, Daniel Fairbanks, John Kimball, Howard Kutchai, Bill Lassiter, Anne McNabb, Hylan Moises, Dave Morton, and Ian Tizard were among the cadre of exceptional advisors for the book's core chapters in human anatomy and physiology. Their insight and attention to detail are reflected on every page. Robert Lapen's manuscript for our discussion of immunology in Chapter 7 was extraordinary in its clarity and accessibility.

Tyler Miller graciously contributed *Focus on Environment* essays, and he and Steve Wolfe were the guardians of accuracy and teachability in the chapters on chemistry and cell biology.

Jerilyn Emori shepherded the book from its manuscript stage through production with intelligence and a sense of humor when we needed it most. Kristin Milotich managed to combine the best kind of professional support with consistent good cheer. The book is much better for the developmental editing of Mary Arbogast and the copyediting of Alan Titche. We are also grateful for the contributions of photo researcher Stuart Kenter and of all the others listed on the copyright page. Marion Hansen did her reliably superb job in obtaining permissions. Art editors Kevin Berry and Kelly Murphy organized, analyzed, advised, and finally put together the remarkable illustration program you see in the following pages.

This book owes its elegant design and clean page makeup in large measure to Gary Head and Cloyce Wall. Gary also provided computer support that saved our sanity. Cloyce helped keep things moving and was instrumental in helping us realize our vision for the book's stunning cover.

We are grateful to all of the people whose dedication and competence helped make this book happen. In the end, however, the credit goes to Jack Carey. Jack perceived the need for this book, brought us together with a single vision of what it could be and should be, and then provided unfailing advice and support for our ideas. Jack gives the most, the best, all the time. No authors could ask for more.

Supplements for Students and Instructors

Full-color transparencies and *35mm slides* of many of the book illustrations are labeled with large, boldface type. Diagrams are on *CD-ROM*. A *Test Items* booklet has more than 2,300 questions by outstanding test writers. Questions are available in electronic form on *IBM* and *Macintosh*.

An *Instructor's Resource Manual* has, for each chapter, an outline, objectives, list of boldface and italic terms, and a detailed lecture outline. It includes suggestions for lecture presentations, classroom and laboratory demonstrations, suggested discussion questions, research paper topics, and annotations for filmstrips and videos. *Lecture outlines* in the Instructor's Resource Manual are available on a *Mac* or *IBM disk* for those who wish to modify the material.

Collaborating with Carolina Biological Supply and Bill Surver of Clemson University, Wadsworth developed *The Wadsworth Biology Videodisc,* which contains many illustrations and photographs for the chapters in *Human Biology.* Many illustrations are broken down into a step-by-step format or animated in full motion, and many are available with fill-in-the-blank labels for testing purposes. A directory and bar code guide comes with the disc, and *HyperCard Stacks for the Mac* and *Toolbook for the IBM* are available.

An activity-oriented *Study Guide and Workbook* asks students to respond in writing to almost all questions. Questions are arranged by chapter section. *Chapter objectives* of the Study Guide are available on disk as part of the testing file for those who wish to modify or select portions of the material. An *Electronic Study Guide* consists of multiple choice questions different from those in the Test Items booklet. After responding to each question, an onscreen prompt lets students review their answers and learn why they are right or wrong. A 75-page *Answer Booklet* has answers to the book's end-of-chapter review questions. A special version of *STELLA II,* a software tool for developing critical thinking skills, is available to adopters, together with a workbook. In addition, *Critical Thinking Exercises* contains critical thinking questions for each chapter.

Wadsworth offers eight additional-readings supplements: *Contemporary Readings in Biology • Science and the Human Spirit: Contexts for Writing and Learning • A Beginner's Guide to Scientific Method • The Game of Science • Environment: Problems and Solutions • Green Lives, Green Campuses • Watersheds: Classic Cases in Environmental Ethics • Environmental Ethics.*

A new *Laboratory Manual* by Morton, Perry, and Perry has 20 experiments and exercises, with hundreds of full-color labeled photographs and diagrams. Many experiments are divided into parts that can be assigned individually, depending on time available. All have objectives, discussion (introduction, background, and relevance), a list of materials for each part of an experiment, procedural steps, pre-lab questions, and post-lab questions. An accompanying *Instructor's Manual* covers quantities, procedures for preparing reagents, time requirements for each portion of the exercise, suggestions to make the lab a success, and vendors of materials with item numbers. It offers additional investigative exercises that can be copied for laboratory use.

REVIEWERS

Alcamo, I. Edward, State University of New York at Farmingdale

Alford, Donald K., Metropolitan State College of Denver

Anderson, D. Andy, Utah State University

Armstrong, Peter, University of California, Davis

Bakken, Aimée, University of Washington

Barnum, Susan R., Miami University

Bedecarrax, Edmund E., City College of San Francisco

Bennett, Jack, Northern Illinois University

Booth, Charles E., Eastern Connecticut State University

Brammer, J. D., North Dakota State University

Brown, Melvin K., Erie Community College–City Campus

Burks, Douglas J., Wilmington College

Chow, Victor, City College of San Francisco

Christensen, Ann, Pima Community College

Connell, Joe, Leeward Community College

Coyne, Jerry, University of Chicago

Crosby, Richard M., Treasure Valley Community College

Danko, Lisa, Mercyhurst College

Delcomyn, Fred, University of Illinois

Denner, Melvin, University of Southern Indiana

Denniston, Katherine J., Towson State University

Fairbanks, Daniel J., Brigham Young University

Frey, John E., Mankato State University

Froehlich, Jeffrey, University of New Mexico

Fulford, David E., Edinboro University of Pennsylvania

Gianferrari, Edmund A., Keene State College of the University of New Hampshire

Gordon, Sheldon R., Oakland University

Hahn, Martin, William Paterson College

Hall, N. Gail, Trinity College

Hassan, Aslam S., University of Illinois

Henry, Michael, Santa Rosa Junior College

Huccaby, Perry, Elizabethtown Community College

Hupp, Eugene W., Texas Woman's University

Johns, Mitrick A., Northern Illinois University

Johnson, Vincent A., Saint Cloud State University

Jones, Carolyn K., Vincennes University

Kareiva, Peter, University of Washington

Kaye, Gordon I., Albany Medical College

Kennedy, Kenneth A. R., Cornell University

Keyes, Jack, Linfield College, Portland Campus

Kimball, John W.

Klein, Keith K., Mankato State University

Krumhardt, Barbara, Des Moines Area Community College, Urban Campus

Kutchai, Howard, University of Virginia

Lammert, John M., Gustavus Adolphus College

Lapen, Robert, Central Washington University

Lassiter, William E., University of North Carolina–Chapel Hill School of Medicine

Levy, Matthew N., Mt. Sinai Hospital

Lucas, Cran, Louisiana State University–Shreveport

Mann, Alan, University of Pennsylvania

Mann, Nancy J., Cuesta College

Matthews, Patricia, Grand Valley State University

Mays, Charles, DePauw University

McCue, John F., St. Cloud State University

McMahon, Karen A., University of Tulsa

McNabb, F. M. Anne, Virginia Polytechnic Institute and State University

Miller, G. Tyler

Mitchell, Robert B., Pennsylvania State University

Moises, Hylan, University of Michigan

Mork, David, St. Cloud State University

Morton, David, Frostburg State University

Mote, Michael I., Temple University

Mowbray, Rodney, University of Wisconsin–La Crosse

Murphy, Richard A., University of Virginia Health Sciences Center

Mykles, Donald L., Colorado State University

Norris, David O., University of Colorado at Boulder

Parson, William, University of Washington

Peters, Lewis, Northern Michigan University

Roberts, Jane C., Creighton University

Sapp Olson, Sally

Scala, Andrew M., Dutchess Community College

Sherman, John W., Erie Community College–North

Shippee, Richard H., Vincennes University

Smigel, Barbara W., Community College of Southern Nevada

Smith, Robert L., West Virginia University

Sorochin, Ron, State University of New York, College of Technology at Alfred

Steele, Craig W., Edinboro University of Pennsylvania

Stone, Analee G., Tunxis Community College

Sullivan, Robert J., Marist College

Tizard, Ian, Texas A&M University

Trotter, William, Des Moines Area Community College

Valentine, James W., University of California, Berkeley

Van Dyke, Pete, Walla Walla Community College

Varkey, Alexander, Liberty University

Weisbrot, David R., William Paterson College

Whipp, Brian J., St. George's Hospital Medical School

Williams, Roberta B., University of Nevada, Las Vegas

Wolfe, Stephen L., University of California, Davis

LEARNING GUIDE FOR STUDENTS

You could view human biology as just another course that you must take in order to graduate. Or you could open your mind to a fantastically rewarding way to gain perspective on the world around you—and on your place in it. We hope, of course, that you will choose the latter. If you do, this book will help you in your journey of discovery. Before you begin reading, take a moment to look over the next few pages, which provide an overview of some of the book's built-in learning devices. Becoming familiar with these unique features can make your walk through the material a lot easier.

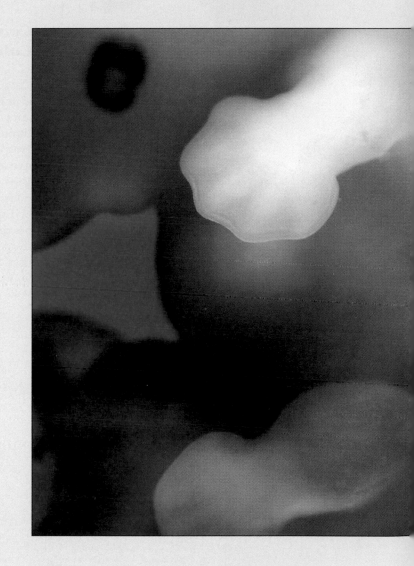

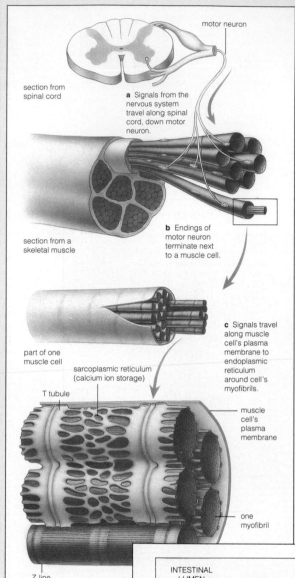

section from
spinal cord

motor neuron

a Signals from the
nervous system
travel along spinal
cord, down motor
neuron.

section from a
skeletal muscle

b Endings of
motor neuron
terminate next
to a muscle cell.

part of one
muscle cell

sarcoplasmic reticulum
(calcium ion storage)

c Signals travel
along muscle
cell's plasma
membrane to
endoplasmic
reticulum
around cell's
myofibrils.

T tubule

muscle
cell's
plasma
membrane

one
myofibril

Z line

Visual Summaries

Who hasn't stared at the words describing a complex topic, stared at illustrations, then stared back at the words, wondering how they all fit together? That's why the visual summaries you'll find in this book can give you perspective on the big picture. These illustrations have detailed, often step-by-step descriptions integrated with the diagrams. When you come across pages with this type of illustration, read the text, then go back and work your way more slowly, step by step, through the illustration. Or use these illustrations as a preview, or advance organizer, for the concept you are learning.

Consistent Color Coding

Certain large molecules are characteristic of life, and they appear repeatedly in the illustrations found in this book. We've assigned a different color to each type. For example, proteins are color-coded green, carbohydrates pink, lipids yellow and gold, DNA blue, and RNA orange. This will remind you of the structural and functional uses of these molecules, when you track them through a metabolic pathway.

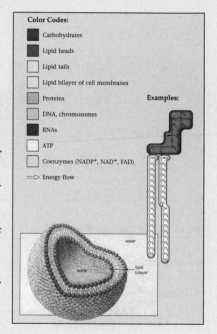

Color Codes:
- Carbohydrates
- Lipid heads
- Lipid tails
- Lipid bilayer of cell membranes
- Proteins
- DNA, chromosomes
- RNAs
- ATP
- Coenzymes (NADP+, NAD+, FAD)
- ⟹ Energy flow

Examples:

water

water

lipid
bilayer

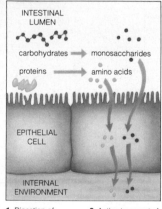

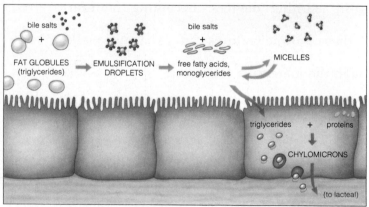

INTESTINAL LUMEN

carbohydrates → monosaccharides

proteins → amino acids

EPITHELIAL CELL

INTERNAL ENVIRONMENT

bile salts

FAT GLOBULES
(triglycerides)

+

EMULSIFICATION
DROPLETS

→

free fatty acids,
monoglycerides

bile salts

+

MICELLES

triglycerides + proteins

CHYLOMICRONS

(to lacteal)

1 Digestion of carbohydrates to monosaccharides and proteins to amino acids completed by pancreatic enzymes.

2 Active transport of monosaccharides and amino acids across plasma membrane of cells, then nutrients move by facilitated diffusion into blood.

1 Emulsification. Wall movements break up fat globules into small droplets. Bile salts keep globules from re-forming. Pancreatic enzymes digest droplets to fatty acids and monoglycerides.

2 Formation of micelles (as bile salts combine with digested products and phospholipids). Products readily slip into and out of micelles.

3 Concentration of monosaccharides and fatty acids in micelles enhance gradients that lead to their diffusion across lipid bilayer of cells' plasma membranes.

4 Reassemby of products into triglycerides inside cells. These join with proteins, then are expelled (by exocytosis) into internal environment.

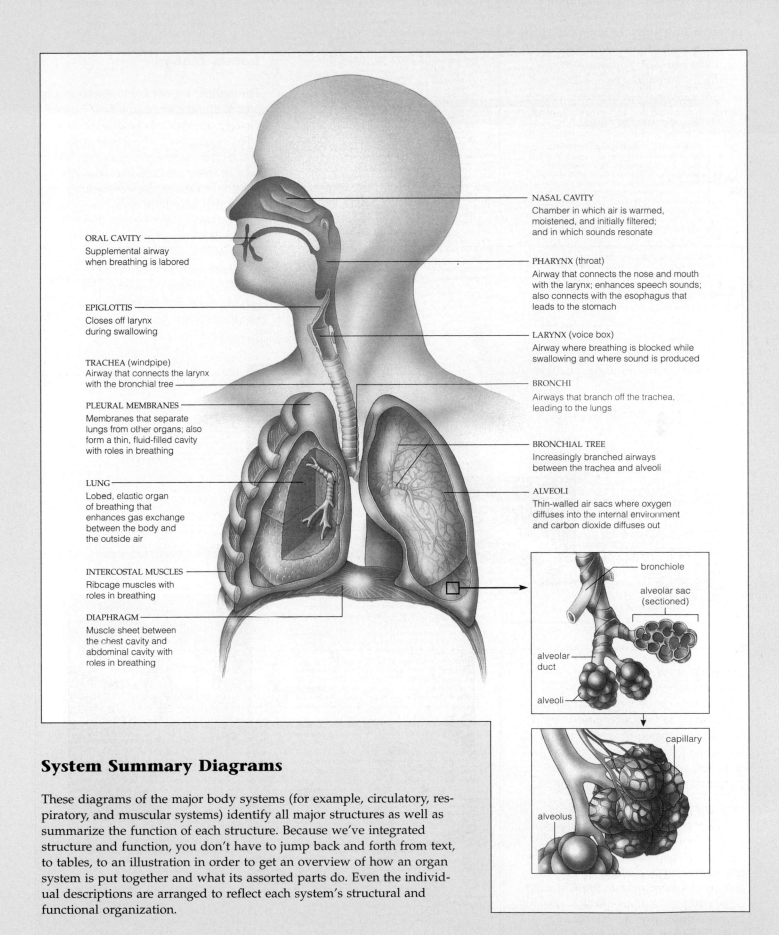

ORAL CAVITY
Supplemental airway
when breathing is labored

EPIGLOTTIS
Closes off larynx
during swallowing

TRACHEA (windpipe)
Airway that connects the larynx
with the bronchial tree

PLEURAL MEMBRANES
Membranes that separate
lungs from other organs; also
form a thin, fluid-filled cavity
with roles in breathing

LUNG
Lobed, elastic organ
of breathing that
enhances gas exchange
between the body and
the outside air

INTERCOSTAL MUSCLES
Ribcage muscles with
roles in breathing

DIAPHRAGM
Muscle sheet between
the chest cavity and
abdominal cavity with
roles in breathing

NASAL CAVITY
Chamber in which air is warmed,
moistened, and initially filtered;
and in which sounds resonate

PHARYNX (throat)
Airway that connects the nose and mouth
with the larynx; enhances speech sounds;
also connects with the esophagus that
leads to the stomach

LARYNX (voice box)
Airway where breathing is blocked while
swallowing and where sound is produced

BRONCHI
Airways that branch off the trachea,
leading to the lungs

BRONCHIAL TREE
Increasingly branched airways
between the trachea and alveoli

ALVEOLI
Thin-walled air sacs where oxygen
diffuses into the internal environment
and carbon dioxide diffuses out

bronchiole

alveolar sac
(sectioned)

alveolar
duct

alveoli

capillary

alveolus

System Summary Diagrams

These diagrams of the major body systems (for example, circulatory, respiratory, and muscular systems) identify all major structures as well as summarize the function of each structure. Because we've integrated structure and function, you don't have to jump back and forth from text, to tables, to an illustration in order to get an overview of how an organ system is put together and what its assorted parts do. Even the individual descriptions are arranged to reflect each system's structural and functional organization.

Making the Most of Muscles

Muscle makes up more than 40 percent of the human body by weight, and most of that is skeletal muscle. Muscle cells adapt to the activity demanded of them. When severe nerve damage or prolonged bed rest prevents a muscle from being used at all, the muscle will rapidly begin to atrophy. Over time, affected muscles can lose up to three-fourths of their mass, with a corresponding loss of strength. More commonly, the skeletal muscles of a sedentary person stay basically healthy but cannot respond to physical demands in the same way that well-worked muscles can.

One of the best ways to maintain or increase muscle efficiency is through regular aerobic exercise—activities such as walking, swimming, biking, organized aerobics classes, and jogging. Aerobic exercise works muscles at a rate at which the body can keep them supplied with oxygen, and it has the following effects on muscle cells:

1. Increase in the number and size of mitochondria, the organelles in which ATP is synthesized.

2. Increase in the number of blood capillaries supplying muscle tissue, and hence increased availability of oxygen and nutrients and more efficient waste removal.

3. Increase in the amount of the oxygen-storing pigment myoglobin in muscle tissues.

Together, such changes result in muscles that are more efficient metabolically and can work longer without becoming fatigued. When muscles also are regularly exerted with high intensity—as a person might do by lifting weights several times a week—individual muscle cells, or fibers, grow larger. Along with other related changes, this translates into whole muscles that are larger and

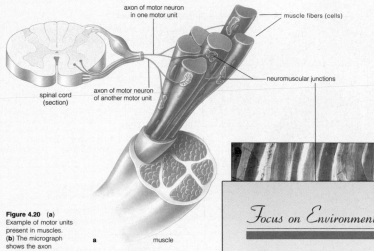

axon of motor neuron in one motor unit

muscle fibers (cells)

neuromuscular junctions

spinal cord (section)

axon of motor neuron of another motor unit

a

muscle

Figure 4.20 (a) Example of motor units present in muscles. (b) The micrograph shows the axon endings of a motor neuron that acts on individual muscle cells in the muscle.

Focus Essays

Throughout the book, numerous examples of applications show how human biology concepts can be used to explain what's going on in the world around you. *Focus* essays are the most prominent of these applications. They highlight links between human biology and issues that you think about every day. *Choices: Biology and Society* essays delve into current questions relating to personal choice and bioethics, such as fetal tissue transplants, screening of embryos for genetic traits, and cigarette smoking. Health concerns such as fad diets and

---

A Planetary Emergency: Loss of Biodiversity

On March 24, 1900, in Ohio, a young boy shot the last known wild passenger pigeon. More recently, fisheries biologist John Musick of the Virginia Institute of Marine Science reported his latest findings on the decline in populations of the dusky shark. Over the past 25 years, Musick has tracked the numbers of the dusky and other shark species that spend much of their lives off the mid-Atlantic coast of the United States and have become popular as human food. Musick's numbers show that, like codfish in parts of the North Atlantic, today the dusky shark has all but disappeared. Based on similar evidence from many scientists, the National Marine Fisheries Service in 1993 imposed catch limits on commercial shark fishing operations. Fleet operators responded with a lawsuit.

Sooner or later all species become extinct, but humans have become a primary factor in the premature extinction of more and more species. Already our actions are leading to the premature extinction of at least six species per hour.

Many biologists consider this epidemic of extinction an even more serious problem than depletion of stratospheric ozone or global warming because it is happening faster and is irreversible. Such rapid extinction cannot be balanced by speciation because it takes many thousands of generations for new species to evolve.

Globally, tropical deforestation (Figure 25.15) is the greatest killer of species, followed by destruction of coral reefs. However, plant extinctions can be more important ecologically than animal extinctions since most animal species depend directly or indirectly on plants for food, and often for shelter as well. Humans also have traditionally depended on plants as sources of medicines. Indeed, 40 percent of prescription drug sales in the United States involve natural plant products. One striking example is the rosy periwinkle of Madagascar (Figure *a*), from which we derive the anticancer drugs vincristine and vinblastine (page 434). As it happens, humans have destroyed 90 percent of the vegetation on Madagascar, so we may never know if it was home to plant sources of other life-saving drugs as well. Meanwhile, in the United States we are going about the business of habitat destruction at a dizzy-

Figure *a* The rosy periwinkle found in the threatened tropical forests of Madagascar.

ing pace. Among other things, we have cut down 90–95 percent of old growth forests and drained half the wetlands, which filter human water supplies and provide homes for waterfowl and juvenile fishes. At least 500 species native to these areas have been driven to extinction, and dozens more are endangered.

The underlying causes of wildlife extinction are human population growth and economic policies that fail to value the environment. Instead, they promote unsustainable exploitation. As our population grows we clear, occupy, and damage more land to supply food, fuel, timber, and other resources. In wealthy countries, affluence leads to greater-than-average resource use per person. In less-affluent nations, the combination of rapid population growth and poverty push the poor to cut forests, grow crops on marginal land, overgraze grasslands, and poach endangered animals.

How can we help stem the tide of extinctions? Among other strategies, individuals can support efforts to reduce deforestation, projected global warming, ozone depletion, and poverty—the greatest threats to earth's wildlife *and* to the human species.

eating disorders, the use of anabolic steroids, and practical strategies for avoiding sexually transmitted diseases are explored in *Focus on Wellness* essays. *Focus on Science* essays will help you see that science is more than test tubes and white-coated technicians—it is a way of knowing about the world. *Focus on the Environment* essays present balanced discussions of environmental concerns, such as pollution of groundwater supplies, noise pollution, and a host of other topics.

Focus on Science

A Weighty Question: How Much Energy Does Your Body Need?

How many kilocalories should you take in each day to maintain what you consider to be an acceptable body weight? To estimate your body's energy requirements, first multiply your desired weight (in pounds) by 10 if you are not very active physically, by 15 if you are moderately active, and by 20 if you are quite active. Then, depending on your age, subtract the following amount from the value obtained in the first step:

Age	Subtract
20–34	0
35–44	100
45–54	200
55–64	300
Over 65	400

For example, if you want to weigh 120 pounds and are quite active, $120 \times 20 = 2,400$ kilocalories. If you are 35 years old and moderately active, then you should take in a total of $1,800 - 100$, or 1,700 kcal a day. Such calculations provide a rough estimate, and factors such as height and gender must be considered as well. Because males tend to have more muscle and so burn more calories, an active woman doesn't need as many kilocalories as an active man of the same height and weight. Nor does she need as many as another active woman who weighs the same but is several inches taller.

Man's Height	Size of Frame		
	Small	Medium	Large
5' 2"	128–134	131–141	138–150
5' 3"	130–136	133–143	140–153
5' 4"	132–138	135–145	142–156
5' 5"	134–140	137–148	144–160
5' 6"	136–142	139–151	146–164
5' 7"	138–145	142–154	149–168
5' 8"	140–148	145–157	152–172
5' 9"	142–151	148–160	155–176
5'10"	144–154	151–163	158–180
5'11"	146–157	154–166	161–184
6' 0"	149–160	157–170	164–188
6' 1"	152–164	160–174	168–192
6' 2"	155–168	164–178	172–197
6' 3"	158–172	167–182	176–202
6' 4"	162–176	171–187	181–207

Woman's Height	Size of Frame		
	Small	Medium	Large
4'10"	102–111	109–121	118–131
4'11"	103–113	111–123	120–134
5' 0"	104–115	113–126	122–137
5' 1"	106–118	115–129	125–140
5' 2"	108–121	118–132	128–143
5' 3"	111–124	121–135	131–147
5' 4"	114–127	124–138	134–151
5' 5"	117–130	127–141	137–155
5' 6"	120–133	130–144	140–159
5' 7"	123–136	133–147	143–163
5' 8"	126–139	136–150	146–167
5' 9"	129–142	139–153	149–170
5'10"	132–145	142–158	152–173
5'11"	135–148	145–159	155–176
6' 0"	138–151	148–162	158–179

Choices: Biology and Society

Issues for a Biotechnological Society

Some people say that no matter what the species of organism, DNA should never be altered. But as we noted earlier, nature alters DNA all the time. The real issue is how to bring about beneficial changes without harming ourselves or the environment.

In the 1980s activists entered a test plot and ripped out strawberry plants that had been sprayed with a bacterium that had been genetically altered to help prevent ice crystals from forming on the plants (Figure *a*). Although previous studies had shown that the "ice-minus" bacteria posed no threat to the environment or human health, many people have voiced similar concerns about releasing recombinant organisms into the environment. Researchers themselves have moved into the vanguard of efforts to develop regulations and guidelines that protect the public interest while permitting beneficial advances. Current concerns include the following:

• Transgenic bacteria or viruses could mutate, possibly becoming new pathogens in the process. However, as described on page 450, regulations require recombinant microbes to be altered in ways that will prevent them from reproducing.

• Bioengineered plants could escape from test plots and pose an ecological risk by becoming "superweeds" resistant to herbicides and other control measures.

• Crop plants with added insect resistance could set in motion an evolutionary see-saw in which new, even more formidable insect pests arise.

• Transgenic fish that feed voraciously and grow rapidly could displace natural species, wrecking the delicate ecological balance in streams and lakes.

• Major ethical problems could arise if people undertake "eugenic engineering"—using biotechnology to insert genes into normal individuals (or sperm or eggs) simply for the purpose of creating individuals that are "more desirable."

• Genetic screening could provide a means of discrimination by insurance companies and others against people

Figure a Spraying an experimental strawberry patch in California with "ice-minus" bacteria. (Government regulations required that the sprayer use elaborate protective gear.)

who carry a gene predisposing them to a disorder, even if such a person shows no sign of ill health.

These examples only touch on some of the profound social and ethical issues raised by recombinant DNA technology and genetic engineering. With new advances coming every day, it may be time for everyone to think about the benefits *and* risks of living in a biotechnological society.

Vignettes

These beautifully illustrated short stories that begin each chapter drive home the relevance of human biology to your own life. For example, the Chapter 7 vignette shown below, "Russian Roulette, Immunological Style," describes how early experiments in providing immunity to smallpox were carried out by ancient Chinese physicians and an intrepid ambassador's wife. In Chapter 13, "Girls and Boys" introduces you to the fascinating topic of how human sex is determined. The Chapter 20 vignette, "The Body Betrayed," sketches the biological basis of cancer and outlines cancer's variable impacts on adults and children, men and women.

Each vignette will lead you, in a nontechnical way, to the chapter's key concepts. Right after the vignette, the concepts are listed as simple summary statements, as indicated below. At the end of each chapter, those same concepts are reinforced, as outlined on the last page of this preview.

8 RESPIRATION

Conquering Chomolungma

Chomolungma is a Himalayan mountain that also goes by the name Everest (Figure 8.1). Its summit soars to 9,700 meters (29,108 feet) above sea level, the highest place on earth. To conquer Chomolungma, climbers must come to terms with blizzards, heart-stopping avalanches, and icy, nearly vertical rock faces. They also must grapple with a life-threatening scarcity of oxygen.

In the air at low elevations, one molecule in five is oxygen. The earth's gravitational pull keeps oxygen and other gas molecules in air concentrated near sea level—or, as we say for gases, under pressure. At altitudes above 3,300 meters (10,000 feet), the breathing game changes. Gravity's pull is weaker, gas molecules spread out—and the pressure decreases. What would be a normal breath at sea level simply will not bring enough oxygen into the lungs. The oxygen deficit can cause "altitude sickness," with symptoms ranging from shortness of breath, headache, and heart palpitations to loss

Figure 8.1 A climber approaching the summit of Chomolungma, where oxygen is dangerously scarce.

7 IMMUNITY

Russian Roulette, Immunological Style

At one time, smallpox swept repeatedly through the world's cities. Some outbreaks were so severe that only half of those stricken survived. No one emerged unscathed. Even survivors ended up with permanent scars. But they did not contract the disease again—they were "immune" to smallpox.

The idea of acquired immunity to smallpox fascinated many people. In Asia, Africa, and then Europe, a few intrepid souls played immunological Russian roulette by allowing themselves to be inoculated with material from the sores of victims. Twelfth-century Chinese inhaled powdered crusts from smallpox sores. In the 1600s the wife of the British ambassador to Turkey injected bits of smallpox scabs into the veins of her children. Others soaked threads in the fluid from smallpox sores, then poked the threads into scratches on the body. Not everyone survived such practices, but those who did acquired immunity to smallpox.

For centuries it was known that humans could catch cowpox, a fairly mild disease, from cattle. Curiously, people who caught cowpox never got smallpox. In 1796, an English physician named Edward Jenner injected material from a cowpox sore into the arm of a boy (Figure 7.1). Six weeks later, after the cowpox reaction subsided, Jenner inoculated the boy with fluid from smallpox sores. He hypothesized that the earlier inoculation would provoke immunity to smallpox, and he was right. The French mocked the procedure, calling it "vaccination" (literally, "encowment"). Much later the French chemist Louis Pasteur devised vaccination procedures for other diseases, and the term became respectable.

By the early 20th century, the fight against infectious disease began in earnest, and it continues today. How the body wages its battle against invaders is the focus of this chapter.

Figure 7.1a Statue honoring Edward Jenner's development of an immunization procedure against smallpox, one of the most dreaded diseases in human history.

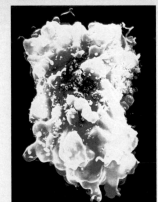

Figure 7.1b Micrograph of a white blood cell being attacked by the virus (blue particles) that causes AIDS. Immunologists are working to develop weapons against this modern-day scourge.

KEY CONCEPTS

1. The human body has physical, chemical, and cellular defenses against invasions by pathogens—viruses, bacteria, and other agents of disease.

2. During early stages of an invasion, white blood cells and plasma proteins take part in an inflammatory response. This counterattack is rapid and *general*; it does not target a specific pathogen.

3. If the invasion persists, certain white blood cells make *specific* immune responses. These cells can recognize molecules that are abnormal or foreign to the body, such as those on the surfaces of bacteria, viruses, cancer cells, and transplanted tissues. If the foreign or abnormal molecule triggers an immune response, it is called an antigen.

4. In one type of immune response, a subgroup of white blood cells produces huge amounts of antibodies. Antibodies are proteins that bind to a specific antigen and tag it for destruction. In another type of immune response, "killer" cells directly destroy body cells that have become abnormal via infection or some other process.

Human Impact Sections

Sometimes key health and social issues warrant fuller treatment than is possible in the context of a text chapter. Where this is the case, we include illustrated "short subjects" entitled *Human Impact*. They cover world hunger and malnutrition; human gene therapy; drug effects and drug abuse; physiological effects of aging; and the global public health crisis related to the spread of infectious disease.

Human Impact

MIND, BODY, AND DRUG ABUSE

A drug is any substance introduced into the body to provoke a specific physiological response. For thousands of years people in nearly every culture have used drugs as medicines, to alter mental states as part of religious or social rituals, or simply for individual effects. **Drug abuse** can be defined as use of a drug in a way that harms health or interferes with a person's ability to function in society.

In general, drug abuse involves **psychoactive drugs**, which act on the central nervous system by binding to receptors in the neuron plasma membrane. Such receptors normally bind neurotransmitter molecules, which transmit chemical messages among neurons (page 234). When drug molecules bind such receptors, the result is a change in the chemical messages neurons send or receive—and in associated mental and physical states.

Drug Effects

Psychoactive drugs exert their effects on brain regions that govern states of consciousness and behavior. Various drugs also influence physiological events, such as heart rate, respiration, sensory processing, and muscle coordination. Many affect a pleasure center in the hypothalamus and artificially fan the sense of pleasure we associate with eating, sexual activity, or other behaviors.

Often the body develops *tolerance* to a drug, meaning it takes larger or more frequent drug doses to produce the same effect. Tolerance reflects *physical drug dependence*. As Chapter 5 described, the liver produces enzymes that detoxify drugs circulating in the bloodstream. Tolerance develops when the level of detoxifying liver enzymes increases in response to the consistent presence of the drug in the blood. Hence a user must increase his or her intake to keep one step ahead of the liver's increasing ability (up to a point) to break down the drug.

Psychological drug dependence, or *habituation*, develops

include substances that can injure health and in some cases can lead to widespread harm in other ways as well. Table I lists some warning signs of potentially serious drug dependence.

Drug Action and Interactions Not surprisingly, a drug's effects depend on the dose, and typically the effects of a higher dose are more intense than those of a smaller dose. For example, smoking a little marijuana may pleasurably intensify sensory perceptions (such as listening to music), but a higher dose may provoke wildly exaggerated, even paranoid perceptions of events.

Different psychoactive drugs used simultaneously can interact, sometimes in ways that are extremely dangerous. In a *synergistic* interaction, two drugs used

Table I   Warning Signs of Drug Dependence
There may be cause for concern about drug dependency if a person consistently or often shows three or more of the following characteristics with respect to a substance:
1. Noticeable tolerance to a substance, marked by significant increase in the amount needed to produce the desired effect
2. Using a substance regularly for an extended period of time
3. Difficulty stopping or cutting down, even when a person consistently wishes to do so
4. Use of a substance to avoid withdrawal symptoms and feel "normal"
5. Concealing substance use from others
6. Taking extreme or dangerous measures to get the drug or to ensure its supply. (Examples: stealing money to buy cocaine; drinking on the job in spite of severe penalties for doing so; going from doctor to doctor until finding one who will prescribe a particular tranquilizer)

Human Impact

INFECTION AND HUMAN DISEASE

The human body is continually under siege from pathogens, including viruses, bacteria, fungi, protozoans, and parasitic worms. "Pathogen" means that these agents are capable of *infection*—of invading and then multiplying in the cells and tissues of a host organism. *Disease* may be the outcome of an infection. It results whenever the body cannot mobilize its defenses quickly enough to prevent a pathogen's activities from interfering with normal body functions. Worldwide, more people die of infectious diseases than from any other cause. The accompanying chart lists the top killers according to 1992 estimates of the World Health Organization.

Modes of Transmission

Infectious diseases are usually transmitted in the following ways:

1. Direct contact, as by touching open sores or other lesions on an infected person. (This is where "contagious" comes from; the Latin *contagio* means touch or contact.) Infected people also can transfer pathogens from such lesions to their own hands or mouth and so infect someone else through a handshake or a kiss. HIV (human immunodeficiency virus), gonorrhea, syphilis, hepatitis B, and other sexually transmissible diseases

described in Chapter 15 are spread primarily by sexual contact.

2. Indirect contact, as by touching doorknobs, food, diapers, hypodermic needles (as used by drug abusers), or other objects that were previously in contact with an infected person. Food that is moist, not refrigerated, and not too acidic can be contaminated by many different pathogens, including ones responsible for amoebic dysentery, bacterial dysentery (*Salmonella* and *Shigella*), and typhoid fever.

3. Inhaling pathogens that have been ejected into the air, as by coughs and sneezes (Figure *a*). Cold and influenza viruses are usually spread this way; the bacteria that cause tuberculosis are virtually always inhaled, which is why tuberculosis is so contagious.

4. Encounters with biological *vectors*, including mosquitoes, flies, fleas, ticks, and other arthropods. Vectors transport pathogens from infected people or contaminated material to new hosts. Many pathogens depend on vectors as *intermediate hosts*. That is, a portion of the pathogen's life cycle must take place inside the vector. Mosquitoes, for instance, are intermediate hosts for the parasite *Plasmodium falciparum*, which causes malaria.

a full-blown sneeze.

Patterns of Occurrence

Infectious diseases often are described in terms of their patterns of occurrence. As the name suggests, *sporadic* diseases break out irregularly and affect only a few people. Whooping cough is an example. *Endemic* diseases occur more or less continuously but they are localized to a relatively small portion of the population. Leprosy and Lyme disease (transmitted by tick bites) are examples; so is impetigo, a highly contagious bacterial infection that is endemic to many day-care centers.

During an *epidemic*, a disease abruptly spreads through large portions of the population for a limited period. When cholera broke out all through Peru in 1991, this was an epidemic. The epidemic of bubonic plague in 14th-century Europe killed 25 million people. When epidemics break out in several countries around the world in a given time span, they collectively are called a *pandemic*. AIDS, which is caused by HIV, is now pandemic; as many as 40 million people may be infected by 2000 (pages 341–346). The single most lethal pandemic in human history (so far) occurred at the end of World War I in 1918, when viral influenza killed 21 million people around the globe.

Virulence

Virulence, which refers to the damage that a pathogen can do to a susceptible host, depends on how easily the pathogen can avoid detection and destruction within the body. For example, a certain virus may cause a mild "24-hour flu," whereas toxins produced by the bacterium *Clostridium botulinum* may kill a person within hours. HIV is apparently difficult to transmit, but is highly virulent because it eventually kills its victims.

Individuals respond differently to pathogens. Most important is the overall state of the body's immune system. As Chapter 7 describes, the immune system can be weakened by factors such as chronic fatigue or stress. In some cases, as with AIDS, it can be essentially shut down.

What Antibiotics Can and Cannot Do

Antibiotics are substances that destroy or inhibit the growth of bacteria and certain other microorganisms. They are produced mainly by bacteria and fungi. Antibiotics such as penicillins, tetracyclines, and streptomycins interfere with a variety of different metabolic functions. Streptomycins block the synthesis of proteins, and peni-

Infectious Diseases' Global Threats to Human Health*		
Disease	Type of Pathogen	Estimated Deaths
Tuberculosis	Bacterium	3.3 million
Malaria	Parasitic protozoan	1–2 million
Hepatitis (includes A, B, non-A/non-B, and delta hepatitis)	Virus	1–2 million
Various respiratory infections (pneumonia, viral influenza, diphtheria, streptococcus)	Virus, bacterium	6.9 million
Diarrheas (includes amoebic dysentery, cryptosporidiosis, and gastroenteritis)	Parasitic, virus, and bacterium	4.2 million
Measles	Virus	220,000
Schistosomiasis	Parasitic flatworm	200,000
Whooping cough	Bacterium	100,000
Hookworm	Parasite	50,000+
Amoebiasis	Parasitic protozoan	40,000+

* Does not include AIDS-related deaths. Recent statistics on worldwide HIV infections and deaths due to AIDS are presented in Chapter 15.

Built-In Study Aids

Besides the visual summaries, in-text summary statements, and other features shown in earlier examples, each chapter concludes with the following study aids:

Critical Thinking Questions

emphasize practical applications of text concepts. They ask you to analyze and pull together different facets of what you have learned and expand your thinking beyond the text material itself. Appendix IV contains a key suggesting an approach you can use to think through plausible answers for each question.

Key Terms

consist of a list of the boldfaced terms defined in the chapter. See if you can give definitions for all of them. The italic page numbers will provide fast entry back into the chapter for the ones you aren't sure of. All terms are also defined in the book's glossary.

Summaries

in list format reinforce the main chapter concepts.

SUMMARY

1. Genes, the units of instruction for heritable traits, are arranged linearly on a chromosome.

2. Human diploid cells have two chromosomes of each type (one from the mother, one from the father). The two are homologues; they pair at meiosis. The sex chromosomes (X and Y) differ from each other in size, shape, and the genes they carry. All other pairs of chromosomes are autosomes; they are roughly the same in size and shape and carry genes for the same traits.

3. Genes on the same chromosome tend to be linked—they tend to stay together during meiosis and end up in the same gamete. But crossing over can disrupt such linkages. The farther apart two genes are on a chromosome, the greater will be the frequency of crossing over and recombination between them.

4. A family pedigree is a chart of genetic relationships of individuals. It helps establish inheritance patterns and track genetic abnormalities through several generations.

5. In disorders involving autosomal recessive inheritance, a person who is homozygous for a recessive allele shows the recessive phenotype. Heterozygotes generally have no symptoms. Cystic fibrosis and Tay-Sachs disease are examples.

6. In autosomal dominant inheritance, a dominant allele usually is expressed to some extent. Disorders include Huntington's disease and achondroplasia.

7. A large number of genetic disorders are X-linked—the mutated gene occurs on the X chromosome. Males, who inherit only one X chromosome, typically are affected. Examples include hemophilia A and B and Duchenne muscular dystrophy.

8. Chromosome structure can be altered by deletions, duplications, inversions, or translocations. Chromosome number can be altered by nondisjunction (the failure of chromosomes to separate during meiosis or mitosis), so that gametes end up with one extra or one missing chromosome. Gene mutations, changes in chromosome structure, and changes in chromosome number tend to cause many genetic disorders.

Review Questions

1. What are the main differences between X and Y chromosomes? 393

2. Why do harmful genes (alleles) persist in a population? 398

3. What factors indicate that a trait is coded by a dominant allele on an autosome? 398

4. What is the difference between an X-linked trait and a sex-influenced trait? 399

5. Describe the general procedures of genetic counseling. 405

Critical Thinking: You Decide (Key in Appendix IV)

1. A female athlete is disqualified from competition because chromosome testing shows that this individual is XY. Later, a medical exam reveals "androgen insensitivity," an abnormality in which an embryo's cells lack receptor proteins for the male hormone testosterone. As a result, female sex characteristics develop, and indeed the person clearly appears to be feminine. Knowing this, do you agree or disagree with the disqualification, and why?

2. Referring to question 1, would you consider androgen insensitivity a genetic disorder? Why or why not?

Self-Quiz (Answers in Appendix III)

1. _____ segregate during _____.

   a. Homologues; mitosis
   b. Genes on one chromosome; meiosis
   c. Homologues; meiosis
   d. Genes on one chromosome; mitosis

2. The two alleles of a gene on homologous chromosomes end up in separate _____.

   a. body cells
   b. gametes
   c. nonhomologous chromosomes
   d. offspring
   e. both b and d are possible

3. Genes on the same chromosome tend to remain together during _____ and end up in the same _____.

   a. mitosis; body cell
   b. mitosis; gamete
   c. meiosis; body cell
   d. meiosis; gamete
   e. both a and d

4. The probability of a crossover occurring between two genes on the same chromosome is _____.

   a. unrelated to the distance between them
   b. increased if they are closer together on the chromosome
   c. increased if they are farther apart on the chromosome
   d. impossible

5. Chromosome structure can be altered by _____.

   a. deletions
   b. duplications
   c. inversions

6. Nondisjunction can be caused by _____.

   a. crossing over in meiosis
   b. segregation in meiosis
   c. failure of chromosomes to separate during meiosis
   d. multiple independent assortments

7. A gamete affected by nondisjunction would have _____.

   a. a change from the normal chromosome number
   b. one extra or one missing chromosome
   c. the potential for a genetic disorder
   d. all of the above

8. Genetic disorders can be caused by _____.

   a. gene mutations
   b. changes in chromosome structure
   c. changes in chromosome number
   d. all of the above

9. Match the following chromosome terms appropriately.
   _____ crossing overa. a chemical change in DNA that may
   _____ deletion   affect genotype and phenotype
   _____ nondisjunction b. movement of a chromosome
   _____ translocation   segment to a nonhomologous
   _____ gene mutation   chromosome
       c. disrupts gene linkages during meiosis
       d. causes gametes to have abnormal
       chromosome numbers
       e. loss of a chromosome segment

Genetics Problems (Answers in Appendix II)

1. If a couple has a large family of six boys, what is the probability that a seventh child will be a girl?

2. Answer the following questions, given that human sex chromosomes are XX for females and XY for males.
   a. Does a male child inherit his X chromosome from his mother or father?
   b. With respect to an X-linked gene, how many different types of gametes can a male produce?
   c. If a female is homozygous for an X-linked gene, how many different types of gametes can she produce with respect to this gene?
   d. If a female is heterozygous for an X-linked gene, how many different types of gametes can she produce with respect to this gene?

3. Suppose that you have two linked genes with alleles A,a and B,b, respectively. An individual is heterozygous for both genes, as in the following:

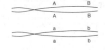

If the crossover frequency between these two genes is 0 percent, what genotypes would be expected among gametes from this individual?

Key Terms

consist of a list of the boldfaced terms defined in the chapter. See if you can give definitions for all of them. The italic page numbers will provide fast entry back into the chapter for the ones you aren't sure of. All terms are also defined in the book's glossary.

4. Individuals affected by Down syndrome typically have an extra chromosome 21, so their cells have a total of 47 chromosomes. However, in a few cases of Down syndrome, 46 chromosomes are present. Included in this total are two normal-appearing chromosomes 21 and a longer-than-normal chromosome 14. Interpret this observation, and indicate how these few individuals can have a normal chromosome number.

5. The mugwump, a space alien, has a reversed sex-chromosome condition. The male is XX and the female is XY. However, perfectly good sex-linked genes are found to have the same effect as in humans. For example, a recessive, X-linked allele c produces red/green color blindness. If a normal female mugwump mates with a phenotypically normal male mugwump whose mother was color blind, what is the probability that a son from that mating will be color blind? A daughter?

6. A normal woman whose father had hemophilia A marries a man who also has hemophilia A. What is the chance their son will have the disorder?

7. Both parents in a couple are carriers for the recessive allele p that codes for the defective protein responsible for PKU. What is the symbol for the dominant allele? What is the probability that any of their four children will have PKU? Now, assume that this probable outcome actually occurs in the family, and mark the following pedigree chart accordingly:

How many of the children are carriers for PKU?

Key Terms

Readings

Cummings, M. 1991. *Human Heredity: Principles and Issues.* St. Paul, Minn.: West.

Elmer-Dewitt, P. October 5, 1992. "Catching a Bad Gene in the Tiniest of Embryos." *Time.*

"Huntington's Gene Finally Found." April 1993. *Science.*

Review Questions

are keyed to italicized and boldfaced sentences to further reinforce main concepts. These conceptual questions are cross-referenced to page numbers in the text. Detailed answers to all review questions are available in a separate booklet.

Self-Quizzes

are a quick way to test your recall of important definitions and key concepts with the help of objective questions. Appendix III lists the answers.

Genetics Problems

are available in the unit on inheritance. They give you a better sense of patterns of inheritance. Appendix II gives detailed solutions to assist understanding.

Recommended Readings

provide you with a starting point for exploring a topic in more depth.

HUMAN BIOLOGY

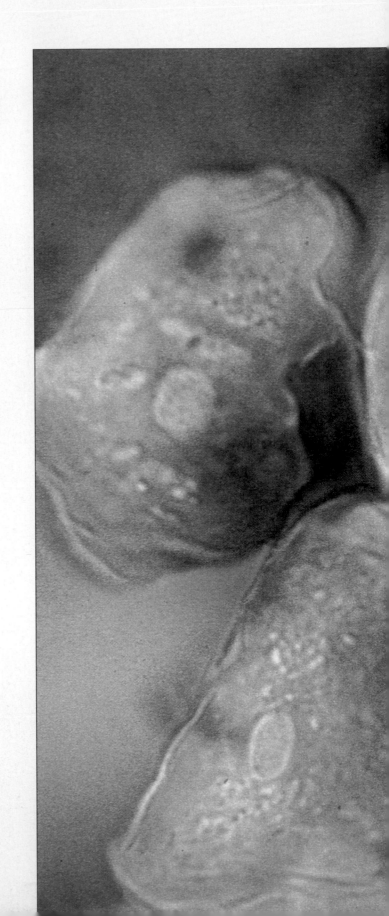

Human cheek cells magnified 200 times.
Each cell is a tiny but highly ordered unit,
the fundamental unit of life.

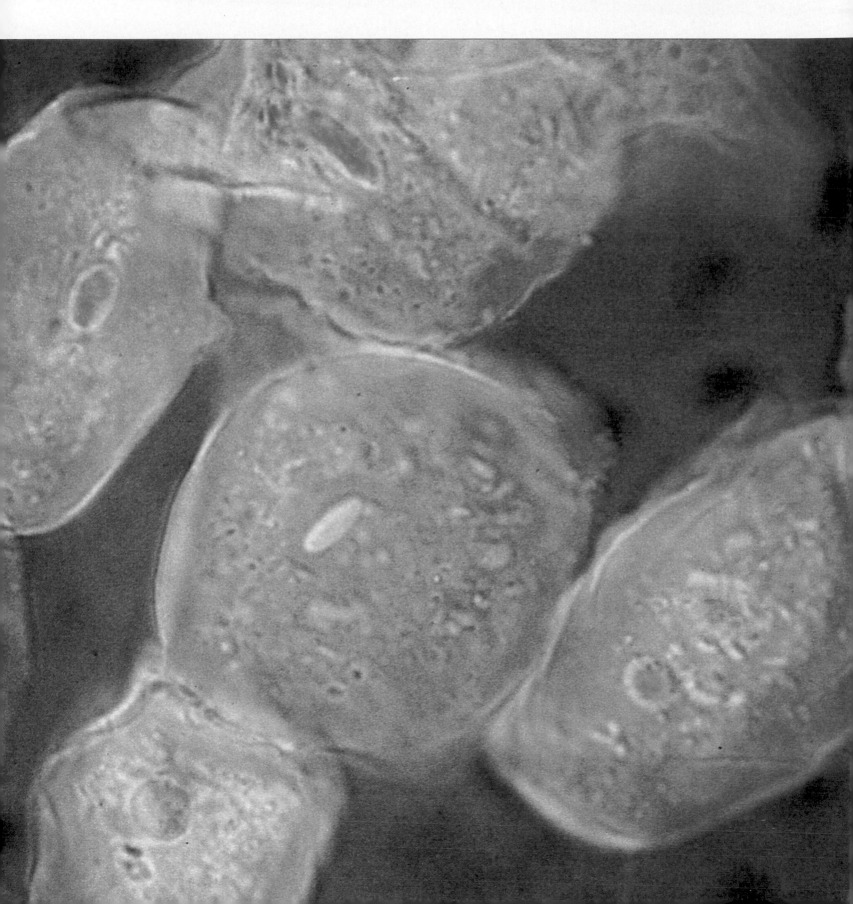

Introduction CONCEPTS IN HUMAN BIOLOGY

Human Biology Revisited

Buried somewhere in your brain are memories of discovering your own hands and feet, your family, the change of seasons, the smell of grass. The memories include early introductions to a great disorganized parade of insects, flowers, frogs, and furred things—mostly living, sometimes dead. There are memories of questions: "What is life?" and "Where do I fit in the world around me?" There are memories of answers, some satisfying, others less so.

By observing, asking questions, and accumulating answers, you have built up a store of knowledge about the world of life (Figure I.1). Experience and education have been refining your questions, and no doubt some answers are difficult to come by. Think of a young man whose brain is functionally dead as a result of a motorcycle accident. If breathing and other basic functions proceed only as long as he remains hooked up to mechanical support systems, is he no longer "alive"? Think of an embryo growing inside a pregnant woman, but currently no more than a cluster of tiny cells. At what point in its development is it a definably "human" life? If questions like these cross your mind, your thoughts about life obviously run deep.

The point is, this book isn't your introduction to human biology, for you have been studying yourself, others, and the world around you ever since information began penetrating your brain. This book simply is human biology *revisited*, in ways that may help carry your thoughts to more organized levels of understanding.

To biologists, the question "What is life?" opens up a story that has been unfolding in countless directions for several billion years. To begin with, "life" is an outcome of ancient events by which nonliving materials became assembled into the first living cells. With the emergence of living cells came biological **evolution**—change in details of the body plan and functions of organisms through successive generations. In the course of evolution, broad groups of life forms, including animals, emerged.

Humans, apes, and some other closely related animal species are primates, and as primates we are also mammals. All mammals, including humans, are *vertebrates*—animals with "backbones." Ultimately, we share our planet with millions of other animal species, as well as with plants, fungi, bacteria, and other organisms. Figure I.2 provides a general picture of the place our human species, *Homo sapiens*, occupies in the living world.

"Life" is also a way of capturing and using energy and materials. "Life" is a way of sensing and responding to specific changes in the environment. "Life" is a

Figure I.1 Think back on all you have known and seen; this is the foundation for your deeper study of human biology.

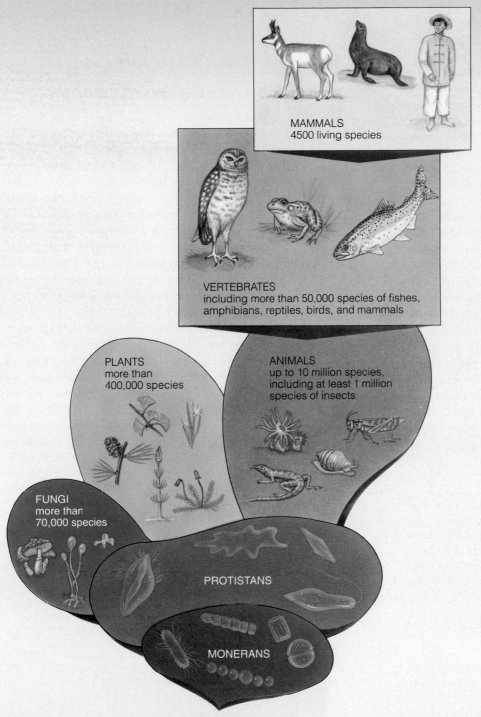

MAMMALS
4500 living species

VERTEBRATES
including more than 50,000 species of fishes,
amphibians, reptiles, birds, and mammals

PLANTS
more than
400,000 species

ANIMALS
up to 10 million species,
including at least 1 million
species of insects

FUNGI
more than
70,000 species

PROTISTANS

MONERANS

FIVE KINGDOMS OF LIFE

Figure I.2 Human beings are only one species among millions
inhabiting the earth. Our species, *Homo sapiens*, is a subgroup
of the class Mammalia, which in turn is classified as a subgroup
of the phylum Vertebrata. The vast majority of animals on earth
are not vertebrates; they are invertebrates, such as insects, and
lack a "backbone." Among the organisms with which we share
our world are at least 1 million species of insects, more than
400,000 species of plants, and vast numbers of single-celled
organisms, including bacteria.

capacity to grow, develop, and reproduce. In fact, much of this text explores aspects of anatomy and functioning through which your body maintains the living state. A partial list of those impressive life support systems includes your brain and an elaborate network of nerve cells, a digestive system that can extract nutrients from thousands of foods, respiratory and circulatory systems that keep your body supplied with oxygen, a urinary system, an immune system, a hormone-based communication system, a skeleton and muscles for support and movement, and organs for reproduction. And this short description only hints at the intricate workings of the human body.

This chapter introduces some basic concepts that provide a foundation for understanding the material to come. Here, too, we introduce several themes that are cornerstones of the study of human biology. One theme is our evolutionary heritage, which includes cultural as well as biological evolution. Another is the state of internal constancy called *homeostasis*. Living cells can only function properly within narrow environmental limits, and each of the life support systems we will study makes an essential contribution to maintaining a constant internal environment. Finally, throughout the text you will also find *Focus* and *Choices* features that relate basic biological concepts to health topics, environmental concerns, and social issues.

KEY CONCEPTS

1. Humans exhibit basic characteristics of life. First, the structural organization and functions of the human body depend on certain properties of matter and energy. Second, humans obtain and use energy and materials from their environment. Third, they make controlled responses to changing conditions. Fourth, they grow and reproduce, based on instructions contained in DNA.

2. Life processes depend on a stable internal state called homeostasis.

3. Human anatomy, physiology, and possibly some aspects of behavior have been shaped by processes of evolution.

4. Biology, like other branches of science, is based on systematic observations, hypotheses, predictions, and relentless testing. The external world, not internal conviction, is the testing ground for scientific theories.

Figure I.3 A climber cautiously chooses her route up a vertical rock face. Although she is alive and the rock is not, both are composed of the same fundamental particles: protons, electrons, and neutrons.

CHARACTERISTICS OF LIFE

Energy, DNA, and Biological Organization

Picture a climber on a rock face, inching cautiously toward the safety of the top (Figure I.3). Without even thinking about it, you know the climber is alive and the rock is not. Yet the climber, the rock, and all other things are composed of the same particles (protons, electrons, and neutrons). The particles are organized into atoms according to fundamental physical laws. At the heart of those laws is something called **energy**—a capacity to make things happen, to do work. Energetic interactions bind atom to atom in predictable patterns, giving rise to the structured bits of matter we call molecules. Energetic interactions among molecules hold a rock together—and they hold a human together.

It takes a special molecule called deoxyribonucleic acid, or **DNA**, to set living things apart from the nonliving world. No chunk of granite or quartz has it. DNA molecules in sperm and eggs contain the instructions for assembling each new human being from carbon, hydrogen, and a few other kinds of "lifeless" substances. By analogy, think of what you can do with a little effort and just two kinds of ceramic tiles in a crafts kit. Following the kit's directions, you can arrange the tiles to form organized patterns (Figure I.4). Similarly, *life emerges from lifeless matter with DNA's "directions," some raw materials, and inputs of energy.*

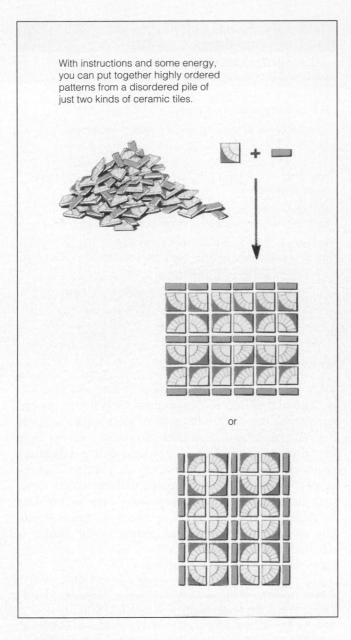

With instructions and some energy, you can put together highly ordered patterns from a disordered pile of just two kinds of ceramic tiles.

or

Figure I.4 Emergence of organized patterns from disorganized beginnings. Two ceramic tile patterns are shown here. You probably can visualize other possible patterns using the same two kinds of tiles. Similarly, the organizational characteristic of life emerges from pools of simple building blocks, given energy sources and specific DNA "blueprints."

Look carefully at Figure I.5, which outlines the levels of organization in nature. The properties of life emerge at the level of the cell. A **cell** is an organized unit that has the capacity to survive and reproduce on its own, given DNA instructions and appropriate sources of energy and raw materials. In other words, the cell is the basic *living* unit.

Biosphere
Those regions of the earth's waters, crust, and atmosphere in which organisms can exist

Ecosystem
A community and its physical environment

Community
The populations of *all* species occupying the same area

Population
A group of individuals of the same kind (that is, the same species) occupying a given area at the same time

Multicellular Organism
An individual composed of specialized, interdependent cells arrayed in tissues, organs, and often organ systems. Humans are complex multicellular animals.

Organ System
Two or more organs interacting chemically, physically, or both in ways that contribute to the survival of the whole organism

Organ
A structural unit in which tissues are combined in specific amounts and patterns that allow them to perform a common task

Tissue
A group of cells and surrounding substances, functioning together in a specialized activity

Cell
Smallest *living* unit; may live independently or may be part of a multicellular organism

Organelle
Sacs or other compartments that separate different activities inside the cell

Molecule
A unit of two or more atoms of the same or different elements bonded together

Atom
Smallest unit of an element that still retains the properties of that element

Subatomic Particle
An electron, proton, or neutron; one of the three major particles of which atoms are composed

Figure I.5 Levels of organization in nature, starting with the subatomic particles that serve as the fundamental building blocks of all organisms, including human beings.

Figure I.6 This child is exhibiting one of the characteristics of life: taking in energy from food. Other characteristics of life include metabolic activity and the ability to sense and respond to change in the environment. Food energy in simple sugar molecules is transferred to molecules of ATP, which transfer it in turn to other molecules that perform the metabolic functions required to maintain cells in the living state. The body constantly monitors the changing level of simple sugar molecules in the blood, which are prime energy sources.

Figure I.5 also shows more inclusive levels of organization, from populations on through communities, ecosystems, and the biosphere. (*Biosphere* refers to all regions of the earth's waters, crust, and atmosphere in which organisms live.) We will return to each of these concepts later in the book, when we consider some basic principles of ecology and human impacts on ecosystems.

The structure and organization of nonliving and living things arise from the properties of matter and energy.

The structure and organization unique to living things begin with instructions contained in DNA molecules.

Table I.1   Characteristics of Life

1. Complex structural organization based on instructions contained in DNA molecules.

2. Direct or indirect dependence on other organisms for energy and material resources.

3. Metabolic activity by the single cell or multiple cells composing the body.

4. Use of homeostatic controls that maintain favorable operating conditions in the body despite changing conditions in the environment.

5. Reproductive capacity, by which the instructions for heritable traits are passed from parents to offspring.

6. Diversity in body form, in the functions of various body parts, and in behavior. Such traits are adaptations to changing conditions in the environment.

7. The capacity to evolve, based ultimately on variations in traits that arise through mutations in DNA.

Metabolism

Table I.1 lists basic characteristics of life, and you will notice that these include metabolism. Only living cells can accomplish this extraordinary feat. **Metabolism** refers to the cell's capacity to (1) extract and convert energy from its surroundings and (2) use energy to maintain itself, grow, and reproduce. Molecules of **ATP**, an "energy carrier," transfer energy to other molecules that either perform metabolic work (enzymes), serve as building blocks, or serve as energy reserves. In humans (Figure I.6), as in most animals and plants, stored energy is released and transferred to ATP by way of *aerobic respiration*, another metabolic process.

Living things show metabolic activity. Their cells acquire and use energy to stockpile, tear down, build, and eliminate materials in ways that promote survival and reproduction.

Interdependency Among Organisms

With few exceptions, a flow of energy from the sun maintains the great pattern of organization in nature. Plants and some other organisms that capture solar energy through photosynthesis are the entry point for this flow. They are food producers for the living world. Animals, including humans, are consumers; directly or indirectly, they feed on energy stored in plant parts. Thus you tap directly into the stored energy when you eat a banana, and you tap into it indirectly when you eat hamburger made from a steer that fed on grass or grain. Bacteria and fungi are decomposers. When they feed on tissues or

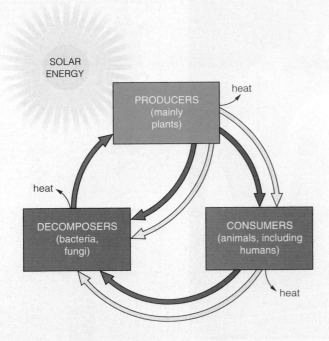

SOLAR ENERGY

heat

PRODUCERS (mainly plants)

heat

DECOMPOSERS (bacteria, fungi)

CONSUMERS (animals, including humans)

heat

Figure I.7 Energy flow and the cycling of materials in the biosphere. Humans are consumers and are inextricably linked with both producers and decomposers.

remains of other organisms, they break down sugars and other biological molecules to simple raw materials, which can be recycled back to producers.

The point here is simple: Every part of the living world is ultimately linked to every other part. Organisms, including humans, are interdependent. There is a one-way flow of energy through them and a cycling of materials among them (Figure I.7). Interactions among organisms influence the cycling of carbon and other substances on a global scale; such interactions even influence the earth's energy "budget." Understand the extent of these interactions—and how they are affected by the activities of 5.6 billion human beings—and you will gain insight into the thinning of the ozone layer, acid rain, and many other modern-day problems.

All organisms, including humans, are part of webs of organization in nature, in that they depend directly or indirectly on one another for energy and raw materials.

Sensing and Responding to Change

Cells of the human body can sense changes in the environment and make controlled responses to such changes. They manage this monitoring task with the help of receptors, which are molecules and structures that can detect specific information about the environment. When cells receive signals from receptors, they adjust their activities in ways that bring about an appropriate response.

For example, your body can withstand only so much heat or cold. It must rid itself of harmful substances, and certain nutrients must be available in certain amounts. Yet temperatures shift, harmful substances may be encountered, and you may sometimes gorge on junk food or fruit or skip a meal entirely.

Suppose you skip breakfast, then lunch, and the level of the sugar glucose in your blood—"blood sugar"—falls. A hormone then signals liver cells to dig into their stores of energy-rich molecules. Those complex molecules are broken down to glucose, which is released into the bloodstream, and your blood-sugar level returns to normal. After you eat, glucose enters your bloodstream, blood sugar rises, and liver cells step up their secretion of another hormone, insulin. Most cells in your body have receptors for insulin, which induces them to take up glucose. With so many cells taking up glucose, the blood-sugar level returns to normal.

Blood is part of the "internal environment," the fluid environment bathing your cells. Usually, the internal environment of your body is kept fairly constant. When the internal environment is maintained within tolerable ranges, a state we call **homeostasis** exists.

Cells sense and respond to changes in the environment. The responses help maintain the stable internal state called homeostasis.

Homeostasis results in favorable operating conditions inside the cell or the body as a whole.

Reproduction

We humans tend to think we enter the world rather abruptly and are destined to leave it the same way. Yet we and all other organisms are part of an immense, ongoing journey that began billions of years ago. Think of the first cell produced when a human egg and sperm join. The cell would not even exist if the sperm and egg had not formed earlier according to DNA instructions that were passed down through countless generations. With time-tested DNA instructions, a new human body develops in ways that will prepare it, ultimately, for helping to produce individuals of a new generation. With **reproduction**—that is, the production of offspring by parents—the journey of life continues.

Every human being is part of a reproductive continuum that extends back through countless generations.

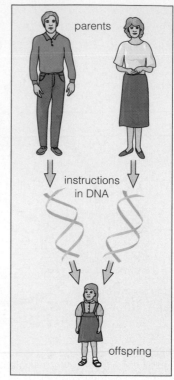

Figure I.8 Reproduction is another life characteristic. Instructions in DNA assure that offspring will resemble parents—and they also permit variations in the details of traits, as the photo demonstrates.

Inheritance and Variations in Heritable Traits

Reproduction involves **inheritance** (Figure I.8). The word means that parents transmit to their offspring instructions for duplicating their body form and other traits. DNA molecules contain the required instructions.

DNA instructions assure that offspring will resemble parents, and they also permit *variations* in the details of traits. For example, although having five fingers on each hand is a human trait, some humans are born with six fingers on each hand instead of five! Variations in traits arise through **mutations**, which are heritable changes in the structure or number of DNA molecules.

Many mutations are harmful, for the separate bits of information in DNA are part of a coordinated whole. For example, a single mutation in a tiny bit of human DNA may lead to hemophilia. In individuals with this genetic disorder, blood cannot clot properly after the body is cut or bruised (page 402).

Adaptation Some mutations may prove to be harmless, or even beneficial, under prevailing conditions. For example, the disease sickle-cell anemia is caused by a DNA change that results in a defective form of the blood protein hemoglobin. The sickle-cell trait is most prevalent in parts of the world where malaria is common. People who inherit the mutation from both parents suffer the debilitating disease. But, for complex reasons, people who inherit the mutation from only one parent are resistant to malaria. For them, the mutation is adaptive; it increases their chances of survival.

An **adaptive trait** is one that helps an organism survive and reproduce under a given set of environmental conditions. In the long course of human evolution, countless DNA mutations, tested in the environments of our ancestors, have given rise to an elaborate nervous system, efficient mechanisms for taking in and distributing oxygen and food molecules, and other characteristics that enable each of us to live the biologically complex life of a human being. Later in the book we will consider the actual mechanisms by which evolution occurs.

DNA is the molecule of inheritance. Its instructions for reproducing traits are passed on from parents to offspring.

Mutations introduce variations in heritable traits.

Although most mutations are harmful, some give rise to variations in form, function, or behavior that are adaptive under prevailing conditions.

THE NATURE OF SCIENTIFIC INQUIRY

The Inquiring Mind: Scientific Methods

Biology, like science generally, is an ongoing record of discoveries arising from methodical inquiries into the natural world. Human biology focuses, naturally enough, on the workings of the human body and closely related topics.

Our fascination with ourselves is probably as ancient as our species's beginnings. Thinkers such as Hippocrates, Leonardo da Vinci, and Charles Darwin pursued their curiosity under the umbrella of "natural history." Modern biologists are just as curious, if not more so, but they now pursue complex topics ranging from the molecular structure of HIV, the virus responsible for AIDS, to the impacts on human health of a hole in the stratospheric ozone. In fact, the range of possible specialization within "human biology" is so broad that no single "scientific method" can be used to approach all the relevant topics and issues.

Even so, scientists everywhere still have practices in common. *Scientists ask questions, make educated guesses about possible answers, and then devise ways to test their predictions, which will hold true if their guesses are good ones.* The following list describes what scientists generally do when they proceed with an investigation:

1. Ask a question or identify a problem.

2. Develop one or more **hypotheses**, or educated guesses, about what the answer (or solution) might be. This might involve sorting through what has been learned already about related phenomena.

3. Using each hypothesis as a guide, make a **prediction**—that is, a statement of what you should be able to observe, if you were to go looking for it. This is often called the "if-then" process. (*If* cigarette smoke is a cancer-causing agent in human lungs, *then* we should be able to detect a higher rate of lung cancer among smokers than among nonsmokers.)

4. Devise ways to *test* the accuracy of predictions. You might do so by making observations, developing models, and doing experiments.

5. If the tests do not turn out as expected, check to see what might have gone wrong. (Maybe you overlooked something, or maybe the hypothesis isn't a good one.)

6. Repeat or devise new tests—the more the better. Hypotheses supported by many different tests are more likely to be correct.

7. Objectively report the test results and conclusions drawn from them.

In broad outline, a scientific approach to studying nature is that simple. You can use this approach to pick your way logically through environmental, medical, and social issues of the sort described later in the book. And understanding how good science operates will help you evaluate media accounts of discoveries, advances, and research in progress. The *Focus on Science* (pages 10–11) gives one example.

About the Word *Theory*

What is the difference between a hypothesis and a theory? In science, a **theory** is a related set of hypotheses which, taken together, form a broad-ranging explanation about some fundamental aspect of the natural world. A scientific theory differs from a scientific hypothesis in its *breadth of application*. Charles Darwin's theory about the evolution of species fits this description: It is a broad, encompassing "Aha!" explanation that, in a few intellectual strokes, makes sense of a huge number of observable phenomena.

Yet is any theory an "absolute truth" in science? Ultimately, no. Why? It is impossible to perform the infinite number of tests required to show that a theory holds true under *all* possible conditions! Objective scientists say only that they are *relatively* certain that a theory is (or is not) correct. Even so, such "relative certainty" can be extremely impressive. Especially after exhaustive tests by many scientists, a theory may be as close to the "truth" as we can get.

Scientists must keep asking themselves: "Will some other evidence show my idea to be incorrect?" They are expected to put aside pride or bias by testing their ideas, even in ways that might prove them wrong. Even if an individual doesn't (or won't) do this, *others will*—for science is conducted in a community that is both cooperative and competitive. Ideas are shared and examined with the understanding that it is just as important to expose errors as it is to applaud insights. Individuals can change their minds when presented with new evidence, and this is a strength of science, not a weakness.

A scientific theory is an explanation about the cause or causes of a broad range of related phenomena. Like hypotheses, theories are open to testing, revision, and tentative acceptance or rejection.

The Limits of Science

There are limits on the kinds of studies scientists can carry out. The reason is simple: Some questions lie beyond the realm of scientific analysis, and lie instead in the realm of personal belief, values, morality, philosophy, or ethics. Why do we exist, for what purpose? Why does any one of us have to die at a particular moment and not another? Should one person aid another incurably ill person in the

Focus on Science

Scientific Sleuthing in Cancer Research

Making Observations and Asking Questions   Put yourself in the shoes of Michael Pariza, a biochemist who is probing the effects of different forms of linoleic acid (LA, for short) on cancer. You're interested in this subject because LA is a "Jekyll-and-Hyde" substance: The human body requires it in small amounts as a building block for fats, but studies on mice and rats show that in large doses it is associated with the development of certain types of cancer. There is also an alternate form of the compound, called CLA—and here the plot thickens. When you paint CLA on the skin of mice that have been exposed to a potent carcinogen (cancer-causing agent), those mice develop only half as many skin cancers as mice painted with LA. A second experiment shows that force-feeding CLA to mice a few days before administering a carcinogen inhibits cancerous stomach tumors. Now you are getting excited, because you know that CLA occurs in some common human foods, including grilled ground beef and some cheese products. Could CLA inhibit cancer when it is consumed as an additive to food?

Moving From Hypothesis to Prediction   Pariza, Clement Ip, and several colleagues agreed on a hypothesis: Based on their existing knowledge, they reasoned that CLA consumed as a regular part of the diet *can* help prevent certain cancerous tumors from developing. Notice that at least one alternative hypothesis would also have been reasonable; for example, they might have proposed that CLA exerts its effects only when it is administered immediately before exposure to a carcinogen. In science, *alternative hypotheses are the rule, not the exception*.

The researchers then made a prediction they (and other scientists) could readily test—that rats fed certain doses of CLA would develop fewer cancers than rats in a control group, when all the rats were exposed to a carcinogen known to cause tumors in mammary glands.

Testing a Prediction: The Role of Experiments Some hypotheses can be tested by direct observation. (You can watch birds visiting a feeder to see whether one type of bird feeds more aggressively, for example.) In this

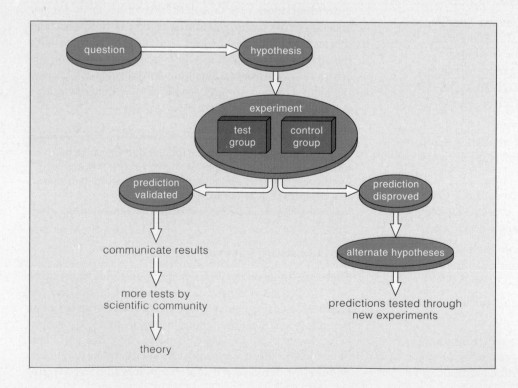

A scientific method.

instance, Pariza and his coworkers conducted a series of experiments. An **experiment** is a test in which nature is manipulated into revealing one of its secrets. An essential part of any experiment is something called a **control group**. Control groups are used to evaluate possible side effects of a test being performed on an experimental group. If an experiment involves laboratory rats, then the control group will be rats; if it involves college students, the controls will be students, and so on. Ideally, members of a control group should be *identical* to those of an experimental group in every respect—*except* for the key factor, or **variable**, under study. Both groups also must be large enough so the results won't be due to chance alone. Generally, experiments are devised to *disprove* a hypothesis. Why? It is impossible to prove beyond a shadow of a doubt that a hypothesis is correct, for it takes an infinite number of experiments to demonstrate that it holds under *all* possible conditions.

Pariza's experiment began with five groups of 30 healthy rats. One group, the control, received a normal diet; the other four groups were fed the same diet, *except* that each group's food contained a certain amount of CLA, which was added for several weeks before all the animals were exposed to the carcinogen. What was the outcome? The team recently reported in the prestigious journal *Cancer Research* that, under the conditions of their experiments, regular feeding of CLA in the diet "is effective in cancer prevention."

Since then, no doubt, other researchers have repeated this same experiment to verify the findings and perhaps add new insights. Pariza, Ip, and the rest of the team are now investigating exactly *how* CLA deters cancer. Based on knowledge they and others have gained, one hypothesis is that CLA may interfere with steps in a pathway that promotes the runaway multiplication of cancer cells.

No one yet knows whether the results of these experiments will apply to humans. If it turns out that they do, the day could come when, thanks to some scientific sleuthing, a CLA-like substance is a routine additive to a wide range of human foods.

act of suicide? Answers to such questions are *subjective*; that is, they come from within us, as an outcome of our individual experiences, religious beliefs, ethical training, and similar factors. As you probably already recognize, subjective answers do not readily lend themselves to scientific analysis. Today, modern medical technology can sometimes create serious ethical dilemmas when it blurs what once seemed to be a clear distinction between life and death (see *Choices*, page 12).

This is not to say that subjective answers are without value. No human society can function without a shared commitment to standards for making judgments, even if those judgments are subjective. Although moral, aesthetic, economic, and philosophical standards vary from one society to the next, all guide their members in deciding what is important and good, and what is not. All attempt to give meaning to what we do. But *the external world, not internal conviction, must be the testing ground for scientific beliefs*.

SUMMARY

1. As living organisms, humans have the following characteristics:

   a. Their structure, organization, and interactions arise from the basic properties of matter and energy.

   b. Processes of metabolism and homeostasis maintain the living state.

   c. They have the capacity for growth, development, and reproduction, based on instructions contained in their DNA.

2. Diversity in features and characteristics arises through mutation. Mutations introduce changes in the DNA. The changes may lead to variations in the form, functioning, or behavior of individual offspring.

3. Individuals vary in their heritable traits (the traits that parents transmit to offspring). Such variations influence an organism's ability to survive and reproduce. Under prevailing conditions, some varieties of a given trait may be more adaptive than others; such traits will be "selected," whereas others will be eliminated through successive generations. Thus the population changes over time; it evolves. These points are central to the theory of evolution by natural selection.

4. There are many specialized scientific methods, corresponding to many different fields of inquiry. The following key terms are important in all of those fields:

   a. Theory: an explanation of a broad range of related phenomena. An example is the theory of evolution by natural selection.

Defining Death

In an austere hospital room, a young mother and father face the most anguished moment of their lives. Their baby has been born with no brain, except for a small portion of brain stem. For this child, none of the qualities that we associate with human life—the potential to think, feel joy or pain, learn, or speak—will ever be possible. For the moment, however, that bit of brain tissue controls lung and heart functions, so the heart beats sporadically and the lungs occasionally take in air. Within hours or days, even those halting functions will cease.

In the United States, about one baby in a thousand is born with this condition, called *anencephaly*. For parents and physicians alike, the situation is agonizing. During the short period while the heart and lungs minimally function, other organs (such as the liver and kidneys) receive enough oxygen-carrying blood to keep them reasonably healthy. If the child is declared legally dead during that time, those organs can be transplanted and can bring the gift of life to others. If doctors wait until the nubbin of brain stem gives out, potentially transplantable organs will be irreversibly damaged by lack of oxygen and will be useless.

If this were your child, what course would you follow? You might not have a choice, if the matter went to court.

In recent years many states have adopted a strict legal standard for such cases. A person may be declared legally dead only if *all* of the brain, including the brain stem, no longer functions. Some ethicists prefer this approach because it does not put society in the position of determining that some parts of the brain—but not others—make a person human and truly alive. Other people disagree strongly in cases in which the patient is in a "persistent vegetative state" (that is, has no higher brain function) and there is no hope of recovery. In particular, advocates for a less strict definition of brain death point to the serious shortage of organs for transplants and the potential for saving other lives.

As individuals, we deal with this kind of issue in many ways. Some people hold religious beliefs that require that individuals receive care as long as the heart can beat. Others draw up "living wills" designed to convey their wishes that life-support equipment, such as respirators, heart bypass pumps, and other high-tech machinery, not be used to prolong life artificially. Many people ignore the issue altogether, hoping that they or their loved ones will never have to grapple with it. Unfortunately life—and death—are not always so simple.

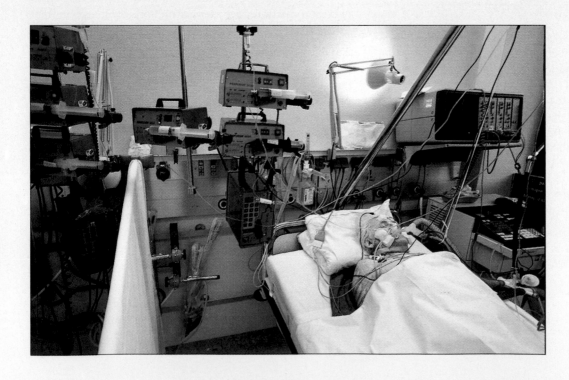

b. Hypothesis: a possible explanation of a specific phenomenon; sometimes called an "educated guess."

c. Prediction: a claim about what an observer can expect to see in nature if a theory or hypothesis is correct.

d. Test: an effort to gather actual observations that may (or may not) match predicted or expected observations.

e. Conclusion: a statement about whether a hypothesis (or theory) should be accepted, rejected, or modified, based on tests of the predictions derived from it.

5. Scientific theories are based on systematic observations, hypotheses, predictions, and tests. The external world, not internal conviction, is the testing ground for scientific theories.

Review Questions

1. For this and subsequent chapters, make a list of the boldface terms that occur in the text. Write a definition next to each, and then check it against the one in the glossary.

2. Why is it difficult to give a simple definition of life? (For this and subsequent chapters, *italic numbers* following review questions indicate the pages on which the answers may be found.) *2*

3. As living organisms, what characteristics do humans exhibit? *4*

4. What is energy? What is DNA? *4*

5. Define metabolic activity, and briefly describe a metabolic event. *6*

6. Describe the one-way flow of energy and the cycling of materials through the biosphere. *7*

7. What is mutation? What role does it play in evolutionary change? *8*

8. Design a test (or series of tests) to support or refute the following hypothesis: A diet high in salt is associated with hypertension (high blood pressure), but hypertension is more common in people who have a family history of the condition.

Critical Thinking: You Decide *(Key in Appendix IV)*

1. Witnesses in a court of law are asked to swear "to tell the truth, the whole truth, and nothing but the truth." What are some of the problems inherent in the question? Can you think of a better alternative?

Self-Quiz *(Answers in Appendix III)*

1. The complex patterns of structural organization characteristic of life are based on instructions contained in _____ .

2. _____ is the ability of cells to extract and transform energy from the environment and use it to maintain themselves, grow, and reproduce.

3. _____ is a state in which the body's internal environment is being maintained within a tolerable range. This state depends on _____ , which are cells or structures that detect specific aspects of the environment.

4. Diverse structural, functional, and behavioral traits are considered to be _____ to changing conditions in the environment.

5. The capacity to evolve is based on variations in traits, which originally arise through _____ .

6. Each of us has some number of traits that also were present in our great-great-great-great-grandmothers and grandfathers. This is an example of _____ .
a. metabolism   c. a control group
b. homeostasis   d. inheritance

7. A scientific approach to explaining some aspect of the natural world includes all of the following except _____ .
a. hypothesis   c. faith and simple consensus
b. testing   d. systematic observations

8. A related set of hypotheses that collectively explain some aspect of the natural world is a scientific _____ .
a. prediction   d. authority
b. test   e. observation
c. theory

Key Terms

adaptive trait *8*	hypothesis *9*
ATP *6*	inheritance *8*
cell *5*	metabolism *6*
control group *11*	mutation *8*
DNA *4*	prediction *9*
energy *4*	reproduction *7*
evolution *2*	theory *9*
experiment *11*	variable *11*
homeostasis *7*	

Readings

Committee on the Conduct of Science. 1989. *On Being a Scientist*. Washington, D.C.: National Academy of Sciences. Paperback.

Larkin, Tim. June 1985. "Evidence vs. Nonsense: A Guide to the Scientific Method." *FDA Consumer*.

Rosenthal, Elizabeth. October 1992. "Dead Complicated." *Discover*. Is a person with no functional brain, maintained by life support machinery, alive or dead? The author, a physician, discusses this and other questions society faces as a result of modern, high-technology medical practices.

1 CHEMICAL FOUNDATIONS FOR CELLS

The Chemistry In and Around You

You live in a chemical world. You, and every other human being, would quickly die if it weren't for countless chemical reactions proceeding within your brain, muscles, lungs, intestines, and other body parts. We also have become major "chemistry consumers." Researchers have learned (or can figure out) what substances are made of, how they can be transformed into *different* substances, and what it takes to accomplish the transformations. These discoveries have led to synthetic fabrics, vaccines, antibiotics, and the plastic components of refrigerators, computers, and cars. They have yielded artificial sweeteners and "fake" fats for diet foods. They have produced the fertilizers, herbicides, and insecticides that help maintain food supplies for the 5.6 billion people on earth.

As a society, we've also discovered that our decisions about chemicals can cause monumental problems. Pesticides are just one example. These powerful chemicals may kill unwanted insects and vegetation, but humans also can inhale them, ingest trace amounts with food, or absorb them through the skin (Figure 1.1). Exposure to high doses of certain pesticides can trigger hives, rashes, asthma, and other moderate allergic reactions, or, in some people, life-threatening immune responses. We do not know for sure the effects of long-term exposure to the trace amounts in and on foods such as tomatoes, oranges, and beef (to name just a few), although a recent study by the National Academy of Sciences reported that chemical components of most fungicides, more than half of all herbicides, and nearly a third of all insecticides may cause cancer in humans.

Figure 1.1a Cropduster with its rain of pesticides. About two-thirds of the pesticides used on U.S. croplands are applied this way.

Toxic chemical by-products of industrial processes, including heavy metals such as lead and mercury and compounds known as dioxins, can seriously pollute water and soil and can poison people exposed to them.

Most of the chemical reactions that keep you alive involve four types of biologically active molecules: proteins, lipids (fats and their relatives), carbohydrates, and nucleic acids (components of DNA and some other key substances). Those molecules are the foundation for the structure and function of each of your cells. The body uses them as building materials, as catalysts that speed up chemical reactions, and as energy storehouses. Each type of biological molecule is built upon a skeleton of linked carbon atoms.

The basic concepts in this chapter will help you understand processes that go on in your own body, every minute of your life. They provide a foundation of knowledge for understanding mechanisms of homeostasis that will be discussed in subsequent chapters. These concepts will also help you, as a "chemistry consumer," assess the benefits and risks of chemical decisions you and others choose to make.

KEY CONCEPTS

1. All matter in the universe consists of atoms. Atoms, in turn, consist of protons, electrons, and neutrons. Most atoms can lose, gain, and share electrons with one another. These interactions result in chemical bonds that hold atoms together. They are the basis for the structural organization and activities of all living things, including human beings.

2. Oxygen, carbon, hydrogen, and nitrogen are the most abundant kinds of atoms in organisms. Ionic, hydrogen, and covalent bonds are the main chemical bonds between these and other atoms.

3. Complex carbohydrates, lipids, proteins, and nucleic acids are the large "molecules of life." Cells assemble and use them as structural materials and energy stores, as molecules that speed up chemical reactions (enzymes), and as libraries of hereditary information.

4. The molecules of life are assembled from four kinds of smaller building blocks: simple sugars such as glucose, fatty acids, amino acids, and nucleotides.

Figure 1.1b Pesticides, fungicides, herbicides, and other chemicals make their way to consumers through a variety of routes, including the supermarket.

ORGANIZATION OF MATTER

Matter is anything that occupies space and has mass ("weight"). All the solids, liquids, and gases within and around you are forms of matter that consist of one or more elements. An *element* is any substance that cannot be broken down to a different substance, at least by ordinary means. Ninety-two elements occur naturally on earth. Of these, just four—oxygen, carbon, hydrogen, and nitrogen—make up most of the human body (Figure 1.2). These and all other known elements are organized into groups with similar chemical properties. They are listed in the periodic table in Appendix A.

An **atom** is the smallest unit of matter that is unique to a particular element. (A single atom of gold is still gold, but the component parts of that atom—electrons, protons, and neutrons—are *not* gold.) A **molecule** consists of two or more atoms joined together. For instance, the oxygen you breathe is a molecule of two oxygen atoms. Chemical shorthand for this arrangement is O_2. A molecule of pure water consists of two hydrogen atoms and one oxygen and is represented by the chemical formula H_2O.

Water is an example of a **compound**, a substance in which atoms of *different* elements are always present in the same proportions.

The Structure of Atoms

Atoms are almost unimaginably small, yet they consist of even smaller subatomic particles (Figure 1.3). Most of an atom consists of open space around a tiny core region, or *nucleus*. The nucleus consists of **protons** and **neutrons**. (Hydrogen, the simplest atom, is an exception and has no neutrons.) Neutrons and protons account for the atom's mass. Neutrons carry no electrical charge, but protons carry a positive charge (p^+). Moving around the nucleus are one or more **electrons**, which carry a negative charge (e^-). An atom has an equal number of electrons and protons. This means that the atom has no net charge overall because the positive and negative charges cancel each other out.

An atom's *atomic number* is the number of protons in the nucleus. This number differs for each element. For example, an atom of hydrogen—and only hydrogen—has one proton; its atomic number is 1. A carbon atom has six protons; its atomic number is 6. *Mass number* or *atomic weight* is the total number of protons and neutrons in the nucleus. Thus the mass number of a carbon atom with six protons and six neutrons is 12.

If we know the atomic number of an atom—that is, the number of protons in the nucleus—then we also know how many electrons it has. In a subsequent section you will see that this information suggests whether a given type of atom can lose, gain, or share electrons.

EARTH'S CRUST		HUMAN	
Oxygen	46.6	Oxygen	65
Silicon	27.7	Carbon	18
Aluminum	8.1	Hydrogen	10
Iron	5.0	Nitrogen	3
Calcium	3.6	Calcium	2
Sodium	2.8	Phosphorus	1.1
Potassium	2.6	Potassium	0.35
Magnesium	2.1	Sulfur	0.25
Other		Sodium	0.15
elements:	1.5	Chlorine	0.15
		Magnesium	0.05
		Iron	0.004
		Iodine	0.0004

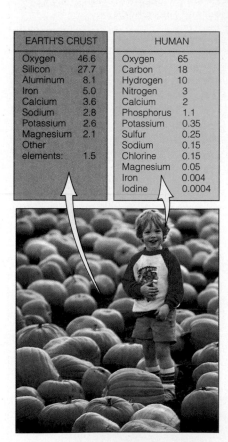

Figure 1.2 Proportions of different elements in the earth's crust and in the human body, as percentages of the total weight of each.

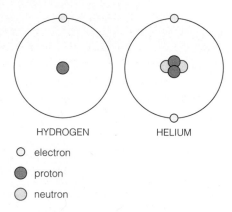

HYDROGEN   HELIUM

○ electron
● proton
○ neutron

Figure 1.3 Model of atomic structure, using a hydrogen atom and a helium atom as examples. For all elements except hydrogen, the nucleus has one or more neutrons as well as protons. Electrons occupy orbitals around the nucleus. At this scale, the nucleus actually would be an invisible speck at the atom's center.

Isotopes: Varying Forms of Atoms

All atoms of an element have the same number of protons (atomic number), but they may vary in how many neutrons they contain. Such varying atoms of an element have different mass numbers and are called **isotopes**. Thus "a carbon atom" might be carbon 12 (six protons, six neutrons), carbon 13 (six protons, seven neutrons), or carbon 14 (six protons, eight neutrons). These can be written as ^{12}C, ^{13}C, and ^{14}C. Radioactive isotopes, or *radioisotopes*, are unstable and tend to spontaneously emit subatomic particles or energy (called gamma rays) until a stable isotope is formed. *Focus on Science* describes some uses of radioisotopes in research and medicine.

All atoms of an element have the same number of electrons and protons, but they can vary in the number of neutrons. Such variant forms of atoms of the same element are called isotopes.

BONDS BETWEEN ATOMS

We turn now to interactions between atoms. Figure 1.4 summarizes a few conventions used to describe these events.

The Nature of Chemical Bonds

A chemical bond is an attraction between the electron structures of two (or more) atoms. It is the force that holds atoms together in molecules. Whether an atom will bond with another depends on how many electrons it has and how those electrons are arranged in the atom.

Electrons, with their negative charges, are attracted to an atom's protons, which carry positive charges. Electrons are also repelled by other electrons that may be present, just as like poles of a magnet repel each other. They move rapidly within *orbitals*, which are regions of space around the nucleus in which electrons are likely to be at any instant. In fact, orbitals occupy most of the atom's space. Each orbital has room for two electrons at most.

It is convenient to think of orbitals as occurring in a series of hollow *shells* around the nucleus (Figure 1.5). The orbital in the first shell is ball-shaped, and it can hold a maximum of two electrons. Moving outward, the next shell has room for four orbitals, and so it can hold up to

We use symbols for elements when writing *formulas*, which identify the composition of compounds. For example, water has the formula H_2O. Symbols and formulas are used in *chemical equations*, which are representations of reactions among atoms and molecules.

In written chemical reactions, an arrow means "yields." Substances entering a reaction (reactants) are to the left of the arrow. Reaction products are to the right. For example, the reaction between hydrogen and oxygen that yields water is summarized this way:

$$2H_2 \quad + \quad O_2 \quad \longrightarrow \quad 2\,H_2O$$

4 hydrogens   2 oxygens   4 hydrogens, 2 oxygens

Note that there are as many atoms of each element to the right of the arrow as there are to the left. Although atoms are combined in different forms, none was consumed or destroyed. The total mass of all products of any chemical reaction equals the total mass of all its reactants. All equations used to represent chemical reactions, including reactions in cells, must be balanced this way.

Figure 1.4 Chemical bookkeeping.

Figure 1.5 How electrons are arranged in atoms. One or at most two electrons occupy a ball-shaped volume of space (an orbital) close to the nucleus. This orbital occupies the space within a "shell." In the next shell, there can be up to eight more electrons (two in each of four orbitals). Atoms with high atomic weights, and hence with many protons and electrons, may have many shells, some with more than four orbitals. However, the outermost shell of an atom is always complete with four orbitals and eight electrons.

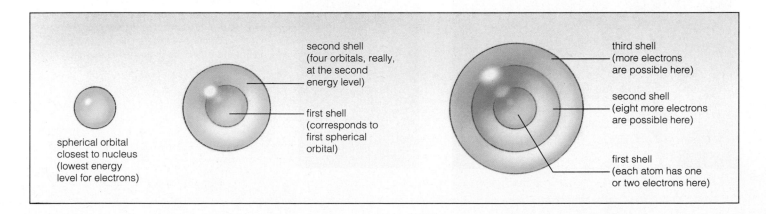

Tracking Chemicals and Saving Lives—Some Uses of Radioisotopes

Radioisotopes are unstable atoms with dissimilar numbers of protons or neutrons. They can capture or emit electrons, subatomic particles, or energy (as gamma rays). This spontaneous process, called radioactive decay, continues until the isotope changes to a new, stable one that is not radioactive.

Tracking Chemicals   All isotopes of an element have the same number of electrons, so they all interact with other atoms in the same way. Accordingly, cells can use any isotope of carbon for a given metabolic reaction. Scintillation counters and other devices can detect emissions from radioisotopes. Thus, with carefully controlled dosages, isotopes can be used as *tracers*. A tracer can help identify the pathways or destinations of a substance that has entered the human body. Think about the body's thyroid, the only gland that takes up iodine. After a tiny amount of ^{123}iodine can be injected into a patient's bloodstream, the thyroid can be scanned with a scintillation counter. Figure *a* shows examples of what such radioisotope scans may reveal.

Saving Lives   In nuclear medicine, radioisotopes are used under carefully controlled conditions to diagnose and treat diseases. Some cancer treatments rely on the cell-destroying capacities of radioisotopes. In radiation therapy, localized cancers are deliberately bombarded with ^{226}radium or ^{60}cobalt. Patients with irregular heartbeats use pacemakers, which are powered by energy emitted from ^{238}plutonium. (This dangerous radioisotope is sealed in a case to prevent emissions from damaging body tissues.) With positron-emission tomography (PET), radioisotopes provide information about abnormalities in the metabolic functions of specific tissues. The radioisotopes, once attached to glucose or some other biological molecule, are injected into a patient, who is moved into a PET scanner (Figure *b*). When cells in certain tissues absorb glucose, the radioisotopes give off energy that can be used to produce a vivid image of variations in metabolic activity.

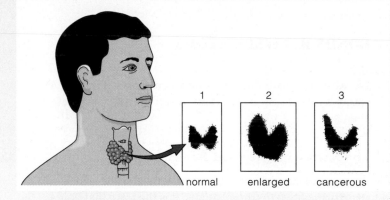

Figure *a* Scans of human thyroid glands after ^{123}iodine was injected into the bloodstream. The thyroid normally takes up iodine (including radioisotopes) and uses it in hormone production. (1) Uptake by a normal gland. (2) Enlarged gland of a patient with a thyroid disorder. (3) Cancerous thyroid gland.

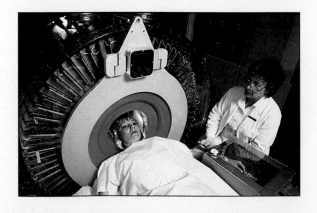

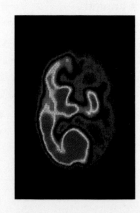

Figure *b* Patient being moved into a PET scanner. The image at right shows a vivid image of a brain scan of a child with a severe neurological disorder. The different colors signify differences in metabolic activity in one half of the brain; the other half shows little activity.

Element	Symbol	Atomic Number	Most Common Mass Number	Number of Electrons in Outermost Shell
Hydrogen	H	1	1	1
Carbon	C	6	12	4
Nitrogen	N	7	14	5
Oxygen	O	8	16	6
Sodium	Na	11	23	1
Magnesium	Mg	12	24	2
Phosphorus	P	15	31	5
Sulfur	S	16	32	6
Chlorine	Cl	17	35	7
Potassium	K	19	39	1

hydrogen
atom

sodium
atom

Figure 1.6 Distribution of electrons (yellow dots) in hydrogen and sodium atoms. Each atom has a lone electron (and room for more) in its outermost shell. Atoms having such partly filled shells tend to enter into reactions with other atoms.

eight electrons. An atom will have as many shells as required to accommodate all of its electrons, but the maximum number of electrons in an outermost shell is always eight.

Many atoms, including most biologically important types, do not have enough electrons to completely fill the outermost shell. For instance, atoms of hydrogen, oxygen, carbon, and nitrogen are this way (Table 1.1). Having a filled outer shell is the most stable state for atoms. Thus, *atoms with an unfilled outer shell tend to form chemical bonds with other atoms and so fill their outer shell* (Figure 1.6). The single shell of hydrogen and helium atoms is full when the orbital within it contains two electrons. Other types of atoms that have unfilled outer shells follow the *octet rule*—they participate in chemical bonds that result in a filled outer shell containing eight electrons. As we shall see next, in different types of bonds electrons are either added to an atom's outer shell, removed from it, or shared between atoms.

Hydrogen, oxygen, carbon, and other biologically important atoms have unfilled orbitals in their outermost shell. To fill the outermost shell (two or eight electrons), atoms tend to form chemical bonds with other atoms. They do this by gaining, losing, or sharing one or more of their electrons.

Ionic Bonding: Electrons Gained or Lost

When an atom (or group of atoms) loses or gains one or more electrons, it becomes positively or negatively charged. A charged atom is called an **ion**.

Atoms lose or gain electrons when another nearby atom can accept or donate electrons. Because one loses and one gains, *both* atoms become ionized. An **ionic bond** is a strong attraction between two oppositely charged ions. Table salt, NaCl, consists of ions of sodium (Na^+) and chloride (Cl^-) held together by the force of this

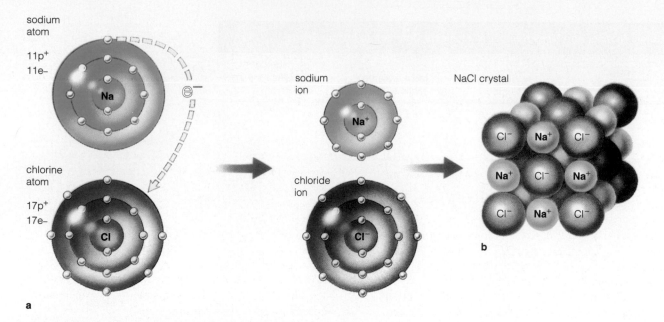

sodium
atom

11p⁺
11e–

Na

chlorine
atom

17p⁺
17e–

Cl

a

sodium
ion

Na⁺

chloride
ion

Cl⁻

NaCl crystal

Cl⁻ Na⁺ Cl⁻

Na⁺ Cl⁻ Na⁺

Cl⁻ Na⁺ Cl⁻

b

Figure 1.7 Ionic bonding in sodium chloride (NaCl). A crystal of table salt is held together by many ionic bonds. (**a**) A sodium atom (with a lone electron in its outermost shell) reacts with a chlorine atom (with an electron vacancy in its outermost shell). Both atoms become ionized by the electron transfer. (**b**) The attraction of their opposite charges holds many ions of sodium (Na⁺) and chloride (Cl⁻) together in crystals of table salt (NaCl).

attraction (Figure 1.7). The sodium fluoride in "anti-cavity" toothpastes is another example of a compound held together by ionic bonds.

Whenever a substance taking part in a chemical reaction gives up electrons, it is said to be *oxidized*. As we will see in Chapter 2, many vital chemical reactions in cells involve this type of ionic bonding. Oxidation reactions can also be extremely detrimental to cells, as this chapter's *Choices* essay describes.

An ion is an atom that has gained or lost one or more electrons. It has an overall negative charge if an electron is gained, and an overall positive charge if an electron is lost.

In an ionic bond, a positive and a negative ion are linked by the mutual attraction of opposite charges.

Covalent Bonding: Electrons Shared

Sometimes one atom cannot pull electrons completely away from another, and the two atoms end up *sharing* electrons. A pair of electrons shared between two atoms is a **covalent bond**. In depicting bonds, a single line represents a single covalent bond between two atoms, as in H—H. In a double covalent bond, two atoms share two pairs of electrons. This happens in an O_2 molecule, expressed as O==O. In a triple covalent bond (such as N≡N), two atoms share three pairs of electrons.

There are two kinds of covalent bonds: nonpolar or polar. In a *nonpolar* covalent bond, both atoms exert the same pull on shared electrons, and there is no difference in charge at the two ends of the bond. The H—H molecule shows this kind of balance. Its two hydrogen atoms, with one proton each, attract the shared electrons equally.

A *polar* covalent bond involves atoms of different elements. In this type of covalent bond, one atom exerts a stronger pull on shared electrons. Why? An atom's power to attract electrons is called *electronegativity*. When a more electronegative atom is joined with one that is less electronegative, part of the molecule comes to have a slightly negative charge, whereas the other part has a slightly positive charge. The molecule has no overall net charge, but the charge is distributed unevenly between the two ends of the polar bond.

A water molecule has two polar covalent bonds (Figure 1.8). Its electrons are more attracted to the oxygen because oxygen has more protons than hydrogen does. The molecule carries no net charge, yet because of its polarity it can weakly attract other atoms.

"Radical" Research—The Facts About Antioxidants

Almost daily, reports in newspapers and magazines and on television proclaim that this or that food, vitamin, or drug will benefit—or harm—our health. As consumers in a chemical world, people are faced with the daunting task of sorting out facts and reasonable possibilities from fiction. You may have read that certain substances, including the yellow-red pigment beta carotene and vitamins E and C, are "antioxidants" that help protect the body against cancer and heart disease and may even slow the aging process. Let's consider the biological facts behind these reports.

An *antioxidant* is any substance that thwarts oxidation; that is, it prevents an atom or molecule from losing one or more electrons to another atom or molecule (page 19). Researchers are in fact probing the human health impacts of antioxidants, which can help counteract the harmful effects of molecules called *free radicals*. Free radicals result from normal cell operations; they are also produced by the effects of cigarette smoke, ultraviolet radiation in sunlight, and other chemical assaults on the body.

How does a free radical do its dirty work? Free radicals contain oxygen atoms that lack a full complement of electrons in their outer shell. As a result, a free radical is extremely reactive: It will oxidize another, stable molecule, "stealing" an electron to fill the empty slot—and disrupt or destroy that molecule in the process. Potential targets include DNA molecules and proteins and lipids that are the main components of cell membranes. Antioxidants curb this destructive process by combining with free radicals before they do damage. Known antioxidants include beta carotene (found in carrots and leafy green vegetables, among other foods) and vitamins C and E.

Researchers are investigating whether free-radical activity may account for changes associated with lung damage from cigarette smoke and smog, with certain degenerative diseases such as rheumatoid arthritis, with the coronary artery disease called atherosclerosis (page 160), with damage to human sperm cells, and even with changes in cell membranes thought to be linked to aging.

Figure 1.8 A view you will see only on planet earth: an ocean of water, a substance that is absolutely vital for life. (**a**) Polarity of a water molecule. (**b**) Hydrogen bonds between water molecules in liquid water, signified by dashed lines.

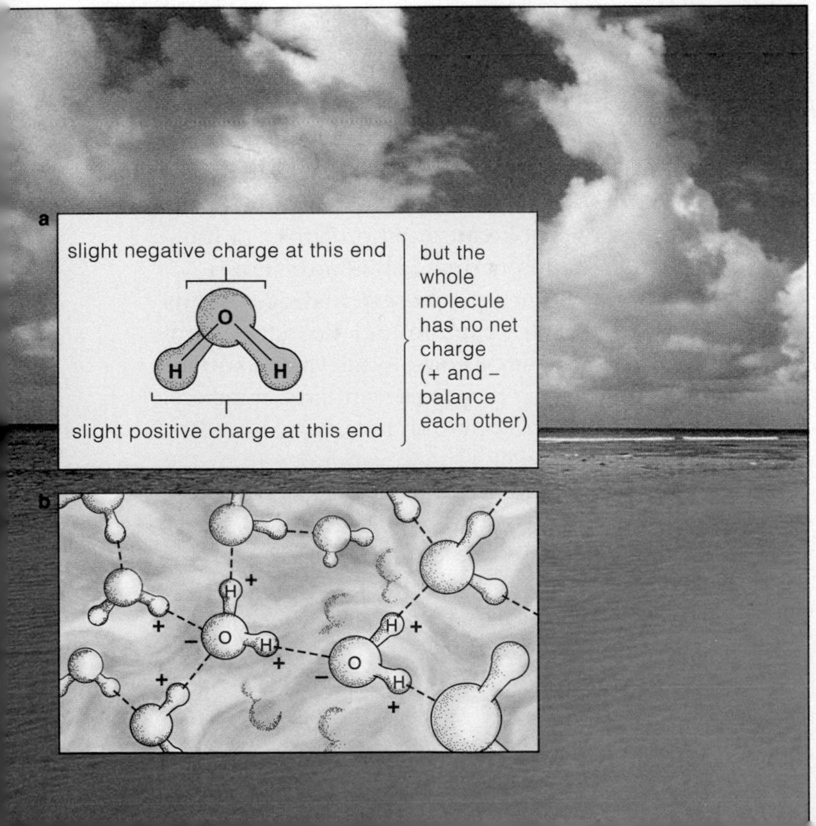

a
slight negative charge at this end | but the whole molecule has no net charge (+ and – balance each other)
slight positive charge at this end

b

In a covalent bond, atoms share a pair of electrons.

If electrons are shared equally between the two atoms, the bond is nonpolar. If they are not shared equally, the bond is polar (slightly positive at one end and slightly negative at the other).

Hydrogen Bonding

In a **hydrogen bond**, a slightly negative atom of a polar molecule interacts weakly with a neighboring hydrogen atom that is already part of a polar covalent bond. The hydrogen, with its slight positive charge, is attracted to the other atom's slight negative charge.

Hydrogen bonds are common in cells. For example, DNA, the genetic material, consists of two parallel strands of chemical units. Hydrogen bonds link the strands, and even though each hydrogen bond is weak, collectively the bonds help stabilize DNA molecules. Hydrogen bonds also hold water molecules together (Figure 1.8) and give

Hydrogen as a Nonpolluting Fuel

Beyond its roles as a component of water and other compounds, hydrogen holds major promise as a nonpolluting
fuel in an energy-hungry world.

Hydrogen is produced easily by passing an electrical
current through water, which breaks down into oxygen
and hydrogen gases. It is much easier to store than electricity. It can be stored in a pressurized tank or in metal
hydrides—metal powders that absorb gaseous hydrogen
and release it when heated for use as a fuel in a car, furnace, or an electricity-producing fuel cell. Unlike gasoline,
solid metallic hydrogen compounds will not explode or
burn if a vehicle's tank is ruptured in an accident. By
weight, hydrogen has about 2.5 times the energy of gasoline, making it an especially attractive aircraft fuel. A test
fleet of 200 cars, a bus, and even a large jet airplane have
been running on hydrogen fuel for several years in the former Soviet Union.

Although the production of hydrogen gas requires
another form of energy (such as electricity) to break down
water molecules, advocates say that this energy input does
not outweigh the benefits of using hydrogen as a fuel, in
part because of the clear environmental gains. Hydrogen is
a clean-burning fuel. Its large-scale use would eliminate air
and water pollution caused by extracting, transporting,
and burning fossil fuels and would reduce the threat of
global warming (page 504). It would also reduce the threat
of wars over dwindling oil supplies. Nuclear power plants,
which produce large amounts of radioactive wastes, could
be phased out.

Proponents say that the key to the "hydrogen revolution" is the development of affordable methods of using
solar energy to produce electricity that in turn can be used
to make hydrogen gas. If this can be accomplished, the
United States could probably make a full transition to a
solar-hydrogen economy by 2045. Many U.S. oil companies, electric utilities, and automobile manufacturers currently view hydrogen fuel as a threat to their short-term
economic interests. Meanwhile, in 1992 a Japanese
automaker unveiled a prototype car that runs on hydrogen
released slowly from metal hydrides heated by the car's
radiator coolant. In Germany, researchers are developing
small hydrogen-producing systems that could be used by
individuals for home heating, cooking, and other power
needs. Germany and Saudi Arabia have built a large solar-
hydrogen plant, and other developments are in the works.

water many of the life-sustaining properties described in
the next section. *Focus on Environment* describes another
potential role for hydrogen.

**In a hydrogen bond, a polar atom or molecule interacts
weakly with a neighboring hydrogen atom that is already
part of a polar covalent bond.**

PROPERTIES OF WATER

By weight, your body is about two-thirds water, which is
essential to maintaining the shape and internal structure
of cells. Water also makes up over 90 percent of the fluid
in your blood. Moreover, many of the chemical reactions
that sustain life require water as a reactant or can occur
only after other substances dissolve in water. Not surprisingly, water's central role in life processes is linked to
properties of the water molecule.

The polarity of water molecules allows them to
hydrogen-bond with one another and with other polar
substances. All polar molecules are attracted to water;
that is, they are **hydrophilic** ("water-loving"). By contrast,
water's polarity repels oil and other nonpolar substances,
which are **hydrophobic** ("water-fearing"). As we shall see
in Chapter 2, hydrophobic interactions help organize the
rather oily, sheetlike layers of cell membranes.

Hydrogen bonds between water molecules also
enable water to absorb a great deal of heat energy before
it heats up significantly or evaporates. This is because, in
a given volume of water, multiple hydrogen bonds link
individual H_2O molecules, and a large amount of heat
must be applied to stretch or break the bonds in large
numbers. Water can be used to cool a hot automobile
engine for the same reason. Water's ability to absorb
much heat before itself becoming hot also helps stabilize
temperatures in cells, which are mostly water. The chemical reactions in cells constantly produce heat, yet cells
must stay relatively cool because cell proteins can function properly only within narrow temperature limits.

When enough heat energy is present, hydrogen bonds
between water molecules rupture and do not reform.
Then water evaporates—molecules at the water's surface
escape into the air. When they depart in large numbers,
heat energy is lost. You cool off on hot, dry days when
sweat, which is 99 percent water, evaporates from the
more than 2.5 million sweat glands in the skin covering
your body.

Water is also a superb *solvent*, which means ions and
polar molecules readily dissolve in it. Dissolved substances are called **solutes**. A substance "dissolves" as

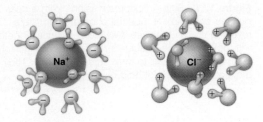

Figure 1.9 Clusters of water molecules around charged ions.

clusters of water molecules form around its individual ions or molecules and break the bonds between them (Figure 1.9). This happens to solutes in cells, between cells, in blood, and in all other body fluids. Most chemical reactions in the body occur in water-based solutions.

ACIDS, BASES, AND SALTS

Acids and Bases

Some substances release protons when they dissolve in water. Free (unbound) protons are called **hydrogen ions**, symbolized as H^+. A substance that releases H^+ in water is an **acid**.

When you consume food, stomach cells are stimulated to secrete hydrochloric acid (HCl), which separates into H^+ and Cl^- in water. These ions make the fluid in your stomach more acidic, and the increased acidity switches on enzymes that can break down food particles. It also helps kill bacteria that may be present. However, if you eat too much of certain kinds of foods during a meal, you may end up with an "acid stomach" and need to reach for an antacid.

Milk of magnesia is one kind of antacid. When dissolved, it releases magnesium ions (Mg^{+2}) and hydroxide ions (OH^-). Hydroxide ions may then combine with excess hydrogen ions in your stomach fluid, neutralizing the acid. Any substance that releases hydroxide ions in water is a **base**.

The pH Scale

The **pH scale** is used to measure the concentration of unbound hydrogen ions in blood, water, and other solutions. Figure 1.10 shows the scale, which ranges from 0 (most acidic) to 14 (most basic). A solution with a pH value of 7 is neutral; it has as many hydrogen ions (H^+) as hydroxide ions (OH^-). An acid solution has more hydrogen ions than hydroxide ions, whereas a basic solution has proportionately more hydroxide ions. A neutral solution becomes more acidic when H^+ ions are added to it and more basic when H^+ ions are removed from it.

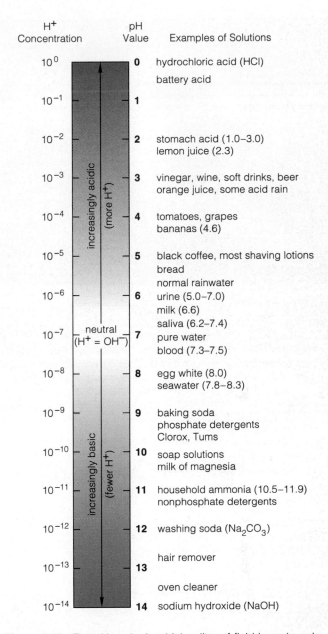

Figure 1.10 The pH scale, in which a liter of fluid is assigned a number according to the number of hydrogen ions in it. The scale ranges from 0 (most acidic) to 14 (most basic). A change of only one unit on the pH scale means a tenfold change in hydrogen ion concentration. For example, gastric fluid in your stomach is ten times more acidic than vinegar, and vinegar is ten times more acidic than tomatoes.

As Figure 1.10 suggests, each unit of change on the pH scale corresponds to a tenfold increase or decrease in H^+ concentration. The pH of human blood and most tissue fluids ranges between 7.35 and 7.45. The normal pH inside cells ranges from about 7.0 to 7.2, or from neutral to slightly basic.

Compared to cells, the pH of an environment may be much higher or lower. River water ranges between 6.8

Figure 1.11 Sulfur dioxide emissions from a coal-burning power plant. Special photographic filters revealed these otherwise invisible emissions. Together with other airborne pollutants, sulfur dioxides dissolve in atmospheric water to form acidic solutions. They are a major component of acid rain.

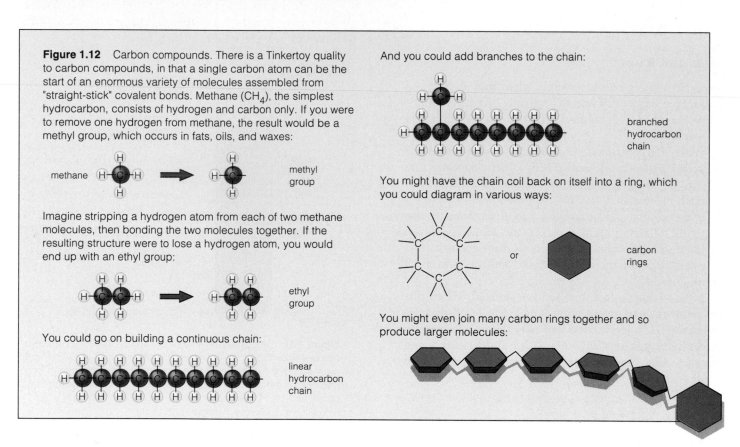

Figure 1.12 Carbon compounds. There is a Tinkertoy quality to carbon compounds, in that a single carbon atom can be the start of an enormous variety of molecules assembled from "straight-stick" covalent bonds. Methane (CH_4), the simplest hydrocarbon, consists of hydrogen and carbon only. If you were to remove one hydrogen from methane, the result would be a methyl group, which occurs in fats, oils, and waxes:

methane   methyl group

Imagine stripping a hydrogen atom from each of two methane molecules, then bonding the two molecules together. If the resulting structure were to lose a hydrogen atom, you would end up with an ethyl group:

ethyl group

You could go on building a continuous chain:

linear hydrocarbon chain

And you could add branches to the chain:

branched hydrocarbon chain

You might have the chain coil back on itself into a ring, which you could diagram in various ways:

or   carbon rings

You might even join many carbon rings together and so produce larger molecules:

and 8.6. Airborne industrial wastes, such as the sulfur dioxide (SO_2) emissions pictured in Figure 1.11, can react with water vapor and lower its pH significantly. One result is acid rain, which causes serious environmental damage (Chapter 25).

Dissolved Salts

Inside organisms, acids commonly combine with bases. The results are ionic compounds called **salts**. Salts often dissolve and form again, depending on pH. Sodium chloride (table salt), which is only one of many salts, forms and dissolves this way:

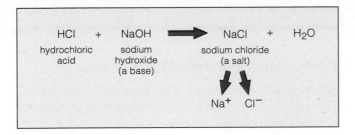

HCl + NaOH $\longrightarrow$ NaCl + H₂O
hydrochloric
acid
sodium
hydroxide
(a base)
sodium chloride
(a salt)

$\downarrow$ $\downarrow$
Na⁺ Cl⁻

Many other salts also dissolve into ions, and such ions serve important functions in cells. For instance, sodium ions (Na^+) and potassium ions (K^+) are involved in the transmission of "messages" through the nervous system. They are components of *electrolytes*, substances that dissociate into ions in solution and that can conduct an electric current.

Buffers

Chemical reactions in cells are sensitive to even slight shifts in pH. Yet many reactions continually use and produce hydrogen ions. Helping to maintain internal pH within narrow limits are compounds that act as buffers. A **buffer** is any molecule that can combine with hydrogen ions, release them, or both. It responds to acid conditions by taking up H^+, and to basic conditions by releasing H^+. Bicarbonate (HCO_3^-) is an important buffer in blood. When the blood becomes too acidic, bicarbonate combines with H^+ to form carbonic acid (H_2CO_3):

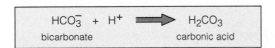

HCO_3^- + H^+ $\longrightarrow$ H_2CO_3
bicarbonate   carbonic acid

Conversely, bicarbonate releases H^+ ions when blood is not acid enough. This buffering action is crucial. For instance, some lung diseases interfere with carbon dioxide elimination, so the blood level of carbonic acid (hence of H^+) rises. This abnormal condition, a form of *acidosis*, makes breathing difficult and weakens the body.

CARBON COMPOUNDS

By far, oxygen, hydrogen, and carbon are the most abundant elements in living things. Much of the oxygen and hydrogen is linked together in water molecules. Notable amounts also are bonded to carbon, the most important structural element in the body.

A carbon atom can bond covalently to as many as four other atoms (Figure 1.12). Commonly, carbon atoms are joined one after another into a chain or ring, and most have hydrogen atoms attached to them. Because the carbons share pairs of electrons equally, the chains and rings are stable backbones for molecules. Any molecule with a carbon backbone is an *organic compound*. Such backbones are not present in water, carbon dioxide, and most other simple, *inorganic compounds*.

Functional Groups

Other atoms besides hydrogen also bond to carbon backbones. Such atoms are **functional groups** (Figure 1.13), which influence the behavior of organic compounds. For example, sugars have hydroxyl groups (—OH). Water reacts with hydroxyl groups—it hydrogen-bonds with them—so sugars dissolve in water.

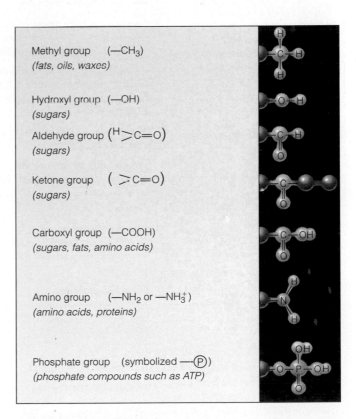

Methyl group (—CH₃)
(fats, oils, waxes)

Hydroxyl group (—OH)
(sugars)

Aldehyde group (^{H}$>$C=O)
(sugars)

Ketone group ($>$C=O)
(sugars)

Carboxyl group (—COOH)
(sugars, fats, amino acids)

Amino group (—NH₂ or —NH₃⁺)
(amino acids, proteins)

Phosphate group (symbolized —Ⓟ)
(phosphate compounds such as ATP)

Figure 1.13 Major functional groups that confer distinctive properties upon carbon compounds.

Families of Small Organic Compounds

The large organic compounds in cells—the "molecules of life"—are complex carbohydrates, lipids, proteins, and nucleic acids (Table 1.2). All are assembled from four families of small organic compounds, each with no more than about 20 carbon atoms. These are the simple sugars, fatty acids, amino acids, and nucleotides we will consider shortly.

The reactions that assemble and disassemble large biological molecules occur with the help of enzymes. **Enzymes** are a class of proteins that catalyze, or speed up, reactions between specific substances.

Simple organic compounds serve as energy sources and building blocks for the large organic compounds in cells, including complex carbohydrates, lipids, proteins, and nucleic acids.

Condensation and Hydrolysis Reactions

Small organic compounds' molecules are "put together" into larger ones by **condensation** reactions that produce covalent bonds between the molecules. Enzymes catalyze these reactions, in which one molecule loses a hydrogen

Table 1.2 Summary of the Main Carbon Compounds In Living Things		
Category	Main Subcategories	Some Examples and Their Functions
Carbohydrates *contain an aldehyde or a ketone group and one or more hydroxyl groups*	**Monosaccharides (simple sugars)**	Glucose — Structural roles, energy source
	Oligosaccharides	Sucrose (a disaccharide) — Form of sugar transported in plants
	Polysaccharides (complex carbohydrates)	Starch — Energy storage Cellulose — Structural roles
Lipids *are largely hydrocarbon, generally do not dissolve in water but dissolve in nonpolar substances*	**Lipids with fatty acids:** *Glycerides:* one, two, or three fatty acid tails attached to glycerol backbone	Fats (e.g., butter) Oils (e.g., corn oil) — Energy storage
	Phospholipids: phosphate group, another polar group, and (often) two fatty acids attached to glycerol backbone	Phosphatidylcholine — Key component of cell membranes
	Waxes: long-chain fatty acid tails attached to alcohol	Waxes in cutin — Water retention by plants
	Lipids with no fatty acids: *Steroids:* four carbon rings; the number, position, and type of functional groups vary	Cholesterol — Component of animal cell membranes, can be rearranged into other steroids (e.g., vitamin D, sex hormones)
Proteins *are polypeptides (up to several thousand amino acids, covalently linked)*	**Fibrous proteins:** Individual polypeptide chains, often linked into tough, water-insoluble molecules	Keratin — Structural element of hair, nails Collagen — Structural element of bones and cartilage
	Globular proteins: One or more polypeptide chains folded and linked into globular shapes; many roles in cell activities	Enzymes — Increase in rates of reactions Hemoglobin — Oxygen transport Insulin — Control of glucose metabolism Antibodies — Tissue defense
Nucleic Acids (and Nucleotides) *are chains of units (or individual units) that each consist of a five-carbon sugar, phosphate, and a nitrogen-containing base*	**Adenosine phosphates**	ATP — Energy carrier
	Nucleotide coenzymes	NAD^+, $NADP^+$ — Transport of protons (H^+) and electrons from one reaction site to another
	Nucleic acids: Chains of thousands to millions of nucleotides	DNA, RNAs — Storage, transmission, translation of genetic information

Condensation

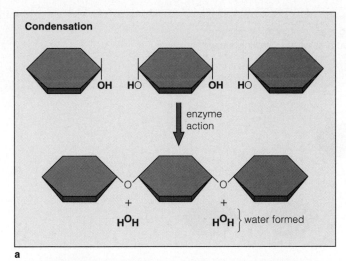

a

Hydrolysis

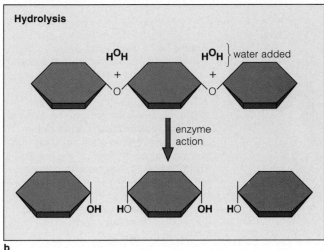

b

Figure 1.14 (**a**) Condensation. In this generalized example, three molecules covalently bond into a larger molecule and two water molecules are formed. (**b**) Hydrolysis. Two covalent bonds of a molecule are split, and H⁺ and OH⁻ derived from water molecules become bonded to the molecular fragments.

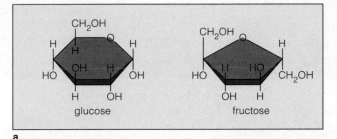

a

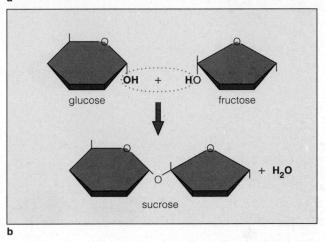

b

Figure 1.15 (**a**) Glucose and fructose, two monosaccharides. (**b**) Condensation of glucose and fructose into sucrose, a disaccharide.

atom and another molecule loses an —OH group; a bond then forms at the exposed sites (Figure 1.14a). The unbound parts, now H⁺ and OH⁻ ions, may combine to form a water molecule (H_2O). Because this kind of reaction often forms water as a by-product, condensation is sometimes called *dehydration synthesis*.

A **hydrolysis** ("water-splitting") reaction is like condensation in reverse. It splits a molecule into two or more parts by breaking covalent bonds. At the same time, H⁺ and OH⁻ derived from water become attached to the exposed sites (Figure 1.14b). Enzymes commonly catalyze hydrolysis reactions, which break apart all of the compounds we consider next.

Carbohydrates

A **carbohydrate** is a simple sugar or a larger molecule composed of sugar units. Cells use carbohydrates as structural materials and to store or transport energy.

The term *saccharide* comes from a Greek word meaning sugar. A *monosaccharide*, or one sugar unit, is the simplest carbohydrate. Sugars are soluble in water, most are sweet-tasting, and the most common ones have five or six carbon atoms. Ribose and deoxyribose (present in RNA and DNA, respectively) are examples of five-carbon sugars. Glucose (Figure 1.15a) is a six-carbon sugar you will encounter repeatedly in this book because it is the preferred energy source for cells. Glucose also is the precursor (parent molecule) of many complex carbohydrates.

An *oligosaccharide* is a short chain of two or more covalently bonded sugar units. Those with two linked sugar molecules—*disaccharides*—include sucrose, lactose, and maltose. Sucrose, probably the most plentiful sugar in nature, forms from a glucose unit and a fructose unit (Figure 1.15b). Table sugar is a crystallized form of sucrose extracted from sugarcane and other plants. Lactose (a glucose unit and a galactose unit) occurs in milk. Maltose (two glucose units) occurs in many seeds and is used in the manufacture of beer.

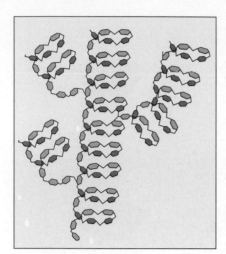

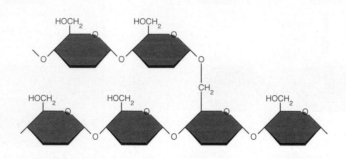

Figure 1.16 Glycogen, a form in which sugars are stored in some animal tissues, including human muscle. A glycogen molecule has a branched structure, as this drawing shows.

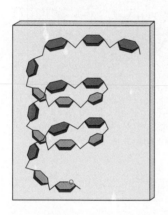

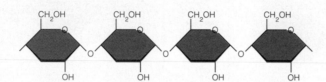

a Starch

Figure 1.17 Starch and cellulose. Both polysaccharides consist only of glucose units, but they have different properties. (**a**) Starch is a sugar storage form in plants. In amylose, the form of starch shown here, oxygen bridges link glucose subunits. (**b**) Cellulose is an insoluble structural material in plant cell walls. It is a major source of "fiber" in the human diet (Chapter 5). Neighboring cellulose molecules link together at —OH groups to form a fine strand. Such strands may be twisted together and then coiled to form cellulose threads, as the micrograph shows.

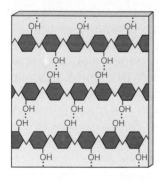

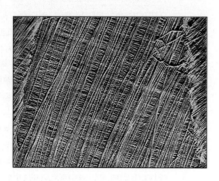

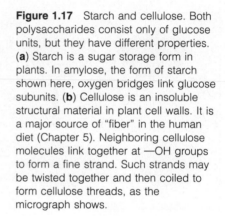

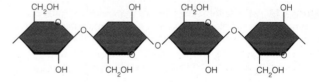

b Cellulose

A *polysaccharide* is a straight or branched chain of hundreds or even thousands of sugar units of the same or different kinds. The most common ones—glycogen, starch, and cellulose—consist entirely of glucose units. Glycogen is a storage form of glucose in animals, including humans (Figure 1.16). Foods such as potatoes, rice, and corn are all rich in starch (Figure 1.17a), which is a storage form of glucose in plants. Cellulose is a tough, insoluble structural material in plants (Figure 1.17b). Humans do not have enzymes that can break down (hydrolyze) the cellulose in vegetables, whole grains, and other plant tissues. We benefit from it, however, as "fiber" that adds bulk and helps move wastes through the lower digestive tract (colon).

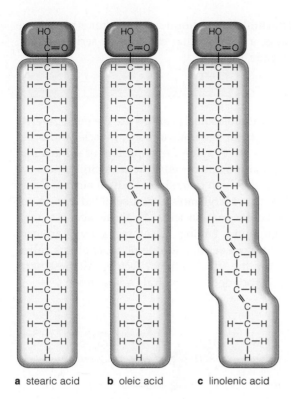

a stearic acid   **b** oleic acid   **c** linolenic acid

Figure 1.18 Structural formulas for three fatty acids. (**a**) Stearic acid's carbon backbone is fully saturated with hydrogen atoms. (**b**) Oleic acid, with its double bond in the carbon backbone, is unsaturated. (**c**) Linolenic acid, with three double bonds, is a "polyunsaturated" fatty acid.

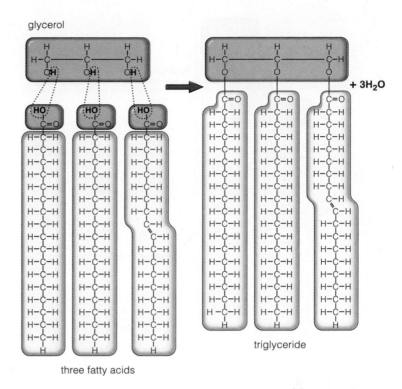

Figure 1.19 Condensation of fatty acids and glycerol into a triglyceride. In this diagram, the R signifies the "rest" of the carbon chain in each fatty acid molecule.

Lipids

Lipids are greasy or oily compounds that show little tendency to dissolve in water, but they do dissolve in nonpolar solvents (such as ether). Like polysaccharides and proteins, lipids can be broken down by hydrolysis reactions. Some lipids store energy. Others are structural materials in membranes, coatings, and other cell structures. Here we will focus on lipids of two types: those with and those without fatty acid components.

Lipids With Fatty Acids   A **fatty acid** is a flexible, tail-like hydrocarbon chain with a —COOH group at one end (Figure 1.18). Fatty acids are largely insoluble in water. Three common lipids having fatty acid tails are glycerides, phospholipids, and waxes.

Glycerides are the body's most abundant lipids and its richest source of stored energy. A glyceride molecule has fatty acid tails attached to a glycerol backbone (Figure 1.19). *Monoglycerides* have one tail, *diglycerides* have two, and *triglycerides* have three. Many fats and oils are composed of such molecules.

Saturated fats, including many fats from animal sources such as butter and lard, tend to be solids at room temperature. "Saturated" means the fatty acid tails have single C—C bonds only, and as many hydrogen atoms as chemically possible are linked to the carbon atoms. In other words, the carbons are "saturated" with hydrogens. In saturated fats the chains of adjacent molecules pack together in a parallel arrangement.

Unsaturated fats, or oils, tend to be liquid at room temperature. Plant-derived fats such as corn oil, canola oil, and olive oil are examples. They have one or more double bonds between carbon atoms in their fatty acid tails (Figure 1.19). Oils are liquid because the double bonds create kinks that disrupt packing between chains.

Phospholipids are the main structural component of cell membranes. A phospholipid has a glycerol backbone, two fatty acid tails, and a hydrophilic "head" that includes a charged phosphate group and other chemical groups (Figure 1.20). This polar part of the molecule is soluble in water. By contrast, the fatty acid tails are hydrophobic. They are not charged, and therefore they are not polar and are insoluble in water.

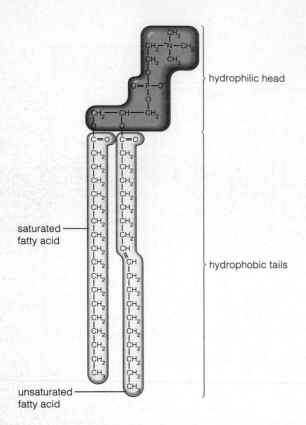

saturated fatty acid

unsaturated fatty acid

hydrophilic head

hydrophobic tails

Figure 1.20 Structural formula of a typical phospholipid found in animal cell membranes. The hydrophilic head is shaded orange, and the hydrophobic tail is in gold.

Waxes also have fatty acid components. Wax coatings such as ear wax protect tissues from drying out.

Lipids Without Fatty Acids Many lipids that have no fatty acid tails are important in membrane structure and in metabolism. They include the steroids. Steroids differ in their functional groups, but they all have the same kind of backbone (four carbon rings):

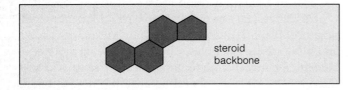

steroid backbone

The steroid cholesterol is a component of animal cell membranes. Vitamin D (essential for bone and tooth development) cannot be synthesized without it. But excess cholesterol in the blood can contribute to the artery disorder called atherosclerosis (see *Focus on Wellness* on page 161). Steroid hormones influence growth, development, reproduction, and everyday functions. Some body-

builders and athletes use hormonelike steroids to increase muscle mass, a practice that can lead to serious health problems (page 106).

Proteins

Of all biological molecules, **proteins** are the most diverse. Some are structural materials in bone, muscle, and other tissues. Others are enzymes, transporters of signals and substances, and movers of cell structures. Still others are antibodies, which help defend the body against disease.

All proteins are large molecules assembled from about 20 different kinds of amino acids. An **amino acid** is a small organic compound having an amino group, an acid group, a hydrogen atom, and one or more atoms called its R group (for the "rest" of the molecule). All these parts are covalently bonded to a central carbon atom:

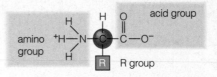

amino group

acid group

R group

Protein Structure The amino acids of proteins are joined together, one after the other, by *peptide bonds*. During protein synthesis, these covalent bonds form between the *amino* group of one amino acid and the *acid* (carboxyl) group of another (Figure 1.21). Three or more joined amino acids are a **polypeptide chain**.

Various kinds of amino acids follow one another in a chain. Consider part of the sequence for insulin, a protein hormone that prompts cells to take up glucose:

glycine — isoleucine — valine — glutamate — glycine — ...

The complete sequence for insulin is unique; no other protein has it. The same is true for each kind of protein. A protein's unique overall sequence is called its *primary structure*. That structure influences a protein's shape, what its function will be, and how it will interact with other substances. It does so in two major ways.

First, oxygen and other atoms of different amino acids in the sequence take part in hydrogen bonds that keep the chain coiled or extended (Figure 1.22). This is a protein's *secondary structure*. *Disulfide bonds* (—S—S—) may form cross-links between adjacent regions of a protein's polypeptide chains, adding structural support.

Second, different R groups in the sequence interact and dictate how the chain bends and twists into its three-dimensional shape. *Tertiary structure* is the protein shape

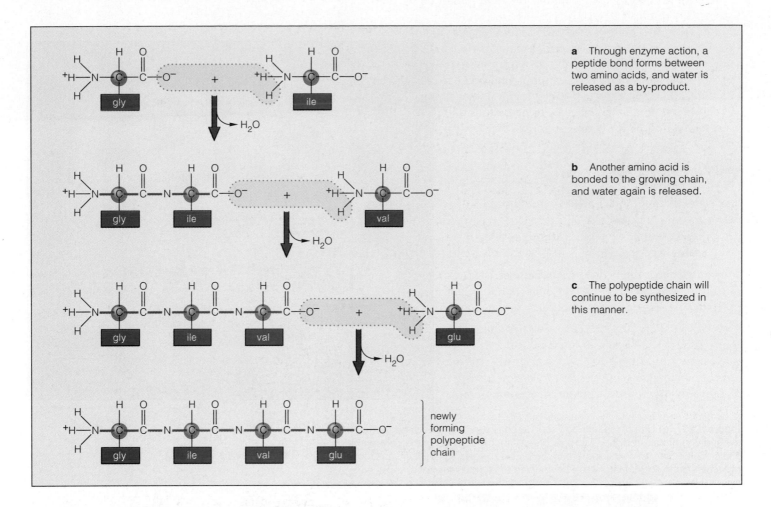

a Through enzyme action, a peptide bond forms between two amino acids, and water is released as a by-product.

b Another amino acid is bonded to the growing chain, and water again is released.

c The polypeptide chain will continue to be synthesized in this manner.

newly forming polypeptide chain

Figure 1.21 Formation of peptide bonds during protein synthesis. This is a condensation reaction.

resulting from R-group interactions. Figure 1.23a provides an example.

Some proteins have *quaternary structure*, meaning they incorporate two or more polypeptide chains. The resulting protein is globular, fiberlike, or a combination of the two. The blood protein hemoglobin (Figure 1.23b) is an example of a globular protein.

You can almost literally run your fingers through the protein keratin, the main component of hair (Figure 1.24). A permanent wave curls straight hair by way of a chemical treatment that disrupts R-group links in keratin, followed by treatment with another chemical that causes new R-group interactions while the hair is twisted around rollers.

The amino acid sequence in polypeptide chains gives rise to the three-dimensional structure of proteins. That structure governs a protein's chemical behavior.

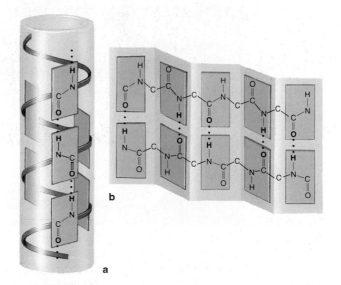

Figure 1.22 Hydrogen bonds (dotted lines) in a polypeptide chain. Such bonds can give rise to a coiled chain (**a**) or a sheetlike array of chains (**b**).

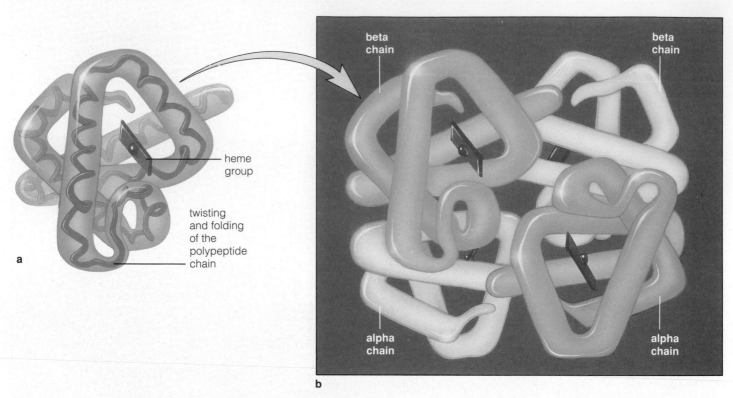

Figure 1.23 (**a**) One of the four polypeptide chains in hemoglobin, an oxygen-transporting protein in blood. The dark green "ribbon" represents the polypeptide chain. Heme, an iron-containing group, binds the oxygen. (**b**) Quaternary structure of human hemoglobin. Many weak bonds hold the four chains tightly together.

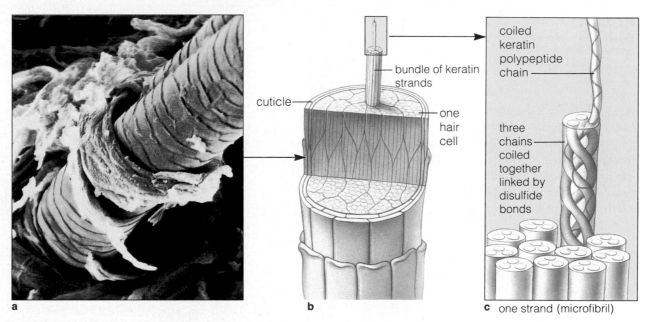

Figure 1.24 Structure of hair. (**a**) Scanning electron micrograph of a hair shaft. Cells in each of the shaft's three regions (**b**) contain the protein keratin. Polypeptide chains of keratin are synthesized inside the hair cells. Groups of three chains become linked by disulfide bonds, then eleven of such groups become coiled into fibrous strands called microfibrils (**c**). The microfibrils in turn become bundled together into larger, cablelike strands called macrofibrils. Each cell in a hair shaft contains large numbers of keratin macrofibrils. Keratin is most densely packed in the tough, outer cuticle—a single layer of dead, overlapping cells.

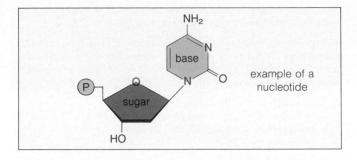

Figure 1.25 Structure of a nucleotide.

Protein Denaturation The loss of any molecule's three-dimensional shape following disruption of bonds is called **denaturation**. Because they are weak and sensitive to changes in temperature and pH, the hydrogen bonds (and other bonds) that help hold a protein in its normal, three-dimensional shape can be disrupted fairly easily. When that happens, polypeptide chains unwind or change shape, and the protein no longer can function.

The protein albumin is concentrated in the "egg white" of uncooked chicken eggs. When you cook an egg, the heat does not affect the strong covalent bonds of albumin's primary structure, but it does destroy weaker bonds holding albumin in its three-dimensional shape. Although denaturation can be reversed for some kinds of proteins when normal conditions are restored, albumin isn't one of them; there is no way to uncook a cooked egg.

Nucleotides and Nucleic Acids

Nucleotides are the fundamental components of hereditary material and help regulate various chemical reactions in cells. Each nucleotide has three parts: a five-carbon sugar (ribose or deoxyribose), one or more phosphate groups, and a nitrogen-containing base that has either a single-ring or a double-ring structure (Figure 1.25).

One nucleotide, **ATP** (adenosine triphosphate), carries energy from one site to another in cells. Still others function as parts of **coenzymes**, molecules that accept hydrogen atoms and electrons from various molecules and transfer them elsewhere. Examples include certain vitamins and compounds abbreviated NAD^+ and FAD.

Nucleic acids are polynucleotides; they consist of four different kinds of nucleotides strung together, forming large single- or double-stranded molecules. Each strand's backbone consists of alternating phosphate and sugar units. The nucleotide bases stick out to the side (Figure 1.26a). Different nucleic acids have their nucleotide bases arranged in different sequences.

Two major nucleic acids are **RNA** (ribonucleic acid) and **DNA** (deoxyribonucleic acid). RNA is usually a sin-

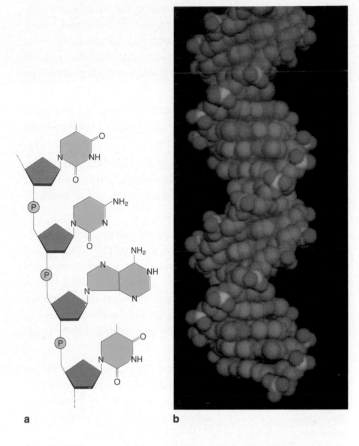

a   **b**

Figure 1.26 (**a**) Examples of bonds between nucleotides in a nucleic acid molecule. (**b**) Model of DNA, a molecule that is central to maintaining and reproducing the cell. The molecule consists of two strands of nucleotides joined by hydrogen bonds and twisted into a double helix. The nucleotide bases are shown in blue.

gle nucleotide strand. As we will see in a subsequent chapter, it functions in the processes by which genetic instructions are used to build proteins. DNA is usually a double-stranded molecule that twists helically, like a spiral staircase (Figure 1.26b). Hydrogen bonds hold the two strands together, and genetic instructions are encoded in the sequence of bases in DNA . You will be reading a good deal more about these molecules in chapters to come.

SUMMARY

1. Protons, neutrons, and electrons are the basic constituents of atoms. All atoms of an element have the same number of protons and electrons. Interaction between atoms depends on the number and arrangement of their electrons (especially the outermost ones), which carry a negative charge.

2. An atom has no net charge because the number of positively charged protons in its nucleus is balanced by an equal number of negatively charged electrons outside its nucleus. An ion is an atom or compound that has gained or lost one or more electrons; it has an overall positive or negative charge.

3. Hydrogen, oxygen, carbon, and nitrogen are the most abundant elements in living things. They all tend to form bonds with other atoms.

4. Most atoms form chemical bonds with other atoms or ions in order to achieve a stable electron structure—that is, two or eight electrons in the atom's outermost electron shell.

5. In ionic bonds, positive and negative ions remain together by mutual attraction of their opposite electrical charges. In covalent bonds, atoms share one or more pairs of electrons. In hydrogen bonds, a hydrogen atom that is already part of a polar covalent bond, and that carries a slight positive charge, interacts weakly with an atom at the more negative region of the same or another polar molecule.

6. Acids release hydrogen ions (H^+) in water. Bases release hydroxide ions (OH^-) that can combine with H^+. At pH 7, the H^+ and OH^- concentrations in a solution are equal. Buffers and other mechanisms maintain cellular pH.

7. Water has roles in reactions and helps give cells shape and internal organization. Because of hydrogen bonds between its molecules, water resists temperature changes and can dissolve other polar substances.

8. Carbon atoms covalently bonded into chains and rings serve as the backbone of organic compounds. The chemical and physical properties of many of those compounds depends largely on functional groups (atoms attached to the backbone).

9. Within cells, small organic molecules (including simple sugars, fatty acids, amino acids, and nucleotides) serve as energy sources or building blocks for larger "molecules of life"—the complex carbohydrates, lipids, proteins, and nucleic acids.

Review Questions

1. Cells use four main families of small organic molecules to assemble carbohydrates, lipids, proteins, and nucleic acids (the large biological molecules). What are they? *26*

2. Which of the following is the carbohydrate, the fatty acid, the amino acid, and the polypeptide? *27–30*
   a. $^+NH_3$—CHR—COO^- c. (glycine)$_{20}$
   b. $C_6H_{12}O_6$ d. $CH_3(CH_2)_{16}COOH$

3. Is the following statement true or false? Most enzymes are proteins, but only some proteins are enzymes. *26*

4. Describe the four levels of protein structure. How do a protein's side groups influence its interactions with other substances? Give an example of what happens when weak bonds holding a protein in its three-dimensional shape are disrupted. *30*

5. Distinguish among the following:
   a. monosaccharide, polysaccharide, disaccharide *27*
   b. peptide bond, polypeptide *30*
   c. glycerol, fatty acid *29*
   d. nucleotide, nucleic acid *33*

Critical Thinking: You Decide *(Key in Appendix IV)*

1. In a magazine you read about a new high-carbohydrate diet, which claims to help you lose weight by ridding the body of its entire supply of sugars. Without access to sugars, the diet says, your body will burn away fat. How would you critique this claim, given what you know of biological molecules?

2. Black coffee has a pH of 5, whereas milk of magnesia has a pH of 10. Is coffee twice as acidic as milk of magnesia?

3. After a dessert party, your best cotton shirt has stains on it from whipped cream and strawberry syrup. When you take the shirt to the cleaners, you are told that two separate cleaning agents will be required to remove the stains. Explain why two agents are needed and what different chemical characteristic each would have.

Self-Quiz *(Answers in Appendix III)*

1. The backbone of organic compounds forms when _____ atoms are covalently bonded into chains and rings.

2. Each carbon atom can form up to _____ bonds with other atoms.
   a. four   c. eight
   b. six   d. sixteen

3. The four categories of large biological molecules are _____, _____, _____, and _____.

4. All of the following *except* _____ are small organic molecules that serve as the main building blocks or energy sources in cells.
   a. fatty acids   d. nucleotides
   b. simple sugars   e. amino acids
   c. lipids

5. Which of the following is *not* a carbohydrate?
   a. glucose molecule   c. fat
   b. simple sugar   d. polysaccharide

6. _____, a class of proteins, make metabolic reactions proceed much faster than they would on their own.
   a. DNA   c. Fatty acids
   b. Amino acids   d. Enzymes

7. Examples of nucleic acids, the basis of inheritance, are _____.
   a. polysaccharides   c. proteins
   b. DNA and RNA   d. simple sugars

8. Which of the following best describes the role of functional groups?
   a. assembling large organic compounds
   b. influencing the behavior of organic compounds
   c. splitting molecules into two or more parts
   d. speeding up metabolic reactions

9. In _____ reactions, small molecules become covalently linked, and water can also form.
   a. symbiotic   c. condensation
   b. hydrolysis   d. ionic

10. Match each type of molecule with the correct description.

   _____ chain of amino acids   a. carbohydrate
   _____ energy carrier   b. phospholipid
   _____ glycerol, fatty acids,   c. protein
         phosphate   d. DNA
   _____ chain of nucleotides   e. ATP
   _____ one or more sugar units

Readings

Goodsell, D. September–October 1992. "A Look Inside the Living Cell." *American Scientist*. Current models of biological molecules.

Lehninger, A. 1982. *Principles of Biochemistry*. New York: Worth. Classic reference in the field.

"The Molecules of Life." October 1985. *Scientific American*. Entire issue devoted to biological molecules.

Zimmer, C. October 1992. "Wet, Wild, and Weird." *Discover*. Water is nature's "hardest" liquid, and other recent, intriguing discoveries about this fluid crucial for life.

2 CELL STRUCTURE AND FUNCTION

A Risky Business

For every living cell in the body, survival is a risky business. The risk is easy to see in parts of Asia, Africa, and the Americas, where cholera is a severe health threat to inhabitants and visitors alike.

Caused by a bacterium, *Vibrio cholerae*, cholera spreads by way of water or food contaminated by human sewage (Figure 2.1). Its most devastating symp-

tom is severe diarrhea. Over a period of several days, body tissues rapidly lose water and essential salts. Treatment involves giving fluids that replenish these substances, as well as antibiotics to kill the microbe. Sometimes, however, the body loses so much water so rapidly that blood pressure plunges and the volume of blood in blood vessels falls drastically. The cells that

Figure 2.1 In many places where public sanitation is primitive or lacking altogether, people run the risk of contracting cholera because drinking water and some food sources (especially fish) are contaminated by human sewage that carries the *Vibrio cholerae* bacterium. Cholera can be lethal if it so rapidly depletes the body of water and needed solutes that the internal balance we call homeostasis is fatally disrupted.

make up tissues become starved for oxygen and nutrients because too little blood is available to deliver those substances. Other complications can occur as well.

As this example suggests, a cell is a highly organized bit of life in a world that is usually less organized and sometimes harsh. In this chapter you will discover that cells generally are built in such a way that they can bring in certain substances, release or keep out others, and conduct their internal activities with great precision—all at a breakneck pace. Every cell also is adapted to function under particular environmental conditions, which the body's homeostatic mechanisms must maintain.

The precision and control necessary to keep each cell alive begins at the plasma membrane, a seemingly flimsy surface layer of little more than lipids and proteins. Across this membrane, water molecules and other materials are selectively exchanged between the cell's surroundings and its interior. Further exchanges are made across internal cell membranes, which form compartments called organelles. Organelles perform a variety of functions that help cells survive and reproduce. The sum total of these activities is called *metabolism*, and we'll also consider here how cells obtain the energy that keeps metabolic fires burning.

KEY CONCEPTS

1. Cells are the smallest units that still retain the characteristics of life, including complex organization, metabolic activity, and reproductive behavior.

2. All cells have an outermost plasma membrane that separates their interior from the surroundings. They have a region of DNA. They have cytoplasm, a region that is structurally and functionally organized for energy conversions, protein synthesis, cell movements, and other activities necessary for survival.

3. All cell membranes are composed of phospholipids and proteins. The phospholipids form a double layer. This "bilayer" gives membranes their basic structure and serves as a barrier to water-soluble substances. The proteins carry out most other membrane functions.

4. Human cells contain a nucleus and other organelles (compartments bounded by internal cell membranes). Organelles separate different chemical reactions and allow them to proceed in an orderly fashion.

5. The three main regions characteristic of human cells are arranged in this general fashion:

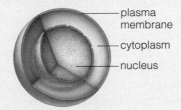

plasma membrane
cytoplasm
nucleus

6. Cells harness and use energy for building, storing, breaking apart, and eliminating substances in ways that help them survive and reproduce. These activities are called metabolism.

7. Enzymes increase the rate of chemical reactions. They take part in nearly all metabolic pathways and function with great specificity; each enzyme can interact with only one or a few particular substrate molecules.

8. ATP is an organic compound that transfers energy needed for life processes from one reaction site to another within cells.

THE NATURE OF CELLS

Basic Cell Structure and Function

Cells are amazingly diverse in structure and function. A single-celled bacterium is a complete organism unto itself. One of your liver cells is not only more complex structurally than a bacterium, but it is also highly specialized for a specific function within one of the most complex life forms on earth. Yet all cells share some basic features.

All have an outer membrane, and all have the genetic material DNA concentrated somewhere in their interior. Bacteria are *prokaryotic*, a term loosely meaning "before nucleus." Nothing separates their DNA from other internal cell parts. Cells more complex than bacteria are *eukaryotic* ("true nucleus"); their DNA is enclosed by a system of membranes. For eukaryotic cells, including human cells, the basic regions and their functions can be divided into three parts:

1. **Plasma membrane**. This outermost membrane allows metabolic events to proceed inside the cell in organized, controlled ways, apart from the environment. It does not isolate the interior, for many substances and signals can move across the membrane.

2. **Nucleus**. This membrane-bound compartment within the cell contains the DNA, along with other molecules that can copy or read its hereditary instructions.

3. **Cytoplasm**. This region is everything enclosed by the plasma membrane, *except* for the nucleus. It consists of a semifluid substance in which particles, filaments, and often membranous parts are organized.

Cell Size and Shape

Can any cell be seen with the unaided human eye? There are a few, including the "yolks" of bird eggs, the fish eggs we call caviar, and even one type of bacterium that lives in the intestines of certain fishes. Generally, however, cells are too small to be observed without microscopes (see *Focus on Science*). To give you a sense of the sizes involved, each of your red blood cells is about 8 *millionths* of a meter in diameter. You could fit a string of about 2,000 of them across your thumbnail!

If a cell isn't small, it probably is long and thin or has outfoldings and infoldings that increase its surface area relative to its volume. The smaller or more elongated or frilly-surfaced the cell, the more efficiently materials can cross its surface and become distributed throughout the interior. Figure 2.2 shows some of the many shapes of human cells.

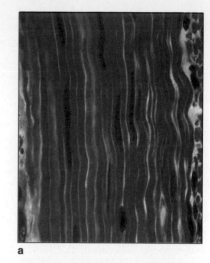

a

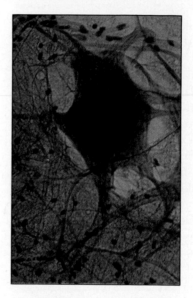

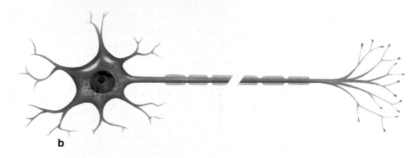

b

Figure 2.2 Examples of different types and shapes of cells of the human body. (**a**) Smooth muscle cells, found in the walls of hollow organs such as the stomach, have an elongated oval shape. (**b**) A motor neuron, a type of nerve cell, has a blob-like cell body with threadlike extensions. The photographs in *Focus on Science* on the facing page show differing images of red blood cells, which are disk-shaped.

Microscopy: It's A Small World

In the 17th century Robert Hooke looked at a thin slice of cork through his microscope and saw tiny, empty compartments (Figure *a*). He gave them the name *cellulae* (meaning "small rooms")—and this was the origin of the biological term *cell*. Today, researchers use much more sophisticated instruments to view cells (Figure *b*).

The micrographs in Figures *c* through *e* compare the type of detail revealed by modern microscopes. (A *micrograph* is simply a photograph of an image formed with a microscope.) The specimen is a red blood cell.

a

b

Microscopy in the 17th century and now. (**a**) Robert Hooke's compound microscope and his drawing of dead cork cells. (**b**) An electron microscope in a modern laboratory.

(**c-e**) Comparison of how different types of microscopes reveal cellular details. The cells are human red blood cells. (**c**) Red blood cells flowing through a capillary as revealed by a light microscope. (**d**) Scanning electron micrograph (SEM) with color added clearly shows the "concave disk" shape of red blood cells. (**e**) Transmission electron micrograph.

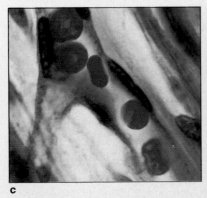

c

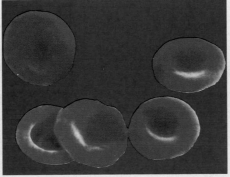

d

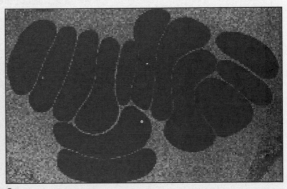

e

The *compound light microscope* has glass lenses that bend incoming light rays to form an enlarged image of a specimen. If you wish to observe a living cell, it must be small or thin enough for light to pass through. Also, to be visible cell parts must differ in color or optical density from their surroundings—but most are nearly colorless and uniform in density. Cells are thus stained (exposed to dyes that react with some parts but not others), but staining can alter the parts and kill the cell. Dead cells begin to break down at once, so they are sometimes preserved before staining.

No matter how good a glass lens system is, when the diameter of the object being viewed is magnified by 2,000 times or more, cell parts appear larger but are not clearer. The properties of light waves passing through glass lenses limit the resolution of smaller details.

Transmission electron microscopes have great magnifying power because they use magnetic lenses to focus electrons (which respond to magnetic force). Electrons are particles, but they also behave like waves—and their wavelengths are much shorter than those of visible light. Details of a cell's internal structure show up best with these microscopes.

With a *scanning electron microscope,* a narrow electron beam is directed back and forth across a specimen's surface, which has been coated with a thin metal layer. The energy in the electron beam triggers the emission of electrons in the metal. The emission patterns can be used to form an image. Scanning electron microscopy provides a three-dimensional view of surface features.

CELL MEMBRANES

Membrane Structure and Function

A cell's organization and its activities depend on membranes. The main components of cell membranes are phospholipids and proteins. A phospholipid, recall, has a hydrophilic ("water-loving") head and two fatty-acid tails, which are hydrophobic ("water-fearing"):

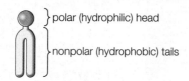
polar (hydrophilic) head

nonpolar (hydrophobic) tails

When phospholipids are immersed in water, the molecules cluster together into layers, with all the fatty-acid tails sandwiched between the hydrophilic heads. This arrangement—a **lipid bilayer**—is the structural basis of cell membranes:

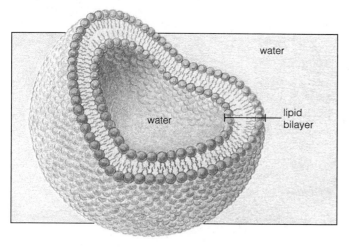
water

water

lipid bilayer

Figure 2.3 shows the **fluid mosaic model** of cell membranes. The bilayer represents the "fluid" aspect of the model, for both the lipid and the protein molecules show quite a bit of movement. The lipids move sideways and their tails flex back and forth, so they cannot become packed into a solid layer. Some lipids in the bilayer have short tails or unsaturated (kinked) tails, which also disrupt the packing. The membrane is said to have a "mosaic" quality because it is a composite of individual lipids and proteins. The proteins carry out most membrane functions. Most are *glycoproteins*, meaning they have carbohydrates covalently bonded to them. The carbohydrate chains always extend outward, never into the cytoplasm.

Some membrane proteins are *channel proteins* through which water-soluble substances enter cells. Some *carrier proteins* act like pumps to transport substances across the lipid bilayer. Still others carry electrons. *Recognition proteins* of the plasma membrane are like molecular fingerprints; they identify a cell as being of a certain type. The carbohydrate chains of certain recognition proteins serve as "self" markers that help cells distinguish between body cells and foreign ones such as bacteria. Other recognition proteins promote adhesion between cells in a tissue. *Receptor proteins* bind hormones and other substances that trigger alterations in metabolism or cell behavior. For example, the machinery for cell growth and division switches on when human growth hormone binds to cell receptors.

The features common to all cell membranes can be summarized as follows:

1. Cell membranes are composed of lipids (especially phospholipids) and proteins.

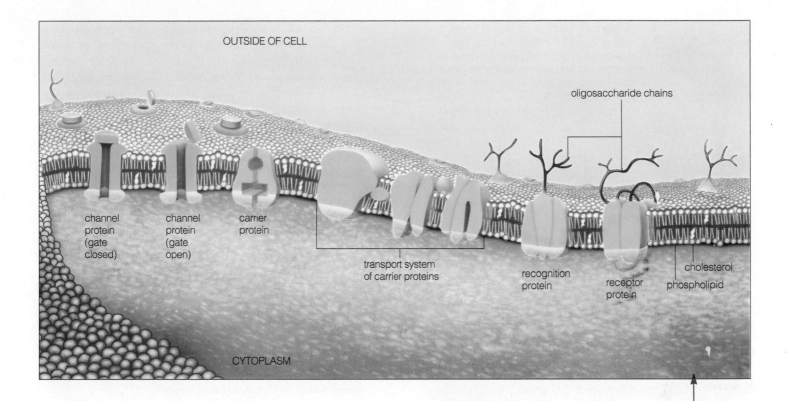

2. The lipid molecules have their hydrophilic heads at the two outer faces of a bilayer and their fatty acid tails sandwiched in between.

3. A lipid bilayer imparts structure to cell membranes and serves as a hydrophobic barrier between two solutions. (A *plasma membrane* separates fluids inside and outside the cell. *Internal cell membranes* separate different fluids within the space of the cytoplasm.)

4. Proteins embedded in the bilayer or positioned at its surfaces carry out most membrane functions.

Diffusion

Cellular life depends on the energy inherent in molecules (or ions), which keeps them in constant motion; it also depends on **concentration gradients**. "*Concentration*" refers to the number of molecules of a substance in a specified volume of fluid. "*Gradient*" implies that one region of the fluid contains more molecules than a neighboring region.

In the absence of other forces, molecules move down their concentration gradient—that is, *from a region where they are more concentrated to a region where they are less concentrated.* They do so because they constantly collide with one another, millions of times a second. Random collisions send the molecules back and forth, but the *net* movement is away from the place of greater concentration. When the difference in concentration is large, molecules

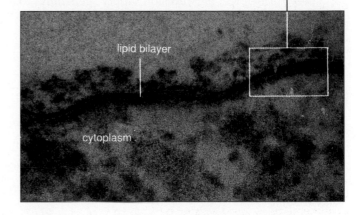

Figure 2.3 Views of the plasma membrane, based on the fluid mosaic model of membrane structure. The micrograph shows the plasma membrane of a eukaryotic cell.

move outward rapidly. When the difference is not so great, the net rate of movement is slower. Even when the gradient no longer exists, individual molecules are still in motion; then, however, the net rate of movement in any direction is zero.

The net movement of solutes from a region of higher concentration to a region of lower concentration is called **diffusion**. Diffusion is a key factor in the movement of substances across cell membranes and through fluid regions of the cytoplasm.

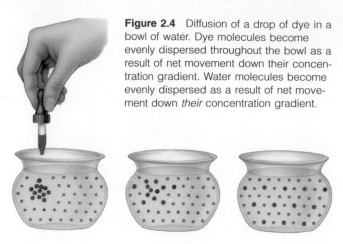

Figure 2.4 Diffusion of a drop of dye in a bowl of water. Dye molecules become evenly dispersed throughout the bowl as a result of net movement down their concentration gradient. Water molecules become evenly dispersed as a result of net movement down *their* concentration gradient.

If more than one solute is present in a solution, each solute still diffuses according to its *own* concentration gradient. Put a drop of dye in one side of a bowl of water, and the dye molecules diffuse in one direction: to the region where they are less concentrated. And the water molecules move in the opposite direction, to the region where *they* are less concentrated (Figure 2.4). Ultimately both are distributed equally throughout the solution.

These principles of diffusion apply generally to uncharged solutes. When a solute is charged, as ions are, it may diffuse down an *electrical* gradient—and this movement can occur *against* its concentration gradient. As we will see in Chapter 10, every moment of your life you depend on electrical gradients that are associated largely with differences in the concentration of sodium ions inside and outside the plasma membranes of your nerve cells.

Diffusion is the net movement of like molecules or ions from a region of higher concentration to a region of lower concentration.

Charged solutes such as ions also can diffuse down electrical gradients.

Water Concentration and Cell Membranes

Cell membranes are *selectively permeable*. Some small molecules can diffuse through the lipid bilayer of a cell membrane or through open channels across it. Glucose and other molecules cannot do this; membrane proteins must assist them across. Because the membrane shows this selective permeability, the concentrations of dissolved substances (that is, solutes) can increase on one side of the membrane and not the other. The resulting solute concentration gradients affect the movement of water into and out of cells.

When solute concentrations are equal on both sides of a cell membrane, there is no net movement of water in either direction; the two fluids are said to be *isotonic* ("iso-" means same). When solute concentrations are not equal, one fluid is *hypotonic* (has fewer solutes) and the other is *hypertonic* (has more solutes). Because water moves down its concentration gradient, it tends to move from a hypotonic solution to a hypertonic one. If cells did not have mechanisms for adjusting to such differences, they would shrivel or burst, as Figure 2.5 illustrates.

The diffusion of water across membranes in response to concentration gradients is called **osmosis**. Osmotic water movements in the kidneys are key to maintaining proper water balance in the body as a whole. *Focus on Environment* explores issues of water balance on a more global scale.

Osmosis is the diffusion of water across a selectively permeable membrane in response to its concentration gradient.

How Solutes Cross Membranes

Lipid-soluble molecules (such as steroids) and small molecules with no net charge (such as oxygen, carbon dioxide, and ethanol) readily diffuse across the membrane's lipid bilayer:

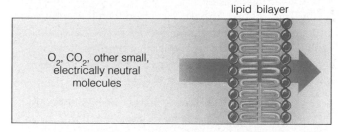

By contrast, glucose and other larger, water-soluble molecules almost never diffuse freely across the lipid bilayer. Neither do positive or negative ions:

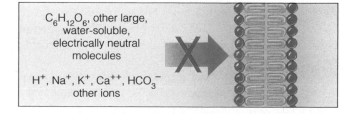

Instead, these substances cross the membrane by means of proteins embedded in the bilayer. Different proteins selectively move them across by active and passive transport mechanisms.

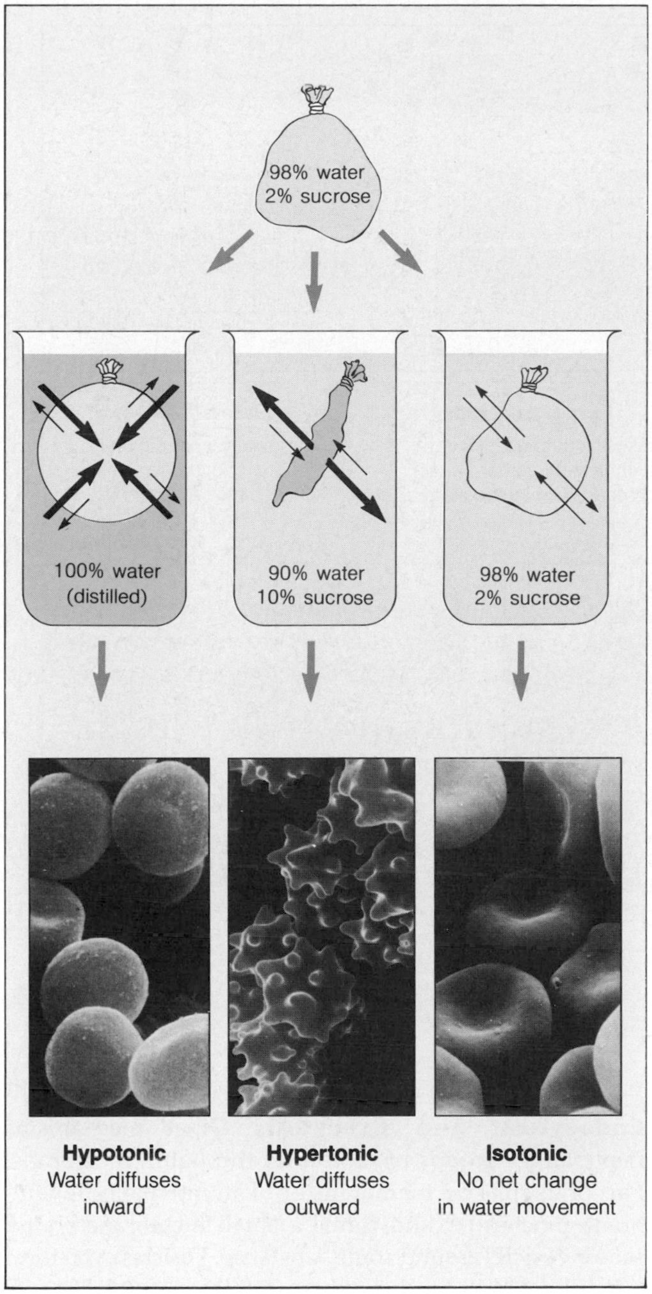

Figure 2.5 Osmosis in different environments. In the sketches, membranous bags through which water but not sucrose can move are placed in hypotonic, hypertonic, and isotonic solutions in three containers. In each container, arrow width represents the relative amount of water movement. The sketches explain what happens to red blood cells that are placed in comparable solutions. Red blood cells have no mechanisms to actively take in or expel water.

Labels within figure:

98% water
2% sucrose

100% water
(distilled)

90% water
10% sucrose

98% water
2% sucrose

Hypotonic
Water diffuses
inward

Hypertonic
Water diffuses
outward

Isotonic
No net change
in water movement

Focus on Environment G. Tyler Miller, Jr.

Conserving Precious Water

High-quality water is a necessity of life—and a commodity that is easily taken for granted. Increasingly, various parts of the world are plagued by periodic drought, pollution of groundwater supplies, and water-wasting practices. Consider, for instance, that in the arid western United States and in parts of Australia, lawn and garden watering accounts for 80 percent of an average household's daily usage. In the United States generally, much of the water used by individuals is wasted. According to the American Water Works Association, you may use two gallons of water per day simply by brushing your teeth with the water running, and an estimated 180 gallons washing your car with the hose running. Some 25–50 gallons of water go down the drain during a ten-minute shower. A toilet flush uses 5–7 gallons. In addition, leaky pipes, water mains, toilets, bathtubs, and faucets waste 20–35 percent of water withdrawn from public supplies in the United States. Interestingly, one-fifth of all U.S. public water systems charge a single low rate (less than $100 a year for an average family) for virtually unlimited use of high-quality water. In those systems and elsewhere, water users pay little or no penalty for wasting such an essential commodity.

Water-wasting habits and population growth are now making water recycling and conservation necessities in some areas. At least 15 states now require that all new toilets use no more than 6 liters (1.6 gallons) per flush. In California, one utility company promoted water-saving toilets and shower heads, and in one year cut per capita water use of its customers by 40 percent. A New Jersey office complex has cut water consumption 62 percent with an on-site treatment and reuse system. Such systems are also being implemented in Japan, Mexico, and other nations with large populations and increasingly limited water supplies.

Conservation and recycling efforts are vital to promote sustainable use of earth's water resources. As Chapter 25 describes, we must also face up to the consequences of practices that seriously pollute the water we depend on.

Figure 2.6 Overview of mechanisms by which substances move across cell membranes. Note that exocytosis and endocytosis occur at the plasma membrane only, not at internal cell membranes.

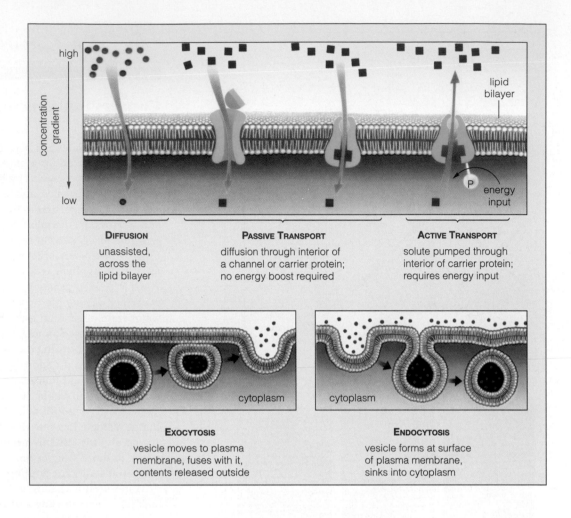

concentration gradient

high

low

lipid bilayer

energy input

DIFFUSION
unassisted, across the lipid bilayer

PASSIVE TRANSPORT
diffusion through interior of a channel or carrier protein; no energy boost required

ACTIVE TRANSPORT
solute pumped through interior of carrier protein; requires energy input

cytoplasm

cytoplasm

EXOCYTOSIS
vesicle moves to plasma membrane, fuses with it, contents released outside

ENDOCYTOSIS
vesicle forms at surface of plasma membrane, sinks into cytoplasm

Active and Passive Transport In **active transport**, a membrane protein undergoes a series of changes in shape after receiving an energy boost (usually from ATP). The changes cause the protein to *pump* specific solutes through its interior and across the membrane. Active transport is necessary whenever a cell must move molecules against their concentration gradient. For example, a cell will continue to stockpile essential amino acids even though the concentrations of such molecules may already be greater within the cell than in the surrounding tissue fluid. Active transport carries amino acids into the cell against their respective concentration gradients—and so increases the cell's reserves for building proteins.

In **passive transport**, a protein serves as a channel or carrier for a specific solute, which moves *only* in the direction that diffusion would take it (down its concentration gradient). Hence this transport mode is also called "facilitated diffusion."

Endocytosis and Exocytosis Other mechanisms move larger groups of molecules through a membrane. Part of the plasma membrane or of an internal cell membrane pinches off and forms a small membrane-bound sac, or **vesicle**, around some substance. Vesicles even form around tiny cells (such as a bacterium) and fluids. In **exocytosis**, vesicles form inside the cytoplasm and then move to the plasma membrane and fuse with it, so that their contents are moved to the outside. In **endocytosis**, a patch of plasma membrane encloses material at the cell surface. Then it sinks in and pinches off, forming a vesicle that either transports the material into the cytoplasm or stores it there. Endocytosis that brings organic matter into the cell is sometimes called *phagocytosis* ("cell eating").

Figure 2.6 illustrates the mechanisms by which substances move across cell membranes. Through such mechanisms, cells take in materials and get rid of wastes at controlled rates. Transport mechanisms also help maintain a cell's volume and pH.

HUMAN CELLS

Typical Components: An Overview

Eukaryotic cells have many **organelles**, which are compartmented portions of the cytoplasm. The most conspicuous organelle is the nucleus.

No chemical apparatus in the world can match the cell for the sheer number of chemical reactions that go on simultaneously in so small a space. Many of the reactions would be incompatible if they occurred within the same compartment of the cell. For example, a molecule of fat can be put together by some reactions and taken apart by others, but a cell would gain nothing if both sets of reactions proceeded at the same time on the same molecule.

Organelles physically separate chemical reactions (many of which are incompatible) into different regions of the cytoplasm.

Organelles separate different reactions in time, as when molecules are produced in one organelle and then used later in other reactions.

Our discussion here will focus on several organelles and other structures found in most types of human cells. The following list identifies them and their functions:

nucleus	physical isolation and organization of DNA
ribosomes	synthesis of polypeptide chains (proteins)
endoplasmic reticulum (ER)	in rough ER, initial modification of polypeptide chains; lipid synthesis in smooth ER
Golgi bodies	further modification of polypeptide chains into mature proteins; sorting and shipping of proteins and lipids for secretion or for use in cell
lysosomes	breakdown of large molecules
diverse vesicles	transport or storage of substances; other functions
mitochondria	ATP formation
cytoskeleton	cell movements, shape, and internal organization

Figures 2.7 and 2.8 show where organelles and structures might be located in a typical animal cell. Keep in

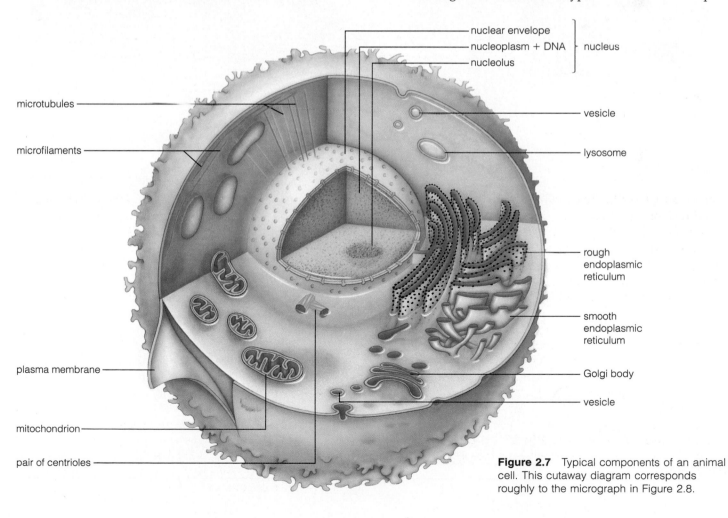

Figure 2.7 Typical components of an animal cell. This cutaway diagram corresponds roughly to the micrograph in Figure 2.8.

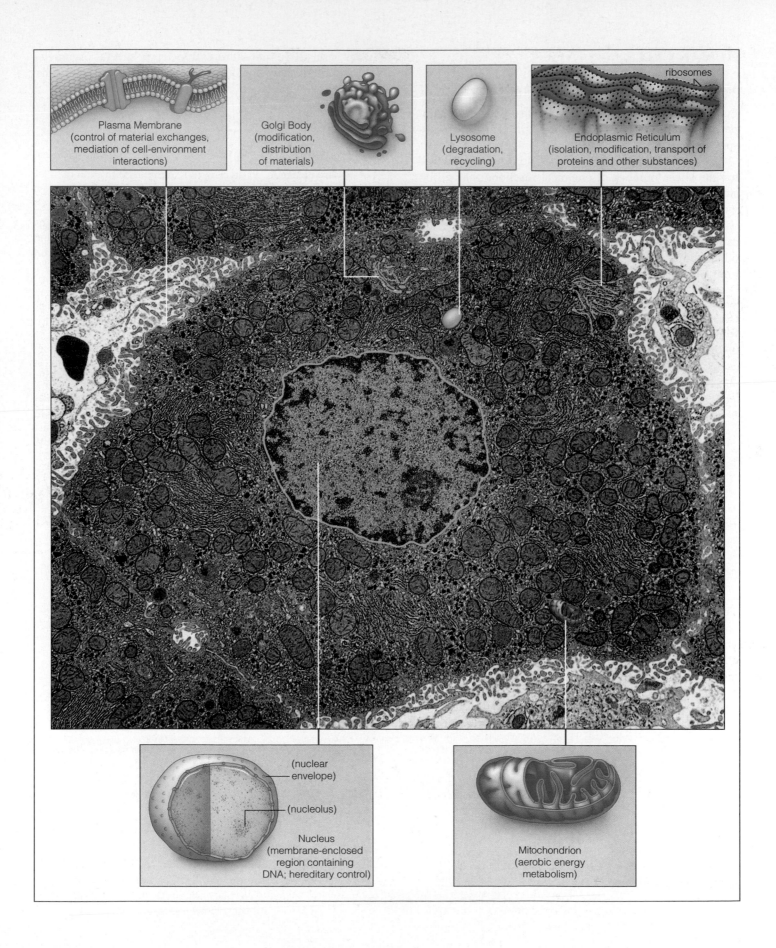

Plasma Membrane
(control of material exchanges,
mediation of cell-environment
interactions)

Golgi Body
(modification,
distribution
of materials)

Lysosome
(degradation,
recycling)

ribosomes

Endoplasmic Reticulum
(isolation, modification, transport of
proteins and other substances)

(nuclear
envelope)

(nucleolus)

Nucleus
(membrane-enclosed
region containing
DNA; hereditary control)

Mitochondrion
(aerobic energy
metabolism)

mind, though, that calling a cell "typical" is about as meaningful as calling a squid a "typical animal." As we shall discover in subsequent chapters, cells with different functions may have dramatic differences in their structures as well.

The Nucleus

The **nucleus** encloses the DNA in a cell. DNA contains instructions for building all of a cell's proteins and, through those proteins, for determining a cell's structure and function. Figure 2.9 shows the structure of the nucleus, and Table 2.1 lists its components. The nucleus serves two main functions: First, it helps control access to hereditary instructions contained in DNA. Second, it separates DNA from the substances and chemical activity in the cytoplasm. When the time comes for a cell to divide, this arrangement makes it easier to package the DNA into parcels for the two daughter cells that form.

Nuclear Envelope The outermost part of the nucleus, the **nuclear envelope**, is two membranes thick. It is continuous with the endoplasmic reticulum, discussed shortly. Each membrane is a lipid bilayer, studded with a variety of proteins. Pores composed of proteins span both bilayers. The pores are passageways where specific substances can move into and out of the nucleus in controlled ways.

Nucleolus As a cell grows, one or more dense masses of irregular size and shape appear in the nucleus. Each mass is a **nucleolus** (plural: *nucleoli*). Nucleoli are sites where the protein and RNA subunits of ribosomes are assembled. The ribosomal subunits are shipped from the nucleus into the cytoplasm, where they join together into intact, functional ribosomes. A nucleolus is visible in Figure 2.9.

Chromosomes Between cell divisions, DNA is threadlike, with many enzymes and other proteins attached to it like beads on a string. Prior to division, however, DNA molecules are duplicated so that each new cell will get a set of hereditary instructions. Before the DNA molecules are sorted into two sets, they fold and twist into condensed structures, proteins and all.

Early microscopists named the threadlike material *chromatin* and the condensed structures *chromosomes*.

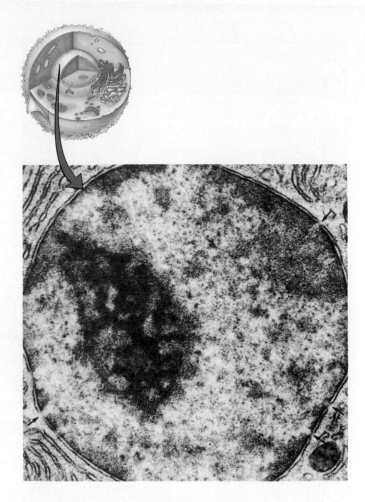

Figure 2.9 Animal cell nucleus, thin section. Arrow points to pores in its nuclear envelope.

Table 2.1	Components of the Nucleus
Nuclear envelope	Double-membraned, pore-riddled boundary between cytoplasm and interior of nucleus
Nucleolus	Dense cluster of the RNA and proteins used to assemble ribosomal subunits
Nucleoplasm	Fluid portion of the nuclear interior
Chromosomes	DNA molecules and numerous proteins attached to them

Figure 2.8 Transmission electron micrograph of a liver cell in cross-section. Although this is a rat liver cell, it is representative of liver cells in other mammals, including humans.

Today, we call a DNA molecule and its associated proteins a **chromosome** regardless of whether it is in a threadlike or condensed form. When you read subsequent chapters, keep in mind that the appearance of chromosomes changes at different times during the life of a cell.

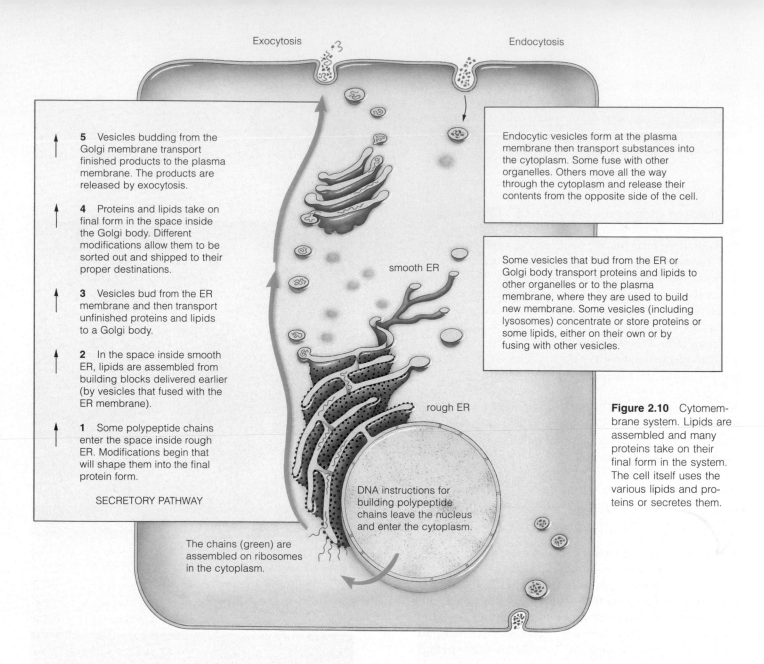

Exocytosis

Endocytosis

5 Vesicles budding from the Golgi membrane transport finished products to the plasma membrane. The products are released by exocytosis.

4 Proteins and lipids take on final form in the space inside the Golgi body. Different modifications allow them to be sorted out and shipped to their proper destinations.

3 Vesicles bud from the ER membrane and then transport unfinished proteins and lipids to a Golgi body.

2 In the space inside smooth ER, lipids are assembled from building blocks delivered earlier (by vesicles that fused with the ER membrane).

1 Some polypeptide chains enter the space inside rough ER. Modifications begin that will shape them into the final protein form.

SECRETORY PATHWAY

Endocytic vesicles form at the plasma membrane then transport substances into the cytoplasm. Some fuse with other organelles. Others move all the way through the cytoplasm and release their contents from the opposite side of the cell.

Some vesicles that bud from the ER or Golgi body transport proteins and lipids to other organelles or to the plasma membrane, where they are used to build new membrane. Some vesicles (including lysosomes) concentrate or store proteins or some lipids, either on their own or by fusing with other vesicles.

smooth ER

rough ER

DNA instructions for building polypeptide chains leave the nucleus and enter the cytoplasm.

The chains (green) are assembled on ribosomes in the cytoplasm.

Figure 2.10 Cytomembrane system. Lipids are assembled and many proteins take on their final form in the system. The cell itself uses the various lipids and proteins or secretes them.

Ribosomes

A ribosome consists of two subunits, each composed of RNA and many different protein molecules. **Ribosomes** are the sites where proteins are synthesized. Each new protein consists of one or more polypeptide chains.

Cytomembrane System

Polypeptide chains are assembled in the cytoplasm on many thousands of ribosomes. Many of the new chains become stockpiled in the cytoplasm; others pass through a series of organelles called the **cytomembrane system**. These organelles include the endoplasmic reticulum, Golgi bodies, and certain vesicles.

In this system, lipids as well as proteins take final form and are then packaged in vesicles. Some vesicles deliver proteins and lipids to regions where new membrane must be built. Others accumulate and store proteins or lipids for specific uses. Still other vesicles are part of a "secretory pathway" (Figure 2.10). They move to the plasma membrane, fuse with it, and release their contents to the surroundings.

Endoplasmic Reticulum Endoplasmic reticulum, or **ER**, is a membrane system that is continuous with the nuclear membrane and curves through the cytoplasm. It has rough and smooth regions, due largely to the presence or absence of ribosomes on the side facing the cytoplasm.

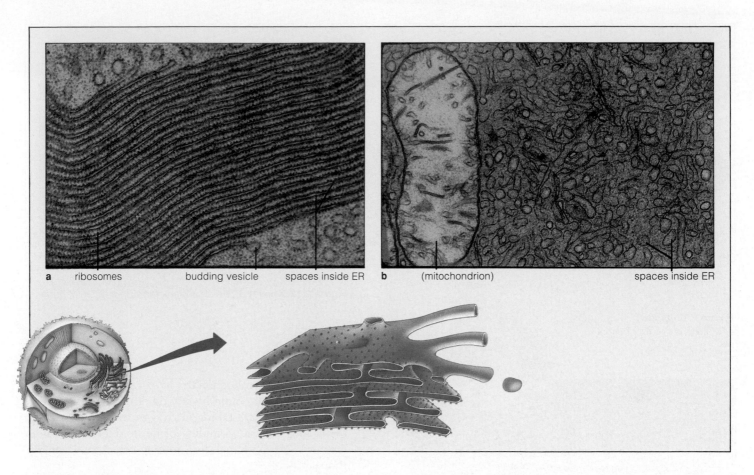

Figure 2.11 Endoplasmic reticulum. (**a**) Rough ER, showing how the membrane surface facing the cytoplasm is studded with ribosomes. (**b**) Smooth ER.

a ribosomes   budding vesicle   spaces inside ER   **b** (mitochondrion)   spaces inside ER

Rough ER often is arranged as stacked, flattened sacs with many ribosomes attached (Figure 2.11a). Polypeptide chains enter spaces inside rough ER, where some chains acquire side chains. Rough ER is abundant in cells that specialize in secreting digestive enzymes and other proteins.

Smooth ER is free of ribosomes and curves through the cytoplasm like connecting pipes (Figure 2.11b). It is the main site of lipid synthesis in many cells. The smooth ER of liver cells inactivates certain drugs and harmful byproducts of metabolism. Sarcoplasmic reticulum, a type of smooth ER in skeletal muscle cells, stores, releases, and re-stores calcium ions for muscle contraction.

Peroxisomes Peroxisomes are membrane-bound sacs of enzymes that break down fatty acids and amino acids. The reactions produce hydrogen peroxide, a potentially harmful substance. But before hydrogen peroxide can injure the cell, another enzyme within peroxisomes converts it to water and oxygen or uses it to break down alcohol. If you drink alcohol, nearly half of it is broken down in peroxisomes of your liver and kidney cells.

Golgi Bodies and Lysosomes A **Golgi body** somewhat resembles a stack of pancakes (Figure 2.12). The "pancakes" actually are flattened sacs in which lipids and proteins are modified. Specific modifications allow them to be sorted out and packaged in vesicles for transport to specific locations.

Vesicles form when portions of the topmost Golgi membrane bulge out and then break away. Among the vesicles are **lysosomes**, the cell's main organelles of digestion. Lysosomes fuse with other vesicles that contain cell substances or foreign particles brought into the cell by endocytosis. Different enzymes in lysosomes can break down virtually all proteins, complex sugars, nucleic acids, and some lipids.

Mitochondria

The energy of ATP drives nearly all cell activities. Reactions in **mitochondria** (singular: *mitochondrion*) form most of the body's ATP supply. Glucose, fatty acids, and other organic substances serve as raw materials for these reactions, which, as we will see shortly, require oxygen. The

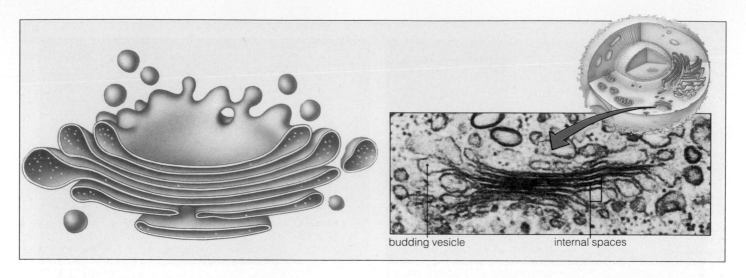

Figure 2.12 (above)   Sketch and micrograph of a Golgi body.

Figure 2.13 (below)   Micrograph and sketch of a mitochondrion.

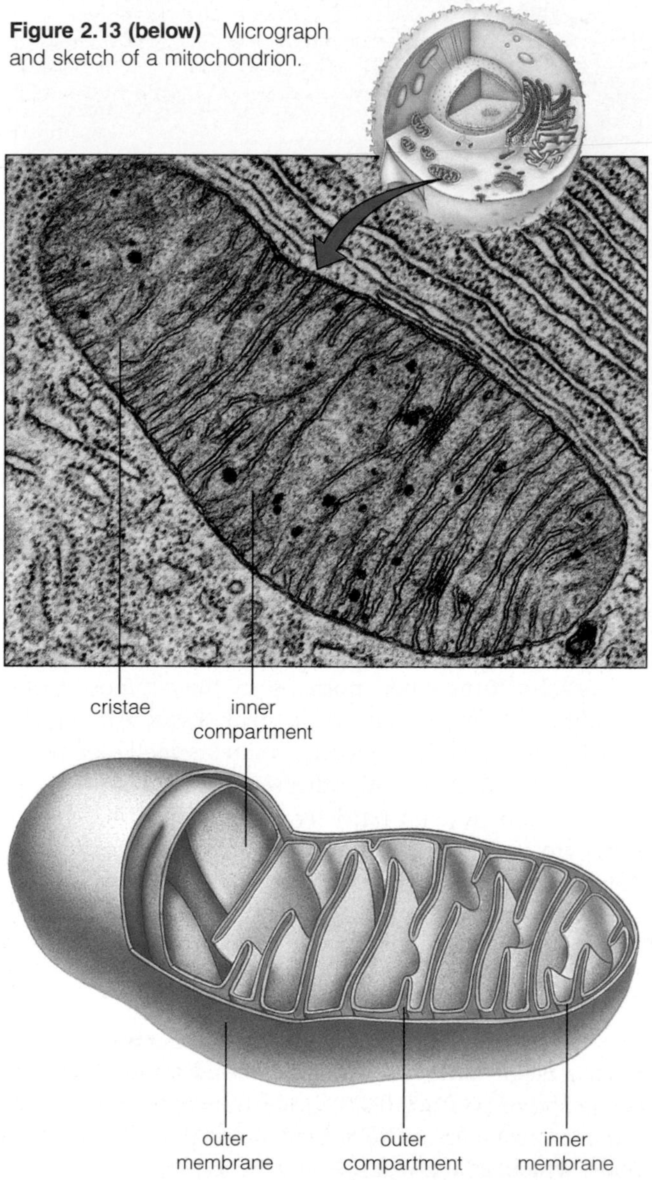

cristae   inner compartment

outer membrane   outer compartment   inner membrane

reactions enable mitochondria to extract far more energy than the cell can obtain by any other means. (When you breathe in, you are taking in oxygen for mitochondria.) Some cells have a sprinkling of these organelles; others (such as energy-demanding muscle cells), may contain thousands.

Each mitochondrion has two membranes: an outer membrane separating the organelle from the cytoplasm, and an inner membrane, usually with many deep, inward folds called *cristae* (Figure 2.13). The double-membrane system creates two compartments that are used in ATP formation.

Mitochondria have piqued the interest of biologists because they resemble bacteria. They also have their own DNA and a few ribosomes, and they divide independently of the cell. For these and other reasons, some researchers hypothesize that mitochondria may have evolved from bacteria that were consumed by an ancient cell, survived in that new environment, and later became modified into ATP-producing organelles.

The Cytoskeleton

An interconnected system of bundled fibers, slender threads, and lattices extends from the nucleus all the way to the plasma membrane. This is the **cytoskeleton**, which gives cells their shape and internal organization. In human cells its key components are tubelike **microtubules**, fine fibers called **intermediate filaments**, and **microfilaments**. All three components (Figure 2.14) are assembled from protein subunits and act like girders to support the cytoplasm.

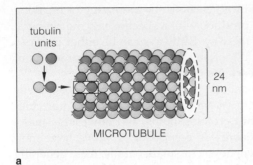 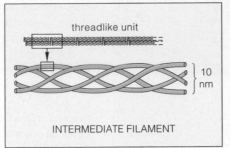 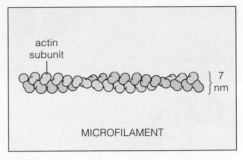

a

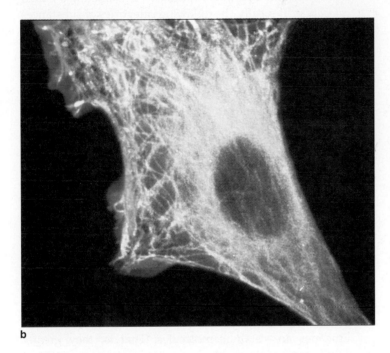

b

Figure 2.14 Cytoskeleton. Its components function in the internal organization, shape, and motion of eukaryotic cells. (**a**) Three classes of protein fibers in the cytoskeleton. *Microtubules* consist of protein subunits (tubulins) linked in parallel rows. *Intermediate filaments* consist of different proteins in different cell types. *Microfilaments* contain actin subunits. Microfilaments and thick filaments of myosin are the basis of muscle contraction. (**b**) Cytoskeleton of a fibroblast, a type of cell that gives rise to certain tissues in humans and other animals. In this composite of three images, microtubules are green and microfilaments are blue and red.

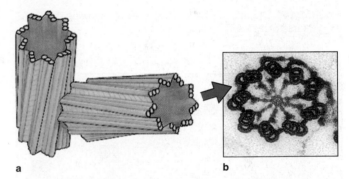

a   b

Figure 2.15 (**a**) A pair of centrioles, which occur near the nucleus of many cells. Centrioles apparently help organize flagella. (**b**) Electron micrograph of a basal body.

Some parts of the cytoskeleton appear only at certain times in a cell's life. Before a cell divides, for instance, some fibers assemble into a "spindle" structure that moves chromosomes; the fibers disassemble when the task is done. Each new cell that arises in the body develops a scaffold of microtubules that will dictate its shape. In some types of cells the scaffold is prominent; in other cell types it is less so. The scaffold's exact structure is determined by several factors, including small collections of proteins and other substances in the cytoplasm that form a "microtubule organizing center" (MTOC) or *centrosome*.

Flagella and Cilia   Many eukaryotic cells, including human sperm, have tail-like structures called **flagella** (singular: *flagellum*) that function in movement. Certain other cells have **cilia** (singular: *cilium*), which are shorter than flagella and more abundant at the cell surface. Cilia and flagella arise from **centrioles** (Figure 2.15a), which

are cylinders composed of several triplet microtubules. After the cilium or flagellum has been completed, the centriole remains attached to it as a *basal body* (Figure 2.15b). Both cilia and flagella have nine pairs of microtubules in a ring around two central microtubules, in a *9 + 2 array* (Figure 2.16).

In your own body, thousands of ciliated cells whisk mucus laden with dust and other undesirable material out of the respiratory tract. Cilia in the female reproductive tract waft the fertilized egg to the uterus, where it may implant and grow.

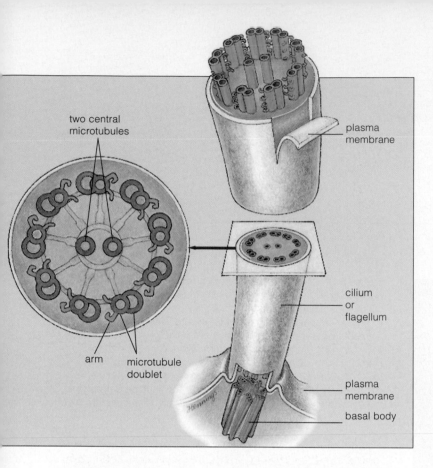

Figure 2.16 Structure of a cilium or flagellum, showing the 9 + 2 array of microtubules.

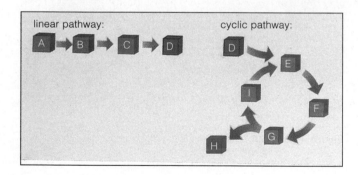

Figure 2.17 Metabolic pathways can be linear or cyclic.

LIFE AND ENERGY: METABOLISM

Metabolic Pathways

The 65 trillion living cells in your body have the capacity to acquire energy and use it to build, store, break apart, and eliminate substances in controlled ways. This dynamic activity is called **metabolism**.

Cells normally maintain, increase, and decrease the concentrations of substances by coordinating different metabolic pathways. A **metabolic pathway** is an orderly sequence of chemical reactions for producing or breaking down a specific product, with a specific enzyme acting at each step along the way. Pathways can be either linear or cyclic (Figure 2.17); often they connect, with products of one becoming reactants for others.

The main metabolic pathways are either biosynthetic or degradative. In *biosynthetic pathways*, small molecules are assembled into proteins, lipids, and other large molecules of higher energy. In *degradative pathways*, large molecules are broken down to smaller molecules, generally releasing energy. Biosynthetic activity is sometimes called *anabolism*, and degradative activity is sometimes called *catabolism*. Typically, the following categories of substances participate in a metabolic pathway:

reactants	substances able to enter a reaction; also called substrates or precursors
intermediates	compounds formed between the start and end of a metabolic pathway
enzymes	proteins that catalyze (speed up) reactions
cofactors	small molecules and metal ions that help enzymes or that carry atoms or electrons from one reaction site to another
energy carriers	mainly ATP, which donates energy to reactions
end products	substances present at the end of a metabolic pathway

Enzymes

Enzymes are catalytic molecules. That is, they greatly speed up the rate at which specific reactions occur. Most are proteins, and they have four features in common: (1) Enzymes do not make anything happen that could not happen on its own. Usually, however, they make it happen at least a million times faster. (2) Enzymes are not permanently altered or used up in the reaction; they can be used over and over again. (3) The same enzyme works for both the forward and reverse directions of most reactions. (4) Each enzyme works only with one or a few specific substrates.

Substrates are molecules that a particular enzyme can chemically recognize, bind, and modify in a specific way. For example, thrombin, an enzyme involved in blood clotting, recognizes and splits only one bond, the peptide bond between the amino acids arginine and glycine:

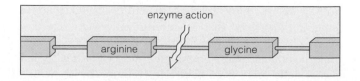

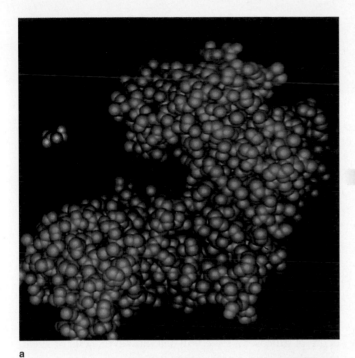

a

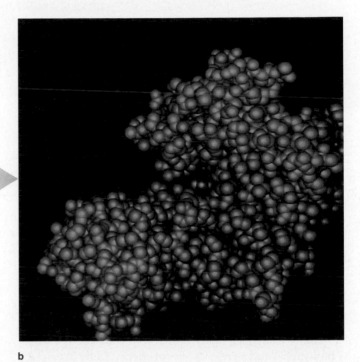

b

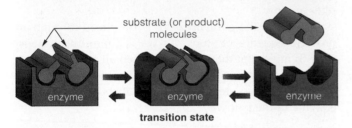

substrate (or product) molecules

enzyme   enzyme   enzyme

transition state

Figure 2.18 An enzyme at work. (**a**) Model of an enzyme (hexokinase) and its substrate (a glucose molecule, shown here in red). The cleft into which the glucose is heading is the enzyme's active site. (**b**) When the substrate makes contact with this active site, the enzyme temporarily changes shape. The upper and lower parts close in around the substrate, making a closer fit.

The most precise fit between enzyme and substrate occurs during a transition state that precedes the reaction. The enzyme-substrate complex is short-lived, partly because it is usually held together by weak bonds. The enzyme returns to its prebinding shape as a product molecule is released.

An enzyme has one or more **active sites** where substrates can interact with it. Many active sites are located in crevices in the enzyme's surface (Figure 2.18).

Enzymes function best within certain ranges of temperature and pH. Exposure to high temperatures often denatures and so destroys enzymes, thereby altering metabolism. Even high fevers can alter enzyme structure significantly enough to impair function. Humans usually die when body temperature reaches 44°C (112°F). Most enzymes in the human body function best at pH 7.35 to 7.4. Higher or lower pH values generally disrupt enzyme structure and activity (see page 25).

Cells maintain, increase, and decrease concentrations of substances by coordinating different metabolic pathways, and they do this mostly by controlling enzymes. Some controls govern enzyme synthesis and so reduce or increase the number of enzyme molecules that are operating at any time at a key step in a pathway. Other controls stimulate or inhibit the rate of activity of enzymes that are already formed. For example, when the body's supply of glucose runs low, a hormone called glucagon indirectly stimulates the key enzyme in the pathway by which glycogen in the liver is degaded to its glucose subunits, and it also inhibits the enzyme that catalyzes the formation of glycogen.

An enzyme speeds up the rate at which a reaction occurs.

Enzymes act only on specific substrates and function best within certain ranges of temperature and pH.

Enzyme Cofactors

During metabolic reactions, certain enzymes speed the transfer of electrons, atoms, or functional groups from one substrate to another. "Cofactors" help catalyze the reaction or serve briefly as transfer agents.

Cofactors include **coenzymes**, which are large organic molecules; **NAD⁺** (nicotinamide adenine dinucleotide) and **FAD** (flavin adenine dinucleotide) are examples. During glucose breakdown, they pick up unbound protons (H^+) and electrons and then transfer them to other reaction sites. **NADP⁺** (nicotinamide adenine dinucleotide phosphate), another coenzyme, transfers protons and electrons within other pathways. Many coenzymes are derived partly from vitamins. Niacin, for instance, is a precursor for NAD⁺ and NADP⁺.

Some metal ions serve as cofactors. For example, ferrous iron (Fe^{++}) assists certain membrane proteins (cytochromes) in mitochondria.

ATP: The Main Energy Carrier

Even though glucose and fatty acids are the basic fuels for metabolism, no cell can directly use the energy released as these molecules are broken down. First that energy must be converted to a usable form, ATP. ATP is a nucleotide, and it consists of adenine (a nitrogen-containing compound), ribose (a five-carbon sugar), and a triphosphate (three phosphate groups), all linked together by covalent bonds.

Many enzymes easily split off the outermost phosphate group and so release useful energy. Cells use energy from ATP for diverse tasks, as when they construct or break apart molecules, actively transport substances across membranes, or activate the contractile proteins in muscle cells. ATP molecules are like coins of a nation—they are the cell's common energy currency.

Cells have mechanisms for renewing ATP after it gives up phosphate. In the **ATP/ADP cycle**, an energy input drives the linkage of ADP (adenosine diphosphate) and a phosphate group (or unbound phosphate) into ATP. Then the ATP can donate a phosphate group elsewhere and revert back to ADP:

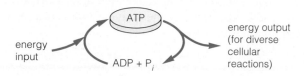

The ATP turnover is continuous. Even if you were bedridden for 24 hours, your cells would use and resynthesize about 40 kilograms (88 pounds) of ATP molecules simply for routine maintenance!

ATP directly or indirectly delivers energy to almost all metabolic pathways.

"Redox" Reactions

Many metabolic pathways involve *oxidation-reduction reactions*, or "redox" reactions for short. Whenever a substance taking part in a chemical reaction gives up electrons, it is said to be *oxidized*. Conversely, when a substance accepts electrons, it is said to be *reduced*. In general, "oxidation-reduction reaction" merely means an electron transfer. In the next section, we see how energy released during electron transfers in mitochondria is captured in ATP.

CELLULAR RESPIRATION

In human cells, *cellular respiration* releases energy from energy-rich organic molecules and changes ADP into ATP. The main ATP-producing pathway is **aerobic respiration**, and as the word *aerobic* suggests, this pathway uses oxygen.

Other ATP-producing pathways are *anaerobic*—they can be carried out without using oxygen. These pathways are much less efficient than aerobic respiration. Being large and highly active, you depend absolutely on the aerobic route, which produces much more ATP per molecule of glucose. As we will see, however, cells of skeletal muscle and some other tissues also can use an anaerobic pathway for short periods when oxygen supplies are low.

Whatever the pathway, ATP is produced through an orderly sequence of steps. As we examine these steps, keep in mind that they do not proceed all by themselves. Enzymes catalyze each one, and the intermediate produced at a given step serves as a substrate for the next enzyme in the pathway.

Aerobic Respiration: A High ATP Yield

Figure 2.19 presents the reactions of aerobic respiration, which take place in three stages. The first stage, *glycolysis*, occurs in the cell cytoplasm outside mitochondria. Its product molecules then enter a mitochondrion. Stage two, the *Krebs cycle*, occurs in the mitochondrion's inner compartment. Neither of these first two stages produces much ATP. But hydrogen and electrons are stripped from intermediates during both stages, and coenzymes deliver these to an electron transport system. Such systems, along with neighboring enzymes, serve as machinery for **oxidative phosphorylation**. This third and final stage, which takes place in the inner mitochondrial membrane (cristae), yields many ATP molecules. As it draws to a close, oxygen accepts the "spent" electrons from the transport system.

The different stages of aerobic respiration go on simultaneously. The following equation summarizes what the substances are at the start and end of the pathway:

$$C_6H_{12}O_6 + 6O_2 \longrightarrow 6CO_2 + 6H_2O$$

glucose

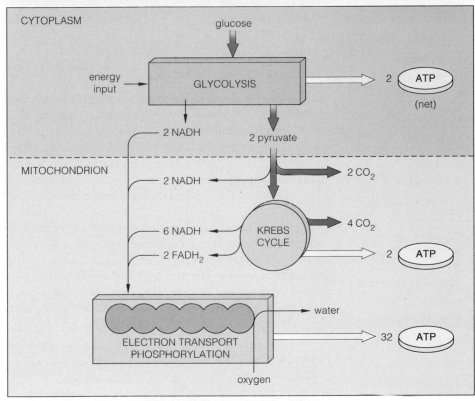

Figure 2.19 Overview of the three stages of aerobic respiration, the main energy-releasing pathway. Only this pathway delivers enough ATP for sustained, strenuous activity. In the first stage—glycolysis—glucose is partially broken down to pyruvate. In the second stage, which includes the Krebs cycle, pyruvate is completely broken down to carbon dioxide. During these two stages, hydrogen and electrons stripped from intermediates are loaded onto coenzymes (NAD^+ and FAD). In the final stage, electron transport and oxidative phosphorylation, the loaded coenzymes (NADH and $FADH_2$) give up hydrogen and electrons. As the electrons pass through transport systems, energy is released and drive ATP formation. Oxygen accepts electrons at the end of the transport system. From start (glycolysis) to finish, the aerobic pathway typically has a net energy yield of 36 ATP.

Typical Energy Yield: 36 ATP

Glucose and oxygen enter the pathway, and carbon dioxide and water are end products. Along the way, ATP is formed. To grasp where and how this all happens, we can focus on each stage separately.

Glycolysis

Recall that glucose is a simple sugar with a backbone of six carbon atoms. In **glycolysis**, glucose (or some other carbohydrate) present in the cytoplasm is partially broken down to **pyruvate**, a molecule with a backbone of three carbon atoms.

We can think about the glucose backbone in this simplified way:

The first steps of glycolysis are *energy-requiring*. They proceed only when two ATP molecules each donate energy to the glucose backbone by transferring a phosphate group to it. (Such transfers are called phosphorylations.) The attachments cause the backbone to split apart into two molecules of **PGAL** (phosphoglyceraldehyde).

PGAL formation marks the start of the *energy-releasing* steps of glycolysis. Each PGAL is converted to an unstable intermediate that gives up a phosphate group to ADP, forming ATP. The next intermediate in the sequence does the same thing. Thus, a total of four ATP molecules form by **substrate-level phosphorylation**—the direct transfer of a phosphate group *from a substrate of the reactions* to ADP.

Remember, though, that two ATP were invested to start the reactions. So the net energy yield is only two ATP.

Meanwhile, hydrogen and electrons that were released from each PGAL are transferred to the coenzyme NAD^+. Like other coenzymes, NAD^+ is "reusable." It picks up hydrogen and electrons stripped from a substrate and so becomes NADH. When it gives them up at a different site, it becomes NAD^+ again.

In sum, glycolysis converts a bit of glucose's stored energy to ATP energy. Hydrogen and electrons stripped from glucose have been loaded onto a coenzyme, and these have roles in the next stage of reactions. So do the end products of glycolysis—two molecules of pyruvate for each glucose molecule that entered the reactions.

Glycolysis is the partial breakdown of a glucose molecule. Two NADH and four ATP form during the reactions, which end with two pyruvate molecules.

When you subtract the two ATP required to start the reactions, the *net* energy yield of glycolysis is only two ATP.

Krebs Cycle

After pyruvate forms in the cytoplasm, it may enter a mitochondrion. The second and third stages of aerobic respiration proceed *only* in this organelle.

The second-stage reactions occur in the mitochondrion's inner compartment (see Figure 2.13). A little more ATP forms. All of pyruvate's carbon and oxygen atoms end up in carbon dioxide and water. And hydrogen and electrons stripped from it are loaded onto coenzymes.

During a few preparatory steps, a carbon atom is stripped from each pyruvate molecule, leaving an acetyl group that gets picked up by a coenzyme (forming acetyl-CoA). The acetyl group becomes attached to oxaloacetate, the point of entry into a cyclic pathway called the **Krebs cycle** (Figure 2.20). Three carbon atoms enter the second stage of reactions (as a pyruvate backbone) and three leave as part of three carbon dioxide molecules. (CO_2 produced during cellular respiration is transported in blood to the lungs, where it is exhaled.)

The Krebs cycle has three main functions. First, hydrogen and electrons from substrates are transferred to NAD^+ and FAD (another coenzyme), which thereby become NADH and $FADH_2$. Second, two substrate-level phosphorylations produce two ATP. Third, intermediates

are juggled back to the form of oxaloacetate. Cells have only so much oxaloacetate, and it must be regenerated to keep the cyclic reactions going.

The second stage adds only two more ATP to the small yield from glycolysis. But it also loads up many coenzymes with hydrogen and electrons that can be used during the third stage:

Glycolysis:		2NADH
Pyruvate conversion preceding Krebs cycle:		2NADH
Krebs cycle:	$2FADH_2$	6NADH
Total coenzymes sent to third stage:	$2FADH_2$ +	10NADH

Electron Transport and Oxidative Phosphorylation

ATP production goes into high gear during the third stage of the aerobic pathway. Electron transport systems and neighboring channel proteins are the production mechanism. These are embedded in the mitochondrion's inner membrane. The actual ATP-making mechanism, called **oxidative phosphorylation**, runs on the electrons and hydrogen ions (H^+) picked up by coenzymes in the first two stages.

Electron transport systems consist of organized sequences of enzymes and other molecules. As coenzymes deliver their hydrogen and electron cargoes, the electrons move steadily through the systems. Some of their energy is used to pump H^+ from the mitochondrion's inner compartment to the outer compartment, through enzymelike channel proteins:

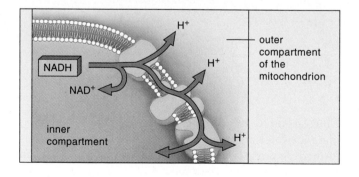

This pumping sets up a H^+ concentration gradient across the membrane. It is large enough that (following the gradient) H^+ flow *back* into the inner compartment through another channel protein, called ATP synthase. As its name suggests, this enzyme makes ATP. Using energy from the return flow of hydrogen ions, ATP synthase catalyzes a reaction that attaches another phosphate to ADP, forming ATP:

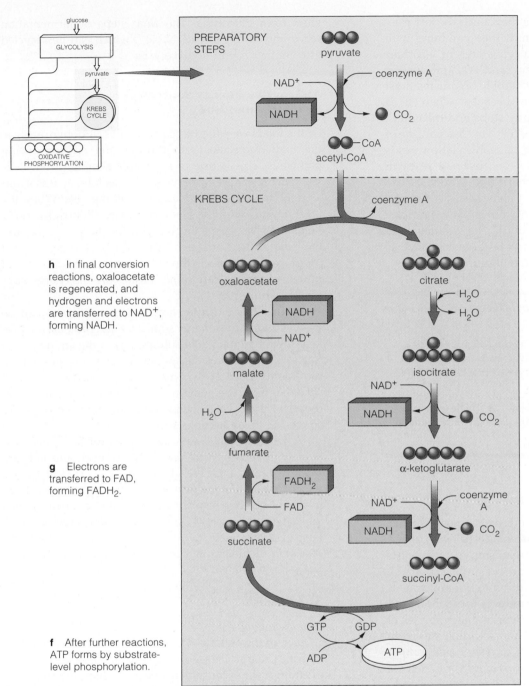

PREPARATORY STEPS

pyruvate

NAD⁺ → coenzyme A

NADH

CO₂

acetyl-CoA — CoA

KREBS CYCLE

coenzyme A

oxaloacetate

citrate

H₂O
H₂O

NADH

NAD⁺

malate

isocitrate

NAD⁺

NADH → CO₂

H₂O

fumarate

α-ketoglutarate

FADH₂

FAD

coenzyme A

NAD⁺

NADH

succinate

NADH → CO₂

succinyl-CoA

GTP   GDP

ADP   ATP

a A pyruvate molecule enters a mitochondrion. It undergoes preparatory conversions before entering cyclic reactions (Krebs cycle).

b First, the pyruvate is stripped of a functional group (COO^-), which departs as CO_2. Next, it gives up hydrogen and electrons to NAD^+, forming NADH. A coenzyme joins with the two-carbon fragment, forming acetyl-CoA.

c The acetyl-CoA is transferred to oxaloacetate, a four-carbon compound that is the point of entry into the Krebs cycle. The result is citrate, with a six-carbon backbone.

d Citrate enters conversion reactions in which a COO^- group departs (as CO_2). Also, hydrogen and electrons are transferred to NAD^+, forming NADH.

e Another COO^- group departs (as CO_2) and another NADH forms. *At this point, three carbon atoms have been released, balancing out the three that entered the mitochondrion (in pyruvate).*

h In final conversion reactions, oxaloacetate is regenerated, and hydrogen and electrons are transferred to NAD^+, forming NADH.

g Electrons are transferred to FAD, forming $FADH_2$.

f After further reactions, ATP forms by substrate-level phosphorylation.

glucose

GLYCOLYSIS

pyruvate

KREBS CYCLE

OXIDATIVE PHOSPHORYLATION

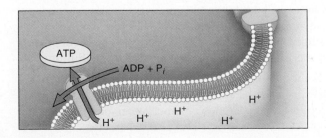

ATP

ADP + P$_i$

H⁺   H⁺   H⁺   H⁺

Figure 2.20 The Krebs cycle and the preparatory steps preceding it. For the sake of completeness, here we give the names of the various intermediate compounds that are produced with each turn of the cycle. However, the most important thing to notice is that for each three-carbon pyruvate molecule that enters a mitochondrion, three CO_2, one ATP, four NADH, and one $FADH_2$ are formed. The steps shown occur twice because the initial glucose molecule was broken down to two pyruvate molecules.

Oxygen picks up electrons as they reach the last reaction site in the transport system. It then combines with H^+ to form water. As long as oxygen is present, electrons flow steadily through the system, and ATP production keeps going. Humans and other aerobic organisms must continually breathe in oxygen because when it is depleted, ATP production abruptly stops and cells die.

The Krebs cycle yields two ATP. The mechanism of electron transport and oxidative phosphorylation yields roughly 32 ATP. We say "roughly" because new research on cellular respiration shows that many variables can affect the final ATP tally, and in some cases it may be lower. As a general rule of thumb, however, we can say that with the two ATP from glycolysis, the total net yield of the aerobic pathway in eukaryotic cells can total as much as 36 ATP for each molecule of glucose.

In aerobic respiration, glucose is completely broken down to carbon dioxide and water.

Coenzymes accept hydrogen and electrons stripped from substrates of the reactions and deliver them to an electron transport system. Oxygen is the final acceptor of those electrons.

From start (glycolysis in the cytoplasm) to finish (electron transport in the mitochondrion), this pathway may yield 36 ATP for every glucose molecule.

Anaerobic Routes: Making ATP Without Oxygen

Alcoholic beverages, soy sauce, sour cream, yeast breads, and a number of other foods are made by way of fermentation carried out by microorganisms such as yeasts and bacteria. **Fermentation** is an anaerobic route by which cells can make ATP without relying on oxygen as an electron acceptor. It rapidly produces a small amount of ATP. Human cells—in the hardworking leg muscles of a weightlifter or tennis player, say—call upon the pathway either for quick, short energy bursts or as a temporary backup when supplies of oxygen for aerobic respiration run low.

Most of the sequence of chemical steps in fermentation is simply glycolysis. In *lactate fermentation*, glycolysis produces two pyruvate and two NADH molecules, and two ATP form. Pyruvate itself then becomes the final acceptor of electrons from NADH and so is converted to the end product lactate. (This is the "fermenting" step; sometimes its product is called lactic acid.) Lactate is a waste product in cells; if not promptly removed, an excess can build up and cause muscle cramps.

Fermentation pathways do not break down glucose completely, so no more ATP is produced beyond the small yield from glycolysis. The final steps of fermentation serve only to regenerate NAD^+, which can be recycled into another round of glycolysis.

Alternative Energy Sources in the Human Body

Of the foods we eat, glucose and other carbohydrates are the main molecules sent through the ATP-producing pathways. This is true of mammals generally. Our cells use glucose for ATP production for as long as it is available. Brain cells can use almost nothing else. When the level of glucose in the blood decreases slightly, the body breaks into its stores of glycogen (in the liver), the animal's equivalent of starch. If the body becomes starved for glucose, it begins to tap into its fat reserves. Structural proteins in muscle and other tissues are the energy supply of last resort.

Figure 2.21 indicates how proteins and fats can be used as alternative energy sources. First, proteins are broken down to amino acid subunits; then the amino group (NH_3) of each subunit is split away. One way or another, the fragments enter the Krebs cycle, where hydrogen is stripped from the remaining carbon atoms and transferred to the cycle's coenzymes. The amino groups are converted to ammonia, a waste product.

A fat molecule, recall, has a glycerol head and one, two, or three fatty-acid tails. First the head and tails are separated, and then the glycerol is converted to PGAL, which enters the reaction sequence of glycolysis. Each fatty acid's carbon backbone is split into fragments, which are converted to acetyl-CoA and enter the Krebs cycle. A fatty acid has many more carbon atoms than does glucose, and its breakdown produces much more ATP.

SUMMARY

1. Each living cell has a plasma membrane surrounding an inner region called the cytoplasm. Eukaryotic cells, including cells of the human body, also have internal compartments called organelles. The largest organelle is the nucleus.

2. Cell membranes consist of phospholipids and proteins. A phospholipid bilayer gives the membrane its structure. Proteins in the bilayer or at one of its surfaces carry out most membrane functions. Properties of membrane phospholipids and proteins make the cell membrane only semipermeable to substances outside and within the cell.

3. Some membrane proteins actively or passively transport specific substances across the membrane. Others are receptors for hormones or other substances that trigger

alterations in cell behavior. Still other proteins have short carbohydrate chains that function in cell-to-cell recognition.

4. Substances move across cell membranes by several transport mechanisms. Simple diffusion is the random, unassisted movement of solutes from a region of higher to lower concentration. Osmosis is the movement of water across a membrane in response to concentration gradients. Facilitated diffusion is movement of a substance through the interior of a membrane protein, in the same direction that simple diffusion would take it. In active transport, a solute is pumped against its concentration gradient through the interior of a membrane protein. This operation requires an energy boost from ATP.

5. A metabolic pathway is a stepwise sequence of reactions in cells to produce or break down a given product. In biosynthetic pathways, large molecules are assembled and energy becomes stored in them. In degradative pathways, large molecules are broken down to smaller ones and energy is released. Enzymes serve as catalysts for reactions; most enzymes are proteins. Cofactors, such as the

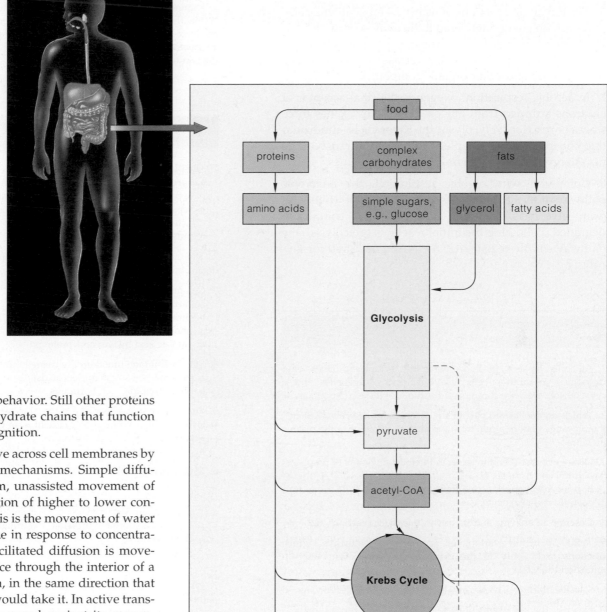

Figure 2.21 Points of entry into the aerobic pathway for proteins, complex carbohydrates, and fats consumed in the diet.

coenzyme NAD^+, help catalyze reactions or carry electrons. ATP is the main energy carrier in cells.

6. The metabolic reactions of nearly all biosynthetic pathways run on energy delivered to them by ATP molecules. ATP can be produced by aerobic respiration, fermentation, and other pathways that release chemical energy from glucose and other organic compounds.

7. In aerobic respiration, oxygen is the final acceptor of electrons stripped from glucose. The pathway has three stages of reactions: glycolysis, the Krebs cycle, and oxidative phosphorylation coupled with electron transport. Its net energy yield is commonly 36 ATP molecules.

8. Compared with aerobic respiration, the anaerobic pathway of fermentation has a small net yield (two ATP, from glycolysis) because glucose is not completely degraded. Following the initial reactions (glycolysis), the pathway simply regenerates the NAD^+ required for glycolysis.

Review Questions

1. All cells have a plasma membrane, cytoplasm, DNA, and ribosomes. Eukaryotic cells also have many organelles and a cytoskeleton. Describe the general functions of these components. *38*

2. Which organelles are part of the cytomembrane system? Sketch their arrangement in a cell, from the nuclear envelope to the plasma membrane. *40*

3. Distinguish between the terms in the following pairs of processes:
   a. diffusion; osmosis *41*
   b. passive transport; active transport *44*
   c. endocytosis; exocytosis *44*

4. Describe an enzyme and its role in metabolic reactions. *52*

5. Think about the various energy-releasing pathways. Which reactions occur only in the cytoplasm? Which occur only in mitochondria? *52*

6. State the function of coenzymes in the energy-releasing pathways. *53*

7. Describe the key events of each stage of aerobic respiration:
   a. glycolysis *55*
   b. the Krebs cycle (and preparatory conversions) *56*
   c. oxidative phosphorylation/electron transport *56*

8. Describe the functional zones of a mitochondrion. Explain where the electron transport systems and carrier proteins required for ATP formation are located. *55*

Critical Thinking: You Decide *(Key in Appendix IV)*

1. Using the *Focus on Science* as a reference, suppose you want to observe the surface of a microscopic section of bone. Would you

benefit most from using a compound light microscope, a transmission electron microscope, or a scanning electron microscope?

2. Jogging is considered aerobic exercise because the cardiovascular system (heart and lungs) can adjust its activity to supply the oxygen needs of working cells. In contrast, sprinting the 100-meter dash might be called "anaerobic" exercise, and golf "nonaerobic" exercise. Explain these last two observations.

Self-Quiz *(Answers in Appendix III)*

1. The plasma membrane _____.
   a. surrounds an inner region of cytoplasm
   b. separates the nucleus from the cytoplasm
   c. separates cell interior from the environment
   d. acts as a nucleus in prokaryotic cells
   e. only a and c are correct

2. The _____ is responsible for cell shape, internal structural organization, and cell movement.

3. Cell membranes consist mainly of a _____.
   a. carbohydrate bilayer and proteins
   b. protein bilayer and phospholipids
   c. phospholipid bilayer and proteins
   d. nucleic acid bilayer and proteins

4. Most membrane functions are carried out by _____.
   a. proteins   c. nucleic acids
   b. phospholipids   d. hormones

5. The passive movement of a solute through a membrane protein as it follows its concentration gradient is an example of _____.
   a. osmosis   c. diffusion
   b. active transport   d. facilitated diffusion

6. Match each organelle with its correct function.
   ____ protein synthesis   a. mitochondrion
   ____ movement   b. ribosome
   ____ intracellular digestion   c. smooth ER
   ____ modification of new proteins   d. rough ER
   ____ lipid synthesis   e. nucleolus
   ____ ATP formation   f. lysosome
   ____ ribosome subunit assembly   g. flagellum

7. Which of the following statements is *not* true? Metabolic pathways _____.
   a. occur in stepwise series of chemical reactions
   b. are speeded up by enzyme activity
   c. may degrade or assemble molecules
   d. always produce energy (i.e., ATP)

8. Enzymes _____.
   a. enhance reaction rates   c. act on specific substrates
   b. are affected by pH   d. all of the above are correct

9. Match each substance with its correct description.
   ____ a coenzyme or metal ion   a. reactant
   ____ substance formed at end of a   b. enzyme
        metabolic pathway   c. cofactor
   ____ mainly ATP   d. energy carrier
   ____ substance entering a reaction   e. end product
   ____ protein that catalyzes a
        reaction

10. Match each type of metabolic reaction with its function:

_____ glycolysis   a. ATP, NADH, and CO_2 form

_____ Krebs cycle   b. glucose to two pyruvate molecules

_____ oxidative   c. H^+ flows through channel
       phosphorylation   proteins, ATP forms

Key Terms

active site 53
active transport 44
aerobic respiration 54
ATP/ADP cycle 54
centriole 51
chromosome 47
cilium 51
coenzyme 53
concentration gradient 41
cytomembrane system 48
cytoplasm 38
cytoskeleton 50
diffusion 41
endocytosis 44
endoplasmic reticulum (ER) 48
exocytosis 44

FAD 53
fermentation 58
flagellum 51
fluid mosaic model 40
glycolysis 55
Golgi body 49
intermediate filament 50
Krebs cycle 56
lipid bilayer 40
lysosome 49
metabolic pathway 52
metabolism 52
microfilament 50
microtubule 50
mitochondrion 49
NAD^+ 53

$NADP^+$ 54
nuclear envelope 47
nucleolus 47
nucleus 38
organelle 45
osmosis 42
oxidation-reduction reaction 54
oxidative phosphorylation 56

passive transport 44
PGAL 55
plasma membrane 38
pyruvate 55
ribosome 48
substrate 52
substrate-level
    phosphorylation 56
vesicle 44

Readings

Kedersha, N. March 1992. "Psychedelic Cells." *Discover*. Text plus vivid color photographs of different types of human cells, obtained using the technique of immunofluorescence.

"Motor Molecules on the Move." 26 June 1992. *Science*. Summary of the latest findings on how elements of the cytoskeleton move cell structures such as organelles and chromosomes.

Sharon, N. and H. Lis. January 1993. "Carbohydrates in Cell Recognition." *Scientific American*.

Trefil, James. June 1992. "Seeing Atoms." *Discover*. How researchers use modern microscopes to study cell organelles, molecules, and even atoms.

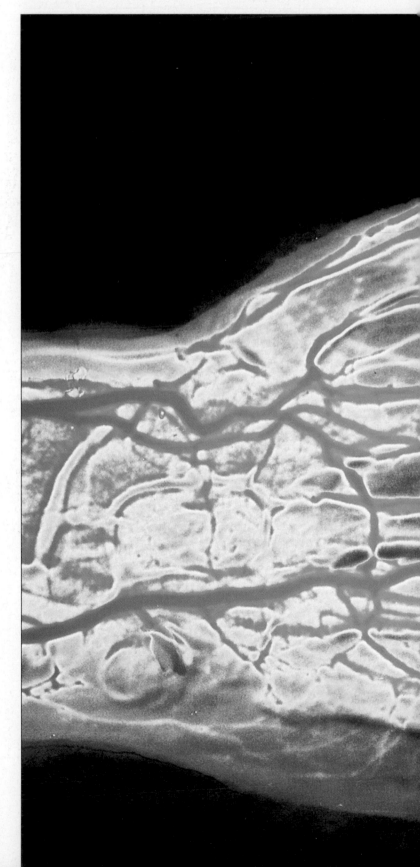

Intricate network of blood vessels in a human hand.

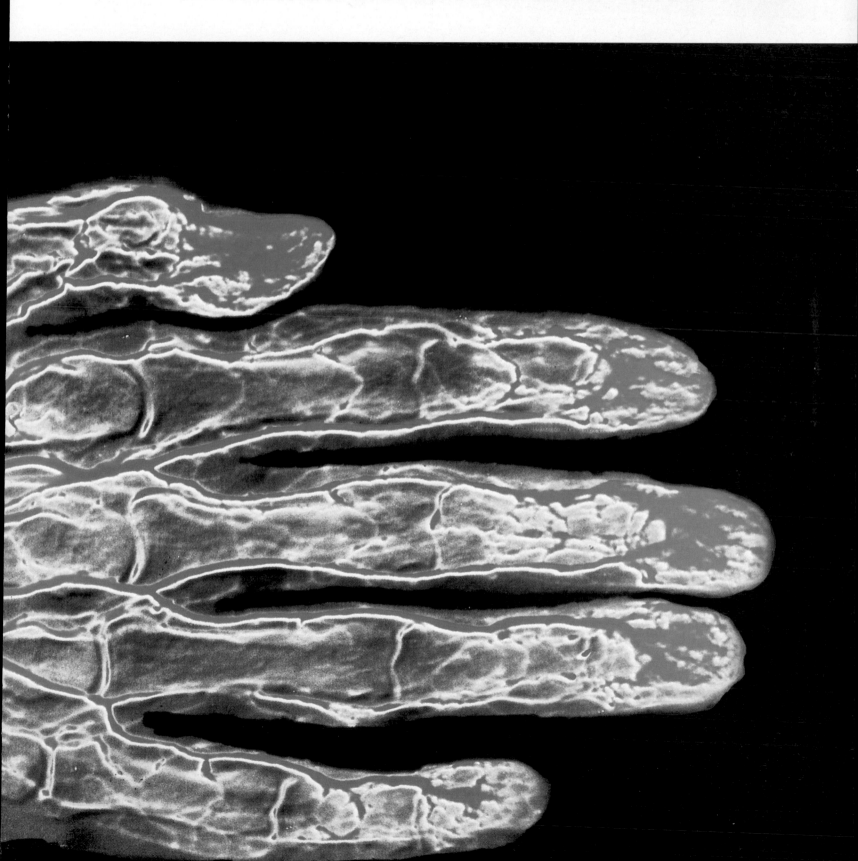

3 TISSUES, ORGAN SYSTEMS, AND HOMEOSTASIS

Go Ahead, Sweat

Let's conduct a little demonstration of the body's adaptive powers. Put a drop of food coloring into a few tablespoons of clear oil, then dab a bit of the mixture on your hand. Next, drink a cup or two of hot cocoa. Within minutes, your internal body temperature will begin to rise, and you'll probably begin to do something humans do remarkably well—you'll sweat. For proof positive, examine your hand under a dissection microscope. Like tiny glistening pearls under the tinted oil, hundreds of droplets of whitish, watery fluid will be rounding up from sweat glands. A square inch of human skin includes roughly 650 sweat glands; over the entire body of an average-sized adult, there are over 2 *million* sweat glands!

These numbers suggest that sweating is a vital adaptation for survival. In fact, it's a key element in *thermoregulation*—a process involving mechanisms such as sweating and shivering that help the body maintain internal body temperature within narrow limits (Figure 3.1). For humans that range is about 35°C to 37.2°C (95°F to 99°F). At temperatures much above or below it, cells may die and organs cease to function.

Figure 3.1 Working up a sweat. As we'll see in this chapter, sweating is just one of the ways the body helps maintain homeostasis, keeping internal operating conditions within life-sustaining limits.

But body functioning depends on more than sweat glands and a mechanism for heat loss. Any function begins with structure—the specializations of different types of cells, tissues, organs, and organ systems. For instance, the digestive system brings nutrients into the body, and the circulatory system transports them to body cells. Together, the circulatory and respiratory systems both supply cells with oxygen needed for energy metabolism and ATP production and relieve them of carbon dioxide wastes. The urinary system helps maintain the volume and composition of the vital fluids bathing body cells. Nervous and endocrine systems together convey information necessary to coordinate the entire operation and to maintain internal conditions that enable life to go on.

The central topics of this unit are the structure of the human body (its *anatomy*) and the mechanisms by which the body functions in the environment (its *physiology*). This chapter provides an overview of the tissues and organ systems that make up the human body. It also surveys the kinds of *homeostatic mechanisms* by which cells, tissues, organs, and organ systems work together in ways that maintain a stable environment inside the body.

KEY CONCEPTS

1. Cells interact at three levels of organization: in *tissues*, many of which are combined in *organs*, which are components of *organ systems*.

2. Organs are constructed of four major classes of tissues: epithelial, connective, nervous, and muscle tissues.

3. Each cell engages in basic metabolic activities that assure its own survival. At the same time, cells of a tissue or organ perform activities that contribute to the survival of the body as a whole.

4. The combined contributions of cells, tissues, organs, and organ systems help maintain homeostasis. This stable internal environment is required for individual cell survival.

THE HUMAN BODY: STRUCTURE AND FUNCTION

As with all other animals on earth, the human body is structurally and physiologically adapted to perform the following specialized tasks:

1. Maintain internal "operating conditions" within some tolerable range, even though external conditions change.

2. Locate and take in nutrients and other raw materials, distribute them through the body, and dispose of wastes.

3. Protect itself against injury or attack from viruses, bacteria, and other foreign agents.

4. Reproduce and often help nourish and protect new individuals during their early development.

Even though humans are extraordinarily complex animals, our organs are constructed of only four basic types of tissues. These are called epithelial, connective, muscle, and nervous tissues. A **tissue** is a group of cells and intercellular substances that function together in one or more specialized tasks just listed. An **organ**, such as the heart, consists of different tissues organized in specific proportions and patterns. An **organ system** consists of two or more organs that interact chemically, physically, or both in performing a common task, such as blood circulation. Cells, tissues, organs, and organ systems "divide the labor," so to speak, in ways that contribute to the survival of the body as a whole.

TISSUES

Tissues begin to form when an embryo embarks on its course of development. That journey starts when one of the father's sperm cells fertilizes one of the mother's eggs. The result is a zygote, which undergoes cell divisions that produce the early embryo. Chapter 14 describes the developmental mechanisms involved.

Cells in the early human embryo soon become arranged as three "primary tissues": ectoderm, mesoderm, and endoderm. These are the embryonic forerunners of all the specialized tissues of the body. *Ectoderm* gives rise to the skin's outer layer and to tissues of the nervous system. *Mesoderm* gives rise to muscle; to the organs of circulation, reproduction, and excretion; to most of the internal skeleton; and to connective tissue layers of the gut and skin. *Endoderm* gives rise to the lining of the gastrointestinal tract and the organs derived from it.

Let's now consider some of the features of the four types of specialized tissues present in the adult human body.

Epithelial Tissue

An epithelial tissue also is called an **epithelium** (plural: *epithelia*), and different types serve different functions. The epidermis of skin, for example, is an epithelium that forms a protective covering for the body's exterior surface. Epithelia that line parts of the gut and other internal cavities function in secreting or absorbing substances. Figure 3.2 shows a few examples of simple epithelial tissues.

Epithelial cells lie close together, with little intervening material. Specialized junctions provide structural and functional links between the cells, which are organized as one or more layers. Epithelium always has one free surface. In many cases, cells at the free surface have cilia or *microvilli* (singular: *microvillus*), which are fingerlike protrusions of cytoplasm. These surface modifications have special roles, as you will see in subsequent chapters. The other surface of epithelium adheres to a **basement membrane**—a noncellular layer, rich with proteins and polysaccharides, that lies between the epithelium and underlying connective tissue (Figure 3.3a).

Simple epithelium functions as a lining for body cavities, ducts, and tubes. It consists of a single layer of cells, which typically have roles in the diffusion, secretion, absorption, or filtering of substances across the layer. *Stratified epithelium* has two or more cell layers (Figure 3.3b); its typical function is protection, as at the skin surface. *Pseudostratified epithelium* consists of a single cell layer that looks like a double layer in cross section because the nuclei of adjacent cells are not aligned. Most of the cells are ciliated. Pseudostratified epithelium occurs in the throat, nasal passages, and other parts of the respiratory and reproductive tracts where the action of cilia sweeps mucus or some other fluid across the epithelial surface.

The different types of epithelium are subdivided into several categories, depending on the shape of cells at the free surface. A *squamous epithelium* has flattened cells, a *cuboidal epithelium* has cube-shaped cells, and a *columnar epithelium* has elongated cells. For example, oxygen and carbon dioxide diffuse readily across the simple squamous epithelium making up the wall of fine blood ves-

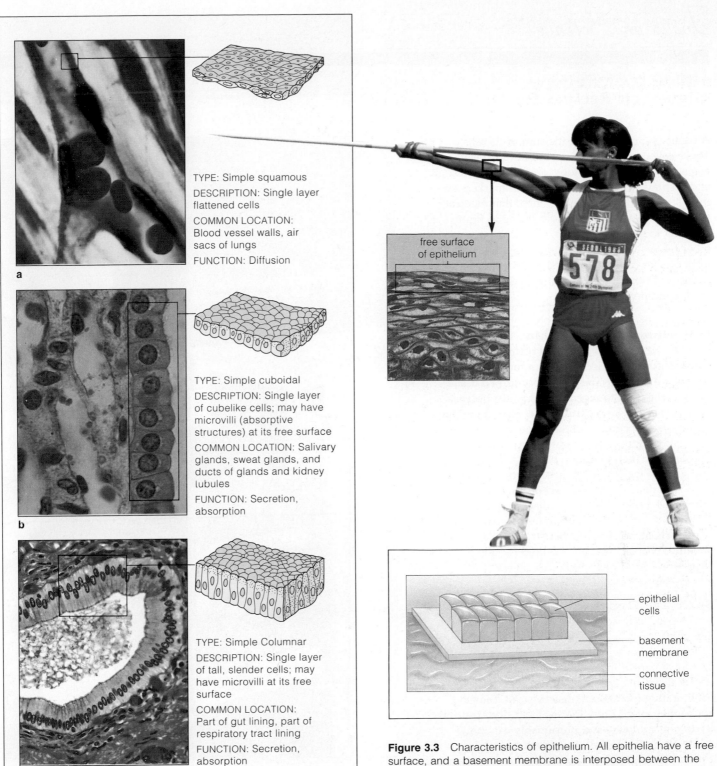

TYPE: Simple squamous

DESCRIPTION: Single layer flattened cells

COMMON LOCATION: Blood vessel walls, air sacs of lungs

FUNCTION: Diffusion

a

TYPE: Simple cuboidal

DESCRIPTION: Single layer of cubelike cells; may have microvilli (absorptive structures) at its free surface

COMMON LOCATION: Salivary glands, sweat glands, and ducts of glands and kidney tubules

FUNCTION: Secretion, absorption

b

TYPE: Simple Columnar

DESCRIPTION: Single layer of tall, slender cells; may have microvilli at its free surface

COMMON LOCATION: Part of gut lining, part of respiratory tract lining

FUNCTION: Secretion, absorption

c

Figure 3.2 Examples of simple epithelium, showing the three basic cell shapes in this type of tissue.

free surface of epithelium

578

epithelial cells

basement membrane

connective tissue

Figure 3.3 Characteristics of epithelium. All epithelia have a free surface, and a basement membrane is interposed between the opposite surface and an underlying connective tissue. The diagram shows simple epithelium, which consists of a single layer of cells. The micrograph shows the upper portion of stratified epithelium, which has more than one layer of cells. The cells become more flattened toward the surface.

High-Tech Recipes for Making Tissues

A tissue is more than just the sum of its cells. Those cells line up with one another in specific ways, form interconnections, produce enzymes, and display various kinds of protein markers on their surfaces. It's no wonder, then, that medical researchers have spent decades trying to figure out how to grow epithelium, muscle, bone, and other tissues in the laboratory. If successful, such technologies could help physicians treat severe burns, arthritis, muscle-wasting diseases, and other conditions.

Epithelium: Replacing Skin   Researchers have already devised a method for growing whole sheets of epidermis from a few of the patient's own cells. The laboratory-grown tissue can then be used to regenerate skin lost because of burns, ulcers, or other damage. First, a postage-stamp–sized section of undamaged epidermis is removed and its cells separated by enzyme action. The cells are then placed in bottles with nutrients and growth factors. In about a month, they proliferate into delicate, pink sheets of "custom-made" epidermis. When the layer is placed over a wound, it apparently orchestrates (probably by way of chemical cues) further structural and biochemical changes that are required to fully regenerate skin's multiple, specialized layers.

Space Age Tissues   A "bioreactor" device developed for Space Shuttle missions (Figure *a*) can grow cartilage, liver and lung cells, and other cell types, all of which behave much as they would as part of intact tissues. The device reduces physical "shear" stresses that can separate cells proliferating in a culture dish. Instead, cells in such a bioreactor grow in extremely close proximity to one another—and presumably can more easily exchange chemical signals that play key roles in tissue activity. Researchers are already using bioreactor cell cultures in attempts to develop treatments for a type of brain cancer and to study the interactions between tissues and disease-causing viruses.

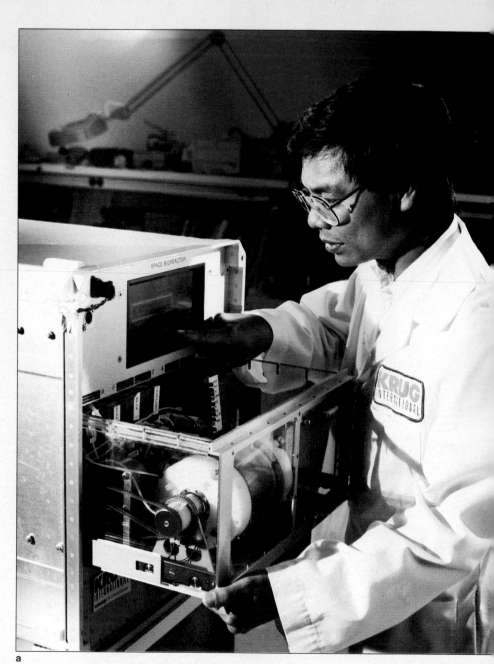

a

Rotating cell culture device developed by NASA researchers allows cells to differentiate into specialized forms.

Table 3.1 Major Types of Epithelium

Type	Shape	Typical Locations
Simple (one layer)	Squamous	Linings of blood vessels, lung alveoli (sites of gas exchange)
	Cuboidal	Glands and their ducts, surface of ovaries, retina of the eye
	Columnar	Stomach, intestines, uterus
Stratified (two or more layers)	Squamous	Skin (keratinized), mouth, throat, esophagus, vagina (nonkeratinized)
	Cuboidal	Ducts of sweat glands
	Columnar	Male urethra, ducts of mammary glands, throat (ciliated)
Pseudostratified	Columnar	Throat, nasal passages, sinuses trachea, male genital ducts

sels, as shown in Figure 3.2a. Table 3.1 summarizes the various types of epithelium and their roles. *Focus on Science* describes experimental efforts to grow epithelium and other tissues outside the body.

Glandular Epithelium Glands are single cells or multicellular secretory structures that are derived from epithelium and are often connected to it. Mucus-secreting *goblet cells*, for instance, are embedded in pseudostratified epithelium that lines the trachea and other tubes leading to the lungs. Stomach epithelium is a sheet of glandular cells that secrete mucus and other substances.

Glands are generally classified according to how their products reach the site where they are used. **Exocrine glands** secrete products onto a free epithelial surface, usually through ducts or tubes. The ducts have only a few branches in *simple* exocrine glands, whereas in *compound* glands the ducts have many branches. Exocrine products include substances such as mucus, saliva, earwax, oil, milk, and digestive enzymes.

Exocrine glands are sometimes subdivided into groups according to the makeup of their secretions. The secretions of *apocrine* glands (including some sweat glands and epithelial tissue in milk-secreting mammary glands) include bits of gland cells that produced them. In *holocrine glands*, such as sebaceous (oil) glands in skin, whole cells full of the material to be secreted are actually shed into the duct, where they burst and release the secretion. Most exocrine glands are *merocrine* glands; their secretions do not include elements of gland cells. Salivary glands and most sweat glands are like this.

Endocrine glands, such as the endocrine islets of the pancreas and the thyroid, have no ducts. As Chapter 12 describes more fully, their products—hormones—are secreted directly into the connective tissue underlying the gland cells and then are picked up and distributed by the bloodstream.

Cell-to-Cell Contacts in Epithelium With few exceptions, epithelial cells adhere strongly to one another by means of specialized attachment sites. These sites of cell-to-cell contact are especially close when substances must not leak from one body compartment to another.

Tight junctions between cells of epithelial tissues form gasketlike seals that keep molecules from freely crossing the epithelium and entering deeper tissues. This allows cells of the epithelium to control what enters the body. For example, during food digestion, various types of nutrient molecules can diffuse into epithelial cells or enter them by active transport, but tight junctions keep out other material.

Maculae adherens junctions or *desmosomes* are like spot welds at the plasma membranes of two adjacent cells. They are anchored to the cytoskeleton in each cell and help hold cells together in tissues that are subject to stretching, such as epithelium of the skin, heart, and stomach.

Epithelia also include *gap junctions*, which contain small, open channels that directly link the cytoplasm of adjacent cells. In heart muscle and smooth muscle, gap junctions provide rapid communication between cells. In liver and other tissues they allow small mole-

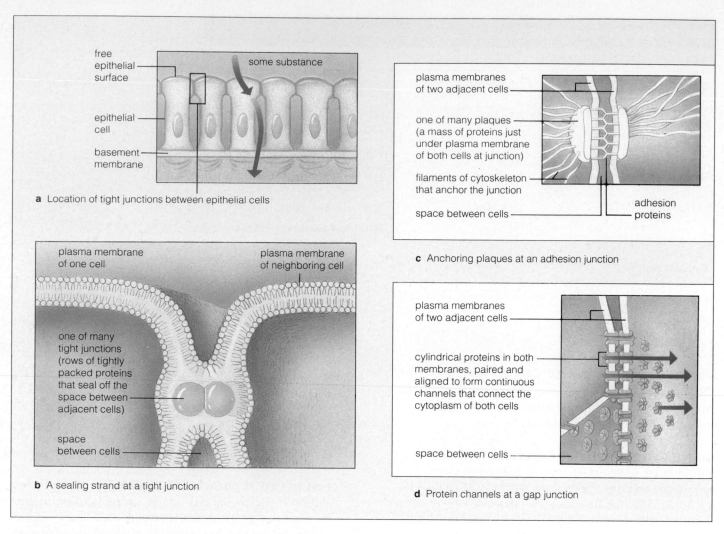

a Location of tight junctions between epithelial cells

b A sealing strand at a tight junction

c Anchoring plaques at an adhesion junction

d Protein channels at a gap junction

Figure 3.4 Examples of cell junctions. (**a**,**b**) In some epithelia, protein strands form tight seals that ring each cell and seal it to its neighbors. The seals prevent substances from leaking across the free epithelial surface. The only way they can reach tissues below is to pass *through* the epithelial cells, which have built-in mechanisms that can control their passage. (**c**) Adhesion junctions (desmosomes) are like spot welds that "cement" cells of epithelium (and all other tissues) together so that they function as a unit. They are abundant in the skin's surface layer and other tissues subjected to abrasion. (**d**) Gap junctions promote diffusion of ions and small molecules from cell to cell. They are abundant in the heart and other organs in which cell activities must be rapidly coordinated.

cules and ions to pass directly from one cell to the next, (Figure 3.4).

Connective Tissue

Name a body part or region, and one or more types of connective tissue will be found there. All are structurally specialized to serve many functions, as Table 3.2 shows. They range from those classified as "connective tissue proper" to cartilage, bone, adipose tissue, and blood.

Connective tissue differs from epithelium in several respects. In all types except blood, fibroblasts and certain other cells secrete fibers that serve as structural elements.

The fibers contain the protein collagen, which makes them strong, or elastin, which makes them elastic. (Plastic surgeons inject collagen to "plump up" skin wrinkles and sunken acne scars and to create fuller lips.) Fibroblasts also secrete a jellylike *ground substance* of proteins and polysaccharides. Together, the ground substance and fibers make up an *extracellular matrix*. The fiber-reinforced matrix fills spaces between connective tissue cells and helps connective tissues withstand physical stresses.

Connective Tissue Proper Tissues in this category have many of the same components, but in different proportions. *Loose connective tissue* (Figure 3.5a) has more types of cells and fewer, thinner fibers. These include loosely arranged collagen and elastin fibers, as well as fibroblasts and macrophages and other cells that migrate through tissues or take up residence in them. Macrophages perform housekeeping tasks and function as part of the immune system to help protect the body against invasion. Those in connective tissue underlying the skin, intestine, and respiratory and urinary tracts serve as a first line of defense when bacteria and other agents enter the body through cuts, abrasions, and other wounds. Loose connective tissue supports epithelia and many organs, and it surrounds blood vessels and nerves.

Dense, irregular connective tissue has thicker fibers and more of them, but far fewer types of cells. Its fibers interweave with one another, but not in a regular orientation. Dense connective tissues occur in parts of organs that are not subjected to continual stretching, such as the dermis, the layer of skin below the surface.

Specialized Connective Tissue Unlike the tissues just described, *dense, regular connective tissue* has its fibers in a parallel orientation (Figure 3.5b). In this arrangement, the collagen fibers strongly resist being pulled along the axis of their parallel orientation whenever the tissue is stretched. Often, parallel bundles of fibers have rows of fibroblasts between them; this is the arrangement in elastic ligaments (which attach bone to bone) and collagen-containing tendons (which attach muscle to bone).

Several types of **cartilage**, another specialized connective tissue, cushion and help maintain the shape of body parts and serve as a transition tissue in growing bones. Cartilage has a solid yet somewhat pliable matrix through which substances diffuse to and from nearby blood vessels. This pliability allows cartilage to resist compression and stay resilient, like a piece of solid rubber. However, the lack of blood vessels in cartilage makes it difficult for this tissue to heal when it is injured. Cells

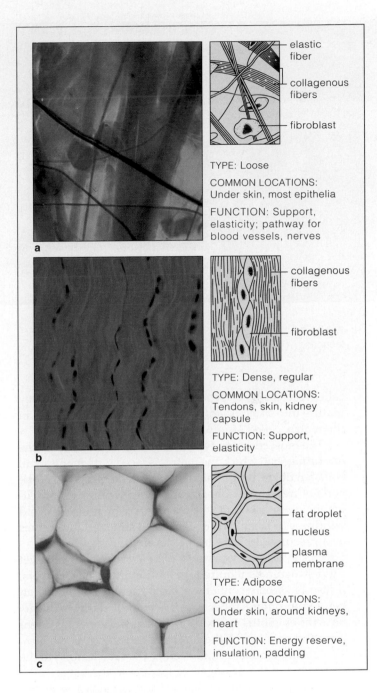

Figure 3.5 Examples of loose connective tissue, dense connective tissue, and adipose tissue.

Table 3.2 Types of Connective Tissue
Connective tissue proper:
Loose connective tissue
Dense, irregular connective tissue
Specialized connective tissue:
Dense, regular connective tissue (ligaments, tendons)
Cartilage
Bone
Adipose tissue
Blood

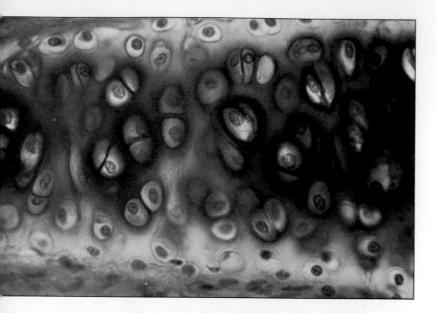

Figure 3.6 Cartilage cells (chondrocytes) inside and outside of lacunae.

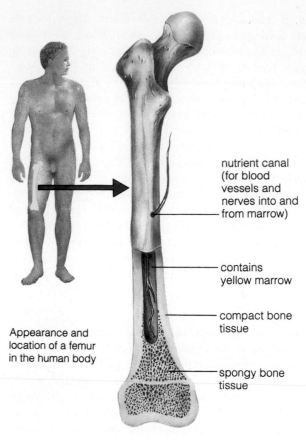

Appearance and location of a femur in the human body

nutrient canal (for blood vessels and nerves into and from marrow)

contains yellow marrow

compact bone tissue

spongy bone tissue

Figure 3.7 Two types of bone tissue in a human leg bone. Spongy bone tissue has tiny, slab-shaped hard parts with spaces between them. Compact bone tissue is much more dense.

called *chondroblasts* produce the cartilage matrix. They mature into *chondrocytes*, which occur in small cavities in the matrix called *lacunae* (Figure 3.6).

The most common type of cartilage in the body is *hyaline cartilage*, in which the matrix is laced with many small collagen fibers. It provides a friction-reducing cover at the ends of freely movable mature bones where they articulate in joints, and it makes up parts of the nose, ribs, and windpipe (trachea). In an embryo, hyaline cartilage serves as a precursor to bone in the developing skeleton.

Elastic cartilage occurs in places where a flexible yet rigid structure is required. In addition to collagen fibers, it also has fibers containing the protein elastin. You can bend your outer ear because it is built of elastic cartilage. So is the epiglottis, a flexible flap of tissue that folds down over the opening of the larynx whenever you swallow. As a result, food and liquids enter the tube leading to the stomach, not the passageway leading to the lungs.

Sturdy and resilient *fibrocartilage* is packed with collagen fibers arranged in thick bundles. It can withstand tremendous pressure, and it forms the cartilage "cushions" in joints such as the knee (page 98) and in the disks that separate vertebrae of the spinal column.

Bone is the weight-bearing tissue of the skeleton. It is the predominant tissue in bone organs, which support or protect softer body structures. Limb bones interact with attached muscles to bring about movement. Bones also function in storing calcium, and some produce red blood cells.

Unlike all other connective tissues, the bone matrix is mineralized and hardened; its collagen fibers and ground substance are loaded with calcium salts. Bone tissue is not completely solid, however; as with cartilage, lacunae occur within the ground substance. Some lacunae harbor living bone cells called *osteocytes* (Figure 4.3). Unlike cartilage, it has other spaces that contain blood vessels.

Figure 3.7 illustrates the two basic types of bone tissue. Dense **compact bone** makes up the shafts of long bones and the outer regions of all bones. To the unaided eye, compact bone appears solid. Using a light microscope, you can see that it is laced with tunnel-like channels that contain blood vessels and nerves. In contrast, the *spongy bone* inside the ends of long bones (and in the interior of bones of the skull, pelvis, and breastbone) looks a little like Swiss cheese. Hard, platelike struts separate large, marrow-filled spaces. This structure provides strength without the weight of compact bone. Chapter 4 considers bone structure and functions more fully.

The connective tissue called **adipose tissue** has large, densely clustered cells that are specialized for fat storage (Figure 3.5c). The animal body can store limited amounts of carbohydrate and can only store protein in muscle cells.

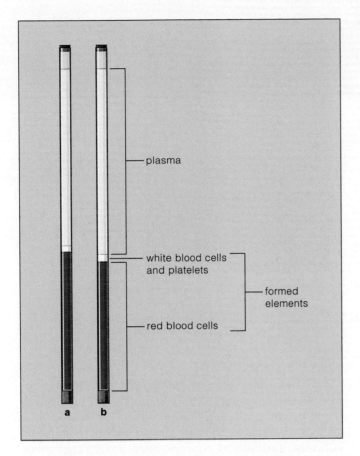

Figure 3.8 The distributions of plasma and formed elements in blood samples from a male (**a**) and a female (**b**). The relative percentage of formed elements in a given volume of blood is termed the *hematocrit*.

labels in figure: plasma; white blood cells and platelets; formed elements; red blood cells; a; b

Focus on Wellness

Future Fats

The rising popularity of low-fat diets reflects the fact that fats consumed in the diet add unwanted and sometimes unhealthy pounds in the form of excess adipose tissue. Animal fats, which contain cholesterol, also increase the risk of artery-clogging cholesterol deposits and heart disease. Yet, as described elsewhere in this text, the human body needs fat—as insulation, emergency energy reserves, and most of all as a basic component of cell membranes and a "solvent" for fat-soluble vitamins.

To satisfy the body's natural craving for fat but devise healthier prepared foods, food chemists have begun to tinker with fats and even create new ones. Some synthetic fats consist of laboratory-created molecules called sucrose polyesters. Their fatty acid units, derived from vegetable oils, are linked by chemical bonds that human digestive enzymes cannot break down. As a result, sucrose polyester passes through the digestive tract without adding calories or fat to the bloodstream. Clinical trials show that this fat substitute is safe for human consumption. However, because its chemical structure differs from that of natural fats, there is some concern that sucrose polyester can react with fat-soluble vitamins and therapeutic drugs in a way that prevents such needed substances from being absorbed. Products containing sucrose polyesters—marketed as "olestra"—might have to be vitamin-fortified to prevent vitamin deficiencies in people who eat foods containing it, without taking in sufficient quantities of "real" fats.

Excess amounts of carbohydrates and proteins are converted to storage fats that are tucked away in adipose cells. Adipose tissue has a rich supply of blood vessels, which serve as easily accessible "highways" for the movement of fats (or their components) to and from individual adipose cells. *Focus on Wellness* surveys new research aimed at developing edible, "nonfattening" fats that may help some people keep their excess adipose tissue to a minimum.

Blood is derived primarily from connective tissue and so is included in this category. It transports oxygen and nutrients to cells and carries wastes away from them; it also transports hormones and enzymes. Blood has several components. Its fluid portion, called *plasma*, serves as this tissue's matrix. The nonfluid portion of blood consists of cells and cell fragments collectively called *formed elements*. The formed elements consist mostly of red blood cells, which carry oxygen. About 5 percent consists of white blood cells, which function in immunity, and platelets, which have a major role in blood clotting (page 157). Platelets are cell fragments that have broken off from large cells in bone marrow. Figure 3.8 shows the relative amounts of different blood components in samples from a male and a female. Chapter 6 discusses blood and its complex functions in detail.

Muscle Tissue

All types of **muscle tissue** contain specialized cells that contract (shorten) in response to stimulation, then lengthen and so return to their original state. Muscle tissue helps move the whole body and its individual parts. There are three types of muscle tissue: skeletal, smooth, and cardiac.

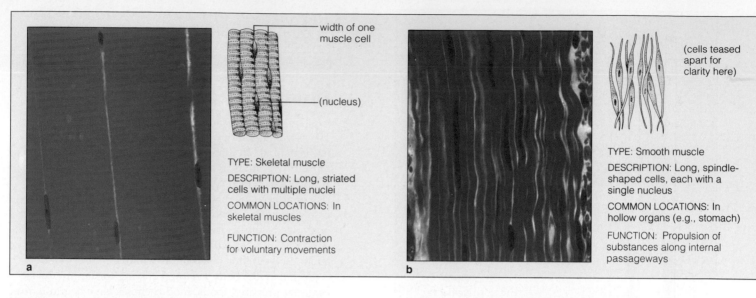

width of one muscle cell

(nucleus)

TYPE: Skeletal muscle

DESCRIPTION: Long, striated cells with multiple nuclei

COMMON LOCATIONS: In skeletal muscles

FUNCTION: Contraction for voluntary movements

a

(cells teased apart for clarity here)

TYPE: Smooth muscle

DESCRIPTION: Long, spindle-shaped cells, each with a single nucleus

COMMON LOCATIONS: In hollow organs (e.g., stomach)

FUNCTION: Propulsion of substances along internal passageways

b

Figure 3.9 Examples of skeletal, smooth, and cardiac muscle tissues.

Skeletal muscle is attached to the body's bones and functions in moving body parts. This tissue is made up of many long, cylindrical cells (Figure 3.9a) sometimes called *muscle fibers*. Each elongated cell has several nuclei, which are located not in the center of the cell but near the outer edges. This multinuclear arrangement comes about when several precursor cells fuse into a single cell, which then matures into a muscle fiber. As the fiber develops, contractile proteins (actin and myosin) accumulate within it. Actin and myosin eventually become organized into a banded pattern that makes skeletal muscle appear *striated* when it is viewed under the microscope.

Typically, skeletal muscle cells form bundles (called fascicles). Several bundles are then enclosed in a tough connective tissue sheath to form "a muscle." Chapter 4 describes how skeletal muscle tissue functions.

The contractile cells of *smooth muscle* are smaller than skeletal muscle fibers, and each has a single nucleus in the center of the cell. There are fewer contractile proteins and they are organized differently than in skeletal muscle, so smooth muscle does not appear striated. For the same reasons, it also does not contract as rapidly as skeletal muscle, but it can maintain constant tension over longer periods of time. The individual cells are tapered at both ends, and cell junctions hold them together (Figure 3.9b) and allow coordinated contraction. They are enclosed in connective tissue. Smooth muscle tissue occurs in walls of blood vessels, the stomach, the bladder, the uterus, and other internal organs. It is said to be "involuntary" because a person usually cannot directly control its contraction.

Cardiac muscle is the contractile tissue of the heart (Figure 3.9c). It is striated like skeletal muscle, but each cell has only a single nucleus. The plasma membranes of adjacent cardiac muscle cells are bound tightly together by adherens junctions at regions called intercalated disks. Gap junctions in the disks are passageways for signals that stimulate contraction, allowing individual heart muscle cells to contract as a unit. When one muscle cell receives a signal to contract, its neighbors are also stimulated into contracting.

Nervous Tissue

Nervous tissue contains cells called *neurons* that serve as lines of communication and control, as well as many accessory cells called *neuroglia* (or simply glia). Tens of thousands of neurons are centrally located in the brain, whereas others extend throughout the body. Each neuron consists of a cell body (which contains the nucleus and cytoplasm) and two types of cell processes. Branched processes called *dendrites* pick up incoming chemical messages (neurotransmitters) released from other neurons. Outgoing messages, called action potentials, are conducted by an *axon*. Depending on the type of neuron, its axon may extend outward a few millimeters or more than a meter. A *nerve* is a cluster of processes from several neurons. Nerves conduct messages from the central nervous system (CNS, consisting of brain and spinal cord) to muscles and glands, and from sensory receptors back to the CNS. Various types of accessory cells physically support or insulate neurons, help shuttle nutrients to them,

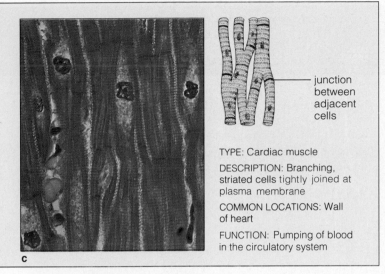

junction
between
adjacent
cells

TYPE: Cardiac muscle

DESCRIPTION: Branching, striated cells tightly joined at plasma membrane

COMMON LOCATIONS: Wall of heart

FUNCTION: Pumping of blood in the circulatory system

c

Figure 3.10 A sampling of the millions of neurons that form communication lines within and between different regions of the human body. Shown here are motor neurons, which relay signals from the brain or spinal cord to muscles and glands. Collectively, the body's neurons sense environmental change, integrate signals about those changes, and initiate appropriate responses.

and scavenge debris, microorganisms, or other foreign matter.

Some types of neurons detect specific changes in environmental conditions. Others coordinate the body's immediate and long-term responses to change. Still others relay signals to muscles and glands that can carry out those responses (Figure 3.10). Neuron structure and function are key topics in Chapter 10.

Some types of neurons detect specific changes in both internal and external environmental conditions. For example, as Chapter 11 describes, millions of specialized sensory neurons in epithelium lining the upper reaches of the nose respond to odor molecules. They are key to our senses of smell and taste. Sensory neurons in the retina of the eye are linked with receptors that respond to light, while different types in skin and muscle tissue are linked to receptors that detect pressure, stretching, and other stimuli.

Other neurons receive sensory input, integrate it with other information, and then trigger responses by influencing the activity of other neurons. In this way, they coordinate the body's immediate and long-term responses to change. Such cells occur mainly in the brain and spinal cord, and make up the central nervous system. Still other neurons are part of a peripheral nervous system that relays signals to muscles and glands that can carry out responses (Figure 3.10). The constant flow of information coordinates and regulates the activities of the body's billions of cells and the different tissues and organs they make up. Neuron structure and functioning will be key topics in Chapter 10.

Tissues begin to form when an embryo embarks on its course of development.

The four basic types of tissues in the human body are epithelial tissue, connective tissue, muscle tissue, and nervous tissue.

Epithelia have different functions, including protection and lining of body cavities and tubes. Connective tissues mainly provide structural support; blood has transport functions and other roles. Muscle tissue contains specialized cells that contract in response to stimulation. Nervous tissue contains cells that serve as lines of communication and control.

ORGAN SYSTEMS

As a human embryo develops, epithelial tissues, connective tissues, nervous tissue, and muscle tissue begin to form. The various tissues eventually become organized into organs, and then into organ systems, that are much

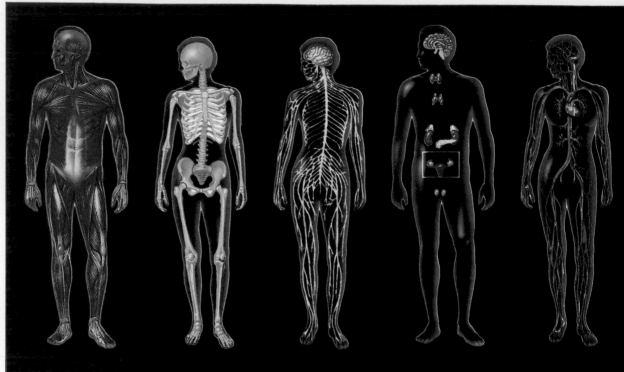

INTEGUMENTARY SYSTEM	MUSCULAR SYSTEM	SKELETAL SYSTEM	NERVOUS SYSTEM	ENDOCRINE SYSTEM	CIRCULATORY SYSTEM
Protection from injury and dehydration: body temperature control: excretion of some wastes; reception of external stimuli; defense against microbes	Movement of internal body parts: movement of whole body: maintenance of posture; heat production.	Support, protection of body parts: sites of muscular attachment, blood cell production, and calcium and phosphate storage.	Detection of external and internal stimuli: control and coordination of responses to stimuli; integration of activities of all organ systems.	Hormonal control of body functioning: works with nervous system in integrative tasks.	Rapid internal transport of many materials to and from cells; helps stabilize internal temperature and pH.

Figure 3.11 Organ systems of the human body. All vertebrates have the same types of systems serving similar functions.

the same in all vertebrates. Figure 3.11 shows the general arrangement of these systems in humans. In the next section we'll consider the integumentary system as an example of an organ system. Other organ systems will be our focus in subsequent chapters.

It might seem like an exaggeration to say that each organ system contributes to the survival of all living cells in the body. After all, what could the body's skeleton and musculature have to do with the life of a tiny cell in the brain or lungs? Yet interactions between the skeletal and muscular systems allow us to move about—toward sources of nutrients and water, for example. Parts of those systems help circulate blood through the body, as when leg muscle contractions help move blood in veins back to the heart. The bloodstream transports nutrients and other

substances to individual cells and carries secreted products and wastes away from them.

Throughout this unit we will be using some standard terms for describing the location of organs and organ systems in the vertebrate body. Take a moment to study Figure 3.12a, which shows the location of some major body cavities in which organs occur. The *cranial cavity* and *spinal cavity* house the central nervous system. Your heart and lungs reside in the *thoracic cavity*. A muscular diaphragm separates the thoracic cavity from the *abdominal cavity*, which holds the stomach, liver, most of the intestine, and other organs. Your reproductive organs, bladder, and rectum are located within the *pelvic cavity*.

Figure 3.12b defines some other anatomical terms that apply to most animals, including humans.

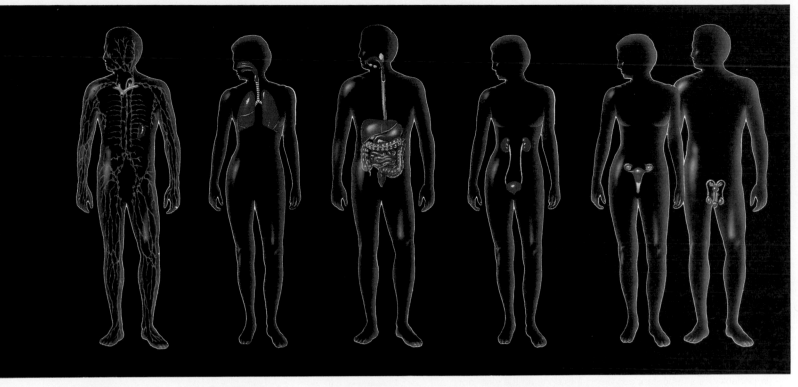

LYMPHATIC SYSTEM	RESPIRATORY SYSTEM	DIGESTIVE SYSTEM	URINARY SYSTEM	REPRODUCTIVE SYSTEM
Return of some tissue fluid to blood; roles in immunity (defense against specific invaders of the body).	Provisioning of cells with oxygen; removal of carbon dioxide wastes produced by cells; pH regulation.	Ingestion of food, water; preparation of food molecules for absorption; elimination of food residues from the body.	Maintenance of the volume and composition of extracellular fluid. Excretion of blood-borne wastes.	Male: production and transfer of sperm to the female. Female: production of eggs; provision of a protected, nutritive environment for developing embryo and fetus. Both systems have hormonal influences on other organ systems.

Figure 3.12 (**a**) Major cavities in the human body. (**b**) Directional terms and planes of symmetry for the human body. The midsagittal plane divides the body into right and left halves. The transverse plane divides it into superior (top) and inferior (bottom) parts. The frontal plane divides it into anterior (front) and posterior (back) parts.

cranial cavity

spinal cavity

thoracic cavity

diaphragm

abdominal cavity

pelvic cavity

a

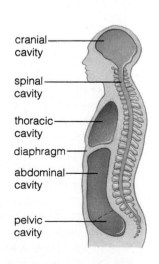

SUPERIOR (of two body parts, the one closer to head)

distal (farthest from trunk or from point of origin of a body part)

frontal plane (aqua)

midsagittal plane (green)

ANTERIOR (at or near front of body)

proximal (closest to trunk or to point of origin of a body part)

POSTERIOR (at or near back of body)

transverse plane (yellow)

INFERIOR (of two body parts, the one farthest from head)

b

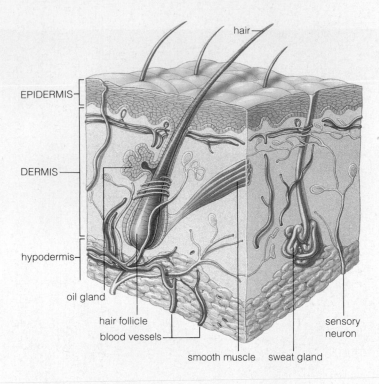

hair

EPIDERMIS

DERMIS

hypodermis

oil gland

hair follicle

blood vessels

smooth muscle   sweat gland

sensory
neuron

Figure 3.13 Two-layered structure of human skin. The hypodermis is a subcutaneous layer; it is beneath the skin, not part of it.

Membranes

Various types of membranes cover or line organs. All are thin and sheetlike and consist of epithelium together with underlying connective tissue. Some, like the *mucous membranes* that line the tubes and cavities of the digestive, respiratory, and reproductive systems, contain glands, including mucous glands that secrete mucus. In chapters to come we will see many examples of how mucous membranes offer protection and secrete or absorb various substances. *Serous membranes*, such as those that line the thoracic cavity and enclose the heart and lungs, do not contain glands. Among other functions, they help hold internal organs in place and provide lubricated smooth surfaces to prevent chafing or abrasion between adjacent organs.

INTEGUMENT: CASE STUDY OF AN ORGAN SYSTEM

Technically, your body's outer cover is an **integument** (after the Latin *integere*, meaning "to cover"). The integument is tough yet pliable, a barrier against a great variety of environmental insults. It includes the skin and the structures derived from epidermal cells of its outer lay-

ers of tissue. By weight and size, the integument is the largest organ of the body, and it is an excellent example of an organ that performs a wide range of functions.

Functions of Skin

No garment ever made approaches the qualities of the one covering your body—your skin. What besides skin maintains its shape in spite of repeated stretchings and washings, kills many bacteria on contact, screens out harmful rays from the sun, is waterproof, repairs small cuts and burns on its own, and, with a little care, will last as long as you do?

Skin does more than protect the rest of the body from dehydration, abrasion, and bacterial attack. It helps control the body's internal temperature. It has so many small blood vessels that it serves as one of the reservoirs for blood. The reservoirs can be tapped and shunted to metabolically active regions, such as leg muscles during strenuous exercise. Skin produces vitamin D, which is required for calcium absorption and metabolism. And signals from sensory receptors in skin help the brain assess what is happening in the outside world.

Structure of Skin

Assuming you are an average-sized adult, your skin weighs about 9 pounds. Stretched out, it would have a surface area of 15-to-20 square feet. Except for places subjected to regular pounding or abrasion (such as the palms of the hands and soles of the feet), human skin is generally not much thicker than a paper towel.

Skin has two distinct regions (Figure 3.13). The outermost region is the **epidermis**; the underlying region is the **dermis**. Beneath the dermis is a subcutaneous ("under the skin") layer, the *hypodermis*, a loose connective tissue that anchors the skin and yet allows it some freedom of movement. Fat stored in the hypodermis helps insulate the body against cold.

Epidermis The epidermis consists mostly of stratified squamous epithelium. Abundant cell junctions knit the epithelial cells together. The cells arise deep within the epidermis and are pushed toward its free surface as cell divisions produce new cells beneath them.

Most epidermal cells are keratinocytes. Each is a tiny factory for manufacturing *keratin*, a tough, water-insoluble protein. Keratinocytes start producing keratin when they are in mid-epidermal regions. By the time they reach the skin surface, they are dead and flattened. All that remain are fibers of keratin packed inside plasma membranes. This is the composition of the outermost layer of skin—the tough, waterproof *stratum corneum* (Figure

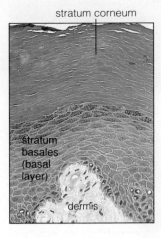
stratum corneum

stratum basales (basal layer)

dermis

Figure 3.14 Section through human skin, showing the uppermost layer of epidermis (stratum corneum); deeper epidermal layers, including the stratum basales (basal layer); and the underlying dermis. The dark spots in the basal layer are melanocytes. The most common skin cancer, basal cell carcinoma, arises in cells in this region. Figure 20.11 on page 441 shows examples of skin cancer lesions.

dead, flattened cells around a developing hair shaft

Figure 3.15 Close look at a hair. This scanning electron micrograph shows overlapping cells of the outer layers of a hair shaft, here emerging from the surface of skin.

3.14). Millions of the flattened keratin packages are worn off daily, but cell divisions continuously push up replacements. The rapid divisions also contribute to skin's capacity to mend itself quickly after cuts or burns.

In the deepest epidermal layer, cells called melanocytes produce *melanin*, a brownish-black pigment. This pigment is transferred to keratin-producing cells and accumulates inside them, forming a shield against ultraviolet radiation. Melanin also contributes to skin color. Skin color variations among humans are due to genetic differences in the distribution of and activity in melanocytes. For instance, the melanocytes of people with black or brown skin produce more and darker melanin than do melanocytes of fair-skinned people. Suntanning also increases melanocyte activity (see *Focus on Environment*, page 80). Skin color is influenced by *hemoglobin* (the oxygen-carrying pigment of red blood cells) and *carotene* (a yellow-orange pigment) in the dermal layer. For example, pale skin does not have much melanin, so the presence of hemoglobin is not masked and the skin looks pinkish.

With its multiple layers of keratinized and melanin-producing epidermal cells, skin helps the body conserve water, avoid damage from ultraviolet radiation, and resist mechanical stress.

Rapid, continuous cell divisions in deep epidermal layers enable skin to heal itself after being abraded, burned, or cut.

Dermis Dense connective tissue makes up most of the dermis, and it fends off damage from everyday stretching and other mechanical insults. There are limits to this protection. For example, the dermis tears when skin over the abdomen is stretched too much during pregnancy, leaving white scars ("stretch marks"). If persistent abrasion occurs on, say, a finger, the epidermis separates from

the dermis, the gap fills with a watery fluid, and a "blister" is the result.

Blood vessels, lymph vessels, and the receptor endings of sensory nerves thread throughout the dermis. Nutrients from the bloodstream reach epidermal cells by diffusing through the dermal tissue. Sweat glands, oil glands, and the husklike structures called hair follicles reside mostly in the dermis, even though they are derived from epidermal tissue. So are fingernails and toenails.

The fluid secreted from *sweat glands* is 99 percent water; it also contains dissolved salts, traces of ammonia and other metabolic wastes, vitamin C, and other substances. As the chapter introduction noted, the body has about 2.5 million sweat glands, which are controlled by sympathetic nerves. One type, which abounds in the palms of the hands, soles of the feet, forehead, and armpits, functions mainly in temperature regulation. These glands also function in "cold sweats," one of the responses a person makes when frightened, nervous, or merely embarrassed. Another type of sweat gland prevails in skin around the sex organs. Their secretion steps up during stress, pain, and sexual foreplay.

Oil glands (also called sebaceous glands) are everywhere except on the palms and soles of the feet. They function to soften and lubricate both the hair and the skin and to kill surface bacteria. *Acne* is a skin condition in which the ducts of oil glands have become infected by bacteria, and the glands have become inflamed.

Hairs are flexible structures composed mostly of keratinized cells. Each has a root embedded in skin and a shaft that projects above the skin's surface. As living cells divide near the base of the root, older cells are pushed upward, then flatten and die. The outermost layer of the shaft consists of flattened cells that overlap one another like roof shingles (Figure 3.15). The most abused of these cells tend to frizz out near the end of the hair shaft, causing "split ends."

Sunlight, the Ozone Layer, and the Skin

Melanin-producing cells of the epidermis are stimulated by exposure to ultraviolet (UV) radiation. With prolonged sun exposure, melanin levels increase and light-skinned people become tanned. Tanning provides some protection against UV radiation, but prolonged exposure can damage the skin. Over the years, tanning causes elastin in the dermis to clump together. The skin loses its resiliency and begins to look like old leather.

Prolonged exposure to UV radiation also suppresses the immune system. Certain specialized cells in the epidermis defend the body against specific viruses and bacteria. A sunburn interferes with the functioning of these cells. This may be why sunburns can trigger the small, painful blisters called "cold sores," which are a symptom of a viral infection. Nearly everyone harbors this virus (Herpes simplex); usually it becomes localized in a nerve ending near the skin surface, where it remains dormant. Stress factors—including sunburn—can activate the virus and trigger the skin eruptions. Sunscreen can help prevent eruptions in people harboring the virus.

Ultraviolet radiation from sunlight or from the lamps of tanning salons also can activate proto-oncogenes in skin cells (page 430). These snippets of DNA can trigger cancer, especially when some factor alters their structure (and hence their function).

Most of the UV radiation to which we are exposed comes from the sun. Until fairly recently, a layer of ozone in the stratosphere has intercepted much of the potentially damaging UV radiation (specifically ultraviolet-B). Today, however—due in large part to substances resulting from human activities described in Chapter 25—the ozone layer over ever larger regions of the globe is being destroyed faster than natural processes can replace it. The rate of skin cancers now is rapidly increasing. As you will read in Chapter 20, skin cancers are among the easiest cancers to cure—but only if they are surgically removed in time. As for the protective ozone layer, experts estimate that even if all ozone-depleting substances were banned tomorrow, it would take about 100 years for the planet to recover.

The average scalp has about 100,000 hairs, but genes, nutrition, and hormones influence hair growth and density. Protein deficiency causes hair to thin, for hair cannot grow without the amino acids required for keratin synthesis. Severe fever, emotional stress, and excessive vitamin A intake also cause hair thinning. Excessive hairiness (hirsutism) may result when the body produces abnormal amounts of testosterone. This hormone influences patterns of hair growth and other secondary sexual traits.

As we age, epidermal cells divide less often, and the skin becomes thinner and more susceptible to injury. Glandular secretions that once kept the skin soft and moistened start dwindling. Collagen and elastin fibers in the dermis break down and become sparser, so the skin loses its elasticity and wrinkles deepen. Excessive tanning, prolonged exposure to drying winds, and smoking accelerate skin aging processes. The *Human Impact* essay that begins on page 331 discusses the aging process in more detail.

Hair follicles, oil glands, sweat glands, and other structures associated with skin are derived from epidermal cells, but they are largely embedded in skin's underlying region, the dermis.

The dermis also contains blood and lymph vessels as well as receptor endings of sensory neurons.

HOMEOSTASIS AND SYSTEMS CONTROL

The Internal Environment

To stay alive, body cells must be continuously bathed in fluid that supplies them with nutrients and carries away metabolic wastes. In this they are no different from an amoeba or any other free-living, single-celled organism. However, trillions of cells are crowded together in your body, and all of them must draw nutrients from and dump wastes into the same 15 liters of fluid. That is less than four gallons!

The fluid outside of cells is called **extracellular fluid**. Much of it is *interstitial*, meaning it occupies the spaces between cells and tissues. The rest is blood plasma. Interstitial fluid exchanges substances with blood and with the cells it bathes.

Drastic changes in the composition and volume of extracellular fluid have drastic effects on cell activities. Its concentrations of hydrogen, potassium, calcium, and other ions are especially vital and must be maintained at levels that are compatible with cell survival. Otherwise, the body as a whole cannot survive.

This brings us to a concept that is central to understanding the structure and function of the human body: *The component parts of the body work together to maintain the stable fluid environment required by its living cells*. The three basic elements of this concept are:

1. Each body cell engages in metabolic activities that promote its own survival.

2. Concurrently, the cells of a given tissue typically perform one or more activities that contribute to the survival of the whole organism.

3. The combined contributions of individual cells, organs, and organ systems help maintain the extracellular fluid. In so doing, they sustain a stable internal environment that permits individual cells to survive.

Mechanisms of Homeostasis

Homeostasis literally means "staying the same." It refers to stable operating conditions in the body's internal environment. Three components—sensory receptors, integrators, and effectors—interact to maintain this state. **Sensory receptors** are cells or parts of cells that can detect a specific change in the environment. For example, when someone kisses you, there is a change in pressure on your lips. Receptors in the skin of your lips translate the stimulus energy into a signal, which can be sent to the brain. The brain is an **integrator**, a control point where different bits of information are pulled together in the selection of a response. It can send signals to your muscles or glands (or both), which are **effectors** that carry out the response. In this case, the response might include flushing with pleasure and kissing the person back. You cannot engage in a kiss indefinitely, however, because eventually you must eat and perform other tasks that maintain your body's operating conditions.

How does the brain reverse the physiological changes induced by the kiss? Receptors only provide it with information about how things *are* operating. The brain also maintains information about how things *should be* operating—that is, information from "set points." When physical or chemical conditions deviate significantly from a set point, the brain functions to bring them back to within an effective operating range. It does this by way of signals that cause specific muscles and glands to increase or decrease their activity. Set points are key elements of many physiological mechanisms, including those that influence eating behavior, breathing, thirst, and urination, to name a few.

The controls that help keep internal conditions within tolerable ranges include feedback mechanisms. In a **negative feedback mechanism** (Figure 3.16), an activity alters a condition in the internal environment, and this triggers a response that reverses the altered condition. To grasp how the mechanism operates, think of a home furnace with a thermostat. The thermostat senses the air temperature and mechanically "compares" it to a preset point on a thermometer built into the furnace control system. When the temperature falls below the preset point, the thermostat signals a switching mechanism that turns on the heating unit. When the air becomes heated enough to match the prescribed level, the thermostat signals the switching mechanism, which shuts off the heating unit.

Similarly, humans (and many other animals) rely on feedback mechanisms to raise or lower body temperature so that it is maintained within a normal range, even dur-

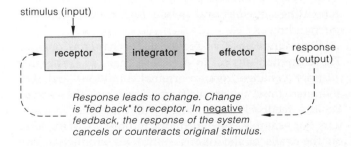

Figure 3.16 Components necessary for negative feedback at the organ level.

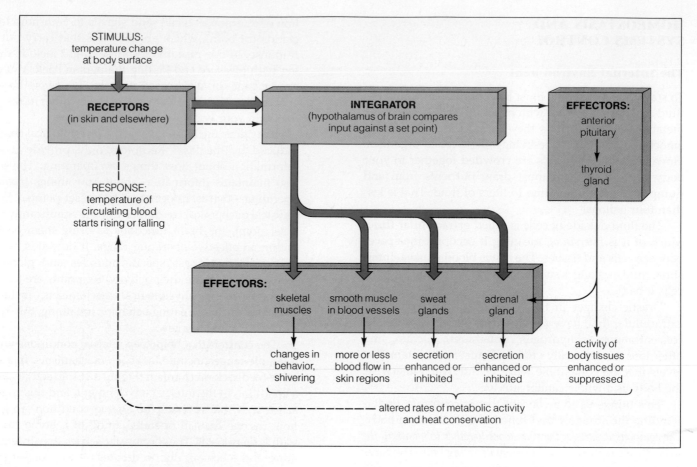

Figure 3.17 Homeostatic controls over the internal temperature of the human body. The dashed line shows how the feedback loop is completed. The blue arrows indicate the main control pathways.

ing extremely hot or cold weather or under other conditions that alter temperature (Figure 3.17). For example, when the body senses that its skin is getting too hot while you work outside in the summer sun, mechanisms are set in motion that slow down metabolic and overall body activity. You move more slowly and may seek the shelter of a shade tree. At the same time, blood flow to the skin increases and your sweat glands are prodded to secrete larger quantities of sweat. As water in sweat evaporates, more heat is lost from the body. These and other mechanisms counter overheating by curbing the body's heat-generating activities and giving up excess heat to the surroundings.

Sometimes **positive feedback mechanisms** operate. These mechanisms set in motion a chain of events that *intensify* a change from an original condition—and after a limited time, the intensification reverses the change. Positive feedback is associated with instability in a system. For example, during childbirth pressure of the fetus on the walls of the uterus stimulates production and secretion of the hormone oxytocin. Oxytocin causes muscles in the walls to contract and exert pressure on the

fetus, which increases pressure on the uterine wall, and so on until the fetus is expelled.

Homeostatic control mechanisms maintain physical and chemical aspects of the internal environment within ranges that are most favorable for cell activities.

What we have been describing here is a general pattern of monitoring and responding to a constant flow of information about the external and internal environments. During this activity, organ systems operate together in coordinated ways. Throughout this unit we will be asking the following questions about organ systems:

1. What physical or chemical aspect of the internal environment are organ systems working to maintain as conditions change?

2. How are organ systems kept informed of change?

3. How do they process incoming information?

4. What mechanisms are set in motion in response?

As you will see in the chapters to follow, all organ systems operate under neural and endocrine control.

SUMMARY

1. A tissue is a group of cells and intercellular material that together perform a specialized activity. An organ is a structural unit in which tissues are combined in definite proportions and patterns that allow them to perform a common task. In an organ system, two or more organs interact chemically, physically, or both in ways that contribute to the survival of the animal as a whole.

2. Epithelial tissues cover external body surfaces and line internal cavities and tubes. Each epithelial tissue has one or more layers of closely adhering cells with no intercellular material between. Epithelium has one free surface exposed to body fluids or the external environment; the opposite surface rests on a basement membrane that intervenes between it and an underlying connective tissue.

3. Connective tissues bind together other tissues or provide them with mechanical or metabolic support. They include connective tissue proper and specialized connective tissues (such as cartilage, bone, and blood).

4. Muscle tissue has contractile cells. It functions in moving the body or its individual parts. Nervous tissue detects, transmits, and coordinates information about change in the internal and external environments, and it controls responses to change.

5. Tissues, organs, and organ systems work together in ways that help maintain a stable internal environment (the extracellular fluid) required for individual cell survival. At homeostasis, conditions in the internal environment are most favorable for cell activities.

6. An example of an organ is skin (integumentary system), which protects the rest of the body from abrasion, bacterial attack, ultraviolet radiation, and dehydration. It helps control internal temperature, and it serves as a blood reservoir for the rest of the body. Its receptors are essential in detecting environmental stimuli.

7. Feedback controls help maintain internal conditions. In negative feedback (the most common mechanism), a change in some condition triggers a response that reverses the change. In positive feedback, a response reverses a change in some condition by intensifying the change for a limited time.

8. Homeostasis depends on receptors, integrators, and effectors. Receptors detect stimuli, which are specific changes in the environment. Integrating centers (such as the brain) process the information and direct muscles and glands—the body's effectors—to carry out responses.

Review Questions

1. Identify the following tissues:

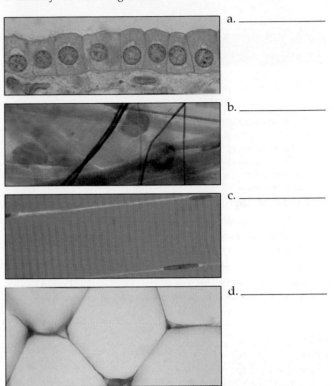

a. _____

b. _____

c. _____

d. _____

2. What is a tissue? An organ? An organ system? List the major organ systems of the human body, along with their functions. *67–76*

3. Define extracellular fluid and interstitial fluid. *81*

4. Epithelial tissue and connective tissue differ from each other in overall structure and function. Describe how. *70*

5. State the overall functions of (a) muscle tissue and (b) nervous tissue. *74*

6. What are some of the functions of skin? List some derivatives of epidermis. *78*

7. A major concept in human physiology relates the functioning of cells, organs, and organ systems to the internal environment. Can you state the three main points of this concept? *81*

8. Define homeostasis. What are the three components necessary for homeostatic control over the internal environment? *81*

9. What are the differences between negative feedback and positive feedback mechanisms? *81–82*

1. A third-degree burn destroys both the dermis and epidermis layers of skin. What kinds of hazards to homeostasis are posed by a serious burn over a large percentage of the body?

2. The disease called scurvy results from a deficiency of vitamin C, which the body requires for collagen synthesis. Explain why (among other symptoms) scurvy sufferers tend to lose teeth, and why any wounds heal much more slowly than normal.

Self-Quiz *(Answers in Appendix III)*

1. The four main types of tissues in the human body are _____, _____, _____, and _____.

2. The human body is structurally and functionally adapted for which of the following activities:
   a. maintenance of the internal environment
   b. nutrient acquisition, processing, and distribution
   c. self-protection against injury or attack
   d. reproduction
   e. waste disposal

3. _____ tissues cover external body surfaces, line internal cavities and tubes, and some form the secretory portions of glands.
   a. Muscle   c. Connective
   b. Nervous   d. Epithelial

4. Most _____ tissues bind or mechanically support other tissues, but one type provides physiological support of other tissues.
   a. muscle   c. connective
   b. nervous   d. epithelial

5. Which of the following is *not* a function of the integumentary system?
   a. protect the body from abrasion
   b. protect the body from dehydration
   c. detect environmental stimuli
   d. bring about body movements
   e. serve as a blood reservoir for the rest of the body

6. _____ tissues detect and coordinate information about environmental changes and control responses to those changes.
   a. Muscle   c. Connective
   b. Nervous   d. Epithelial

7. _____ tissues contract and make possible internal body movements as well as movements through the external environment.
   a. Muscle   c. Connective
   b. Nervous   d. Epithelial

8. Cells in the human body _____.
   a. engage in metabolic activities that ensure their survival
   b. perform activities that contribute to the survival of the body as a whole

   c. contribute to maintaining the extracellular fluid
   d. digest nutrients

9. In a state of _____, physical and chemical aspects of the body are being kept within tolerable ranges by controlling mechanisms.
   a. positive feedback   c. homeostasis
   b. negative feedback   d. metastasis

10. In negative feedback mechanisms, _____.
    a. a detected change brings about a response that tends to return internal operating conditions to the original state
    b. a detected change suppresses internal operating conditions to levels below the set point
    c. a detected change raises internal operating conditions to levels above the set point
    d. fewer solutes are fed back to the affected cells

11. Fill in the blanks: _____ detect specific environmental changes, an _____ pulls different bits of information together in the selection of a response, and _____ carry out the response.

12. Match the concepts:
    ____ muscles and glands   a. integrating center
    ____ positive feedback   b. the most common homeostatic mechanism
    ____ sites of body receptors   c. eyes and ears
    ____ negative feedback   d. effectors
    ____ brain   e. chain of events that intensifies the original condition

Key Terms

adipose tissue *72*
basement membrane *66*
blood *73*
bone *72*
cartilage *71*
compact bone *72*
connective tissue *70*
dermis *78*
effector *81*
endocrine gland *69*
epidermis *78*
epithelium *66*
exocrine gland *69*

extracellular fluid *81*
gland *69*
homeostasis *81*
integrator *81*
integument *78*
muscle tissue *73*
negative feedback mechanism *81*
nervous tissue *74*
organ *66*
organ system *66*
positive feedback mechanism *82*
sensory receptor *81*
tissue *66*

Readings

Bloom, W. and D. W. Fawcett. 1986. *A Textbook Of Histology,* 11th ed. Philadelphia: Saunders. Outstanding reference text.

Green, H. November 1991. "Cultured Cells for the Treatment of Disease." *Scientific American.*

Human Impact

INFECTION AND HUMAN DISEASE

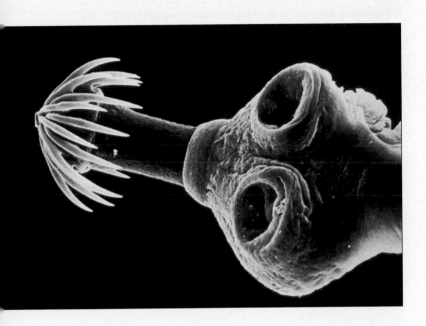

The human body is continually under siege from pathogens, including viruses, bacteria, fungi, protozoans, and parasitic worms. "Pathogen" means that these agents are capable of *infection*—of invading and then multiplying in the cells and tissues of a host organism. *Disease* may be the outcome of an infection. It results whenever the body cannot mobilize its defenses quickly enough to prevent a pathogen's activities from interfering with normal body functions. Worldwide, more people die of infectious diseases than from any other cause. The accompanying chart lists the top killers according to 1992 estimates of the World Health Organization.

Modes of Transmission

Infectious diseases are usually transmitted in the following ways:

1. Direct contact, as by touching open sores or other lesions on an infected person. (This is where "contagious" comes from; the Latin *contagio* means touch or contact.) Infected people also can transfer pathogens from such lesions to their own hands or mouth and so infect someone else through a handshake or a kiss. HIV (human immunodeficiency virus), gonorrhea, syphilis, hepatitis B, and other sexually transmissible diseases described in Chapter 15 are spread primarily by sexual contact.

2. Indirect contact, as by touching doorknobs, food, diapers, hypodermic needles (as used by drug abusers), or other objects that were previously in contact with an infected person. Food that is moist, not refrigerated, and not too acidic can be contaminated by many different pathogens, including ones responsible for amoebic dysentery, bacterial dysentery (*Salmonella* and *Shigella*), and typhoid fever.

3. Inhaling pathogens that have been ejected into the air, as by coughs and sneezes (Figure *a*). Cold and influenza viruses are usually spread this way; the bacteria that cause tuberculosis are virtually always inhaled, which is why tuberculosis is so contagious.

4. Encounters with biological *vectors*, including mosquitoes, flies, fleas, ticks, and other arthropods. Vectors transport pathogens from infected people or contaminated material to new hosts. Many pathogens depend on vectors as *intermediate hosts*. That is, a portion of the pathogen's life cycle must take place inside the vector. Mosquitoes, for instance, are intermediate hosts for the parasite *Plasmodium falciparum*, which causes malaria.

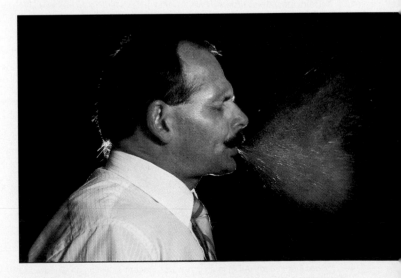

a full-blown sneeze.

Patterns of Occurrence

Infectious diseases often are described in terms of their patterns of occurrence. As the name suggests, *sporadic* diseases break out irregularly and affect only a few people. Whooping cough is an example. *Endemic* diseases occur more or less continuously but they are localized to a relatively small portion of the population. Leprosy and Lyme disease (transmitted by tick bites) are examples; so is impetigo, a highly contagious bacterial infection that is endemic to many day-care centers.

During an *epidemic*, a disease abruptly spreads through large portions of the population for a limited period. When cholera broke out all through Peru in 1991, this was an epidemic. The epidemic of bubonic plague in 14th-century Europe killed 25 million people. When epidemics break out in several countries around the world in a given time span, they collectively are called a *pandemic*. AIDS, which is caused by HIV, is now pandemic; as many as 40 million people may be infected by 2000 (pages 341–346). The single most lethal pandemic in human history (so far) occurred at the end of World War I in 1918, when viral influenza killed 21 million people around the globe.

Virulence

Virulence, which refers to the damage that a pathogen can do to a susceptible host, depends on how easily the pathogen can avoid detection and destruction within the body. For example, a certain virus may cause a mild "24-hour flu," whereas toxins produced by the bacterium *Clostridium botulinum* may kill a person within hours. HIV is apparently difficult to transmit, but is highly virulent because it eventually kills its victims.

Individuals respond differently to pathogens. Most important is the overall state of the body's immune system. As Chapter 7 describes, the immune system can be weakened by factors such as chronic fatigue or stress. In some cases, as with AIDS, it can be essentially shut down.

What Antibiotics Can and Cannot Do

Antibiotics are substances that destroy or inhibit the growth of bacteria and certain other microorganisms. They are produced mainly by bacteria and fungi. Antibiotics such as penicillins, tetracyclines, and streptomycins interfere with a variety of different metabolic functions. Streptomycins block the synthesis of proteins, and peni-

---

Infectious Diseases: Global Threats to Human Health*

Disease	Type of Pathogen	Estimated Deaths
Tuberculosis	Bacterium	3.3 million
Malaria	Parasitic protozoan	1–2 million
Hepatitis (includes A, B, non-A/non-B, and delta hepatitis)	Virus	1–2 million
Various respiratory infections (pneumonia, viral influenza, diphtheria, streptococcus)	Virus, bacterium	6.9 million
Diarrheas (includes amoebic dysentery, cryptosporidiosis, and gastroenteritis)	Parasitic, virus, and bacterium	4.2 million
Measles	Virus	220,000
Schistosomiasis	Parasitic flatworm	200,000
Whooping cough	Bacterium	100,000
Hookworm	Parasite	50,000+
Amoebiasis	Parasitic protozoan	40,000+

* Does not include AIDS-related deaths. Recent statistics on worldwide HIV infections and deaths due to AIDS are presented in Chapter 15.

cillins disrupt the formation of bonds necessary to hold molecules together in the cell walls of certain bacteria. Today there are more than 160 commonly prescribed antibiotics; as described in the next section, microbial resistance to some of these drugs is becoming an extremely serious threat to public health.

Many antibiotics can have potent side effects on the body. Penicillins, tetracyclines, and other drugs inhibit or destroy normal intestinal bacteria as well as the targeted microbe, so they can cause digestive upsets. Tetracyclines can cause yellow-tan staining of children's teeth. Women who take antibiotics often must simultaneously use an antifungal drug to combat yeast infections. This is advisable because antibiotics also kill beneficial bacteria in the vaginal tract, and when the bacteria are absent yeast tend to "overgrow" and produce large amounts of irritating secretions.

Antibiotics do not work against viruses, although some fairly new *antiviral drugs* show promise. Acyclovir, for example, is used to treat cold sores and genital herpes. It is not a cure, but it can ease a patient's symptoms during an outbreak. AZT and a few other antiviral drugs at least temporarily help some AIDS patients.

Antibiotic-Resistant Microbes

Bacterial resistance to antibiotics is rapidly becoming a human health crisis. The problem began in the 1950s, when antibiotic use increased dramatically. In many countries antibiotics are available as over-the-counter drugs, so that many lay people "self-prescribe" potent medication whenever they don't feel well. In the United States antibiotics are prescription drugs, but many people fail to take the full recommended dose. In doing so, they unknowingly kill off only the most susceptible bacteria, creating a situation that actually favors the microbes that are naturally more resistant. When doctors prescribe antibiotics unnecessarily, or when they order a broad-spectrum antibiotic (usually effective against many types of microbes) when a more specific one would do, they compound the problem.

Today, the list of microbes that have developed resistance to one or more drugs includes those responsible for tuberculosis, gonorrhea, malaria, urinary tract infections, bacterial dysentery, pneumonias, and many surgical wound infections. Researchers are forced continually to seek new natural or synthetic drugs, and infectious disease specialists have proposed a worldwide surveillance system to identify new resistant strains before they can become established.

Case Study: Tuberculosis

Globally, tuberculosis kills more humans than any other single infectious disease. Roughly one-third of the world's population is infected with *Mycobacterium tuberculosis*, the bacillus-type bacterium that causes TB. Africa, China, India, and other tropical or Asian countries are hit hardest. The bacteria are transmitted in airborne droplets produced by coughing or sneezing. In most cases, cells of the immune system kill invading bacilli. If small lesions (tubercles) have formed in the lungs, their healing will leave a scar visible on a chest X ray. A person who is disease-free may learn that he or she has been exposed to the bacillus only when a routine TB skin test registers positive. (This result only indicates past exposure to the bacterium, not active disease.) Although TB once was a major health problem in the United States, improved hygiene, nutrition, and medical care—including effective antibiotics—caused a steady decline in new cases for most of the 20th century. Since 1985, however, TB has begun to make a comeback.

A large number of recent cases have developed in people infected with HIV, which weakens and eventually destroys the immune system; an HIV-infected person who is exposed to *M. tuberculosis* cannot fight off the microbe before disease develops. And unlike HIV, TB *can* spread from an infected individual to others through inhaling air contaminated with bacilli.

Intravenous drug abusers are at high risk for HIV. They are also more likely to wind up housed in the close quarters of jails, prisons, and shelters for the homeless, and are less likely to receive adequate health care. Health officials cite these interacting factors to explain why active TB is becoming increasingly common among drug abusers and in correctional facilities and public shelters. In some states, TB is arriving along with immigrants from poor countries, who unknowingly carry the microbe. Also, resistant new strains of the TB bacterium have evolved. So far, drug-resistant TB has turned up in about 25 states, and even with the best treatments available it kills 50–80 percent of people diagnosed with the disease. It is entirely possible that drug-resistant TB could slowly spread to the general population. To combat the scourge, our only recourse as a society is to institute widespread TB screening programs and hope that researchers can develop new anti-TB drugs.

4 MUSCULOSKELETAL SYSTEM: SUPPORT AND MOVEMENT

Challenge of the Iditarod

"All right—GO!"

Susan Butcher just might be the finest long-distance sled-dog racer of all time. In the past 10 years she has competed in a string of Iditarod races, commanding a 200-pound sled and a dog team (Figure 4.1) over 1,157 frozen miles between Anchorage and Nome, Alaska. Butcher has won the grueling race three times.

Unlike one of Susan Butcher's well-tended Alaskan huskies, we humans do not have a skeleton and mus-

cles adapted for hauling loads over great distances. Through hard work and ingenuity, however, we *can* maximize the potential of our own remarkably versatile adaptations for body movement and support.

For example, long before each race Susan Butcher followed a marathoner's regimen of diet and exercise to put her arm and leg muscles in peak condition. The physical workouts increased the capacity of her musculoskeletal system to support and help move her body.

Figure 4.1 Susan Butcher, illustrating her body's systems of support and movement along the Iditarod Trail.

How? Exercise increases bone mass, and muscles respond to rigorous training by becoming larger and loaded with mitochondria. Additional blood vessels develop, increasing blood flow for the transport of nutrients and wastes. The results can be amazing increases in strength and endurance. As we'll see in this chapter, structural support and body movement each involve specific tissues that are functionally linked in key ways.

The body's softer parts require a supporting framework—one service, among several, that our bony skeleton provides. The skeleton's bones also serve as rigid parts against which muscles can work and so translate the pull of contracting cells into body movement. And whereas long bones of your legs, for instance, are quite sturdy, much of the structural material inside them is surprisingly lightweight. This minimizes the energy "cost" of movement. In muscles, cells and tissues are organized to work with one another as well as with the skeleton. In tandem with the nervous system, this fine-tuning enables a person to execute the movements and positional changes required for the full range of human activities—from mushing a sled to turning the pages of a textbook.

KEY CONCEPTS

1. Bones function in body movement, in protection and support for soft organs, in mineral storage, and in blood cell formation.

2. The body contains three kinds of muscle tissues: smooth, cardiac, and skeletal muscle. The cells of all three types of muscle tissue can contract (shorten) and then return to the resting position.

3. Skeletal muscle contracts in response to stimulation by the nervous system. Cardiac muscle can contract intrinsically—that is, without nerve stimulation. Smooth muscle responds to stimulation by nerves and hormones, and many smooth muscles also can contract intrinsically.

4. The force of skeletal muscle contraction acts against a skeleton of bone and cartilage. Interactions of muscles and the skeleton bring about movements.

As is true for all vertebrates, two organ systems are directly responsible for your body's shape and capacity for movement. These are the skeletal and muscular systems. Figure 4.2 gives a general picture of these systems in the human body.

BONE

Bone Function

Just as skin is more than a body covering, so is the skeletal system more than a frame for muscles and other organs. Its major parts, **bones**, are complex organs composed of a number of tissues. Bones serve the following functions:

1. Movement: By interacting with skeletal muscle, bones help maintain or change the position of body parts.

2. Protection: Bones form hard compartments that enclose and protect the brain, lungs, and other organs.

3. Support: Bones support and anchor muscles.

4. Mineral storage: Bone tissue is a "bank" for the deposit and withdrawal of mineral ions required in body fluids and for many metabolic activities. In this and other ways, bone contributes to homeostasis.

5. Blood cell formation: Blood cells are produced in parts of some bones.

Bone Structure

In size and shape, human bones range from tiny ear bones to clublike thighbones. With respect to shape, bones are classified as long, short (or cubelike), flat, or irregular. All incorporate other tissues as well as bone tissue.

The main component of bones is *bone tissue*, a connective tissue that has both living and nonliving elements. The living cells are **osteocytes**. As a bone develops, precursors of osteocytes called *osteoblasts* secrete collagen fibers and a ground substance of proteins and carbohydrates. Eventually, the osteocytes reside within spaces in the ground substance called *lacunae*. The ground substance becomes hardened (mineralized) by deposits of calcium salts.

Figure 4.3 shows the internal organization of a typical long bone, the thighbone (femur). The tissue that forms the bone's shaft and the outer portion of its two ends is dense *compact bone*. Such tissue forms the shaft of all long bones. The collagen fibers within compact bone

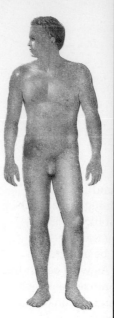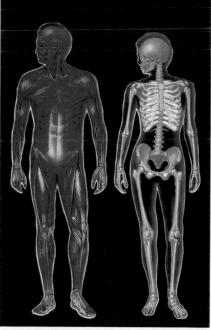

Figure 4.2 Starting point for a tour of the body's muscles and its skeletal system.

give long bones the strength to withstand mechanical stresses. An outer membrane, the periosteum, covers the bone shaft. Compact bone tissue is organized as thin, concentric layers around small canals. Each set of circular layers is an **osteon** (also called a *Haversian system*), and the canals—known as *Haversian canals*—are interconnected channels for blood vessels and nerves that service the osteocytes in compact bone tissue. Osteocytes in different lacunae extend slender cell processes into narrow channels between lacunae, called *canalculi*. By way of canalculi, nutrients can readily move through the hardened ground substance from osteocyte to osteocyte. Wastes can exit by the same route.

The bone tissue inside the shaft and the ends has a spongelike appearance. Tiny, flattened struts are fused together in a latticework to make up this *spongy bone tissue*, which actually is quite firm and strong. In some bones, **red marrow** fills the spaces between the struts. Red marrow is a major site of blood cell formation. As a human grows older, the red marrow in the shafts of most long bones becomes replaced with **yellow marrow**. Yellow marrow is mostly fat. If the need arises, it can convert to red marrow and produce red blood cells.

Bones are complex organs composed of living cells, a nonliving mineralized matrix of collagen fibers and ground substance, and an outer covering of dense connective tissue.

How Bones Develop Before a long bone forms in a growing embryo, a cartilage "model" forms for it (Figure 4.4). Once the outer covering (the periosteum) is in place

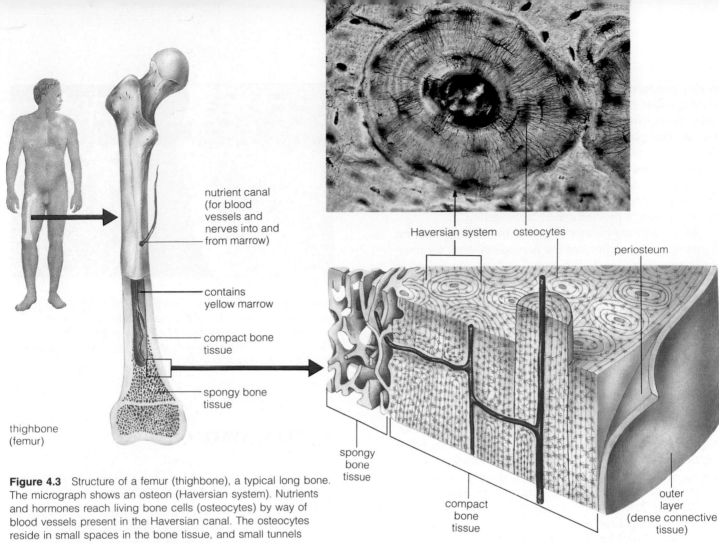

nutrient canal
(for blood
vessels and
nerves into and
from marrow)

contains
yellow marrow

compact bone
tissue

spongy bone
tissue

thighbone
(femur)

Haversian system   osteocytes

periosteum

spongy
bone
tissue

compact
bone
tissue

outer
layer
(dense connective
tissue)

Figure 4.3 Structure of a femur (thighbone), a typical long bone. The micrograph shows an osteon (Haversian system). Nutrients and hormones reach living bone cells (osteocytes) by way of blood vessels present in the Haversian canal. The osteocytes reside in small spaces in the bone tissue, and small tunnels (caniculi) connect neighboring spaces.

Figure 4.4 Long bone formation, starting with osteoblast activity in a cartilage model (here, already formed in the embryo). Bone-forming cells are active first in the shaft region, then at the knobby ends. In time, cartilage is left only in the epiphyseal plates at the ends of the shaft.

on the model, it gives rise to osteoblasts, the bone-form-ing cells. A supportive bony "collar" forms around the cartilage shaft, then the cartilage within the shaft breaks down and the marrow cavity forms. An artery, a vein, and other elements (including some osteoblasts) then infiltrate the cavity. The osteoblasts continue to secrete matrix materials, which gradually become mineralized. Eventually the osteoblasts become surrounded by their own secretions. They are now osteocytes (see Figure 4.3), mature living bone cells that maintain mature bones.

The two expanded ends of a long bone are called *epiphyses,* and in growing children and young adults they are separated from the bone shaft by an **epiphyseal plate** of cartilage. The plate develops and is maintained under the influence of growth hormone (GH) produced by the pituitary gland, which secretes GH in response to hor-

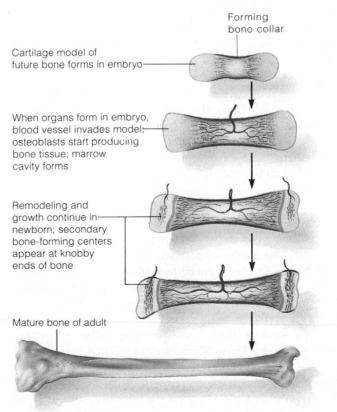

Forming
bone collar

Cartilage model of
future bone forms in embryo

When organs form in embryo,
blood vessel invades model;
osteoblasts start producing
bone tissue; marrow
cavity forms

Remodeling and
growth continue in
newborn; secondary
bone-forming centers
appear at knobby
ends of bone

Mature bone of adult

mones produced by the thyroid gland. As long as the plate consists of cartilage, the bone can lengthen as an individual grows. When growth stops in late adolescence, the cartilage plates are replaced by bone.

Bone Tissue Turnover Minerals are constantly deposited in and withdrawn from bone tissue. This turnover occurs as osteoblasts deposit bone and *osteoclasts* break it down, a process called "remodeling."

As we will see in later chapters, the mineral calcium is essential for functions of the nervous system, for muscle contraction, and for many other physiological processes. When the level of calcium in the blood falls below a certain point, the hormone PTH (parathyroid hormone) stimulates osteoclasts to secrete enzymes that break down bone tissue. As the component minerals dissolve, the released calcium enters interstitial fluid; from there, it is taken up by the blood. If the amount of calcium in blood is greater than required, another hormone, calcitonin, stimulates osteoblasts to take up calcium from the interstitial fluid and use it to produce new bone. Chapter 12 discusses these and other hormonal controls.

Bone deposits increase relative to withdrawals when mature bone is subjected to mechanical stress, partly by way of muscles that pull on the bone during body movements. Hence brisk walking, jogging, "pumping iron," and similar activities cause affected bones to become denser—and stronger—as more bone tissue is created. Remodeling also occurs after breaks or other bone injuries.

Before adulthood, bones grow by way of remodeling. During this time the body requires ample calcium to meet the combined demands of bone growth and other metabolic needs. Turnover is also especially important. For example, the diameter of the thighbone increases as osteocytes (derived from the periosteum) deposit calcium phosphate at the surface of the shaft. At the same time, osteoclasts destroy a small amount of bone tissue *inside* the shaft. Thus the thighbone becomes thicker and stronger, but not too heavy.

As a person ages, bone may break down faster than it is renewed, especially in older women. The backbone, pelvis (hip bones), and other bones decrease in mass. This progressive bone deterioration is called *osteoporosis* (Figure 4.5). The spine can collapse and curve so much that the rib cage lowers (a condition called *lordosis*), crowding internal organs. Studies suggest that deficiencies of calcium and sex hormones, excess protein intake, smoking, and reduced physical activity may contribute to osteoporosis. On the other hand, lifelong habits of getting plenty of exercise (to stimulate deposits of bone tissue) and taking in enough dietary calcium may help minimize bone loss later in life.

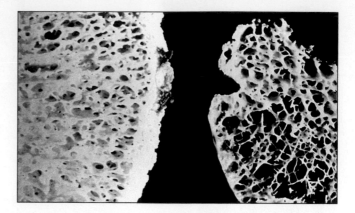

Figure 4.5 Osteoporosis. In normal tissue (left), mineral deposits continually replenish mineral withdrawals. After the onset of osteoporosis (right), replacements can't keep pace with withdrawals. The tissue gradually erodes, and bones become progressively hollow and brittle.

SKELETAL SYSTEM

The human body has 206 bones in two skeletal divisions: the **axial skeleton** and the **appendicular skeleton** (Figure 4.6). Straps of dense, regular connective tissue called **ligaments** connect the bones at joints. As befits this function, the connective tissue of ligaments contains elastin and thus is flexible and resilient. **Tendons** attach muscles to bones (or sometimes to other muscles). In contrast to ligaments, they are cords or straps of dense, regular connective tissue that contains collagen, which increases the tissue's strength.

Axial Skeleton

The axial skeleton is made up of bones that roughly form the body's vertical axis. These include the skull, vertebral column (backbone), ribs, and sternum (the breastbone).

The Skull We begin with the skull, which consists of more than two dozen bones that are divided into several groupings. Although many bones are traditionally called by complex-sounding names derived from Latin, their roles are much simpler to grasp. For example, one grouping, the "cranial vault" or **brain case**, includes eight bones that together surround and protect the brain. As Figure 4.7a shows, the *frontal bone* makes up the forehead and upper ridges of the eye sockets. It contains air spaces called **sinuses**, which lighten the skull and are lined with mucous membrane. Sinuses drain into the upper respiratory tract, a frequent source of misery for anyone who

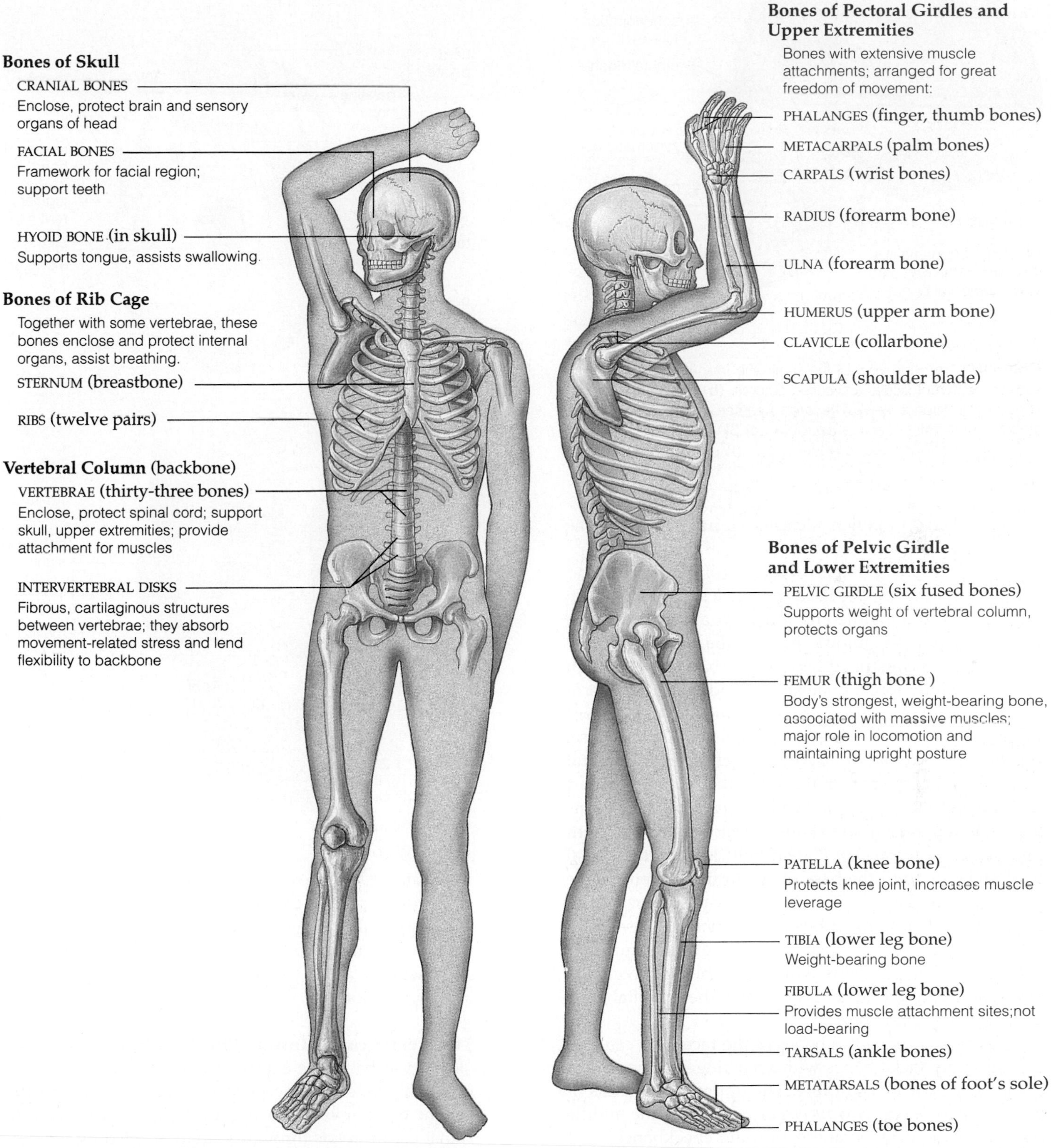

Bones of Skull

CRANIAL BONES
Enclose, protect brain and sensory organs of head

FACIAL BONES
Framework for facial region; support teeth

HYOID BONE (in skull)
Supports tongue, assists swallowing.

Bones of Rib Cage

Together with some vertebrae, these bones enclose and protect internal organs, assist breathing.

STERNUM (breastbone)

RIBS (twelve pairs)

Vertebral Column (backbone)

VERTEBRAE (thirty-three bones)
Enclose, protect spinal cord; support skull, upper extremities; provide attachment for muscles

INTERVERTEBRAL DISKS
Fibrous, cartilaginous structures between vertebrae; they absorb movement-related stress and lend flexibility to backbone

Bones of Pectoral Girdles and Upper Extremities

Bones with extensive muscle attachments; arranged for great freedom of movement:

PHALANGES (finger, thumb bones)

METACARPALS (palm bones)

CARPALS (wrist bones)

RADIUS (forearm bone)

ULNA (forearm bone)

HUMERUS (upper arm bone)

CLAVICLE (collarbone)

SCAPULA (shoulder blade)

Bones of Pelvic Girdle and Lower Extremities

PELVIC GIRDLE (six fused bones)
Supports weight of vertebral column, protects organs

FEMUR (thigh bone)
Body's strongest, weight-bearing bone, associated with massive muscles; major role in locomotion and maintaining upright posture

PATELLA (knee bone)
Protects knee joint, increases muscle leverage

TIBIA (lower leg bone)
Weight-bearing bone

FIBULA (lower leg bone)
Provides muscle attachment sites;not load-bearing

TARSALS (ankle bones)

METATARSALS (bones of foot's sole)

PHALANGES (toe bones)

Figure 4.6 Human skeletal system. Major bones of its axial division are listed to the left. Major bones of its appendicular division are listed to the right.

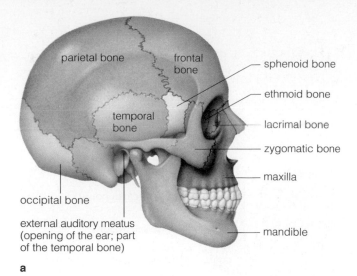

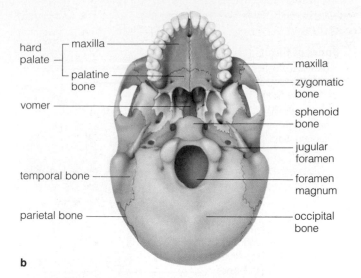

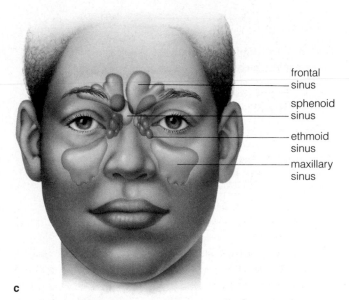

Figure 4.7 (**a**) Side view of the skull. The irregular junctions between different bones are called sutures. (**b**) The skull from underneath (inferior view). The large foramen magnum is situated atop the uppermost cervical vertebra. (**c**) Sinuses in bones associated with the nasal cavity.

has a head cold or pollen allergies. Figure 4.7c shows the locations of sinuses in cranial and facial bones.

Temporal bones form the lower sides of the cranium and surround the ear canals. Each canal is a tunnel-like space that leads to the middle and inner ear and opens to the outside; within the middle ear are the tiny bones that function in hearing (page 259). A *sphenoid bone* lies just in front of each temporal bone and extends inward to form part of the inner eye socket. The *ethmoid bone* also contributes to the inner socket and helps support the nose. A pair of *parietal bones* above and behind the temporal bones form a large part of the skull; they sweep upward and meet at the top of the head. An *occipital bone* forms the back and base of the skull; it encloses a large opening, the *foramen magnum* ("great hole"). Here, the spinal cord emerges from the base of the brain and enters the spinal column (Figure 4.7b). Quite a few passageways run through and between various skull bones for nerves and blood vessels, especially at the base of the skull. For instance, the jugular veins, which carry blood leaving the brain, pass through openings between the occipital bone and each temporal bone.

Figure 4.7 also shows bones of the face. You can easily feel many facial bones with your fingers. The largest is the lower jaw or **mandible**. The upper jaw consists of two *maxillary bones*. Two *zygomatic bones* form the middle portion of the protuberances we call "cheekbones" and the outer parts of the eye sockets. A small, flattened *lacrimal bone* fills out the inner eye socket. Tear ducts pass between this bone and the maxillary bones and drain into the nasal cavity. The upper and lower jaws also contain the teeth in *tooth sockets*.

Palatine bones make up part of the floor and side wall of the nasal cavity. (Extensions of these bones, together with the maxillary bones, form the back of the hard palate, the "roof" of the mouth.) A *vomer bone* forms part of the nasal septum, a thin "wall" that divides the nasal cavity into two sections.

The Vertebral Column: The Backbone The flexible, curved backbone extends from the base of the skull to the hipbones (pelvic girdle), where it transmits the weight of your torso to the lower limbs. The delicate spinal cord threads through a cavity formed by the vertebrae, which are arranged one above the other. As Figure 4.8 shows, humans have 24 vertebrae, including seven in the neck region (cervical), 12 in the chest area (thoracic), and five in the lower back (lumbar), plus a sacrum and a coccyx (the "tailbone").

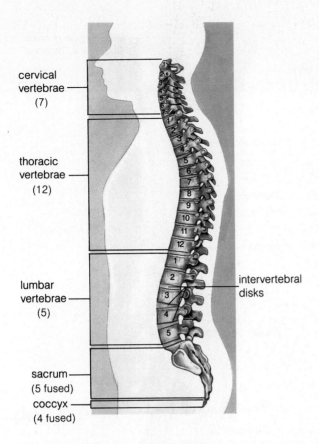

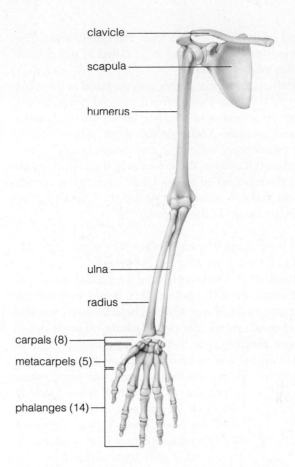

Figure 4.8 Side view of the vertebral column or backbone. The cranium balances on the column's uppermost vertebra.

Figure 4.9 Bones of the pectoral girdle, the arm, and the hand.

Intervertebral disks, which contain cartilage, occur between the vertebrae. They serve as shock absorbers and flex points. However, severe or rapid shocks may cause the core of a disk to *herniate*. If the disk ruptures, the jellylike core material may squeeze out and press against neighboring nerves or the spinal cord. Surgery may be required to ease the excruciating pain.

Ribs and Sternum In addition to protecting the spinal cord, absorbing shocks, and providing flexibility, the vertebral column also serves as an attachment point for 12 pairs of **ribs**, which serve as a scaffold for the body cavity of the upper torso. The upper ribs also attach to the **sternum** (Figure 4.6). As you will see in later chapters, this "rib cage" helps protect the lungs, heart, and other internal organs and is vitally important in breathing.

Appendicular Skeleton

"Append" means to hang, and the appendicular skeleton includes the bones of body parts that we sometimes think of as dangling from the main body frame: arms, hands,

legs, and feet. It also includes a pectoral girdle at each shoulder and the pelvic girdle at the hips.

Pectoral Girdle and Upper Limbs Each **pectoral girdle** (Figure 4.9) has a large, flat shoulder blade—a **scapula**—and a long, slender collarbone or **clavicle** that connects to the breastbone (sternum). The rounded shoulder end of the **humerus**, the long bone of the upper arm, fits into an open socket in the scapula. The human arm is capable of remarkably versatile movements; it can swing in wide circles and back and forth, lift objects, or tug on a rope. We owe this freedom of movement to the fact that the pectoral girdles and upper limbs are loosely attached to the rest of the body by muscles. Although this arrangement is sturdy enough under normal conditions, it is vulnerable to strong blows. Fall on an outstretched arm and you might fracture a clavicle or dislocate a shoulder. The collarbone is the bone most frequently broken, perhaps because it is most exposed and least protected.

The humerus connects with two bones of the forearm, the **radius** (on the thumb side) and the **ulna** (on the "pinky finger" side). The upper end of the ulna joins the lower end of the humerus to form the elbow joint. The prominence sometimes (mistakenly) called a "wristbone" is the lower end of the ulna.

The radius and ulna together join the hand at the wrist joint, where they meet eight small, curved *carpal* bones. Ligaments attach them to the long bones. Blood vessels, nerves, and tendons pass over the wrist; when a blow, constant pressure, or repetitive movement (such as prolonged typing) damages these tendons, the result can be a painful disorder called carpal tunnel syndrome. Bones of the hand, the five *metacarpals*, end at the knuckles. *Phalanges* are the bones of the fingers.

Pelvic Girdle and Lower Limbs   For most of us, the shoulders and arms are much more flexible than are our hips and legs. Why? Although there are similarities in the basic "design" of both girdles, this lower part of the appendicular skeleton is adapted to bear the body's entire weight when we stand. The **pelvic girdle** (Figure 4.10) is much more massive than the pectoral girdles, and it is attached to the axial skeleton by exceptionally strong ligaments. It forms an open basin: A pair of *coxal bones* attach to the lower spine (sacrum) in back, then curve forward and meet at the *pubic arch*. ("Hipbones" are actually the upper *iliac* regions of the coxal bones.) This combined structure is the *pelvis*. In females the pelvis is broader than in males, and it shows other structural differences that are evolutionary adaptations for the function of child-bearing.

The legs contain the body's largest bones. In terms of length, the thigh bone or **femur** ranks number one. It is also extremely strong. When you run or jump, your femurs routinely withstand stresses of several tons per square inch (aided by contracting leg muscles). The femur's ball-like upper end fits snugly into a deep socket in the coxal (hip) bone. The other end connects with one of the bones of the lower leg, the thick, load-bearing *tibia* on the inner (big toe) side. A slender *fibula* parallels the tibia on the outer (little toe) side. The tibia is your shinbone. A triangular kneecap, the *patella*, helps protect the knee joint.

Bones of the ankle and foot correspond closely with those of the wrist and hand. *Tarsal* bones make up the ankle and heel, and the foot contains five long bones, the *metatarsals*. The largest metatarsal, leading to the big toe, supports a great deal of body weight and is thicker and stronger than the others. Like fingers, toes contain phalanges.

Joints

"Joints" are areas of contact or near-contact between bones. In the most common type, the **synovial joint**, adjoining bones are separated by a cavity. Such joints are freely movable and are stabilized in part by straplike liga-

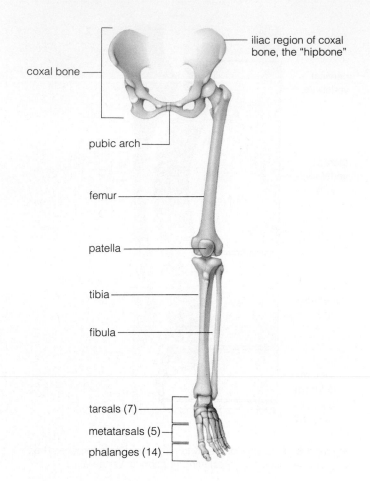

Figure 4.10   Bones of the pelvic girdle, the leg, and the foot.

ments. A capsule of dense connective tissue surrounds the bones of a synovial joint. Cells that line the interior of the capsule secrete a lubricating *synovial fluid* into the joint cavity.

Synovial joints include the *ball-and-socket* joints at the hips. These types of joints are capable of a wide range of movements, including rotation and movements in different planes (up-down, side-to-side, and so on). *Hingelike* synovial joints such as the knee and elbow move in one plane only; that is, they are limited to simple flexing and extending (straightening), like a door hinge. Figure 4.11 shows various ways body parts can move at joints.

As a person ages, the cartilage covering the bone ends of freely movable joints may degenerate, a condition called *osteoarthritis*. In contrast, *rheumatoid arthritis* is a degenerative disorder that results from an immune system malfunction. The synovial membrane becomes inflamed and thickened, cartilage is eroded away, and the bones fall out of their proper alignment. The bone ends may eventually fuse together. *Focus on Science* (page 98) describes some other ailments that can disrupt the func-

flexion

extension

flexion

extension

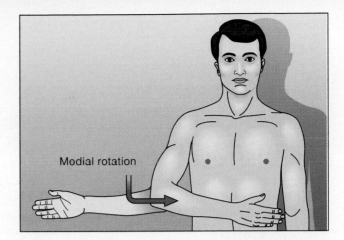

Modial rotation

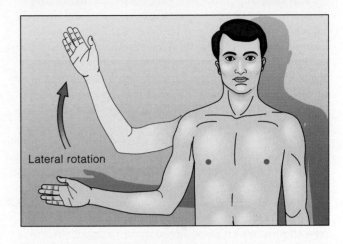

Lateral rotation

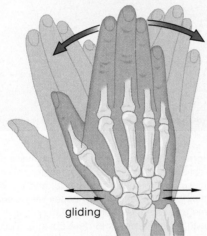

gliding

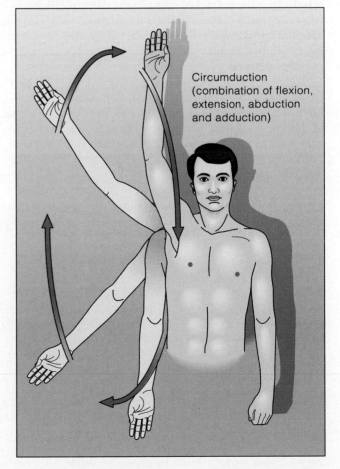

Circumduction (combination of flexion, extension, abduction and adduction)

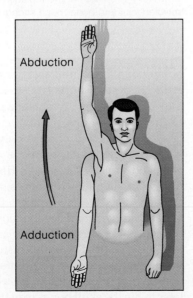

Abduction

Adduction

Figure 4.11 Examples of ways body parts can move at joints. The synovial joint at the shoulder permits the greatest range of movement.

The Vulnerable Knee

Joggers, skiers, and players of football, baseball, and other sports have learned the hard way that mobility—even a career—can hinge on the health of their knees.

The knee joint links two long bones, the femur and tibia. Those bones are separated by a cavity and are held together by ligaments and tendons that form a capsule around the joint (Figure *a*). A membrane that lines the capsule produces fluid that lubricates the joint; where the bone ends meet, they are capped with a cushioning layer of cartilage. A wedge of fibrocartilage between the femur and tibia called a *meniscus* adds stability. Fluid-filled sacs (*bursae*) reduce friction.

When the knee is hit hard or twisted too much, its cartilage can be torn. Often the body cannot repair the damage. If most or all of the damaged tissue is not removed, it can cause arthritis. Each year more than 50,000 pieces of torn cartilage are surgically removed from the knees of football players. A common procedure today is *arthroscopy*, an operation in which the surgeon makes a small slit and inserts an instrument that carries a miniaturized light source and lens that are hooked up to a television camera. The apparatus enables the physician to see and correct the injury while minimizing the scarring and other side effects of major surgery.

Ligaments strap the femur and tibia together, but blows to the knee can tear them apart, causing a *sprain*. A ligament is composed of connective tissue fibers. If only some of the fibers are torn, it may heal itself. A completely severed ligament must be surgically repaired, and quickly. If not, scavenger cells in the synovial fluid will destroy the injured tissue.

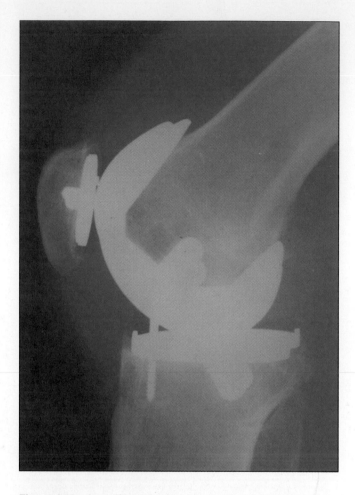

Figure 4.12 An artificial knee joint made of stainless steel alloy. Most artificial joints today are constructed of metal or a combination of metal and plastic, and the whole assembly is attached to the articulating bones with a special adhesive. With time and use, this arrangement may loosen. Researchers are having some success in developing artificial joints made of porous materials (such as ceramics); the patient's bone may grow into tiny cavities in the replacement joint, making the bond between joint and bone much stronger.

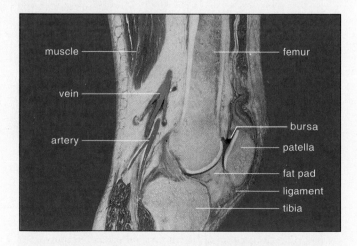

a Longitudinal section through the knee joint. In this view, the lateral and medial menisci are not readily visible.

tioning of a freely movable joint. Surgeons can now use artificial joints (Figure 4.12) to replace some joints when they become seriously damaged.

In **cartilaginous joints**, cartilage fills the space between bones and permits only slight movement. Such joints occur between vertebrae and between the breastbone and some of the ribs. In **fibrous joints**, fibrous connective tissue unites the bones, and no cavity is present. Fibrous joints loosely connect the flat skull bones of a fetus. During childbirth, the loose connections allow the bones to slide over each other and so prevent skull fractures. The skull of a newborn still has fibrous joints, and membranous areas called fontanels ("soft spots"). The fibrous joints harden into *sutures* during childhood. Much later in life the skull bones may become completely fused.

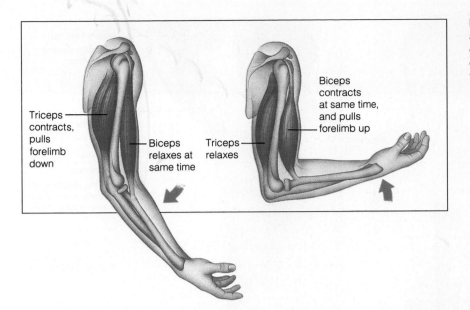

In the figure, the labels read:

- Triceps contracts, pulls forelimb down
- Biceps relaxes at same time
- Triceps relaxes
- Biceps contracts at same time, and pulls forelimb up

MUSCULAR SYSTEM

How Muscles and Bones Interact

The human body has more than 600 muscles, arranged as pairs or groups. Some muscles work together (*synergistically*) to promote the same movement. Others work in opposition (*antagonistically*), so that the action of one opposes or reverses the action of another. When muscles contract, they pull on the bones to which they are attached. In general, one end of a muscle, the **origin**, is attached to a bone that stays relatively motionless during a movement. The other end of the muscle, called the **insertion**, is attached to the bone that moves most. Together, the skeleton and the muscles attached to it are like a system of levers in which rigid rods (bones) move about at fixed points (the joints). Most joints have muscle attachments nearby. This means a muscle must contract only a small distance to produce a large movement of some body part.

Figure 4.13 shows an antagonistic pair of muscles, the biceps and triceps of a human arm. When the biceps contracts, the elbow joint flexes (bends). As it relaxes and its partner (the triceps) contracts, the limb extends and straightens. Such coordinated action results partly from *reciprocal innervation* by nerves from the spinal cord. By this mechanism, when one muscle group is stimulated, signals sent to the opposing group prevent it from contracting. In addition, the nervous system uses signals from stretch receptors in tendons or muscles to coordinate the contractions.

Skeletal Muscle

We turn now to skeletal muscle, the functional partner of bone. Recall from Chapter 3 that there are three types of muscle tissues: skeletal, cardiac, and smooth. Although skeletal and cardiac muscle are quite different in appearance from smooth muscle, all muscle cells are specialized to generate force by contracting—that is, by becoming shorter. Muscle cells also can relax; after contracting, they return to their original resting position.

Skeletal muscle contracts in response to stimulation by the nervous system. Cardiac muscle can respond to nervous stimulation, but it also contracts intrinsically, without outside stimulation. Smooth muscle responds to a variety of control systems, including nerves, hormones, and intrinsic mechanisms.

Skeletal muscle is the only type of muscle tissue that interacts with the skeleton to bring about movement. As Chapter 3 described, smooth muscle occurs mostly in the walls of internal organs, where it helps maintain organ shape. It also serves other functions; for example, smooth muscle contractions in the stomach and intestinal walls help propel substances through the digestive tract. Cardiac muscle occurs only in the heart, and its pumping action is described in Chapter 6.

The cells of skeletal, cardiac, and smooth muscle tissue all generate force by contracting. After contracting they return to their resting position.

Skeletal muscle responds to stimulation by the nervous system. Cardiac muscle can contract intrinsically—that is, without nerve stimulation. Smooth muscle responds to stimulation by nerves and hormones, and many smooth muscles also can contract intrinsically.

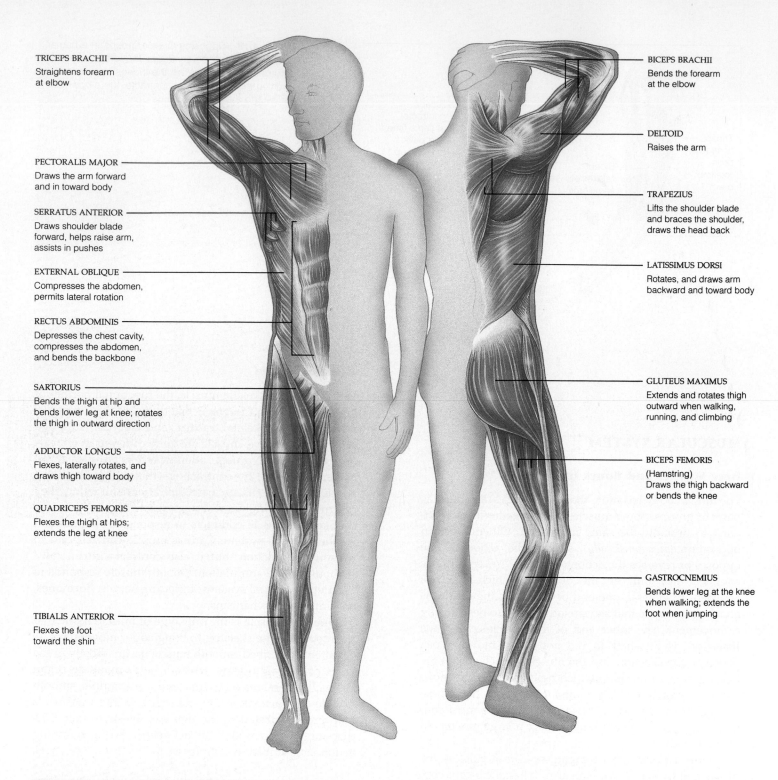

TRICEPS BRACHII
Straightens forearm
at elbow

PECTORALIS MAJOR
Draws the arm forward
and in toward body

SERRATUS ANTERIOR
Draws shoulder blade
forward, helps raise arm,
assists in pushes

EXTERNAL OBLIQUE
Compresses the abdomen,
permits lateral rotation

RECTUS ABDOMINIS
Depresses the chest cavity,
compresses the abdomen,
and bends the backbone

SARTORIUS
Bends the thigh at hip and
bends lower leg at knee; rotates
the thigh in outward direction

ADDUCTOR LONGUS
Flexes, laterally rotates, and
draws thigh toward body

QUADRICEPS FEMORIS
Flexes the thigh at hips;
extends the leg at knee

TIBIALIS ANTERIOR
Flexes the foot
toward the shin

BICEPS BRACHII
Bends the forearm
at the elbow

DELTOID
Raises the arm

TRAPEZIUS
Lifts the shoulder blade
and braces the shoulder,
draws the head back

LATISSIMUS DORSI
Rotates, and draws arm
backward and toward body

GLUTEUS MAXIMUS
Extends and rotates thigh
outward when walking,
running, and climbing

BICEPS FEMORIS
(Hamstring)
Draws the thigh backward
or bends the knee

GASTROCNEMIUS
Bends lower leg at the knee
when walking; extends the
foot when jumping

Figure 4.14 Some of the major skeletal
muscles of the human musculoskeletal
system.

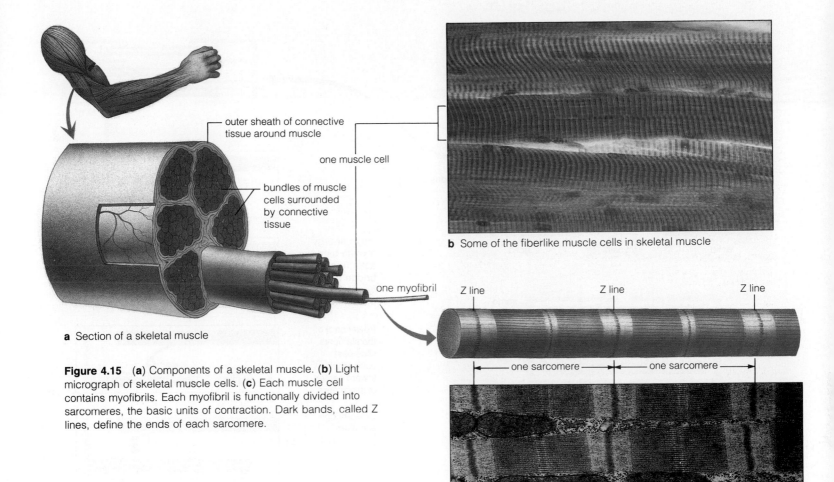

outer sheath of connective tissue around muscle

one muscle cell

bundles of muscle cells surrounded by connective tissue

one myofibril

a Section of a skeletal muscle

b Some of the fiberlike muscle cells in skeletal muscle

Z line   Z line   Z line

one sarcomere   one sarcomere

Figure 4.15 (**a**) Components of a skeletal muscle. (**b**) Light micrograph of skeletal muscle cells. (**c**) Each muscle cell contains myofibrils. Each myofibril is functionally divided into sarcomeres, the basic units of contraction. Dark bands, called Z lines, define the ends of each sarcomere.

c Two of the myofibrils inside a muscle cell. In each myofibril, the contractile units called sarcomeres are arranged one after the other. The oval-shaped organelle is a mitochondrion.

Functional Organization of a Skeletal Muscle

Figure 4.14 shows the major skeletal muscles of the human body. Each is composed of a few hundred to many thousands of muscle cells ("muscle fibers"). Fibrous connective tissue encloses the muscle cells and blends with the tendons that attach the muscle to bone.

Chapter 3 described various structural features of skeletal muscle cells (page 74). Recall that each muscle cell contains threadlike structures packed together in a parallel array. You can see these structures, called **myofibrils**, in Figure 4.15. Each myofibril is divided into sarcomeres, which occur one after another along its length. **Sarcomeres** are the basic units of muscle cell (or "fiber") contraction.

Figure 4.15c shows that a sarcomere contains many filaments lying side by side. Some of the filaments are thin, others are thick. Each *thin* filament is actually two beaded strands twisted together. The "beads" are ball-shaped molecules of the globular protein **actin**:

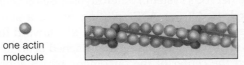

one actin molecule

one actin filament

Each *thick* filament consists of molecules of **myosin**, a protein with a head and a long tail. The myosin tails are packed together in parallel. The heads, which look like double-headed golf clubs, stick out to the sides:

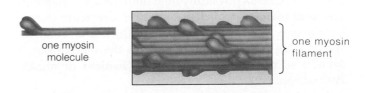

one myosin molecule

one myosin filament

Skeletal and cardiac muscle owe their striped (striated) appearance to the highly ordered organization of actin and myosin filaments in the sarcomeres of myofibrils.

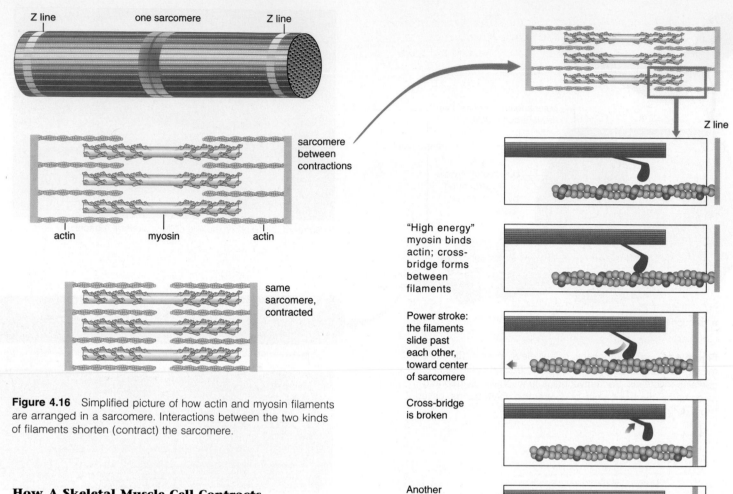

Figure 4.16 Simplified picture of how actin and myosin filaments are arranged in a sarcomere. Interactions between the two kinds of filaments shorten (contract) the sarcomere.

Figure 4.17 Sliding-filament model of muscle contraction in the sarcomeres of muscle cells. For simplicity, the action of only one myosin head is shown.

How A Skeletal Muscle Cell Contracts

Skeletal muscles never push on bones; they always pull. Why? The only way that skeletal muscles can move the body parts to which they are attached is to shorten. When a skeletal muscle shortens, its cells are shortening. And when a muscle cell shortens, its component sarcomeres are shortening. The combined decreases in length of the individual sarcomeres account for contraction of the whole muscle.

Figure 4.16 shows the arrangement of actin and myosin in sarcomeres. How does a sarcomere contract? Myosin filaments physically bind onto the actin filaments and pull them toward the center of a sarcomere during contraction. In this movement, the thick and thin filaments slide over each other, and so the mechanism is known as the **sliding-filament mechanism** of muscle contraction.

The interactions between myosin and actin filaments take place in a cyclic fashion. In each cycle, myosin heads (called "cross-bridges") become linked with actin in the thin filament, the linked myosin and actin filament move, and then the myosin heads detach from the actin (Figure 4.17). We can analyze the cycle beginning at a point when enzyme action splits an ATP molecule into ADP and

phosphate, temporarily creating a "high-energy" form of myosin. In this state, the myosin binds actin. (As we will see shortly, the attachment of myosin to actin also relies on calcium ions, which enter a muscle cell when it is stimulated by the nervous system.) After myosin binds actin, the energy released from ATP splitting drives a short power stroke by the myosin head, which tilts toward the center of the sarcomere. This pulls the attached actin filaments toward the center of the sarcomere, which shortens. Now the cycle repeats. An energy input from ATP causes each myosin head to detach, the attachment steps occur with the next actin binding site in line, and the actin filaments move a bit more. A single contraction takes a whole series of power strokes by myosin heads in each sarcomere.

In the absence of ATP, the myosin cross-bridges cannot detach. Following death, for instance, ATP production stops along with other metabolic activities. Cross-bridges remain locked in place, and all skeletal muscles in the body become rigid. This condition, *rigor mortis*, lasts up to 60 hours after death. It ends as biological molecules, including actin and myosin, gradually break down.

In skeletal muscle, contraction occurs in sarcomeres, which are contractile units organized one after another in the myofibrils of muscle cells. Each sarcomere contains parallel arrays of actin filaments and myosin filaments.

Each sarcomere shortens when linked actin and myosin filaments slide past each other.

The combined decrease in length of individual sarcomeres accounts for contraction of a muscle.

Neural Control of Skeletal Muscle Cell Contraction

Skeletal muscle cells contract under commands from the nervous system. *Motor neurons* of the nervous system deliver signals that stimulate contraction. We consider the nature of these signals, called action potentials, in Chapter 10. Here, it is enough to understand what happens when commands from the nervous system stimulate a muscle cell. Figure 4.18 summarizes the pathway.

From the point of stimulation, action potentials spread and rapidly reach small, tubular extensions of the plasma membrane. The small tubes connect with a system of membrane-bound chambers that lace around the cell's myofibrils. That system, the **sarcoplasmic reticulum (SR)**, takes up, stores, and releases calcium ions in controlled ways. The arrival of stimulation from the nervous system causes calcium ions within the SR to be released. The released ions then diffuse through the cytoplasm and bind to a regulatory protein (troponin). The troponin is asso-

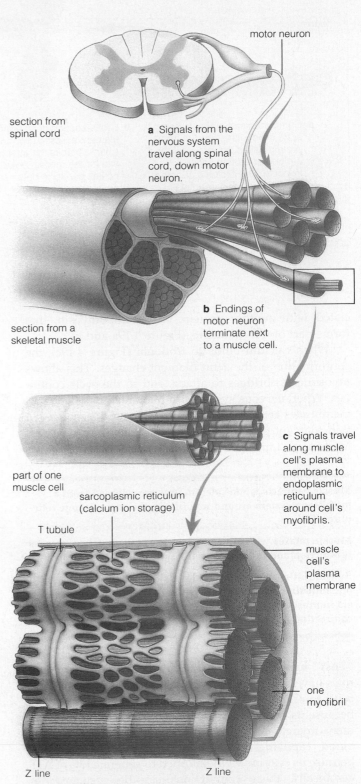

a Signals from the nervous system travel along spinal cord, down motor neuron.

section from spinal cord

motor neuron

section from a skeletal muscle

b Endings of motor neuron terminate next to a muscle cell.

part of one muscle cell

c Signals travel along muscle cell's plasma membrane to endoplasmic reticulum around cell's myofibrils.

sarcoplasmic reticulum (calcium ion storage)

T tubule

muscle cell's plasma membrane

one myofibril

Z line

Z line

d Signals trigger calcium release from sarcoplasmic reticulum lacing around the myofibrils. Calcium allows actin and myosin filaments inside the myofibrils to interact and bring about contraction.

Figure 4.18 Pathway for signals from the nervous system that stimulate contraction of skeletal muscle. The plasma membrane of each muscle cell surrounds myofibrils and connects with inward-threading tubes (T tubules). The membrane-bound tubes are close to the sarcoplasmic reticulum, a calcium-storing system that functions in the control of contraction.

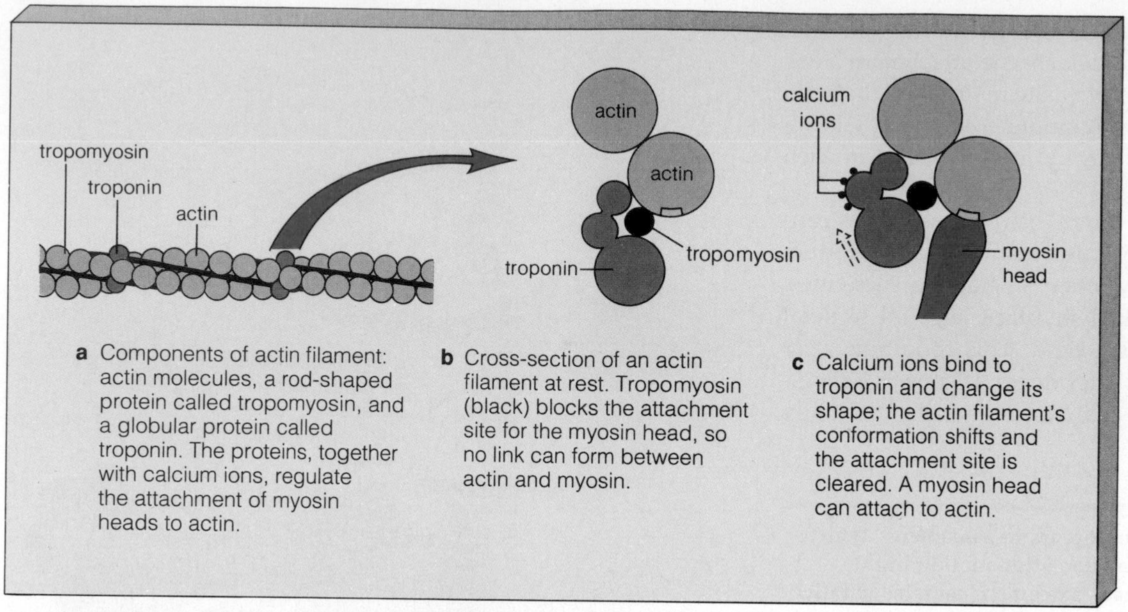

Figure 4.19 Role of calcium in the linkage of myosin heads with actin.

a Components of actin filament: actin molecules, a rod-shaped protein called tropomyosin, and a globular protein called troponin. The proteins, together with calcium ions, regulate the attachment of myosin heads to actin.

b Cross-section of an actin filament at rest. Tropomyosin (black) blocks the attachment site for the myosin head, so no link can form between actin and myosin.

c Calcium ions bind to troponin and change its shape; the actin filament's conformation shifts and the attachment site is cleared. A myosin head can attach to actin.

ciated with another protein, called tropomyosin, that forms part of thin filaments (Figure 4.19a and b).

When calcium binds to troponin (Figure 4.19c), the conformation of the actin filament changes. This allows myosin cross-bridges to attach, and so the cycle continues. When nervous system stimulation shuts off, the SR membranes actively take up calcium, and the changes in thin filament conformation just described are reversed. Myosin can't bind to actin, and the muscle cell relaxes.

Muscle contracts when calcium ions are released from a membrane system around myofibrils, the sarcoplasmic reticulum (SR).

Muscle relaxes when calcium ions are actively taken up after contraction and are stored in the sarcoplasmic reticulum.

By controlling the electrical signals that reach the sarcoplasmic reticulum, the nervous system controls calcium ion levels in muscle tissue—and so exerts control over muscle contraction.

"Fast" and "Slow" Muscle   Humans and other mammals have two general types of skeletal muscle cells. One type, so-called "slow" or "red" muscle, is crimson in color because its cells are packed with the oxygen-storing pigment myoglobin and are served by larger numbers of blood capillaries. It contracts relatively slowly, but because its cells are well-equipped to generate lots of ATP aerobically, those contractions can be sustained for extended periods. For example, certain muscles of the back and leg—called "postural" muscles because they aid body support—must contract for long periods when a

person is standing. Under the microscope, a sample of one of these muscles would reveal a high proportion of red muscle cells. Tissue from the muscles of the hands, on the other hand, will have fewer blood capillaries and a higher proportion of "fast" or "white" muscle cells, which contain fewer mitochondria and less myoglobin. This type of muscle cannot sustain contractions over long periods, but it *can* contract rapidly and powerfully for short periods.

When athletes train rigorously, one goal is to increase the relative size and contractile strength of fast or slow fibers in their muscles. A sprinter will benefit from larger, stronger fast muscle fibers in his or her thighs, whereas a distance swimmer will follow a regimen designed to increase the number of mitochondria in shoulder muscle cells (see *Focus on Wellness*, page 105). Some athletes resort to the synthetic hormones called anabolic steroids to rapidly build muscle mass, a practice we examine in this chapter's *Choices* essay.

CONTRACTION OF A SKELETAL MUSCLE

Ultimately, how strong you are depends on how forcefully your skeletal muscles can contract. That, in turn, depends on the size of a given muscle, how many of its cells are contracting, and the frequency of stimulation by the nervous system. A skeletal muscle contains a large number of cells, but not all of them contract at the same time. Together, a motor neuron and the muscle cells under its control are called a **motor unit** (Figure 4.20). Different muscles have dramatically different numbers of cells in

Focus on Wellness

Making the Most of Muscles

Muscle makes up more than 40 percent of the human body by weight, and most of that is skeletal muscle. Muscle cells adapt to the activity demanded of them. When severe nerve damage or prolonged bed rest prevents a muscle from being used at all, the muscle will rapidly begin to atrophy. Over time, affected muscles can lose up to three-fourths of their mass, with a corresponding loss of strength. More commonly, the skeletal muscles of a sedentary person stay basically healthy but cannot respond to physical demands in the same way that well-worked muscles can.

One of the best ways to maintain or increase muscle efficiency is through regular aerobic exercise—activities such as walking, swimming, biking, organized aerobics classes, and jogging. Aerobic exercise works muscles at a rate at which the body can keep them supplied with oxygen, and it has the following effects on muscle cells:

1.   Increase in the number and size of mitochondria, the organelles in which ATP is synthesized.

2.   Increase in the number of blood capillaries supplying muscle tissue, and hence increased availability of oxygen and nutrients and more efficient waste removal.

3.   Increase in the amount of the oxygen-storing pigment myoglobin in muscle tissues.

Together, such changes result in muscles that are more efficient metabolically and can work longer without becoming fatigued. When muscles also are regularly exerted with high intensity—as a person might do by lifting weights several times a week—individual muscle cells, or fibers, grow larger. Along with other related changes, this translates into whole muscles that are larger and stronger.

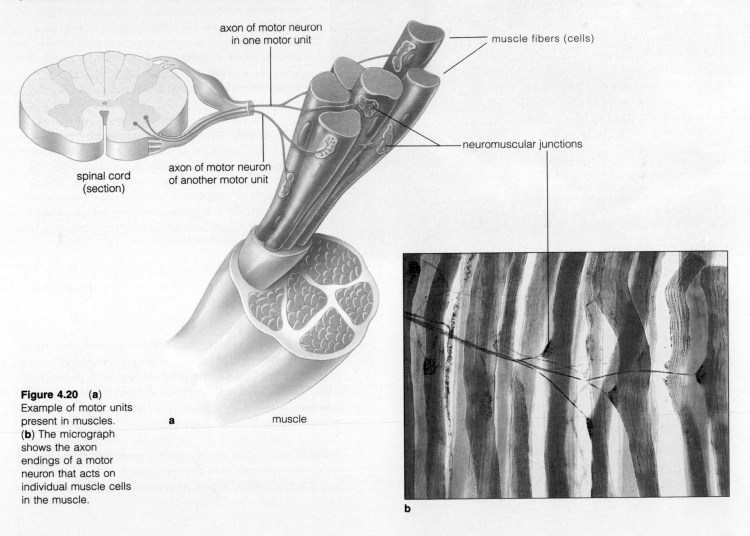

axon of motor neuron in one motor unit

muscle fibers (cells)

neuromuscular junctions

spinal cord (section)

axon of motor neuron of another motor unit

Figure 4.20 **(a)** Example of motor units present in muscles. **(b)** The micrograph shows the axon endings of a motor neuron that acts on individual muscle cells in the muscle.

a

muscle

b

Are Anabolic Steroids Worth the Gamble?

Aerobic exercise may replace fat with muscle, but it does not build significantly larger skeletal muscles. Only *resistance* exercise, such as weightlifting, does that. Some people, especially competitive athletes, desire significantly larger (and stronger) muscles, and they want them *now*. In spite of legal sanctions and persistent warnings from health professionals, they choose to use anabolic steroids.

Anabolic steroids are synthetic hormones, and as with other potent drugs, only prescription use is legal. The roughly 20 varieties of anabolic steroids mimic the effects

of a sex hormone, testosterone, and stimulate the synthesis of protein molecules, including muscle proteins. Using anabolic steroids while engaged in a weight-training exercise program can lead to rapid gains in lean muscle mass and strength. On the other hand, physicians, researchers, and athletes themselves report a long list of minor and major side effects of steroid use. In men, acne, baldness, shrinking testes, and infertility are the first signs of toxicity. The drugs may be linked to early onset of a cardiovascular disease, atherosclerosis. There is some evidence that consistent use contributes to kidney damage and to various types of cancer. In women, anabolic steroids trigger the development of a deep voice, pronounced facial hair, and irregular menstrual periods. Breasts may shrink and the clitoris may become grossly enlarged.

Some steroid-using men experience irritability and increased aggressiveness. Many competitive athletes look upon the added aggressiveness as a plus. Other men, however, experience uncontrollable aggression, delusions, and wildly manic behavior. In 1988, one steroid-using athlete purposely drove his car into a tree at 35 miles per hour.

With all of the suspected dangers associated with anabolic steroids, some may wonder why anyone would place his or her body and future in such jeopardy. Not everyone may be convinced that the drugs do enough damage to outweigh the "edge" they give in competition. What should a competitor do in a world that accords winning athletes wealth and the status of hero while relegating others to the pile of also-rans? What would *you* do?

their motor units—three or four in the motor unit of an eye muscle, and several hundred in motor units of a leg muscle. Hence we have much more precise control over eye movements than we do over leg movements.

Control of Contraction

Acetylcholine (ACh) is a *neurotransmitter*, a substance released by neurons that can have excitatory or inhibitory effects on the cells of muscles and glands throughout the

body. ACh also acts on certain cells in the brain and spinal cord. **Neuromuscular junctions** are synapses between a motor neuron and muscle cells. At this type of junction, the branched endings of certain extensions of the motor neuron (axons) are positioned on the muscle cell membranes (Figure 4.18b). When the neuron is stimulated, calcium channels open in the plasma membrane of the endings, and calcium ions (Ca^{++}) from the extracellular fluid flow inward. This causes vesicles in the axon ending to release ACh. When ACh binds to receptors on the muscle cell membrane, it has an excitatory effect that may set in motion events within the muscle cells that lead to contraction. The poison curare, extracted from a South American tree frog, blocks the binding of ACh by muscle cells and so prevents muscles—including those required for breathing—from contracting.

Physiologists can artificially stimulate a motor neuron with electrical impulses and make recordings of the changes in muscle contraction. When a single, brief stimulus of a given strength activates a certain number of motor units, the muscle contracts briefly, then relaxes. This response is a **muscle twitch** (Figure 4.21). When a motor unit is stimulated again before a twitch response is completed, it twitches again. The strength of the contraction depends on how far the twitch response has proceeded by the time the second signal arrives. The "pull" of the new contraction is added to that of the contraction already underway, a phenomenon called **temporal summation** (Figure 4.21c). As a result, with additional stimuli the muscle shortens even more.

Our muscles normally operate near or at maximum temporal summation, a condition called **tetany.** (In the disease *tetanus,* muscles remain contracted, possibly fatally, due to the effects of a bacterial toxin.) In order to postpone *muscle fatigue,* in which hard-working muscle cells outstrip their supply of ATP and temporarily become unable to respond to stimulation, sets of motor units take turns sustaining the contraction of a muscle.

Individual cells in a motor unit always contract according to an **all-or-none principle:** They either contract fully in response to stimulation, or they do not respond at all. If a muscle is contracting only weakly—as might happen in your forearm muscles when you pick up a pencil—it is because the nervous system is activating only a small number of motor units. In a stronger contraction—when you heft a stack of books, say—a larger number of motor units is activated, at a high frequency.

Even when muscles are relaxed, some of their motor units are contracted. This steady, low-level contracted state is called *muscle tone.* Muscle tone helps stabilize joints and maintain general muscle health.

Exercise Levels and Energy for Contraction

A resting muscle cell has a small reservoir of ATP. When the cell is called upon to contract, its demand for ATP suddenly surges. During the onset of contraction, a chemical reaction transfers phosphate from the compound creatine phosphate to ADP, forming ATP. Although cell stores of creatine phosphate are also limited, this fleeting reaction generates enough ATP to power contraction until other ATP-producing pathways begin to operate.

During prolonged, moderate exercise, the oxygen-using reactions of aerobic respiration provide most of the

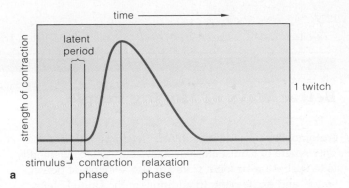

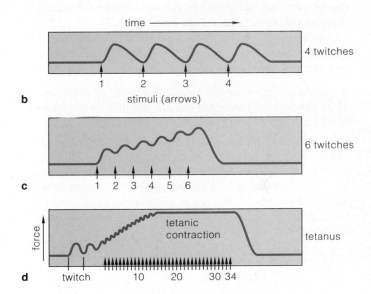

Figure 4.21 Recordings of twitches in artificially stimulated muscles. (**a**) A single twitch. (**b**) Two stimulations per second cause a series of twitches. (**c**) Six per second cause a summation of twitches. (**d**) About 20 per second cause a tetanic contraction.

ATP required for contraction. When the demand for muscle action is intense (say, a 100-meter sprint), it may exceed the body's capacity to produce enough ATP. Then, the cells produce more and more ATP by lactate fermentation. In this anaerobic pathway, glycogen in cells is broken down to its component glucose molecules, which are then converted to lactate (the ionized form of lactic acid). The ATP yield is small, and as lactate builds up in muscles it contributes to fatigue.

The interaction of muscles and bones to bring about body movements has been a central focus of this chapter. Before continuing on, take a moment to consider some larger issues of "people moving," the topic of this chapter's *Focus on Environment* essay.

Better Ways of Moving People

Automobiles provide us with the freedom to move not only where our musculoskeletal systems will conveniently take us, but far beyond. Cars are also symbols of power, status, and excitement. In addition, in the United States one of every six nonfarm jobs is connected to automobile manufacturing or related industries.

Unfortunately, motor vehicles have many harmful effects on air, water, and other resources. Consider just a few of these effects:

• Motor vehicles are the largest source of air pollution. They produce 50 percent of the air pollution in the United States, in spite of strict emissions standards.

• Worldwide, motor vehicles account for 13 percent of the input of the primary greenhouse gas, carbon dioxide, into the atmosphere. In the United States they account for almost 25 percent of CO_2 emissions and 13 percent of the chlorofluorocarbons that act as greenhouse gases (page 504), and they also deplete life-sustaining ozone in the stratosphere.

• Motor vehicle use contributes directly and indirectly to water pollution from oil spills, gasoline spills, leakage and dumping of used engine oil and battery fluids, and contamination of underground drinking water from leaky oil and gasoline storage tanks.

Currently, efforts are underway to develop cars that run on electricity or nonpolluting hydrogen fuels (see page 22). It may be decades before such new technologies are affordable and routinely available, however. Meanwhile, the most constructive approach for average citizens is to minimize unnecessary personal automobile use and to support policies that encourage the development and use of mass transit systems (including economic incentives such as higher gasoline taxes). For many of us, such changes will no doubt take some time to get used to. However, they are rapidly becoming essential to sustaining a livable world.

SUMMARY

1. Bones are the structural elements of the human skeleton. They function in movement (by interacting with skeletal muscles to which they are attached), protection and support of other body parts, mineral storage, and blood cell formation.

2. The skeleton has an axial portion (skull, backbone, ribs, and breastbone) and an appendicular portion (limb bones, pelvic girdle, and pectoral girdles). Intervertebral disks are shock pads and flex points in the backbone.

3. Smooth, cardiac, and skeletal muscle tissue all contract (shorten) when stimulated. Only skeletal muscle interacts with the skeleton to bring about movement of the body or its parts.

4. Skeletal and cardiac muscle cells contain many threadlike myofibrils, which contain actin (thin) and myosin (thick) filaments. The filaments are organized in orderly arrays in sarcomeres (the basic units of contraction).

5. Sarcomeres contract when nerve stimulation triggers the release of calcium ions from a membrane system (sarcoplasmic reticulum) in the muscle cell. Calcium binding alters the actin filaments so that the heads of adjacent myosin filaments can bind to them. ATP provides the energy to drive the cross-bridge power strokes that cause actin filaments to slide past the myosin filaments and so shorten the sarcomere.

6. In combination with skeletal muscles, the skeleton works like a system of levers in which rigid rods (bones) move about at fixed points (joints). A limb can be moved and rotated around a joint because of the way pairs or groups of muscles are arranged relative to joints.

Review Questions

1. What are some of the functions of bone tissue? *90*

2. Discuss the properties of the three types of muscle tissue. *99–100*

3. Look at Figure 4.15. Then, on your own, sketch and label the fine structure of a muscle, down to one of its individual myofibrils. Can you identify the basic unit of contraction in a myofibril? *101*

4. How do actin and myosin interact in a sarcomere to bring about muscle contraction? What role does ATP play? What role does calcium play? *102*

5. What is a motor unit? Why does a rapid-fire sequence of muscle twitches generate a stronger overall contraction than a single twitch? *105*

Critical Thinking: You Decide *(Key in Appendix IV)*

1. You are training young athletes for the 100-meter dash. They need muscles specialized for speed and strength, rather than for endurance. What kinds of muscle characteristics would your training regimen aim to develop? How would you alter it to train marathoners?

2. Growth hormone, or GH, is used clinically to spur growth in children who are unusually short because they have a GH deficiency. However, it is useless for a short but otherwise normal 25-year-old to request GH treatment from a physician. Why?

Self-Quiz *(Answers in Appendix III)*

1. _____ and _____ systems work together to move the body and specific body parts.

2. The three types of muscle tissue are _____, _____, and _____.

3. Which of the following serve as shock pads and flex points in the human backbone?
a. vertebrae   c. lumbar bones
b. cervical bones   d. intervertebral disks

4. Which structure stores calcium ions necessary for muscle contraction?
a. plasma membrane   c. sarcoplasmic reticulum
b. motor neuron   d. T tubule

5. The smallest unit of contraction in skeletal muscle cells is the _____.
a. myofibril   c. muscle fiber
b. sarcomere   d. myosin filament

6. Muscle contraction will not occur _____.
a. in the absence of calcium ions   c. both a and b
b. in the absence of ATP   d. neither a nor b

7. Match the following terms concerning muscle structure and function.
____ myofibrils
____ sarcoplasmic reticulum
____ sarcomere
____ muscle cell
____ ATP

a. contains many myofibrils
b. the contractile unit of muscle
c. indirectly provides energy to drive the power stroke that slides actin filaments past myosin filaments
d. composed of actin and myosin filaments
e. calcium ion storage site in a muscle

Key Terms

actin *101*
all-or-none principle *107*
appendicular skeleton *92*
axial skeleton *92*
bone *90*
brain case *92*
cartilaginous joint *98*
clavicle *95*
cross-bridge formation *102*
epiphyseal plate *91*
femur *96*
fibrous joint *98*
humerus *95*
insertion *99*
intervertebral disk *95*
ligament *92*
mandible *94*
motor unit *104*
muscle twitch *107*
myofibril *101*
myosin *101*
neuromuscular junction *106*

origin *99*
osteocyte *90*
osteon *90*
pectoral girdle *95*
pelvic girdle *96*
radius *95*
red marrow *90*
ribs *95*
sarcomere *101*
sarcoplasmic reticulum *103*
scapula *95*
sinus *92*
skeletal muscle *99*
skull *92*
sliding-filament mechanism *102*
sternum *95*
synovial joint *96*
temporal summation *107*
tendon *92*
tetany *107*
ulna *95*
yellow marrow *90*

Readings

Diamond, J. May 1993. "Building to Code." *Discover.* How an engineer might analyze the functional design of bones and other organs.

Lewis, R. July–August 1991. "Arthritis: Modern Treatment for that Old Pain in the Joints." *FDA Consumer.*

Vander, A., J. Sherman, and D. Luciano. 1990. *Human Physiology*, 5th ed. New York: McGraw-Hill.

5 DIGESTION AND NUTRITION

Tales of a Tube

Surrounded by snow and ice, an Aleut hunter slices slabs of raw whale blubber for dinner. In a village in Nepal, a grandmother eats only a bowl of rice. Ten thousand miles away, an American college student downs a bacon cheeseburger, a salad, and coffee. People in various cultures feast on snake meat, insects, hot chilis, salted fish, and cassava tubers (Figure 5.1).

Through it all, our digestive systems receive whatever we choose to swallow and convert that array of foods into carbohydrates, lipids, and proteins the body can use for energy and to build, repair, and maintain its tissues. In this way, the digestive system plays a key role in homeostasis, and thus in survival. The system consists basically of a food-processing tube and some accessory organs.

Many Americans have poor eating habits. They may eat large quantities of snack foods, skip meals, or eat too much and often too fast. Worse yet, the typical American diet tends to be high in fat, cholesterol, and salt. The world distribution of some illnesses suggests that the body may pay a high price for certain dietary habits. For example, in the United States a fat-laden diet is a major contributor to health-threatening obesity and heart disease. Obesity also increases an adult's risk of developing diabetes and contributes to other health problems. In regions where salt-cured and smoke-cured foods are staples, the rates of stomach and esophagus cancer are disturbingly high. Nongenetic colon cancer is much more prevalent among people whose diets lack sufficient fiber, as is common in the United States. Most people of rural Africa and India cannot afford to eat much more than whole grains, which are high in fiber content, and they rarely suffer colon cancer.

In this chapter we'll look at how the components of the digestive system carry out their life-supporting functions and at the roles various nutrients play in human health. As you read, keep in mind the digestive system's overall role in homeostasis. When it has made nutrients available, other activities—including circulation, respiration, and excretion—come into play to help maintain the body in the living state.

KEY CONCEPTS

1. The digestive system has specialized regions, including the mouth, stomach, and large and small intestines. As food moves through the regions in sequence, it is broken down mechanically and chemically, nutrients are absorbed, and residues are eliminated. Accessory organs produce enzymes and other substances that aid in the digestive process.

2. To maintain an acceptable body weight and overall health, energy intake must balance energy output (in the form of resting metabolism, physical activity, and so on).

3. Proper nutrition requires the intake of vitamins, minerals, and carbohydrates, as well as amino acids and fatty acids that the body cannot produce itself.

4. Digestion and absorption of food is an essential aspect of homeostasis. Interactions among the digestive, circulatory, respiratory, and urinary systems supply the body's cells with nutrients and oxygen, dispose of wastes, and maintain the volume and composition of extracellular fluid.

Figure 5.1 (left page) Taking in food, which will be digested into nutrients the body requires to survive.

OVERVIEW OF THE DIGESTIVE SYSTEM

The **digestive system** functions to mechanically and chemically break down food into nutrient molecules that the body uses to help maintain homeostasis. The system is a tube with many specialized regions (Figure 5.2, next page). It extends from the mouth to the anus and is also called the **gastrointestinal (GI) tract**. The space inside the tube, the *lumen*, passes through the body like a tunnel. From that perspective, material in the GI tract is *outside* the body. Stretched out, the GI tract would be 6.5–9 meters (21–30 feet) long in an adult. Its specialized regions are the mouth (oral cavity), pharynx, esophagus, stomach, small intestine, and large intestine. The large intestine terminates in the rectum, anal canal, and anus.

Accessory Organs Various *accessory organs* secrete enzymes and other substances that have essential roles in different aspects of digestion and absorption. These "organs" include glands in the wall of the gastrointestinal tract, the salivary glands, and the liver, pancreas, and gallbladder.

Digestive system operations include four overall functions:

1. **Motility** Muscular movement of the digestive tube wall, leading first to the mechanical breakdown, mixing, and passage of ingested nutrients and then to elimination of undigested and unabsorbed residues.

2. **Secretion** Release of digestive enzymes, fluids, and other substances into the digestive tube lumen.

3. **Digestion** Chemical breakdown of food into particles and then into nutrient molecules small enough to be absorbed.

4. **Absorption** Passage of digested nutrients, fluid, and ions across the tube wall and into the blood or lymph, which will distribute them throughout the body.

The digestive system is the entry point for nutrients into the body, but as Figure 5.3 suggests, this is only a beginning. Digested nutrients enter the internal environment by way of the bloodstream, and the circulatory system typically distributes nutrients to cells throughout the body. The respiratory system helps cells use nutrients by supplying oxygen for aerobic respiration and eliminating carbon dioxide wastes. And even though the kinds and amounts of nutrients being absorbed can vary depending on the diet, the urinary system helps maintain a steady state in the volume and composition of blood and other extracellular fluid.

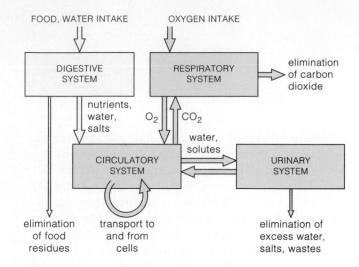

Figure 5.3 Links between the digestive, respiratory, circulatory, and urinary systems. These organ systems work together to supply the body's cells with nutrients and with oxygen for aerobic cell respiration, and to eliminate wastes. This chapter focuses on the digestive system; subsequent chapters address the other systems depicted here.

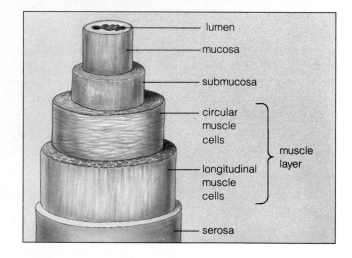

Figure 5.4 The four-layered wall of the gastrointestinal tract. The layers are not drawn to scale.

Gastrointestinal Tract Structure

The entire digestive tube is lined with mucous membrane. From the esophagus onward, the tube wall consists of four basic layers (Figure 5.4). The *mucosa* (the innermost layer of epithelial tissue) faces the lumen, through which food passes. The mucosa is surrounded by the *submucosa*, a connective tissue layer with blood and lymph vessels and nerve plexuses (local networks of nerve cells). The next layer is *smooth muscle*—usually two sublayers, one in a circular orientation and the other oriented lengthwise. An outer layer of connective tissue, the *serosa*, is

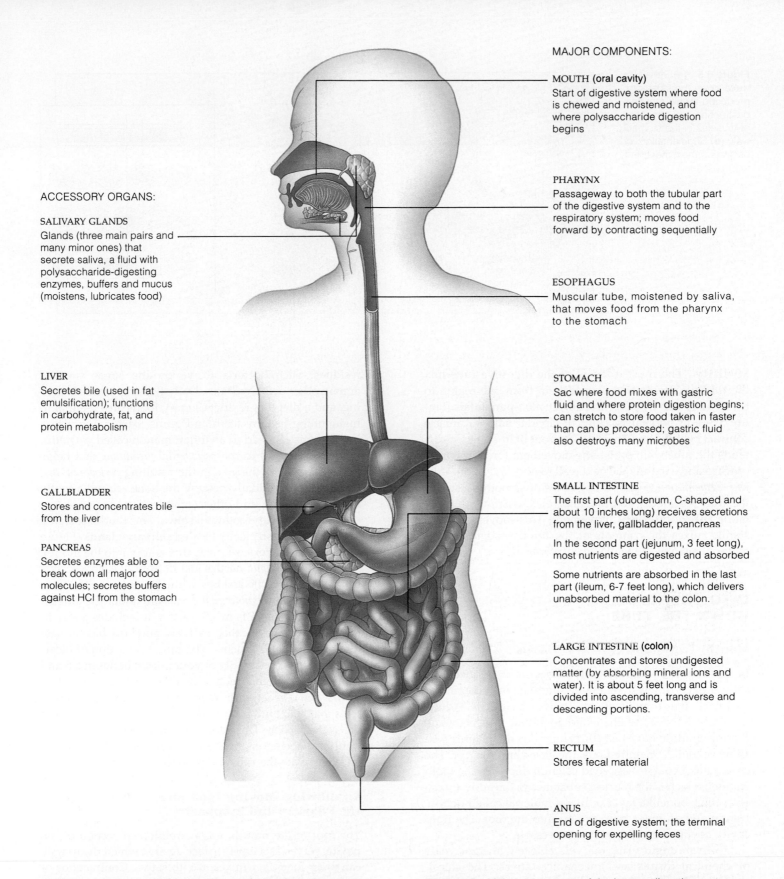

MAJOR COMPONENTS:

MOUTH (oral cavity)
Start of digestive system where food is chewed and moistened, and where polysaccharide digestion begins

PHARYNX
Passageway to both the tubular part of the digestive system and to the respiratory system; moves food forward by contracting sequentially

ESOPHAGUS
Muscular tube, moistened by saliva, that moves food from the pharynx to the stomach

STOMACH
Sac where food mixes with gastric fluid and where protein digestion begins; can stretch to store food taken in faster than can be processed; gastric fluid also destroys many microbes

SMALL INTESTINE
The first part (duodenum, C-shaped and about 10 inches long) receives secretions from the liver, gallbladder, pancreas

In the second part (jejunum, 3 feet long), most nutrients are digested and absorbed

Some nutrients are absorbed in the last part (ileum, 6-7 feet long), which delivers unabsorbed material to the colon.

LARGE INTESTINE (colon)
Concentrates and stores undigested matter (by absorbing mineral ions and water). It is about 5 feet long and is divided into ascending, transverse and descending portions.

RECTUM
Stores fecal material

ANUS
End of digestive system; the terminal opening for expelling feces

ACCESSORY ORGANS:

SALIVARY GLANDS
Glands (three main pairs and many minor ones) that secrete saliva, a fluid with polysaccharide-digesting enzymes, buffers and mucus (moistens, lubricates food)

LIVER
Secretes bile (used in fat emulsification); functions in carbohydrate, fat, and protein metabolism

GALLBLADDER
Stores and concentrates bile from the liver

PANCREAS
Secretes enzymes able to break down all major food molecules; secretes buffers against HCl from the stomach

almost as thin as kitchen plastic wrap. Thickened regions of muscle occur in the wall at either end of the stomach and at other specialized locations. These muscular regions serve as one-way valves called *sphincters,* which help control the forward movement of food and prevent backflow.

Figure 5.2 Major components of the human digestive system and their functions. Organs with accessory roles in digestion are also labeled.

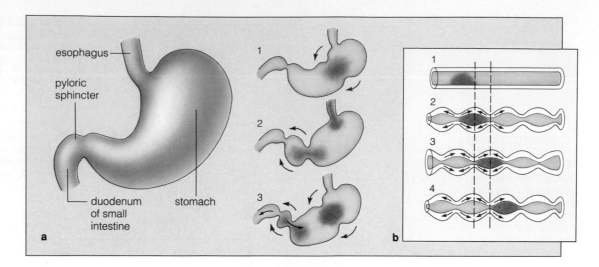

Figure 5.5 (a) Peristaltic wave down the stomach, produced by alternating contraction and relaxation of muscles in the stomach wall. (b) Segmentation, or oscillating movement, in the intestines.

esophagus

pyloric sphincter

duodenum of small intestine

stomach

a

b

Motility The muscle layers of the digestive tube mix the tube's contents and propel them from one region to the next by wavelike contractions called **peristalsis** (Figure 5.5a). In peristalsis, rings of circular smooth muscle contract behind food and relax in front of it. The food distends the tube wall, peristaltic movement forces the food onward and expands the next wall region, and so on. During *segmentation* (Figure 5.5b), rings of smooth muscle in the wall repeatedly contract and relax, creating an oscillating (back-and-forth) movement. This movement constantly mixes the contents of the lumen and forces the material against the wall's absorptive surface.

INTO THE MOUTH, DOWN THE TUBE

Mouth, Teeth, and Salivary Glands

In the mouth or *oral cavity*, food begins to be mechanically broken down by chewing, and enzymes begin the chemical digestion of starch (polysaccharides). Adults normally have 32 teeth to aid in biting and chewing. (Young children have just 20 "primary teeth.") Figure 5.6a shows a tooth's main regions, the *crown* and the *root*. The crown has a coat of hardened calcium deposits, the tooth enamel, which is the hardest substance in the body. It covers a thick bonelike layer of living material called dentin. Dentin and an inner pulp extend into the root. The pulp cavity contains nerves and blood vessels.

Teeth are engineering marvels, able to withstand years of chemical insults and mechanical stress. The chisel-shaped incisors bite off chunks of food, the cone-shaped cuspids (canines) tear it, and the flat-topped molars and premolars grind it (Figure 5.6b). Bacteria that cause tooth decay *(caries)* abound in the mouth, living on food

residues. Such bacteria thrive on the sugar sucrose, among other carbohydrates. Daily flossing, gentle brushing, and a diet that restricts sugary foods are essential to maintaining healthy teeth and gums. Bacterial infection in the gums can lead to an inflammation called *gingivitis*, which can spread to the *periodontal membrane* that helps anchor a tooth in the jaw. In the resulting *periodontal disease*, bacteria gradually destroy the bone around a tooth, the tooth loosens, and other complications can arise.

Chewing mixes food with saliva. This fluid is secreted from several strategically located **salivary glands** (Figure 5.6c) and flows through ducts that empty into the mouth. A large *parotid gland* nestles just in front of each ear. *Submandibular glands* lie just below the lower jaw in the floor of the mouth, and *sublingual glands* are under the tongue.

Although saliva is mostly water, it includes a starch-degrading enzyme, called **salivary amylase**, bicarbonate ions (HCO_3^-), and mucins. The buffering action of bicarbonate ions keeps the pH of your mouth between 6.5 and 7.5, a range within which salivary amylase can function, even when you eat acidic foods. Mucins are modified proteins that help bind bits of food into a softened, lubricated ball called a *bolus*. This allows starch digestion to continue until acid secretions in the stomach penetrate the bolus and inactivate the salivary amylase.

Swallowing: Moving Food into the Pharynx and Esophagus

The roof of the mouth, a bone-reinforced section of the **palate,** serves as a hard surface against which the tongue can press food as it mixes it with saliva. Contractions of the tongue muscle force the food bolus into the **pharynx** (throat). As Figure 5.2 shows, this passageway connects with both the windpipe or *trachea*, which leads to the lungs, and the **esophagus**, a muscular tube that leads to

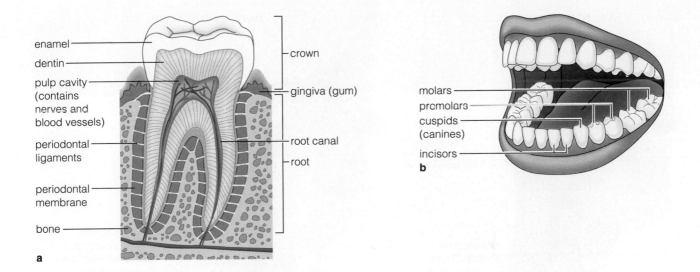

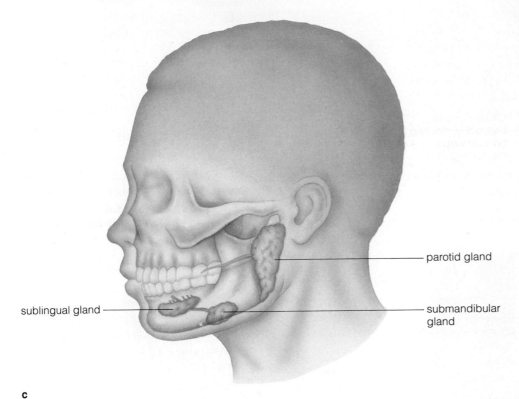

Figure 5.6 (**a**) Anatomy of a human tooth. (**b**) Types and locations of teeth in the human mouth. (**c**) Locations of the salivary glands.

the stomach. The membrane lining the pharynx and esophagus secretes mucus, but these regions do not have roles in digestion. Peristaltic contractions in their walls simply propel food into the stomach.

Swallowing is initiated when an individual pushes a bolus into the pharynx. This stimulates sensory receptors in the pharynx wall, and the receptors trigger involuntary contractions called the "swallowing reflex." This reflex involves a complex series of steps that both prevent food from moving up into the nose and help keep it from entering the trachea. When you swallow a mouthful of food, you normally don't choke on it because the larynx ("voicebox") rises and a flaplike valve, the *epiglottis*, closes off the opening to the respiratory tract and prevents breathing for a split second while food moves into the esophagus. When swallowed food reaches the junc-

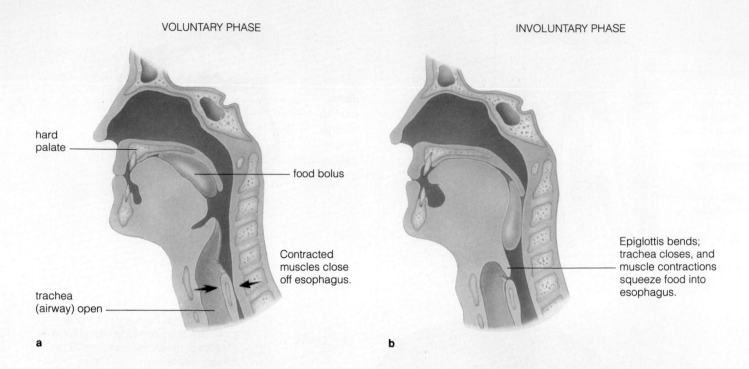

hard
palate

food bolus

Contracted
muscles close
off esophagus.

trachea
(airway) open

Epiglottis bends;
trachea closes, and
muscle contractions
squeeze food into
esophagus.

a

b

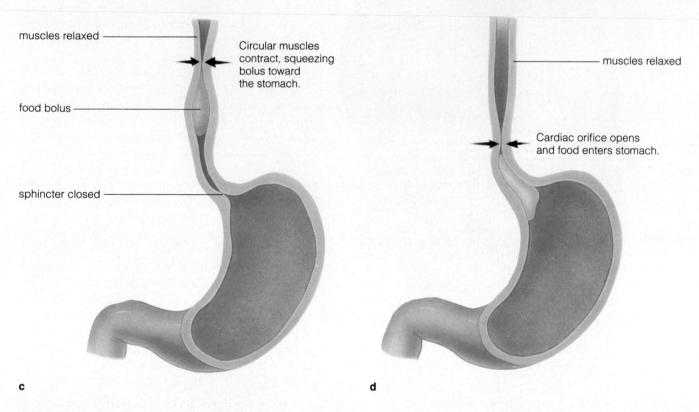

muscles relaxed

Circular muscles
contract, squeezing
bolus toward
the stomach.

muscles relaxed

food bolus

Cardiac orifice opens
and food enters stomach.

sphincter closed

c

d

Figure 5.7 How food reaches the stomach: Swallowing and peristalsis. (**a**) First, contractions of the tongue push the food bolus into the pharynx. (**b**) Next, the larynx rises, the epiglottis bends downward, closing off the trachea, and contractions of throat muscles squeeze the food bolus into the esophagus. (**c**, **d**) Finally, peristalsis in the esophagus moves the bolus through the cardiac orifice into the stomach.

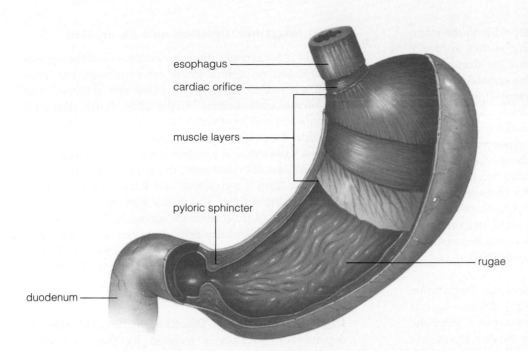

esophagus

cardiac orifice

muscle layers

pyloric sphincter

duodenum

rugae

Figure 5.8 Cutaway view of the stomach. Folds called *rugae* allow this sac to accommodate food by stretching. Because the duodenum is the region of the small intestine closest to the stomach and its acidic secretions, a high percentage of ulcers occur here.

tion of the esophagus with the stomach (Figure 5.7), it enters the stomach through a valvelike *cardiac orifice* (Figure 5.7d).

Stomach: Storage, Digestion, and Control

The **stomach** is a muscular, expandable sac (Figure 5.8). Stretched to the limit, an adult's stomach can hold several liters of food, but when the stomach is empty its walls crumple into folds called *rugae*. The stomach has three main functions: (1) It stores and mixes ingested food, (2) its secretions begin the chemical digestion of food, and (3) it helps control the movement of food into the small intestine.

Stomach Secretions Each day, cells in the stomach lining secrete about two liters of substances, including mucus, hydrochloric acid (HCl), and pepsinogens, which are precursors of digestive enzymes known as **pepsins.** Together with water, these make up the stomach's rather acid *gastric fluid.* The increased acidity helps dissolve bits of food to form a liquid mixture called *chyme;* it also kills many microbes present in food. ("Heartburn" occurs when gastric fluid backs up into the esophagus.) Cells in the stomach lining also secrete *intrinsic factor,* a protein essential for the absorption of vitamin B_{12} in the small intestine.

In response to the sight, aroma, and taste of food, the brain signals secretory cells in the salivary glands and stomach lining to step up their activity. As food enters the

stomach, stretching of the stomach wall activates receptors that give rise to additional signals calling for stepped-up secretions. Secretory cells also are stimulated directly by substances such as partially digested proteins and the caffeine in coffee, tea, chocolate, and colas.

Protein digestion starts in the stomach. The high acidity resulting from HCl secretions denatures proteins and exposes the peptide bonds. It also converts pepsinogens to pepsins, which break the bonds. Accumulating protein fragments trigger secretion of gastrin, a hormone that stimulates the pepsinogen- and HCl-secreting cells even more.

What prevents HCl and pepsin from breaking down the stomach lining itself? Control mechanisms usually assure that enough mucus and bicarbonate are secreted to protect the lining. This protective layer of mucus and bicarbonate is sometimes called the "gastric mucosal barrier." Sometimes, however, the barrier is disrupted. Recent studies indicate that infection by a bacterium *(Helicobacter pylori)* may interfere with mucus production, and this causes the inner surface of the stomach to break down so that hydrogen ions diffuse into the lining. This in turn triggers the release of histamine from tissue cells. Histamine is a chemical that acts on local blood vessels, and it stimulates more HCl secretion, which further damages tissues. The open sore, a *peptic ulcer,* may bleed, and left untreated it can give rise to secondary infections or anemia (from blood loss). Although *H. pylori* infection may be the culprit in many cases, some people appear to have a genetic predisposition to develop ulcers. Chronic stress and smoking are also contributing factors, and drinking alcohol aggravates the condition.

Stomach Emptying Waves of peristalsis in the stomach mix the chyme and build up force as they approach the powerful *pyloric sphincter* between the stomach and the opening of the small intestine. The arrival of a strong contraction closes the sphincter, so most of the chyme is squeezed back. With each contraction, only a small amount of chyme moves into the small intestine.

Depending on the volume and composition of chyme, it can take from two to six hours for the stomach to empty after a meal. Large meals activate more receptors in the stomach wall, these receptors call for stronger contractions, and the stomach empties faster. On the other hand, increased acidity or fat content in the small intestine stimulates hormone secretions that *slow* stomach emptying. Hence food is not moved along faster than it can be processed.

Only alcohol and a few other substances begin to be absorbed across the stomach wall. And whereas liquids imbibed on an empty stomach pass rapidly from the stomach to the small intestine, where further absorption occurs, the rate of gastric emptying slows when food, especially fatty food, is also in the stomach. That is why a person feels the effects of alcohol less quickly when drinking accompanies a meal or high-fat snack such as peanuts.

Small Intestine: Digestion and Absorption

Table 5.1 lists the major digestive enzymes and the regions of the digestive system where carbohydrates, fats, proteins, and nucleic acids are digested and absorbed into the internal environment. As the table shows, digestion is completed and most nutrients are absorbed in the **small intestine.** An adult's small intestine is about an inch and a half in diameter and 6 meters long—some 20 feet! It has three sections: the *duodenum,* the *jejunum,* and the *ileum.* Secretions from the pancreas and liver enter a common duct that empties into the duodenum. In all, each day about 9 liters of fluid from the stomach, liver, pancreas, and intestinal glandular cells enter the small intestine, and 95 percent of that fluid is absorbed across the intestinal lining. Let's examine how this part of the GI tract performs its essential functions.

Digestion Processes Digestion in the small intestine depends on substances secreted by three accessory organs: the pancreas, the liver, and the gallbladder. Chyme entering the duodenum triggers hormonal signals that stimulate a brief flood of digestive enzymes from the **pancreas,** an elongated organ just above and behind the opening to the small intestine. As part of "pancreatic

Table 5.1	Major Enzymes of Digestion			
Enzyme	Source	Where Active	Substrate	Main Breakdown Products*
Carbohydrate Digestion:				
Salivary amylase	Salivary glands	Mouth	Polysaccharides	Disaccharides
Pancreatic amylase	Pancreas	Small intestine	Polysaccharides	Disaccharides
Disaccharidases	Intestinal lining	Small intestine	Disaccharides	Monosaccharides (e.g., glucose)
Protein Digestion:				
Pepsins	Stomach lining	Stomach	Proteins	Protein fragments
Trypsin and chymotrypsin	Pancreas	Small intestine	Proteins	Protein fragments
Carboxypeptidase	Pancreas	Small intestine	Protein fragments	Amino acids
Aminopeptidase	Intestinal lining	Small intestine	Protein fragments	Amino acids
Fat Digestion:				
Lipase	Pancreas	Small intestine	Triglycerides	Free fatty acids, monoglycerides
Nucleic Acid Digestion:				
Pancreatic nucleases	Pancreas	Small intestine	DNA, RNA	Nucleotides
Intestinal nucleases	Intestinal lining	Small intestine	Nucleotides	Nucleotide bases, monosaccharides

*Yellow parts of table identify breakdown products that can be absorbed into the internal environment.

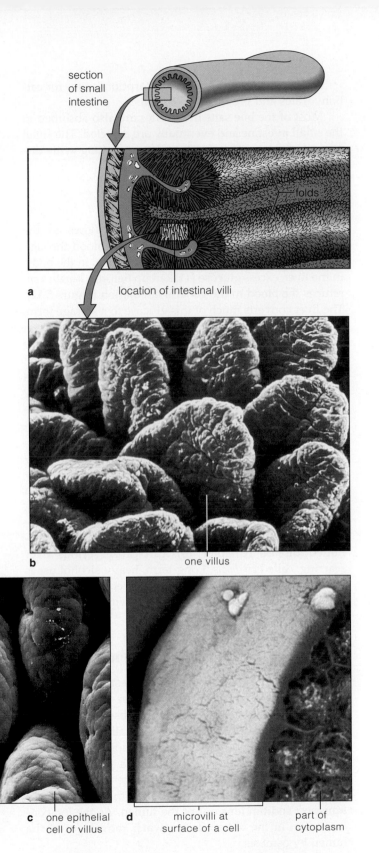

section of small intestine

folds

location of intestinal villi

a

one villus

b

c one epithelial cell of villus

d microvilli at surface of a cell

part of cytoplasm

Figure 5.9 Villi and microvilli in the small intestine. (**a**) Surface view of the deep, permanent folds of the inner layer of the intestinal tube. (**b**) Some of the fingerlike projections (villi) that cover the inner layer. Because the villi are so dense and numerous, they give the surface a velvety appearance. (**c**) Tip of a single villus; individual epithelial cells are visible. (**d**) The dense crown of microvilli at the surface of a single cell.

juice," these enzymes act on carbohydrates, fats, proteins, and nucleic acids (see Table 5.1). For example, like pepsin in the stomach, the pancreatic enzymes *trypsin* and *chymotrypsin* digest the polypeptide chains of proteins into peptide fragments. The fragments are then degraded to amino acids by *carboxypeptidase* from the pancreas and by *aminopeptidase* (present on the surface of the intestinal mucosa). The pancreas also secretes the buffer bicarbonate, which helps neutralize the HCl arriving from the stomach and so maintains an appropriate chemical environment for the functioning of pancreatic enzymes. Two pancreatic hormones, insulin and glucagon, do not function in digestion but do have roles in nutrition.

Fat digestion depends only partly on enzymes. It also depends on *bile*, a mixture of substances secreted continually by the liver. Bile contains bile salts, bile pigments, cholesterol, and lecithin (a phospholipid). It flows through ducts from the liver to the small intestine. When no food is moving through the GI tract, a sphincter closes off the main duct and bile backs up into a saclike storage organ, the **gallbladder.**

By a process called *emulsification,* bile salts speed up fat digestion. Most fats in the human diet are triglycerides, which are insoluble in water and tend to aggregate into large fat globules in the chyme. When movements of the intestinal wall agitate chyme, fat globules break up into small droplets, which become coated with bile salts. Because bile salts carry negative charges, the coated droplets repel each other and stay separated. This suspension of fat droplets is the "emulsion." Emulsion droplets, as compared to fat globules, give fat-digesting enzymes a much greater surface area to act upon. Thus triglycerides can be broken down much more rapidly to fatty acids and monoglycerides.

Bile salts are acids that can break down cholesterol. As we will see shortly, this is how the body excretes cholesterol. When there is chronically more cholesterol than available bile salts can dissolve, the excess may precipitate out. *Gallstones,* which are mostly hard clumps of cholesterol, can develop in the gallbladder and cause severe pain if they become lodged in bile ducts. In severe cases they can result in an inflammation called *cholecystitis.* Treatments include surgery, drugs, and sometimes breaking gallstones into small particles using a laser or ultrasound.

Absorption Processes The lining of the small intestine is densely folded into **villi** (singular: *villus*). Villi are tiny, fingerlike extensions of the mucosa, and individual epithelial cells have a crown of **microvilli** on their surface. Microvilli are threadlike projections of the plasma membrane. Both the villi and microvilli (Figure 5.9) greatly increase the surface area exposed to chyme and available for absorption.

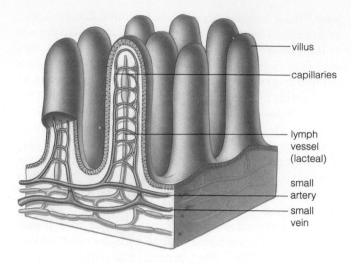

villus

capillaries

lymph vessel (lacteal)

small artery

small vein

Figure 5.10 Location of blood vessels and lymph vessels in intestinal villi. Monosaccharides and most amino acids that have crossed the intestinal lining enter the blood vessels; fats enter the lymph vessels, which drain into the general circulation.

By the time proteins, lipids, and carbohydrates are halfway through the small intestine, most have been broken down into smaller molecules. Some of these molecules—glucose and other monosaccharides, amino acids, and some nucleotides—directly enter the epithelial cells of villi. Some water does also. Mineral ions, too, move across the intestinal lining, actively transported by proteins in the epithelial cell plasma membranes. These nutrients then move through the cells, on through extracellular fluid, and into small blood vessels inside the villus (Figure 5.10).

By contrast, bile salts assist the absorption of fatty acids and monoglycerides that result when lipids are digested. Such molecules have hydrophobic regions and so do not dissolve in watery chyme. Left alone, they would float like a film of oil on a puddle and not gain proximity to the surface of absorptive cells. Instead, however, the molecules clump together with bile salts, a phospholipid called lecithin, cholesterol, and other substances, and form tiny droplets called *micelles*. Nutrient molecules within micelles continuously exchange places with nutrient molecules that are dissolved in the chyme. If concentration gradients are favorable when a micelle comes into close contact with the absorptive cell surface, such molecules diffuse out of the micelle, into the solution bathing the microvilli, and into epithelial cells. There, fatty acids and monoglycerides recombine into triglycerides. The triglycerides combine with proteins into particles, called *chylomicrons,* that leave the cells by exocytosis and enter the internal environment. Once absorbed in a villus, chylomicrons enter a lymph vessel called a **lacteal.** Lacteals drain into other lymphatic vessels, which in turn drain into the blood circulation system.

Figure 5.11 summarizes the absorption routes for carbohydrates and proteins and for fats.

Most of the bile salts in micelles are also absorbed in the small intestine and eventually are recycled. The small quantity that remains in the GI tract—along with dissolved cholesterol—is excreted in feces.

Digestive Functions of the Liver

Nutrient-laden blood in intestinal villi flows to the **hepatic portal vein**. This vein shunts the blood through vessels in the **liver**, one of the largest organs in the body. In the liver, excess glucose is taken up before a *hepatic vein* returns the blood to the general circulation (Figure 5.12). The liver converts much of this glucose to a storage form, glycogen. As we've already noted, the liver's role in digestion is to secrete bile—as much as 1,500 ml every day—which is stored in the gallbladder and released when chyme enters the small intestine. It also processes incoming nutrient molecules into substances the body requires (such as blood plasma proteins) and removes toxins ingested in food or already circulating in the bloodstream.

Liver cells can be injured by viruses that cause *hepatitis.* Without prompt treatment, hepatitis can permanently harm liver function. Type A hepatitis is transmitted via contaminated food or water. Symptoms include nausea, abdominal discomfort, loss of appetite, and jaundice (yellowing of the skin). Type B hepatitis, a sexually transmitted disease (Chapter 15), can cause serious liver scarring, called *cirrhosis.* Long-term, heavy alcohol use is another common cause of cirrhosis. Liver transplants are becoming more successful, but people with advanced cirrhosis still often die of liver failure.

Large Intestine: Concentration of Residues

Material not absorbed in the small intestine moves into the **large intestine** (colon). Here, these "leftovers" are concentrated as water and salts are reabsorbed. The concentrated material is stored and eventually eliminated as *feces,* a mixture of undigested and unabsorbed food material, water, and bacteria. Bile pigments give feces their brown color. The mixture becomes concentrated as water moves across the colon's lining. Cells in the lining actively transport sodium ions out of the lumen. As the ion concentration in the lumen drops, water moves out of the lumen by osmosis.

About 30 percent of the dry weight of feces consists of bacteria. Such microorganisms, including *Escherichia coli,* normally inhabit the intestines and are nourished by the residues there. Their metabolic activity produces useful fatty acids and vitamins (such as vitamin K), some of which also are absorbed into the bloodstream. It also produces a not-so-pleasant "by-product": intestinal gas. Dis-

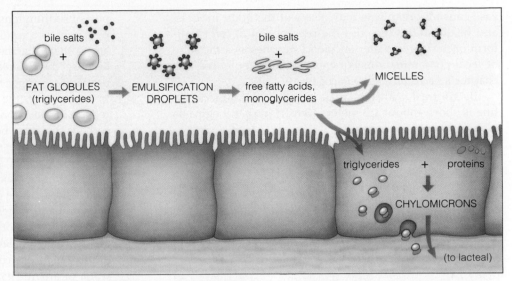

INTESTINAL LUMEN

carbohydrates → monosaccharides

proteins → amino acids

EPITHELIAL CELL

INTERNAL ENVIRONMENT

bile salts

FAT GLOBULES (triglycerides)

EMULSIFICATION DROPLETS

bile salts

free fatty acids, monoglycerides

MICELLES

triglycerides + proteins

CHYLOMICRONS

(to lacteal)

1 Digestion of carbohydrates to monosaccharides and proteins to amino acids completed by pancreatic enzymes.

2 Active transport of monosaccharides and amino acids across plasma membrane of cells, then nutrients move by facilitated diffusion into blood.

1 Emulsification. Wall movements break up fat globules into small droplets. Bile salts keep globules from re-forming. Pancreatic enzymes digest droplets to fatty acids and monoglycerides.

2 Formation of micelles (as bile salts combine with digested products and phospholipids). Products readily slip into and out of micelles.

3 Concentration of monosaccharides and fatty acids in micelles enhance gradients that lead to their diffusion across lipid bilayer of cells' plasma membranes.

4 Reassemby of products into triglycerides inside cells. These join with proteins, then are expelled (by exocytosis) into internal environment.

Figure 5.11 Summary of digestion and absorption processes in the small intestine.

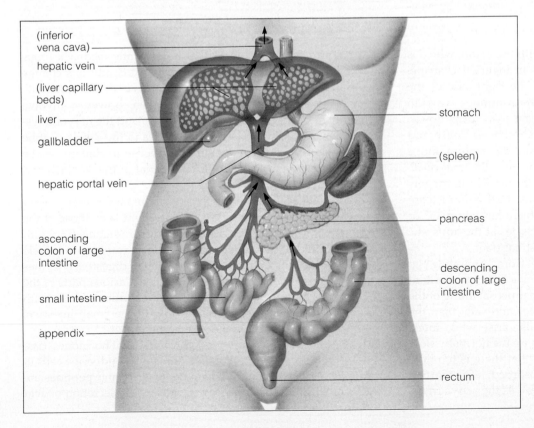

(inferior vena cava)

hepatic vein

(liver capillary beds)

liver

gallbladder

hepatic portal vein

ascending colon of large intestine

small intestine

appendix

stomach

(spleen)

pancreas

descending colon of large intestine

rectum

Figure 5.12 Hepatic portal system. Blood carrying nutrients absorbed in various regions of the gastrointestinal tract flows in many small veins to the larger hepatic portal vein and thence into beds of capillaries (tiny blood vessels) in the liver. There, nutrient molecules, including excess glucose, are taken up by liver cells, and toxins are removed. Blood leaves the liver in hepatic veins, which empty into a large vessel (the inferior vena cava) leading to the heart. Arrows show the direction of blood flow.

ease-causing organisms may also exit the body in feces, and health authorities use the presence of *E. coli* ("coliform bacteria") in water and food supplies as a measure of fecal contamination. For more on this topic, see this chapter's *Focus on Environment* essay.

By contrast with the small intestine, the large intestine is short—about 1.2 meters (5 feet) long. It begins as a blind pouch, the *cecum:*

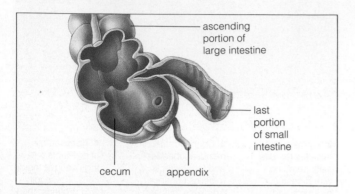

The *appendix,* a slender projection from the pouch, has no known digestive function, but it probably plays a role in immunity. *Appendicitis* is an inflammation of the appendix caused by bacterial infection. Symptoms include pain near the navel or right side of the abdomen, nausea, vomiting, and loss of appetite. Suspected appendicitis requires urgent medical attention. If the infected appendix bursts, bacteria are spewed into the abdominal (peritoneal) cavity, and the victim can rapidly develop *peritonitis,* a potentially life-threatening infection.

The Colon   The cecum merges with the **colon**, which is subdivided into several regions in an inverted U-shape. The *ascending colon* travels up on the right side of the abdominal cavity, the *transverse colon* continues across to the left side, and the *descending colon* then turns downward. The *sigmoid colon* makes an S-shaped curve and connects with the **rectum**. Feces in the rectum move toward the outside of the body through the *anal canal.* When feces distend the rectal wall, the stretching triggers the expulsion of feces from the body. Defecation is controlled by the nervous system, which can stimulate or inhibit contractions of sphincter muscles at the **anus**, the terminal opening of the gastrointestinal tract.

Colon problems can range from annoying to extremely serious. *Diarrhea* can occur when an irritant (such as a bacterial toxin) causes the mucosal lining of the small intestine to secrete more water and salts than the large intestine can absorb. It can also arise when infections, stress, and other factors speed up the normally slow peristalsis in the large intestine, so that there is too little time for sufficient water to be absorbed. (The normal feces-concentrating mechanism involves the active trans-

port of sodium ions across the colon lining; water follows passively as a result.) Recall from Chapter 2 that cholera and other diarrheal diseases are dangerous largely because they deplete the body of the water and salts needed for proper functioning of nerve and muscle cells. In *constipation*, the problem is reversed: Food residues remain in the colon for too long, too much water is reabsorbed, the feces become dry and hard, and defecation becomes difficult.

Stress, lack of exercise, and other factors can cause constipation; a major cause is a lack of "bulk" in the diet. *Bulk* is the volume of fiber (mainly cellulose from plants) and other undigested food material that cannot be decreased by absorption in the colon. There are two categories of fiber. *Soluble fiber* consists of plant carbohydrates (such as fruit pectins) that swell or dissolve in water or are nutrient sources for beneficial intestinal bacteria. *Insoluble fiber* such as cellulose and other plant structural compounds does not easily dissolve in water and cannot be digested by human intestinal bacteria. Wheat bran is an example of insoluble fiber, but virtually all plant foods contain both fiber types. People whose diets lack sufficient fiber tend to produce a low volume of feces, and because of the low volume, feces move more slowly through the colon. Some studies suggest that prolonged contact with feces not only irritates the colon, but over time may trigger changes that lead to colon cancer. Constipation is one cause of the enlarged rectal blood vessels called hemorrhoids.

Control of the Digestive System

Recall from Chapter 3 that homeostatic feedback loops operate when physical and chemical conditions change in the *internal* environment. Controls over the "external" environment of the GI tract lumen, however, involve interactions of the endocrine system and the nervous system, including local nerve plexuses in the digestive tube wall. The controls operate in response to changes in the volume and composition of material in the stomach and intestines.

For instance, in the first few hours after a meal, food distends the walls of the stomach, and later those of the small intestine. This distension triggers signals from sensory receptors in the tube wall, which can lead to muscle contractions in the wall or secretion of digestive enzymes and other substances into the lumen. Various parts of the brain monitor such activities and coordinate them with other events, such as blood flow to the small intestine, where nutrients are being absorbed.

There are a number of gastrointestinal hormones. *Gastrin* is secreted into the bloodstream by endocrine cells in the stomach's lining when amino acids and peptides are in the stomach. It mainly stimulates the secretion of acid

Food That Makes You Sick

At the Centers for Disease Control and Prevention (CDC) in Atlanta, Georgia, scientists are tracking a growing rogue's gallery of microorganisms that can contaminate human food. In the United States, more than 6 million people become sick with food-borne diseases each year, and more than 9,000 of those people die.

The root of the problem is simple. Disease-causing microbes are present in food or water that have been contaminated with feces from infected animals. Typically, an offending pathogen takes hold in the GI tract, rapidly multiplies, and releases toxins or otherwise causes tissue damage so that the tract cannot function normally.

Usually, food on the grocer's shelf becomes tainted when processing or handling procedures for meat, dairy products, or raw fruits or vegetables do not kill the offending bacteria. Sometimes contaminated water is the source. For example, in 1993 more than 240,000 people in Milwaukee developed diarrhea and other intestinal problems when the municipal water supply became contaminated with agricultural runoff carrying *Cryptosporidium* bacteria from infected dairy animals.

Listeria bacteria consumed in improperly pasteurized dairy products or infected, improperly cooked meat or shellfish can cause *listeriosis*. In a healthy adult, the symptoms may only be a brief bout of "flu," but elderly people, pregnant women, infants, and people with impaired immune systems can become deathly ill with meningitis (inflammation of membranes covering the brain and spinal cord) and other complications. Recent outbreaks of severe *gastroenteritis*, an inflammation of the stomach and intestines, have been traced to restaurant food such as mayonnaise and undercooked hamburgers contaminated with *E. coli* 0157:H7. Other common culprits are *Salmonella* and *Campylobacter* bacteria on or in red meat, poultry, pork, raw (unpasteurized) milk or other dairy products—

even on melons that have been shipped in ice made from contaminated water! Symptoms may include stomach cramps, nausea or vomiting, and diarrhea.

How can we counter the threat from food-borne microbes? One obvious tack is to improve government inspection methods so that tainted products are detected before being distributed. Another, more controversial strategy is to expose foods to ionizing radiation (page 413) and thereby kill any bacteria present. At home, consumers should carefully wash raw produce and meats, cook meats and eggs thoroughly, and wash utensils and cutting boards with soapy water after each use. It may also be a good idea to switch to using wooden cutting boards. Recent studies show that nasty bacteria cannot grow as well on wood as they do on plastic.

into the stomach. Acid secretion can be inhibited by *somatostatin*, which is produced by another type of endocrine cell in the stomach lining. Endocrine cells in the lining of the small intestine secrete other hormones. As acid arrives in the small intestine, *secretin*, a peptide hormone, stimulates the pancreas to secrete bicarbonate (page 25). *Cholecystokinin* (CCK) is released in response to fat in the small intestine, and it enhances the actions of secretin and stimulates gallbladder contractions. Secretin

and CCK also slow the rate of stomach emptying. *Gastric inhibitory peptide* (GIP) is released in response to the presence of glucose and fat in the small intestine. It slows both the secretions of gastric glands and the rate of gastric emptying. GIP also stimulates the release of insulin, which is necessary for glucose uptake by cells or storage (as glycogen) by the liver. For this reason, its abbreviation is sometimes used to stand for *glucose insulinotropic peptide.*

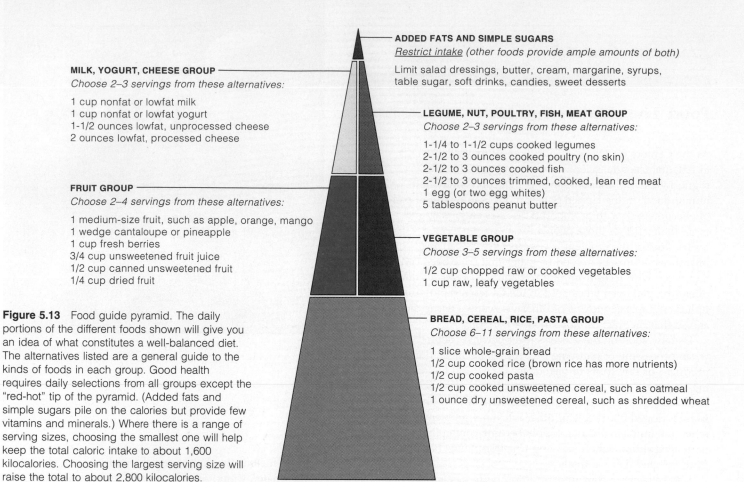

MILK, YOGURT, CHEESE GROUP

Choose 2–3 servings from these alternatives:

1 cup nonfat or lowfat milk
1 cup nonfat or lowfat yogurt
1-1/2 ounces lowfat, unprocessed cheese
2 ounces lowfat, processed cheese

FRUIT GROUP

Choose 2–4 servings from these alternatives:

1 medium-size fruit, such as apple, orange, mango
1 wedge cantaloupe or pineapple
1 cup fresh berries
3/4 cup unsweetened fruit juice
1/2 cup canned unsweetened fruit
1/4 cup dried fruit

ADDED FATS AND SIMPLE SUGARS
Restrict intake (other foods provide ample amounts of both)

Limit salad dressings, butter, cream, margarine, syrups, table sugar, soft drinks, candies, sweet desserts

LEGUME, NUT, POULTRY, FISH, MEAT GROUP

Choose 2–3 servings from these alternatives:

1-1/4 to 1-1/2 cups cooked legumes
2-1/2 to 3 ounces cooked poultry (no skin)
2-1/2 to 3 ounces cooked fish
2-1/2 to 3 ounces trimmed, cooked, lean red meat
1 egg (or two egg whites)
5 tablespoons peanut butter

VEGETABLE GROUP

Choose 3–5 servings from these alternatives:

1/2 cup chopped raw or cooked vegetables
1 cup raw, leafy vegetables

BREAD, CEREAL, RICE, PASTA GROUP

Choose 6–11 servings from these alternatives:

1 slice whole-grain bread
1/2 cup cooked rice (brown rice has more nutrients)
1/2 cup cooked pasta
1/2 cup cooked unsweetened cereal, such as oatmeal
1 ounce dry unsweetened cereal, such as shredded wheat

Figure 5.13 Food guide pyramid. The daily portions of the different foods shown will give you an idea of what constitutes a well-balanced diet. The alternatives listed are a general guide to the kinds of foods in each group. Good health requires daily selections from all groups except the "red-hot" tip of the pyramid. (Added fats and simple sugars pile on the calories but provide few vitamins and minerals.) Where there is a range of serving sizes, choosing the smallest one will help keep the total caloric intake to about 1,600 kilocalories. Choosing the largest serving size will raise the total to about 2,800 kilocalories.

The nervous system and the endocrine system exert control over conditions in the digestive system in response to the volume and composition of material in the digestive tract lumen.

NUTRITION

What is "eating properly"? In 1979, the U. S. Surgeon General released a report representing a medical consensus on how to promote health and avoid ailments such as high blood pressure, heart disorders, colon cancer, and tooth decay. The report advised Americans to eat "less saturated fat and cholesterol; less salt; less sugar; relatively more complex carbohydrates such as whole grains, cereals, fruits, and vegetables; and relatively more fish, poultry, legumes (such as peas, beans, and peanuts); and less red meat." With some refinements, these guidelines are still good advice today. The food guide pyramid in Figure 5.13 represents the most recent government recommendations for healthy eating.

Energy Needs and Body Weight

Nutritionists measure energy in units called "Calories," which is a sometimes confusing shorthand for "kilocalories." A **kilocalorie** is 1,000 calories of energy, and that is the term we will use here.

To maintain an acceptable weight and keep the body functioning normally, caloric intake must be balanced with energy output (see *Focus on Science*, page 125). Physiologists define **basal metabolic rate** (BMR) as the amount of energy it takes to sustain the body when a person is resting, awake, and has not eaten for 12–18 hours. BMR varies from person to person; it is generally higher in males. Other factors that influence a person's energy output include differences in physical activity, age, hormone activity, and emotional state.

In most adults, energy input balances output, so body weight remains much the same over long periods. As any dieter knows, it is as if the body has a "set point" for its weight and works to counteract deviations from its set point. **Obesity** is an excess of body fat caused by imbalances between caloric intake and energy output. Yet, what is too fat or too thin? What is a person's "ideal weight"? Many charts have been developed (see Figure 5.14),

A Weighty Question: How Much Energy Does Your Body Need?

How many kilocalories should you take in each day to maintain what you consider to be an acceptable body weight? To estimate your body's energy requirements, first multiply your desired weight (in pounds) by 10 if you are not very active physically, by 15 if you are moderately active, and by 20 if you are quite active. Then, depending on your age, subtract the following amount from the value obtained in the first step:

Age	Subtract
20–34	0
35–44	100
45–54	200
55–64	300
Over 65	400

For example, if you want to weigh 120 pounds and are quite active, 120 × 20 = 2,400 kilocalories. If you are 35 years old and moderately active, then you should take in a total of 1,800 − 100, or 1,700 kcal a day. Such calculations provide a rough estimate, and factors such as height and gender must be considered as well. Because males tend to have more muscle and so burn more calories, an active woman doesn't need as many kilocalories as an active man of the same height and weight. Nor does she need as many as another active woman who weighs the same but is several inches taller.

Man's Height	Size of Frame		
	Small	Medium	Large
5′ 2″	128–134	131–141	138–150
5′ 3″	130–136	133–143	140–153
5′ 4″	132–138	135–145	142–156
5′ 5″	134–140	137–148	144–160
5′ 6″	136–142	139–151	146–164
5′ 7″	138–145	142–154	149–168
5′ 8″	140–148	145–157	152–172
5′ 9″	142–151	148–160	155–176
5′10″	144–154	151–163	158–180
5′11″	146–157	154–166	161–184
6′ 0″	149–160	157–170	164–188
6′ 1″	152–164	160–174	168–192
6′ 2″	155–168	164–178	172–197
6′ 3″	158–172	167–182	176–202
6′ 4″	162–176	171–187	181–207

Woman's Height	Size of Frame		
	Small	Medium	Large
4′10″	102–111	109–121	118–131
4′11″	103–113	111–123	120–134
5′ 0″	104–115	113–126	122–137
5′ 1″	106–118	115–129	125–140
5′ 2″	108–121	118–132	128–143
5′ 3″	111–124	121–135	131–147
5′ 4″	114–127	124–138	134–151
5′ 5″	117–130	127–141	137–155
5′ 6″	120–133	130–144	140–159
5′ 7″	123–136	133–147	143–163
5′ 8″	126–139	136–150	146–167
5′ 9″	129–142	139–153	149–170
5′10″	132–145	142–158	152–173
5′11″	135–148	145–159	155–176
6′ 0″	138–151	148–162	158–179

Figure 5.14 (**above**) "Ideal" weights for adults according to one insurance company (compiled in 1983). Values shown are for people ages 25 to 59 wearing shoes with 1-inch heels and 3 pounds of clothing (for women) or 5 pounds (for men). (**left**) Extreme obesity puts severe strain on the circulatory system. The body produces many more small blood vessels (capillaries) to service the increased tissue masses. The heart becomes more stressed because it must pump harder to keep blood circulating.

mostly by insurance companies that want to identify overweight people who are considered to be insurance risks. In general, people who are 20 percent heavier than the "ideal" are considered obese.

Some researchers who study causes of death suspect that a person's "ideal" weight actually may be 10–15 pounds heavier than the charts indicate. On the other hand, some nutritionists are convinced the charts' values should be reduced. Whatever the ideal range may be, serious disorders do arise with extremes at either end of that range (see *Focus on Wellness*, to the right).

To maintain an acceptable body weight, energy input (caloric intake) must be balanced with energy output (in the form of metabolic activity and exercise).

A Well-Balanced Diet Many nutritionists now contend that food in the proportions shown in Figure 5.13 and in the following list constitutes a well-balanced diet. The percentages represent the percent of total caloric intake:

Complex carbohydrates:	60–80%
Fats and other lipids:	10–30%
Proteins:	9% or less

These are the basic types of nutrients we consider next.

Types of Nutrients

Carbohydrates Complex carbohydrates are the body's preferred energy sources. As we have seen, they can be readily broken down into glucose units (see Chapter 2). Glucose is the primary energy source for your brain, muscles, and other body tissues. Moreover, foods containing complex carbohydrates typically also include fiber that helps add bulk to the diet. According to many nutritionists, fruits, cereal grains, legumes, and other foods rich in fibrous carbohydrates should make up *at least* 50 to 60 percent of daily caloric intake. Studies show that many people get a large proportion of their calories from foods that are high in fat and contain little complex carbohydrate and little fiber.

Proteins When proteins are digested and absorbed, their amino acids become available for the body's own protein-building programs. Of the 20 common amino acids, eight are **essential amino acids**. Our cells cannot synthesize these amino acids; they must obtain them from food. The eight are isoleucine, leucine, lysine, methionine

When Does a Diet Become a Disorder?

How much of our total tissue mass should be fat? According to current standards, it should be not more than 18–24 percent for a female who is under 30 years old. For a male under 30, body fat should be no more than 12–18 percent of tissue mass. Yet an estimated 34 million Americans are overweight to the point of obesity. Although many sincerely attempt to shed the excess fat by dieting, most eventually regain what they lost—and often put on more.

Dieting and Exercise Why is it so difficult to lose weight permanently? The difficulty is partly a result of our evolutionary heritage. The fat-storing cells of adipose tissue are an adaptation for survival. Collectively, they represent an energy reserve that may carry an animal through times when food isn't available. Remember this when you start putting on unwanted fat. Once a fat-storing cell is added to your body, food intake can affect how empty or full it is, but no amount of dieting can get rid of the cell.

Research suggests that the body interprets dieting as "starvation," triggering changes in metabolism. With these changes, food is used more conservatively, so that *fewer* calories are burned. Meanwhile, the dieter's appetite surges, and the "starved" fat cells quickly refill when a diet ends.

In "yo-yo" dieting, a person repeatedly gains and loses weight. This may alter cell metabolism in ways that make it more difficult to shed weight each time a new round of dieting begins. Frequent changes in body weight may also increase the risk of heart disorders.

Some people opt to shed pounds by exercising, only to discover that the going is slow indeed. Losing just one pound of fat requires energy expenditure of about 3,500 kilocalories. The accompanying table shows the caloric cost of some common activities. For example, a 120-pound female must play almost 10 hours of tennis to burn just one pound of fat. A 160-pound male must play tennis for nearly 7 and 1/2 hours. *The only real solution to permanent weight loss is to combine a moderate reduction in caloric intake with increased physical activity.* This strategy may minimize the "starvation" response and increase the rate at which the body burns calories. Exercise also increases muscle mass, and even at rest muscle burns more calories than other types of tissues. To keep off unwanted weight, you must continue to eat in moderation and maintain a program of regular exercise.

Anorexia Nervosa When dieting becomes obsessive, one result can be a potentially fatal eating disorder called

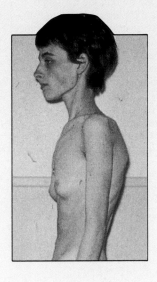

Calories Expended in Some Common Activities

Activity	Kcal/hour per pound of body weight	Hrs. needed to lose 1 lb. fat		
		120 lbs	**155 lbs**	**185 lbs**
Basketball	3.78	7.7	6.0	5.0
Cycling (9 mph)	2.70	10.8	8.4	7.0
Football	3.60	8.1	6.3	5.2
Hiking	2.52	11.6	8.9	7.5
Housecleaning	1.62	18.0	13.9	11.7
Golf	2.34	12.4	9.6	8.1
Jogging	4.15	7.0	5.4	4.5
Mowing lawn (push mower)	3.06	9.5	7.4	6.2
Playing music (seated)	1.08	26.9	20.9	17.5
Racquetball	3.90	7.5	5.8	4.8
Running (9-minute mile)	5.28	5.5	4.3	3.6
Snow skiing (cross-country)	4.43	6.6	5.1	4.3
Swimming (slow crawl)	3.48	8.4	6.5	5.4
Tennis	3.00	9.7	7.5	6.3
Walking (moderate pace)	2.16	13.5	10.4	8.7

To calculate these values for your own body weight, first multiply your weight by the kcal/hour expended for an activity to determine total kcal you use during one hour of the activity. Then divide that number into 3,500 (kcal in a pound of fat) to obtain the number of hours you must perform the activity to burn a pound of body fat.

anorexia nervosa. Most anorexics are women in their teens or early twenties, and the disorder is increasingly common.

People with anorexia nervosa have a skewed perception of their body weight. They have an overwhelming fear of being fat and so begin to starve themselves and, frequently, overexercise. Contributing emotional factors may include fear of growing up in general and maturing sexually in particular. Menstrual cycles stop when a female's body fat falls below a certain percentage.

Bulimia   At least 20 percent of college-age women now suffer to various degrees from *bulimia* ("an oxlike appetite"). Outwardly they may look healthy, but their food intake is out of control. In extreme cases, a bulimic may take in more than 50,000 kilocalories in an hour and

then vomit or use laxatives to purge the body. This binge-purge routine may occur once a month or several times a day. Repeated vomiting, which brings stomach acids into the mouth, can erode teeth to stubs. Extreme bulimics can die of heart failure, stomach rupture, or kidney failure.

Some women start the routine because it seems like a simple way to lose weight. Others have emotional problems. Often they are well-educated and accomplished, but they strive for perfection and may have problems with control exerted by other family members. Some counselors believe that eating may actually be an unpleasant event for bulimics, but the purging (which they themselves control) relieves them of anger and frustration.

Treatment for both anorexia nervosa and bulimia usually includes long-term psychotherapy as well as diet counseling. Severe cases may require hospitalization.

COMPLETE PROTEINS	INCOMPLETE PROTEINS		
No Limiting Amino Acid	Low in Lysine	Low in Methionine, Other Sulfur-Containing Amino Acids	Low in Tryptophan

legumes:
  soybean
  tofu
  soy milk
 	
cereal grains:
  wheat germ
 	
milk
cheeses (except
  cream cheese)
yogurt
eggs
meats
	legumes:
	  peanuts
	
	cereal grains:
	  barley
	  buckwheat
	  corn meal
	  oats
	  rice
	  rye
	  wheat
	
	nuts, seeds:
	  almonds
	  cashews
	  coconut
	  English walnuts
	  hazelnuts
	  pecans
	  pumpkin seeds
	  sunflower seeds
		legumes:
		  beans (dried)
		  black-eyed peas
		  garbanzos
		  lentils
		  lima beans
		  mung beans
		  peanuts
		
		nuts:
		  hazelnuts
		
		fresh vegetables:
		  asparagus
		  broccoli
		  green peas
		  mushrooms
		  parsley
		  potatoes
		  soybeans
		  Swiss chard
			legumes:
			  beans (dried)
			  garbanzos
			  lima beans
			  mung beans
			  peanuts
			
			cereal grains:
			  corn meal
			
			nuts:
			  almonds
			  English walnuts
			
			fresh vegetables:
			  corn
			  green peas
			  mushrooms
			  Swiss chard

Figure 5.15 Essential amino acids—a small portion of the total protein intake. All eight must be available at the same time, in certain amounts, if cells are to build their own proteins. Milk and eggs have high amounts of all eight in proportions that humans require; they are among the complete proteins. Nearly all plant proteins are incomplete, so vegetarians should construct their diet carefully to avoid protein deficiency. For example, they can combine different foods from the three columns of incomplete proteins shown. Also, vegetarians who avoid dairy products and eggs should take vitamin B_{12} and B_2 (riboflavin) supplements. Animal protein is a luxury in most traditional societies, yet good combinations of plant proteins are worked into their cuisines— including rice/beans, chili/cornbread, tofu/rice, and lentils/wheat bread.

total protein intake

isoleucine
leucine
lysine
methionine
phenylalanine
threonine
tryptophan
valine

essential amino acids

(or cysteine), phenylalanine (or tyrosine), threonine, tryptophan, and valine (Figure 5.15a).

Most animal proteins are "complete," meaning they contain high amounts of all essential amino acids. Most plant proteins are "incomplete." They may have all the essential amino acids, but not in the amounts required for proper human nutrition. To get adequate amounts of needed amino acids, strict vegetarians must eat certain combinations of different plants (Figure 5.15b). Nutritionists use a measure called "net protein utilization" (NPU) to compare proteins from different sources. NPU values range from 100 (all essential amino acids present in ideal proportions) to 0 (one or more absent).

Because enzymes and other proteins are vital for the body's structure and function, it makes sense that a protein-deficient diet is no joking matter. Protein deficiency is most damaging among the young, for rapid brain growth and development occur early in life. Even mild protein starvation during the mother's pregnancy or for some months after a child is born can retard the child's growth and affect mental and physical performance. The *Human Impact* essay following this chapter describes some of the severe impacts of undernutrition, which is a critical problem in the poorest developing nations.

Lipids Fats and other lipids have fundamental roles in the body. For example, lecithin (a phospholipid) and cholesterol are essential components of animal cell membranes. Fat deposits serve as energy reserves, cushion many organs (such as the eyes and kidneys), and provide insulation beneath the skin. They also help the body store fat-soluble vitamins.

The liver can synthesize most of the fats the body needs, including cholesterol, from protein and carbohydrates. The ones it cannot produce are called **essential fatty acids**. Linoleic acid is one example. One teaspoon a day of corn oil, olive oil, or some other polyunsaturated fat in food provides enough of it.

Today, 40 percent of the kilocalories in the average diet in the United States comes from fats. Most health authorities agree that this figure should be less than 30 percent. Aside from questions of weight control, studies show that a diet high in fats increases the risk of colon cancer and heart disease, among other disorders. (Some studies have suggested a relationship between a high-fat diet and breast cancer. However, other recent findings do not support this link, and the issue is not yet resolved.) Butter and other animal fats are largely saturated fats, which tend to raise the level of cholesterol in the blood. Cholesterol is necessary in the synthesis of bile acids and steroid hormones, but for many people too much cholesterol may damage the circulatory system (page 161). Work is underway to develop edible but nondigestible oils, such as sucrose polyester, as fat substitutes for people who have trouble cutting back on their fat intake.

Vitamins and Minerals Normal metabolism depends on small amounts of more than a dozen organic substances called **vitamins**. Most plant cells synthesize nearly all of these substances, but in the course of evolution animal cells have lost this ability. Hence animals must obtain vitamins from food. Human cells need at least 13 different vitamins, each with specific metabolic roles (Table 5.2). Many metabolic reactions depend on several different vitamins, and the absence of one vitamin can affect the functions of the others.

Table 5.2 Vitamins: Sources, Functions, and Effects of Deficiencies or Excesses*

Vitamin	Common Sources	Main Functions	Signs of Severe Long-Term Deficiency	Signs of Extreme Excess
Fat-Soluble Vitamins:				
A	Its precursor comes from beta-carotene in yellow fruits, yellow or green leafy vegetables; also in fortified milk, egg yolk, fish liver	Used in synthesis of visual pigments, bone, teeth; maintains epithelia	Dry, scaly skin; lowered resistance to infections; night blindness; permanent blindness	Malformed fetuses; hair loss; changes in skin; liver and bone damage; bone pain
D	D_3 formed in skin and in fish liver oils, egg yolk, (fortified milk); converted to active form elsewhere	Promotes bone growth and mineralization; enhances calcium absorption	Bone deformities (rickets) in children; bone softening in adults	Retarded growth; kidney damage; calcium deposits in soft tissues
E	Whole grains, dark-green vegetables, vegetable oils	Possibly inhibits effects of free radicals; helps maintain cell membranes; blocks breakdown of vitamins A and C in gut	Lysis of red blood cells; nerve damage	Muscle weakness, fatigue, headaches, nausea
K	Colon bacteria form most of it; also in green leafy vegetables, cabbage	Blood clotting; ATP formation via electron transport	Abnormal blood clotting, severe bleeding (hemorrhaging)	Anemia; liver damage and jaundice
Water-Soluble Vitamins:				
B_1 (thiamin)	Whole grains, green leafy vegetables, legumes, lean meats, eggs	Connective tissue formation; folate utilization; coenzyme action	Water retention in tissues; tingling sensations; heart changes; poor coordination	None reported from food; possible shock reaction from repeated injections
B_2 (riboflavin)	Whole grains, poultry, fish, egg white, milk	Coenzyme action	Skin lesions	None reported
Niacin	Green leafy vegetables, potatoes, peanuts, poultry, fish, pork, beef	Coenzyme action	Contributes to pellagra (damage to skin, gut, nervous system, etc.)	Skin flushing; possible liver damage
B_6	Spinach, tomatoes, potatoes, meats	Coenzyme in amino acid metabolism	Skin, muscle, and nerve damage; anemia	Impaired coordination; numbness in feet
Pantothenic acid	In many foods (meats, yeast, egg yolk especially)	Coenzyme in glucose metabolism, fatty acid and steroid synthesis	Fatigue, tingling in hands, headaches, nausea	None reported; may cause diarrhea occasionally
Folate (folic acid)	Dark green vegetables, whole grains, yeast, lean meats; colon bacteria produce some	Coenzyme in nucleic acid and amino acid metabolism	A type of anemia; inflamed tongue; diarrhea; impaired growth; mental disorders	Masks vitamin B_{12} deficiency
B_{12}	Poultry, fish, red meat, dairy foods (not butter)	Coenzyme in nucleic acid metabolism	A type of anemia; impaired nerve function	None reported
Biotin	Legumes, egg yolk; colon bacteria produce some	Coenzyme in fat, glycogen formation and in amino acid metabolism	Scaly skin (dermatitis), sore tongue, depression, anemia	None reported
C (ascorbic acid)	Fruits and vegetables, especially citrus, berries, cantaloupe, cabbage, broccoli, green pepper	Collagen synthesis; possibly inhibits effects of free radicals; structural role in bone, cartilage, and teeth; role in carbohydrate metabolism	Scurvy, poor wound healing, impaired immunity	Diarrhea, other digestive upsets; may alter results of some diagnostic tests

*The guidelines for appropriate daily intakes are being worked out by the Federal Food and Drug Administration.

Metabolic activity also requires certain inorganic substances called **minerals** (Table 5.3). For example, most cells use calcium and magnesium in many different reactions. Potassium is essential to the functioning of muscle and nerve cells. Cytochrome molecules, which are components of electron transport chains in aerobic cell respiration, are built partly with iron. Red blood cells contain iron in hemoglobin, the oxygen-carrying pigment in blood. Iron deficiency is a common dietary problem worldwide. It is especially acute in developing countries, where sufferers are also more vulnerable to parasites that rob the body of iron and other nutrients.

People who are in good health can get all the vitamins and minerals they need from a balanced diet of whole foods. Generally, specific vitamin and mineral supplements are necessary only for strict vegetarians, newborns and the elderly, and individuals suffering from a chronic illness or taking medication that affects the body's use of specific nutrients. For example, research shows that vitamin K supplements may help older women retain calcium and so diminish bone loss (osteoporosis) by 30 percent. Other studies suggest that folic acid, a B vitamin, may help reduce the buildup of fatty deposits in the blood vessels of people at risk of atherosclerosis (page 161).

Table 5.3	Major Minerals: Sources, Functions, and Effects of Deficiencies or Excesses*			
Mineral	Common Sources	Main Functions	Signs of Severe Long-Term Deficiency	Signs of Extreme Excess
Calcium	Dairy products, dark green vegetables, dried legumes	Bone, tooth formation; blood clotting; neural and muscle action	Stunted growth; possibly diminished bone mass (osteoporosis)	Impaired absorption of other minerals; kidney stones in susceptible people
Chloride	Table salt (usually too much in diet)	HCl formation in stomach; contributes to body's acid-base balance; neural action	Muscle cramps; impaired growth; poor appetite	Contributes to high blood pressure in susceptible people
Copper	Nuts, legumes, seafood, drinking water	Used in synthesis of melanin, hemoglobin, and some transport chain components	Anemia; changes in bone and blood vessels	Nausea; liver damage
Fluorine	Fluoridated water, tea, seafood	Bone, tooth maintenance	Tooth decay	Digestive upsets; mottled teeth and deformed skeleton in chronic cases
Iodine	Marine fish, shellfish, iodized salt, dairy products	Thyroid hormone formation	Enlarged thyroid (goiter) with metabolic disorders	Goiter
Iron	Whole grains, green leafy vegetables, legumes, nuts, eggs, lean meat, molasses, dried fruit, shellfish	Formation of hemoglobin and cytochrome (transport chain component)	Iron-deficiency anemia; impaired immune function	Liver damage, shock, heart failure
Magnesium	Whole grains, legumes, nuts, dairy products	Coenzyme role in ATP-ADP cycle; roles in muscle, nerve function	Weak, sore muscles; impaired neural function	Impaired neural function
Phosphorus	Whole grains, poultry, red meat	Component of bone, teeth, nucleic acids, ATP, phospholipids	Muscular weakness; loss of minerals from bone	Impaired absorption of minerals into bone
Potassium	Diet provides ample amounts	Muscle and neural function; roles in protein synthesis and body's acid-base balance	Muscular weakness	Muscular weakness, paralysis, heart failure
Sodium	Table salt; diet provides ample to excessive amounts	Key role in body's acid-base balance; roles in muscle and neural function	Muscle cramps	High blood pressure in susceptible people
Sulfur	Proteins in diet	Component of body proteins	None reported	None likely
Zinc	Whole grains, legumes, nuts, meats, seafood	Component of digestive enzymes; roles in normal growth, wound healing, sperm formation, and taste and smell	Impaired growth, scaly skin, impaired immune function	Nausea, vomiting, diarrhea; impaired immune function and anemia

*The guidelines for appropriate daily intakes are being worked out by the Federal Food and Drug Administration.

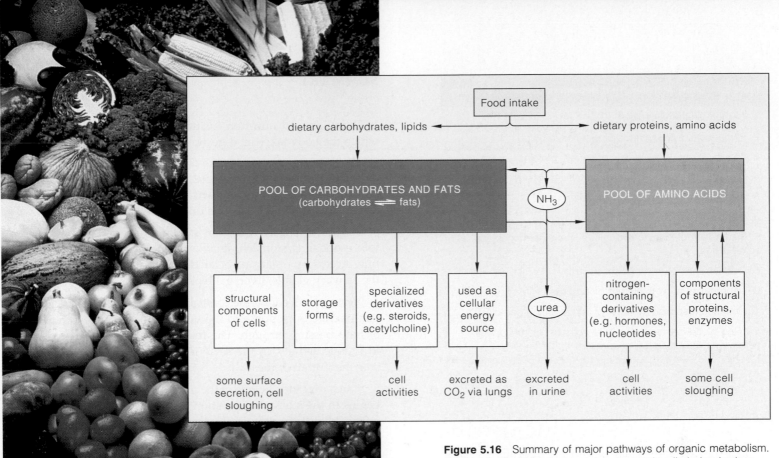

Figure 5.16 Summary of major pathways of organic metabolism. Carbohydrates, fats, and proteins are continually being broken down and resynthesized. Urea forms mainly in the liver.

There is strong evidence that vitamins E and C and beta carotene (a pigment in carrots, some fruits, and leafy green vegetables) act as *antioxidants*. Recall from Chapter 2 that antioxidants combine with the highly reactive molecules called free radicals and counteract their destructive effects on DNA and cell membranes. A recent study of 120,000 men and women found that subjects who had higher daily intakes of vitamin E *may* have lower risk of serious coronary artery disease. Other research is focusing on possible cancer-thwarting properties of cruciferous vegetables—members of the mustard family such as broccoli, cabbage, and kale.

No one should take massive doses of any vitamin or mineral supplement except under medical supervision. As Tables 5.2 and 5.3 indicate, excess amounts of many vitamins and minerals are harmful. For example, large doses of several fat-soluble vitamins, especially vitamins A and D, can cause serious disorders. As with all fat-soluble vitamins, excess amounts can accumulate in tissues, especially in the liver, and interfere with normal metabolic function. Too much iron has been implicated in cardiovascular damage.

Severe shortages or self-prescribed massive excesses of vitamins and minerals can disturb the delicate balances in body function that promote health.

Storage and Metabolic Interconversion of Nutrients

Figure 5.16 summarizes the main routes by which nutrient molecules are shuffled and reshuffled once they have been absorbed into the body. Most carbohydrates, lipids, and proteins are broken down continually, with their component parts picked up and used again in new molecules.

When you eat, your body builds up its pools of organic molecules. Excess carbohydrates and other dietary molecules are transformed mostly into fats, which are stored in adipose tissue. Some are also converted to glycogen in the liver and in muscle tissue. While organic molecules are being absorbed and stored, most cells use glucose as their main energy source.

Table 5.4 Some Activities That Depend on Liver Functioning-
1. Carbohydrate metabolism
2. Partial control of synthesis of proteins in blood
3. Assembly and disassembly of certain proteins
4. Urea formation from nitrogen-containing wastes
5. Assembly and storage of some fats
6. Fat digestion (bile is formed by the liver)
7. Inactivation of many chemicals (such as hormones and some drugs)
8. Detoxification of many poisons
9. Degradation of worn-out red blood cells
10. Immune response (removal of some foreign particles)
11. Red blood cell formation (liver absorbs, stores factors needed for red blood cell maturation)

In a fasting individual or during rigorous physical activity, glucose supplies become depleted. Then, the body taps into its fat stores for energy. Fat stored mainly in adipose tissue is broken down to glycerol and fatty acids, which are released into blood. The glycerol is converted to glucose in the liver. Cells can take up the circulating fatty acids and use them for ATP production.

As described in the previous section, the nervous system and endocrine system interact to control these and other aspects of organic metabolism.

During a meal and for several hours thereafter, glucose moves into cells, where it will be used for energy.

When glucose supplies run low, the body metabolizes fat molecules as the main energy source.

Although the liver is central to the storage and interconversion of absorbed carbohydrates, lipids, and proteins, it has other vital roles (Table 5.4). For example, it helps maintain the concentrations of blood's organic substances and removes many toxic substances from blood. The liver inactivates most hormone molecules and sends them on to the kidneys for excretion (in urine). Also, ammonia (NH_3) is produced when cells break down amino acids, and it can be toxic to cells. The circulatory system carries ammonia to the liver, where it is converted to a much less toxic waste product, urea. Urea leaves the body via the kidneys, in urine.

SUMMARY

1. Digestion and nutrition include all the processes by which the body takes in, digests, absorbs, and uses food.

2. The digestive system breaks down food molecules by mechanical and chemical means. It also enhances absorption of the breakdown products into the internal environment, and it eliminates the unabsorbed residues.

3. The human digestive system includes the mouth, pharynx, esophagus, stomach, small intestine, and large intestine. Accessory organs associated with digestion include the salivary glands, liver, gallbladder, and pancreas.

4. Nervous and endocrine controls over the digestive system operate in response to the volume and composition of food passing through. The response can be a change in muscle activity, in the secretion rate of hormones or enzymes, or all of these.

5. Starch digestion begins in the mouth; protein digestion begins in the stomach. Digestion is completed and most nutrients are absorbed in the small intestine. Following absorption, monosaccharides (including glucose) and most amino acids are sent directly to the liver. Fatty acids and monoglycerides enter lymph vessels and then the general circulation.

6. To maintain acceptable weight and overall health, caloric intake must balance energy output. Complex carbohydrates are the body's preferred energy source. The body converts excess carbohydrates and proteins into fat and stores it in fat cells. Eight essential amino acids, a few essential fatty acids, vitamins, and minerals must be provided by the diet.

Review Questions

1. To answer this question, study Figure 5.3. Then, on your own, diagram the connections between metabolism and the digestive, circulatory, urinary, and respiratory systems. *112*

2. What are the main functions of the stomach? The small intestine? The large intestine? *113*

3. Name four kinds of breakdown products that are small enough to be absorbed across the intestinal lining and into the internal environment. *118*

4. A glass of whole milk contains lactose, protein, triglycerides (in butterfat), vitamins, and minerals. Explain what happens to each component when it passes through your digestive tract. *118–119*

5. Describe some of the reasons why each of the following is nutritionally important: carbohydrates, proteins, fats, vitamins, and minerals. *126–130*

Critical Thinking: You Decide (Key in Appendix IV)

1. Some nutritionists claim that the secret to long life is to be slightly underweight as an adult. Given that a person's weight is related partly to diet, partly to activity level, and partly to genetics, what underlying factors could be at work to generate statistics in support of this claim?

2. As a person ages, the number of cells in the body steadily decreases and their energy needs decline. If you were planning an older person's diet, what kind(s) of nutrients would you emphasize, and why? Which ones would you recommend less of? Include vitamins and minerals in your answer.

3. Along the lines of question 2, formulate a healthy diet for an actively growing seven-year-old.

Self-Quiz (Answers in Appendix III)

1. The _____, _____, _____, and _____ systems interact in supplying body cells with raw materials, disposing of wastes, and maintaining the volume and composition of extracellular fluid.

2. Various specialized regions of the digestive system function in _____ and _____ food and in _____ unabsorbed food residues.

3. Maintaining good health and normal body weight requires that _____ intake be balanced by _____ output.

4. The preferred energy sources for the body are complex _____.

5. The human body cannot produce its own vitamins or minerals, and it also cannot produce certain _____ and _____.

6. Which of the following structures is *not* associated with digestion?
   a. salivary glands   d. gallbladder
   b. thymus gland   e. pancreas
   c. liver

7. Digestion is completed and breakdown products are absorbed in the _____.
   a. mouth   c. small intestine
   b. stomach   d. large intestine

8. After absorption, fatty acids and monoglycerides move into the _____.
   a. bloodstream   c. liver
   b. intestinal cells d. lacteals

9. Excess carbohydrates and proteins are converted to _____ for storage.
   a. amino acids   c. fats
   b. starches   d. monosaccharides

10. Match each of the following digestive system components with its description.
   ____ liver
   ____ small intestine
   ____ human digestive system
   ____ nutrition
   ____ digestive system controls

   a. begins at mouth, ends at anus
   b. operate in response to food volume and composition
   c. functions are digestion, absorption, use of food
   d. where most digestion is completed
   e. receives monosaccharides and amino acids

Key Terms

absorption 112
anus 122
basal metabolic rate 124
colon 122
digestion 112
digestive system 112
essential amino acid 126
essential fatty acid 128
esophagus 114
gallbladder 119
gastrointestinal (GI) tract 112
hepatic portal vein 120
kilocalorie 124
lacteal 120
large intestine 120
liver 120
microvillus 119

mineral 130
motility 112
obesity 124
palate 114
pancreas 118
pepsin 117
peristalsis 114
pharynx 114
rectum 122
salivary amylase 114
salivary gland 114
secretion 112
small intestine 118
stomach 117
villus 119
vitamin 129

Readings

"Ulcers as an Infectious Disease." 9 April 1993. *Science*.

Wardlaw, G. M., P. M. Insel, and M. F. Seyler. 1992. *Contemporary Nutrition: Issues and Insights*. St. Louis: Mosby Year Book.

Weiss, P. 1 September 1990. "Fat and Fiction." *Science News*.

WORLD HUNGER AND MALNUTRITION

If you are reading this book, you probably enjoy the luxury of a varied diet and adequate food supplies. Like millions of affluent people, you may carefully tailor your meals, picking and choosing certain foods that optimize your intake of certain vitamins, antioxidants, proteins, minerals, fats, and other substances. You may even be in the enviable position of needing to limit your food intake and exercise more because you have access to much more food energy than your body needs. In short, where food and nutrition are concerned, you are incredibly lucky.

Today, anyone who watches television or reads the newspaper realizes that many of our fellow humans are less fortunate. What you may not realize is that over a lifetime, an average person born in a country such as the United States, Italy, or Australia will consume *20 or 30 times* as much food as an individual born in poor undeveloped and developing nations of Africa, Asia, and South America. For the not-so-lucky, malnutrition, and specifically undernutrition, may rule and even end their lives.

Malnutrition is a state in which body functions or development suffer due to inadequate or unbalanced food intake. A malnourished person may suffer from *undernutrition*—that is, the individual lacks food and thus does not obtain sufficient kilocalories or nutrients to sustain proper growth, body functioning, and development. Another word for undernutrition is *starvation*, and globally it is the most common form of malnutrition. Less often, a malnourished person consumes plenty of food from the standpoint of kilocalories but has an unbalanced diet that lacks essential vitamins or minerals.

Global Undernutrition

At least half a billion people in the world go hungry every day. Virtually all of them are too poor to obtain adequate food. Over the long term, lack of food energy or essential nutrients disrupts homeostasis and weakens the body's immune defenses, among other effects. Every year up to 20 million people, three-fourths of them children, die of starvation or from diseases that easily ravage a chronically undernourished body. Many other poorly nourished people are chronically weak, lethargic, prone to illness, or mentally impaired. Let's consider some of the most serious effects of severe nutritional deficiencies.

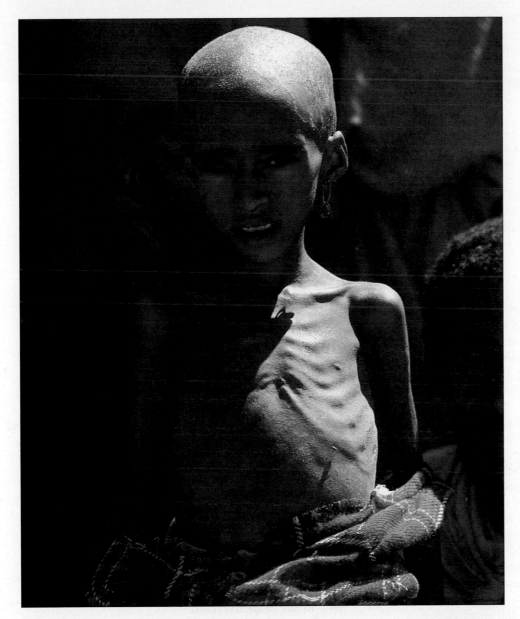

Figure a Child with kwashiorkor.

Protein Deficiency Many undernourished people take in not only too few kilocalories but especially too little protein. The result, *protein-energy malnutrition,* causes weakness, weight loss, impaired immunity, and other symptoms in adults, but it is especially devastating to infants and children. The child in Figure *a* shows the classic signs of *kwashiorkor,* a condition in which a child who may consume near-normal amounts of calories fails to grow and gain weight and shows other evidence of chronic protein deficiency. The swollen abdomen is due to edema (fluid retention) because there are too few proteins in the child's blood plasma. The osmotic imbalance causes water to leave the bloodstream and collect in tissue spaces. Proteins are also essential for proper functioning of the immune system, so affected children are sickly and susceptible to infection. Impaired brain development often leads to permanent mental retardation. Kwashiorkor typically develops when a child is weaned and its diet shifts from protein-rich breast milk to low-protein starchy foods.

Many of the gruesome photographs of skeletal infants in Africa and elsewhere show the body-wasting disease called *marasmus.* Usually marasmus affects very young children who lack both protein and food calories. It is tragically common among bottle-fed infants of poor families in which parents dilute prepared formula, essentially "nourishing" their children mainly with water. (In addition, the local water supply may be contaminated, so that bottle-fed youngsters become infected with waterborne diseases.) Symptoms include low weight and muscle wasting, dry hair and skin, and retarded growth and mental development.

Some effects of kwashiorkor and marasmus can be halted or reversed if the child's diet is improved before the damage is too severe. As a practical matter, however, relief efforts are often too little and too late to save many affected children from death. Those that survive are likely to function at abnormally low levels both physiologically and intellectually.

Other Nutrient Deficiencies People in many underdeveloped countries are malnourished because their food sources are extremely limited. A starchy food, such as rice, corn, millet, or cassava, may make up the lion's share of most meals. This lack of variety translates into deficiencies of key vitamin and mineral nutrients. For example, nutritionists estimate that one-sixth of all humans, almost 1 billion people, are anemic because their diet lacks iron. Iron is a component of hemoglobin, the oxygen-transporting protein in blood (page 142). Major sources of iron include meat and green, leafy vegetables. Iron deficiency impairs oxygen delivery to cells, and it also limits the ability of cells to extract energy from carbohydrates. In addition, it impairs the functioning of the immune system. All these effects further strain the physical resources of a person who may have a parasitic blood infection. Authorities cite a lack of dietary iron as a major factor contributing to high death rates from intestinal and respiratory infections in many countries (page 85). Interestingly, it is also the most common nutritional deficiency of children in the United States.

Lack of vitamin A is responsible for *xerophthalmia*, a form of blindness that annually afflicts a quarter of a million children. Globally, xerophthalmia is the leading cause of preventable blindness, and it is irreversible. Children are also the chief victims of iodide deficiency. This mineral, an ionic form of iodine, is a key component of thyroid hormones that help govern normal growth and metabolism. In adults, low iodide causes goiter (overgrowth of the thyroid gland); in children it can irreversibly retard both physical and mental development. Other common ailments associated with undernutrition in poor nations include scurvy (vitamin C deficiency), beriberi (vitamin B_1 deficiency), and rickets (vitamin D deficiency). Such illnesses occur in addition to the general weakness and apathy typical of an individual whose body lacks sufficient calories.

Groups at Highest Risk

As you have surely gathered by now, children are the main victims of malnutrition. The damage can begin during fetal development, when a pregnant mother does not have access to sufficient protein and other nutrients. Babies born to undernourished mothers may be one to three weeks premature; they may have a low birth weight, poorly developed respiratory systems, and other potentially life-threatening complications. As surviving children move through infancy and childhood, they may be subject to the nutrient deficiencies already described, and research shows that if undernutrition persists they will likely have problems with learning. In the long run, undernourishment of pregnant mothers and their children may result in millions of people who, for want of adequate food, are unable to participate in and contribute fully to society.

Research shows that pregnancy also poses much greater health risks for malnourished women. One study found that the combination of a high birth rate and chronic nutrient deficiencies in Africa resulted in women there having *300 times* the risk of dying from complications of pregnancy as compared to women in North America.

Even in prosperous nations, elderly individuals are also at greater risk of undernutrition. The causes may be economic, social, or psychological; for example, elderly people living alone may eat less food or lack a varied diet because they become depressed or find little pleasure in having solitary meals. Over time, failure to take in adequate nutrients has the predictable effects of declining immunity and other physiological functions.

Social and Political Factors in Global Malnutrition

The United Nations and relief organizations estimate that current global food supplies are adequate to meet the minimum nutritional needs of all or most of the earth's human population. Why then do people starve in places like Somalia and Sudan? Although the situation is complex, part of the answer is that world food supplies and human populations are not equally distributed. In general, people in developed nations have access to more food—and have more money to pay for it—than do people in poorer developing nations. Most of the world's grain and other basic foods are also produced in relatively wealthy countries, whereas 75 percent of the human population lives in developing regions such as Africa. Efforts to increase food production in poor countries are complicated by the high costs of fuel, fertilizers, and other inputs to agriculture, and by the fact that arable land is being converted to other uses as the human population expands (page 514). In addition, governments often encourage farmers to grow crops (such as coffee, sugar, and cocoa) that can be sold internationally to bring cash into a nation's economy but that do not provide nutritious food to local people.

Politics can also get in the way of food distribution. For example, in recent years in Sudan, Ethiopia, Somalia, and the former Czechoslovakia, food earmarked for the general public has been stolen by warring factions and then sold to raise funds, or has rotted in storage because warlords wish to prevent it from feeding their enemies.

Although humanitarian agencies attempt to redistribute food when famine occurs in some part of the world, in the final analysis the solutions to global undernutrition may lie elsewhere. As Chapter 25 describes, within the next three decades the human population is projected to double from its current 5.6 billion to more than 11 billon. Given the recurring food crises we now face, it is difficult to imagine how we can adequately feed so many people; clearly, it is crucial that the human population stabilize at a number the planet's resources can sustain over the long term. In nations in which food is now plentiful, governments and individuals will have to find the will and the means to adequately feed those less fortunate.

6 CIRCULATION AND BLOOD

Heartworks

For Augustus Waller, Jimmie the bulldog was no ordinary pooch. Connected to wires and soaked to his ankles in buckets of salty water, Jimmie was a four-footed window into the workings of the heart.

Waller and other early physiologists suspected that every heartbeat might produce a pattern of electrical currents that could be recorded painlessly at the body surface. That is where Jimmie and the buckets came in. Saltwater is an efficient conductor of electricity—so efficient that it carried faint signals from Jimmie's beating heart, through the skin of his legs, to a crude monitoring device. With this device, in the late 1880s Waller made one of the world's first recordings of a beating heart—an *electrocardiogram* or ECG (Figure 6.1).

Feel the repeated thumpings of your own heart at the chest wall. The pattern of electrical activity of your heart began when you were an embryo, as patches of newly formed cardiac muscle started contracting. Eventually, one patch took the lead. Ever since, it has been the primary pacemaker for your heart's activity. The pacemaker faithfully sets, adjusts, and resets the rate at which blood is pumped from your heart, through a vast network of blood vessels, then back to the heart. At rest your heart rate will be moderate, somewhere around 70 beats a minute. Play a fast game of volleyball and the demands by your muscles for blood-borne oxygen and glucose will escalate. Then, your heart may start pounding 150 times a minute or more to deliver sufficient blood to them.

Figure 6.1 History in the making—Dr. Augustus Waller's pet bulldog, Jimmie, taking part in a painless experiment (**a**) that yielded one of the world's first electrocardiograms (**b**).

a

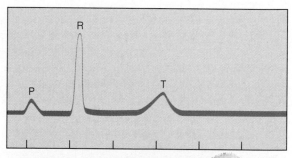

b

We have come a long way from Jimmie in monitoring the heart. Sensors now detect the faintest signals of an impending heart attack. Computers analyze a patient's beating heart and build images of it on a video screen. Surgeons substitute battery-powered pacemakers for malfunctioning natural ones.

With this chapter we turn to the human circulatory system, the means by which substances are rapidly moved to and from all living cells. As you will see, the system is absolutely central to the body's ability to maintain stable operating conditions in the internal environment.

KEY CONCEPTS

1. Cells survive by exchanging substances with their surroundings. In humans, a circulatory system allows rapid movement of substances to and from all living cells. The system consists of a heart, blood, and blood vessels, which are supplemented by a lymphatic system.

2. Blood flows in two circuits. In the pulmonary circuit, the heart pumps deoxygenated blood to the lungs, where it releases carbon dioxide and picks up oxygen. Then blood flows back to the heart. In the systemic circuit, the heart pumps oxygenated blood to all body regions. After giving up oxygen in those regions, blood flows back to the heart. Blood also transports carbon dioxide, plasma proteins, vitamins, hormones, lipids, and other solutes.

3. Arteries and veins are large-diameter transport tubes. Capillaries are narrow diameter tubes for diffusion. Venules collect blood from capillaries and channel it to veins. Arterioles have adjustable diameters; they are control points for the distribution of blood to different regions of the body.

CIRCULATORY SYSTEM: AN OVERVIEW

Like the boulevards and freeways crisscrossing a metropolis, the **circulatory system** is the body's internal rapid-transport system. It is the highway by which cells receive such essential substances as oxygen, nutrients, and secretions (such as hormones) from other cells. It is also the disposal route for waste products of metabolism. As we will see, blood serves as the "carrier" for these various substances, among its other roles.

The basic task of the circulatory system is to circulate blood to the immediate neighborhood of every living cell in the body. This task is vital because the various types of differentiated cells in the human body have few mechanisms for adjusting to drastic changes in the composition, volume, and temperature of interstitial fluid. As Chapter 3 described, interstitial fluid is the "external environment" within which cells live. To stay alive, cells depend on circulating blood to make constant "pickups" and "deliveries" of substances and so maintain stable operating conditions—homeostasis—in the interstitial fluid. The circulatory system functions together with the other organ systems shown in Figure 6.2.

Components of the Circulatory System

The human circulatory system has three main components: (1) blood, a fluid connective tissue composed of water, solutes, and formed elements (blood cells and platelets); (2) the heart, a muscular pump that generates the pressure required to move blood throughout the body; and (3) blood vessels, which are blood transport tubes of different diameters.

The heart pumps constantly, so the volume of flow through the entire system is equal to the volume of blood returned to the heart. Yet as conditions vary, the rate and volume of blood flow must be adjusted. Blood flows rapidly through large-diameter vessels. Elsewhere in the system it must flow slowly, so that there is enough time for substances to be exchanged with interstitial fluid. As you will see, the required slowdown occurs in *capillary beds*, where blood flow fans out through vast numbers of small-diameter blood vessels called capillaries. By dividing up the blood flow, capillaries handle the same total volume of flow as the large-diameter vessels, but at a more leisurely pace.

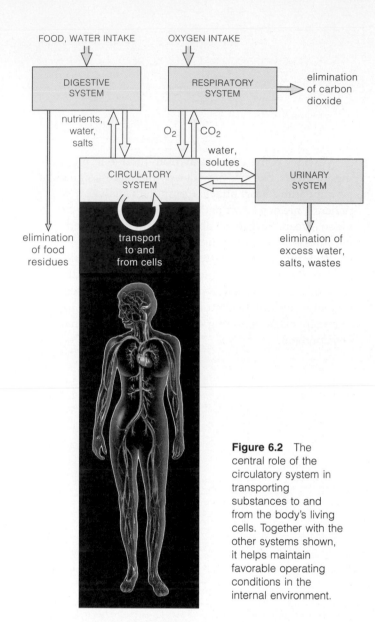

Figure 6.2 The central role of the circulatory system in transporting substances to and from the body's living cells. Together with the other systems shown, it helps maintain favorable operating conditions in the internal environment.

Functional Links with the Lymphatic System

The heart's pumping action puts pressure on blood flowing through the circulatory system. Partly because of this pressure, small amounts of water and some proteins dissolved in blood are forced out of the capillaries and become part of interstitial fluid. However, an elaborate network of drainage vessels picks up excess interstitial fluid and reclaimable solutes and then returns them to the circulatory system. This network is part of the *lymphatic system*. Later you will see how other parts of the lymphatic system help cleanse disease-causing agents from fluid being returned to the blood.

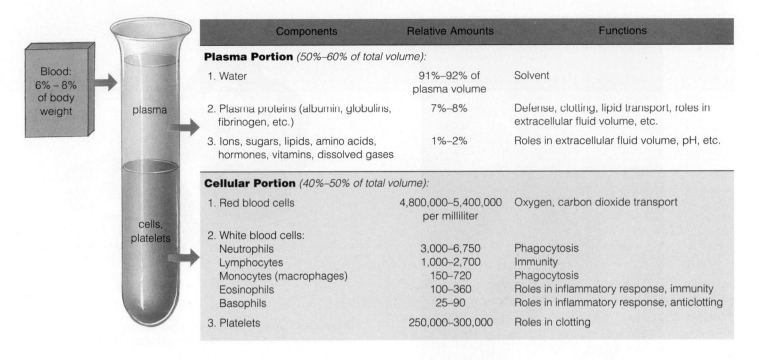

Components	Relative Amounts	Functions
Plasma Portion (50%–60% of total volume):		
1. Water	91%–92% of plasma volume	Solvent
2. Plasma proteins (albumin, globulins, fibrinogen, etc.)	7%–8%	Defense, clotting, lipid transport, roles in extracellular fluid volume, etc.
3. Ions, sugars, lipids, amino acids, hormones, vitamins, dissolved gases	1%–2%	Roles in extracellular fluid volume, pH, etc.
Cellular Portion (40%–50% of total volume):		
1. Red blood cells	4,800,000–5,400,000 per milliliter	Oxygen, carbon dioxide transport
2. White blood cells:		
Neutrophils	3,000–6,750	Phagocytosis
Lymphocytes	1,000–2,700	Immunity
Monocytes (macrophages)	150–720	Phagocytosis
Eosinophils	100–360	Roles in inflammatory response, immunity
Basophils	25–90	Roles in inflammatory response, anticlotting
3. Platelets	250,000–300,000	Roles in clotting

Blood: 6% – 8% of body weight

plasma

cells, platelets

Figure 6.3 Components of blood. If a blood sample placed in a test tube is prevented from clotting, it will separate into a layer of straw-colored liquid, the plasma, that floats over the darker-colored cellular portion of blood.

BLOOD

Blood serves a remarkable array of functions. It brings oxygen and nutrients to cells, transports cell secretions such as hormones, and carries away metabolic wastes. Blood helps stabilize internal pH, and it serves as a highway for phagocytic cells that scavenge tissue debris and fight infections. It also helps equalize body temperature by carrying excess heat from regions of high metabolic activity (such as skeletal muscles) to the skin, where heat can be dissipated.

Blood carries nutrients and raw materials to cells, carries wastes from them, and performs other functions that help maintain a favorable environment for cell activities.

The volume of blood in the body depends on body size and on the concentrations of water and solutes. For average-sized adults, blood volume is generally about 4–5 liters, which is about 7 percent of body weight. As in the old saying, blood *is* thicker than water, and it flows more slowly. This is because blood is a rather viscous fluid consisting of plasma and a cellular portion including red blood cells, white blood cells, and platelets (Figure 6.3). Let's consider each of these components in turn.

Plasma

The straw-colored plasma is mostly water. Plasma makes up about 55 percent of whole blood and serves as a transport medium for blood cells and platelets. The water in plasma also serves as a solvent for ions and molecules, including hundreds of different *plasma proteins*. Among these proteins are albumins, globulins, and fibrinogen. The concentrations of plasma proteins influence the osmotic movement of water between blood and interstitial fluid—and hence the blood's fluid volume. Albumin is important in this water-balancing act, for it represents nearly two-thirds of the total amount of plasma proteins. Albumin also carries a variety of chemicals through the system, from metabolic wastes to therapeutic drugs. Some alpha and beta globulins transport lipids (including cholesterol) and fat-soluble vitamins. As you will see later in this chapter and the next, gamma globulins function in immune responses, and fibrinogen serves in blood clotting.

Plasma also contains ions, glucose and other simple sugars, lipids, amino acids, vitamins, hormones and other communication molecules, and dissolved gases (mostly oxygen, carbon dioxide, and nitrogen). The ions (such as Na^+, Cl^-, and K^+) help maintain extracellular pH and

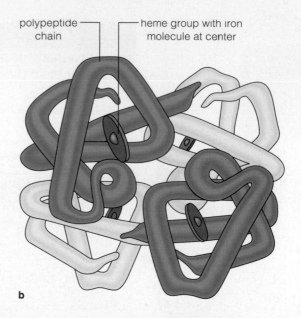

polypeptide chain

heme group with iron molecule at center

b

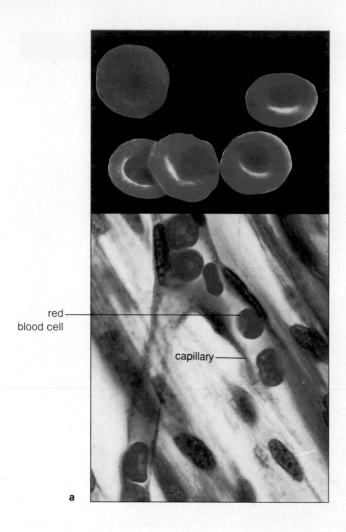

red blood cell

capillary

a

Figure 6.4 (**a**) Red blood cells, which are biconcave disks, viewed alone and within capillaries. (**b**) Structure of a hemoglobin molecule.

fluid volume. The lipids include fats, phospholipids, and cholesterol. Lipids being transported from the liver to different regions generally are bound with proteins, forming lipoproteins.

Red Blood Cells

About 45 percent of whole blood consists of erythrocytes, or **red blood cells**. Each is a biconcave disk, rather like a thick pancake with an indentation on each side (Figure 6.4a). Their red color comes from the iron-containing protein hemoglobin, which transports oxygen. Figure 6.4b (and Figure 1.23 on page 32) shows a hemoglobin molecule's structure. It has two components: the protein *globin* and nonprotein *heme groups* that contain iron. Globin is built of four linked polypeptide chains, and a heme group is associated with each chain. It is the iron molecule at the center of each heme group that binds oxygen.

Oxygen in the lungs diffuses into the blood plasma, then into individual red blood cells, where it binds with the iron in hemoglobin. This oxygenated hemoglobin, or *oxyhemoglobin*, is bright red. Hemoglobin depleted of oxygen appears bluer, especially when it is observed through blood vessel walls and skin.

Hemoglobin and Oxygen Transport If you were to analyze a liter of blood drawn from an artery, you would find only a small amount of oxygen dissolved in it—just 3 milliliters. Yet, like all large, active, warm-blooded animals, humans require a great deal of oxygen to maintain the metabolic activity of their cells. Hemoglobin meets this need. In addition to 3 milliliters of dissolved oxygen, a liter of arterial blood generally carries roughly 194 milliliters of oxygen bound to heme groups of hemoglobin molecules.

As conditions vary in different tissues and organs, so does the tendency of hemoglobin to bind with and hold onto oxygen. Several factors influence this process, which is vital in helping to maintain homeostasis. The most important factor is the amount of oxygen present relative to the amount of carbon dioxide. Other factors are the temperature and the acidity of tissues. Hemoglobin binds oxygen most readily where blood plasma contains a relatively large amount of oxygen, where the temperature is relatively cool, and where the pH is roughly neutral. This is exactly the environment in the lungs, where the body's blood supply must become oxygenated. By contrast,

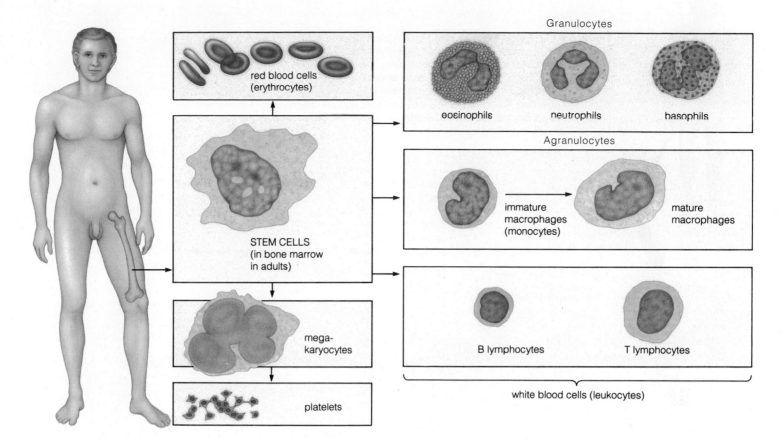

Figure 6.5 Cellular components of blood.

metabolic activity requires oxygen, and it increases both the temperature and the acidity (lowers pH) of tissues. Under those conditions, the oxyhemoglobin of red blood cells arriving in tissue capillaries tends to give up oxygen, which then becomes available to metabolically active cells. We can summarize these events this way:

Hemoglobin also transports some of the carbon dioxide wastes of aerobic metabolism and carries hydrogen ions that help control the pH of body fluids (page 23). We will return to hemoglobin's functions in Chapter 8, which considers the array of interacting elements that enable the respiratory system to efficiently transport gases to and from body cells.

The Life Cycle of Red Blood Cells As Figure 6.5 shows, red blood cells are derived from stem cells in red bone marrow, which occurs in certain bones of the skull,

the vertebrae, the sternum (breastbone), ribs, pelvis, and ends of long bones. (In general, a **stem cell** is unspecialized. It can give rise to descendants that differentiate into specialized cells.) The kidneys produce a hormone, *erythropoietin*, that stimulates a subset of stem cells to produce red blood cells. It has been estimated that roughly 3 million new red blood cells enter the circulation each second! As red blood cells mature (Figure 6.6), they lose their nucleus, ribosomes, and other vital structures. They do not divide or manufacture new proteins, but they have enough enzymes and other proteins to function for about 120 days.

Athletes sometimes resort to "blood doping." First, some of their blood is withdrawn and stored. In response to the loss of red blood cells, erythropoietin stimulates the production of replacements. The stored blood is then reinjected several days prior to an athletic event, so that the athlete has a greater-than-normal number of red blood cells to carry oxygen to muscles—and a temporary, if unethical, competitive advantage.

As red blood cells near the end of their useful life, die, or become damaged or abnormal, phagocytes called *macrophages* ("big eaters") remove them from the blood. As a macrophage dismantles a hemoglobin molecule,

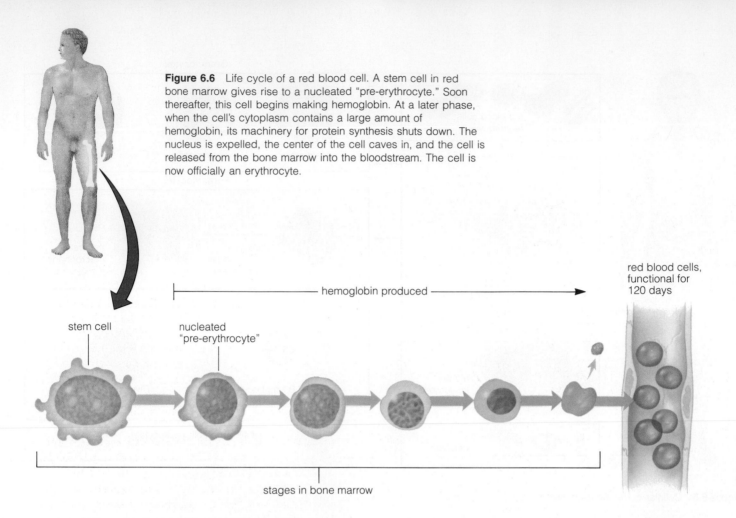

Figure 6.6 Life cycle of a red blood cell. A stem cell in red bone marrow gives rise to a nucleated "pre-erythrocyte." Soon thereafter, this cell begins making hemoglobin. At a later phase, when the cell's cytoplasm contains a large amount of hemoglobin, its machinery for protein synthesis shuts down. The nucleus is expelled, the center of the cell caves in, and the cell is released from the bone marrow into the bloodstream. The cell is now officially an erythrocyte.

hemoglobin produced

red blood cells, functional for 120 days

stem cell

nucleated "pre-erythrocyte"

stages in bone marrow

amino acids from its globin proteins are returned to the bloodstream, and the iron in its heme groups is returned to the red bone marrow, where it may be stored or recycled in new red blood cells. The rest of the heme group becomes converted to the reddish pigment bilirubin, which travels to the liver and is mixed with bile secreted during digestion.

Ongoing replacements from stem cells keep the red blood cell count fairly constant. A *cell count* is the number of cells in a milliliter of blood. The red blood cell count averages 5.4 million in males, 4.8 million in females.

Normally, feedback mechanisms help stabilize the red blood cell count. Suppose you have just started a ski trip. Air contains less oxygen per unit volume at high altitudes, so your body must work harder there to obtain the required oxygen. First your kidneys secrete erythropoietin, which stimulates stem cells in bone marrow to step up production of red blood cells. New oxygen-carrying red blood cells enter the bloodstream, increasing the oxygen–carrying capacity of the blood. Within a few days the oxygen level rises in the blood and in tissues. This information feeds back to the kidneys, erythropoeitin production falls, and red blood cell production drops.

Anemias A condition called *anemia* develops when for one reason or another the blood chronically delivers too little oxygen to tissues. Affected people tend to feel tired and cold and may look pale and be short of breath. In *iron-deficiency anemia*, red blood cells contain a less-than-normal amount of hemoglobin. This disorder generally results from an iron-poor diet and is treatable with a mineral supplement. *Pernicious anemia* is caused by a deficiency of vitamin B_{12}. It usually reflects an underlying disorder in which the intestines are unable to absorb that vitamin from food. The usual treatment involves regular injections of vitamin B_{12}. Other types of anemia develop when a person has too few red blood cells, as when a hemorrhage causes severe blood loss or disease or radiation therapy destroys stem cells in bone marrow.

Two serious human genetic diseases, *sickle cell anemia* (page 382) and *thalassemia*, result from abnormal hemoglobin molecules that in turn cause red blood cells to become deformed or prone to rupture. Chapter 17 describes the genetic origins and devastating consequences of sickle cell anemia. This chapter's *Focus on Science* explores some strategies for replacing blood or its elements when a medical need arises.

White Blood Cells

Even though leukocytes, or **white blood cells**, make up a tiny fraction of whole blood, they have vital functions in day-to-day housekeeping and defense. Some scavenge dead or worn-out cells, as well as anything that is chemically perceived as foreign to the body. Others target or destroy specific bacteria, viruses, or other agents of disease. All types of leukocytes arise from stem cells in bone marrow (Figure 6.5). They circulate in blood, but most go to work after they squeeze out of blood capillaries and enter tissues. White blood cell counts vary, depending on whether an individual is sedentary or highly active, healthy or battling an infection.

There are five types of white blood cells, divided into two major classes (Figure 6.5). The first class, **granulocytes**, includes *neutrophils*, *eosinophils*, and *basophils*. All have a lobed nucleus, and when the cell is stained various types of granules are visible in the cytoplasm. About two-thirds of all leukocytes are neutrophils, "search and destroy" cells that follow chemical trails to inflamed tissues. Eosinophils participate in allergic responses and attacks on invading parasites. Basophils are thought to have roles in inflammation and allergic responses (Chapter 7).

Leukocytes in the second class of white blood cells do not have visible granules in the cytoplasm, and so they are termed **agranulocytes**. One type, *monocytes*, differentiate into wandering macrophages ("big eaters") that engulf invaders and cellular debris. Another type, *lymphocytes* (the B cells and T cells), carry out specific immune responses, and we consider them in detail in Chapter 7.

Life for most white blood cells is challenging and short. The life expectancy of a white cell is measured in days or, during a major infection, a few hours.

Platelets

Some stem cells in bone marrow develop into "giant" cells (megakaryocytes). These cells shed fragments of cytoplasm enclosed in a bit of plasma membrane. The fragments, called **platelets**, are oval or rounded disks about 2–4 micrometers across. Each platelet lasts only five to nine days, but millions are always circulating in blood. Substances associated with platelets contribute to a chain of reactions leading to blood clotting, as will be described shortly.

CARDIOVASCULAR SYSTEM

"Cardiovascular" comes from the Greek *kardia*, meaning heart, and the Latin *vasculum*, meaning vessel. In the human cardiovascular system, a heart pumps blood into

Focus on Science

New Blood

Suppose you are rushed to the hospital with severe internal bleeding. You may need a blood transfusion—a serious emergency in any case, even more serious if you happen to have a rare blood type. Or suppose you have been diagnosed with a cancer for which the most effective treatment involves transfusing you with platelets—the cell fragments in blood that have roles in blood-clotting and are now becoming weapons in various advanced cancer therapies. Just a few years ago, your life and health might have depended precariously on whether a hospital or blood bank could quickly obtain a quantity of blood cells, platelets, or fresh whole blood of your blood type. The predicament is real: Each year, hospital patients in the United States require more than 20 million units of red blood cell products and platelets, and many more units of other blood products.

Platelets, red blood cells, and other blood elements are highly perishable. Once obtained from a donor, fresh blood products must be used within a few days. Blood and blood elements can be frozen, but at a biological price. As water molecules form ice, the resulting crystals can rupture the plasma membrane and other membranes within a cell. Using traditional methods, freezing platelets generally destroys at least half of them. Red blood cells must be stored in costly specialized freezers and thawed by way of complex procedures. These and other problems often create life-threatening shortages of blood products for patients who need them.

Not long ago, a small California research firm announced that it had succeeded in experimentally freezing platelets in a way that limited damage from ice crystals. The platelets are extracted from whole blood, then mixed with a carbohydrate-rich fluid that replaces much of the water inside and surrounding the platelets. The mixture can be stored in a simple home freezer. When it is thawed, it is spun at high speed in a centrifuge along with purification chemicals. The 15-minute procedure separates the platelets from the other solutions. About 75 percent of the platelets survive the freezing, thawing, and purification, and are then ready for use.

Similar methods for freezing red blood cells, other blood elements, and bone marrow are now under development. Like the platelet-freezing technique, all must undergo careful testing on human volunteers before they are approved for medical use. It may not be long, however, before obtaining platelets for a cancer treatment, whole blood for an accident victim, or red blood cells for an anemia sufferer is as easy as pulling a container from a well-stocked hospital freezer.

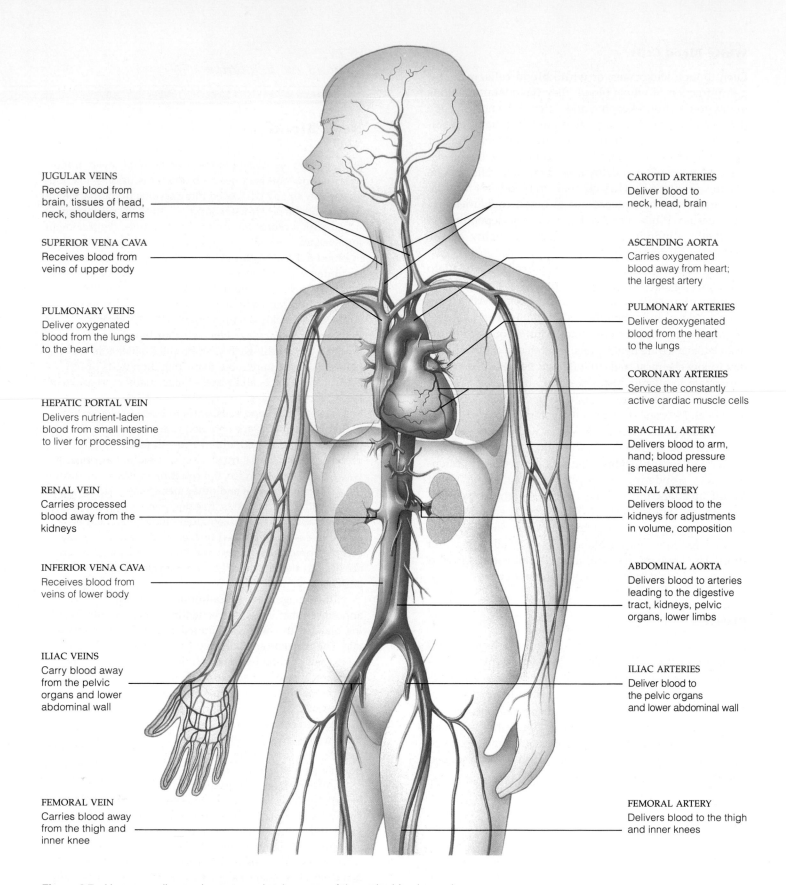

JUGULAR VEINS
Receive blood from
brain, tissues of head,
neck, shoulders, arms

SUPERIOR VENA CAVA
Receives blood from
veins of upper body

PULMONARY VEINS
Deliver oxygenated
blood from the lungs
to the heart

HEPATIC PORTAL VEIN
Delivers nutrient-laden
blood from small intestine
to liver for processing

RENAL VEIN
Carries processed
blood away from the
kidneys

INFERIOR VENA CAVA
Receives blood from
veins of lower body

ILIAC VEINS
Carry blood away
from the pelvic
organs and lower
abdominal wall

FEMORAL VEIN
Carries blood away
from the thigh and
inner knee

CAROTID ARTERIES
Deliver blood to
neck, head, brain

ASCENDING AORTA
Carries oxygenated
blood away from heart;
the largest artery

PULMONARY ARTERIES
Deliver deoxygenated
blood from the heart
to the lungs

CORONARY ARTERIES
Service the constantly
active cardiac muscle cells

BRACHIAL ARTERY
Delivers blood to arm,
hand; blood pressure
is measured here

RENAL ARTERY
Delivers blood to the
kidneys for adjustments
in volume, composition

ABDOMINAL AORTA
Delivers blood to arteries
leading to the digestive
tract, kidneys, pelvic
organs, lower limbs

ILIAC ARTERIES
Deliver blood to
the pelvic organs
and lower abdominal wall

FEMORAL ARTERY
Delivers blood to the thigh
and inner knees

Figure 6.7 Human cardiovascular system, showing some of the major blood vessels.
Arteries, which carry oxygenated blood to tissues, are shaded red. Veins, which carry
deoxygenated blood away from tissues, are shaded blue. Notice, however, that for the
pulmonary arteries and veins the roles are reversed.

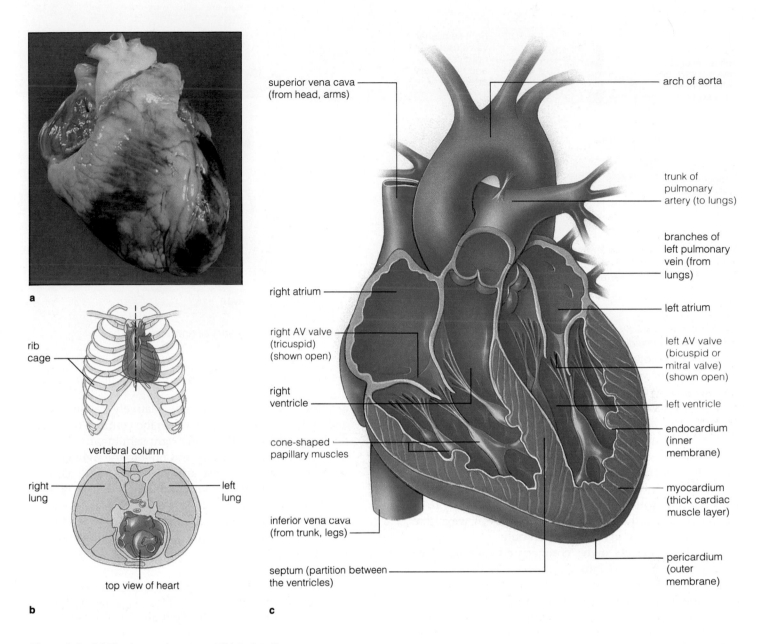

Figure 6.8 (**a**) The human heart, and (**b**) its location.
(**c**) Cutaway view showing the heart's internal organization.

large-diameter arteries. From there the blood flows into small muscular arterioles, which branch into tiny capillaries. Blood flows from capillaries into small venules, then into large veins, which return blood to the heart. Figure 6.7 gives an overall picture of the cardiovascular system.

The Heart

Heart Structure In a lifetime of 70 years, the human **heart** beats some 2.5 billion times. The heart's structure reflects its role as a durable pump. The heart is mostly

cardiac muscle tissue, the **myocardium**. It is surrounded, protected, and lubricated by a tough, fibrous sac, the *pericardium*. Its inner chambers have a smooth lining (*endocardium*) composed of connective tissue and a single layer of epithelial cells. The epithelial cell layer is known as *endothelium*. It lines the inside of blood vessels as well as the heart.

A thick wall, the **septum**, divides the heart into two halves, right and left (Figure 6.8). Each half has two chambers: an **atrium** (plural: *atria*) located above a **ventricle**. Flaps of membrane separate the two chambers and serve as a one-way flow valve, called an **atrioventricular valve** (AV valve). The AV valve in the right half of the heart is

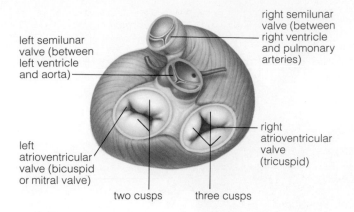

left semilunar valve (between left ventricle and aorta)

right semilunar valve (between right ventricle and pulmonary arteries)

left atrioventricular valve (bicuspid or mitral valve)

right atrioventricular valve (tricuspid)

two cusps

three cusps

Figure 6.9 The valves of the human heart. In this drawing, you are looking down at the heart, and the atria have been removed so that the atrioventricular and semilunar valves are visible.

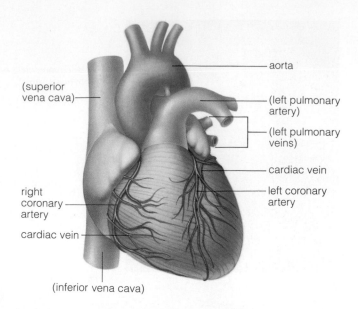

(superior vena cava)

aorta

(left pulmonary artery)

(left pulmonary veins)

cardiac vein

left coronary artery

right coronary artery

cardiac vein

(inferior vena cava)

Figure 6.10 Coronary arteries and veins.

termed a *tricuspid valve* because its three flaps come together in pointed cusps (see Figure 6.9). In the heart's left half the AV valve consists of just two flaps; it is called the *bicuspid valve* or *mitral valve*. Tough, collagen-reinforced strands (*chordae tendineae* or "heart strings") connect the AV valve flaps to cone-shaped *papillary muscles* that extend out from the ventricle wall. When a blood-filled ventricle contracts, this arrangement prevents the flaps from opening backwards into the atrium.

Each half of the heart also has a **semilunar valve** between the ventricle and the arteries leading away from it. During a heartbeat, it opens and closes in ways that keep blood moving in one direction only.

The heart has its own "coronary circulation." Two **coronary arteries** lead into a capillary bed that services cardiac muscle cells (Figure 6.10). They branch off the **aorta**, the major artery carrying oxygenated blood away from the heart. Coronary arteries become dangerously clogged in some cardiovascular disorders described later in this chapter.

Blood Circulation Routes

Each half of the heart (atrium and ventricle) functions as a pump. Together, these side-by-side pumps are the basis of two cardiovascular circuits through the body, the pulmonary and systemic circuits (Figure 6.11). The **pulmonary circuit** receives blood from body tissues and circulates it through the lungs for gas exchange. The **systemic circuit** transports blood to and from tissues.

Pulmonary Circuit The pulmonary circuit begins as blood from body tissues enters the right atrium, then moves through the atrioventricular valve into the right ventricle. As the ventricle fills, the atrium contracts. Blood

arriving in the right ventricle is relatively low in oxygen and high in carbon dioxide. When the ventricle contracts, the blood moves through the right semilunar valve into the main pulmonary artery and on into the right and left pulmonary arteries. These arteries transport the blood to the two lungs, where (in lung capillaries) it picks up oxygen and gives up carbon dioxide that will be exhaled. The freshly oxygenated blood returns through two sets of pulmonary veins to the heart's left atrium, completing the circuit.

Systemic Circuit In the systemic circuit, oxygenated blood pumped by the left half of the heart moves through the body and returns to the right atrium. The left atrium receives blood from pulmonary veins. The blood moves through an atrioventricular (bicuspid) valve to the left ventricle. This heart chamber then contracts forcefully to send blood coursing through a semilunar valve into the aorta.

Figure 6.7 shows that the aorta descends quite far down into the body torso. Along its length major arteries branch off, funneling blood to organs and tissues. After passing through tissue regions (where oxygen is used and carbon dioxide is produced), the deoxygenated blood returns to the right half of the heart, where it enters the pulmonary circuit once again. Notice that in both the pulmonary and systemic circuits, blood travels through arteries, arterioles, capillaries, venules, and finally returns to the heart in veins. Blood from the head, arms, and chest arrives through the *superior vena cava,* and the *inferior vena cava* collects blood from lower body regions.

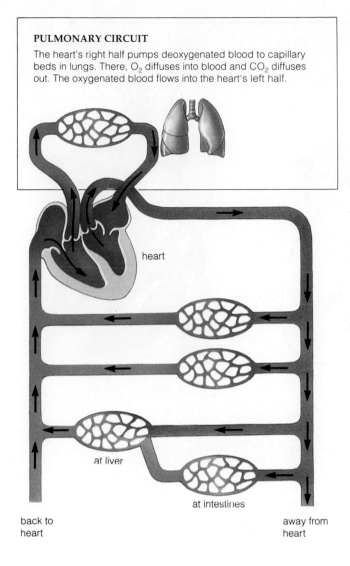

PULMONARY CIRCUIT

The heart's right half pumps deoxygenated blood to capillary beds in lungs. There, O_2 diffuses into blood and CO_2 diffuses out. The oxygenated blood flows into the heart's left half.

heart

at liver

at intestines

back to
heart

away from
heart

SYSTEMIC CIRCUIT

The heart's left half pumps the oxygen-rich blood to capillary beds at organs and tissues in all body regions. There, O_2 diffuses out of blood and CO_2 diffuses in. Then the blood, now oxygen-poor, flows into the heart's right half.

In most cases, a given volume of blood passes through only one capillary bed when it makes the trip away from and back to the heart. There are exceptions, as when blood passes through a capillary bed at the intestines, then is shunted through a capillary bed at the liver before returning to the heart.

In terms of exchange of substances between the blood and tissues, **capillary beds** are where the real action is. Most capillary beds have two types of vessels: true capillaries where exchanges between blood and tissues take place, and so-called "thoroughfare channels" that serve as connecting pipes between arterioles and venules. Blood flow into true capillaries is controlled by *precapillary sphincters*, collarlike arrangements of smooth muscle cells at points where the capillaries branch from thoroughfare channels.

Arteries branch off the aorta, carrying blood directly to capillary beds in specific tissues and organs. For example, in a person at rest about once every 60 seconds roughly one-fourth of the blood pumped into the systemic circulation enters the kidneys via *renal arteries*. Chapter 9 details kidney functions, which include filtering fluid from the blood, removing metabolic wastes and excess ions, and returning needed substances to the blood.

As Figure 6.11 suggests, after a meal the blood passing through capillary beds in the digestive tract (where it picks up glucose and other nutrients) detours through the *hepatic portal vein* to another capillary bed in the liver, which has a key role in nutrition (page 120). The slow flow of blood through this second bed gives the liver time to remove impurities and other substances. Blood leaving the bed enters the general circulation through a *hepatic vein*. (The liver also receives its own supply of oxygenated blood through the *hepatic artery*.)

Cardiac Cycle: Heartbeat

Blood is pumped each time the heart beats. A "heartbeat" is actually a sequence of events in which the heart's chambers contract and relax. The sequence takes place almost simultaneously in both sides of the heart. The contraction phase is called **systole** (SISS-ta-lee) and the relaxation phase is called **diastole** (dye-ASS-ta-lee). This sequence

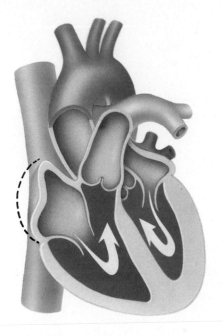
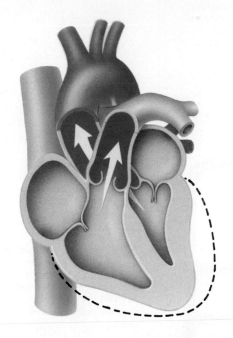
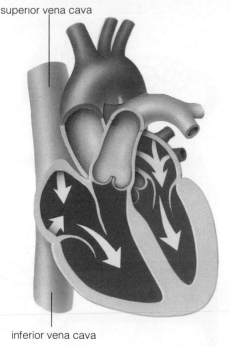

superior vena cava

inferior vena cava

Atria contract (systole) and blood enters ventricles through AV valves. Semilunar valves are closed. Ventricles are in diastole.

Atria relax (diastole) and ventricles contract (systole). Blood from right ventricle moves via right semilunar valve to pulmonary artery. Blood from left ventricle moves via left semilunar valve to aorta.

Semilunar valves close. Atria and ventricles are briefly in diastole. Blood is entering right heart chambers through the vena cavae, and filling begins again.

Figure 6.12 Blood flow through the heart during part of a cardiac cycle. The closing of heart valves sets up vibrations, producing a "lub-dup" sound that can be heard at the chest wall. At each "lub," AV valves are closing as the ventricles contract. At each "dup," semilunar valves are closing as the ventricles relax.

of muscle contraction and relaxation is a **cardiac cycle** (Figure 6.12). Each heartbeat lasts about eight-tenths of a second.

During the cycle, the ventricles relax well before the atria contract, and the ventricles contract when the atria relax. When the relaxed atria are filling with blood, fluid pressure inside them increases and the AV valves open. Blood flows into the ventricles, which are 80 percent filled by the time the atria contract. As the filled ventricles begin to contract, fluid pressure inside *them* increases, and the AV valves snap shut. As the ventricles continue contracting, ventricular pressure rises above that in blood vessels leading away from the heart. The increased pressure forces the semilunar valves open, and blood flows out of the heart and into the aorta and pulmonary artery. After blood has been ejected, the ventricles relax, and the semilunar valves close. For about half a second the atria and ventricles are all in diastole, then the blood-filled atria contract, and the cycle repeats.

Heart Sounds The blood and heart movements during the cardiac cycle generate a "lub-dup" sound that reflect the forceful closing of the heart's one-way valves. You can hear the sound through a stethoscope positioned against the chest wall. At each "lub," the AV valves are closing as the ventricles contract. At each "dup" the semilunar valves are closing as the ventricles relax.

During a cardiac cycle, contraction of the atria helps fill the ventricles. Contraction of the ventricles pumps blood into the systemic circulation.

How Heart Muscle Contracts

The heart muscle contains subsets of noncontractile cells that make up the **cardiac conduction system**. The system includes pacemaker cells that are *self-excitatory*. That is,

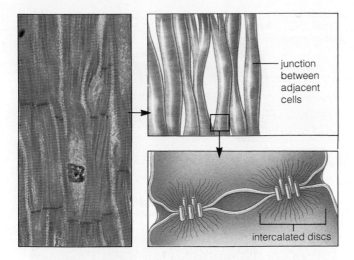

Figure 6.13 Intercalated discs containing communication junctions at the ends of adjacent cardiac muscle cells. Signals travel rapidly across the junctions and cause cells to contract nearly in unison.

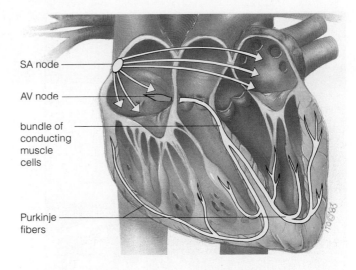

Figure 6.14 Location of cardiac muscle cells that conduct signals for contraction through the heart.

they spontaneously produce and conduct the electrical events (action potentials) that stimulate contractions in the heart's contractile cells. Even if all nerves leading to the heart are severed, the heart will keep on beating!

Recall that in skeletal muscle the ends of individual cells (fibers) are attached to bones or tendons. In cardiac muscle, however, the cells branch and then connect with one another at their endings (Figure 6.13). Communication junctions (called *intercalated discs*) occur where the plasma membranes of neighboring cells are joined together. With each heartbeat, signals calling for contraction spread so rapidly across the junctions that the cardiac muscle cells contract together as if they were a single unit.

Normally, excitation begins with a small mass of cells in the upper wall of the right atrium. This **sinoatrial (SA) node** is the *cardiac pacemaker*. The SA node generates one wave of excitation after another, usually 70 or 80 times a minute. Each wave spreads over both atria, causing them to contract. The wave rapidly reaches the **atrioventricular node** (AV node) in the septum dividing the two atria. Notice in Figure 6.14 that bundles of conducting fibers extend from the AV node to each ventricle. At intervals along each bundle, conducting cells called *Purkinje fibers* branch off and make contact with contractile muscle cells in the ventricles. When a stimulus reaches the AV node, there is a brief pause, then the stimulus passes along the bundles to Purkinje fibers and on to contractile muscle fibers in each ventricle. The pause gives the atria time to finish contracting before the wave of excitation spreads to the ventricles.

Of all cells of the cardiac conducting system, the SA node fires off impulses at the highest frequency and is the first region to respond in each cardiac cycle. Thus, the SA node is the **cardiac pacemaker** mentioned at the start of this chapter. Its rhythmic firing is the basis for the normal rate of heartbeat. The electrocardiogram in Figure 6.25 traces a normal heartbeat sequence. People whose SA node chronically malfunctions may have an artificial pacemaker implanted to provide a regular stimulus for heart contraction.

The SA node is the cardiac pacemaker that establishes a regular heartbeat. Its spontaneous, repetitive excitation spreads along a system of muscle cells that stimulate contractile tissue in the atria, then the ventricles, in a rhythmic cycle.

Nervous Controls over Heart Rate   Whereas the nervous system triggers the contraction of skeletal muscle, it can only *adjust* the rate and strength of cardiac muscle contraction. Stimulation by *sympathetic nerves* of the autonomic nervous system increases the force and rate of heart contractions, while conversely *parasympathetic nerves* can slow heart activity. Centers for nervous control of heart functions lie in the spinal cord, the brain's medulla oblongata, and other brain regions (Chapter 10).

Vascular System and Blood Pressure

A healthy heart pumps blood into the systemic circuit with a great deal of force. The force blood exerts against

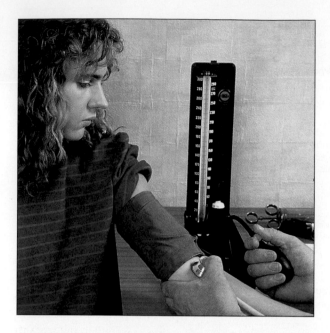

Figure 6.15 Measuring blood pressure with a device called a sphygmomanometer. A hollow cuff attached to a pressure gauge is wrapped around the upper arm. Then it is inflated with air to a pressure above the highest pressure of the cardiac cycle (at systole, when the ventricles contract). Above the systolic pressure, no sounds are heard through a stethoscope positioned above the artery (because no blood is flowing through it).

Air in the cuff is slowly released, so some blood flows into the artery. The turbulent flow causes soft tapping sounds, and when this first occurs, the value on the gauge is the systolic pressure—about 120mm mercury (Hg) in young adults at rest. (This means the measured pressure would force mercury to move upward 120 millimeters in a narrow glass column.)

More air is released from the cuff. Just after the sounds become dull and muffled, blood flows continuously, and so the turbulence and tapping sounds stop. The silence corresponds to the diastolic pressure at the end of a cardiac cycle, just before the heart pumps out blood. Generally the reading is about 80mm Hg. In this example, the *pulse* pressure (the difference between the highest and lowest pressure readings) is 120 − 80, or 40mm Hg.

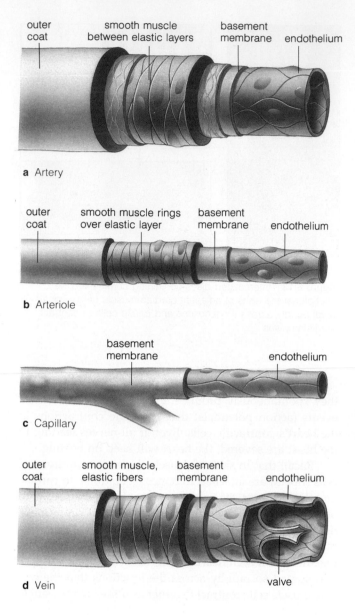

a Artery

b Arteriole

c Capillary

d Vein

Figure 6.16 Structure of blood vessels. The basement membrane contains a special form of collagen.

vessel walls is measurable as **blood pressure**. Pressure is highest in the aorta, then drops along the systemic circuit away from and back to the heart.

Measurements of blood pressure (Figure 6.15) are written as a fraction. For an adult, 120/80 is in the normal range. The top number is *systolic pressure*, the peak of pressure in the aorta while the heart's left ventricle contracts and pushes blood into it. The bottom number is *diastolic pressure*. It measures the lowest blood pressure in the aorta, when the heart is relaxed and blood is flowing out of the aorta. Values for systolic and diastolic pressure provide information about a person's cardiovascular

health, including whether blood is flowing normally through vessels.

Resistance When a fluid moves through a tube, including a blood vessel, friction and other factors combine to create resistance to the movement of molecules and particles in the fluid. When friction occurs, energy is lost (as heat). In the systemic circulation, there is increasing resistance to flowing blood as it moves from arteries to arterioles, through capillaries, and into venules and veins. As resistance causes circulating blood to lose energy, blood pressure drops. Figure 6.16 depicts blood vessel struc-

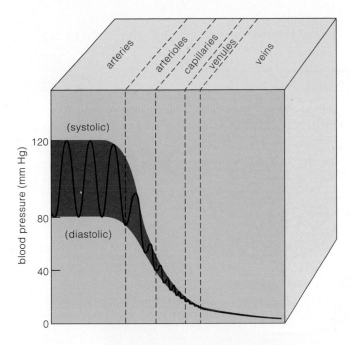

Figure 6.17 Blood pressure. This diagram shows the drop in fluid pressure for a volume of blood making a trip along the systemic circuit.

ture—an important factor, as we will see, in blood pressure changes at different points in the system.

Velocity The speed with which blood travels also declines as blood flows away from the heart. Vigorously pumped by the muscular left ventricle, blood entering the systemic circulation is moving rather rapidly when it leaves the heart in the aorta. Just as a river flows more slowly when it becomes divided into many small channels, the flow of blood slows as the total cross-sectional area of vessels increases. Its velocity is greatest in the aorta, decreases in the more numerous arterioles, and slows to a crawl (relatively speaking) in countless slender capillaries. Hence, there is sufficient time for materials to diffuse between capillaries and tissues. Velocity increases once again as blood moves into veins for the return trip to the heart.

Arteries In the systemic circulation, **arteries** conduct oxygenated blood to all body tissues. (Pulmonary arteries, recall, carry deoxygenated blood.) As Figure 6.16a shows, artery walls consist of several layers of tissues, including a thick central layer of smooth muscle sandwiched between thinner layers containing elastin. The thick, elastic wall of a large artery bulges somewhat under the pressure surge created during each cardiac cycle when a ventricle contracts. In arteries lying near the body surface, as in the wrist, you can feel this surge as a **pulse**.

The bulging of large arteries helps keep blood flowing continuously through capillaries. Some of the blood volume pumped during the systole phase of each cardiac cycle is momentarily stored in the "bulge"; the elastic recoil of the artery then forces that stored blood through the capillaries during diastole, when heart chambers are relaxed.

Arteries present little resistance to blood flow because their diameters are relatively large. As a result, pressure does not drop much in the arterial portion of the blood circuits.

Arterioles Arteries branch into vessels of smaller diameter, the **arterioles**. The greatest drop in blood pressure occurs at arterioles (Figure 6.17). This is because, for reasons given shortly, arterioles offer the greatest resistance to blood flow.

The wall of an arteriole has rings of smooth muscle over a single layer of elastic fibers (Figure 6.16b). This enables arterioles to enlarge in diameter when the smooth muscle relaxes, or shrink in diameter when the smooth muscle contracts. Adjustments are made in response to hormones and to signals from the nervous system, as well as to changes in local chemical conditions. Blood is directed to a region of high metabolic activity when the diameters of arterioles in those regions enlarge. Blood is directed away from less active regions when the diameters of arterioles in those regions constrict.

Capillaries Capillary beds are *diffusion zones* for exchanges of gases and other substances between blood and interstitial fluid. A **capillary** has the thinnest wall of any blood vessel. Its wall consists of a layer of flat endothelial cells, separated from one another by narrow spaces (Figure 6.16c).

Most capillaries have such a small diameter that red blood cells must squeeze through them single file (Figure 6.4a). As you might guess, each capillary presents high resistance to blood flow. Yet because there are so many of them in a capillary bed, their combined diameters are greater than the combined diameters of arterioles leading into them. Said another way, because capillaries branch so extensively, in cross-section the area of a capillary bed is much greater than the area of its associated arterioles (Figure 6.18). Thus a capillary bed presents less *total* resistance to flow than do the arterioles leading into it, and so blood pressure drops less steeply.

Capillaries thread through nearly every tissue in the body, coming within .01 millimeter of almost every living cell. Most solutes, including oxygen and carbon dioxide, diffuse across the capillary wall. Some proteins cross it by endocytosis (followed by exocytosis on the other side), and certain ions probably pass through pores in capillary walls and spaces between endothelial cells.

Figure 6.18 Comparison of cross-sectional areas represented by arteries, arterioles, and capillaries in the circulatory system.

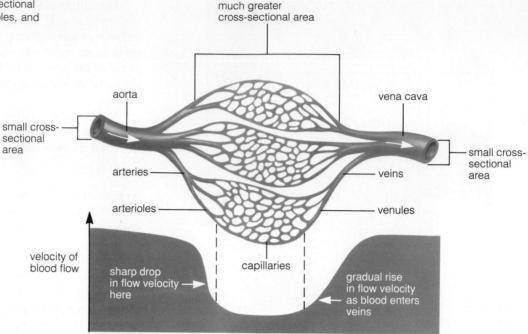

much greater cross-sectional area

aorta

vena cava

small cross-sectional area

arteries

veins

arterioles

venules

small cross-sectional area

velocity of blood flow

sharp drop in flow velocity here

capillaries

gradual rise in flow velocity as blood enters veins

Fluid also enters and leaves capillaries in response to various types of pressure (Figure 6.19). The force a fluid exerts against a surface is called *hydrostatic pressure*. In capillaries, this is the same as blood pressure, and it forces some water out of capillaries, especially at the arterial end where pressure is greatest. Normally, water also moves *into* capillaries in response to *osmotic pressure*, that is, it follows its concentration gradient into capillaries as the concentration of proteins and other solutes there rises. On balance, however, more water tends to leave capillaries than to enter them. As water moves in either direction, some solutes follow along.

The fluid and solute movements we have been describing help maintain the proper fluid balance between the bloodstream and the surrounding tissues. They are vital to maintaining the blood volume needed for adequate blood pressure.

Arteries are large-diameter transport vessels. Their large diameters offer low resistance to blood flow, so there is little drop in blood pressure in arteries.

Arterioles are control points where adjustments can be made in the volume of blood flow to be delivered to different capillary beds. They offer great resistance to flow, so there is a major drop in pressure in arterioles.

Capillary beds are diffusion zones for exchanges between blood and interstitial fluid. Collectively, they have a greater cross-sectional area than that of arterioles leading into the beds, so they present less total resistance to flow. There is some drop in pressure here.

Venules and Veins   Capillaries merge into **venules**. The venule wall is only a bit thicker than that of a capillary, so some diffusion occurs across it. Venules merge into large-diameter **veins**, the transport tubes leading back to the heart. Veins are major blood volume reservoirs, for they contain 50–60 percent of the total blood volume. Although a vein wall is thin and not as resilient as an artery wall, it does contain some smooth muscle (Figure 6.16d).

As Figure 6.17 shows, blood pressure in the venous system is very low. Several mechanisms assist blood movement back to the heart. Skeletal muscle attached to moving limbs bulges against adjacent veins (Figure 6.20); this increases venous pressure and helps drive blood to the heart. Also, some veins (mainly in the limbs) have valves. When blood starts moving backward because of gravity, it pushes the valves into a closed position, preventing backflow. Inherited defects, obesity, pregnancy, and other factors can result in weakened venous valves. The walls of a *varicose vein* have become overstretched because weak valves have chronically allowed blood to pool in the vein.

Venous blood pressure also is influenced by how rapidly a person breathes. When inhaled air pushes down on internal organs, it changes the pressure gradient between the heart and veins. The veins expand, and the volume of blood flowing in them increases.

Veins are blood volume reservoirs as well as transport tubes. Blood pressure in the venous system is low, and mechanisms such as skeletal muscle movements and one-way valves help return venous blood to the heart.

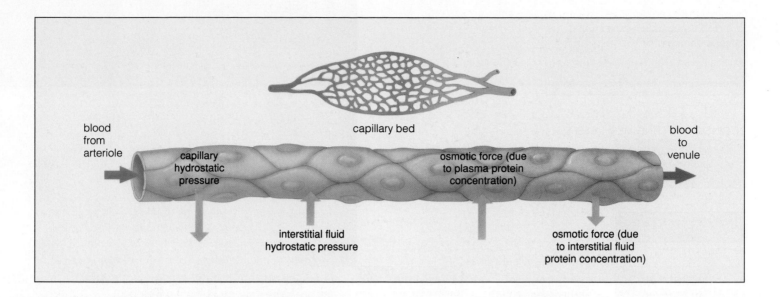

blood
from
arteriole

capillary bed

capillary
hydrostatic
pressure

osmotic force (due
to plasma protein
concentration)

blood
to
venule

interstitial fluid
hydrostatic pressure

osmotic force (due
to interstitial fluid
protein concentration)

Figure 6.19 Fluid movements in an idealized capillary bed. The movements are important in maintaining the distribution of extracellular fluid between the bloodstream and interstitial fluid.

At the arteriole end of a capillary, the difference between capillary blood pressure and interstitial fluid pressure causes some water (but very few plasma proteins) to leave the capillary.

The difference in water concentration between plasma and interstitial fluid causes water to enter a capillary by osmosis, following its concentration gradient. (Plasma has a greater solute concentration, with its protein components, and therefore a lower water concentration.)

Fluid loss at the arteriole end of a capillary bed tends to be balanced by fluid intake at the venule end.

Edema is a condition in which excess fluid accumulates in interstitial spaces. This happens to some extent during exercise. As arterioles dilate in local tissue regions, capillary blood pressure increases and more fluid is forced out of capillaries into tissues. Edema also results from an obstructed vein or from heart failure.

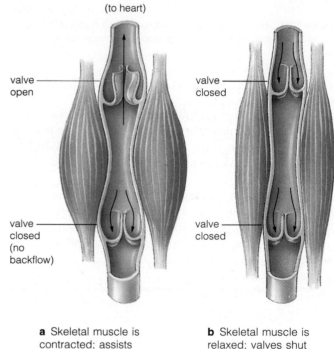

(to heart)

valve
open

valve
closed

valve
closed
(no
backflow)

valve
closed

a Skeletal muscle is contracted; assists blood flow to heart

b Skeletal muscle is relaxed; valves shut and prevent backflow

Figure 6.20 (at right) Structure of valves in veins. Skeletal muscle contractions and venous valves are important in returning blood to the heart.

Homeostatic Controls over Circulation

Maintaining Blood Pressure Various mechanisms work to maintain adequate blood pressure over time and so ensure adequate blood flow to all regions of the body. Blood pressure is controlled by the centers in the medulla oblongata and other brain regions, as mentioned earlier. The centers integrate information from sensory receptors in cardiac muscle tissue and in certain arteries, such as the aorta and the carotid arteries in the neck. They use this information to coordinate the rate and strength of heartbeats with changes in the diameter of arterioles and, to some extent, of veins.

When an abnormal increase in blood pressure is detected, the medulla commands the heart to beat more slowly and contract less forcefully. It also sends signals to smooth muscle cells in the wall of arterioles. The cells relax, and the outcome is **vasodilation**—an enlargement (dilation) of arteriole diameter. With less resistance to blood flow, blood pressure falls. Conversely, when an abnormal decrease in blood pressure is detected, the medulla commands the heart to beat faster and contract

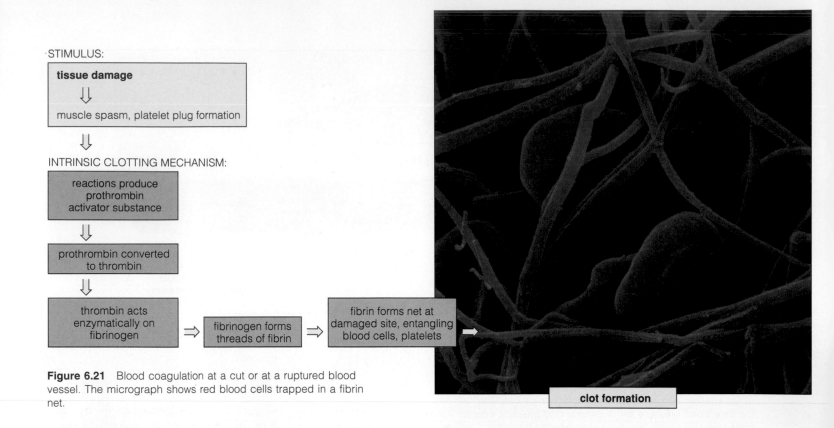

STIMULUS:

tissue damage
⇩
muscle spasm, platelet plug formation

⇩

INTRINSIC CLOTTING MECHANISM:

reactions produce prothrombin activator substance
⇩
prothrombin converted to thrombin
⇩
thrombin acts enzymatically on fibrinogen ⇒ fibrinogen forms threads of fibrin ⇒ fibrin forms net at damaged site, entangling blood cells, platelets ⇒

clot formation

Figure 6.21 Blood coagulation at a cut or at a ruptured blood vessel. The micrograph shows red blood cells trapped in a fibrin net.

more forcefully. It also stimulates the smooth muscle cells of arterioles to contract. In this case, the outcome is **vasoconstriction**—a decrease in arteriole diameter. With the resulting increased resistance to blood flow, pressure rises.

Arterioles in various regions have different receptors that can be activated by epinephrine, angiotensin, and other hormones. Epinephrine triggers vasoconstriction or vasodilation. Angiotensin triggers widespread vasoconstriction.

Integrating centers in the brain function to keep the resting level of blood pressure fairly constant over time.

Control of Blood Distribution   The nervous system, endocrine system, and changes in local chemical conditions interact to assure that blood circulation meets the metabolic demands of various tissues. For instance, after you eat a large meal, more blood is diverted to your digestive system, which swings into full gear as other systems more or less idle. When the body is exposed to cold wind or snow for an extended time, blood is diverted away from the skin to deeper tissue regions, so that the metabolically generated heat that warmed the blood in the first place can be conserved.

Sympathetic nerves as well as hormones can stimulate vasoconstriction in many tissues, including the kid-

neys and muscle tissues that are not being called upon to contract. Thus they can cause blood to be diverted away from areas that—for a time, at least—do not require a large supply.

During strenuous exertion such as running, the oxygen level in skeletal muscle falls and the levels of carbon dioxide, hydrogen ions, potassium ions, and other substances rise. The changes in local chemical conditions cause vasodilation in arterioles. Now more blood flows to the active muscles, delivering more raw materials and carrying away cell products and wastes. At the same time, vasoconstriction is occuring in arterioles in tissues of the digestive tract and kidneys.

Adjustments in the distribution of blood flow are made in response to signals from the nervous system and the endocrine system, and to changes in local chemical conditions.

Hemostasis: Preventing Blood Loss

Breaks or cuts even in small blood vessels are potentially disastrous to the human body. In **hemostasis**, a constellation of mechanisms can stop bleeding. These mechanisms include blood vessel spasm, platelet plug formation, and blood coagulation.

First, smooth muscle in a damaged vessel wall contracts in an automatic response called a spasm. The blood

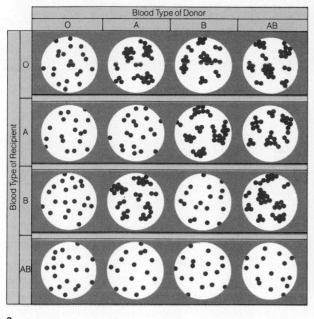

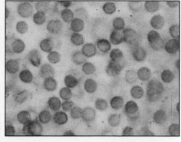

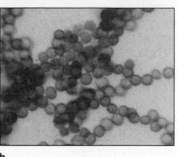

Figure 6.22 (**a**) Agglutination responses in blood types O, A, B, and AB when mixed with blood samples of the same and different types. (**b**) Micrographs showing the absence of agglutination in a mixture of two different but compatible types (above) and agglutination in a mixture of incompatible blood types (below).

vessel constricts, so blood flow through it temporarily stops. Next, platelets clump together, temporarily plugging the rupture. They also release the hormone serotonin and other chemicals that help prolong the spasm and attract more platelets. Finally, blood converts to a gel, or *coagulates*, through one of two responses.

In the *intrinsic clotting mechanism*, a plasma protein becomes activated and triggers reactions that lead to the formation of an enzyme (thrombin), which acts on a large, rod-shaped plasma protein (fibrinogen). The rods adhere to one another, forming long, insoluble threads (fibrin) that stick to one another. The result is a net in which blood cells and platelets become entangled (Figure 6.21). The entire mass is a blood clot. The clot retracts into a compact mass, drawing the walls of the vessel together.

Blood also can coagulate through an *extrinsic clotting mechanism*. "Extrinsic" means that the series of reactions leading to blood clotting is triggered by the release of enzymes and other substances *outside* of the blood itself (that is, from damaged blood vessels or from the surrounding tissues). The substances lead to thrombin formation, and the remaining steps parallel those shown in Figure 6.21.

Blood Typing

Genetically determined surface proteins mark each of your body cells as "self." The cells, including red blood cells, belong to you and no one else. Immune system proteins called *antibodies* recognize and organize an attack on foreign cells (including disease-causing bacteria) because such cells have "nonself" proteins. When the blood of two

people mixes, as during a transfusion, antibodies act against blood cells bearing "nonself" markers. The same thing happens during pregnancy, if antibodies chance to diffuse from the mother's circulatory system to that of her unborn child.

ABO Blood Typing One kind of self marker on human red blood cells has variant forms known as A, B, and O. This group is of particular interest because severe immune responses result when incompatible types are mixed (Figure 6.22). In type A blood, red blood cells bear A markers. Type B blood has B markers, type AB has both A and B, and type O has neither. If you are type A, you do not carry antibodies against A markers, but you do have antibodies against B markers. Hence, if you receive a transfusion of type B blood, an **agglutination** response will occur. In this response, antibodies act against the "nonself" red blood cells and cause them to clump. The clumps can clog small blood vessels, causing severe tissue damage and even death. If you are type B, you cannot safely receive type A blood. If you are type AB, your body will tolerate donations of type A, B, or AB blood. If you are type O, you have antibodies against A *and* B markers. The only "safe" blood for a type O person is type O.

Rh Blood Typing Other markers on red blood cells also can trigger agglutination responses. For example, *Rh blood typing* is based on the presence or absence of an Rh antigen (so named because it was first identified in the blood of *rh*esus monkeys). Rh^+ (positive) individuals have blood cells with this antigen; Rh^- (negative) individuals do not. Ordinarily, people do not have antibodies that act against Rh antigen. However, if an Rh⁻ person has been given a transfusion of Rh⁺ blood, antibodies will be produced against it and will continue circulating in the bloodstream.

If an Rh⁻ woman becomes pregnant by an Rh⁺ man, there is a chance the fetus will be Rh⁺. During pregnancy or childbirth, some red blood cells of the fetus may leak into the mother's bloodstream. If they do, they will stimulate her body to produce antibodies against the Rh markers (Figure 6.23). If the woman becomes pregnant *again*, Rh antibodies will enter the fetal bloodstream. If this second fetus happens to have Rh⁺ blood, the antibodies will cause the fetus's red blood cells to swell and rupture.

In extreme cases of this disorder, called *hemolytic disease of the newborn*, too many cells are destroyed and the fetus dies before birth. If it is born alive, the baby's blood can be replaced with blood free of Rh antibodies. Currently, a known Rh⁻ woman can be treated after her first pregnancy with a drug that inactivates Rh⁺ fetal blood cells circulating in the mother's bloodstream. This prevents her body from again producing anti-Rh⁺ antibodies if she becomes pregnant with another Rh⁺ fetus.

CARDIOVASCULAR DISORDERS

More than 40 million Americans have cardiovascular disorders, which claim about 750,000 lives every year. The most common cardiovascular disorders are *hypertension* (sustained high blood pressure) and *atherosclerosis* (a progressive narrowing of the arterial lumen). Atherosclerosis is the major cause of most *heart attacks*—damage or death of heart muscle due to an interruption of its blood supply. It also can cause *stroke*, damage to the brain due to an interruption of blood circulation to it.

Most heart attacks bring a "crushing" pain behind the breastbone that lasts a half-hour or more. Frequently, the pain radiates into the left arm, shoulder, or neck. The pain can be mild but usually is excruciating. Often it is accompanied by sweating, nausea, vomiting, and dizziness or loss of consciousness.

Table 6.1 lists some other diseases and disorders that can affect cardiovascular functioning.

Hypertension

In hypertension, blood pressure is sustained at elevated levels even when the person is at rest. Heredity may be a factor here; the disorder tends to run in families. Diet also is a factor. For example, high salt intake can raise the blood pressure in people predisposed to the disorder. High blood pressure increases the workload of the heart,

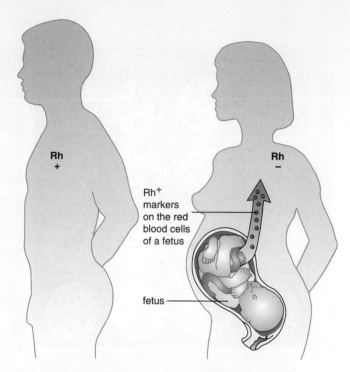

a A forthcoming child of an Rh⁻ woman and Rh⁺ man inherits the genetic instructions for the Rh⁺ marker. The placenta, a complex tissue that forms during pregnancy, allows the mother's blood to intermingle with that of the growing fetus. So fetal blood cells bearing the Rh⁺ markers enter her bloodstream.

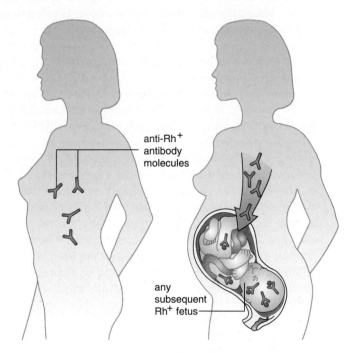

b The foreign markers in the mother's body stimulate production of antibody molecules. Suppose the woman becomes pregnant agian. If the new fetus (or any other one) inherits instructions for the Rh⁺ marker, the anti-Rh⁺ antibodies will act against them.

Figure 6.23 Development of antibodies in response to Rh⁺ blood.

Table 6.1 The Cardiovascular System Under Attack

Cause/Risk Factors	Major Effects	Symptoms
Heart/Vascular System		
Coronary heart disease Blockage of coronary arteries, usually by fatty deposits and/or blood clots. Risk factors include heredity, hypertension, smoking, obesity, lack of exercise, and elevated blood cholesterol.	Reduced supply of blood and hence oxygen to the heart muscle.	Heart attack; pain in the chest, arm, or neck (angina pectoris).
Angina pectoris Pain caused by reduced blood supply to the heart due to atherosclerosis, anemia, arrhythmia, or narrowing of the aortic valve.	Lack of oxygen for heart muscle cells.	Pain in the chest, arms, or jaw.
Myocardial infarction (heart attack) Unhealthy diet, smoking, hypertension, stress, diabetes mellitus, age, heredity.	Sudden death of part of the heart muscle; may lead to heart failure (reduced pumping efficiency).	Chest pain, shortness of breath, clammy skin, nausea, fainting.
Myocarditis and endocarditis Usually triggered by viral or bacterial infection.	Inflammation of the heart muscle and heart valves, respectively.	Irregular heart beat, breathlessness, chest pain, heart failure.
Circulatory shock Reduced blood volume and/or blood pressure arising from heart attack, blood or fluid loss due to injury, illness, or burns, or poisoning or spinal injury.	Severe reduction of blood flow throughout body tissues.	Rapid, shallow breathing and rapid, weak pulse; dizziness; clammy skin, fainting.
Blood		
Anemia Iron deficiency, failure of bone marrow to produce stem cells, destruction of red blood cells, vitamin deficiencies that result in RBCs with reduced capacity to transport oxygen.	Reduced concentration of hemoglobin in the blood, hence reduced oxygen transport.	Tiredness, lethargy, headache, dizziness, pale skin; in severe cases, angina pectoris.
Leukemias Cancers in which stem cells in bone marrow overproduce white blood cells. May be chronic or acute.	Brain, liver, and other organs begin to fail due to invasion of cancerous white blood cells. Bone marrow becomes choked with the abnormal cells, with resulting failure to produce normal blood cells of all types.	Enlarged lymph nodes, liver, and spleen; headache, bruising, fever, night sweats, repeated infections, anemia.
Infectious mononucleosis Epstein-Barr virus or cytomegalovirus.	Viral infection of lymphocytes; most common in teenagers and young adults.	Fever, headache, severe sore throat; swollen lymph glands in armpits, neck, and groin.

which can become enlarged in a way that hampers blood pumping. High blood pressure also can cause arterial walls to "harden" and so influence the delivery of oxygen to the brain, heart, kidneys, and other vital organs. It is a significant cause of *stroke*, a ruptured blood vessel in the brain.

Hypertension has been called the silent killer because affected people may show no outward symptoms; they

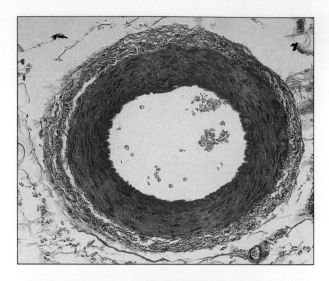

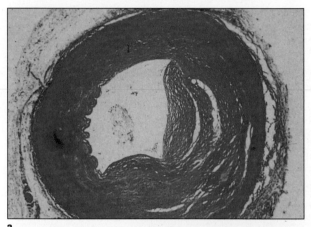

a

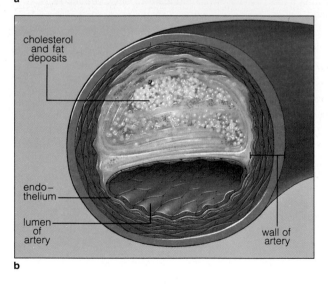

b

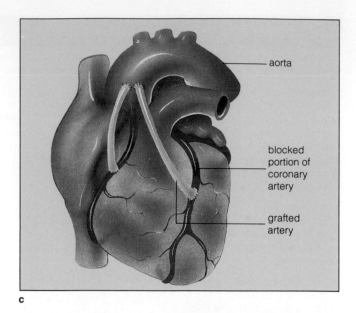

c

Figure 6.24 (**a**) Cross-section of a normal artery (above) and a partially obstructed one (below). (**b**) Diagram of an artherosclerotic plaque. (**c**) Two coronary bypasses (green).

Atherosclerosis

Arteriosclerosis refers to a condition in which arteries thicken and lose their elasticity. In *atherosclerosis*, conditions worsen because lipid deposits also build up in the arterial walls and shrink the diameter of the arterial lumen.

Recall that lipids such as fats and cholesterol are insoluble in water (page 29). Lipids absorbed from the digestive tract are picked up by lymph vessels that empty into the bloodstream. In the lymph vessels the lipids become bound to protein carriers that keep them suspended in the blood plasma. In atherosclerosis, abnormal smooth muscle cells have multiplied, and connective tissue components have increased in arterial walls. Lipids have been deposited within cells and extracellular spaces of the wall's endothelial lining. Calcium salts have been deposited in and around the lipids, and a fibrous net has formed over the whole mass. This *atherosclerotic plaque* sticks out into the lumen of the artery (Figure 6.24a and b).

Plaque formation is related to cholesterol intake (see *Focus on Wellness*), but other factors also contribute. When cholesterol is transported through the bloodstream, it is bound to protein carrier molecules called lipoproteins. These include *high-density lipoproteins* (HDL), *low-density lipoproteins* (LDL), and *very low density lipoproteins* (VLDL). High levels of LDL are related to a tendency toward heart trouble. LDLs, with their cholesterol cargo, tend to infiltrate arterial walls. In contrast, HDLs seem to attract cho-

often believe (mistakenly) that they are in the best of health. Of 23 million Americans who are hypertensive, most are not undergoing treatment. About 180,000 will die each year.

Maintaining Cardiovascular Health

Cardiovascular disorders are the leading cause of death in the United States. Many factors associated with those disorders have been identified. They include cholesterol deposits in arteries, and a range of other factors.

Cholesterol   Normally, the liver produces enough cholesterol to satisfy the body's needs. Together with the liver's output, cholesterol from the diet also ends up circulating in the blood. If you habitually eat cholesterol-rich food, you may end up with a high blood level of cholesterol. If you have a genetic disorder called *familial hypercholesterolemia*, the same thing might happen no matter what kinds of food you eat.

When circulating in blood, cholesterol is bound to *low-density lipoproteins* (LDLs), compounds that the liver manufactures. LDLs can bind to receptors on cells throughout the body. Cells take up LDLs and their cholesterol cargo for use in cell activities.

In some people, not enough LDL is removed from the blood, possibly because cells may not have enough LDL receptors. As the blood level of LDL increases, so does the risk of atherosclerosis. LDLs—with their bound cholesterol—can enter arterial walls. Abnormal cells and cell products multiply at the entry sites. Then calcium salts and a fibrous net form over the mass, forming an atherosclerotic plaque or "hardening of the arteries" (Figure *a*). Blood clots may form at plaques and narrow or block the arteries. A heart attack may follow.

By contrast, HDLs attract cholesterol *out* of arterial walls and transport it to the liver, where it can be broken down. Atherosclerosis is rare in rats, which have mostly HDLs. It is common in humans, who in general have mostly LDLs.

People who want to reduce LDL levels are usually advised to restrict their intake of cholesterol and saturated fats. Some research suggests that HDL levels are higher in people who exercise regularly. They also appear to be higher in people who do not use tobacco in any form and who drink moderate amounts of alcohol.

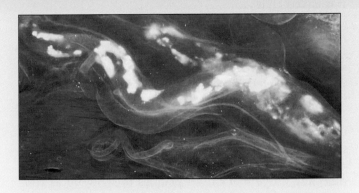

a   Plaques in blood vessels that service the heart. Such plaques form when fats and cholesterol become deposited in the blood vessel wall; then calcium salts and a fibrous net form over the abnormal mass. Blood clots may form here and narrow or block the arteries, leading to a heart attack.

Other "Healthy Heart" Issues   Beyond cholesterol, a variety of other factors and life-style choices directly affect the heart's health. The more body fat you carry, the more your body develops additional blood capillaries to service the increased number of cells and the harder the heart has to work to pump blood through the increasingly divided vascular circuit. Smokers should be aware that the nicotine in tobacco stimulates the adrenal glands to secrete epinephrine, which constricts blood vessels and so triggers an accelerated heartbeat and a rise in blood pressure. Carcinogens in cigarette smoke may also contribute to development of plaques. The carbon monoxide present in cigarette smoke has a greater affinity for binding sites on hemoglobin than does oxygen, and its action means that the heart has to pump harder to deliver oxygen to tissues. In short, smoking can destroy not only your lungs but also your heart.

Other risk factors for heart disease include hypertension (high blood pressure), lack of exercise, diabetes mellitus, a family history of heart disorders, and increasing age. Until age 50, males are at much greater risk than are females.

lesterol out of the walls and transport it to the liver, where it can be metabolized. In addition, it appears that unsaturated fats, including olive oil and fish oil, can reduce the level of LDLs in the blood. Smoking and a diet high in saturated fats appear to aggravate plaque formation.

Researchers are also probing the possible roles of certain bacteria and viruses in some cases of atherosclerosis.

Sometimes platelets become caught on the rough edges of a plaque and are stimulated into secreting some of their chemicals. When they do, a clot forms. As the clot and plaque grow, the artery narrows. Blood flow to the tissue that the artery supplies diminishes or may be blocked entirely. A clot that stays in place is called a *thrombus*. If it becomes dislodged and travels the bloodstream, it is called an *embolus*.

With their narrow diameters, the coronary arteries and their branches are highly susceptible to clogging through plaque formation or occlusion by a clot. When such an artery becomes narrowed to one-quarter of its former diameter, the resulting symptoms can range from mild chest pain (*angina pectoris*) to a full-scale heart attack.

Atherosclerosis can be diagnosed on the basis of several procedures. These include stress electrocardiograms (recording the electrical activity of the cardiac cycle while a person is exercising on a treadmill) and *angiography* (injecting a dye that causes plaques to show up as contrasting masses on X rays; see page 161). Treatments of serious blockages include *coronary bypass surgery*. There are several variations on this operation. Often, a section of a vein taken from the arm or leg is stitched to the aorta and to the coronary artery below the affected region (Figure 6.24c). In *laser angioplasty*, laser beams are used to vaporize the atherosclerotic plaques. In *balloon angioplasty*, a small balloon is inflated within a blocked artery to flatten a plaque and so increase the arterial diameter. Such procedures only buy time; they do not cure the underlying cardiovascular problem.

Up to about age 55, women generally are at less risk of heart disease, probably because the female hormone estrogen helps protect against the formation of atherosclerotic plaques. However, when a woman reaches menopause (usually in her early 50s) her estrogen level plummets and her risk of heart disease rises to equal that of a man.

Arrhythmias

An **arrhythmia** is an irregular or abnormal heart rhythm. Arrhythmias can be detected by an ECG (Figure 6.25), and some are normal. For example, the resting cardiac rate of many athletes who are trained for endurance is lower than average, a condition called *bradycardia*. Inhibition of their cardiac pacemaker by parasympathetic signals has increased as an adaptive response to ongoing strenuous exercise. There is more time for the ventricles to fill, so each contraction pumps blood more efficiently.

A cardiac rate above 100 beats per minute, *tachycardia*, occurs normally during exercise or stressful situations.

Serious tachycardia can be triggered by drugs (including caffeine, nicotine, alcohol, and cocaine), hyperthyroidism, and other factors. Coronary artery disease also can cause arrhythmias. *Fibrillation* is an extreme medical emergency in which contractions of the heart chambers,

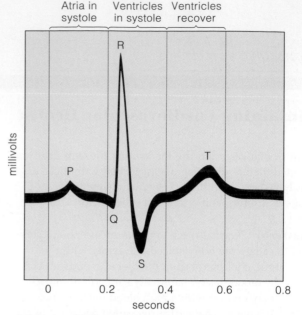

a ECG of a single, normal heartbeat

Bradycardia (here, 46 beats per minute):

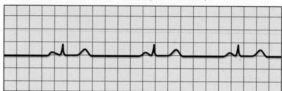

Tachycardia (here, 136 beats per minute):

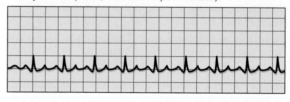

Ventricular fibrillation:

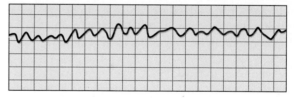

b Examples of ECG readings

Figure 6.25 Examples of ECG readings. (**a**) A single, normal heartbeat. The P wave is generated by electrical signals from the SA node that stimulate contraction of the atria. As the stimulus moves over the cardiac muscle of the ventricles by way of Purkinje fibers to the ventricles, it is recorded as the QRS wave complex. After the ventricles contract they go through a brief period of recovery. Electrical activity during this period is marked by the T wave. (There is also an atrial recovery period "hidden" in the QRS complex.) (**b**) ECG readings for bradycardia, tachycardia, and ventricular fibrillation.

especially the ventricles, become so uncoordinated that blood pumping may stop. A heart attack, drug overdose, or other injuries to the heart can trigger fibrillation. If the patient is lucky, electrical shocks to the heart or defibrillating drugs may restore a normal rhythm in time to avoid brain damage or death.

LYMPHATIC SYSTEM

We conclude this chapter with a brief section on the **lymphatic system**, which supplements the circulatory system by returning excess tissue fluid to the bloodstream. But think of this section as a bridge to the next chapter, on immunity, for the lymphatic system is also vital to the body's defenses against injury and attack. The system's components include transport vessels and lymphoid organs (Figure 6.26). Tissue fluid that has moved into the transport vessels of the lymphatic system is called **lymph**.

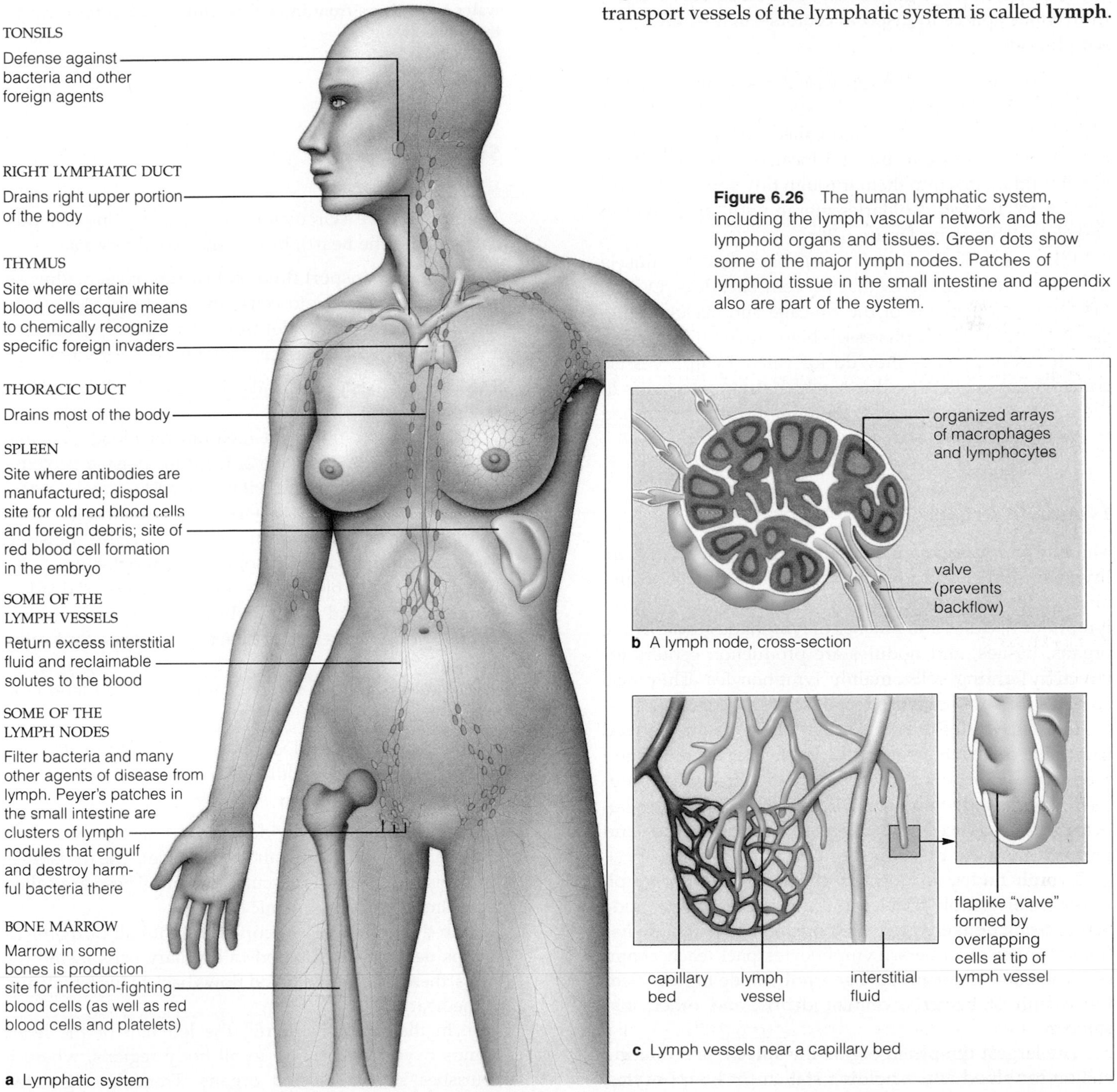

TONSILS

Defense against bacteria and other foreign agents

RIGHT LYMPHATIC DUCT

Drains right upper portion of the body

THYMUS

Site where certain white blood cells acquire means to chemically recognize specific foreign invaders

THORACIC DUCT

Drains most of the body

SPLEEN

Site where antibodies are manufactured; disposal site for old red blood cells and foreign debris; site of red blood cell formation in the embryo

SOME OF THE LYMPH VESSELS

Return excess interstitial fluid and reclaimable solutes to the blood

SOME OF THE LYMPH NODES

Filter bacteria and many other agents of disease from lymph. Peyer's patches in the small intestine are clusters of lymph nodules that engulf and destroy harmful bacteria there

BONE MARROW

Marrow in some bones is production site for infection-fighting blood cells (as well as red blood cells and platelets)

a Lymphatic system

Figure 6.26 The human lymphatic system, including the lymph vascular network and the lymphoid organs and tissues. Green dots show some of the major lymph nodes. Patches of lymphoid tissue in the small intestine and appendix also are part of the system.

organized arrays of macrophages and lymphocytes

valve (prevents backflow)

b A lymph node, cross-section

flaplike "valve" formed by overlapping cells at tip of lymph vessel

capillary bed

lymph vessel

interstitial fluid

c Lymph vessels near a capillary bed

Lymph Vascular System

The **lymph vascular system** includes lymph capillaries, lymph vessels, and ducts. Collectively, these transport vessels serve the following functions:

1. Return of excess filtered fluid to the blood

2. Return of small amounts of proteins that leave the capillaries

3. Transport of fats absorbed from the digestive tract

4. Transport of foreign particles and cellular debris from tissue spaces to disposal centers—that is, the lymph nodes

At one end of the lymph vascular system are *lymph capillaries*. They occur in the tissues of almost all organs and serve as "blind-end" fluid collection tubes. There is no obvious entrance at the end located in tissues; interstitial fluid moves into them through tiny gaps in the capillary wall. Lymph capillaries merge with larger lymph vessels, as shown by Figure 6.26c.

Like veins, larger *lymph vessels* have smooth muscle in their walls and valves that prevent backflow. During breathing, movements of the rib cage and skeletal muscle adjacent to the lymph vessels help move fluid through lymph vessels, just as they do for veins. Lymph vessels converge into collecting ducts, which drain into veins in the lower neck. In this way, the lymph fluid is returned to the blood circulation.

Lymphoid Organs

The **lymphoid organs** include the lymph nodes, spleen, thymus, tonsils, and appendix. The small intestine and airways leading to the lungs (the bronchi) also contain lymphatic nodules and patches of lymphatic tissue. These organs, tissues, and nodules are production centers for infection-fighting cells, mainly lymphocytes. They also are sites for some defense responses (Chapter 7).

Like all white blood cells, lymphocytes are derived from stem cells in bone marrow. They enter the blood and take up residence in lymphoid organs. Properly stimulated, they divide by mitosis. In fact, most new lymphocytes are produced by divisions in the blood and lymphoid organs, not in bone marrow.

Lymph nodes are located at intervals along lymph vessels. All lymph trickles through at least one node before being delivered to the bloodstream. Each node has several inner chambers. Lymphocytes pack each chamber. There are also macrophages in the node to help clear the lymph of bacteria, cellular debris, and other substances.

The largest lymphoid organ, the **spleen**, is a filtering station for blood and a holding station for lymphocytes.

The spleen's inner chambers contain large stores of red blood cells and macrophages. Red blood cells are produced here in developing human embryos.

In the **thymus**, lymphocytes multiply, differentiate, and mature into fighters of specific types of disease agents. The thymus is central to immunity, the focus of Chapter 7.

The lymphatic system works to prevent the spread of disease agents through the circulatory system. It also reclaims water and solutes from interstitial fluid and returns them to the bloodstream.

SUMMARY

1. Humans have a circulatory system consisting of a muscular pump (the heart), blood, and blood vessels.

2. Blood is a transport fluid that carries oxygen, glucose, and other substances to cells, and products and wastes (including carbon dioxide) from them. It helps maintain an internal environment favorable for cell activities.

3. Blood consists of red and white blood cells, platelets, and plasma.

   a. Plasma is a transport medium for blood cells and platelets. Plasma water is a solvent for plasma proteins, ions, nutrients, hormones, vitamins, and several gases.

   b. Red blood cells transport oxygen (bound to hemoglobin) between the lungs and cells. They also transport some carbon dioxide, also bound to hemoglobin.

   c. Some white blood cells are scavengers of dead or worn-out cells and other debris. Others serve in the defense of the body against bacteria, viruses, and other foreign agents.

   d. Platelets are cell fragments involved in blood clotting.

4. The heart pumps blood into arteries. From there it flows into arterioles, capillaries, venules, veins, and back to the heart.

5. The human heart is divided into two halves, each with two chambers (an atrium and a ventricle). The division is the basis of two cardiovascular circuits. These are called the pulmonary and systemic circuits.

   a. In the pulmonary circuit, the right half of the heart pumps deoxygenated blood to capillary beds inside the lungs, then oxygenated blood flows back to the left atrium of the heart.

   b. In the systemic circuit, the left half of the heart pumps oxygenated blood to all body regions, where it nourishes all tissues and organs. Then deoxygenated blood flows from those regions back to the heart's right atrium.

6. Heart contractions (specifically, the contracting ventricles) are the driving force for blood circulation. Fluid pressure is highest at the start of a circuit, then progressively drops in arteries, arterioles, capillaries, then veins. It is lowest in the relaxed atria.

7. Blood moves through several types of blood vessels in both the pulmonary and systemic circuits.

   a. Arteries are elastic conduits that smooth out pulsations in blood pressure caused by heart contractions.

   b. Arterioles are pressure regulators and control points for the variable distribution of blood to different body regions.

   c. Beds of capillaries are diffusion zones between the blood, interstitial fluid, and cells.

   d. Veins are blood-volume reservoirs and help adjust volume flow back to the heart.

8. The lymphatic system supplements the circulatory system by returning excess fluid that seeps out of blood capillaries back to the circulation. Some of its components have major roles in immune responses.

1. What are the functions of blood? *140*

2. Describe the cellular components of blood. Describe the plasma portion of blood. *111–112*

3. Define the functions of the following: heart, cardiovascular system, and lymphatic system. *145–147; 163*

4. Distinguish between the following:
   a. systemic and pulmonary circuits *148*
   b. lymph vascular system and lymphoid organs *163*

5. Label the component parts of the human heart:

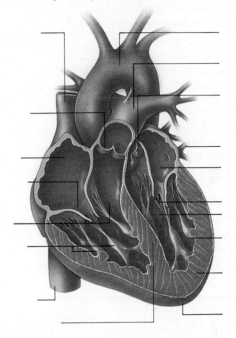

6. Explain how the medulla oblongata of the brain helps regulate blood flow to different body regions. *155*

7. State the main function of blood capillaries. What drives water and solutes out of and into capillaries in capillary beds? *153–154*

8. State the main function of venules and veins. What forces work together in returning venous blood to the heart? *154*

Critical Thinking: You Decide *(Key in Appendix IV)*

1. A patient suffering from hypertension may receive drugs that decrease the heart's output, dilate arterioles, or increase urine production. In each case, how would the drug treatment help relieve hypertension?

2. Aplastic anemia develops when certain drugs or radiation destroy red bone marrow, including the stem cells that give rise to red and white blood cells and platelets. Predict some symptoms a person with aplastic anemia would be likely to develop. Include at least one symptom related to each type of formed element in blood.

1. A _____ system functions in the rapid exchange of substances to and from all living cells, and usually it is supplemented by a _____ system.

2. _____ and _____ are large-diameter blood vessels for fluid transport; _____ and _____ are fine-diameter, thin-walled blood vessels for diffusion; and _____ serve as control points over the variable distribution of blood to different body regions.

3. Which of the following are *not* components of blood?
   a. red and white blood cells
   b. platelets and plasma
   c. assorted solutes and dissolved gases
   d. all of the above are components of blood

4. Red blood cells are produced in the _____ and function in transporting _____ and some _____.
   a. liver; oxygen; mineral ions
   b. liver; oxygen; carbon dioxide
   c. bone marrow; oxygen; hormones
   d. bone marrow; oxygen; carbon dioxide

5. White blood cells (other than lymphocytes) are produced in the _____ and function in both _____ and _____.
   a. liver; oxygen transport; defense
   b. lymph glands; oxygen transport; pH stabilization
   c. bone marrow; day-to-day housekeeping; defense
   d. bone marrow; pH stabilization; defense

6. In the pulmonary circuit, the _____ half of the heart pumps _____ blood to capillary beds inside the lungs, then _____ blood flows back to the heart.
   a. left; deoxygenated; oxygenated
   b. right; deoxygenated; oxygenated
   c. left; oxygenated; deoxygenated
   d. right; oxygenated; deoxygenated

7. In the systemic circuit, the _____ half of the heart pumps _____ blood to all body regions, then _____ blood flows back to the heart.
   a. left; deoxygenated; oxygenated
   b. right; deoxygenated; oxygenated
   c. left; oxygenated; deoxygenated
   d. right; oxygenated; deoxygenated

8. Fluid pressure in the circulatory system is _____ at the beginning of a circuit, then _____ in arteries, arterioles, capillaries, and then veins. It is _____ in the relaxed atria.
   a. low; rises; highest   c. low; drops; lowest
   b. high; drops; lowest   d. high; rises; highest

9. Match the type of blood vessel with its major function.
   ____ arteries   a. diffusion
   ____ arterioles   b. control of blood distribution
   ____ capillaries   c. transport, blood volume reservoirs
   ____ veins   d. blood transport and pressure regulators

10. Match the following circulation components with their descriptions.
   ____ capillary beds   a. two atria, two ventricles
   ____ lymph vascular system   b. driving force for blood
   ____ heart chambers   c. zones of diffusion
   ____ heart contractions   d. return of interstitial fluid to blood

Key Terms

agglutination *157*
agranulocyte *145*
aorta *148*
arrythmia *162*
arteriole *153*
artery *153*
atrioventricular node *151*
atrioventricular valve *147*
atrium *147*
blood *141*
blood pressure *151*
capillary *153*
capillary bed *149*
cardiac conduction system *150*
cardiac cycle *150*
circulatory system *140*
coronary artery *148*
diastole *149*
granulocyte *145*
heart *147*
hemostasis *156*
lymph *163*
lymphatic system *163*

lymph node *164*
lymphoid organ *164*
lymph vascular system *164*
myocardium *147*
platelet *145*
pulmonary circuit *148*
pulse *153*
red blood cell (erythrocyte) *142*
semilunar valve *148*
septum *147*
sinoatrial node *151*
spleen *164*
stem cell *143*
systemic circuit *148*
systole *149*
thymus *164*
vasoconstriction *156*
vasodilation *155*
vein *154*
ventricle *147*
venule *154*
white blood cell (leukocyte) *145*

Readings

"Heart Disease: Women at Risk." May 1993. *Consumer Reports*.

Golde, D. December 1991. "The Stem Cell." *Scientific American*. Description of the "master cell" that gives rise to red blood cells, white cells, and other cellular components of human blood.

Rundle, R. 1 September 1993. "Fresh Ideas on Frozen Blood Spur Firm." *Wall Street Journal*.

Vogel, S. 1992. *Vital Circuits: On Pumps, Pipes, and the Workings of the Circulatory System*. New York: Oxford University Press.

7 IMMUNITY

Russian Roulette, Immunological Style

At one time, smallpox swept repeatedly through the world's cities. Some outbreaks were so severe that only half of those stricken survived. No one emerged unscathed. Even survivors ended up with permanent scars. But they did not contract the disease again—they were "immune" to smallpox.

The idea of acquired immunity to smallpox fascinated many people. In Asia, Africa, and then Europe, a few intrepid souls played immunological Russian roulette by allowing themselves to be inoculated with material from the sores of victims. Twelfth-century Chinese inhaled powdered crusts from smallpox sores. In the 1600s the wife of the British ambassador to Turkey injected bits of smallpox scabs into the veins of her children. Others soaked threads in the fluid from smallpox sores, then poked the threads into scratches on the body. Not everyone survived such practices, but those who did acquired immunity to smallpox.

For centuries it was known that humans could catch cowpox, a fairly mild disease, from cattle. Curiously, people who caught cowpox never got smallpox. In 1796, an English physician named Edward Jenner injected material from a cowpox sore into the arm of a boy (Figure 7.1). Six weeks later, after the cowpox reaction subsided, Jenner inoculated the boy with fluid from smallpox sores. He hypothesized that the earlier inoculation would provoke immunity to smallpox, and he was right. The French mocked the procedure, calling it "vaccination" (literally, "encowment"). Much later the French chemist Louis Pasteur devised vaccination procedures for other diseases, and the term became respectable.

By the early 20th century, the fight against infectious disease began in earnest, and it continues today. How the body wages its battle against invaders is the focus of this chapter.

Figure 7.1a Statue honoring Edward Jenner's development of an immunization procedure against smallpox, one of the most dreaded diseases in human history.

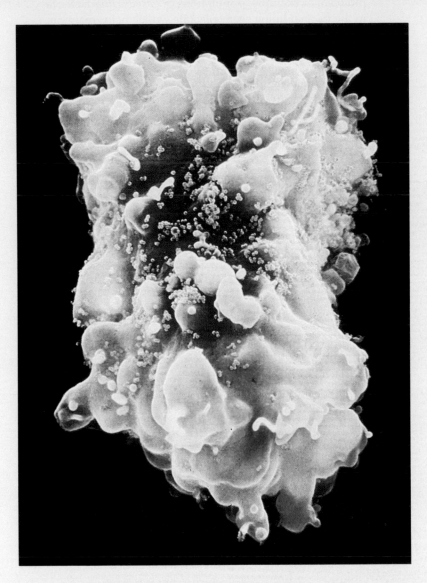

Figure 7.1b Micrograph of a white blood cell being attacked by the virus (blue particles) that causes AIDS. Immunologists are working to develop weapons against this modern-day scourge.

KEY CONCEPTS

1. The human body has physical, chemical, and cellular defenses against invasions by pathogens—viruses, bacteria, and other agents of disease.

2. During early stages of an invasion, white blood cells and plasma proteins take part in an inflammatory response. This counterattack is rapid and *general*; it does not target a specific pathogen.

3. If the invasion persists, certain white blood cells make *specific* immune responses. These cells can recognize molecules that are abnormal or foreign to the body, such as those on the surfaces of bacteria, viruses, cancer cells, and transplanted tissues. If the foreign or abnormal molecule triggers an immune response, it is called an antigen.

4. In one type of immune response, a subgroup of white blood cells produces huge amounts of antibodies. Antibodies are proteins that bind to a specific antigen and tag it for destruction. In another type of immune response, "killer" cells directly destroy body cells that have become abnormal via infection or some other process.

Throughout your life you are attacked by a huge assortment of pathogens. We use the term **pathogen** for viruses, bacteria, fungi, protozoa, and parasitic worms that cause disease. Table 7.1 lists the body's three lines of defense against pathogen attack. Next we consider each line of defense in turn.

FIRST LINE OF DEFENSE: SURFACE BARRIERS TO INVASION

The first line of defense involves physical barriers and chemical defenses at the body's surface. Usually pathogens cannot get past skin or the mucous membranes that line the digestive tract, the respiratory tract, and the urinary and reproductive tracts. Skin, for example, is relatively dry, has a low pH, and has thick layers of dead cells. Many kinds of harmless bacteria are adapted to life under these conditions, but few pathogens can survive in such an environment. Problems arise when conditions are altered. Fungi that cause *athlete's foot* take hold when skin between the toes stays moist in the warm, dark environment of shoes. HIV, the virus that causes

AIDS, can enter the body through cuts or abrasions in mucous membranes (among other routes).

Bacteria that normally inhabit mucous membranes help keep pathogens in check. They do so by outcompeting other microbes for nutrients and/or by secreting metabolic by-products that make the environment inhospitable. For example, populations of *Lactobacillus* bacteria in the vaginal lining secrete lactate, which helps maintain a low vaginal pH that most bacteria and fungi cannot tolerate. When a woman takes an antibiotic to cure a bacterial infection, she may develop a vaginal yeast infection because the drug also kills *Lactobacillus*.

In respiratory airways, mucus contains **lysozyme**, an enzyme that attacks and destroys many bacteria. Protec-

---

Table 7.1   The Human Body's Three Lines of Defense Against Pathogens

Barriers at Body Surfaces (*nonspecific* targets):

1. Intact skin; mucous membranes at other body surfaces

2. Infection-fighting substances in tears, saliva, and so on

3. Normally harmless bacterial inhabitants of body surfaces that outcompete pathogenic visitors

4. Flushing effect of urination and diarrhea; sneezing, coughing, tears

Nonspecific Responses (*nonspecific* targets):

1. Inflammation
   a. Fast-acting white blood cells (neutrophils, eosinophils)
   b. Macrophages (also take part in immune responses)
   c. Complement proteins, blood-clotting proteins, other infection-fighting substances

2. Phagocytic functions of macrophages in lymph nodes, spleen, other organs

Immune Responses (*specific* targets):

1. White blood cells (macrophages, T cells, B cells)

2. Communication signals (e.g., interleukins) and chemical weapons (e.g., antibodies, complement proteins)

Table 7.2   Major White Blood Cells and Their Roles in Defense

Cell Type	Main Characteristics
Macrophage	Phagocyte; takes part in specific and nonspecific defense responses; presents antigen to helper T cells; cleans up damaged tissue
Neutrophil	Fast-acting phagocyte; takes part in inflammation but not sustained responses
Eosinophil	Secretes enzymes that attack parasitic worms; helps reduce inflammation of allergic responses
Basophil and mast cell	Secrete histamine, prostaglandins, other substances that act on small blood vessels during inflammation; also roles in allergies

Lymphocytes (all take part in most immune responses):

1. B cell	Only cell that produces antibodies and positions them at its surface. When sensitized by exposure to antigen, a B cell produces clonal populations of antibody-secreting effector cells and memory cells
2. Helper T cell	Recognizes antigen-presenting cells; secretes various interleukins that stimulate rapid division of B cells and T cells
3. Cytotoxic T cell	Kills infected, abnormal, or foreign cells with a single lethal hit; forms clonal populations
4. Natural killer (NK) cell	Kills infected, abnormal, or foreign cells with a single lethal hit
5. Memory cell	Reserve B or T cell; part of a clonal population that continues circulating and is activated rapidly during a secondary immune response

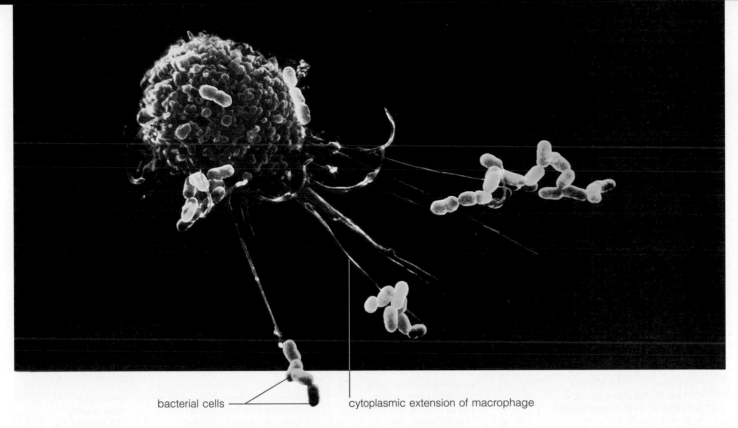

bacterial cells ———   cytoplasmic extension of macrophage

Figure 7.2 Scanning electron micrograph of a macrophage that is probing its surroundings with cytoplasmic extensions. The macrophage engulfs bacterial cells that come in contact with it.

tion also comes from lysozyme and other enzymes in tears, saliva, and gastric and intestinal fluid. Urine, with its low pH and flushing action, helps keep most pathogens from surviving in the urinary tract.

Skin and the mucous membranes lining body surfaces are the first line of defense against pathogens.

Bacteria that inhabit skin and mucous membranes help check invasion by pathogens. Lysozyme and other enzymes in body fluids also offer protection.

SECOND LINE OF DEFENSE: NONSPECIFIC RESPONSES

Certain types of white blood cells (leukocytes) and plasma proteins serve as a second line of defense if a pathogen breaches the surface barriers. These defenders mount a *nonspecific response*—they are adapted to repel any invasion, regardless of the particular pathogen involved. Nonspecific responses are usually triggered by tissue damage. By contrast, a *specific* response is triggered by a molecular configuration that occurs on only *one kind* of pathogen. Specific responses, which we consider later, occur whether tissues are damaged or not.

Phagocytes and Related White Blood Cells

Recall from Chapter 6 that all white blood cells arise from stem cells in bone marrow. Many white blood cells circulate in blood, then enter damaged or invaded tissues by squeezing between endothelial cells that make up the walls of capillaries. Many others move to lymph nodes and the spleen. (You may wish to review Figure 6.26 on page 163, which shows the lymphatic system.) Still others reside in the connective tissue beneath skin and mucous membranes, as well as in the liver, kidneys, lungs, and joints.

Three kinds of white blood cells live only for a few hours or days and act swiftly against danger in general. *Neutrophils* are phagocytes ("eating cells") that ingest and digest bacteria. *Eosinophils* secrete enzymes that attack parasitic worms; they also phagocytize foreign proteins and help control allergic responses (page 184). *Basophils* are not phagocytes, but instead secrete histamine. Histamine has a key role in inflammation (described shortly) and serves as a chemical trail that attracts other white blood cells.

Although slower to act, **macrophages** (Figure 7.2) can live in the body for months. A macrophage ("big eater") is something like a cellular vacuum cleaner—it engulfs and digests nearly any foreign agent and bits of damaged tissue. Immature macrophages circulating in blood are called *monocytes*. Table 7.2 summarizes white blood cells and their roles in defense and immunity.

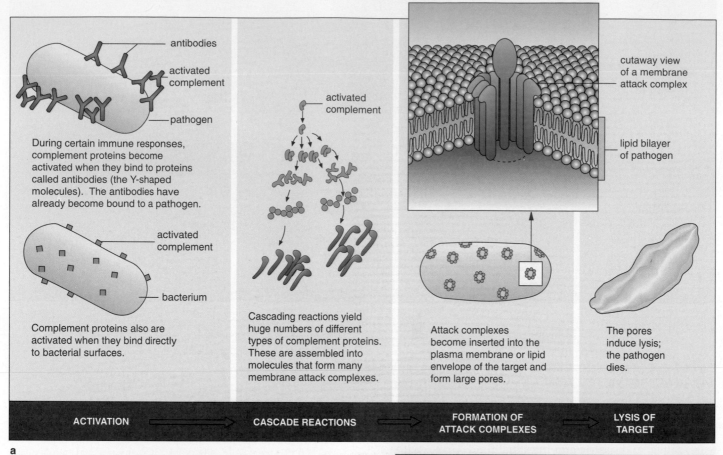

During certain immune responses, complement proteins become activated when they bind to proteins called antibodies (the Y-shaped molecules). The antibodies have already become bound to a pathogen.

Complement proteins also are activated when they bind directly to bacterial surfaces.

Cascading reactions yield huge numbers of different types of complement proteins. These are assembled into molecules that form many membrane attack complexes.

Attack complexes become inserted into the plasma membrane or lipid envelope of the target and form large pores.

The pores induce lysis; the pathogen dies.

ACTIVATION ⟹ CASCADE REACTIONS ⟹ FORMATION OF ATTACK COMPLEXES ⟹ LYSIS OF TARGET

a

Figure 7.3 How complement proteins form membrane attack complexes. (**a**) One reaction pathway begins when complement binds to certain bacterial surfaces. Another operates during immune responses to specific invaders, as described later in the chapter. Both pathways produce membrane pore complexes that induce lysis (cell bursting) in the target pathogen. Complement activity protects against many bacteria and some parasitic protists, as well as viruses with lipid envelopes, which are derived from the plasma membrane of a previously infected host cell. (**b**) What membrane attack complexes look like.

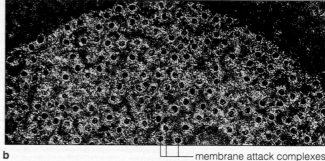

b — membrane attack complexes

Complement Proteins

A set of plasma proteins called the **complement system**, or simply "complement," have roles in both nonspecific and specific defenses. About 20 different complement proteins circulate in the blood in inactive form. Complement can be activated in one of two ways. A complement protein called C1 can bind to a "complex" that consists of a defender protein (an *antibody*) that is already bound to all or part of an invader (an *antigen*). Alternatively, a different complement protein can interact with carbohydrate molecules that are present on the surfaces of some microorganisms. If even a few molecules of one of these complement proteins are activated, a huge "cascade" of reactions begins. The activated molecules in turn activate many molecules of another complement protein, each of which activates many molecules of yet another protein at the next reaction step, and so on. Thus great numbers of molecules are deployed, with the following effects.

Some complement proteins join together to form pore complexes. As Figure 7.3 shows, these are structures with an interior channel. The pore complexes become inserted into the plasma membrane of many pathogens and induce lysis. **Lysis**, recall, is "cell bursting." It follows gross structural disruption of the plasma membrane, and it leads to cell death. Pore complexes also become inserted

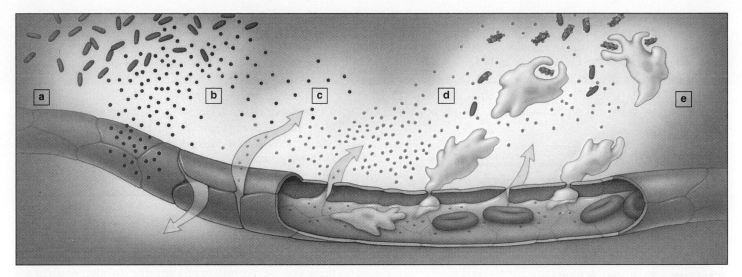

a Bacteria invade a tissue. They kill cells or release harmful metabolic by-products.

b The substances released by bacteria and by damaged or killed body cells accumulate in the tissue.

c The substances make the tissue's small blood vessels more permeable. Plasma fluid and various plasma proteins escape into the tissue.

d Some plasma proteins attack bacteria. Others create chemical gradients that facilitate migration of phagocytes to the tissue. Still others repair tissue damage (as by clotting mechanisms).

e Phagocytic white blood cells engulf bacteria.

Figure 7.4 Acute inflammation. In this example, a bacterial invasion induces chemical changes in a local tissue, and those changes trigger increased blood flow to the area. Small blood vessels become more permeable. Plasma fluid, certain plasma proteins, and phagocytic white blood cells leave the blood and enter the tissue. The invaders are killed and the tissue can then be repaired.

into the lipid coat surrounding some bacteria. Lysozyme molecules diffuse through the pores and, when they reach the bacterial cell wall, destroy it.

Some activated complement proteins promote inflammation. With their huge cascades, these proteins create concentration gradients that attract phagocytes to irritated or damaged tissue. In addition, the surface of many invaders has binding sites for a complement protein, and so the invader ends up with a complement "coat." Phagocytes bind to the coat and proceed to destroy the cell "wearing" it.

Inflammation

By a mechanism called **acute inflammation**, neutrophils, complement proteins, and other plasma proteins can escape the bloodstream and enter a damaged tissue (Figure 7.4). Localized redness, swelling (*edema*), heat, and pain are the classic signs of acute inflammation. They result from changes in capillaries and other small blood vessels in the affected area.

An inflammatory response develops when a powerful blow or burn, an infection, or some other trauma damages or kills a tissue's cells. *Mast cells*, a type of tissue-dwelling basophil, release histamine and other substances that promote dilation of small blood vessels. The

vessels become engorged with blood, so the tissue reddens and becomes warmer. (Blood, recall, carries metabolic heat.) Dilation also causes cells that make up the blood vessel wall to pull apart slightly and become "leaky" to plasma. Plasma, with its infection-fighting weapons (including complement proteins), enters the tissue. Swelling and pain follow. Voluntary movements that aggravate the pain tend to be avoided—and this promotes tissue repair.

Within a few hours, neutrophils squeeze out through gaps in the blood vessel walls and swiftly go to work. Monocytes arrive later, differentiate into macrophages, and engage in more sustained action.

While macrophages are engulfing pathogens, they secrete **interleukins**. Interleukins are proteins that regulate interactions among white blood cells and certain other cells. One type, interleukin-1, also signals the hypothalamus, a brain region that controls body temperature (among other functions), and induces it to secrete a prostaglandin, a signaling chemical that can reset the body's thermostat and raise the "set point." A **fever** is a body temperature that has climbed to a higher set point. Aspirin, ibuprofen, and acetaminophen reduce fever by inhibiting production of prostaglandins. A fever of about 39°C (100°F) is actually a helpful mechanism that promotes an increase in body defense activities. It also raises body temperature to a level that is too hot for the functioning of most pathogens.

Interleukin-1 also induces drowsiness during a fever. Drowsiness reduces the body's demands for energy, so that more can be diverted to defense and repair tasks. Macrophages also engulf dead cells and other debris, and blood-clotting proteins repair damaged blood vessels (page 157).

In acute inflammation, phagocytes and plasma proteins (including complement) leave the bloodstream and then defend and repair an invaded or damaged tissue.

THIRD LINE OF DEFENSE: THE IMMUNE SYSTEM

Overview of Immune Responses

Sometimes physical barriers and inflammation are not enough to overwhelm an invader, and some aspect of body functioning begins to break down. In other words, you get sick. When that happens, armies of white blood cells called **B lymphocytes** and **T lymphocytes** join the battle. (B cells mature in the *bone* marrow, T cells in the *thymus*.) Lymphocytes are central to the body's third line of defense, the **immune system**. Although the word *system* conjures up an image of various organs with linked roles in a particular physiological function (such as respiration or circulation), the immune system consists of lymphocytes and other specialized cells in the blood and in other body tissues and organs, mainly those of the lymphatic system. This cell-based system has two defining features. One is immunological *specificity*: Lymphocytes can recognize and eliminate specific pathogens. The second feature is immunological *memory*: During a first-time confrontation with a pathogen, some lymphocytes are produced and set aside. These lymphocytes can mount a rapid attack if the same type of invader returns.

Self and Nonself   How do lymphocytes identify an invader? Recall that all cells have various proteins as part of the plasma membrane. These proteins include **MHC** (major histocompatibility complex) **markers**. Nearly all body cells display MHC markers, which arise during fetal development. Except for identical twins, no two people have exactly the same markers. MHC proteins thus serve as "self" markers. A person's white blood cells recognize these identity tags and will not attack cells that bear them. A pathogenic virus, bacterium, or other disease agent also bears molecular markers on its surface that give it a unique identity. B lymphocytes can distinguish between

such *nonself* (foreign) markers and self markers, and can mount an immune response against anything foreign. As we will see, T lymphocytes mount a response when they encounter a cell bearing a complex of self markers and a foreign marker.

When an invader's nonself marker is detected, B and T lymphocytes are stimulated to divide repeatedly, and huge populations form. Simultaneously, different subpopulations become specialized to respond to the invader in different ways. Some are composed of **effector cells**, which are fully differentiated cells that engage and destroy the enemy. Other subpopulations are composed of **memory cells**, and these enter a resting phase. Memory cells will "remember" the specific invader and undertake a larger, more rapid response if it ever returns.

Thus immunological specificity and memory involve three events: *recognition* of a specific invader, *repeated cell divisions* to form huge populations of lymphocytes, and *differentiation* of lymphocytes into subpopulations of specialized effector cells and memory cells.

Any nonself marker that triggers the formation of large numbers of lymphocytes is called an **antigen**. Most antigens are protein molecules on infectious agents or on tumor cells, and each has a unique three-dimensional shape. As you will see, lymphocytes bear receptor molecules that can bind to molecules having those shapes. This is how lymphocytes "recognize" their targets. **Antibodies** are the antigen-binding receptors synthesized by B cells. T cells have antigen-binding receptors of their own.

An antigen is any nonself marker that triggers an immune response. Lymphocytes recognize antigens by way of special receptors, such as antibodies.

Types of Defenders   During all immune responses, four kinds of white blood cells are called into action. Here are their names and functions:

1. **Macrophages**. Besides engulfing anything detected as foreign, these phagocytes can present antigens to lymphocytes for recognition.

2. **Helper T cells**. When activated, these T lymphocytes produce and secrete chemicals that promote formation of large effector cell and memory cell populations.

3. **Cytotoxic T cells**. These T lymphocytes eliminate infected body cells and tumor cells by a direct, lethal attack.

4. **B cells**. These lymphocytes produce antibodies, then either position them at the B cell surface or secrete them. Secreted antibodies circulate in the blood.

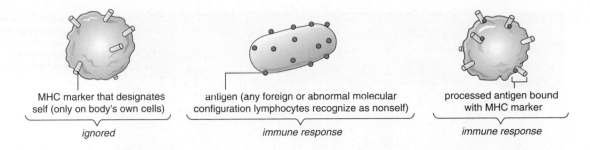

Figure 7.5 Molecular cues that stimulate lymphocytes to begin an immune response.

MHC marker that designates self (only on body's own cells)

ignored

antigen (any foreign or abnormal molecular configuration lymphocytes recognize as nonself)

immune response

processed antigen bound with MHC marker

immune response

Immune Functions of Macrophages

Like all your body cells, each of your macrophages and lymphocytes has a set of MHC markers incorporated into its plasma membrane. However, certain MHC proteins occur *only* on macrophages and lymphocytes. These special MHC markers enable cells of the immune system to recognize one another and also play a direct role in immune functions.

Suppose a wood splinter rips into one of your fingers and bacteria enter the punctured tissue. The bacteria multiply, inflammation follows, and phagocytes start engulfing and destroying them. Some of the phagocytes are macrophages, and digestive enzymes inside them destroy the bacterial cells—but not the bacterial antigen. Instead, the antigen molecules are cleaved into fragments, which bind to MHC proteins to form an **antigen-MHC complex**. The complexes are moved to the cell surface, where they are displayed. Any cell that does this—in most cases, a macrophage—is an **antigen-presenting cell**.

When a processed antigen is bound to an appropriate MHC molecule and displayed, lymphocytes respond to the combination (Figure 7.5). This is the antigen recognition that promotes cell divisions leading to huge numbers of lymphocytes.

Macrophages can function as antigen-presenting cells. They combine antigen fragments with special MHC molecules, then display the complex at their surface. Such complexes are the signal that starts an immune response.

Let's turn now to the functions of the immune system's two branches: the cell-mediated immune responses and antibody-mediated immune responses (Figure 7.6).

Cell-Mediated Immune Responses: Helper T Cells and Cytotoxic T Cells

A *cell-mediated immune response* is directed against cells that are infected by a virus or bacterium, or against abnormal cells such as those that have become cancerous. Other targets include foreign cells of tissue grafts or organ transplants. This type of immune response relies on T cells.

After the precursors of T cells arise in bone marrow, they start down a pathway toward a fully differentiated state. Each cell travels through the bloodstream to the thymus. As T cells mature in this organ, they become split into distinct groups. One group, mature **helper T cells**, are sometimes called CD4 cells. A second group, mature cytotoxic T cells, are called CD8 cells. (The terms refer to different surface proteins displayed by each type of T cell.) There may also be a third category of T cell that functions to shut down an immune response. Immunologists differ on this point. Some recognize a separate group of *suppressor T cells*, whereas others believe that subsets of helper T cells secrete chemical signals that dampen an immune response when it is no longer required.

As a helper T cell or a cytotoxic T cell matures, it acquires receptors for a specific antigen. When this process is complete, the fully developed cells leave the thymus and circulate in the blood as "virgin" T cells (Figure 7.6). A key point to remember is that *each virgin T cell displays receptors for only one kind of antigen*. As long as it does not encounter that antigen, the T cell will remain inactive. It will become activated only if the antigen is presented under the proper circumstances.

A virgin T cell ignores MHC markers on the body's own cells. It also ignores free-floating antigen molecules. However, if (1) the T cell encounters the kind of antigen that it has receptors for, and (2) the antigen is presented by an antigen-presenting cell *in an antigen-MHC complex*, then the T cell will be stimulated to divide and give rise to a clone. (A *clone* is a population of genetically identical cells.) The T cell's descendants differentiate into subpopulations of effector cells and memory cells, each with receptors for the antigen. As a result, many more T cells become available to join the battle against the invader. Circulating effector helper T cells secrete various proteins, including interleukin-2, which enhances the division and differentiation of cells. Effector cytotoxic T cells recognize the antigen-MHC complexes that form on infected body cells (or on tumor cells).

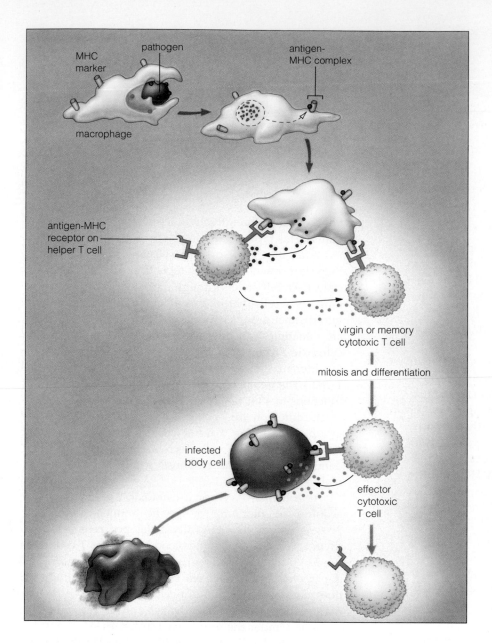

a A macrophage engulfs and digests a pathogen, then cleaves its antigen into fragments that bind to MHC markers. The macrophage becomes an antigen-presenting cell; it displays processed antigen-MHC complexes at its surface.

b Receptors on T cells bind to the complexes. Binding stimulates the macrophage to secrete interleukin-1 (pink dots). This stimulates helper T cells to secrete other interleukins (green dots). These stimulate virgin or memory cytotoxic T cells to divide and differentiate into large populations of effector and memory cells. Only the effector cytotoxic T cells have cell-killing abilities.

c An effector encounters a target: an infected body cell that has the processed antigens bound with MHC markers at its surface. The effector delivers a lethal hit. It releases perforins and toxic substances (green dots) onto its target and so programs it for death.

d The effector disengages from the doomed cell and reconnoiters for new targets. Meanwhile, perforins make holes in the target's plasma membrane. Toxins move into the cell, disrupt organelles, and make the DNA disassemble. The infected cell dies.

Figure 7.6 Example of a cell-mediated immune response, as carried out by activated cytotoxic T cells. An antigen-presenting macrophage provides the signal that starts this response.

Activated cytotoxic T cells destroy their targets with a single "lethal hit," a chemical barrage that directly kills the target (Figure 7.6). First they secrete **perforins**, proteins that form doughnut-shaped pores in a target's plasma membrane. They also secrete toxins that enter the cell and disrupt its organelles and DNA. Having made its attack, the cytotoxic effector quickly disengages and moves on to new targets.

Other cytotoxic cells, such as **natural killer cells** (NK cells), also arise from stem cells in bone marrow. Although they are not well understood, they appear to be lymphocytes (but not T or B cells). They need not encounter an antigen-MHC complex to take action. NK cells patrol for and kill (by lysis) tumor cells and infected cells bearing what probably are odd molecular configurations on their surfaces.

Effector helper T cells secrete interleukins that promote the cell divisions and differentiation required for an immune response. Effector cytotoxic T cells destroy infected cells, tumor cells, or foreign cells with a lethal chemical attack.

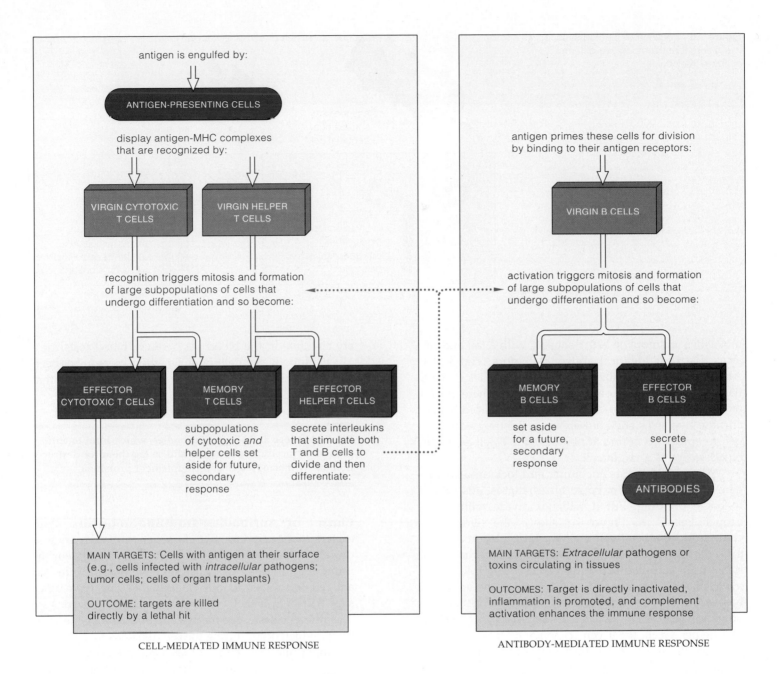

Figure 7.7 Overview of the key interactions among B and T lymphocytes during an immune response. Most often, both types of white blood cells are activated when a specific antigen has been detected. Antigen is any large molecule that lymphocytes recognize as not being "self" (normal body cells). A first-time encounter with an antigen elicits a *primary* immune response. A subsequent encounter with the same type of antigen elicits a *secondary* immune response, which is larger and more rapid. Memory cells that formed but were not used during the first response can immediately engage in the second one.

Before an organ is transplanted, the MHC markers of a potential donor are analyzed to determine how closely they match the patient's MHC markers. The very real danger is that donor MHC markers will be different enough to be recognized as antigen and thus will trigger an immune response against donor cells—causing the transplant to be rejected.

Antibody-Mediated Immune Responses: B Cells

Figure 7.7 summarizes the interactions between B and T cells. Like T cells, B cells also arise from stem cells in bone marrow and start down a pathway that culminates in full differentiation. As they mature in bone marrow, B cells start synthesizing many copies of a single kind of antibody molecule.

All antibodies are proteins, but each kind has binding sites that match up to only a single antigen. Many of the

Figure 7.8 Location of binding sites for two different antigens on two different antibody molecules.

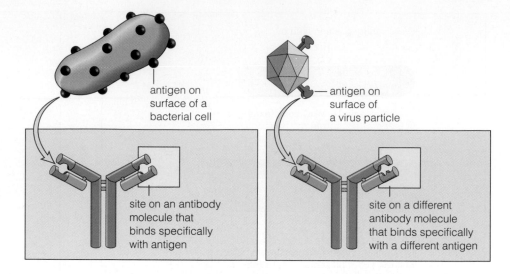

antigen on surface of a bacterial cell

antigen on surface of a virus particle

site on an antibody molecule that binds specifically with antigen

site on a different antibody molecule that binds specifically with a different antigen

molecules are more or less Y-shaped, with a tail and two arms that bear identical antigen receptors (Figure 7.8). After they are synthesized, antibody molecules move to the plasma membrane of a maturing B cell. There, the tail of each molecule becomes embedded in the lipid bilayer, and the two arms stick out above it. Bristling with antigen receptors (its bound antibodies), the B cell enters the bloodstream as a virgin cell.

When the receptors encounter and lock onto an antigen, the B cell is said to be *sensitized*; that is, it is primed to begin dividing. But it will not divide without the proper signals. As Figure 7.9 shows, the signal must come from a helper T cell *already activated* by an antigen-presenting cell. In a kind of immunological tit-for-tat, the B cell itself can play this antigen-presenting role. Like a macrophage, a sensitized B cell processes the antigen it has encountered and displays an antigen-MHC complex on its surface. When the B cell now presents this complex to a helper T cell, the T cell responds by secreting interleukins—the signal the B cell requires to begin dividing. In the presence of interleukins secreted by the helper T cell, the sensitized B cell and its descendants undergo repeated cell divisions.

The resulting clonal B cell population differentiates into effector cells and memory B cells. The effector cells (formerly called plasma cells) produce and secrete huge numbers of antibody molecules. When an antibody binds antigen, it tags the invader for destruction, as by phagocytes and complement proteins (Figure 7.3a).

The main targets of antibody-mediated responses are *extracellular* pathogens and toxins, which are freely circulating in tissues or body fluids. Antibodies cannot bind to pathogens or toxins hidden inside a host cell.

Not all the antibody molecules formed during each immune response are used up. At any time, a great variety circulate in the blood. As a result, blood tests for a given antibody can determine whether a person has been exposed to a specific pathogen, such as the viruses that cause hepatitis and AIDS.

B cells produce and secrete antibodies, which bind to antigens of extracellular pathogens and so tag them for destruction (as by macrophages and complement proteins).

Classes of Antibodies: Immunoglobulins With each immune response, five classes of antibodies are produced. Collectively, the five classes are known as **immunoglobulins** (Igs). They all have binding sites for antigen, but the antibodies in each class also have other sites that permit them to function in a specialized way. The different classes are the protein products of gene shufflings, which occur during the cell divisions that give rise to subpopulations of effector cells and memory B cells.

IgM antibodies are the first to be secreted during immune responses. After binding antigen, they set the complement cascade in motion. They also bind invaders to one another, and the clumped invaders are more readily eliminated by phagocytes.

IgG antibodies activate complement proteins and neutralize many toxins. The IgGs are long-lasting. They are the only antibodies to cross the placenta during pregnancy, so they protect a fetus with the mother's acquired immunities. They also are secreted into the first milk produced by mammary glands and are then absorbed into the suckling newborn's bloodstream.

IgA antibodies enter mucus-coated body surfaces, including those in the respiratory, digestive, and reproductive tracts. There they neutralize infectious agents. A

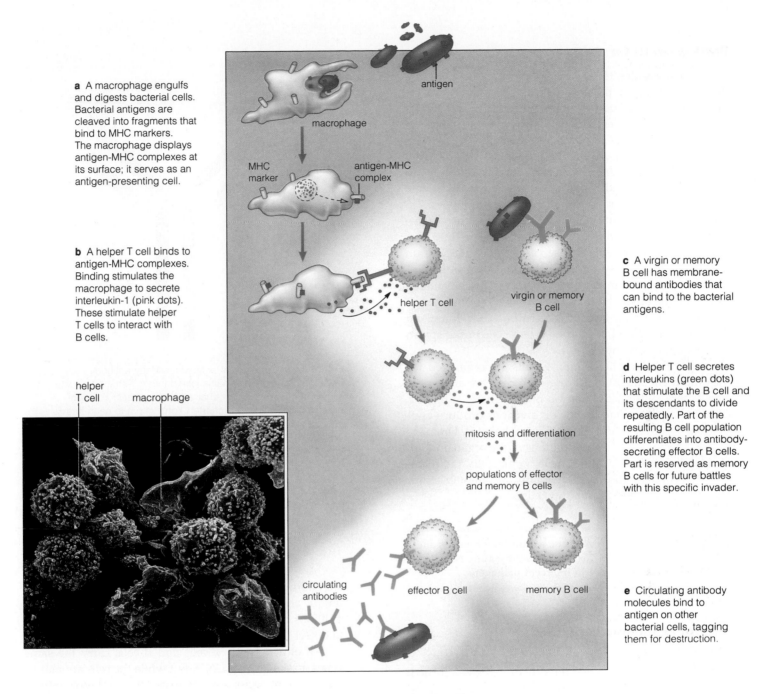

a A macrophage engulfs and digests bacterial cells. Bacterial antigens are cleaved into fragments that bind to MHC markers. The macrophage displays antigen-MHC complexes at its surface; it serves as an antigen-presenting cell.

b A helper T cell binds to antigen-MHC complexes. Binding stimulates the macrophage to secrete interleukin-1 (pink dots). These stimulate helper T cells to interact with B cells.

antigen

macrophage

MHC marker

antigen-MHC complex

helper T cell

virgin or memory B cell

mitosis and differentiation

populations of effector and memory B cells

circulating antibodies

effector B cell

memory B cell

helper T cell

macrophage

c A virgin or memory B cell has membrane-bound antibodies that can bind to the bacterial antigens.

d Helper T cell secretes interleukins (green dots) that stimulate the B cell and its descendants to divide repeatedly. Part of the resulting B cell population differentiates into antibody-secreting effector B cells. Part is reserved as memory B cells for future battles with this specific invader.

e Circulating antibody molecules bind to antigen on other bacterial cells, tagging them for destruction.

Figure 7.9 Example of an antibody-mediated immune response to a bacterial invasion. The micrograph shows helper T cells being activated by an antigen-presenting macrophage.

mother's first milk delivers them to the mucous lining of her newborn's GI tract.

The tails of *IgE* antibodies bind to basophils and mast cells, and the antigen receptors face outward. When the receptors bind antigen, the basophils and mast cells release substances (such as histamine) that promote inflammation.

Researchers do not fully understand the function of *IgD* antibodies, which occur mainly on the surfaces of

mature B cells. They probably act as antigen receptors and have a role in the events leading up to the differentiation of sensitized B cells into effector B cells and memory B cells.

Antibodies bind to antigens of extracellular pathogens. Binding tags an antigen for destruction by macrophages and complement proteins.

Battlegrounds for the Immune System

So far, we have described some key immune system battles—but where are the battlegrounds? Antigen-presenting cells and lymphocytes interact in organs that actually promote immune responses. These are the lymphoid organs described in Chapter 6 and shown in Figure 6.26. They include tonsils, adenoids, and most other lymph nodes just beneath the mucous membranes of the respiratory, digestive, and reproductive systems. Such locations allow antigen-presenting cells and lymphocytes to intercept invaders that have just penetrated the surface barriers. Similarly, antigen is present in tissue fluid that enters lymphatic vessels. These vessels eventually drain into the circulatory system, which could distribute antigen to every tissue in the body. However, before antigen can reach the bloodstream it must trickle through lymph nodes, which are packed with defending cells. If antigen does enter the blood, immune system cells in the spleen and bone marrow intercept it.

Inside the lymphoid organs, cells are organized for maximum effect. Antigen-presenting cells are arrayed at sites where substances enter an organ. There, antigen is engulfed and presented, and cell divisions produce populations of effector and memory lymphocytes. Drainage moves effector cells to other regions of the organ and out to other body parts. All the while, virgin cells and memory cells circulate through the organs, reconnoitering for antigen and antigen-presenting cells.

Immune responses are carried out within lymph nodes and other lymphoid organs, where antigen-presenting cells and lymphocytes are organized to intercept pathogens.

Control of Immune Responses

Antigen provokes an immune response, and removal of antigen stops it. Thus when the tide of battle has turned, effector cells have destroyed most of the antigen-bearing pathogens in the body. With fewer antigen molecules present, stimulation of the response declines and finally stops. Inhibitory signals from suppressor T cells may also slow down antibody production and other aspects of the response.

The Basis of Immunological Specificity and Memory

Immunological Specificity Contact with people, food, water, air, soil, pets, and just about everything else in your surroundings also brings you into contact with

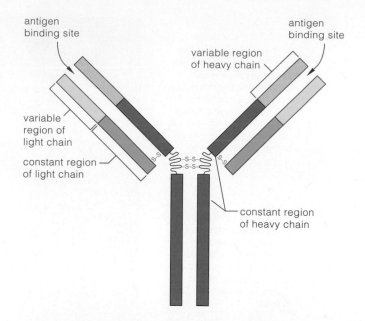

Figure 7.10 Structure of antibodies. Many antibody molecules have four polypeptide chains joined into a Y-shaped structure, as shown here. Two "heavy" chains form the basic Y shape; disulfide bonds attach a shorter "light" chain to each heavy chain. Light and heavy chains each have constant regions that are similar in all antibody molecules. But they also have variable regions where the configuration is unique in each kind of antibody; this is the binding site for one kind of antigen. The antigen fits both into and onto the grooves and protrusions of the binding site.

an enormous variety of pathogens, each with many unique antigens. How do lymphocytes come to have the millions of different antigen-specific receptors required to detect those threats?

All of a person's T cells and B cells carry the same genes, in the form of DNA. But while the cells are maturing, different regions of the genes that code for antigen receptors are shuffled at random into one of millions of possible combinations. In each T and B cell, this random process of *recombination* produces a gene that codes for one of millions of possible antigen receptors.

Figure 7.10 shows the Y-shaped structure of an antibody. Each arm of the molecule consists of two polypeptide chains. In each B cell, the established gene sequence was translated into an amino acid sequence, which dictated the folding of those chains (page 32). In each arm, the result was a particular configuration of grooves and bumps; this configuration is an antigen binding site. An antibody can bind only an antigen that has the appropriate shape.

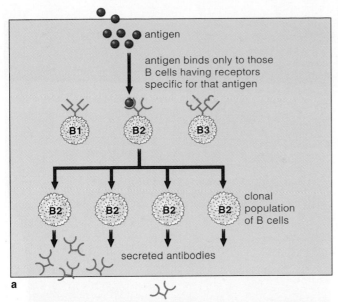

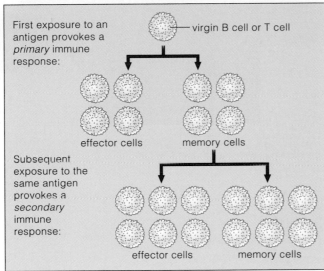

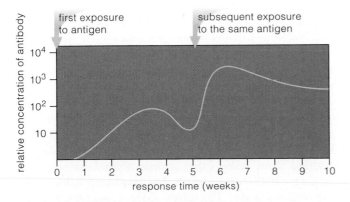

Figure 7.11 (a) Clonal selection of a B cell that produced the specific antibody that can combine with a specific antigen. Only antigen-selected B cells (and T cells) are activated and give rise to a population of immunologically identical clones. (b) Immunological memory. Not all B and T cells are used in a primary immune response to an antigen. Many continue to circulate as memory cells, which become activated during a secondary immune response.

DNA recombinations help explain the **clonal selection theory** first proposed by Macfarlane Burnet. All the clonal descendants of a virgin lymphocyte have the same gene sequence. That sequence codes for the one receptor configuration that can bind to the specific antigen that "selected" the parent cell (Figure 7.11a).

Immunological Memory Many people have a positive reaction to the common skin test for tuberculosis. In such individuals, a hard, red swelling develops at the test site, usually within 48 hours of the test. This positive response in a person who has no medical history of the disease commonly indicates that at some time the individual's immune system successfully fought off an infection by the tuberculosis bacterium. It is unmistakable evidence of the body's "immunological memory" of a *primary* (first-time) *immune response* to an invasion. The body's response to any subsequent encounter with the same antigen, called a *secondary immune response*, can develop more quickly than the primary response, last longer, and be greater in magnitude (Figure 7.12). Why? Memory cells that form during a primary response do not engage in battle. Instead, they continue to circulate for years, even decades in some cases. When a memory cell encounters the same antigen, it rapidly gives rise to large clones—and even more memory cells.

From the preceding discussion we can see that the makeup of a person's memory cell populations depends on the particular antigens to which that individual is exposed. *Focus on Environment* describes one practical consequence of this fact.

PRACTICAL APPLICATIONS OF IMMUNE SYSTEM COMPONENTS

Immunization

Immunization refers to various methods that promote increased immunity against specific diseases. In *active immunization*, an antigen-containing preparation called a

Figure 7.12 Differences in magnitude and duration between a primary and a secondary immune response to the same antigen. (The secondary response starts at week 5.)

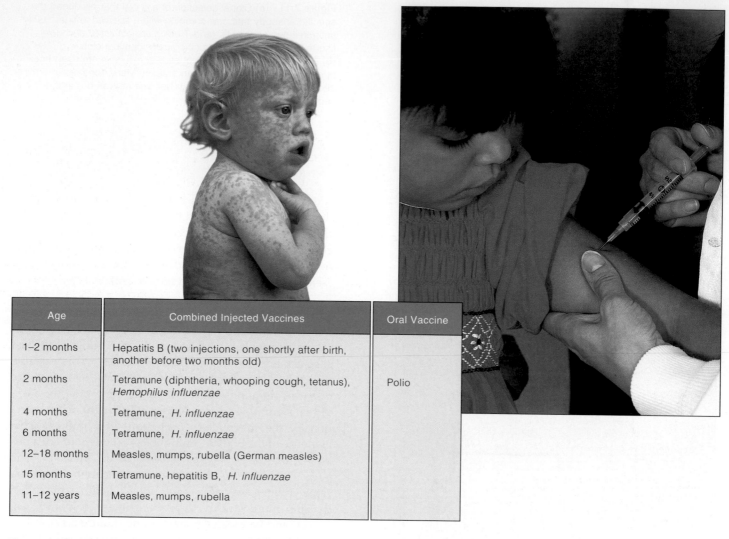

Age	Combined Injected Vaccines	Oral Vaccine
1–2 months	Hepatitis B (two injections, one shortly after birth, another before two months old)	
2 months	Tetramune (diphtheria, whooping cough, tetanus), *Hemophilus influenzae*	Polio
4 months	Tetramune, *H. influenzae*	
6 months	Tetramune, *H. influenzae*	
12–18 months	Measles, mumps, rubella (German measles)	
15 months	Tetramune, hepatitis B, *H. influenzae*	
11–12 years	Measles, mumps, rubella	

Figure 7.13 Immunization schedule for children in the United States. Pediatricians routinely immunize infants and children during office visits. Low-cost or free vaccinations also are available at many community clinics.

vaccine is injected into the body or taken orally, sometimes according to a schedule (Figure 7.13). A first injection elicits a primary immune response. Another injection (booster) elicits a secondary response, which produces more effector cells and memory cells that can provide long-lasting protection against the disease.

Many vaccines are made from killed or weakened pathogens. Sabin polio vaccine, for example, is made from a weakened polio virus. Other vaccines are based on inactivated forms of naturally occurring toxins, such as the bacterial toxin that causes tetanus. Still other experimental vaccines have been made from harmless genetically engineered viruses (Chapter 21). These viruses incorporate genes from three or more different viruses in their genetic material. After an engineered virus is introduced into a host, the introduced genes are expressed, antigens are produced, and immunity is established.

Passive immunization is often used to help people already infected with pathogens, including the ones that cause diphtheria, tetanus, measles, and hepatitis B (page 351). A person receives injections of antibody molecules purified from some other source. (The best source is another person who already has produced a lot of the required antibody.) The effects do not last long because the person's own B cells are not producing memory cells. However, injected antibody molecules may help counter the immediate attack.

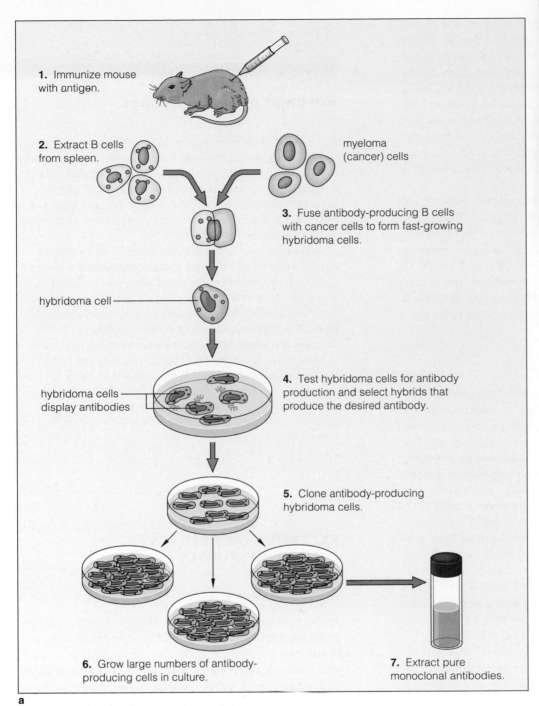

1. Immunize mouse with antigen.

2. Extract B cells from spleen.

myeloma (cancer) cells

3. Fuse antibody-producing B cells with cancer cells to form fast-growing hybridoma cells.

hybridoma cell

hybridoma cells display antibodies

4. Test hybridoma cells for antibody production and select hybrids that produce the desired antibody.

5. Clone antibody-producing hybridoma cells.

6. Grow large numbers of antibody-producing cells in culture.

7. Extract pure monoclonal antibodies.

a

b

Figure 7.14 (**a**) Technique for preparing monoclonal antibodies. (**b**) This woman is using a home pregnancy test that relies on monoclonal antibodies. As soon as a woman becomes pregnant, a membrane called the chorion surrounds the developing embryo and begins secreting the hormone HCG (human chorionic gonadotropin). The test kit's monoclonal antibodies will bind with HCG in the woman's urine, causing the material in the vial to change color.

Immunization with a vaccine can confer active immunity to a pathogen. One or more injections trigger immune responses and the production of antibodies and memory cells.

Injection of antibodies provides short-term passive immunity.

Monoclonal Antibodies

The antibodies produced by B cells make such a powerful contribution to immune responses because of their remarkable specificity. Commercially prepared **monoclonal antibodies** harness this power for medical and research uses. Monoclonal antibodies are obtained by techniques that produce large quantities of a desired antibody derived from B cells (Figure 7.14a). Commonly, a mouse is immunized with a specific antigen, and then B cells are extracted from its spleen and fused with cancerous cells known as *myeloma cells*. The new hybrid is called a *hybridoma cell*. Like cancer cells, the hybridoma cells multiply rapidly (page 428). Some produce the antibody of interest. A clonal population of an antibody-producing hybridoma cell can be maintained indefinitely in the laboratory. The name *monoclonal antibodies* comes from the fact that the antibody molecules are produced by a population of genetically identical cells that is a clone of a single antibody-producing cell.

Monoclonal antibodies are being put to various uses, including home pregnancy tests (Figure 7.14b) and screening for prostate cancer and some sexually transmitted diseases. They are also used for passive immunization against a variety of diseases. And as Chapter 20 describes, they also have a range of potential uses in cancer treatment, including acting as highly specific "magic bullets" to deliver lethal drugs to cancer cells only, sparing the body's healthy tissues.

Applications of Lymphokines

Signaling molecules such as the interleukins produced by lymphocytes are sometimes called *lymphokines*. Along with monoclonal antibodies, interleukins and other lymphokines are becoming tools of medical research. For example, *tumor necrosis factor* (TNF) secreted by cytotoxic T cells can kill cancerous cells. In an experimental treatment of patients with the deadly skin cancer malignant melanoma, researchers extracted cytotoxic T cells from the patients' tumors and, using recombinant DNA techniques, altered the T cells genetically so that they produced much more TNF than normal. Doses of the altered cells were then administered to the patients, along with interleukin-2 to further stimulate T cell activity. The experimental therapy is continuing in the hope that the "souped-up" cytotoxic T cells will be able to kill cancer cells with a level of success that can save lives.

T cells also produce proteins known as **interferons**. Interferons are aptly named because they interfere with the replication of viruses. Activated T cells secrete gamma interferon, which both suppresses viral replication and stimulates the activity of macrophages that kill tumor cells. Gamma interferon is now being made in large quantities by genetic engineering methods, and as you might guess, researchers are actively exploring its potential uses in treating viral diseases, including virus-triggered cancers. To date, however, results have been mixed. Beta interferon was recently approved for treatment of multiple sclerosis, an *autoimmune disease* in which the immune system attacks portions of the nervous system (page 185). This disease is just one example of an abnormal immune response, a phenomenon we consider next.

Revenge of the Microbes

For T and B cells, immunological experience is what counts. A child's thymus and bone marrow actively crank out T and B cells with specificities for many thousands of antigens. The composition of these T and B cell populations isn't static, however. They tend to become lopsided over time as a person—or a whole population—encounters some antigens and not others.

Exposure to antigens that are common in a person's environment results in expanded populations of memory T and B cells primed to battle those particular invaders. If you grew up in North America in the late 20th century, you almost certainly have hosts of memory cells that can fight off microbes common in your environment. Those defenses are so efficient, you may not even know a battle is being waged. On the other hand, you may *not* have ready contingents of T and B cells primed to battle bacteria present in the water of Acapulco or New Delhi—which is why wise tourists travel prepared to cope, at least temporarily, with intestinal problems caused by unfamiliar waterborne bacteria.

EXCESSIVE OR DEFICIENT IMMUNE RESPONSES

Allergies

Millions of people suffer from **allergy**, in which a normally harmless substance provokes an immune response. A substance that triggers an allergic reaction is an *allergen*; common allergens include pollen, a variety of foods and drugs, dust mites, fungal spores, insect venom, drugs, and cosmetics. Some responses to allergens start within minutes of exposure; others are delayed. Either way, they cause moderate to severe inflammation.

Some people are genetically inclined to allergies. In addition, infections, emotional stress, and even changes in air temperature can trigger reactions that otherwise might not occur. Every time allergic individuals are exposed to certain antigens, IgE antibodies are secreted and bind to mast cells. When IgE on a mast cell then binds to antigen, the mast cell is stimulated to secrete histamine, prostaglandins, and other substances that enhance the inflammatory response. These substances also stimulate secretion of mucus and cause smooth muscle in airways to constrict. In asthma and hay fever, the symptoms of the allergic response are congestion, sneezing, a drippy nose, and labored breathing.

In rare cases, inflammatory reactions spread through the body and trigger a life-threatening condition called *anaphylactic shock*. For example, a person who is allergic to wasp or bee venom can die within minutes of a single sting. Air passages leading to the lungs undergo massive constriction, and fluid escapes too rapidly from dilated, extremely leaky capillaries. Blood pressure plummets and can lead to circulatory collapse.

Antihistamines (anti-inflammatory drugs) are often used to relieve the short-term symptoms of allergies, and desensitization programs may help over the long term. In such programs, skin tests are used to identify the particular allergens that bother a patient. Inflammatory responses to some of them can be blocked if the patient's body can be stimulated to make IgG instead of IgE. Over an extended period, larger and larger doses of specific allergens are administered. Each time, the person's body produces more circulating IgG and IgG memory cells. IgG binds with the allergen and blocks its attachment to IgE— and blocks inflammation.

Autoimmune Disorders

In an **autoimmune response**, the powerful weapons of the immune system are unleashed against normal body cells. An example is *rheumatoid arthritis*, in which skeletal joints are chronically inflamed. Often, affected people have high levels of an antibody (rheumatoid factor) that locks onto IgG molecules as if they were antigens and then deposits them on membranes at joints. The deposits trigger complement cascades and inflammation. Fluid leaking into the joints lifts the membrane from underlying tissues, and membrane cells divide repeatedly in response. Over time, the joint becomes filled with these cells and is eventually immobilized.

The autoimmune disease *systemic lupus erythematosus* (SLE) primarily affects younger women, but males can develop it as well. For unknown reasons, patients develop antibodies to their own nuclear DNA, and DNA-antibody complexes accumulate in joints, blood vessels, and the kidney glomeruli. Symptoms include fatigue, painful arthritis, iron-deficiency anemia, and in some cases major kidney dysfunction. Various drug treatments can help relieve many symptoms, but there is currently no cure for SLE. Table 7.3 lists some other immune system disorders.

Table 7.3 The Immune System Under Attack		
Cause/Risk Factors	Major Effects	Symptoms
AIDS Human immunodeficiency virus (HIV), transmitted in body fluids such as blood, semen, breast milk (page 343)	Destroys CD4 (helper T) cells, thereby disrupting immune responses	Early stage: Unexplained weight loss, fatigue, enlarged lymph glands, fever, diarrhea, oral yeast infection. Full-blown AIDS is marked by the appearance of infections and tumors
Chronic fatigue syndrome Unknown; possibly a virus	Overproduction of interleukin 2 and cytotoxic T cells; abnormally low levels of interferons	Persistent deep fatigue; fever, chills, muscle and joint pain, headache, sore throat, swollen glands
Immunosuppression (non-HIV) Genetic disorder; radiation exposure; various drugs used during cancer therapy, organ transplants, and other medical procedures; protein-deficient diet; stress	Destruction of lymphoid tissues (radiation) or impairment of growth and division of lymphocytes; immunosuppressive effects of stress are strongly suspected but not confirmed	Extreme vulnerability to infection, tumor development, cancer
Lymphoma Any of a group of cancers of lymphoid tissue; sometimes associated with immune system suppression by drugs administered during an organ transplant. Some lymphomas may be a result of HIV infection; *Burkitt's lymphoma* may be caused by the Epstein-Barr virus	Unchecked growth of cells of the spleen and/or lymph nodes, leading to severe impairment or destruction of those tissues	Swelling of lymph nodes in the groin or neck; enlarged liver and spleen. Skin ulcers, headache, fever; complications of cancer metastases

Deficient Immune Responses

When the body has too few lymphocytes, immune responses are not effective. This condition can be inherited or acquired. Either way, a person whose immune responses are deficient becomes highly vulnerable to infections that are not life-threatening to the general population. This is what happens in **AIDS** (acquired immune deficiency syndrome).

AIDS is caused by the human immunodeficiency virus (HIV). At present, AIDS is incurable and eventually fatal. HIV is transmitted when bodily fluids of an infected person enter another person's tissues. The virus cripples the immune system by attacking helper T cells and macrophages. This leaves the body dangerously susceptible to opportunistic infections and some otherwise rare forms of cancer. Chapter 15 on sexually transmissible diseases describes various aspects of HIV infection, including how the virus replicates inside a human host and what the prospects are for treating or curing HIV-infected people.

SUMMARY

1. The body protects itself from pathogens (infection-causing agents) with three lines of defense:

   a. Physical and chemical barriers at body surfaces.

   b. Nonspecific responses to tissue irritation or damage. These include inflammation and the participation of organs with phagocytic functions, such as the liver and spleen.

   c. Immune responses to specific pathogens, foreign cells, or abnormal body cells. The white blood cells listed in Table 7.2 have central roles in the body's defense.

2. Intact skin and mucous membranes lining various body surfaces are physical barriers to infection. Glandular secretions (in tears, saliva, and gastric fluid) are examples of chemical barriers. So are metabolic products of resident bacteria on body surfaces.

3. Tissue redness, warmth, swelling, and pain are signs of inflammation. The inflammatory response begins with changes in blood flow to a besieged tissue. Substances released by pathogens and by dead or damaged body cells make small blood vessels leaky. Circulating white blood cells, including macrophages, enter the tissue and destroy invaders. Plasma proteins also enter the tissue. The complement proteins induce lysis of pathogens and attract phagocytes. Blood-clotting proteins help repair damaged blood vessels.

4. Immune responses have the following characteristics:

   a. Each response is triggered by antigen, a unique molecular configuration that lymphocytes recognize as foreign (nonself). An example is a protein at the surface of a particular virus.

   b. Each immune response has specificity, meaning it is directed against one antigen alone.

   c. Each immune response has memory. A subsequent encounter with the same antigen will trigger a more rapid, secondary response of greater magnitude.

   d. An immune response normally is not made against the body's own self-proteins.

5. Macrophages and other antigen-presenting cells process and display fragments of antigen with their own MHC markers. T lymphocytes have receptors that can bind to antigen-MHC complexes. Binding is the start signal for an immune response.

6. Immune responses begin with recognition of antigen, proceed through repeated cell divisions that form clonal populations of lymphocytes, then to differentiation into subpopulations of effector cells and memory cells. Interleukins and other signals among white blood cells drive the responses, which are carried out by effector B cells and antibodies, helper T cells, and effector cytotoxic T cells. Memory cells are set aside for secondary responses.

7. Activated cytotoxic T cells monitor body tissues for any cell bearing antigen-MHC complexes. They directly destroy cells infected with intracellular pathogens, tumor cells, and cells of tissue or organ transplants.

8. Antibodies are Y-shaped protein molecules, each with binding sites for only one kind of antigen. Only B cells produce them. When antibody binds to antigen, toxins may become neutralized, pathogens may be tagged for destruction, or pathogens may be prevented from attaching to body cells.

9. In active immunization, vaccines provoke an immune response and production of memory cells. In passive immunization, injections of antibodies help patients combat an infection that is already under way.

10. An allergy is an immune response to a generally harmless substance. An autoimmune response is an attack by lymphocytes on the body's normal cells. An immune deficiency is a weakened or nonexistent capacity to mount an immune response.

Review Questions

1. While jogging barefoot in the surf, your toes accidentally land on a jellyfish. Soon the bottoms of your toes are swollen, red, and warm to the touch. Describe the events that result in these signs of inflammation. *173*

2. The immune system is characterized by specificity and memory. Describe what these terms mean. *174*

3. Distinguish between the following pairs:
   a. neutrophil and macrophage

b. cytotoxic T cell and natural killer cell
c. effector cell and memory cell
d. antigen and antibody

4. Are phagocytes deployed during nonspecific defense responses, immune responses, or both? *174*

5. Antibodies and interleukins are central to immune responses. Define them and state their functions. *173–174*

6. What is immunization? What is a vaccine? *181–182*

7. What is the difference between an allergy and an autoimmune response? AIDS is neither of these; what is it? *184–185*

Critical Thinking: You Decide   *(Key in Appendix IV)*

1. HIV wreaks havoc with the immune system by attacking certain T cells and macrophages. Would the impact be altered if the virus attacked macrophages only? Explain.

2. Given what you now know about how foreign invaders trigger immune responses, explain why mutated forms of viruses, which have altered surface proteins, pose a monitoring problem for memory cells.

Self-Quiz   *(Answers in Appendix III)*

1. _____ are barriers to pathogens at body surfaces.
   a. Intact skin and mucous membranes   d. Urine
   b. Tears, saliva, and gastric fluid   e. all are correct
   c. Resident bacteria

2. Macrophages are derived from white blood cell precursors called _____.
   a. lymphocytes   d. monocytes
   b. basophils   e. neutrophils
   c. eosinophils

3. Activated complement functions in defense by _____.
   a. neutralizing toxins   c. promoting inflammation
   b. enhancing resident bacteria   d. forming holes in memory lymphocyte membranes

4. _____ are large molecules that lymphocytes recognize as foreign and that elicit an immune response.
   a. Interleukins   c. Immunoglobulins   e. Histamines
   b. Antibodies   d. Antigens

5. Immunoglobulins designated _____ increase antimicrobial activity in mucus.
   a. IgA   c. IgG   e. IgD
   b. IgE   d. IgM

6. Antibody-mediated responses work best against _____.
   a. intracellular pathogens   c. toxins   e. all are correct
   b. extracellular pathogens   d. b and c are correct

7. Antigens (nonself molecular markers) are _____.
   a. carbohydrates   c. steroids   e. nucleotides
   b. tryglycerides   d. proteins

8. _____ would be a target of an effector cytotoxic T cell.
   a. Extracellular virus particles in blood
   b. Cervical tumor cells
   c. Parasitic flukes in the liver
   d. Bacterial cells in pus
   e. Pollen grains in nasal mucus

9. _____ destroy infected cells with a lethal hit.
   a. Complement proteins   d. Natural killer cells
   b. Antibody molecules   e. b and c are correct
   c. Cytotoxic T cells

10. Match the following immunity concepts:
    _____inflammation   a. neutrophil
    _____antibody secretion   b. effector B cell
    _____phagocyte   c. nonspecific response
    _____immunological memory d. deliberately provoking
    _____vaccination   memory cell production
    _____allergy   e. basis of secondary
                                 immune response
                              f. immune response to a
                                 normally harmless substance

Key Terms

acute inflammation *173*	immunization *181*
AIDS *186*	immunoglobulin *178*
allergy *184*	interferon *184*
antibody *174*	interleukin *173*
antigen *174*	lymphoid organ *180*
antigen-MHC complex *175*	lysis *172*
antigen-presenting cell *175*	lysozyme *170*
autoimmune response *185*	macrophage *171*
B lymphocyte (B cell) *174*	memory cell *174*
clonal selection theory *181*	MHC marker *174*
complement system *172*	monoclonal antibodies *183*
cytotoxic T cell *174*	natural killer cell *176*
effector cell *174*	perforin *176*
fever *173*	pathogen *170*
helper T cell *175*	T lymphocyte (T cell) *174*
immune system *174*	vaccine *182*

Readings

Caldwell, M. August 1993. "The Long Shot." *Discover*. Discussion of the problems researchers face in trying to develop a vaccine against HIV.

"Determining What Immune Cells See." 31 January, 1992. *Science*.

"Life, Death, and the Immune System." September 1993. *Scientific American*. Special issue devoted to immunity, including articles on AIDS, allergies, cancer, and immune therapies.

8 RESPIRATION

Conquering Chomolungma

Chomolungma is a Himalayan mountain that also goes by the name Everest (Figure 8.1). Its summit soars to 9,700 meters (29,108 feet) above sea level, the highest place on earth. To conquer Chomolungma, climbers must come to terms with blizzards, heart-stopping avalanches, and icy, nearly vertical rock faces. They also must grapple with a life-threatening scarcity of oxygen.

In the air at low elevations, one molecule in five is oxygen. The earth's gravitational pull keeps oxygen and other gas molecules in air concentrated near sea level—or, as we say for gases, under pressure. At altitudes above 3,300 meters (10,000 feet), the breathing game changes. Gravity's pull is weaker, gas molecules spread out—and the pressure decreases. What would be a normal breath at sea level simply will not bring enough oxygen into the lungs. The oxygen deficit can cause "altitude sickness," with symptoms ranging from shortness of breath, headache, and heart palpitations to loss of appetite, nausea, and vomiting.

Figure 8.1 A climber approaching the summit of Chomolungma, where oxygen is dangerously scarce.

The Chomolungma base camp is 6,300 meters (19,000 feet) above sea level. When climbers reach 7,000 meters (23,000 feet), conditions start to get lethal. The very low air pressure and oxygen scarcity combine to make many blood vessels leaky. Brain and lung tissues swell with plasma fluid, a condition called *edema*. Climbers with severe edema may fall into a coma and die.

Given the risks, experienced climbers keep themselves in top physical condition and live for a time at high elevations before their assault on the summit. They know that "thinner air" triggers the formation of billions of additional red blood cells. It also stimulates the production of more blood capillaries, mitochondria, and other temporary physiological changes that improve the odds of survival. Most climbers also carry bottled oxygen.

Few of us will ever find ourselves on Chomolungma, pushing to the limit our ability to obtain oxygen. Here in the lowlands, disease, smoking, and other environmental insults strain it in more ordinary ways—although the risks can be just as great, as you will see by this chapter's end.

KEY CONCEPTS

1. Humans are highly active animals. The energy to drive their activities comes mainly from aerobic metabolism, which uses oxygen and produces carbon dioxide as a waste product. In a process called respiration, oxygen moves into the internal environment and carbon dioxide moves to the external environment.

2. Oxygen diffuses into the body as a result of a pressure gradient. Its pressure is higher in air than it is in metabolically active tissues, where cells rapidly use oxygen. Carbon dioxide follows a gradient in the reverse direction. Its pressure is higher in tissues (where CO_2 is a by-product of metabolism) than it is in air.

3. The respiratory system includes a pair of lungs, which provide an efficient respiratory surface—a thin, moist layer of epithelium. Blood flowing through the body's circulatory system picks up oxygen and gives up carbon dioxide at this respiratory surface.

4. Controls over breathing operate mainly in response to changing carbon dioxide levels in the blood.

RESPIRATION AND GAS EXCHANGE

Like other animals, humans take in oxygen for aerobic respiration inside cells. Respiring cells also give up carbon dioxide by-products of metabolism. In a process called **respiration**, oxygen moves into the internal environment of the body and carbon dioxide is released to the external environment. By interacting with other organ systems (Figure 8.2), the respiratory system helps maintain stable operating conditions for body cells.

Factors That Affect Gas Exchange

All respiratory systems rely on diffusion of gases. Like other substances, oxygen and carbon dioxide tend to diffuse down a concentration gradient—or, as we say for gases, down a *pressure gradient*. When molecules of either gas are more concentrated outside the body, more pressure is available to drive them into the body, and vice versa.

Oxygen and carbon dioxide are not the only gases in air. At sea level, a given volume of dry air is approximately 78 percent nitrogen, 21 percent oxygen, 0.04 percent carbon dioxide, and 0.96 percent other gases. Each gas exerts only part of the total pressure exerted by the whole mix of gases. Said another way, each exerts a "partial pressure." At sea level, atmospheric pressure, which can be measured by a mercury barometer (Figure 8.3), is about 760mm Hg. Thus, the partial pressure of oxygen is 21 percent of 760, or about 160mm Hg. The partial pressure of carbon dioxide is about 0.3mm Hg.

The respiratory system works with partial pressure gradients that exist between the internal and external environments. As we'll see shortly, gases diffuse across a thin, moist membrane in the lungs that acts as a **respiratory surface**. How rapidly oxygen or carbon dioxide diffuses across the respiratory surface depends on the surface area of the membrane, its thickness, and the difference in partial pressure across it. The greater the surface area and the larger the partial pressure gradient, the faster diffusion occurs. Given the human body's rather high level of metabolic activity, it's not surprising that the lungs provide a huge surface area for gas exchange. In addition, there are normally steep gas pressure gradients between the lungs and body tissues.

Transport Pigments

Hemoglobin and other transport pigments associated with the circulatory system have vital effects on pressure gradients. Each hemoglobin molecule

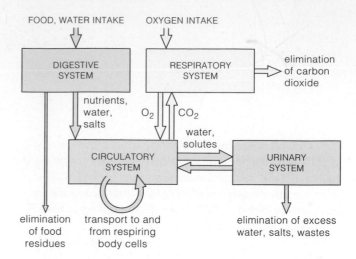

Figure 8.2 Interactions between the respiratory system and other organ systems. The respiratory system takes in oxygen required for aerobic respiration and rids the body of carbon dioxide formed during the reactions.

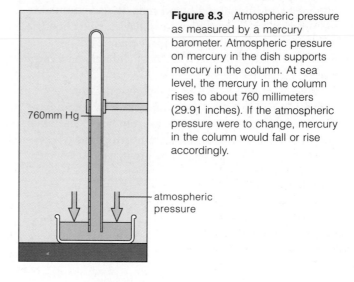

Figure 8.3 Atmospheric pressure as measured by a mercury barometer. Atmospheric pressure on mercury in the dish supports mercury in the column. At sea level, the mercury in the column rises to about 760 millimeters (29.91 inches). If the atmospheric pressure were to change, mercury in the column would fall or rise accordingly.

The respiratory system relies on diffusion of gases (oxygen and carbon dioxide).

The surface area of the respiratory membrane, its thickness, and the steepness of gas partial pressure gradients together determine how much of a given gas can diffuse across the respiratory surface in a given time.

binds loosely with as many as four oxygen molecules in the lungs (where oxygen concentrations are high) and releases them in tissues where oxygen concentrations are low. This binding of oxygen by hemoglobin and its transport away from the respiratory surface help maintain a lower partial pressure of oxygen in blood plasma. The low partial pressure favors oxygen diffusion from the lungs into blood flowing through lung capillaries.

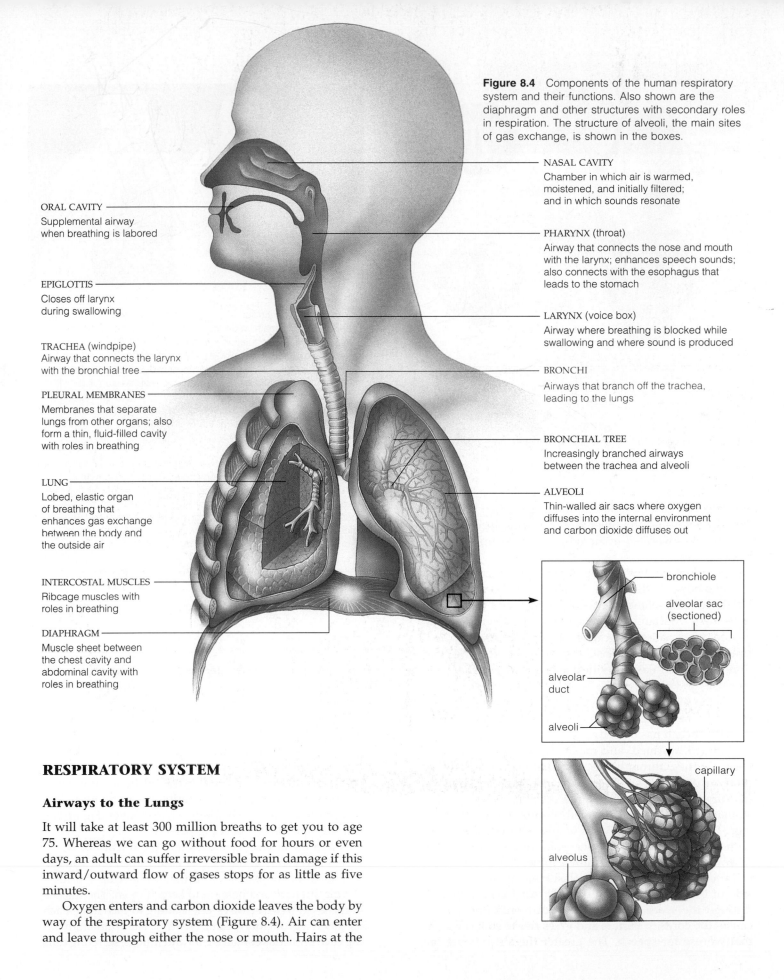

Figure 8.4 Components of the human respiratory system and their functions. Also shown are the diaphragm and other structures with secondary roles in respiration. The structure of alveoli, the main sites of gas exchange, is shown in the boxes.

NASAL CAVITY
Chamber in which air is warmed, moistened, and initially filtered; and in which sounds resonate

ORAL CAVITY
Supplemental airway when breathing is labored

PHARYNX (throat)
Airway that connects the nose and mouth with the larynx; enhances speech sounds; also connects with the esophagus that leads to the stomach

EPIGLOTTIS
Closes off larynx during swallowing

LARYNX (voice box)
Airway where breathing is blocked while swallowing and where sound is produced

TRACHEA (windpipe)
Airway that connects the larynx with the bronchial tree

BRONCHI
Airways that branch off the trachea, leading to the lungs

PLEURAL MEMBRANES
Membranes that separate lungs from other organs; also form a thin, fluid-filled cavity with roles in breathing

BRONCHIAL TREE
Increasingly branched airways between the trachea and alveoli

LUNG
Lobed, elastic organ of breathing that enhances gas exchange between the body and the outside air

ALVEOLI
Thin-walled air sacs where oxygen diffuses into the internal environment and carbon dioxide diffuses out

INTERCOSTAL MUSCLES
Ribcage muscles with roles in breathing

DIAPHRAGM
Muscle sheet between the chest cavity and abdominal cavity with roles in breathing

bronchiole

alveolar sac (sectioned)

alveolar duct

alveoli

capillary

alveolus

RESPIRATORY SYSTEM

Airways to the Lungs

It will take at least 300 million breaths to get you to age 75. Whereas we can go without food for hours or even days, an adult can suffer irreversible brain damage if this inward/outward flow of gases stops for as little as five minutes.

Oxygen enters and carbon dioxide leaves the body by way of the respiratory system (Figure 8.4). Air can enter and leave through either the nose or mouth. Hairs at the

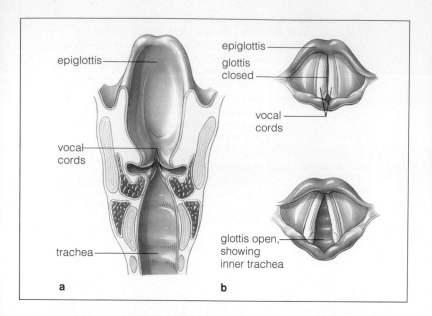

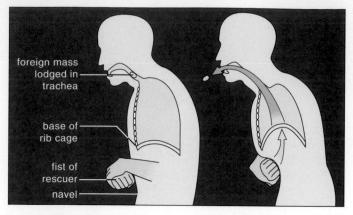

Figure 8.5 The structures where most sounds necessary for speech originate. (**a**) Front view of the larynx, showing the location of the vocal cords. (**b**) The two vocal cords as viewed from above when the glottis (the space between them) is closed or opened.

Figure 8.6 The Heimlich maneuver. Each year, several thousand people choke to death when food enters the trachea instead of the esophagus. Strangulation can result if the air flow is blocked for as little as four or five minutes. The Heimlich maneuver, *an emergency procedure only,* often can dislodge the misdirected chunks of food. The idea is to elevate the diaphragm forcibly, causing a sharp decrease in chest cavity volume and a sudden increase in alveolar pressure. The increased pressure forces air up the trachea—with sufficient force, it is hoped, to dislodge the obstruction.

To perform the Heimlich maneuver, stand behind the victim, make a fist with one hand, and then position the fist, thumb-side in, against the victim's abdomen. The fist must be slightly above the navel and well below the rib cage. Next, press the fist into the abdomen with a sudden upward thrust. Repeat the thrust several times if needed. The maneuver can be performed on someone who is standing, sitting, or lying down.

Once the obstacle is dislodged, a physician should examine the person at once, for an inexperienced rescuer can inadvertently cause internal injuries or crack a rib. The risk may be worth taking, given that the alternative is death.

entrance and ciliated epithelium lining the two *nasal cavities* filter out dust and other large particles. Incoming air also becomes warmed in the nose and picks up moisture from mucus.

The nasal cavities are separated by a *septum* of cartilage and bone. As described in Chapter 4, *paranasal sinuses* lie above and behind the nasal cavities and are linked with them by channels. (Hence, nasal sprays prescribed for colds or allergies can help unclog mucus-filled sinuses.) Tear glands constantly produce moisture that drains into the nasal cavities; when you cry the flow increases and gives you the "sniffles." As we see in Chapter 11, the sense of smell involves a region of specialized sensory cells in the mucous membrane lining the upper reaches of each nasal cavity.

Filtered, warmed, and moistened air flows from the nasal cavities, through the nasopharynx, and into the **pharynx** or throat. This is the entrance to both the **larynx** (an airway) and the esophagus (a tube leading to the stomach). The larynx is formed by nine pieces of cartilage. If viewed from above, it would resemble a triangle with a space in its center. One point of the triangle, the thyroid cartilage, is the "Adam's apple."

Vocal cords are a pair of elastic ligaments on either side of the larynx wall (Figure 8.5). Air forced outward through the space between the vocal cords (the *glottis*) causes the cords to vibrate and gives rise to sound waves that we use for speech. The greater the air pressure on the vocal cords, the louder the sound. The greater the muscle tension on the cords, the higher the pitch of the sound.

For obvious reasons, airways are nearly always open. The flaplike *epiglottis*, attached to the larynx (Figure 8.5), points up during breathing. When you swallow, however, the larynx moves upward so that the epiglottis partly covers the opening of the larynx. In this position, the epiglottis helps prevent food from entering the respiratory tract and causing choking. The Heimlich maneuver (Figure 8.6) can save the life of a choking person.

From the larynx, air moves into the **trachea** or windpipe. Press gently at the base of your throat, and you can feel some of the bands of cartilage that ring the tube, adding strength and helping to keep it open. The trachea branches into two airways, one leading to each lung. Each airway is a **bronchus** (plural: *bronchi*). The epithelium lining each bronchus contains cilia and mucus-secreting cells

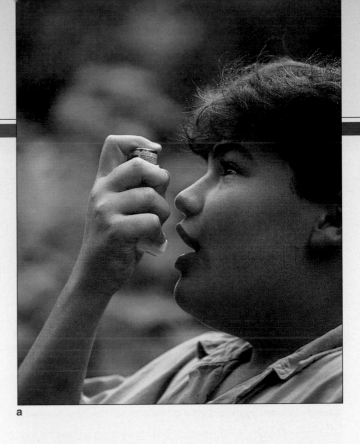

a

Asthma: Life, Breath, and Bronchi

Millions of people know the panicky sensation: Without warning, a vise seems to close around your chest. You break out in a sweat and your pulse begins to race as you wheeze and gasp frantically for a breath of air.

Asthma is a chronic, incurable condition of the respiratory tract in which breathing can become so difficult so quickly that the victim literally feels in imminent danger of suffocating. For reasons that are not yet clear, the incidence of asthma in the United States is growing.

Among other triggers, allergens can set off an asthma attack. Offending substances can include pollen, dairy products, shellfish, flavorings and other foods, pet hairs or dandruff—even the dung of tiny mites in house dust. In susceptible people, attacks can result from noxious fumes, stress, strenuous exercise, or a respiratory infection.

Underlying asthma's frightening symptoms is a basic feature of respiratory anatomy. Stiff rings and plates of cartilage in the walls of bronchi hold the tubes open (as they do the trachea). However, this cartilage gives way to smooth muscle as the bronchi branch into bronchioles deep within the lungs—and therein lies the key to asthma. Factors that trigger an attack actually cause the smooth muscle to contract in strong spasms, so that the bronchi and bronchioles suddenly narrow. In a severe attack, bronchioles may close off completely. At the same time, mucus

pours from bronchial epithelium, further clogging the constricted passages.

Fortunately, a variety of strategies and medications exist to help asthma sufferers. People can identify the sources of allergies and minimize their exposure to those substances. Victims of frequent attacks can take preventative drugs. Aerosol inhalants squirt a fine mist into the airways (Figure *a*). The droplets contain a drug that dilates the bronchial passages and helps restore free breathing.

(Figure 8.7). Bacteria and airborne particles stick in the mucus; then the upward-beating cilia sweep the debris-laden mucus toward the throat. The bronchi narrow as they branch repeatedly. They end in tiny **bronchioles**, which in turn branch into terminal **respiratory bronchioles**.

Flip Figure 8.4 upside down, and you can see why these structures collectively are sometimes called the "respiratory tree." In *asthma*, the tree's finely branched bronchioles can constrict or close off completely, preventing normal breathing (see *Focus on Wellness*).

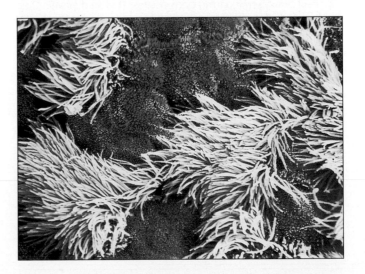

Figure 8.7 Color-enhanced scanning electron micrograph of cilia (gold) in the respiratory tract. Mucus-secreting cells (rust-colored) are interspersed among the ciliated cells. Bacteria and other foreign material stick to mucus-coated microvilli at the free surface of the ciliated cells, then are swept back toward the throat.

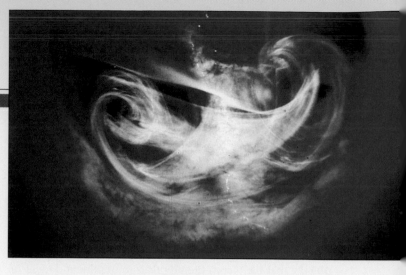

To Smoke or Not to Smoke?

People who now smoke, or who are considering starting, might do well to consider the following information on the risks of smoking and the advantages of quitting.

Cigarette smoke (right) swirling through the human windpipe and down the two bronchial tubes to the lungs.

Risks Associated with Smoking	Benefits of Quitting
Shortened Life Expectancy: Nonsmokers live 8.3 years longer on average than those who smoke two packs daily from the midtwenties on	Cumulative risk reduction; after 10 to 15 years, life expectancy of ex-smokers approaches that of nonsmokers
Chronic Bronchitis, Emphysema: Smokers have 4–25 times more risk of dying from these diseases than do nonsmokers	Greater chance of improving lung function and slowing down rate of deterioration
Lung Cancer: Cigarette smoking is the major cause of lung cancer	After 10 to 15 years, risk approaches that of nonsmokers
Cancer of Mouth: 3–10 times greater risk among smokers	After 10 to 15 years, risk is reduced to that of nonsmokers
Cancer of Larynx: 2.9–17.7 times more frequent among smokers	After 10 years, risk is reduced to that of nonsmokers
Cancer of Esophagus: 2–9 times greater risk of dying from this	Risk proportional to amount smoked; quitting should reduce it
Cancer of Pancreas: 2–5 times greater risk of dying from this	Risk proportional to amount smoked; quitting should reduce it
Cancer of Bladder: 7–10 times greater risk for smokers	Risk decreases gradually over 7 years to that of nonsmokers
Coronary Heart Disease: Cigarette smoking is a major contributing factor	Risk drops sharply after a year; after 10 years, risk reduced to that of nonsmokers
Effects on Offspring: Women who smoke during pregnancy have more stillbirths, and weight of liveborns averages less (hence, babies are more vulnerable to disease, death)	When smoking stops before fourth month of pregnancy, risk of stillbirth and lower birthweight eliminated
Impaired Immune System Function: Increase in allergic responses, destruction of macrophages in respiratory tract	Avoidable by not smoking
Bone Healing: Evidence suggests that surgically cut or broken bones require up to 30 percent longer to heal in smokers, possibly because smoking depletes the body of vitamin C and reduces the amount of oxygen reaching body tissues. Reduced vitamin C and reduced oxygen interfere with production of collagen fibers, a key component of bone. Research in this area is continuing.	Avoidable by not smoking

Information provided by the American Cancer Society.

Air moves to the lungs by way of the nose, pharynx, larynx, trachea, and bronchi.

Except for alveoli, the entire respiratory tract is lined with ciliated epithelium that contains mucus-secreting cells.

Sites of Gas Exchange in the Lungs

The two **lungs** are elastic, cone-shaped organs separated from each other by the heart. Each lung functions as an internal respiratory surface. The left lung has two lobes, the right lung three. The lungs are located within the *rib*

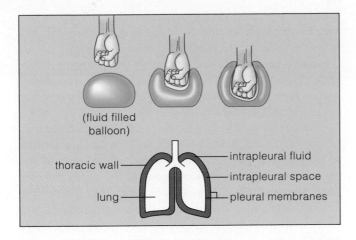

(fluid filled balloon)

thoracic wall — intrapleural fluid
— intrapleural space
lung — pleural membranes

Figure 8.8 Position of the lungs and pleural sac relative to the chest (thoracic) cavity. By analogy, when you push a closed fist into a fluid-filled balloon, the balloon completely surrounds the fist except at your arm. A lung is analogous to the fist; the pleural sac is analogous to the balloon. For clarity, the intrapleural fluid volume is exaggerated.

cage above the **diaphragm**, a sheet of muscle between the thoracic (chest) and abdominal cavities. They are soft and spongy and are not attached directly to the wall of the chest cavity. Instead, each is positioned within a thin, double membrane of epithelium called a **pleura**.

Imagine pushing a closed fist into a fluid-filled balloon (Figure 8.8). A lung occupies the same kind of position as your fist, and the pleural membrane folds back on itself (as does the balloon) to form a closed *pleural sac*. An extremely narrow *intrapleural space* separates the two facing surfaces of the membrane. A thin film of lubricating *intrapleural fluid* fills the space and prevents friction between the pleural membranes. Pneumonia and some other ailments can cause one or both pleurae to become inflamed and swollen. The membranes then rub against one another, making breathing painful—a condition called *pleurisy*.

Inside the lungs, airways become progressively shorter, narrower, and more numerous. The terminal airways, the respiratory bronchioles, have cup-shaped outpouchings from their walls. Each thin, delicate outpouching is an **alveolus** (plural: *alveoli*). Most often, alveoli are clustered together, forming a larger pouch called an **alveolar sac** (Figure 8.4, inset). The alveolar sacs collectively are the major sites of gas exchange.

A dense mesh of blood capillaries surrounds the 150 million or so alveoli in each lung. Together, the alveoli provide a tremendous surface area for exchanging gases with the bloodstream. If they were stretched out as a single layer, they would cover the body several times over—or the floor of a racquetball court!

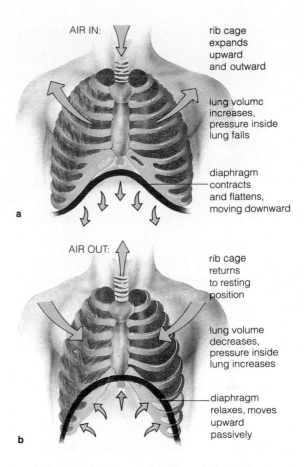

AIR IN:

rib cage expands upward and outward

lung volume increases, pressure inside lung falls

diaphragm contracts and flattens, moving downward

a

AIR OUT:

rib cage returns to resting position

lung volume decreases, pressure inside lung increases

diaphragm relaxes, moves upward passively

b

Figure 8.9 Changes in the size of the chest cavity during breathing. Blue line indicates the position of the diaphragm when you inhale (**a**) and exhale (**b**).

Gases are exchanged in the alveoli of the lungs.

THE MECHANISM OF BREATHING: AIR PRESSURE CHANGES IN THE LUNGS

Ventilation

Every time you take a breath, you are ventilating the respiratory surfaces of your lungs. **Ventilation** has two phases: **Inspiration** (inhaling) draws air into the airways, and **expiration** (exhaling) expels it. The air movements result from rhythmic increases and decreases in the volume of the chest cavity. Every time the volume changes, pressure gradients between the lungs and the air outside the body are reversed. Gases in the respiratory tract follow those gradients.

As you start to inhale, the dome-shaped diaphragm contracts and flattens, and muscles between the ribs contract, pulling the ribs upward and outward (Figure 8.9).

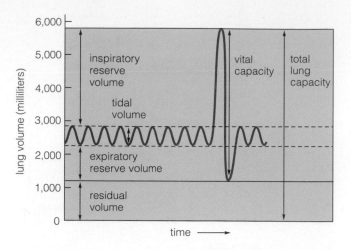

Figure 8.10 Capacities of the human lung. During normal breathing at rest, a tidal volume of air enters and leaves the lungs. Forced inhalation can bring much larger quantities (the inspiratory reserve) into the lungs, and forced exhalation can release some of the air (the expiratory reserve) normally kept in the lungs. A residual volume of gas remains trapped in partially filled alveoli despite the strongest exhalation.

Figure 8.11 Diagram of a section through an alveolus and an adjacent blood capillary. Compared to the red blood cell's diameter, the diffusion distance across the capillary wall, the interstitial fluid, and the alveolar wall is small.

Then, as the chest cavity expands, the rib cage moves away slightly from the lung surface. Pressure in the intrapleural space decreases, becoming lower than it was, compared to atmospheric pressure. The pressure difference causes the lung itself to expand more. Just as air rushes in to fill a vacuum, this expansion causes fresh air to flow down the airways.

When you start to exhale, the elastic lung tissue recoils passively, the diaphragm relaxes, and the rib cage returns to its previous position. As a result, the volume of the chest cavity shrinks. This compresses the air in alveolar sacs. The pressure in the sacs is now greater than the atmospheric pressure, so air follows the gradient and moves out from the lungs. When exercise or some other activity increases the body's demand for oxygen, muscles of the rib cage and abdomen are recruited to hasten airflow. When you sneeze or cough, abdominal muscles contract suddenly, pushing the diaphragm upward. The rapid decrease in the volume of the chest cavity pushes air out the nose and mouth with explosive force.

The air movements of breathing occur because of increases and decreases in the volume of the chest cavity and the corresponding changes in pressure gradients between the lungs and air outside the body.

A lung can "collapse" when, due to an event such as being punctured by a broken rib, air enters the intra-

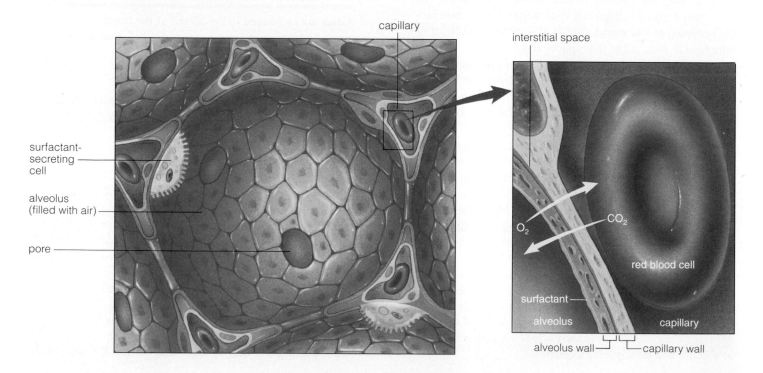

pleural space. Then, because there's no pressure difference with the outside, the lung stays deflated.

Lung Volumes

When you are resting, a **tidal volume** of about 500 milliliters (two cupfuls) of air enters or leaves your lungs in a normal breath. You can increase the volume of air you inhale or exhale considerably, however. In addition to air taken in as part of the tidal volume, a person can forcibly inhale roughly 3,100 milliliters of air, called the *inspiratory reserve volume*. Forcibly exhaling, you can expel a *expiratory reserve volume* of about 1,200 milliliters. **Vital capacity**—the maximum volume of air that can move out of the lungs after a person inhales as deeply as possible—is about 4,800 milliliters for a young man and about 3,800 ml for a young woman. People rarely take in more than half their vital capacity, even when they breathe deeply during strenuous exercise. Even at the end of your deepest exhalation, your lungs still cannot be completely emptied of air; roughly another 1,200 milliliters of *residual volume* remains (Figure 8.10).

How much of the 500 milliliters of inhaled air is actually available for gas exchange? About 150 milliliters of exhaled "dead" air remain in the air-conducting tubes between breaths and never reach the lungs. Thus only 350 (500 − 150) milliliters of fresh air reach the alveoli with each inhalation. If you breathe 10 times a minute, you are supplying your alveoli with (350 × 10) or 3,500 milliliters of fresh air per minute—the same volume contained in three and one-half 1-liter bottles of soda pop.

GAS EXCHANGE AND TRANSPORT

Ventilation brings oxygen into the lungs and moves carbon dioxide out of them. But ventilation is not the same thing as *respiration*. In respiration, gases diffuse between interstitial fluid and individual cells in tissues, or between blood and interstitial fluid. Respiration has external and internal phases. In **external respiration**, oxygen moves from alveoli into the blood, and carbon dioxide moves in the opposite direction. In **internal respiration**, oxygen moves out of the blood into tissues, and carbon dioxide leaves tissues and moves into the blood.

External Respiration: Gas Exchange in Alveoli

Each alveolus (Figure 8.11) is only a single layer of epithelial cells surrounded by a thin basement membrane. A thin film of interstitial fluid separates the walls of alveoli from the walls of lung capillaries. This fluid-filled space

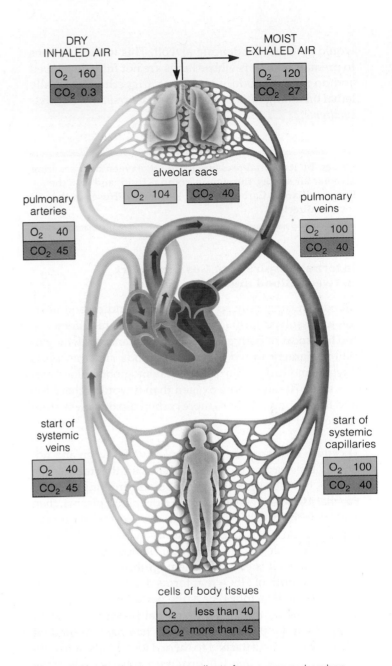

Figure 8.12 Partial pressure gradients for oxygen and carbon dioxide through the respiratory tract. The point to remember about the values shown is that *each gas moves from regions of higher to lower partial pressure.*

is narrow, and gases can diffuse rapidly across it. Figure 8.12 shows the partial pressure gradients for oxygen and carbon dioxide throughout the respiratory system of a person at rest. Oxygen diffuses across the respiratory surface and into the bloodstream, and carbon dioxide diffuses in the reverse direction.

Some of the epithelial cells of alveoli are surfactant-secreting cells (see Figure 8.11). *Pulmonary surfactant* is a substance that reduces the surface tension of the watery fluid film. Without surfactant, the force of surface tension would collapse the delicate alveoli. This in fact happens to premature infants whose lungs are not fully developed and do not yet produce surfactant, triggering potentially lethal breathing difficulties called *infant respiratory distress syndrome.*

Driven by its partial pressure gradient, oxygen diffuses from alveolar air spaces, through interstitial fluid, and into the lung capillaries. Carbon dioxide similarly diffuses in the reverse direction.

Internal Respiration: Gas Transport Between Blood and Other Tissues

Blood can carry only so much oxygen and carbon dioxide in dissolved form. To meet the body's requirements, the transport of both gases must be enhanced. Hemoglobin, a pigment in red blood cells, binds and transports oxygen and carbon dioxide. With hemoglobin, blood carries some 70 times more oxygen than it would otherwise, and it transports 17 times more carbon dioxide away from tissues.

Oxygen Transport   There is plenty of oxygen and relatively little carbon dioxide in inhaled air that reaches the alveoli, but the opposite is true of blood entering the lung capillaries. So oxygen diffuses into the blood plasma, then into red blood cells. Up to four oxygen molecules can rapidly form a weak, reversible bond with each hemoglobin molecule in a red blood cell. Hemoglobin with oxygen bound to it is called **oxyhemoglobin**, or HbO_2.

The amount of HbO_2 that forms depends on the interplay of several factors. One is the partial pressure of oxygen. In general, the higher its partial pressure, the more oxygen will be picked up, until all hemoglobin-binding sites have oxygen attached to them. HbO_2 holds onto oxygen rather weakly and will give it up in tissues where the partial pressure of oxygen is lower than in the blood.

Certain conditions reduce hemoglobin's affinity for binding oxygen. They all occur in tissues with high metabolic activity—and with correspondingly greater oxygen demands. For example, hemoglobin-oxygen binding becomes weaker as temperature rises. It also becomes weaker as pH declines. Several events contribute to a falling pH; the reaction that forms HbO_2 releases hydrogen ions (H^+), and as the level of H^+ increases, the blood becomes more acidic. Blood pH also declines due to an increase in the level of carbon dioxide given off by active cells.

When tissues chronically receive less oxygen than proper functioning requires (see *Focus on Environment*), red blood cells increase their production of a compound called 2,3-diphosphoglycerate (DPG), which reversibly binds hemoglobin. The more DPG bound to hemoglobin, the *less* affinity hemoglobin has for binding oxygen and thus the more oxygen can be available to tissues.

Carbon Dioxide Transport   Cellular repiration produces carbon dioxide as a waste. For this reason, there is relatively more carbon dioxide in metabolically active tissues than in blood flowing through the capillaries threading through them. Carbon dioxide thus diffuses into the bloodstream and is then transported to the lungs in one of three ways. About 7 percent of the carbon dioxide remains dissolved in plasma. About 23 percent binds with hemoglobin, forming **carbaminohemoglobin** ($HbCO_2$). But most of it—approximately 70 percent—is transported in plasma in the form of bicarbonate (HCO_3^-).

Bicarbonate forms when carbon dioxide combines with water in plasma to form carbonic acid (H_2CO_3). The carbonic acid dissociates (separates) into bicarbonate and hydrogen ions:

$$CO_2 + H_2O \rightleftharpoons \underset{\text{carbonic acid}}{H_2CO_3} \rightleftharpoons \underset{\text{bicarbonate}}{HCO_3^-} + H^+$$

The reactions proceed slowly in plasma, converting only 1 in every 1,000 molecules of carbon dioxide. Thus, the reaction rate in plasma is insignificant. However, much of the carbon dioxide diffuses into red blood cells, which contain the enzyme **carbonic anhydrase**. Recall that enzymes speed up chemical reactions in cells; with this enzyme, the reaction rate increases by 250 times. Inside red blood cells, most of the carbon dioxide that is not bound to hemoglobin is converted to carbonic acid and its dissociation products. As a result, the concentration of free carbon dioxide in the blood drops rapidly. Thus, carbonic anhydrase helps maintain the gradient that keeps carbon dioxide diffusing from interstitial fluid into the bloodstream.

The bicarbonate ions formed during the reaction tend to diffuse out of the red blood cells and into the blood plasma. Hemoglobin binds some of the hydrogen ions and thus acts as a buffer. (Recall from Chapter 2 that a *buffer* is any molecule that combines with or releases hydrogen ions in response to changes in cellular pH.) Certain plasma proteins also bind H^+. Such buffering mechanisms help prevent an abnormal decline in blood pH.

The reactions are reversed in the alveoli, where the partial pressure of carbon dioxide is lower than it is in the surrounding capillaries. The water and carbon dioxide that form as a result of the reactions diffuse into the alveolar sacs. From there the carbon dioxide is exhaled from the body. The water helps keep the respiratory tract moistened.

Health Hazards of Breathing Polluted Air

In large cities, in certain occupations, and anyplace near a cigarette smoker, airborne particles and irritating gases put extra workloads on the lungs. Chronically breathing polluted air can result in any of the following disorders:

Bronchitis Bronchitis can be brought on by any form of air pollution that increases mucus secretions and interferes with ciliary action in the lungs. Ciliated epithelium in the bronchioles is especially sensitive to cigarette smoke. Mucus and the particles it traps—including bacteria—accumulate in airways, and coughing starts. The cough persists as long as the irritation does, and the bronchial walls become inflamed. Bacteria or chemical agents start destroying the wall tissue. Cilia are lost from the lining, and mucus-secreting cells multiply as the body fights against the accumulating debris. Eventually scar tissue forms and can block parts of the respiratory tract.

Emphysema Even acute bronchitis is easily treated if the person is otherwise in good health. When inflammation continues, however, scar tissue builds up and the bronchi become clogged with more and more mucus. Enzymes begin to dissolve the elastic tissues of alveoli. As the walls of alveoli break down, stiffer fibrous tissue comes to surround them. Remaining alveoli enlarge and the balance between air flow and blood flow is skewed. The outcome is emphysema, a disorder in which the lungs are so distended and inelastic that gases cannot be exchanged efficiently (compare Figures *a* and *b*). Running, walking, even exhaling can be difficult. Roughly 1.3 million people in the United States have emphysema.

Smoking, poor diet, chronic colds, and other respiratory ailments sometimes make a person susceptible to emphysema later in life. Many sufferers lack a functional gene coding for antitrypsin, a protein that inhibits tissue-destroying enzymes produced by bacteria. Emphysema can develop over 20 or 30 years. By the time it is detected, lung tissue is permanently damaged.

Lung Cancer and Other Perils of Cigarette Smoke
When humans started smoking tobacco, they inadvertently began wreaking havoc with their lungs. Smoke from one cigarette can keep cilia in bronchioles from beating for hours. Noxious particles it contains can stimulate mucus secretions (which in time can clog the airways) and kill the infection-fighting phagocytes that nor-

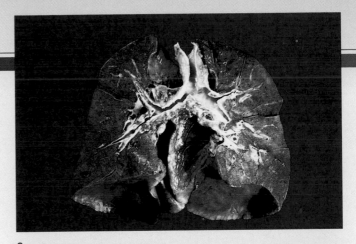

a

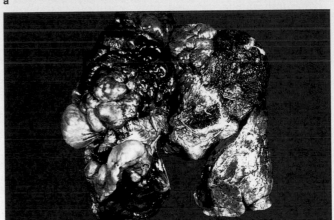

b

(**a**) Normal appearance of human lung tissue. (**b**) Appearance of lung tissue from someone affected by emphysema.

mally patrol the respiratory epithelium. The ensuing "smoker's cough" can pave the way for bronchitis and emphysema. Recent studies strongly suggest that smoking dramatically increases a person's chances of developing cataracts, a disorder in which the lens of the eye clouds over.

Today we know for certain that cigarette smoke—including "secondhand smoke" inhaled by a nonsmoker—causes lung cancer and contributes to emphysema and other ills. In the body, some compounds in coal tar and cigarette smoke are converted to highly reactive forms. These are the real carcinogens; they provoke genetic damage leading to lung cancer. The pain associated with terminal lung cancer is agonizing.

Susceptibility to lung cancer is related to the number of cigarettes smoked per day, and how often and how deeply smoke is inhaled. Cigarette smoking is responsible for at least 80 percent of all lung cancer deaths. As people become educated about the health hazards of smoking, sales of cigarettes are beginning to fall in many countries.

Hemoglobin in red blood cells dramatically increases the oxygen-carrying capacity of blood. Oxygen binding by hemoglobin is influenced by the levels (partial pressures) of oxygen and carbon dioxide in blood and by blood pH and temperature.

Most carbon dioxide is transported as bicarbonate in plasma. Hemoglobin also carries some CO_2, and helps maintain stable blood pH by taking up hydrogen ions liberated when bicarbonate forms.

CONTROLS OVER BREATHING

To maintain satisfactory operating conditions in the body, gas exchange in the respiratory system must be efficient enough to steadily deliver oxygen to respiring cells and to carry away carbon dioxide wastes. Gas exchange is most efficient when breathing is regulated so that the rate of air flow is matched with the rate of blood flow.

Local Controls in the Lungs

Some controls over breathing come into play locally in the lungs. For example, if your heart begins pumping hard and fast but your lungs aren't ventilating at a corresponding pace, blood flows too fast and air movement is too sluggish for the efficient disposal of carbon dioxide. An increase in the blood level of carbon dioxide affects smooth muscle in the bronchiole walls. Bronchioles dilate, enhancing the air flow. Similarly, a decrease in carbon dioxide levels causes the bronchiole walls to constrict, thereby decreasing the air flow.

Local controls also work on blood vessels in the lungs. When air flow is too great relative to blood flow, oxygen levels rise in some parts of the lungs. The increase affects smooth muscle in blood vessel walls. The vessels dilate, so blood flow to the region increases. If there is too little air flow, the vessels constrict and blood flow decreases.

Neural Controls

Physiologists are beginning to understand the neural controls that regulate normal, quiet breathing. Contraction of the diaphragm and the muscles that move the rib cage are under the control of neurons in the reticular formation, a system of nerve cells (neurons) running through the brainstem. One cell cluster of this formation (in the *medulla* region of the brainstem) coordinates the signals calling for inhalation; another coordinates the signals for exhalation. The resulting rhythmic contractions are fine-tuned by respiratory centers in another part of the brain (the *pons*) that can stimulate or inhibit both cell clusters.

As Figure 8.13a suggests, when you inhale, signals from the reticular formation travel nerve pathways to the diaphragm and chest. The signals stimulate the rib muscles and diaphragm to contract, the rib cage expands, and air moves into the lungs. When the diaphragm and chest muscles relax, the rib cage returns to its unexpanded state due to elastic recoil and air in the lungs moves outward. When you breathe rapidly or deeply, stretch receptors in alveoli send signals to brain control centers, which respond by inhibiting contraction of the diaphragm and rib cage muscles—so you exhale.

Monitoring of Carbon Dioxide You might suppose that nervous system controls over breathing mainly involve monitoring the level of oxygen in blood. However, the nervous system is more sensitive to changes in the level of carbon dioxide. Both gases are monitored in arterial blood. When conditions warrant, nervous system signals adjust contractions of the diaphragm and muscles in the chest wall, and so adjust the rate and depth of breathing.

Sensory receptors in the medulla of the brain can detect rising carbon dioxide levels. Corresponding increases in H^+ in the blood (Figure 8.14b) affect the pH of cerebrospinal fluid, which bathes the medulla; the pH shift stimulates the receptors, which signal the change to respiratory centers in the brain. The brain also receives input from sensory receptors, including **carotid bodies**, where the carotid arteries branch to the brain, and **aortic bodies** in arterial walls near the heart. Both types of receptors are sensitive to changes in levels of carbon dioxide and of oxygen in arterial blood, and to changes in blood pH. The brain responds by increasing the rate of ventilation, so more oxygen can be delivered to affected tissues.

Gas exchange mechanisms in the body are like parts of a fine Swiss watch—intricately coordinated and smooth in their operation. *Focus on Environment* outlines some respiratory disorders that can result from breathing polluted air, many of them serious threats to health. Table 8.1 summarizes these and some other respiratory system ailments.

RESPIRATION IN UNUSUAL ENVIRONMENTS

Decompression Sickness

In oceans, the water pressure greatly increases with depth. Pressure in itself is not a problem, so long as it is dealt with. For example, to prevent their lungs from collapsing, divers may rely on tanks of compressed air (that is, air under pressure). Divers also must deal with the additional gaseous nitrogen (N_2) dissolved in their body fluids and tissues, especially in adipose (fat) tissues.

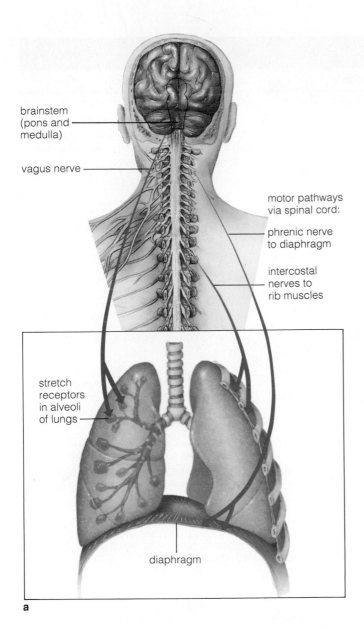

brainstem (pons and medulla)

vagus nerve

motor pathways via spinal cord:

phrenic nerve to diaphragm

intercostal nerves to rib muscles

stretch receptors in alveoli of lungs

diaphragm

a

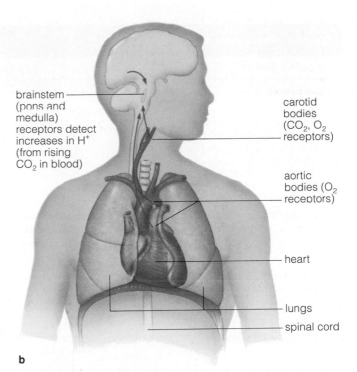

brainstem (pons and medulla) receptors detect increases in H⁺ (from rising CO₂ in blood)

carotid bodies (CO₂, O₂ receptors)

aortic bodies (O₂ receptors)

heart

lungs

spinal cord

b

Figure 8.13 Controls of breathing. (**a**) In normal quiet breathing, cell clusters (reticular formation) in the brainstem coordinate signals to the diaphragm and muscles that move the rib cage, triggering inhalation. When a person breathes deeply or rapidly, another cell cluster receives signals from stretch receptors in the lungs and coordinates signals for exhalation. (**b**) Different sensory receptors detect changes in the concentrations of carbon dioxide and oxygen in the blood.

As a diver ascends, the total pressure of the surrounding water decreases, so N_2 tends to move out of tissues and into the bloodstream. If the ascent is too rapid, the N_2 comes out of solution faster than the lungs can dispose of it. When this happens, bubbles of N_2 form. Too many bubbles cause pain, especially at the joints. Hence the term *the bends* for what is otherwise known as *decompression sickness*. Bubbles that obstruct blood flow to the brain can lead to deafness, impaired vision, and paralysis. At depths of about 150 meters, N_2 poses still another threat, for at high partial pressures it produces feelings of euphoria, even drunkenness (*nitrogen narcosis*), and divers have been known to offer the mouthpiece of their airtank to a fish. To forestall such problems, knowledgeable divers who make deep descents use a gas mixture that does not contain nitrogen.

Hypoxia

As you read at the start of this chapter, the partial pressure of oxygen decreases as altitude increases. Visitors who have not had time to acclimatize to the thinner air at high altitudes can become *hypoxic*—that is, their tissues are chronically short of oxygen. Generally, above 2,400 meters (about 8,000 feet) above sea level, respiratory centers work to compensate for the oxygen deficiency by triggering *hyperventilation* (breathing much more frequently and more deeply than normal). Snorkelers and competitive swimmers sometimes hyperventilate. The strategy doesn't increase the amount of oxygen available to tissues. It does increase pH, and it *decreases* the level of carbon dioxide in the blood. These effects prolong the amount of time before neural controls prompt the swimmer to surface and take a breath of air.

Hypoxia also occurs in *carbon monoxide poisoning*. Carbon monoxide, a colorless, odorless gas, is present in automobile exhaust fumes and in smoke from burning tobacco, coal, charcoal, or wood. It binds to hemoglobin

Table 8.1 Respiratory System Under Attack

Cause/Risk Factors	Major Effects	Symptoms
Lung cancer		
Most cases associated with tobacco smoking, or regular exposure to smoke (*passive smoking*). Other risk factors include inhaling asbestos fibers, exposure to radioactivity, and living where air pollution is high	Tumors develop in lung tissue, with resulting loss of respiratory surface for gas exchange. Tumor spread may trigger pneumonia, lung collapse, or accumulation of fluid between the pleural membranes	Persistent cough or chronic bronchitis, chest pain, shortness of breath, coughing up blood. Weight loss, especially if cancer has spread to other tissues
Infections		
Bronchitis Often associated with colds or flu. Can be *acute* or *chronic;* may be a symptom of lung cancer	Increased mucus secretion in bronchi and trachea, which leads to coughing. Chronic form can result in inflammation of bronchial walls and scarring	Cough, breathlessness, chest pain. Yellow or greenish phlegm; sometimes, fever
Pneumonia Infection by viruses or bacteria; inhaling toxic fumes	Inflammation of lung tissue; fluid buildup in alveoli	Dry cough, fever; shortness of breath. Symptoms sometimes include chills, blood in phlegm, chest pain, blue tinge to skin
Histoplasmosis Fungal infection, most common in southeastern United States	Inflammation and increased mucus secretion in respiratory tract; can spread to retina of the eye and cause blindness	Flulike symptoms, including cough and fever
Influenza Viral infection, usually spread from person to person by contact with infectious particles (coughs, sneezes, discarded tissues)	Initial infection of mucous membranes in nose and throat, spreading into the trachea and lungs. May be followed by bacterial infection, which leads to bronchitis or pneumonia	Chills, fever, headache, muscle pain, sore throat, usually followed by cough, chest pain, and runny nose due to increased mucus secretion
Tuberculosis Infection by the bacterium *Mycobacterium tuberculosis.* TB had become rare in the United States until the advent of HIV and AIDS. Now, this disease is becoming common once again among AIDS sufferers	Destruction of patches of lung tissue; possible spread of infection to other body regions	Flulike illness; if infection progresses, mild fever, night sweats, fatigue, cough and chest pain. Phlegm containing blood and/or pus. Can be fatal if not treated
Allergic rhinitis (hay fever) Airborne pollen, animal hair, feathers, or other irritants	Release of histamine, which triggers inflammation and fluid production in nasal passages, sinuses, eyelids and surface layer of eyes	Sneezing, runny nose, watery, itchy eyes; dry throat, wheezing
Injuries		
Pneumothorax Injury, such as knife or bullet wound or broken rib, that permits air to enter pleural space	Air enters space between pleural membranes, eliminating pressure difference between lungs and external environment	Lungs collapse
Asbestosis and *silicosis* Inhalation of particles or fibers of asbestos and silica, respectively	Irritation of lung tissue followed by development of scar tissue, so lungs become stiff and inflexible. Occasionally, lung cancer	Breathlessness; chronic cough with heavy phlegm. Increased susceptibility to other lung disorders
Emphysema		
Inflammation resulting from chronic bronchitis or severe asthma	Enzymatic destruction of walls of alveoli and deterioration of gas exchange	Difficulty breathing, usually becoming progressively worse. Can lead to heart failure and death
Pulmonary embolism		
Blood clot in artery leading to the lungs	Blockage of blood flow to lungs	Heart failure; general collapse of the circulatory system
Respiratory distress syndrome		
Lack of *surfactant* in the lungs of a premature newborn. Normally, surfactant acts to keep lung alveoli open so gas exchange can proceed	Alveoli begin to close up within hours of birth	Labored, rapid breathing. Can cause death in premature infants. Also affects infants born to diabetic mothers

at least 200 times more tightly than oxygen does. Even very small amounts can tie up half of the body's hemoglobin and seriously disrupt oxygen delivery to tissues.

SUMMARY

1. Human cells rely mainly on aerobic respiration, a pathway that requires oxygen and produces carbon dioxide. The internal process by which the human body as a whole acquires oxygen and disposes of carbon dioxide is called respiration.

2. Air is a mixture of gases, each exerting a partial pressure. Each gas tends to move from areas of higher partial pressure to areas of lower partial pressure. The respiratory system makes use of this tendency.

3. In the lungs, oxygen and carbon dioxide diffuse across a moist, thin layer of epithelium (the respiratory surface). Airways carry gases to and from one side of the respiratory surface, and blood vessels carry gases to and from the other side.

4. The interconnected airways of the human respiratory system are the nasal cavities, pharynx, larynx, trachea, bronchi, and bronchioles. Alveoli, located at the end of the airways, are the sites of primary gas exchange in the system.

5. During inhalation, the chest cavity expands, the pressure in the lungs falls below atmospheric pressure, and air flows into the lungs. During normal exhalation, these processes are reversed.

6. Driven by its partial pressure gradient, oxygen brought into the lungs diffuses from alveolar air spaces into the pulmonary capillaries. Then it diffuses into red blood cells and binds weakly with hemoglobin. When oxygenated blood reaches body tissues, hemoglobin gives up the oxygen, which diffuses out of the capillaries, across interstitial fluid, and into cells.

7. Hemoglobin combines with or releases oxygen in response to shifts in oxygen levels, carbon dioxide levels, pH, and temperature.

8. Driven by its pressure gradient, carbon dioxide diffuses from cells, across interstitial fluid, and into the bloodstream. Most CO_2 reacts with water to form bicarbonate, but the reactions are reversed in the lungs, where carbon dioxide diffuses from capillaries into the air spaces of the alveoli.

9. The nervous system controls respiration mainly by monitoring blood levels of carbon dioxide.

1. Label the component parts of the human respiratory system:

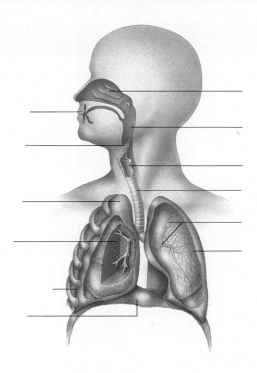

2. What is the main requirement for gas exchange in respiration? *190*

3. What is oxyhemoglobin? Where does it form? *198*

4. What drives oxygen from alveolar air spaces, through interstitial fluid, and across capillary epithelium? What drives carbon dioxide in the reverse direction? *190, 195, 198*

5. How does hemoglobin help maintain the oxygen partial pressure gradient during gas transport in the body? What reactions enhance the transport of carbon dioxide throughout the body? *198*

6. Gas exchange is most efficient when the rates of air flow and blood flow are balanced. Give an example of a local control that comes into play in the lungs when the two rates are imbalanced. Do the same for a neural control. *200*

Critical Thinking: You Decide *(Key in Appendix IV)*

1. Some cigarette manufacturers have conducted public relations campaigns urging their customers to "smoke responsibly." What are some social and biological issues in this controversy? How do these issues apply to the nonsmoking spouse or children of a smoker? To nonsmoking patrons in a restaurant? To the unborn child of a pregnant smoker? In your opinion, what behavior would constitute "responsible smoking"?

2. People occasionally poison themselves with carbon monoxide by building a charcoal fire in an enclosed area. Assuming help arrives in time, what would be the *most* effective treatment: placing the victim outdoors in fresh air, or administering pure oxygen? Why?

Self-Quiz *(Answers in Appendix III)*

1. _____ is used and _____ is produced in aerobic metabolism.

2. Operation of the respiratory system depends on the tendency of a _____ to diffuse down its _____.

3. The internal surface of the lungs consists of thin, moist layers of _____ across which gases can easily _____.

4. Controls over respiration include mechanisms that match _____ flow to _____ flow.

5. Which of the following statements is *not* related to the function of the respiratory system?
   a. Air is a mixture of gases.
   b. Each gas in air exerts a partial pressure.
   c. Each gas in air tends to move from areas of higher partial pressure to areas of lower partial pressure.
   d. All of the above are directly applicable.

6. During inhalation, _____.
   a. the pressure in the thoracic cavity is greater than the pressure within the lungs
   b. the pressure in the thoracic cavity is less than the pressure within the lungs
   c. the diaphragm moves upward and becomes more curved
   d. the chest cavity volume decreases

7. Oxygen diffusing into pulmonary capillaries also diffuses into _____ and binds with _____.
   a. white blood cells; carbon dioxide
   b. red blood cells; carbon dioxide
   c. white blood cells; hemoglobin
   d. red blood cells; hemoglobin

8. Due to its partial pressure gradient, carbon dioxide diffuses from cells, into interstitial fluid, and into the _____; in the lungs, carbon dioxide diffuses into the _____.
   a. alveoli; bronchioles
   b. bloodstream; bronchioles
   c. alveoli; bloodstream
   d. bloodstream; alveoli

9. Hemoglobin performs which of the following respiratory functions:
   a. transports oxygen
   b. transports some carbon dioxide
   c. acts as a buffer to help maintain blood pH
   d. all of the above

10. Match the following respiratory components with their descriptions.
   _____ bronchi   a. microscopically small air sacs where gases are exchanged
   _____ alveoli   b. contains vocal cords
   _____ trachea   c. throat
   _____ larynx   d. air-conducting tube; windpipe
   _____ pharynx   e. connects trachea to lungs

Key Terms

alveolar sac 195	larynx 192
alveolus 195	lung 194
aortic body 200	oxyhemoglobin 198
bronchiole 193	pharynx 192
bronchus 192	pleura 195
carbaminohemoglobin 198	respiration 190
carbonic anhydrase 198	respiratory bronchiole 193
carotid body 200	respiratory surface 190
diaphragm 195	tidal volume 197
expiration 195	trachea 192
external respiration 197	ventilation 195
inspiration 195	vital capacity 197
internal respiration 197	vocal cords 192

Readings

American Cancer Society. *Dangers of Smoking; Benefits of Quitting and Relative Risks of Reduced Exposure,* rev. ed. New York: American Cancer Society.

Fackelmann, K. 2 March 1991. "Respiratory Rescue." *Science News.* Promising new treatments for allergy.

Gorman, C. 22 June 1992. "Asthma: Deadly But Treatable." *Time.*

Spence, A. and E. B. Mason. 1992. *Human Anatomy and Physiology.* St. Paul: West Publishing Company.

9 CONTROL OF SALT-WATER BALANCE AND BODY TEMPERATURE

Survival at Stovepipe Wells

In Death Valley, the sand dunes at Stovepipe Wells cover some 14 square miles. Just south of the dunes are 200 square miles of waterless salt flats. The whole parched region has claimed more than a few human lives.

Without adequate water, a human being can't survive for long anywhere, let alone in Death Valley (Figure 9.1). Homeostasis is a major theme in our study of human anatomy and physiology, and maintaining water balance is a crucial aspect of internal stability. Among other considerations, living cells are mostly water, and water is required for many critical chemical reactions within them. Water in blood and interstitial

fluid acts as a solvent for important salts and other molecules. And most of the potentially toxic nitrogen-containing wastes of protein metabolism are transported out of the body as part of the watery fluid called urine.

Although individuals differ, an adult routinely loses about 2,400 milliliters (two and a half quarts) of water daily, mainly in urine, feces, vapor in exhaled air, and in sweat. In the sweltering Death Valley summer, however, sweating increases dramatically as a mechanism to help dissipate heat. Whereas a person might normally lose about 100 ml of water in sweat a day, in extreme heat (or during heavy exercise) the loss can rise to 900

ml or more *per hour*. The body cannot long tolerate such rapid water loss. So, within the body of a hiker at Stovepipe Wells, a remarkable homeostatic juggling act goes on. In response to changing internal conditions, the brain issues commands you experience as thirst—telling you, in effect, to drink water. Water-conserving mechanisms are also set in motion while other mechanisms operate simultaneously to cool the body, dispose of metabolic wastes, and make other required exchanges with the external environment.

We also may take in *too much* water, eat salty foods that contain large amounts of sodium, or do something else that upsets the optimal balance of water and solutes in blood and body fluids. Fortunately, two fist-sized organs, the kidneys, come to the rescue. Water and solutes from the blood flow continuously through the kidneys, where adjustments occur in how much water and which solutes are reabsorbed into body fluids or disposed of as urine. Those adjustments help our bodies cope with everything from water imbalances to antibiotic drugs, and they will be our main focus in this chapter. We will also consider how homeostatic mechanisms maintain body temperature within limits that support life.

KEY CONCEPTS

1. The body continually gains and loses water, nutrients, and ions. It also continually produces metabolic wastes. Even with all these inputs and outputs, the body's extra-cellular fluid does not change much in its overall volume and composition.

2. The urinary system helps balance the intake and output of water and dissolved substances (solutes). This system eliminates excess water and solutes as a fluid called urine.

3. Urine forms in tubelike nephrons inside a pair of kidneys, which are blood-filtering organs. Nephrons receive water and solutes from a set of blood capillaries. They return most of the filtrate to a second set of blood capillaries; the rest leaves the body as urine.

4. The kidneys precisely adjust how much water and sodium the body excretes or conserves.

5. Body temperature is maintained within a favorable range through controls over metabolic activity and adaptations in structure, physiology, and behavior.

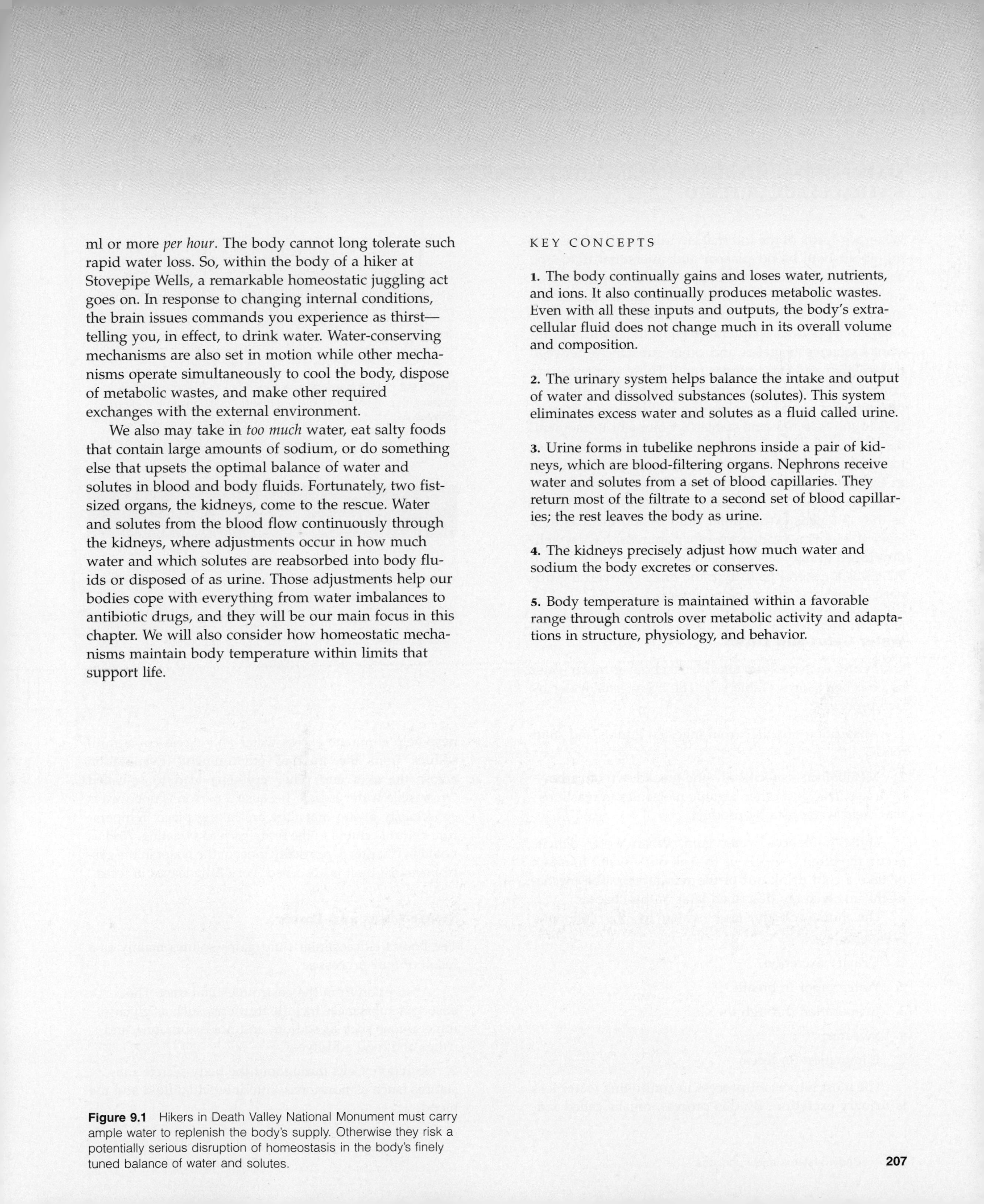

Figure 9.1 Hikers in Death Valley National Monument must carry ample water to replenish the body's supply. Otherwise they risk a potentially serious disruption of homeostasis in the body's finely tuned balance of water and solutes.

MAINTAINING HOMEOSTASIS IN THE EXTRACELLULAR FLUID

When we speak of the internal environment, we are talking about both blood plasma and interstitial fluid, the solute-rich solutions in the body's tissues. Together, they compose the **extracellular fluid** (ECF) that services all living cells of the body. The fluid within cells is *intracellular fluid*. From previous chapters we know that there is a constant exchange of gases and other substances between intracellular and extracellular fluid. Those exchanges are crucial for keeping cells functioning smoothly, and they cannot occur properly unless the volume and composition of the ECF remains stable. Yet moment to moment, the ECF changes as gases, cell secretions, ions, and other materials enter or leave it. To maintain stable conditions in ECF, especially concentrations of water and various ions, there must be mechanisms that remove substances as they enter the extracellular fluid or add them as they leave it. Humans (and many other animals) have a well-developed urinary system that performs this task. Figure 9.2 gives a general picture of the links between the urinary system and other organ systems.

Water Gains and Losses

Ordinarily, each day you take in just about as much water as your body loses (Table 9.1). The body *gains* water by two processes:

1. Absorption of water from ingested liquids and solid foods

2. Metabolism—specifically, the breakdown of carbohydrates, fats, and other organic molecules in reactions that yield water as a by-product.

Thirst influences water gain. When water deficits occur, the brain compels us to seek out a water fountain or take a cold drink out of the refrigerator. The mechanisms involved are described later in this chapter.

The human body *loses* water by the following processes:

1. Urinary excretion

2. Water vapor in breath

3. Evaporation through the skin

4. Sweating

5. Elimination (in feces)

The most important process in controlling water loss is **urinary excretion**. By this process, organs called kid-

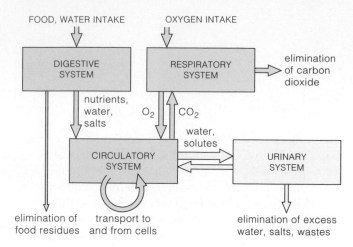

Figure 9.2 Links between the urinary system and other organ systems that maintain stable operating conditions in the internal environment.

Table 9.1	Normal Daily Balance Between Water Gain and Water Loss in Adult Humans		
Water Gain (milliliters)		Water Loss (milliliters)	
Ingested in solids:	850	Urine:	1,500
Ingested as liquids:	1,400	Feces:	200
Metabolically derived:	350	Evaporation:	900
	2,600		2,600

neys help eliminate excess water and excess or harmful solutes from the internal environment. Evaporation across the skin and other epithelia lead to so-called "insensible water losses" because a person is not always consciously aware that they are taking place. Temperature-control centers in the brain govern sweating. And as noted in Chapter 5, normally most of the water in the gastrointestinal tract is absorbed; very little leaves in feces.

Solute Gains and Losses

The body's extracellular fluid gains solutes mainly as a result of four processes:

1. Absorption from the gastrointestinal tract. The absorbed substances include nutrients such as glucose, mineral ions such as sodium and potassium ions, and drugs and food additives.

2. Secretion. Cells throughout the body secrete substances (such as hormones) into interstitial fluid and the blood.

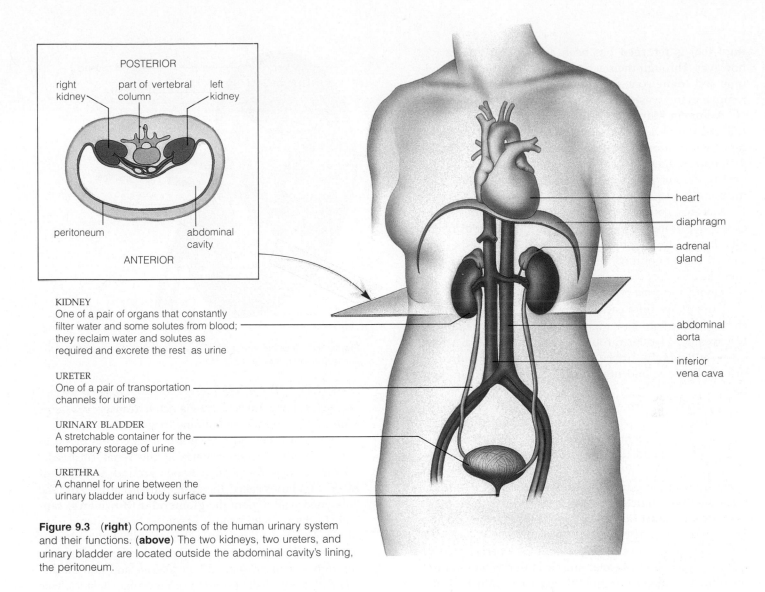

POSTERIOR

right kidney

part of vertebral column

left kidney

peritoneum

abdominal cavity

ANTERIOR

heart

diaphragm

adrenal gland

abdominal aorta

inferior vena cava

KIDNEY
One of a pair of organs that constantly filter water and some solutes from blood; they reclaim water and solutes as required and excrete the rest as urine

URETER
One of a pair of transportation channels for urine

URINARY BLADDER
A stretchable container for the temporary storage of urine

URETHRA
A channel for urine between the urinary bladder and body surface

Figure 9.3 (**right**) Components of the human urinary system and their functions. (**above**) The two kidneys, two ureters, and urinary bladder are located outside the abdominal cavity's lining, the peritoneum.

3. Metabolism, which produces *waste products* such as carbon dioxide and ammonia.

4. Respiration. Oxygen enters the blood via the respiratory system, and carbon dioxide enters the blood from cellular respiration.

Carbon dioxide, the most abundant waste product of metabolism, is exhaled from the lungs. *Uric acid* is formed in reactions that degrade nucleic acids (DNA and RNA). If allowed to accumulate, uric acid can crystallize and collect in the joints, causing *gout*.

Other major metabolic wastes include several by-products of protein metabolism. One of these is *ammonia*, which is formed in "deamination" reactions whereby nitrogen-containing amino groups are stripped from amino acids. If allowed to accumulate in the body, ammonia can be highly toxic.

Urea is produced in the liver in reactions that link two ammonia molecules to carbon dioxide. It is the main nitrogen-containing waste product of protein breakdown.

About 40–60 percent of the urea filtered from blood (in the kidneys) is reabsorbed, and the rest is excreted. Protein breakdown also produces *phosphoric acid, sulfuric acid*, and small amounts of other nitrogen-containing compounds, some of which are highly toxic. These too are excreted, along with creatinine, a by-product of ATP-forming reactions during muscle contraction. A small percentage of nitrogen-containing wastes (ammonia, urea, and uric acid) is excreted in sweat.

Urinary System

Figure 9.3 shows the components of the urinary system and their functions. The paired **kidneys** continuously filter water, mineral ions, organic wastes, and other substances from the blood. Normally only a tiny portion of the water and solutes entering the kidneys leaves as a fluid called urine. In fact, except when you take in a great deal of fluid (without exercise), all but about 1 percent of the water is returned to the blood. The composition of the fluid that is returned has been adjusted in vital ways, however. Through their action, kidneys regulate the volume and solute concentrations of extracellular fluid and remove many waste products.

A human kidney is about the size of a large dinner roll and is divided internally into several lobes. An outer *cortex* wraps around a central region, the *medulla* (Figure 9.4). Each kidney is enclosed within a tough coat of connective tissue, the *renal capsule* (from the Latin *renes*, meaning kidneys).

Kidneys contain blood vessels and slender tubes called **nephrons**. Water and solutes filtering out of the blood enter the nephrons. Most of the filtrate is reabsorbed from nephrons, but some continues on through tubelike *collecting ducts* and drains into the kidney's central cavity, the *renal pelvis*. This fluid is urine.

Urine flows from each kidney into a **ureter**, a tubular channel, then into a storage organ, the **urinary bladder**. It leaves the bladder through a tube, the **urethra**, which opens at the body surface. The two kidneys, two ureters, urinary bladder, and urethra constitute the **urinary system**.

Urination, or urine flow from the body (also called "micturition"), is a reflex response. As the urinary bladder fills, tension increases in its strong, smooth-muscled walls. Where the bladder joins the urethra, smooth muscle acts as an *internal urethral sphincter* that helps prevent urine from flowing into the urethra. As muscle tension increases, the sphincter relaxes; at the same time the bladder walls contract and force fluid through the urethra. The internal sphincter cannot be controlled voluntarily, but a person can exert control over an *external urethral sphincter* formed by skeletal muscle closer to the urethral opening. Learning to control this external sphincter is the basis of urinary "toilet training" in young children.

Kidney stones are deposits of uric acid, calcium salts, and other substances that have settled out of urine and have collected in the renal pelvis. Smaller kidney stones usually pass naturally from the body during urination. Larger ones can become lodged in the renal pelvis or ureter or, on rare occasions, in the bladder or urethra. The blockage can interfere with urine flow and cause intense pain. Large kidney stones must be removed by medical or surgical procedures. In *lithotripsy*, high-energy sound waves literally blast the stone to bits small enough to pass out in the urine.

Nephron Structure

Each human kidney has more than a million nephrons. The nephron tube consists only of a single layer of epithelial cells, but the cells and junctions between them vary in different regions of the tube. Some regions are highly permeable to water and solutes. Other regions prevent

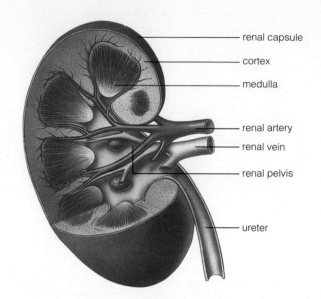

Figure 9.4 Internal structure of the kidney and the major blood vessels leading into and out of it.

the passage of solutes *except* via active transport systems built into the plasma membrane (page 44).

A vessel called an *afferent arteriole* delivers blood to each nephron. Filtration starts at the **renal corpuscle**, where the nephron wall balloons around a cluster of blood capillaries called the **glomerulus** (Figure 9.5). The ballooned wall region, the **glomerular (Bowman's) capsule**, forms a cup for water and solutes being filtered from blood. This fluid flows from the cup into the nephron's *proximal tubule* (closest to the glomerular capsule), then through its hairpin-shaped *loop of Henle* and into the *distal tubule* (most distant from the glomerular capsule). This distal tubule delivers the fluid that enters it to a collecting duct.

Unlike capillaries in most other parts of the body, the glomeruli (capillaries that receive blood inside a glomerular capsule) do not link arterioles and venules and so channel blood directly back to the general circulation. Instead, they converge to form an arteriole (the efferent arteriole) that branches into *another* set of capillaries. These **peritubular capillaries** thread around the rest of the nephron and recapture water and essential solutes (Figure 9.5c). Eventually peritubular capillaries merge into venules, which carry filtered blood out of the kidney.

URINE FORMATION

Urine-Forming Processes

Urine consists of water and solutes that the body does not need to retain in order to maintain the extracellular

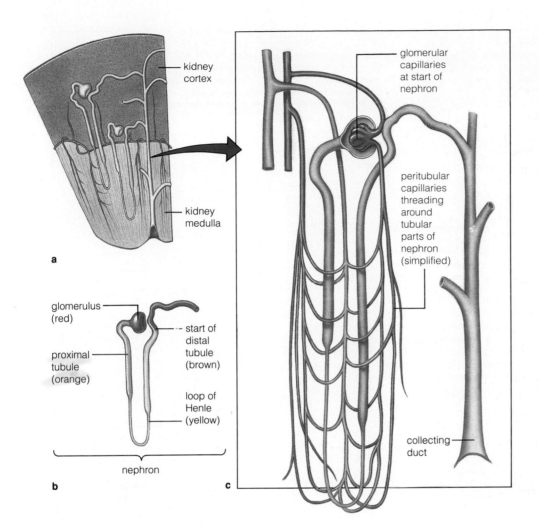

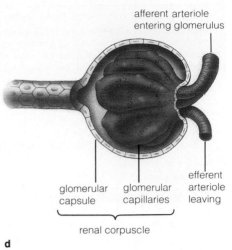

kidney
cortex

kidney
medulla

a

glomerulus
(red)

proximal
tubule
(orange)

start of
distal
tubule
(brown)

loop of
Henle
(yellow)

nephron

b

glomerular
capillaries
at start of
nephron

peritubular
capillaries
threading
around
tubular
parts of
nephron
(simplified)

collecting
duct

c

afferent arteriole
entering glomerulus

glomerular
capsule

glomerular
capillaries

efferent
arteriole
leaving

renal corpuscle

d

Figure 9.5 Diagrams of a nephron and its association with two sets of blood capillaries.

(**a**) Orientation of a nephron relative to the kidney cortex and medulla. (**b**) Functional regions of a nephron. (**c**) The two sets of blood capillaries associated with the nephron. In this sketch, the second set, peritubular capillaries, is highly simplified for clarity; these capillaries actually thread around all tubular parts of the nephron. (**d**) A closer look at the glomerulus, the nephron's blood-filtering unit.

fluid (Table 9.2). Urine forms through a sequence of three processes: filtration, reabsorption, and secretion.

In **filtration**, water and solutes leave the blood in the glomerular capillaries and move into the cupped region inside the glomerular capsule. This *filtrate* is forced out of the capillaries by blood pressure. The blood is "filtered" because blood cells, proteins, and other large solutes are left behind while water and smaller solutes (such as glucose, sodium, and urea) leave the capillaries. The filtrate flows on into the proximal tubule.

Reabsorption proceeds along the nephron's tubular parts. Most of the filtrate's water and solutes—including sodium ions, vitamins, and nutrients such as glucose—move across the tubule wall and *out* of the nephron (by diffusion or active transport), then into adjacent capillaries.

Secretion also occurs across the tubule walls, but in the *opposite* direction. Excess potassium ions in blood plasma and a few other substances are moved *out* of the

Table 9.2	Typical Kinds and Daily Amounts of Solutes in Normal Urine *
Solute	Amount of Solute
Urea	20.0–35.0 grams
Sodium	4.0–6.0 grams
Chloride	6.0 9.0 grams
Potassium	2.5–3.5 grams
Creatinine	1.0–1.5 grams
Calcium	0.01–0.30 grams

capillaries and into cells of the nephron wall. The cells of the nephron wall secrete the substances into the forming urine. Among other functions, this highly regulated process also rids the body of uric acid, some breakdown products of hemoglobin and other proteins, and other

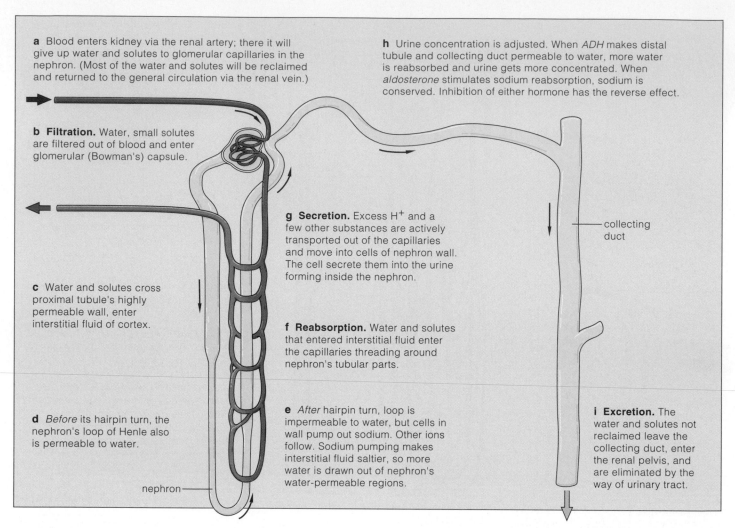

a Blood enters kidney via the renal artery; there it will give up water and solutes to glomerular capillaries in the nephron. (Most of the water and solutes will be reclaimed and returned to the general circulation via the renal vein.)

b Filtration. Water, small solutes are filtered out of blood and enter glomerular (Bowman's) capsule.

c Water and solutes cross proximal tubule's highly permeable wall, enter interstitial fluid of cortex.

d *Before* its hairpin turn, the nephron's loop of Henle also is permeable to water.

nephron

g Secretion. Excess H$^+$ and a few other substances are actively transported out of the capillaries and move into cells of nephron wall. The cell secrete them into the urine forming inside the nephron.

f Reabsorption. Water and solutes that entered interstitial fluid enter the capillaries threading around nephron's tubular parts.

e *After* hairpin turn, loop is impermeable to water, but cells in wall pump out sodium. Other ions follow. Sodium pumping makes interstitial fluid saltier, so more water is drawn out of nephron's water-permeable regions.

h Urine concentration is adjusted. When *ADH* makes distal tubule and collecting duct permeable to water, more water is reabsorbed and urine gets more concentrated. When *aldosterone* stimulates sodium reabsorption, sodium is conserved. Inhibition of either hormone has the reverse effect.

collecting duct

i Excretion. The water and solutes not reclaimed leave the collecting duct, enter the renal pelvis, and are eliminated by the way of urinary tract.

Figure 9.6 Processes involved in urine formation. *Filtration* occurs in the cup-shaped region of a nephron. *Reabsorption* and *secretion* occur at its tubular regions. Excretion occurs via the urethra, at the end of the urinary tract.

metabolic wastes. It also removes drugs, including penicillin, and other water-soluble foreign substances such as pesticides. Drug testing of athletes and employees in certain professions relies on the use of urinalysis to detect drug residues secreted into the urine. *Focus on Wellness* (page 213) describes how urinalysis can help evaluate a person's health.

Figure 9.6 summarizes the processes of urine formation. Table 9.3 lists average daily values for reabsorption of some substances from urine.

Urine forms through the processes of filtration, reabsorption, and secretion. It includes water and solutes not needed to maintain the extracellular fluid, as well as water-soluble wastes.

Table 9.3	Average Daily Reabsorption Values for a Few Substances		
	Amount Filtered	Amount Excreted	Proportion Reabsorbed
Water	180 liters	1.8 liters	99%
Glucose	180 grams	None, normally	100%
Sodium ions	630 grams	3.2 grams	99.5%
Urea	54 grams	30 grams	44%

Factors That Influence Blood Filtration

Each day, more blood flows through the kidneys than through any other organ except the lungs. Each minute, about 1.5 quarts of blood course through them! That is nearly one-fourth of the cardiac output (at rest). How can kidneys handle blood flowing through on such a massive scale? There are two mechanisms.

Avoiding Trouble in the Urinary Tract

Urinary tract infections regularly plague millions of people. Women are especially prone to bladder infections because of their urinary anatomy: The female urethra is short, just a little over an inch long. (In males, the urethra is about 9 inches long.) Hence it is relatively easy for bacteria from outside the body to make their way to a female's bladder and trigger the inflammation known as *cystitis*—or even all the way to the kidneys to cause *pyelonephritis*. An increasingly large number of urinary tract infections in both sexes result from sexually transmitted microbes. The organisms that cause *gonorrhea* and *chlamydia* are major culprits (page 346).

In males, the prostate gland wraps around the urethra. As a man ages, the prostate may swell either occasionally or chronically, narrowing the urethra and preventing urine from draining effectively. Then, bacterial growth can trigger infection. Urinary problems can also be an early sign of prostate cancer.

Urinalysis The procedure called *urinalysis* analyzes the composition of urine and is used to help diagnose illness. For example, the presence of glucose in urine may be a symptom of adult onset diabetes, whereas white blood cells (pus) frequently indicate a urinary tract infection. Red blood cells can reveal bleeding due to infection, kidney stones, cancer, or an injury. High levels of albumin and other proteins in urine may indicate severe hypertension, kidney disease, and some other disorders. Bile pigments enter the urine when liver functions are impaired by cirrhosis and hepatitis.

There are simple measures everyone can take to help keep their urinary tract healthy. Drink plenty of fluids, and practice careful hygiene to minimize the opportunity for bacteria to enter the urethra. People who are prone to bladder infections may also want to limit their intake of alcohol, caffeine, and other acidic or spicy foods, all of which can irritate the bladder.

To begin with, the afferent arterioles delivering blood to a glomerulus have a wider diameter—and less resistance to flow—than most arterioles. The hydrostatic pressure caused by heart contractions therefore does not drop as much when blood enters them. By contrast, the efferent arteriole that receives blood from glomeruli offers *high* resistance to blood flow. Because of the properties of the afferent and efferent arterioles, pressure in the glomerular capillaries is higher than in other capillaries.

Second, glomerular capillaries are highly permeable. They do not allow blood cells or protein molecules to escape, but compared to other capillaries they are 10–100 times more permeable to water and small solutes. Because of the higher hydrostatic pressure and greater capillary permeability, the kidneys can filter an average of 45 gallons (180 liters) per day.

At any given time, the *rate* at which kidneys process a given volume of blood depends on the blood flow to them and on how fast their tubules are reabsorbing water. Reabsorption, as you will see shortly, is partly under hormonal control. Blood flow to the kidneys is influenced by neural controls. For example, conditions elsewhere in the body may trigger neural signals that call for vasoconstriction of the arterioles leading into the kidneys. Blood is then diverted away from the kidneys, and when the flow volume to the kidneys is reduced, the filtration rate decreases.

Local events also influence filtration rates. When blood pressure decreases in the arterioles leading into the glomeruli, the smooth muscle in the arteriole walls relaxes, the arterioles dilate, and more blood flows in. When blood pressure rises, the arterioles constrict and less blood flows in. This helps keep blood flow to the kidneys relatively constant, despite changes in blood pressure.

Reabsorption of Water and Sodium

Your kidneys precisely adjust how much water and sodium your body excretes or conserves. It makes no difference whether you drink too much or too little water at lunch, wolf down salty potato chips, or follow a low-sodium diet—your kidneys will make the necessary adjustments to keep the water and salt content of your body relatively constant. Let's take a look at how this regulation is accomplished.

Proximal Tubule Of all the water and sodium filtered in the kidneys, about two-thirds is promptly reabsorbed at the **proximal tubule**—the part of the nephron closest

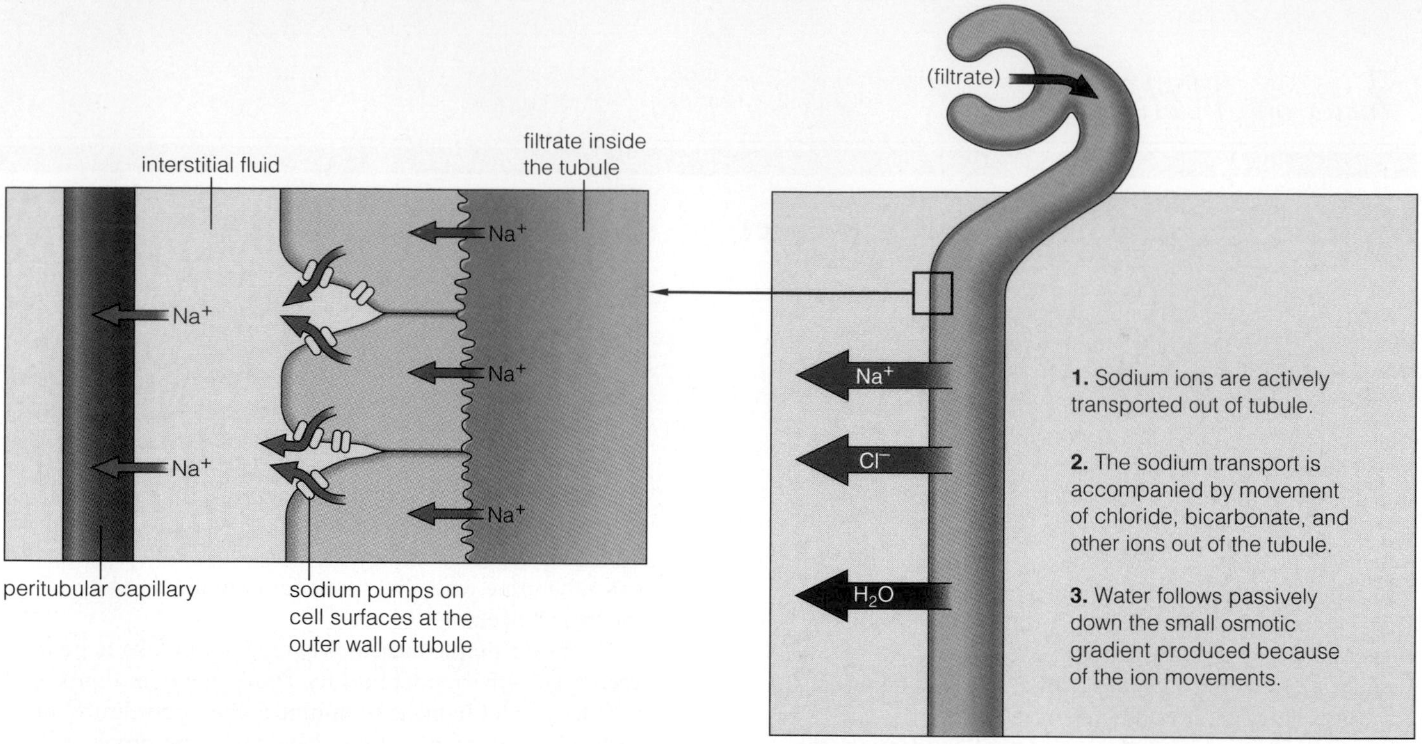
(filtrate)
filtrate inside
the tubule
interstitial fluid
Na+
Na+
Na+
Na+
Na+
peritubular capillary
sodium pumps on
cell surfaces at the
outer wall of tubule
Na+
Cl−
H2O
1. Sodium ions are actively
transported out of tubule.
2. The sodium transport is
accompanied by movement
of chloride, bicarbonate, and
other ions out of the tubule.
3. Water follows passively
down the small osmotic
gradient produced because
of the ion movements.

Figure 9.7 Reabsorption of solutes as a result of the active transport of sodium out of the proximal tubule.

to the glomerulus. As Figure 9.7 shows, epithelial cells of the proximal tubule wall have transport proteins at their outer surface. Nearly all cells have proteins of this sort, which function as sodium "pumps." In this case, the proteins actively transport sodium ions from the filtrate inside the tubule into the interstitial fluid. Sodium ions (Na^+) are positively charged; negatively charged ions including chloride (Cl^-) and bicarbonate (HCO_3^-) follow the sodium. Glucose and amino acids are reabsorbed by other transport mechanisms that also are linked to sodium reabsorption.

This outward movement reduces the solute concentration inside the tubule and increases the solute concentration in the interstitial fluid outside the tubule. The wall of the proximal tubule is quite permeable to water, so water follows the osmotic gradient and moves passively out of the tubule.

In this fashion, the volume of fluid remaining within the proximal tubule decreases greatly, but the total concentration of solutes—sodium especially—changes very little.

Urine Concentration and Dilution The situation changes after filtrate moves on through the proximal tubule and enters the **loop of Henle**. This hairpin-shaped structure descends into the kidney medulla. In the interstitial fluid surrounding the loop, the solute concentration increases progressively with depth within the medulla.

The descending limb of the loop is permeable to water. Water moves out of the descending limb by osmosis, and the solute concentration in the fluid remaining inside increases until it matches that in the interstitial fluid. In the ascending limb, water cannot cross the tubule wall, but sodium is actively transported out. As sodium (and chloride) ions move out of the filtrate, the solute concentration rises outside the tubule and falls inside it. This increase in solute concentration outside the tubule favors the movement of water out of the descending limb.

As solutes leave the tubule, a very high solute concentration develops in the deeper portions of the medulla. Urea contributes to this steep gradient. As water is reabsorbed, the urea left behind in the filtrate becomes concentrated. Some will be excreted in urine, but as filtrate moves into the final portion of the collecting duct, some urea seeps out. It enters the interstitial fluid in the inner medulla, which further increases the concentration of solutes there. Solute concentration is highest in the very deepest parts of the inner medulla.

With so many solutes—but no water—having left the fluid in the ascending limb of the tubule, solutes are not very concentrated there. Hence, the filtrate that finally reaches the distal tubule in the kidney cortex is quite dilute, with a low sodium concentration. As you will see next, the stage is set for the excretion of urine that is either highly dilute or highly concentrated—or anywhere in between.

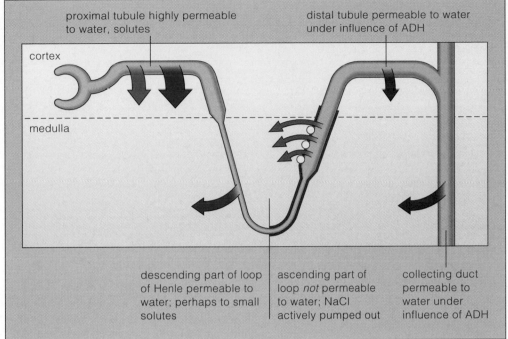

proximal tubule highly permeable to water, solutes

distal tubule permeable to water under influence of ADH

cortex

medulla

isotonic

hypertonic

descending part of loop of Henle permeable to water; perhaps to small solutes

ascending part of loop *not* permeable to water; NaCl actively pumped out

collecting duct permeable to water under influence of ADH

Figure 9.8 *Permeability characteristics of different parts of the nephron. (Both the distal tubule and the collecting duct have very limited permeability to solutes; most of the solute movements in these regions are related directly or indirectly to active transport mechanisms.)*

Hormonal Adjustments of Reabsorption Because so much water and sodium are reabsorbed from the nephron's proximal tubule and loop of Henle, the volume of dilute urine reaching the start of the **distal tubule** (farthest from the glomerulus) has been greatly reduced. Yet if even that reduced volume were excreted without adjustments, the body would rapidly become depleted of both water and sodium. Controlled adjustments are made at cells located in the walls of distal tubules and collecting ducts. Two hormones serve as the agents of control. **ADH** (antidiuretic hormone) influences water reabsorption, and **aldosterone** influences sodium reabsorption.

Let's first consider the role of ADH in adjusting the rate of water reabsorption. The hypothalamus in the brain controls the release of ADH from the posterior pituitary gland. It triggers ADH secretion either when the solute concentration of extracellular fluid rises above a set point (compare page 81) or when blood pressure falls. The solute concentration can increase when a person takes in too little water or becomes dehydrated; severe bleeding (hemorrhage) might cause a rapid decline in blood pressure. ADH acts on distal tubules and collecting ducts, making their walls more permeable to water (Figure 9.8). Thus, in the kidney cortex, water is reabsorbed from the dilute filtrate inside the distal tubules. The volume of fluid inside is now reduced somewhat, and it passes down through the collecting ducts, which plunge down into the medulla. Remember that solute concentrations in

the surrounding interstitial fluid are high in the inner medulla. This encourages the reabsorption of even more water, so only a small volume of very concentrated urine is excreted.

Conversely, when water intake is excessive, the solute concentration in extracellular fluid falls. ADH secretion is inhibited. Without ADH, the walls of the distal tubules and collecting ducts become less permeable to water. Less water is reabsorbed, and a large volume of dilute urine can be excreted. In this way the body rids itself of the excess water.

A *diuretic* is any substance that promotes urine production. Alcohol is a diuretic, which is one reason why drinking beer to replenish body fluids after exercise may be self-defeating. Cold water does the job much more effectively. Caffeine is also a diuretic.

ADH enhances water reabsorption at distal tubules and collecting ducts when the body must conserve water. As a result, the urine is concentrated.

When excess water must be excreted, ADH secretion is inhibited. Less water is reabsorbed, and dilute urine is produced.

Let's now consider how aldosterone helps adjust the rate of sodium reabsorption. When the body loses more sodium than it takes in, the volume of extracellular fluid

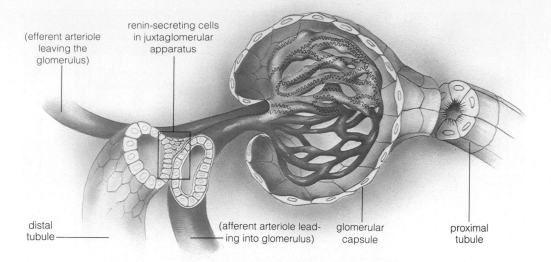

Figure 9.9 Location of juxtaglomerular apparatus and renin-secreting cells that play a role in sodium reabsorption.

(efferent arteriole leaving the glomerulus)

renin-secreting cells in juxtaglomerular apparatus

distal tubule

(afferent arteriole leading into glomerulus)

glomerular capsule

proximal tubule

falls. This is because, as we've just seen, where sodium goes, water follows. Sensory receptors in the walls of blood vessels in the kidney, the heart, and elsewhere detect the decrease, and renin-secreting cells in the kidneys (Figure 9.9) are called into action. Those cells are part of the **juxtaglomerular apparatus**. The name refers to a region of contact between the arterioles of the glomerulus and the distal tubule of the nephron.

Renin acts on molecules of an inactive protein (angiotensinogen) that circulates in the bloodstream. In effect, enzyme action removes part of the molecule, leaving angiotensin I. Subsequent reactions convert angiotensin I to a hormone, angiotensin II. Among other effects, this hormone stimulates cells of the adrenal cortex, the outer portion of a gland perched on top of each kidney (page 277), to secrete aldosterone. Aldosterone causes cells of the distal tubules and collecting ducts to reabsorb sodium faster (by stimulating active transport of Na$^+$), and less sodium is excreted in the urine. Conversely, when the extracellular fluid contains too much sodium, aldosterone secretion is inhibited. Less sodium is reabsorbed, and more is excreted.

In both instances, water "follows the salt." When less sodium is excreted, so is less water, and when more sodium is excreted, more water leaves the body as well.

For various reasons, sodium regulating mechanisms do not operate properly in some people, and the body cannot fully rid itself of excess sodium. Inevitably, tissues retain excess water, which leads to a rise in blood pressure. Abnormally high blood pressure (hypertension) can adversely affect the kidneys as well as the vascular system and brain (page 158). Some hypertensive people can help control their blood pressure by restricting their intake of sodium chloride in table salt and many processed foods.

Aldosterone enhances sodium reabsorption at distal tubules and collecting ducts when the body must conserve sodium.

Salt-Water Balance and Thirst

The body does not rely entirely on events in the kidneys when it needs water. The same stimuli that lead to ADH secretion and increased reabsorption of water in the kidneys also stimulate thirst. Suppose you eat a box of salty popcorn at the movies. Soon the salt is absorbed into your bloodstream. The solute concentration in your extracellular fluid rises, and the hypothalamus detects the increase and prompts ADH secretion. In addition, the **thirst center** is stimulated. Signals from this cluster of nerve cells in the hypothalamus can inhibit saliva production. Your brain interprets the resulting sensation of mouth dryness as "thirst" and leads you to seek fluids.

In fact, a "cottony mouth" is one of the early signs that the body is becoming dehydrated. Dehydration commonly results after severe blood loss, burns, or diarrhea. It also results after profuse sweating, after the body has been deprived of water for a long time, or as a side effect of some medications. Under such circumstances, thirst becomes exceptionally intense.

MAINTAINING ACID-BASE BALANCE

So far, we have focused on the kidney's primary function—that is, how it regulates the total volume of body fluids by controlling water and sodium reabsorption. However, together with the respiratory system and other

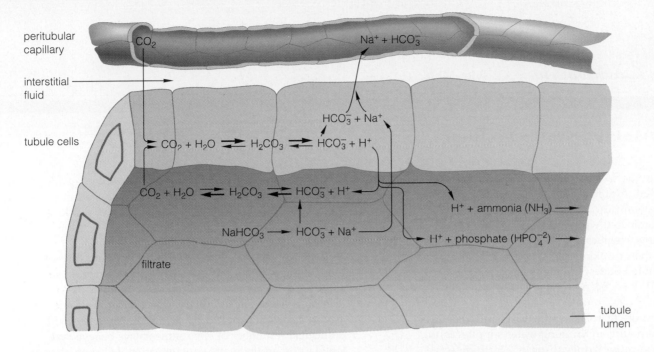

Figure 9.10 How the bicarbonate-carbon dioxide buffer system in the kidneys helps regulate pH.

organ systems, the kidneys also help keep the extracellular fluid from becoming too acidic or too basic (alkaline). In other words, the kidneys help control pH.

The overall acid-base balance of extracellular fluid is maintained by controls over ion concentrations, especially hydrogen ions (H^+). The controls are exerted through (1) buffer systems, (2) respiration, and (3) excretion by way of the kidneys.

The normal extracellular pH for the human body is between 7.34 and 7.45. Maintaining that range involves neutralizing or eliminating a variety of acidic and basic substances entering the blood from the GI tract and from normal metabolism. Recall that acids lower pH and bases raise it. In a person on an ordinary diet, normal cell activities produce excess acids. The acids dissociate, releasing H^+ and lowering the pH. The effect is minimized when excess hydrogen ions react with different kinds of buffers. A *buffer system* consists of substances that combine with H^+ when its concentration rises above a set point and release H^+ when its concentration falls below the set point. The bicarbonate-carbon dioxide buffer system, introduced earlier on page 25, is an example:

$$H^+ + HCO_3^- \rightleftharpoons H_2CO_3 \rightleftharpoons H_2O + CO_2$$
$$\text{bicarbonate} \qquad \text{carbonic acid}$$

Here, the H^+ is neutralized as it combines with bicarbonate to form carbonic acid; the enzyme carbonic anhydrase catalyzes the dehydration of carbonic acid into

carbon dioxide and water. The carbon dioxide that forms during the reactions is exhaled by the lungs.

Buffer systems only neutralize H^+ temporarily. In the kidneys, however, cells of the kidney tubules can secrete hydrogen ions into the filtrate. In this way the urinary system can eliminate excess H^+ and so help regulate the pH of body fluids. It also resupplies bicarbonate to the blood, restoring the bicarbonate-carbon dioxide buffer system. Figure 9.10 will help you keep track of the events described next.

The reactions in the bicarbonate-carbon dioxide buffer system also proceed in reverse in cells of the nephron wall. As Figure 9.10 shows, after bicarbonate (HCO_3^-) forms in those cells, it is moved into interstitial fluid, then into the capillaries around the nephron, then into the general circulation. Sodium accompanies the bicarbonate, and this combination buffers excess acid in the blood just as sodium bicarbonate in antacids combats "acid indigestion." The H^+ that forms in tubule cells is secreted into the nephron. There it can react with bicarbonate to form CO_2, or it can combine with phosphate ions or ammonia before it leaves the body in urine.

Through such mechanisms, the kidneys help maintain health. When the mechanisms fail, serious problems arise (see *Focus on Science*).

A key homeostatic role of the urinary system is to restore the buffers in body fluids and to eliminate excess hydrogen ions, thus helping maintain an optimal pH for metabolic activity.

Is Your Drinking Water Polluted?

In many parts of the world, the quality of drinking water is being degraded. Aquifers used as sources of drinking water in many countries, including the United States, are becoming contaminated with pesticides, fertilizers, and hazardous organic chemicals. In China, for example, 41 large cities get their drinking water from polluted groundwater. In the United States, hundreds of potentially dangerous chemicals have been detected in groundwater supplies.

Some of the more than 700 synthetic organic chemicals found in trace amounts in drinking-water supplies in the United States can cause kidney disorders, birth defects, and various types of cancer in laboratory animals. At high levels, water-soluble inorganic chemicals—including acids, salts, and compounds of toxic metals such as mercury and lead—can also make water unfit to drink, harm aquatic life, and have other negative effects. Excessive nitrates (from agricultural and industrial chemicals) in drinking water can reduce the oxygen-carrying capacity of the blood and even cause stillbirths.

In the United States overall, up to 25 percent (by volume) of the usable groundwater is contaminated. In some areas, up to 75 percent by volume is contaminated—including every major aquifer in New Jersey. In California, at least 1 million people drink water contaminated with pesticides. A number of environmentalists believe that long-lasting groundwater contamination will soon emerge as one of our most serious water resource problems. Do you know where your drinking water comes from—and what is in it?

MAINTAINING BODY TEMPERATURE

Maintaining the volume and composition of the internal environment is serious business; the need to maintain its temperature is equally serious. Temperatures in the environment just outside the body often change quickly, and exercise can send metabolic rates soaring. Such changes trigger slight increases or decreases in the body's normal **core temperature**. "Core" refers to the body's innermost tissues, as opposed to the tissues near its surface. Normal human core temperature is about 37°C (98.6°F).

Heat is an inevitable by-product of metabolic activity. (Even as you sit reading this book, you are producing roughly one kilocalorie of heat per hour per kilogram of body weight.) If that heat were to accumulate internally, your body temperature would steadily rise. Above 41°C (105.8°F), some protein enzymes become denatured and cease to function properly. Likewise, the rate of enzyme activity generally *decreases* by at least half when body temperature drops by 10°F. As body core temperature drops below 35°C (95°F) reduced enzyme functioning causes the heart rate to fall, and heat-generating mechanisms such as shivering stop (Table 9.4). Breathing slows, and a person may lose consciousness. Below 80°F the heart may stop beating entirely. Given these physiological facts, you can see why humans require mechanisms that help maintain body temperature within narrow limits.

Humans are **endotherms**, which means "heat from within." Our body temperature is controlled mainly by (1) metabolic activity and (2) controls over heat conservation and dissipation. We also make behavioral adjustments (such as building a fire or changing clothes) that supplement the physiological controls. Next we consider temperature regulation in greater detail.

Table 9.4 Physiological Responses to Cold Stress	
Core Temperature	**Responses**
36°–34°C (about 95°F)	Shivering response, increase in respiration. Increase in metabolic heat output. Constriction of peripheral blood vessels; blood is routed to deeper regions. Dizziness and nausea set in.
33°–32°C (about 91°F)	Shivering response stops. Metabolic heat output drops.
31°–30°C (about 86°F)	Capacity for voluntary motion is lost. Eye and tendon reflexes inhibited. Consciousness is lost. Cardiac muscle action becomes irregular.
26°–24°C (about 77°F)	Ventricular fibrillation sets in (page 162). Death follows.

Kidney Failure and Dialysis

An estimated 13 million people in the United States have kidneys in which the nephrons have been so damaged (as by diabetes or immune responses) that the filtering of blood and formation of urine are seriously impaired. Control of the volume and composition of the extracellular fluid is disturbed, and toxic by-products of protein breakdown can accumulate in the bloodstream. Patients can suffer nausea, fatigue, and memory loss. In advanced cases, death may result. A *kidney dialysis machine* can restore the proper solute balances. Like the kidney itself, the machine helps maintain extracellular fluid by selectively removing and adding solutes to the bloodstream.

"Dialysis" refers to the exchange of substances across a membrane between solutions of differing compositions. In *hemodialysis,* the dialysis machine is connected to an artery or a vein, and then blood is pumped through tubes made of a material similar to cellophane. The tubes are submerged in a warm-water bath. The precise mix of salts, glucose, and other substances in the bath sets up the correct gradients with the blood. Dialyzed blood is returned to the body.

In *peritoneal dialysis,* exchanges are made several times every day. Fluid of the proper composition is put into the abdominal cavity, left in place for a period of time, and then drained out. Here, the lining of the cavity (the peritoneum) serves as the dialysis membrane.

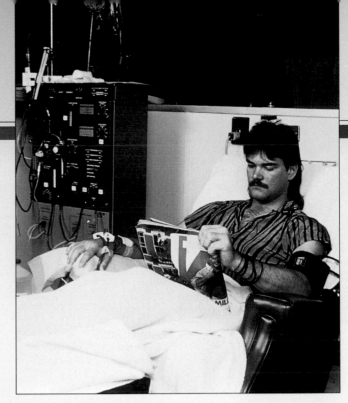

Patient undergoing kidney dialysis.

Hemodialysis generally takes about four hours; blood must circulate repeatedly before solute concentrations in the body are improved. The procedure must be performed three times a week and is used as a temporary measure in patients with reversible kidney disorders. In chronic cases, the procedure must be used for the rest of the patient's life or until a functional kidney is transplanted. With treatment and a controlled diet, many people can pursue a fairly active, close-to-normal life-style.

Temperature Regulation

Responses to Cold Stress Table 9.5 summarizes the major responses to cold stress. They include peripheral vasoconstriction, the pilomotor response, shivering, and nonshivering heat production.

The hypothalamus governs responses to cold stress. In **peripheral vasoconstriction**, thermoreceptors at the body surface detect a decrease in temperature and fire off signals to the hypothalamus. In turn, the hypothalamus sends out commands to smooth muscles in the walls of arterioles in the skin. When the muscles contract, vasoconstriction occurs. Less blood flows to capillaries near the body surface, so body heat is retained. When your fingers or toes become cold, 99 percent of the blood that would otherwise flow to the skin covering them is curtailed.

Table 9.5 Summary of Human Responses to Cold Stress and to Heat Stress		
Environmental Stimulus	Main Responses	Outcome
Drop in temperature	Vasoconstriction of blood vessels in skin; pilomotor response; changes in behavior (e.g., curling up the body to reduce surface area exposed to the environment)	Heat is conserved
	Increased muscle activity; shivering; nonshivering heat production	Heat production increases
Rise in temperature	Vasodilation of blood vessels in skin; sweating; changes in behavior; heavy breathing	Heat is dissipated from body
	Decreased muscle activity	Heat production decreases

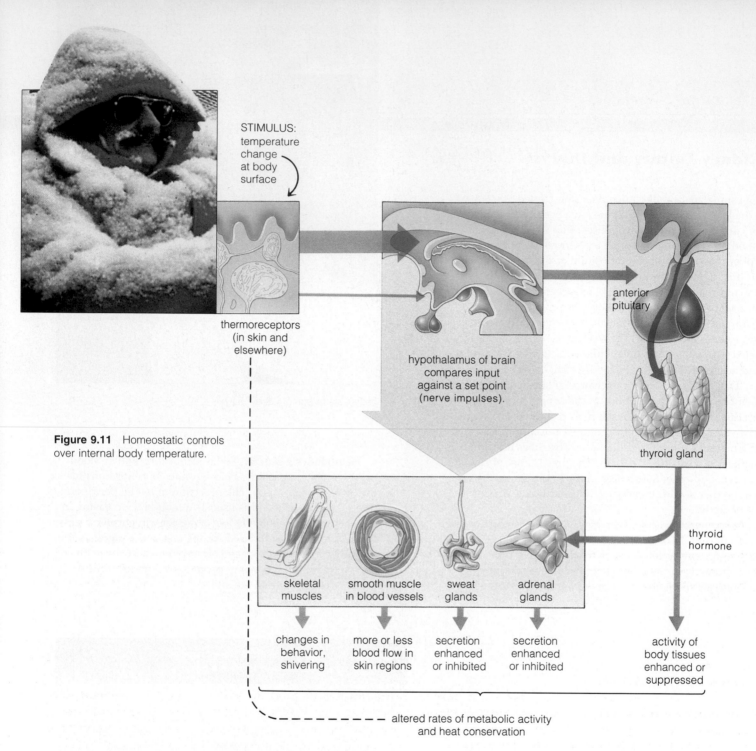

STIMULUS:
temperature
change
at body
surface

thermoreceptors
(in skin and
elsewhere)

hypothalamus of brain
compares input
against a set point
(nerve impulses).

anterior
pituitary

thyroid gland

Figure 9.11 Homeostatic controls over internal body temperature.

skeletal
muscles

smooth muscle
in blood vessels

sweat
glands

adrenal
glands

thyroid
hormone

changes in
behavior,
shivering

more or less
blood flow in
skin regions

secretion
enhanced
or inhibited

secretion
enhanced
or inhibited

activity of
body tissues
enhanced or
suppressed

altered rates of metabolic activity
and heat conservation

In the **pilomotor response** to a drop in outside temperature, smooth muscle controlling the erection of body hair is stimulated to contract. This creates a layer of still air that reduces heat losses from the body. (Of course, this response is much more effective in mammals with more body hair than humans.) Heat loss can be further restricted by behavioral responses that reduce the amount of body surface exposed for heat exchange—as when you hold both arms tightly against your body.

When other responses are not enough to counter cold stress, the hypothalamus calls for an increase in skeletal muscle activity that leads to *shivering*. The word refers to rhythmic tremors in which the muscles contract about 10–20 times per second. Within a short time, heat production throughout the body increases several times over. Figure 9.11 summarizes responses to cold stress.

Heat production also can be increased without shivering. Prolonged or severe exposure to cold can lead to a hormonal response that elevates the rate of metabolism. This *nonshivering heat production* is especially notable in brown adipose tissue (found in the neck, armpits, and near the kidneys), where heat is generated as the lipid molecules are broken down. Human infants have this tissue; adults have very little unless they are cold-adapted.

Figure 9.12 A tragic episode of hypothermia.

In 1912, the ocean liner *Titanic* set out from Europe on her maiden voyage across the cold Atlantic waters to America. In that same year, a huge chunk of the leading edge of a Greenland glacier broke off and began floating out to sea. Late at night on April 14, off the coast of Newfoundland, the iceberg and the *Titanic* made their ill-fated rendezvous. The *Titanic* was the largest ship afloat and was believed to be unsinkable. Survival drills had been neglected, and there were not enough lifeboats to hold even half the 2,200 passengers. The *Titanic* sank in about 2-1/2 hours.

Within two hours rescue ships were on the scene, yet 1,517 bodies were recovered from a calm sea. All were wearing life jackets. None had drowned. Probably every one of those individuals had died from hypothermia, a drop in body temperature below a tolerable level.

Figure 9.13 Evaporative water loss and sweating. Humans and some other mammals have sweat glands that move water and specific solutes through pores to the skin surface. In fact, an average-sized human can produce 1–2 liters of sweat per hour. For every liter of sweat that evaporates, the body loses 600 kilocalories of heat energy. During extreme exercise, this mechanism balances the high rate of heat production in skeletal muscle. In itself, sweat dripping from the skin does not dissipate body heat by evaporation. When you drip with sweat while exercising on a hot, humid day, the rate of evaporation does not keep pace with the rate of sweat secretion—the high water content of the surrounding air slows evaporation.

When defenses against cold are not adequate, the result is *hypothermia*, a condition in which the core temperature falls below normal. In humans, a drop in core temperature of only a few degrees affects brain function and leads to confusion; further cooling can lead to coma and death (Figure 9.12). Some victims of extreme hypothermia, children particularly, have survived prolonged immersion in cold water of even lower temperatures. One reason is that mammals, including humans, have a *dive reflex*. When a mammal is submerged, the heart rate slows and blood is shunted to vital organs such as the brain.

Severe damage to tissue through localized freezing is called *frostbite*. Cells that become frozen may be destroyed unless thawing is precisely controlled, as sometimes can be done in a hospital.

Responses to Heat Stress   Table 9.5 also summarizes the main responses to heat stress. Here again, the hypothalamus has roles in responses to increases in core temperature. In **peripheral vasodilation**, hypothalamic signals cause blood vessels in the skin to dilate. More blood flows from deeper body regions to skin regions, where the excess heat it carries is dissipated.

Evaporative heat loss is another response that can be influenced by the hypothalamus, which can activate sweat glands. Your skin has 2.5 million or more sweat glands, and considerable heat is dissipated when the water they give up to the skin surface evaporates (Figure 9.13). With extreme sweating, as might occur in a marathon race, the body loses important salts—especially sodium chloride—as well as copious amounts of water. Such losses may change the character of the internal environment to the extent that the runner may collapse and faint.

Sometimes peripheral blood flow and evaporative heat loss are not enough to counter heat stress. The result is *hyperthermia*, a condition in which the core temperature increases above normal. If the increase is not too great, a person can suffer *heat exhaustion*. Vasodilation and water losses from heavy sweating cause a drop in blood pressure, the skin feels cold and clammy, and the person may collapse. When heat stress is great enough to completely break down the body's temperature controls, *heat stroke* occurs. Sweating stops, the skin becomes dry, and body temperature rapidly increases to a level that can be lethal.

Fever   During a *fever*, the hypothalamus actually resets the body's "thermostat" that dictates what the core temperature should be. The normal response mechanisms are brought into play, but they are carried out to maintain a higher temperature! At the onset of fever, heat loss decreases and heat production increases. Paradoxically, at the time the person feels chilled. When the fever "breaks," peripheral vasodilation and sweating increase as the body attempts to reduce the core temperature to normal; then, the person feels warm.

Fever may be an important defense mechanism against infections, and perhaps against cancer. Following infection, macrophages and other cells secrete signaling molecules, including interleukin-l and interferons (page 183). The secretions somehow influence the hypothalamus to reset the thermostat. Hormonelike substances called *prostaglandins* (page 287) apparently contribute to the changes. We know, for example, that aspirin and other drugs that reduce fever interfere with the synthesis of prostaglandins. (Prostaglandins injected directly into the hypothalamus can induce fever.)

The controlled increase in body temperature during a fever seems to enhance the body's immune response. Therefore, administering aspirin and other drugs may interfere with these beneficial effects. Even so, fever-reducing drugs are essential when the core temperature approaches dangerous levels.

SUMMARY

1. By balancing water and solute gains with water and solute losses, the body maintains its internal environment.

2. Water is gained by absorption from the gastrointestinal tract and by metabolism. A thirst center in the hypothalamus controls water gain via drinking. Water is lost by evaporation from the lungs and skin, elimination from the gastrointestinal tract, and excretion of urine. Controls over water loss deal mainly with varying the composition and volume of urine.

3. Urine forms in a pair of kidneys. Each human kidney contains about 1 million tubelike blood-filtering units called nephrons. A nephron has a cup-shaped beginning (glomerular capsule), a proximal tubule, a loop of Henle, and a distal tubule that leads into a collecting duct.

4. Kidney function depends on intimate links between the nephron and the bloodstream. Blood flows from an arteriole into a set of capillaries inside the glomerular (Bowman's) capsule, then through a second arteriole into a second set of capillaries that thread around the tubular parts of the nephron, then back to the bloodstream by way of a venule.

5. Each day, massive volumes of fluid are filtered by the glomeruli. Most of the water and solutes filtered are reabsorbed, and the excess is excreted as urine. Urine composition and volume depends on three processes:

a. Filtration of blood at the glomerulus of a nephron, with blood pressure providing the force for filtration.

b. Reabsorption. Water and solutes move out of tubular parts of the nephron and back into adjacent blood capillaries.

c. Secretion. Some ions and a few foreign substances are transported out of those capillaries and into the nephron, so that they are disposed of in urine.

6. Reabsorption of many solutes may occur passively, following concentration gradients. But in other instances, active transport (with its expenditure of energy) is required. Sodium reabsorption is an important example of active transport. Water reabsorption is always passive, occurring along its osmotic gradient.

7. The hormone ADH is secreted when the body must retain water; it acts on the distal nephron walls and makes them permeable to water, so that more filtrate is returned to the blood. When the body must rid itself of excess water, ADH secretion is inhibited. The hormone aldosterone controls sodium reabsorption in the distal part of the nephron, and, indirectly, water reabsorption.

8. Body temperature is determined by the balance between metabolically produced heat and the heat absorbed from and lost to the environment.

9. Body temperature is controlled largely by metabolic activity and by precise controls over heat produced and heat lost.

Review Questions

1. The human body has mechanisms for maintaining body fluid concentration and composition. Which organs cooperate in these tasks? *206*

2. Describe what happens during (a) filtration, (b) reabsorption, and (c) secretion in the kidney's nephron/capillary unit. What do these three processes influence? *209*

3. Which hormone is involved in the control of water reabsorption? Which hormone plays a major role in sodium reabsorption? *213*

4. Which ion is especially important in maintaining the body's acid-base balance? *215*

Critical Thinking: You Decide *(Key in Appendix IV)*

1. Alcohol inhibits production of ADH. What practical consequence does this fact have for drinkers?

2. A urinalysis reveals that the patient's urine contains glucose, urea, hemoglobin, and sodium. Which of these substances are abnormal in urine and why?

3. As a person ages, nephron tubules lose some of their ability to concentrate urine. What is the effect of this change?

Self-Quiz *(Answers in Appendix III)*

1. Maintaining the volume and composition of extracellular fluid depends on three kidney functions: _____, _____, and _____.

2. Urine forms in _____.
   a. glomeruli   c. nephrons
   b. the kidney cortex   d. ureters

3. The body gains water by _____.
   a. gastrointestinal absorption   c. both a and b
   b. metabolism   d. neither a nor b

4. The body loses water by _____.
   a. evaporation from lungs and skin
   b. elimination from the gastrointestinal tract
   c. excretion of urine
   d. all of the above

5. Each human kidney contains about _____ nephrons.
   a. 1,000   c. 100,000
   b. 10,000   d. 1,000,000

6. The processes responsible for urine composition and volume generally occur in which order?
   a. filtration, reabsorption, secretion
   b. secretion, reabsorption, filtration
   c. reabsorption, secretion, filtration
   d. secretion, filtration, reabsorption

7. Which of the following descriptions does *not* match the process named?
   a. *filtration:* water and solutes leave blood plasma and enter the glomerular capsule

   b. *reabsorption:* water and solutes selectively returned to blood capillaries
   c. *secretion:* excess ions and some other substances move from blood capillaries into nephron
   d. all of the above match
   e. none of the above match

8. Concentration gradients between nephrons and surrounding interstitial fluid play central roles in _____.
   a. blood filtration
   b. dialysis
   c. secretion and reabsorption
   d. hypertension

9. The hormone ADH controls _____.
   a. nephron production
   b. sodium reabsorption
   c. secretion
   d. water balance

10. Match the following salt-water balance concepts:
   ____ aldosterone   a. blood filter of a nephron
   ____ nephron   b. controls sodium reabsorption
   ____ thirst mechanism c. occurs at nephron tubules
   ____ reabsorption   d. site of urine formation
   ____ glomerulus   e. controls water gain

Key Terms

ADH *215*	peripheral vasodilation *221*
aldosterone *215*	peritubular capillary *210*
core temperature *218*	pilomotor response *220*
distal tubule *215*	proximal tubule *213*
endotherm *218*	reabsorption *211*
extracellular fluid *208*	renal corpuscle *210*
filtration *211*	secretion *211*
glomerular (Bowman's) capsule *210*	thirst center *216*
glomerulus *210*	ureter *210*
juxtaglomerular apparatus *216*	urethra *210*
kidney *209*	urinary bladder *210*
loop of Henle *214*	urinary excretion *208*
nephron *210*	urinary system *210*
peripheral vasoconstriction *219*	urine *211*

Readings

Flieger, Ken. March 1990. "Kidney Disease: When Those Fabulous Filters Are Foiled." *FDA Consumer.* Excellent, nontechnical survey of kidney functions and treatments for kidney disease.

Vander, A., J. Sherman, and D. Luciano. 1990. "The Kidneys and Regulation of Water and Inorganic Ions" in *Human Physiology,* 5th ed. New York: McGraw-Hill.

10 NERVOUS SYSTEM

Why Crack the System?

Suppose your biology instructor asks you to volunteer for an experiment. A microchip will have to be implanted in your brain, and it will make you feel *really* good. But it may harm your health, take years off your life, and perhaps destroy part of your brain. Your behavior will change for the worse, so you may have trouble completing school, getting or keeping a job, even having a normal family life. The longer the chip is implanted, the less you will want to give it up. You won't get paid to participate in this experiment—*you* pay the experimenter, first at bargain rates, then a little more each week. The chip is illegal. If you get caught using it, you and the experimenter will go to jail.

Sometimes Jim Kalat, a professor at North Carolina State University, proposes this experiment (which of course is hypothetical). Few students volunteer. Then he substitutes *drug* for "microchip" and *dealer* for "experimenter"—and an amazing number of students

come forward! Like 30 million other Americans, the "volunteers" seem ready to engage in self-destructive uses of drugs that alter emotional and behavioral states.

The destruction shows up in unexpected places. Each year, for instance, about 300,000 infants are born addicted to crack cocaine smoked or injected by their mothers. Crack relentlessly stimulates a region deep within the brain. It dampens normal urges to eat and sleep, and blood pressure rises. Elation and sexual desire intensify. In time, brain cells that produce the stimulatory chemicals can't keep up with the incessant demands. The chemical vacuum makes crack users frantic, then profoundly depressed. Only crack makes them "feel good" again.

Figure 10.1 Owners of an evolutionary treasure—a complex nervous system that is the foundation for our memory, for reasoning, for coordinated movement, and for our future.

Many legal drugs, including alcohol, can also disrupt normal functioning of the human nervous system. For example, chronic heavy drinkers typically have problems with short-term memory. A pregnant woman who drinks, especially if she drinks heavily, puts her fetus at risk of developing fetal alcohol syndrome. This devastating disorder of the nervous system is characterized by severe learning disabilities and behavioral problems (page 327).

Directly or indirectly, the nervous system is responsible for our ability to perceive joy and pain, to read a book, and to remember what we've read. It is also responsible for nearly every body movement (Figure 10.1). Constant, coordinated signals that travel rapidly through its communication lines—nerves composed of cells called neurons—control the muscles concerned with facial expressions, food digestion, breathing, and other vital functions. In this regard it may be useful to keep in mind that substance abuse—whether the substance is cocaine or a legal drug such as alcohol—can distort or even eliminate these remarkable possibilities.

KEY CONCEPTS

1. The nervous system detects information about external and internal conditions, integrates information, and calls for responses from the body's muscles and glands. Its communication lines consist of nerve cells called neurons.

2. Due to differences in ion concentrations, there is an uneven distribution of electrical charge across a neuron's plasma membrane. The inside of the cell is slightly negative compared to the fluid outside the neuron. When the neuron is stimulated, this charge difference across the membrane may briefly reverse. Such reversals, called action potentials, are the mechanism by which electrical messages are sent through the nervous system.

3. Action potentials self-propagate along a neuron, but they cannot cross the small gap (synaptic cleft) between neurons. Chemical signals bridge the gaps and either stimulate or inhibit the adjoining neuron.

4. The nervous system includes the brain, spinal cord, and many nerves. The human brain is complex, with centers for receiving, integrating, and responding to information.

OVERVIEW OF THE NERVOUS SYSTEM

The key point about a nervous system is not what it is, but *what it does*. The human body is multicellular and complex. Maintaining homeostasis within it requires continuous, finely tuned adjustments in all sorts of functions—and making adjustments requires efficient, internal communication. Indeed, a constant flow of information coordinates, integrates, and regulates the activities of the body's billions of cells and the different tissues and organs they make up. Much of this communication occurs by way of the nervous system interacting closely with the endocrine system, which we consider in Chapter 12.

The nervous system has two divisions. The *central nervous system* consists of the brain and spinal cord. It receives and integrates signals from the external world and from elsewhere in the body and coordinates responses. Cordlike communication lines called *nerves* make up the *peripheral nervous system*, which carries messages between the brain and spinal cord and other body regions. Two types of cells make up both divisions of the nervous system. One type is specialized for communication, and the other type provides structural and functional support. The remainder of this chapter will flesh out this general picture. Let's begin by examining the structure and functioning of nervous system cells.

CELLS OF THE NERVOUS SYSTEM

The nerve cell, or **neuron**, is the basic unit of communication in the human nervous system. Neurons collectively monitor changing conditions and issue commands for responsive actions that benefit the body as a whole. These tasks involve three different classes of nerve cells: sensory neurons, interneurons, and motor neurons. Each class of nerve cell has its own role in a control scheme by which the nervous system monitors and responds to change.

Sensory neurons act as *receptors* that detect a specific stimulus (light, pressure, sound waves, or some other form of energy). They relay information about the stimulus *to* the spinal cord and brain. There, **interneurons** act as *integrators*; they receive sensory input, integrate it with other information, and then influence the activity of other neurons. **Motor neurons** relay information *away* from the brain and spinal cord to muscles or glands. Muscles and glands, the body's *effectors*, carry out responses:

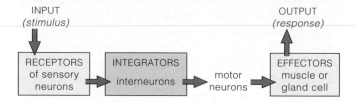

Neurons make up only about half the volume of the human nervous system, and no more than 20 percent of its cells. The rest are **neuroglial cells**, which physically support and protect neurons; they also help maintain proper concentrations of important ions in the fluid around neurons. Some neuroglia impart structure to the brain, much

Figure 10.2 Component parts of a motor neuron. The micrograph shows its cell body and dendrites.

dendrites cell body

INPUT ZONE

axon

CONDUCTING ZONE

OUTPUT ZONE

axon endings

10 μm

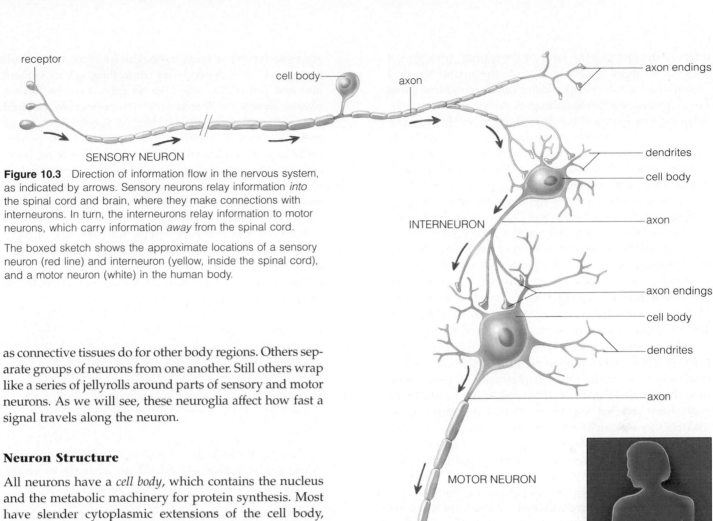

receptor cell body axon axon endings

SENSORY NEURON

dendrites
cell body
axon

INTERNEURON

axon endings
cell body
dendrites
axon

MOTOR NEURON

axon endings
at effector

Figure 10.3 Direction of information flow in the nervous system, as indicated by arrows. Sensory neurons relay information *into* the spinal cord and brain, where they make connections with interneurons. In turn, the interneurons relay information to motor neurons, which carry information *away* from the spinal cord.

The boxed sketch shows the approximate locations of a sensory neuron (red line) and interneuron (yellow, inside the spinal cord), and a motor neuron (white) in the human body.

as connective tissues do for other body regions. Others separate groups of neurons from one another. Still others wrap like a series of jellyrolls around parts of sensory and motor neurons. As we will see, these neuroglia affect how fast a signal travels along the neuron.

Neuron Structure

All neurons have a *cell body*, which contains the nucleus and the metabolic machinery for protein synthesis. Most have slender cytoplasmic extensions of the cell body, called *processes*. The number and length of neuron processes differ enormously—so much, in fact, that there really is no such thing as a "typical" neuron. The neurons described most often are motor neurons of the sort shown in Figure 10.2. A motor neuron has two types of processes. Its many short, slender **dendrites** carry nerve impulses toward the cell body, whereas its single long, cylindrical **axon** carries nerve impulses away from the cell body. The junction between the axon and the cell body is called the *axon hillock*; in motor neurons and interneurons it is here that electrical signals are initiated.

A motor neuron axon has finely branched endings that terminate on muscle cells. Figure 10.3 shows the direction of information flow in the nervous system. As the figure suggests, dendrites and axons also occur on many sensory neurons and interneurons. Generally speaking, we can think of dendrites and the cell body as "input zones," where a neuron receives signals (Figure 10.2). Axon endings are "output zones," where it sends signals to other cells.

How a Neuron Responds to Stimulation

Different types of signals travel through the nervous system. The signals that travel along the plasma membrane of individual neurons are electrical events.

There is a difference in electric charge across the plasma membrane of a neuron that is "resting." The extracellular fluid just outside the membrane tends to be positively charged, whereas the cytoplasmic fluid just inside the membrane is negatively charged. This charge difference is expressed as a difference in electrical voltage, and it is called the **resting membrane potential**. The name refers to the fact that the force of the mutual attraction of opposite charges on either side of the membrane is a potential source of energy. Under certain conditions, that force can increase the likelihood that charge-bearing particles such as ions will move (following an *electrical gradient*). We consider these events in more detail shortly. As we will see, the movement of ions across a neuron's plasma membrane is the source of the signals that travel through the nervous system.

When a neuron receives signals at its input zone, the resting membrane potential there may change temporarily; that is, there may be a change in the voltage difference across the membrane. A very weak disturbance might trigger only slight changes across a small patch of membrane. But a stronger disturbance can trigger an **action potential**: an abrupt, brief *reversal* in the polarity of charge across the plasma membrane. For a fraction of a second, the cytoplasmic side of a patch of membrane—which was negative with respect to the outside—becomes positive. This brief change in membrane potential sets in motion a wave of excitation that is actively propagated from its beginning point on down the entire length of the axon: Each reversal triggers another action potential at the neighboring patch of membrane, this triggers another at the next patch, and so on away from the input zone.

Any cell that can respond to stimulation by producing action potentials is said to have an *excitable membrane*. Most neurons have this kind of plasma membrane.

Action potentials are initiated and end on the same neuron; they are not transferred to neighboring cells. For example, if your hand brushes against a hot iron, that "disturbance" triggers action potentials that quickly reach the axon endings of motor neurons in your hand. The arriving action potentials trigger the release of molecules called *neurotransmitters* that serve as chemical signals to adjacent muscle cells. Muscles in your forearm contract in response to the signals, and you jerk your hand away from the iron.

In a resting neuron, there is a steady voltage difference across the plasma membrane. That is, the inside is more negative than the outside. The energy inherent in that difference is the resting membrane potential.

Most neurons have an excitable plasma membrane. With adequate stimulation, the membrane can produce action potentials.

An action potential is an abrupt, brief reversal in the voltage difference across the membrane: The inside becomes more positive than the outside. Action potentials travel along the cell membrane to the axon terminals.

Neurons "At Rest" In between action potentials, the neuron restores and maintains the voltage difference across each patch of membrane. This involves an ion "balancing act" that depends on controlling ion movement into and out of the cell. Like all cells, neurons have a plasma membrane that is only semipermeable. Its lipid bilayer bars the passage of charged substances such as potassium ions (K^+) and sodium ions (Na^+). The plasma membrane also contains various types of proteins, however, including channel proteins. Channel proteins are hollow, and ions can flow through them from one side of the membrane to the other in controllable ways (Figure 10.4).

For example, assume that a motor neuron has 15 sodium ions inside the membrane for every 150 outside. Assume that it also has 150 potassium ions inside for every 5 on the outside. We can depict each ion's concentration gradient, which runs from the large letters to the small letters, as follows:

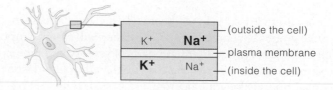

Such concentration gradients dictate the direction in which sodium and potassium ions will diffuse across the membrane through the hollow interior of channel proteins.

Gated Ion Channels Some of the protein channels through which ions cross the plasma membrane are "gated." Like automatic doors in a supermarket, they open and close in response to specific stimuli. They can allow the flow of specific ions across the membrane, or shut it off. Some gated ion channels are *voltage-gated*—they respond to changes in electrical potential. Others are chemically gated and respond to chemical signals. As we will see, chemically gated channels are crucial in the transfer of signals between neurons.

Some channels selectively permit certain ions to "leak" (diffuse) through them, following their concentration gradients. In particular, certain channels permit potassium ions to leak out of the cell more readily than they permit sodium ions to leak inward. As positively charged potassium ions exit the cell, the cytoplasmic side of the membrane becomes more negatively charged than the side facing interstitial fluid. Negatively charged proteins in the cytoplasm add to the charge difference. So, in addition to ion concentration gradients, there is also an *electrical gradient* across the membrane. The voltage difference of the resting membrane potential is the amount of energy inherent in the concentration and electric gradients between the two differently charged regions. For many neurons, it is about −70 millivolts.

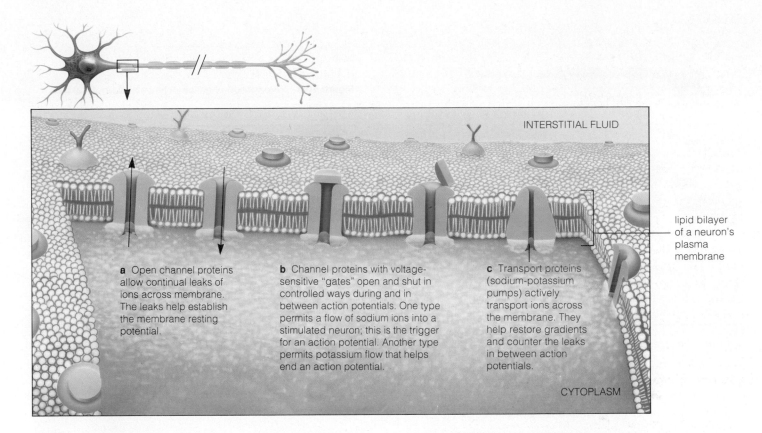

a Open channel proteins allow continual leaks of ions across membrane. The leaks help establish the membrane resting potential.

b Channel proteins with voltage-sensitive "gates" open and shut in controlled ways during and in between action potentials. One type permits a flow of sodium ions into a stimulated neuron; this is the trigger for an action potential. Another type permits potassium flow that helps end an action potential.

c Transport proteins (sodium-potassium pumps) actively transport ions across the membrane. They help restore gradients and counter the leaks in between action potentials.

INTERSTITIAL FLUID

lipid bilayer of a neuron's plasma membrane

CYTOPLASM

Figure 10.4 Pathways for ions across the plasma membrane of a neuron.

Pump Proteins In a resting neuron, most of the gated channels for sodium almost completely are shut. Even so, some sodium does leak into the neuron. Unless something offsets the inward leakage of positive ions, the resting membrane potential eventually will disappear. Transport proteins called **sodium-potassium pumps** prevent this from happening. Using energy from ATP, they actively transport potassium into the neuron, and at the same time they pump sodium ions out. To put this important mechanism in perspective, even when you are relatively inactive an estimated 20–30 percent of your metabolic energy use goes to keeping your sodium-potassium pumps pumping!

Figure 10.5 summarizes the balancing effect of the pumping and leaking mechanisms that maintain membrane conditions between action potentials.

Concentration and electric gradients exist across the plasma membrane of a neuron.

In a resting neuron, active transport maintains the concentration and electrical gradients. Potassium ions are pumped into the neuron and sodium ions are pumped out.

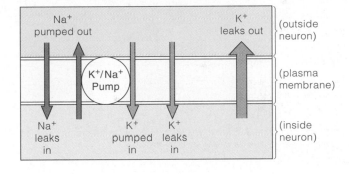

Na^+ pumped out

K^+ leaks out

(outside neuron)

K^+/Na^+ Pump

(plasma membrane)

Na^+ leaks in

K^+ pumped in

K^+ leaks in

(inside neuron)

Figure 10.5 Active pumping and passive leaking processes that maintain the distribution of sodium and potassium ions across the plasma membrane of a neuron at rest. Arrow widths indicate the magnitude of the movements. Relatively more potassium leaks out across the membrane because certain ion channels remain "open" to K^+ all the time. The total inward and outward movements for each kind of ion are balanced in a way that maintains the resting membrane potential.

Local, Graded Signals: Prelude to Action Potentials

In all neurons, stimulation at an input zone produces graded, local signals. *Graded* means the signals can vary in magnitude—they can be small or large—depending on the intensity and duration of the stimulus. *Local* means

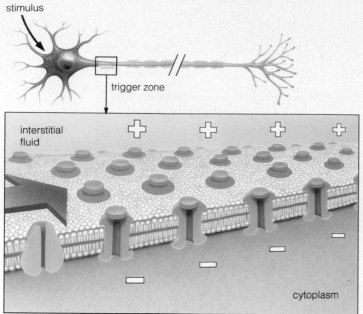

stimulus

trigger zone

Figure 10.6 Propagation of action potentials along an axon.

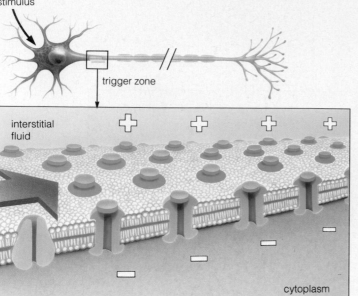

interstitial fluid

cytoplasm

a Membrane at rest (inside negative with respect to the outside). An electrical disturbance (red arrow) spreads from an input zone to an adjacent trigger region of the membrane, which has many gated sodium channels.

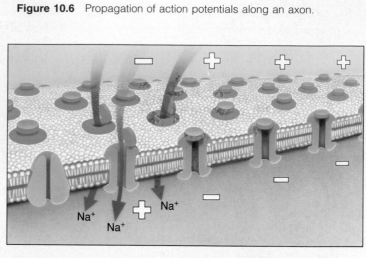

Na⁺ Na⁺ Na⁺

b A strong disturbance initiates an action potential. Sodium gates open, the inflow decreases the negativity inside; this causes more gates to open, and so on, until threshold is reached and the voltage difference across the membrane reverses.

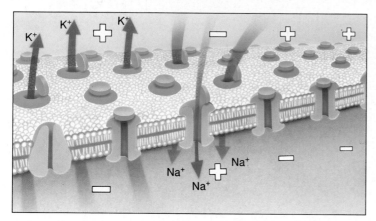

K⁺ K⁺ K⁺
Na⁺ Na⁺ Na⁺

c The reversal causes sodium gates to shut and potassium gates to open (at purple arrows). Potassium follows its gradient (out of the neuron). Voltage is restored. The disturbance produced by the action potential triggers another action potential at the adjacent membrane site, and so on, away from the point of stimulation.

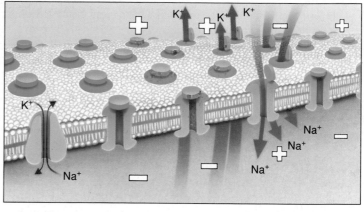

K⁺ K⁺ K⁺
K⁺
Na⁺ Na⁺ Na⁺ Na⁺

d The inside of the membrane becomes negative again following each action potential, but the sodium and potassium concentration gradients are not yet fully restored. Active transport at sodium-potassium pumps restores the gradients.

the signal does not spread far—usually half a millimeter or less. Input zones simply do not have the type of ion channels needed to propagate a signal farther than this. However, when stimulation is intense or prolonged, graded signals can spread into a nearby *trigger zone* of the membrane. This is a region that contains many voltage-gated channels for sodium ions, and it is the site where an action potential can be initiated.

How an Action Potential Is Triggered

Before an action potential occurs, the inside of a neuron at rest is more negative than the outside; that is, the plasma membrane is *polarized*. During an action potential, the membrane is *depolarized* as the inside becomes more positive than the outside. Finally, following an action potential, resting conditions are restored—the membrane is *repolarized*.

We can now describe an action potential as the rapid depolarization of a region of the neuron membrane and the brief reversal of charge across it. An action potential is triggered when a disturbance is strong enough to cause the membrane potential to change by a certain minimum amount, a *threshold* level. The change occurs as increasing numbers of voltage-gated sodium channels open (Figure 10.6). As the positively charged sodium ions flow inward, the inside of the membrane becomes less negative. This electrical stimulus causes even more voltage-gated sodium channels to open, more sodium to enter, and so forth, until the charge difference reverses. The ever-increasing flow of sodium is an example of positive feedback, whereby an event intensifies as it proceeds (page 82):

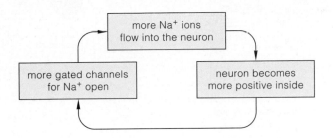

There is no such thing as a "partial" action potential. An action potential is an *all-or-none* event—a stimulus either triggers a complete action potential, or (if the stimulus is too weak) no action potential occurs at all. Once threshold is reached, the continued opening of sodium gates no longer depends on the strength of the stimulus. Because the positive-feedback cycle is now under way, the inward-rushing sodium itself is enough to cause more sodium gates to open. Figure 10.7 includes a recording of the voltage difference across the membrane before, during, and after an action potential. Notice how the membrane potential is recorded as a "spike." All action potentials in a given neuron spike to the same level above threshold.

Duration of an Action Potential An action potential lasts only about a millisecond. At the membrane region where it occurred, sodium gates shut, potassium gates open, and ion movements restore the original membrane potential. Also, active transport by sodium-potassium pumps helps restore the original ion gradients.

After an action potential occurs in a trigger zone, it automatically repeats, or *propagates* itself along the membrane. The electrical disturbance spreads to adjacent patches of membrane, threshold is reached, voltage-gated sodium channels open, and so on away from the stimulation site. Figure 10.6 shows how action potentials travel away from the stimulation site. They cannot move backward because a *refractory period* follows each one. During

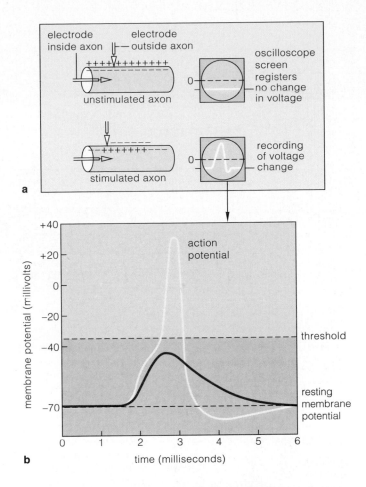

Figure 10.7 Action potentials. (**a**) Electrodes positioned inside and outside an axon can be used to detect voltage changes when the axon is stimulated. Changes show up as deflections in a beam of light on an oscilloscope screen. (**b**) The yellow line is a typical waveform for an action potential. The red line represents a local signal that did not reach the threshold of an action potential, so spiking did not occur. The solid white line is a recording of an action potential.

this brief time, the affected membrane region cannot be depolarized to the required threshold. All these events take place very rapidly; most neurons can produce hundreds of action potentials every second.

Action Potentials Along Sheathed Axons Many axons bundled in nerves throughout the body are surrounded by a whitish **myelin sheath**. Myelin is a fatty substance that protects the axon and insulates one nerve fiber from adjacent ones, so that nerve impulses travel much faster along myelinated axons.

In the peripheral nervous system the myelin sheath consists of the plasma membranes of specialized glial cells

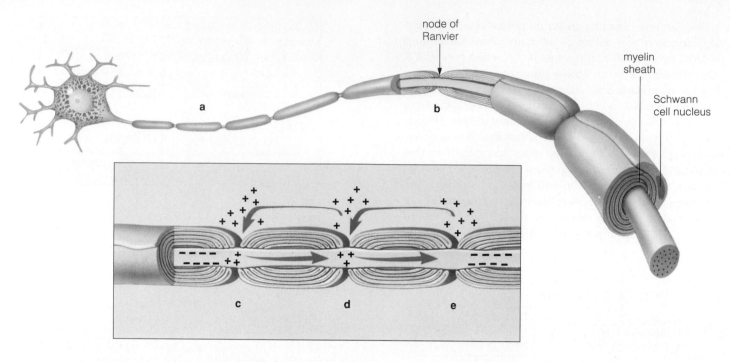

node of
Ranvier

myelin
sheath

Schwann
cell nucleus

a

b

c   d   e

Figure 10.8 Propagation of an action potential along a motor neuron having a myelin sheath. (**a**) The sheath is a series of Schwann cells, each wrapped like a jellyroll around the axon. Ions can cross the membrane only at nodes between Schwann cells (**b**). The unsheathed nodes have very dense arrays of voltage-gated sodium channels (**c, d**). A disturbance caused by an action potential spreads down the axon. When it reaches a node, sodium gates open, sodium ions rush inward, and another action potential results. (**e**) This new disturbance spreads rapidly to the next node and so forth down the line.

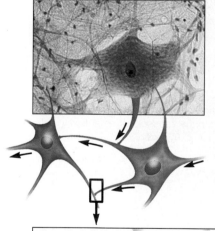

Figure 10.9 Chemical synapses. Typically, action potentials spread along axons, away from the neuron cell body (**a**). In (**b**), the axon terminates next to another neuron; this is a chemical synapse. Information flows from the presynaptic cell to the postsynaptic cell by way of a neurotransmitter (**c**).

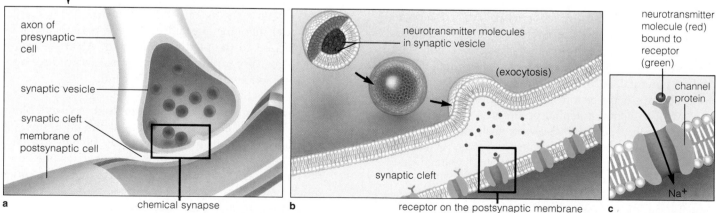

axon of
presynaptic
cell

synaptic vesicle

synaptic cleft

membrane of
postsynaptic cell

a   chemical synapse

neurotransmitter molecules
in synaptic vesicle

(exocytosis)

synaptic cleft

b   receptor on the postsynaptic membrane

neurotransmitter
molecule (red)
bound to
receptor
(green)

channel
protein

Na+

c

called **Schwann cells**, which wrap like jellyrolls around the axon (Figure 10.8). Each Schwann cell is separated from adjacent ones by a small region (a *node of Ranvier*) where the axon membrane is exposed and loaded with gated sodium channels. Action potentials jump from node to node. The sheathed regions hinder the flow of ions across the membrane, which forces the ions to flow along the length of the axon until they can exit at a node and generate a new action potential there. The node-to-node hopping in myelinated neurons is called *saltatory conduction*, after a Latin word meaning "to jump." Saltatory conduction affords the most rapid signal propagation with the least metabolic effort by the cell. In the largest myelinated axon, signals travel 120 meters per second, or 270 miles per hour. In general, myelinated axons can carry signals more than 100 times faster than unmyelinated ones.

There are no Schwann cells in the central nervous system (brain and spinal cord). There, processes from other glial cells (called *oligodendrocytes*) form the sheaths of myelinated axons. In *multiple sclerosis*, patches of the myelin sheath around axons in the brain and spinal cord are slowly destroyed. Depending on where the destruc-

tion occurs, typical symptoms include numbness, muscle weakness or paralysis, and inability to control urination and elimination. The major effects and symptoms of various insults to the nervous system are listed in Table 10.1.

CHEMICAL SYNAPSES

When an action potential reaches the output zone of a neuron, it usually does not proceed further. At the axon ending, an action potential triggers the release of a **neurotransmitter**, a type of signaling molecule, into the junction between the neuron and an adjacent cell. These junctions are called **chemical synapses** (Figure 10.9). Some occur between two neurons, others between a neuron and a muscle cell or gland cell. Although the two cells are firmly anchored together, a small space, the *synaptic cleft*, separates them.

A neuron that releases a neurotransmitter into the cleft is called the *presynaptic cell*. The presynaptic cell contains numerous vesicles filled with neurotransmitters. The cell affected by the transmitter is the *postsynaptic* cell.

Table 10.1	The Nervous System Under Attack	
Cause/Risk Factors	Major Effects	Symptoms
Meningitis Viral or bacterial infection of meninges; often fatal	Inflammation of the membranes covering the brain and/or spinal cord	Severe headache, fever, stiff neck, nausea, vomiting
Encephalitis Usually, infection by a virus, such as Herpes simplex; also, HIV infection	Inflammation of the brain	Fever, headache, mental confusion, memory loss, seizures, gradual loss of consciousness
Epilepsy Brain injury or infection, drug overdose, metabolic imbalance. Often, no specific cause can be identified	Abnormal electrical activity that temporarily alters one or more brain functions. Attacks are sometimes triggered by fatigue, stress, or flashing lights	Seizures of various types. *Grand mal* seizures involve loss of consciousness, jerking body movements. *Petit mal* attacks involve only momentary loss of consciousness
Multiple sclerosis Cause unknown. Possibly an autoimmune disease with viral origins. Some people may have genetic susceptibility	Progressive destruction of myelin sheaths of neurons in brain and spinal cord	Numbness, muscle weakness, fatigue, incontinence; vision, gait, and speech may also be affected. Some victims become severely disabled and bedridden
Alzheimer's disease Cause unknown. Victims have lower than normal levels of acetylcholine in brain tissues; some families show a genetic predisposition to the disorder	Progressive degeneration of neurons and shrinking of the brain. Abnormal buildup of masses of amyloid protein	In three general stages, mild memory losses, severe short-term memory loss and disorientation; sudden mood changes; severe confusion, paranoid delusions, extreme antisocial or childlike behavior
Concussion Violent blow to the head or neck	Disruption of electrical activity of brain neurons	Brief unconsciousness, followed by mental confusion, blurred vision, vomiting
Neuralgia Physical injury or viral infection; specific cause often unknown	Irritation of or damage to a nerve	Usually, intermittent bouts of severe pain; often, a shooting pain along the affected nerve. Nerves of the face and head are commonly affected, as in *migraine headache*

When an action potential arrives at the presynaptic cell membrane facing the cleft, it causes voltage-gated channels for calcium ions to open. Calcium ions are more concentrated outside the cell, and when they move inside (down their gradient), they cause synaptic vesicles filled with a neurotransmitter to fuse with the plasma membrane. The contents of the vesicles are released into the cleft, and they diffuse to the postsynaptic cell. There the neurotransmitter molecules bind briefly to membrane receptor molecules. Depending on the type of channels being opened up, a neurotransmitter may either *excite* or *inhibit* the membrane (help drive its membrane potential toward or away from threshold). After neurotransmitter molecules exert their effects, the presynaptic cell takes them up or enzymes inactivate them.

Effects of Neurotransmitters

Acetylcholine (ACh) is a neurotransmitter that has excitatory or inhibitory effects on the cells of muscles and glands throughout the body. It also acts on certain cells in the brain and spinal cord. Recall from Chapter 4 what ACh does at *neuromuscular junctions* (page 106), which are synapses between a motor neuron's axon endings and muscle cells (Figure 10.10). An action potential traveling down the motor neuron spreads through all the endings and causes the release of ACh into each synaptic cleft. When ACh binds to receptors on the muscle cell membrane, it has an excitatory effect. It may trigger action potentials, which in turn initiate events in the muscle cell that lead to contraction. ACh receptors at some neuromuscular junctions are destroyed in a condition called *myasthenia gravis*, causing drooping eyelids, muscle weakness, and fatigue.

Serotonin, another neurotransmitter, acts on brain cells that govern sleeping, sensory perception, temperature regulation, and emotional states. *Norepinephrine* affects brain regions that apparently are concerned with emotional states, as well as with dreaming and awaking. Other neurotransmitters that act on different parts of the brain are *dopamine* and *GABA* (gamma aminobutyric acid). Two debilitating degenerative diseases underscore how important neurotransmitters are to normal life. *Parkinson's disease*, which involves a progressive, severe loss of muscle control, results from a lack of dopamine in certain parts of the brain. *Choices* (page 235) discusses some of the biological and ethical issues associated with a controversial, potentially powerful new treatment for this disorder. A loss or degeneration of neurons that manufacture and release acetylcholine may play a key role in *Alzheimer's disease*, with its devastating attacks on personality and intellectual functions. The *Human Impact* feature on aging (page 331) describes current research on Alzheimer's disease.

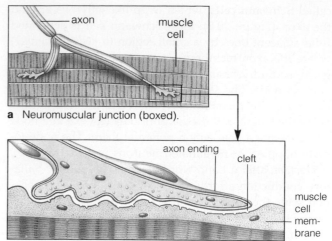

a Neuromuscular junction (boxed).

b Motor end plate (troughs in muscle cell membrane).

Figure 10.10 (**a**) One type of chemical synapse: the neuromuscular junction between a motor neuron and a muscle cell (boxed area). The myelin sheath of the axon stops at the junction, leaving the membranes of the two interacting cells exposed to each other. (**b**) Close-up of the troughs in the muscle cell membrane where the axon endings are positioned.

Neuromodulators

The effects of neurotransmitters are often influenced by signaling molecules known generally as **neuromodulators**. A neuromodulator may make a postsynaptic neuron more or less sensitive to a given neurotransmitter, or it may increase or decrease the synthesis or release of a neurotransmitter from a presynaptic neuron. Neuromodulators include *endorphins*, naturally occurring peptide molecules that are much more potent painkillers than morphine. When runners and other athletes push themselves beyond the point of normal fatigue, they can experience a euphoric "high" as endorphin release increases. Endorphins also may have roles in memory and learning, temperature regulation, and sexual behavior, as well as in emotional depression and other mental disorders.

Synaptic Integration

At any given time, excitatory and inhibitory signals are arriving at a postsynaptic cell. Some signals drive the membrane closer to threshold; others maintain the resting level or drive it away from threshold. An *excitatory postsynaptic potential* (EPSP) brings the membrane closer to threshold; it has a depolarizing effect. An *inhibitory postsynaptic potential* (IPSP) either drives the membrane away from threshold (a *hyperpolarizing* effect) or maintains the membrane potential at the resting level.

All these signals compete for control of the membrane. In a process called **synaptic integration**, the competing

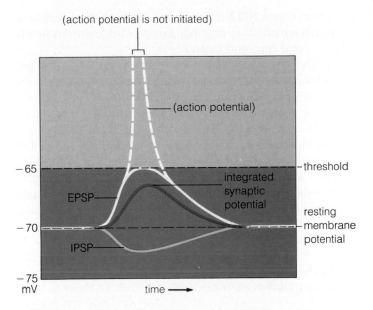

Figure 10.11 Synaptic integration. In this example, an exitatory synapse and an inhibitory synapse nearby are activated at the same time. The IPSP reduces the magnitude of the EPSP from what it could have been, pulling the membrane potential away from threshold. The red line represents the integration of these two synaptic potentials. Threshold is not reached in this case; hence an action potential cannot be initiated.

signals that impinge on the receptive regions of a neuron are added together or *summed* at an integrating zone (usually the initial segment of the axon). If summing results in a signal that reaches the threshold level, the postsynaptic neuron will respond with an action potential. If not, no action potential will be initiated. Through synaptic integration, signals arriving at any given neuron in the body can be reinforced or dampened, sent on or suppressed (Figure 10.11).

Synaptic integration sums up competing excitatory and inhibitory signals acting on a neuron.

PATHS OF INFORMATION FLOW

The destination of a given signal in the nervous system depends on the circuit or pathway in which it occurs. The brain has many "local" circuits in which the interaction of neurons is confined to a single region. In contrast, signals between the brain or spinal cord and other body regions travel via nerves. A **nerve** consists of the axons of many neurons, held together in bundles by connective

Choices: Biology and Society

Fetal Tissue Transplants

Every year 50,000 people in the United States, many of them in the prime of life, are diagnosed with Parkinson's disease. Their hands or feet quiver with muscle tremors while their joints grow increasingly stiff. They walk with a teetering shuffle, sometimes interspersed with rapid, tiny, uncontrollable steps. Sadly, these symptoms may be only the beginning. Many victims become almost completely immobile, seriously depressed, and have their mental functions deteriorate.

Parkinson's disease results when, for unknown reasons, clusters of neurons deep within the brain, called *basal ganglia*, degenerate and stop producing the neurotransmitter dopamine. Normally, dopamine produced by the basal ganglia moderates the flow of nerve impulses that stimulate skeletal muscle contractions. When the neurotransmitter is absent, skeletal muscles contract intensely, causing tremors and restricting voluntary movements. There is no cure. Patients can be treated with drugs such as levodopa, which the brain converts into dopamine. Drugs are often only partly successful in relieving symptoms, however, and become useless with time as the neurons that manufacture dopamine (from the levodopa) degenerate and are lost.

There is another, potentially powerful treatment: transplantation of fetal brain cells into the brain of a Parkinson's sufferer. In theory, if the cells from a young fetus are properly placed in the recipient's brain, they will mature and begin cranking out dopamine. The procedure is still highly experimental, and although several recipients have shown marked improvement in their conditions, to date the operation remains highly controversial.

Transplantable tissue must be in an early stage of development. It can come from miscarried fetuses, or possibly be grown in tissue culture in the laboratory. However, much of what is available comes from aborted fetuses. Opponents of fetal tissue transplantation charge that use of such tissue raises the specter of fetuses being conceived and aborted *solely to provide tissues for transplantation*. People who object to abortions on moral grounds are adamantly opposed to medical use of fetal tissues under any circumstances. On the other hand, proponents say that it is unethical to destroy tissues that could enable Parkinson's victims to lead nearly normal lives. And they point out that society can impose ethical safeguards, such as prohibiting tissue recipients and their doctors from having any contact with women who seek abortions for personal reasons. Who is right? More and more, as technology extends our ability to treat disease using novel methods, individuals will be asking themselves that question.

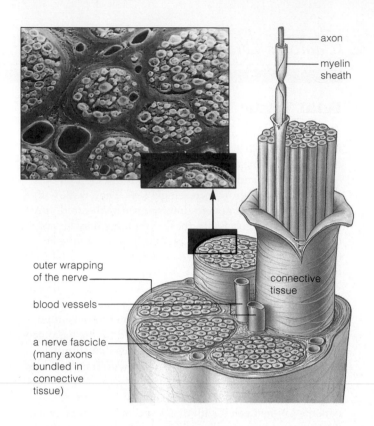

axon

myelin sheath

outer wrapping of the nerve

connective tissue

blood vessels

a nerve fascicle (many axons bundled in connective tissue)

tissue (Figure 10.12). (Nerves are distinct from *nerve tracts*, which are bundled-together axons of interneurons inside the spinal cord and brain.)

One of the simplest nerve pathways is the **reflex arc**, in which sensory neurons directly synapse with motor neurons. A reflex arc controls **reflexes**—simple, stereotyped, movements that occur automatically in response to sensory stimuli. The *stretch reflex* is an example; it works to contract a stretched muscle (Figure 10.13). For instance, without thinking about it you can hold out a bowl and keep it stationary when someone puts fruit in it. As the fruit adds weight to the bowl and your hand starts to drop, the biceps muscle in your arm is stretched.

Figure 10.12   Structure of a nerve. Axons in the nerve are bundled together inside wrappings of connective tissue.

Figure 10.13   The stretch reflex. Inside a skeletal muscle, sensory neurons are located in muscle spindles. Stretching stimulates them and generates action potentials that travel down to the axon endings. These synapse with a motor neuron that has axons leading back to the stretched muscle. The motor neuron can initiate muscle contraction.

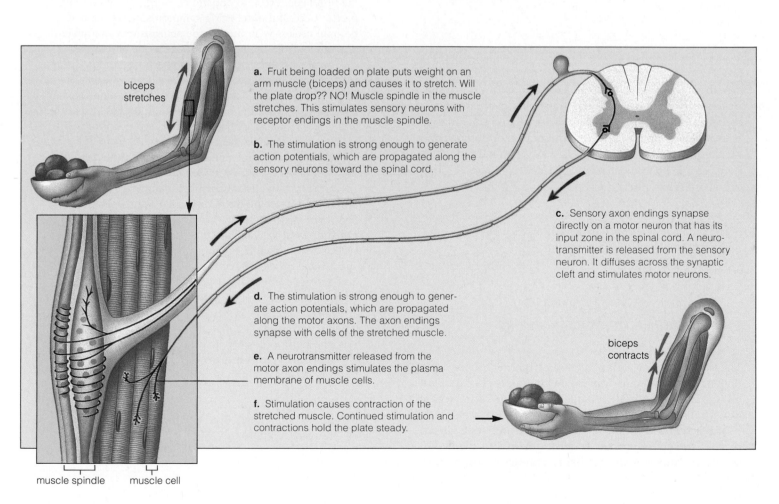

biceps stretches

a. Fruit being loaded on plate puts weight on an arm muscle (biceps) and causes it to stretch. Will the plate drop?? NO! Muscle spindle in the muscle stretches. This stimulates sensory neurons with receptor endings in the muscle spindle.

b. The stimulation is strong enough to generate action potentials, which are propagated along the sensory neurons toward the spinal cord.

c. Sensory axon endings synapse directly on a motor neuron that has its input zone in the spinal cord. A neurotransmitter is released from the sensory neuron. It diffuses across the synaptic cleft and stimulates motor neurons.

d. The stimulation is strong enough to generate action potentials, which are propagated along the motor axons. The axon endings synapse with cells of the stretched muscle.

e. A neurotransmitter released from the motor axon endings stimulates the plasma membrane of muscle cells.

biceps contracts

f. Stimulation causes contraction of the stretched muscle. Continued stimulation and contractions hold the plate steady.

muscle spindle   muscle cell

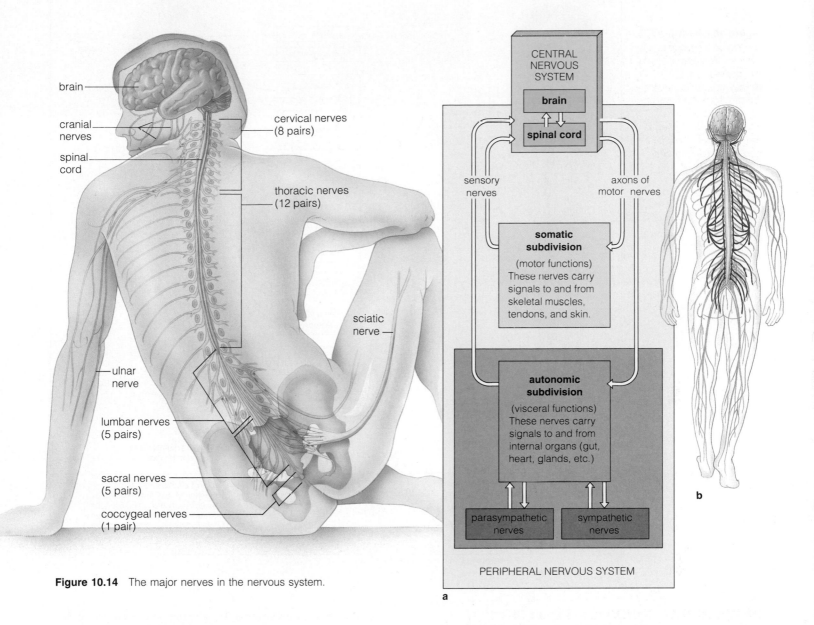

Figure 10.14 The major nerves in the nervous system.

Labels in figure:

brain

cranial nerves

spinal cord

cervical nerves (8 pairs)

thoracic nerves (12 pairs)

sciatic nerve

ulnar nerve

lumbar nerves (5 pairs)

sacral nerves (5 pairs)

coccygeal nerves (1 pair)

CENTRAL NERVOUS SYSTEM

brain

spinal cord

sensory nerves

axons of motor nerves

somatic subdivision

(motor functions) These nerves carry signals to and from skeletal muscles, tendons, and skin.

autonomic subdivision

(visceral functions) These nerves carry signals to and from internal organs (gut, heart, glands, etc.)

parasympathetic nerves

sympathetic nerves

PERIPHERAL NERVOUS SYSTEM

a

b

Figure 10.15 (**a**) Functional divisions of the nervous system. (**b**) The central nervous system is coded blue. For the peripheral nervous system, somatic nerves are coded green, and autonomic nerves are coded red.

The stretching activates receptor endings of certain sensory neurons in the muscle; those sensory neurons synapse with motor neurons in the spinal cord. Axons of the motor neurons lead back to the stretched muscle, and action potentials that reach the axon endings trigger the release of ACh, which initiates contraction. Continued receptor activity excites the motor neurons further, allowing them to maintain your hand's position.

In most reflexes, sensory neurons make connections with interneurons, which then activate or suppress the motor neurons necessary for a coordinated response. An example is the *withdrawal reflex*, a rapid pulling away from a harmful or unpleasant stimulus. If you have ever brushed up against a hot stove burner, you know this reflex action can be completed even before you are conscious it has occurred.

Figure 10.14 shows some of the major nerves in the human body. Figure 10.15 depicts the two divisions of the human nervous system. The **central nervous system** (CNS) includes the brain and spinal cord. The **peripheral nervous system** includes all the nerves that carry signals to and from the brain and spinal cord. At some sites in the peripheral nervous system, cell bodies of several neurons occur in clusters called *ganglia* (singular: ganglion). Both the CNS and PNS also have neuroglial cells, such as the oligodendrocytes (CNS) and Schwann cells (PNS) previously described.

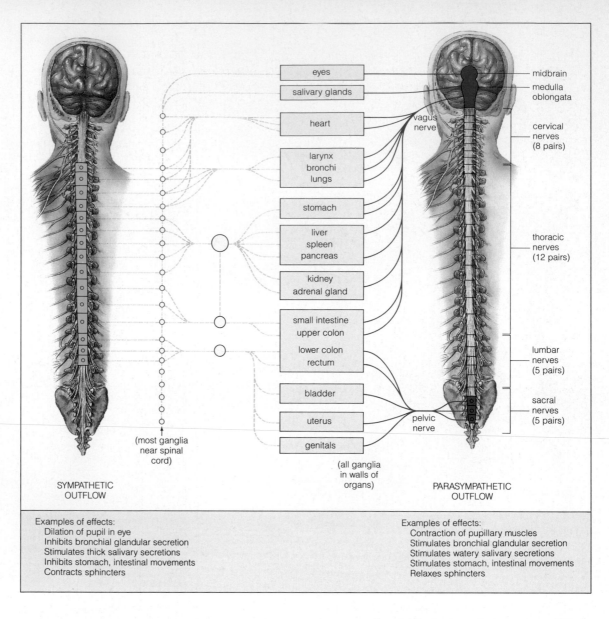

Figure 10.16 The autonomic nervous system. Its pairs of sympathetic and parasympathetic nerves extend from the central nervous system to internal organs. Ganglia are clusters of cell bodies of the neurons that are bundled together in nerves.

eyes
salivary glands
heart
larynx
bronchi
lungs
stomach
liver
spleen
pancreas
kidney
adrenal gland
small intestine
upper colon
lower colon
rectum
bladder
uterus
genitals

vagus nerve
pelvic nerve

midbrain
medulla oblongata
cervical nerves (8 pairs)
thoracic nerves (12 pairs)
lumbar nerves (5 pairs)
sacral nerves (5 pairs)

(most ganglia near spinal cord)

(all ganglia in walls of organs)

SYMPATHETIC OUTFLOW

PARASYMPATHETIC OUTFLOW

Examples of effects:
Dilation of pupil in eye
Inhibits bronchial glandular secretion
Stimulates thick salivary secretions
Inhibits stomach, intestinal movements
Contracts sphincters

Examples of effects:
Contraction of pupillary muscles
Stimulates bronchial glandular secretion
Stimulates watery salivary secretions
Stimulates stomach, intestinal movements
Relaxes sphincters

PERIPHERAL NERVOUS SYSTEM

The peripheral nervous system consists of 31 pairs of spinal nerves (Figure 10.14), which connect with the spinal cord, and 12 pairs of cranial nerves that connect directly with the brain. Some nerves of the peripheral system carry only sensory information. The optic nerves, which carry visual signals from the eyes to the brain, are sensory nerves. Other nerves contain both sensory and motor axons. For example, the musculocutaneous nerve has sensory axons leading from the skin of the forearm to the spinal cord and motor axons leading out to the muscles that flex the forearm.

Somatic and Autonomic Systems

The peripheral nervous system has two subdivisions: the somatic system and the autonomic system (Figure 10.15). The **somatic system** deals with conscious, usually vol-untary body movements. Its sensory axons carry signals inward from receptors in the skin, skeletal muscles, and tendons; its motor axons carry signals out to the body's skeletal muscles. Each of the motor axons extends all the way from the central nervous system to the particular skeletal muscle it innervates. Some of those axons, such as spinal neuron axons that synapse with skeletal muscle that moves your big toe, are several feet long!

Sometimes the nerves carrying sensory input to the central nervous system are said to be *afferent*, a word meaning "to bring to." Nerves carrying motor output away from the central nervous system to muscles and glands are *efferent*, meaning "to carry outward." The reflex pathway shown in Figure 10.13 is a simple example of how somatic nerves work.

The **autonomic system** (ANS) deals with the invol-untary, usually unconscious control of internal ("vis-ceral") organs and structures (Figure 10.16). Autonomic motor axons carry signals to smooth muscle, cardiac

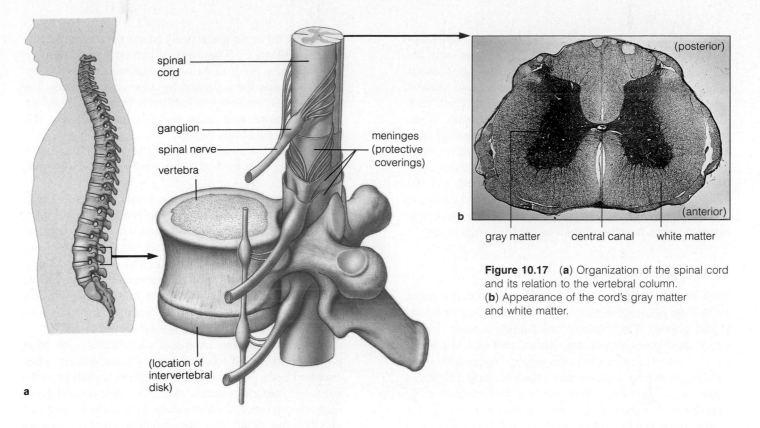

spinal cord

ganglion

spinal nerve

vertebra

meninges (protective coverings)

(location of intervertebral disk)

a

(posterior)

b

(anterior)

gray matter   central canal   white matter

Figure 10.17 (**a**) Organization of the spinal cord and its relation to the vertebral column. (**b**) Appearance of the cord's gray matter and white matter.

(heart) muscle, and glands in different body regions. Unlike somatic neurons, single autonomic neurons don't extend the entire distance between muscles or glands and the central nervous system. Instead, *preganglionic* ("before a ganglion") *neurons* have cell bodies within the spinal cord or brain stem, and their axons travel through nerves to autonomic system ganglia outside the central nervous system. There, the axons synapse with *postganglionic* ("after a ganglion") *neurons*. Axons of this second set of neurons make the actual connection with the body's muscles and glands (effectors).

As described next, autonomic nerves are divided into two subgroups that counteract one another.

Sympathetic and Parasympathetic Nerves There are two subdivisions of autonomic nerves. Signals from one subdivision, the **parasympathetic nerves**, tend to slow down the body overall and support basic "housekeeping" tasks required for homeostasis. For example, parasympathetic pathways act to divert blood flow to the GI tract when food is being digested. (This is one reason why you feel drowsy after a large meal.) Parasympathetic nerve action dominates when the body is not receiving much outside stimulation. In contrast, signals from **sympathetic nerves** tend to suppress housekeeping tasks and increase overall body activities during times of heightened awareness, excitement, danger, or even day-to-day stress. Sympathetic nerves prepare an animal to fight or flee when threatened (the "fight or flight" response), or to frolic intensely—for example, in sexual behavior.

The interplay of parasympathetic and sympathetic autonomic nerves is central to maintaining the body's homeostatic balance. As the foregoing discussion suggests, it brings about constant minor adjustments in internal organs as conditions change. For example, even while sympathetic signals are causing your heart to beat a little faster, parasympathetic signals are opposing this effect. At any given moment, your heart rate is the result of interaction of the opposing signals. In addition to increasing or decreasing digestive secretions at appropriate times, the ANS also acts to raise or lower body temperature, blood pressure, breathing rate, and so forth.

CENTRAL NERVOUS SYSTEM

The Spinal Cord

The **spinal cord** (Figure 10.17a) lies within a closed channel formed by the bones of the vertebral column. The bones and ligaments attached to them protect the cord. So do the *meninges*, a series of three coverings of connective tissue that are layered around the spinal cord and brain. The cord consists partly of bundles of myelinated axons called **nerve tracts**. The myelin sheaths of these axons are white, and the tracts are referred to as *white matter*. The cord also contains dendrites, cell bodies of neurons, interneurons, and neuroglial cells. These form its *gray matter*. In cross-section, the gray matter of the spinal cord looks vaguely like a butterfly (Figure 10.17b).

The spinal cord is a vital expressway that carries signals between the peripheral nervous system and the brain. It is also a reflex control center. Sensory and motor neurons involved in many reflex movements of skeletal muscle make direct connections in the spinal cord; such *spinal reflexes* do not require input from the brain. When you jerk your hand away from a hot stove burner, you are experiencing a spinal reflex in action. Higher brain centers also receive information about the sensory stimulus, and so you become aware of "hot burner!" even as your hand is moving away from it. The spinal cord also contributes to some *autonomic reflexes*, which deal with internal organ functions (such as bladder emptying).

Divisions of the Brain

Your impressively large, intricate **brain** contains about 100 billion neurons and typically weighs about 3 pounds (1,300 grams). The brain is the body's master control panel. It receives, integrates, stores, and retrieves information, and it coordinates appropriate responses by intricately stimulating and inhibiting the activities of different body parts. The brain starts out as a continuation of the anterior end of the spinal cord. Like the spinal cord, it is protected by bones (of the cranium) and by meninges.

As noted earlier, the three **meninges** are membranes of connective tissue that are layered between the skull bones and the brain tissue itself. Meninges cover and protect the delicate CNS neurons and blood vessels that service the tissue. Folds in the tough, outermost one (the *dura mater*) separate the right and left brain hemispheres. The meninges also enclose spaces filled with *cerebrospinal fluid*, which cushions and helps nourish the brain. We will return to its functions shortly.

Table 10.2 summarizes the functions of the three divisions of the brain: the hindbrain, midbrain, and forebrain.

Hindbrain The **hindbrain** consists of the medulla oblongata, cerebellum, and pons. The *medulla oblongata* merges with the spinal cord. Within it are groups of neurons associated with reflexes that regulate respiration, heart rate, swallowing, coughing, and other basic tasks. Signals for voluntary muscle movements travel along nerve tracts that pass from higher brain centers via the medulla to the spinal cord. The medulla also influences brain centers that help govern waking and sleeping.

The *cerebellum* has extensive connections with other parts of the brain. It receives sensory input related to balance and body position, and it integrates signals from the eyes, ears, muscle spindles, skin, and elsewhere. It keeps higher brain centers informed about how the body trunk and limbs are positioned, how much different muscles are contracted or relaxed, and in which direction the body or limbs happen to be moving. The activities of the cerebel-

Table 10.2	Regional Divisions of the Brain	
Divisions	Main Components	Some Functions
FOREBRAIN	Cerebrum	In two cerebral hemispheres, centers for coordinating sensory and motor functions, for memory, and for abstract thought. Most complex coordinating center; intersensory association, memory circuits
	Olfactory lobes	Relaying of sensory input from the nose to olfactory structures of cerebrum
	Limbic system	Via interacting scattered brain centers (including hypothalamus), coordination of skeletal muscle and internal organ activity underlying emotional expression
	Thalamus	Major coordinating center for sensory signals; relay station for most sensory impulses to cerebrum
	Hypothalamus	Neural-endocrine coordination of visceral activities (e.g., water-solute balance, temperature control, carbohydrate metabolism)
	Pituitary gland	"Master" endocrine gland (controlled by hypothalamus). Control of growth, metabolism, etc.
	Pineal gland	Control of some daily (circadian) rhythms
MIDBRAIN	Tectum	Largely reflex coordination of visual, tactile, auditory input; contains nerve tracts ascending to thalamus, descending from cerebrum
HINDBRAIN	Medulla oblongata	Contains tracts extending between pons and spinal cord; reflex centers involved in respiration, cardiovascular function, etc.
	Cerebellum	Unconscious coordination of motor activity underlying limb movements, maintaining posture, spatial orientation
	Pons	"Bridge" of transverse nerve tracts from cerebrum to both sides of cerebellum. Also contains longitudinal tracts connecting forebrain and spinal cord

lum appear to be quite important in coordinating voluntary movements.

The *pons* (meaning "bridge") consists mainly of prominent bands of axons that extend into each side of the cerebellum. These nerve tracts link the cerebellum with the cerebrum and also with the brainstem (described next).

Midbrain   The **midbrain** has a roof of gray matter, the *tectum*, where visual and auditory sensory input converges before being sent on to higher brain centers. The midbrain, pons, and medulla oblongata together represent the *brainstem*. Within the core of the brainstem and extending through its entire length is a major network of interneurons, the **reticular formation**. The reticular formation has connections with neurons in the spinal cord that control skeletal muscle contractions. By way of these links it functions together with the cerebellum in governing muscle activity associated with maintaining balance, posture, and muscle tone. In addition, reticular formation neurons also synapse with neurons of the medulla and of forebrain regions such as the hypothalamus and the cerebral cortex, the "highest" brain center of all, and influence activity in those regions.

Forebrain   The **forebrain** is the most developed portion of the brain in "higher" animals. It includes the **cerebrum**, where information processing occurs and sensory input and motor responses are integrated; a pair of olfactory lobes that deal with the sense of smell; and the *thalamus* and *hypothalamus*. The thalamus mainly serves as a relay switchboard for sensory information; signals arriving via sensory nerve tracts are "projected" onto clusters of neuron cell bodies (called *nuclei*) in the thalamus, and are relayed onward to the cerebrum or elsewhere. The nuclei also process some outgoing motor information. The *hypothalamus* monitors internal organs and influences forms of behavior related to their activities, such as thirst, hunger, and sex.

Cerebrospinal Fluid   To get an idea of how delicate nervous tissue is, try holding a jiggling blob of Jell-O in your hands. The brain and spinal cord would be highly vulnerable to damage if they were not protected by bones and meninges. In addition, both are surrounded by **cerebrospinal fluid**. This clear extracellular fluid helps to cushion the brain and spinal cord from abrupt, jarring movements. It is secreted from specialized capillaries within four ventricles, which are cavities within the brain that connect with one another and with the central canal of the spinal cord. The fluid fills the ventricles, and it also fills the space between the innermost layer of the meninges and the brain itself (Figure 10.18).

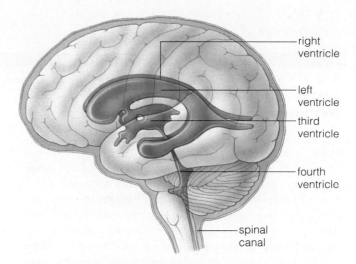

Figure 10.18   Location of the cerebrospinal fluid in the human brain. This extracellular fluid surrounds and cushions the brain and spinal cord. It also fills the four interconnected cavities (cerebral ventricles) within the brain and the central canal of the spinal cord.

The bloodstream exchanges substances with the cerebrospinal fluid, which in turn exchanges substances with neurons. However, the structure of brain capillaries makes those vessels relatively impermeable to many substances. This **blood-brain barrier** helps control *which* blood-borne substances are allowed to reach the neurons. Endothelial cells making up the walls of the capillaries are fused together by continuous tight junctions (Figure 3.4). This means that substances must pass *through* the cells, rather than between them, to reach the brain. Transport proteins embedded in the plasma membrane of those cells selectively transport glucose and other water-soluble substances across the barrier. In addition, lipid-soluble substances quickly diffuse through the lipid bilayer of the plasma membrane. This "loophole" is one reason why caffeine, nicotine, alcohol, barbiturates, heroin, and other lipid-soluble drugs have such rapid effects on brain function. Certain environmental pollutants such as the heavy metal mercury are also lipid-soluble and can cross the blood-brain barrier. As *Focus on Environment* explains, this fact has sobering implications for human health.

The Cerebral Hemispheres

The human cerebrum vaguely resembles the much-folded nut inside a walnut shell. A deep fissure divides the human cerebrum into two parts, the left and right *cerebral*

Environmental Assault on the Nervous System

Have you ever heard someone called "mad as a hatter"? The phrase originated a century ago in Great Britain, when hat makers used the heavy metal mercury to process felt. Traces of that substance entered the bloodstream, crossed the blood-brain barrier, and caused irreparable neurological damage. In an adult, symptoms of mercury poisoning include muscle tremors, impaired vision, difficulty walking and talking, paranoia, and other emotional problems. Victims were often believed to be mentally deranged or "mad."

Mercury can also have other, more insidious effects. They begin when mercury compounds are dumped or washed from the atmosphere into waterways, lakes, and marshlands. There, metabolic processes in bacteria convert it to methyl mercury, which dissolves in fatty tissues in fish and other animals. Consumed in food (especially fish) by a pregnant woman, methyl mercury becomes concentrated in adipose (fat) cells and breast milk—and also wreaks havoc on the developing fetus. In Minimata, Japan, in the 1960s, more than 200 women living near a mercury-discharging factory gave birth to mentally retarded, physically deformed babies. In the late 1980s, authorities in Michigan, Minnesota, and Wisconsin warned nursing mothers and women who were even *considering* pregnancy not to eat *any* fish from many lakes in the region. Those advisories are still in force today. In fact, the U.S. Environmental Protection Agency estimates that mercury contamination in fish is rising by 3–5 percent per year. Ocean fish have also been found to carry large amounts of mercury in their tissues.

From a health standpoint, consumers are well advised to limit their intake of fish taken from waters where mercury pollution may be a problem. Clearly, pregnant women or those considering pregnancy should be especially careful. In the environmental arena, we can all support action on the part of government, industry, and individuals to minimize the entry of mercury-containing pollutants into the environment. Recent studies show that emissions from garbage incinerators and coal-fired power plants may now be even more serious sources of mercury than waste-water discharges. In Sweden, stringent controls over waste-burning have reduced the amount of mercury released into the air by human activities by 90 percent in just the last 10 years.

hemispheres. Other fissures and folds in each hemisphere follow certain patterns and divide it into the frontal, parietal, temporal, and occipital lobes (Figure 10.19).

Much of the gray matter of the cerebral hemispheres is arranged as a thin surface layer, the **cerebral cortex**. The cerebral cortex weighs about a pound, and if you were to stretch it flat, it would cover a surface area of two and a half square feet. The white matter consists of major nerve tracts that keep the hemispheres in communication with each other and with the rest of the body. Some of the tracts originate in the brainstem, where they are aligned close together, then fan out extensively in the cerebral hemispheres. Each hemisphere has its own set of tracts that serve as communication lines among its different regions. A prominent band of 200 million axons, the *corpus callosum*, runs between the two hemispheres and keeps them in communication with each other.

Many experiments have focused on the functioning of the cerebral hemispheres. We know, for example, that each hemisphere can function separately. However, the left cerebral hemisphere responds primarily to signals from the right side of the body. The opposite is true for the right cerebral hemisphere. Signals that travel by way of the corpus callosum coordinate the functioning of both hemispheres.

The main regions responsible for spoken language skills generally reside in the left hemisphere. The main regions responsible for nonverbal skills such as music, mathematics, and other abstract abilities generally reside in the right hemisphere.

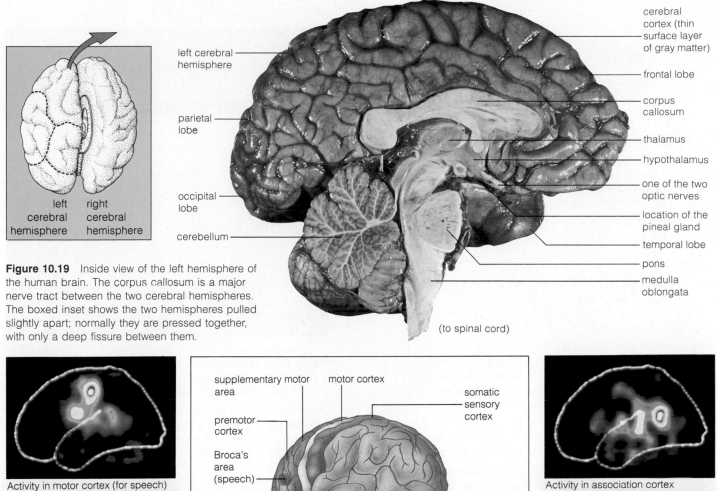

left cerebral hemisphere

parietal lobe

occipital lobe

cerebellum

cerebral cortex (thin surface layer of gray matter)

frontal lobe

corpus callosum

thalamus

hypothalamus

one of the two optic nerves

location of the pineal gland

temporal lobe

pons

medulla oblongata

(to spinal cord)

left cerebral hemisphere   right cerebral hemisphere

Figure 10.19 Inside view of the left hemisphere of the human brain. The corpus callosum is a major nerve tract between the two cerebral hemispheres. The boxed inset shows the two hemispheres pulled slightly apart; normally they are pressed together, with only a deep fissure between them.

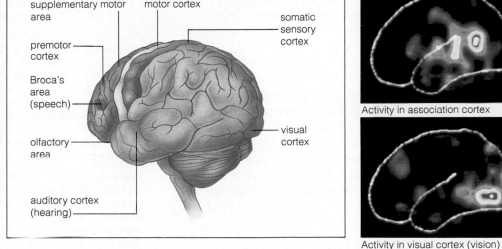

Activity in motor cortex (for speech)

Activity in auditory cortex (hearing)

supplementary motor area

premotor cortex

Broca's area (speech)

olfactory area

auditory cortex (hearing)

motor cortex

somatic sensory cortex

visual cortex

Activity in association cortex

Activity in visual cortex (vision)

Figure 10.20 Primary receiving and association areas for the human cerebral cortex. Signals from receptors near the body surface enter primary receiving areas. Association areas of the cortex coordinate and process various types of sensory input from receptors that encode and transmit information of different kinds. The PET scans (page 18) show which brain regions are active during the performance of specific tasks (speaking, hearing, generating and observing words).

Functions of the Cerebral Cortex

Different regions of the cerebral cortex have different functions. Some receive sensory input from the skin, muscles, and tendons. Others receive sensory input from the nose, eyes, and ears, and so on. Still other regions are concerned with integrating and processing incoming signals and with coordinating appropriate responses (Figure 10.20).

For example, motor centers of the cortex coordinate instructions for motor responses, including voluntary contractions of skeletal muscles. By experimentally stimulating different points on the motor cortex, researchers discovered that they could stimulate the contraction of

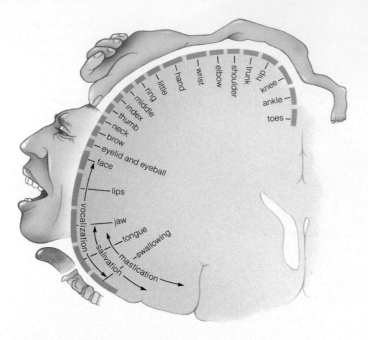

Figure 10.21 Section through the motor cortex of the right hemisphere of someone facing you. Notice the distorted human body draped over the diagram. The distortions show which body regions are controlled by different parts of the motor cortex.

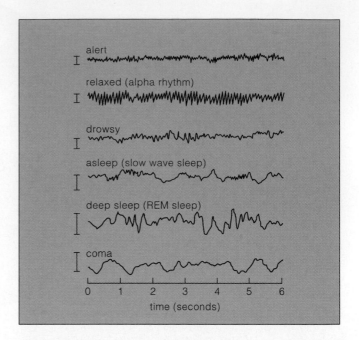

Figure 10.22 EEG patterns. Vertical bars indicate a range of 50 microvolts of electrical activity. The irregular horizontal graph lines indicate the electrical response over time.

different muscles. A strikingly large area of the motor cortex deals with thumb and tongue muscles, indicating the great control required for hand movements and verbal expression (Figure 10.21). A primary center for sensory input from the skin and joints lies just behind the motor cortex. Specialized *association areas* process inputs from different tracts of sensory neurons (including visual, auditory, and other sensory stimuli) and help determine appropriate motor responses. Information from memory stores is also added to the primary sensory information to give it fuller meaning.

States of Consciousness

States of consciousness include sleeping, dozing, daydreaming, and full alertness. The central nervous system governs these states, and psychoactive drugs such as cocaine, alcohol, heroin, and marijuana can alter them. The *Human Impact* essay on drug abuse that follows this chapter describes the effects of these and other psychoactive substances in more detail.

Throughout the spectrum of consciousness, neurons in the brain are constantly chattering among themselves. This neural chatter shows up as wavelike patterns in an *electroencephalogram* (EEG). An EEG is an electrical recording of the frequency and strength of potentials from the brain's surface. Figure 10.22 shows examples of EEG patterns. More recently, PET scans (positron emission tomography, page 18) are providing information on the precise location of brain activity while it is occurring.

The prominent wave pattern for someone who is relaxed, with eyes closed, is an *alpha rhythm*. The EEG waves are recorded in "trains" of one after the other, about 10 per second. Alpha waves predominate during the state of meditation. With a transition to sleep, wave trains gradually become larger, slower, and more erratic. This *slow-wave sleep* pattern shows up about 80 percent of the total sleeping time for adults. It occurs when sensory input is low. Subjects awakened from slow-wave sleep usually report that they were not dreaming. Often they seemed to be mulling over recent, ordinary events.

Slow-wave sleep is punctuated by brief spells of *REM* sleep. *Rapid eye movements* accompany this pattern (the eyes jerk beneath closed lids). So do irregular breathing, faster heartbeat, and twitching fingers. Most people awakened from REM sleep report they were experiencing vivid dreams. The transition from sleep or deep relaxation into alert wakefulness is marked by a shift to low-amplitude, higher-frequency wave trains. Associated with this accelerated brain activity are increased blood flow and oxygen uptake in the cortex. The transition, *EEG arousal*, occurs when conscious effort is made to focus on external stimuli or even on one's own thoughts.

Certain neurons of the reticular formation make up a **reticular activating system (RAS)** that controls the changing levels of consciousness. The RAS sends signals to the spinal cord, cerebellum, and cerebrum as well as back to itself. The flow of signals along these circuits—and the inhibitory or excitatory chemical changes accompanying them—affects whether you stay awake or fall asleep.

Interneurons of one of the "sleep centers" of the reticular formation release serotonin. This neurotransmitter inhibits other neurons that arouse the brain and maintain wakefulness. Thus, high serotonin levels are linked to drowsiness and

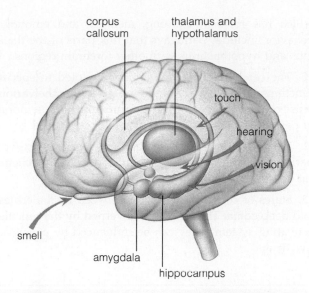

corpus
callosum

thalamus and
hypothalamus

touch

hearing

vision

smell

amygdala

hippocampus

Figure 10.23 The limbic system, our "emotional brain," includes regions called the thalamus, hypothalamus, amygdala, and hippocampus that have key roles in memory. The cerebral cortex processes sensory input, then sends it to parts of the limbic system and forebrain.

sleep. Substances released from another brain center counteract serotonin's effects and bring about wakefulness.

Damage to parts of the RAS circuits can lead to unconsciousness and coma. Among the drugs commonly used in general anesthesia are substances that induce unconsciousness by suppressing the RAS.

Emotional States

Love, anger, sadness, and other emotions are a fundamental aspect of human life, yet we understand very little of their biological basis. We do know that emotions are governed by interactions between the cerebral cortex and other forebrain regions collectively called the **limbic system** (Figure 10.23).

The **hypothalamus** is the gatekeeper of the limbic system. Connections between the cerebral cortex and lower brain centers pass through it. Through these connections, the reasoning possible in the cerebral cortex can dampen rage, hatred, and other "gut reactions." The hypothalamus also monitors internal organs in addition to emotional states.

Memory

Memory is the storage and retrieval of information about previous experiences. It is part of your conscious experience and underlies your capacity for *learning*. Today we know that information becomes stored in stages. "Memory traces," or chemical and structural changes in neuron associations necessary for storage, occur in many differ-

ent brain regions (Figure 10.23). Short-term memory lasts a few seconds to a few hours and seems to be limited to only a few bits of information—words of a sentence, telephone numbers, and so on. Long-term storage lasts more or less permanently, and its capacity seems to be limitless.

Whereas short-term memory is a fleeting stage of neural excitation, long-term memory is thought to depend on more permanent chemical or structural changes in the brain. People who suffer lesions in the hippocampus (part of the limbic system) lose the ability to form new memories. Affected individuals live almost entirely in the immediate present; they have memories only of events that were experienced before the hippocampal damage.

Nerve cells are among the few kinds of cells that are *not* replaced during a person's lifetime. You are born with billions, and as you grow older some 50,000 die off each day. The nerve cells formed during embryonic development are the same ones present, whether damaged or otherwise modified, at the time of death.

The part about being "otherwise modified" is tantalizing. *There is evidence that neuron structure is not static, but rather can be modified in several ways.* Most likely, such modifications depend on electrical and chemical interactions with neighboring neurons. Electron micrographs show that some synapses "wither" as a result of disuse. Such regression weakens or breaks connections between neurons. The visual cortex of mice raised without visual stimulation showed such effects of disuse. Similarly, there is some evidence that intensively stimulated synapses may grow in size or sprout buds or spines to form more connections! The chemical and physical transformations that underlie changes in synaptic connections may correspond to memory storage.

SUMMARY

1. Nervous systems detect, interpret, and respond directly to stimuli. Reflexes (simple, stereotyped movements made in response to sensory stimuli) are the basic operating machinery of nervous systems. In simple reflexes, sensory neurons directly signal motor neurons that act on muscle cells. In more complex reflexes, interneurons coordinate and refine the responses.

2. A neuron can receive and respond to stimuli because of its membrane properties. An unstimulated neuron shows a steady voltage difference across its plasma membrane (the inside is more negative than the outside). The resting neuron maintains concentration gradients of potassium ions, sodium ions, and other charged substances across the membrane.

3. Under adequate stimulation, the voltage difference across the membrane changes dramatically, past a certain minimum amount (the threshold level). Then, gated sodium channels across the membrane open in an accelerating way and suddenly reverse the voltage difference, which registers as a spike (action potential) on recording devices.

4. Action potentials propagate themselves along the neuron membrane until they reach an output zone, where axon endings form a junction (chemical synapse) with another neuron or a muscle or gland cell. The presynaptic cell releases a neurotransmitter into the cleft, and this has an excitatory or inhibitory effect on the postsynaptic cell. Integration is the moment-by-moment combining of all signals—excitatory and inhibitory—acting on a neuron.

5. The direction of information flow through the body depends on the organization of neurons into circuits and pathways. Local circuits are sets of interacting neurons confined to a single region in the brain or spinal cord. Nerve pathways extend from neurons in one body region to neurons or effectors in different regions.

6. The central nervous system consists of the brain and spinal cord. The peripheral nervous system consists of nerves and ganglia in other body regions. Nerves carry signals between the central nervous system and the peripheral nervous system.

7. The peripheral nervous system has a somatic subdivision, which deals with skeletal muscles concerned with voluntary body movements and sensations arising (mainly) from skin. It also has an autonomic subdivision, which deals with the functions of the heart, lungs, glands, and other internal organs.

8. The spinal cord has nerve tracts that carry signals between the brain and the peripheral nervous system. It also is a center for some direct reflex connections that underlie limb movements and internal organ activity.

9. The brain has three regional divisions (hindbrain, midbrain, and forebrain).

a. The hindbrain includes the medulla oblongata, pons, and cerebellum and contains reflex centers for vital functions and muscle coordination.

b. The midbrain functions in coordinating and relaying visual and auditory information. The midbrain, medulla oblongata, and pons constitute the brain stem. The reticular formation, an extensive network of interneurons, extends the length of the brain stem and helps govern activities of the nervous system as a whole.

c. The forebrain includes the cerebrum, thalamus, hypothalamus, and limbic system. The thalamus relays sensory information and helps coordinate motor responses. The hypothalamus monitors internal organs and influences thirst, hunger, sexual activity, and other behaviors related to their functioning. The limbic system, which has roles in learning, memory, and emotional behavior, includes pathways that link parts of the thalamus and hypothalamus with other forebrain regions.

10. The cerebral cortex has regions devoted to specific functions, such as receiving information from the various sense organs, integrating this information with memories of past events, and coordinating motor responses.

11. Memory apparently occurs in two stages: a short-term period and long-term storage, which depends on chemical or structural changes in the brain.

12. States of consciousness vary between total alertness and deep coma. The levels are governed by the reticular activating system. They can be influenced by psychoactive drugs.

Review Questions

1. Define sensory neuron, interneuron, and motor neuron. 226

2. Label the functional zones of a motor neuron: 227

3. Distinguish between a local signal at the input zone of a neuron and an action potential. 229

4. What is a synapse? Explain the difference between an excitatory and an inhibitory synapse. Define neural integration. 233

5. What is a reflex? Describe the events in a stretch reflex. 236

6. What constitutes the central nervous system? The peripheral nervous system? 237

7. Distinguish between the following:
   a. neurons and nerves 226, 235
   b. somatic system and autonomic system 238
   c. parasympathetic and sympathetic systems 239

8. Label the parts of the human brain: 243

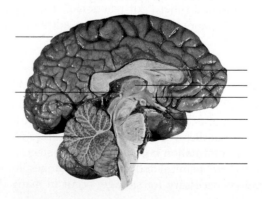

Critical Thinking: You Decide (Key in Appendix IV)

1. In some cases of ADD (attention deficit disorder) the impulsive, erratic behavior typical of so-called "hyperactive" youngsters can be normalized with drugs that *stimulate* the central nervous system. Explain this finding in terms of neurotransmitter activity in the brain.

2. Meningitis is an inflammation of the meninges that cover the brain and spinal cord. Diagnosis involves making a "spinal tap" (lumbar puncture) and analyzing a sample of cerebrospinal fluid for signs of infection. Why analyze this fluid and not blood?

Self-Quiz (Answers in Appendix III)

1. The nervous system senses, interprets, and issues commands for responses to _____. Its communication lines are organized gridworks of nerve cells, or _____.

2. A neuron responds to adequate stimulation with _____, a type of self-propagating signal.

3. When action potentials arrive at a synapse between a neuron and another cell, they stimulate the release of molecules of a _____ that diffuse over to that cell.

4. The moment-by-moment combining of all signals acting on all the different synapses on a neuron is called _____.

5. Interactions among neurons in your body _____.
   a. involve neurotransmitters
   b. are mediated by memory
   c. are mediated by learning and reasoning
   d. all of the above are correct

6. In the simplest kind of reflex, _____ directly signal _____, which act on muscle cells.
   a. sensory neurons; interneurons
   b. interneurons; motor neurons
   c. sensory neurons; motor neurons
   d. motor neurons; sensory neurons

7. The accelerating flow of _____ ions through gated channels across the membrane is the actual trigger for an action potential.
   a. potassium
   b. sodium
   c. hydrogen
   d. a and b are correct

8. The peripheral nervous system consists of the _____; the central nervous system consists of the _____.
   a. nerves and ganglia; brain and spinal cord
   b. brain and spinal cord; nerves and ganglia
   c. spinal cord; brain
   d. nerves and interneurons; brain and spinal cord

9. _____ nerves slow down the body overall and divert energy to basic housekeeping tasks; _____ nerves slow down housekeeping tasks and increase overall activity during times of heightened awareness, excitement, or danger.
   a. Autonomic; somatic

   b. Sympathetic; parasympathetic
   c. Parasympathetic; sympathetic
   d. Peripheral; central

10. Match each of the following central nervous system regions with some of its functions.
   _____ spinal cord
   _____ medulla oblongata
   _____ hypothalamus
   _____ limbic system
   _____ cerebral cortex

   a. receives sensory input, integrates it with stored information, coordinates motor responses
   b. monitors internal organs and related behavior (e.g., thirst, hunger, sex)
   c. governs emotions
   d. coordinates reflexes (e.g., for respiration, blood circulation)
   e. makes reflex connections for limb movements, internal organ activity

Key Terms

action potential 228	nerve 235
autonomic system 238	nerve tract 239
axon 227	neuroglial cell 226
blood-brain barrier 241	neuromodulator 234
brain 240	neuron 226
central nervous system 237	neurotransmitter 233
cerebral cortex 242	parasympathetic nerve 239
cerebrospinal fluid 241	peripheral nervous system 237
cerebrum 241	reflex 236
chemical synapse 233	reflex arc 236
dendrite 227	resting membrane potential 227
forebrain 241	reticular activating system 244
hindbrain 240	reticular formation 241
hypothalamus 245	Schwann cell 233
interneuron 226	sensory neuron 226
limbic system 245	sodium-potassium pump 229
memory 245	somatic system 238
meninges 240	spinal cord 239
midbrain 241	sympathetic nerve 239
motor neuron 226	synaptic integration 234
myelin sheath 231	

Readings

Changeux, J. P. November 1993. "Chemical Signaling in the Brain." *Scientific American*.

Fischbach, G. September 1992. "Mind and Brain." *Scientific American*. This and the following reading are from a special issue devoted to the brain.

Hinton, G. September 1992. "How Neural Networks Learn from Experience." *Scientific American*.

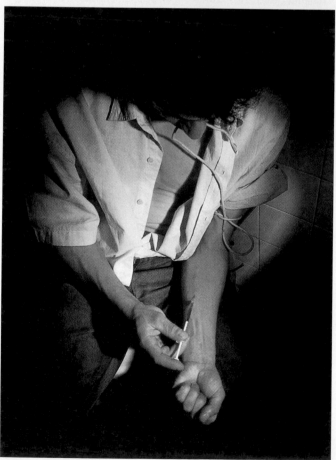

Different ways people abuse drugs

Human Impact

MIND, BODY, AND DRUG ABUSE

A drug is any substance introduced into the body to provoke a specific physiological response. For thousands of years people in nearly every culture have used drugs as medicines, to alter mental states as part of religious or social rituals, or simply for individual effects. **Drug abuse** can be defined as use of a drug in a way that harms health or interferes with a person's ability to function in society.

In general, drug abuse involves **psychoactive drugs**, which act on the central nervous system by binding to receptors in the neuron plasma membrane. Such receptors normally bind neurotransmitter molecules, which transmit chemical messages among neurons (page 234). When drug molecules bind such receptors, the result is a change in the chemical messages neurons send or receive—and in associated mental and physical states.

Drug Effects

Psychoactive drugs exert their effects on brain regions that govern states of consciousness and behavior. Various drugs also influence physiological events, such as heart rate, respiration, sensory processing, and muscle coordination. Many affect a pleasure center in the hypothalamus and artificially fan the sense of pleasure we associate with eating, sexual activity, or other behaviors.

Often the body develops *tolerance* to a drug, meaning it takes larger or more frequent drug doses to produce the same effect. Tolerance reflects *physical drug dependence*. As Chapter 5 described, the liver produces enzymes that detoxify drugs circulating in the bloodstream. Tolerance develops when the level of detoxifying liver enzymes increases in response to the consistent presence of the drug in the blood. Hence a user must increase his or her intake to keep one step ahead of the liver's increasing ability (up to a point) to break down the drug.

Psychological drug dependence, or *habituation*, develops when a user begins to crave the feelings associated with using a particular drug. The inability to "feel good" or function normally without a regular supply of a drug—whether it is caffeine, alcohol, nicotine, or crack cocaine—is a clear indication of habituation.

Habituation and tolerance are evidence of **addiction** to a particular substance. Although all drugs can be misused and abused, the psychoactive drugs described here include substances that can injure health and in some cases can lead to widespread harm in other ways as well. Table I lists some warning signs of potentially serious drug dependence.

Drug Action and Interactions   Not surprisingly, a drug's effects depend on the dose, and typically the effects of a higher dose are more intense than those of a smaller dose. For example, smoking a little marijuana may pleasurably intensify sensory perceptions (such as listening to music), but a higher dose may provoke wildly exaggerated, even paranoid perceptions of events.

Different psychoactive drugs used simultaneously can interact, sometimes in ways that are extremely dangerous. In a *synergistic* interaction, two drugs used together

Table I   Warning Signs of Drug Dependence
There may be cause for concern about drug dependency if a person consistently or often shows three or more of the following characteristics with respect to a substance:
1. Noticeable tolerance to a substance, marked by significant increase in the amount needed to produce the desired effect
2. Using a substance regularly for an extended period of time
3. Difficulty stopping or cutting down, even when a person consistently wishes to do so
4. Use of a substance to avoid withdrawal symptoms and feel "normal"
5. Concealing substance use from others
6. Taking extreme or dangerous measures to get the drug or to ensure its supply. (Examples: stealing money to buy cocaine; drinking on the job in spite of severe penalties for doing so; going from doctor to doctor until finding one who will prescribe a particular tranquilizer)
7. Intoxication or withdrawal symptoms interfere with performance at work or school and/or cause problems in personal relationships
8. Extreme anger and defensiveness if someone else suggests there may be a problem
9. Reducing, avoiding, or giving up customary activities because they have begun to interfere with substance use

have a much more powerful effect than they would have if used separately. Alcohol and barbiturates (such as Seconal and Nembutal) both depress the central nervous system, for example, and used in tandem they can lethally depress respiratory centers in the brain. Some drug combinations are *antagonistic*—one drug blocks the effects of another. Others are *potentiating*, in that one drug enhances the effects of another. For instance, labels on allergy medications often warn against drinking alcohol while using such products because even a little alcohol may deepen the drowsiness antihistamines cause in certain people and seriously impair a person's ability to function normally.

Stimulants

Stimulants include caffeine, nicotine, amphetamines, and cocaine. First a stimulant increases alertness and body activity, then it leads to depression.

Caffeine Coffee, tea, chocolate, and many soft drinks contain caffeine, one of the most widely used stimulants. Low doses of caffeine stimulate neurons of the cerebral cortex first, and cause increased alertness and restlessness. Higher doses act at the medulla oblongata to disrupt motor coordination and transmission of signals from other brain regions governing intellectual functions.

Nicotine Nicotine, a component of tobacco, has powerful effects on the central and peripheral nervous systems. It mimics the neurotransmitter acetylcholine and can directly stimulate a number of sensory receptors. Inhaled in cigarette smoke, nicotine enters lung alveoli and then the bloodstream. Within less than 10 seconds, about 15 percent of inhaled nicotine has reached the brain, where it triggers secretion of adrenal hormones that increase a smoker's heart rate and blood pressure (page 286). These and other metabolic changes are the effects a smoker comes to crave. Other short-term effects can include water retention, irritability, and gastric upsets. An addicted smoker can quickly learn to regulate nicotine intake to maintain the blood level necessary for the nicotine "lift" while avoiding many of the unpleasant side effects. Because people who smoke or chew tobacco also take in tars and other substances that cause lung and oral cancers (page 438), the long-term impact of nicotine addiction can be devastating.

Amphetamines Amphetamines (including "speed") are synthetic chemicals that resemble dopamine and norepinephrine, natural signaling molecules that stimulate the pleasure center. They include methamphetamine, dexedrine, benzedrine, and MDMA or "Ecstasy." Small doses tend to make a person feel alert and interested in surrounding activities. Some people come to rely on them to relieve fatigue or boredom or curb the appetite. Over time, however, the brain responds to amphetamines by producing less and less of its own signaling molecules. It comes to depend on artificial stimulation. As tolerance develops, the user requires increasing doses to achieve the desired effect. Those effects can also wear off quickly, causing the person to suddenly "crash" in exhaustion. Chronic users may become seriously malnourished and develop heart and other circulatory problems. They may also suffer paranoid delusions and exhibit hostile, violent behaviors.

Cocaine Unlike amphetamines, cocaine stimulates the pleasure center by *blocking* the reabsorption of dopamine, norepinephrine, and other signaling molecules that are normally released at synapses. Depending on whether it is smoked or injected, cocaine can act on the brain within 10–20 seconds of entering the bloodstream. Receptor cells are incessantly stimulated over an extended period. Heart rate and blood pressure rise; sexual appetite increases. Then the effects change. The signaling molecules that have accumulated in synaptic clefts diffuse away, but the cells that produce them cannot make up for the extraordinary loss. The sense of pleasure evaporates as the receptor cells (which are now hypersensitive to stimulation) demand stimulation. The cocaine user becomes anxious and depressed. After prolonged, heavy use of cocaine, "pleasure" is impossible to experience. The addict loses weight and cannot sleep restfully. The immune system becomes compromised, and heart abnormalities set in. Otherwise-healthy people have been known to die of heart failure after a single cocaine "experiment."

Granular cocaine, which is inhaled ("snorted") or injected in solution, has been around for some time. Crack cocaine, a cheaper but more potent form, is burned and the smoke inhaled. As society has discovered, crack is incredibly addictive (Table II); its highs are higher, but the crashes are more devastating. Most authorities rank the social ills associated with crack cocaine—from lost jobs and disrupted families to street crime and the tragedy of "crack babies"—as the most serious of all illegal drug problems.

Sedatives, Hypnotics, and Antianxiety Drugs

Drugs that act as sedatives, hypnotics, or antianxiety agents lower the activity in nerves and parts of the brain, so they reduce activity throughout the body. Some act at synapses in the reticular formation and in the thalamus.

Depending on the dose, most of these drugs can produce responses ranging from emotional relief, sedation, sleep (hypnosis), anesthesia and coma, to death. At low doses the person at first feels excited or euphoric. Higher doses suppress excitatory synapses in the CNS, leading to depression.

Alcohol Unlike other drugs in this group, alcohol acts directly on the plasma membrane to alter cell function. It is water-soluble and is absorbed across the mucosal lining of the stomach and small intestine. Alcohol's effects can vary according to a person's gender and other circumstances. In general, alcohol is absorbed more slowly when there is food in the stomach. Because of gender-related differences in enzyme secretions, females tend to become intoxicated more rapidly than do males.

Due to the initial "high," some people think of alcohol as a harmless substance that helps them relax or eases tension. Evidence suggests that in moderate amounts (up to one or two drinks per day) it may not do serious damage in many people. Even so, alcohol is one of the most powerful psychoactive drugs known, and alcohol abuse is a major cause of illness, death, and disruption of personal relationships. There is growing evidence that even small amounts of alcohol can damage a developing fetus (page 327); pregnant women should not drink alcohol. Among adults, under certain circumstances as little as an ounce or two can produce disorientation, uncoordinated

Table II Most Commonly Used—and Abused—Recreational Drugs

This table tallies recreational drug use only. Drugs are ranked by estimated number of users and how the drug ranks in terms of the ease with which a user may become addicted. For the addiction ranking, 1 is most addicting and 10 is least addicting; some substances fall below a rank of 10 (as indicated by dashes), but they can be habit-forming nonetheless. In interpreting the chart, keep in mind that a large number of people regularly use or have used more than one drug.

Drug	Number of Users	Addiction Ranking
Alcohol	102,919,000	7
Nicotine	60,744,000	1
Marijuana, Hashish	10,206,000	—
Cocaine (non-crack)	1,601,000	10
Analgesics (painkillers)	1,536,000	—
Inhalants	1,188,000	—
Stimulants (non-caffeine)	957,000	depends on type: 2 (smoked methamphetamine) 4 (injected methamphetamine) 9 (amphetamine)
Tranquilizers (Valium, etc.)	568,000	5
Sedatives (Quaalude, etc.)	568,000	6
Hallucinogens	553,000	—
Crack cocaine	494,000	3
Heroin	48,000	8

Estimates are based on information published by the National Institute on Drug Abuse (1991) for the population of the United States age 12 and over.

motor functions, and diminished judgment. Long-term addiction destroys nerve cells and causes permanent damage to the brain, heart, and liver. It is the most common cause of the liver scarring known as cirrhosis (page 120) and is a significant factor in more than 50 percent of all fatal traffic accidents.

Blood alcohol concentration (BAC) is a measure of the percentage of alcohol in the blood. In most states, a person with a BAC of 0.08 is considered legally drunk and may not drive. When the BAC reaches 0.15 to 0.4, a person becomes obviously intoxicated and cannot function normally physically or mentally. A BAC greater than 0.4 can be lethal.

Some studies suggest there may be a genetic component to at least some cases of alcoholism, although the relationship has yet to be clearly defined. On the other hand, social factors, such as beginning a pattern of heavy drinking in the teens or twenties, clearly play a role.

Barbiturates and Antianxiety Drugs *Barbiturates* include drugs such as phenobarbitol and secobarbitol. Different ones are prescribed medically to help control epileptic seizures and induce relaxation, and sometimes as general anesthetics. Used abusively as "downers," they can cause extreme drowsiness and markedly impaired judgment. Substances such as Valium (diazepam), Xanax (alprazolam), and Dalmane (fluorazepam) are *benzodiazepines,* drugs that are used medically to combat extreme anxiety, severe insomnia, and panic attacks. Used long-term or at high dosages, they can lead to dependence, in part because withdrawal symptoms (including headaches, anxiety, nausea, vomiting, and even seizures) can be severe.

Another drug in this group is methaqualone, sometimes sold as Quaalude. Although it has legitimate medical uses as a sedative, methaqualone can be habit-forming. At levels that constitute abuse, it causes mental confusion, impaired coordination and judgment, and abusive behavior toward others.

Analgesics

When stress leads to physical or emotional pain, the brain produces its own analgesics, or natural pain relievers. Endorphins and enkephalins are examples. These substances seem to inhibit activity in many parts of the nervous system, including central nervous system centers concerned with emotions and perception of pain.

Narcotic analgesics such as codeine, morphine, and heroin sedate the body and relieve pain. Such substances are *opioids* derived from opium. Although they have medical uses, they are habit-forming, and some can be seri-

ously addictive. Deprivation following massive doses of heroin leads to fever, chills, hyperactivity and anxiety, violent vomiting, cramping, and diarrhea. A synthetic form of heroin known as 3-methylfentanyl or China White is one type of "designer drug"—a laboratory-made chemical that differs from the natural compound but mimics its effects.

Psychedelics and Hallucinogens

These so-called "mind-expanding" drugs alter sensory perception and other aspects of consciousness. Some skew acetylcholine or norepinephrine activity. Others, such as LSD (lysergic acid diethylamide), affect the activity of serotonin, a brain hormone. LSD is extraordinarily potent. Consuming an amount less than the volume of a match head will dramatically warp the user's perceptions. In addition to physical changes such as increases in blood pressure and heart rate, it causes vivid hallucinations that can last up to 12 hours. Some people become psychotic on LSD, seeing visions of horrible insects or writhing snakes; some have died when they became convinced they had superhuman powers and jumped from high windows or stepped in front of speeding vehicles.

Marijuana, made from crushed leaves, flowers, and stems of the plant *Cannabis sativa,* is another hallucinogen. Its main psychoactive ingredient is tetrahydrocannabinol, or THC, which enters the bloodstream when marijuana is smoked or eaten in food. In low doses, THC is like a depressant. It slows down motor activity, relaxes the body, and it elicits mild euphoria. However, it can produce disorientation, anxiety bordering on panic, and paranoid delusions. Chronic marijuana smoking irritates the lungs and throat, and babies born to women who smoked marijuana during pregnancy tend to be irritable and grow poorly during early life. Like alcohol, marijuana can affect a person's ability to perform complex tasks, such as driving a car or studying. Recent studies also point to a link between marijuana smoking and suppression of the immune system and the male reproductive system.

Marijuana is sometimes used medically to help control the nausea that is a side effect of cancer chemotherapy. Some medical research suggests that controlled doses may also help reduce fluid pressure associated with the eye disease glaucoma (page 268).

Deliriants

Certain substances, including many that we do not usually think of as drugs, disrupt brain functioning in ways that lead to *toxic psychosis* or delirium. In general, their effects include mental confusion, skewed perception of the surroundings (including the normal behavior of others), and hallucinations. One type commonly sold as a street drug is the synthetic compound *phencyclidine* or PCP ("angel dust"). PCP also functions as an anesthetic, and impairs proprioception, a person's awareness of the position of different body parts. It may trigger a permanent psychotic state or temporary but dangerous outbursts of violent behavior. The danger is compounded by the fact that the individual often has delusions of being invulnerable and feels no physical pain. Large doses produce additional symptoms ranging from dizziness, memory impairment, and profuse sweating to seizures and heart failure.

Other deliriants are known as *inhalants*. As their name suggests, inhalants produce psychoactive vapors that a user inhales. Commonly abused ones include cleaning fluid and similar solvents, the butane in cigarette lighters, model airplane glue, kerosene, "poppers" (amyl nitrate, a heart medication), and gases contained in certain aerosol products. The "laughing gas" nitrous oxide used by many dentists is an inhalant. Some inhalants work directly on brain neurons. Others interfere with gas exchange in the lungs and only indirectly affect the brain.

Inhalants sometimes become the recreational drug of choice for people, often including teenagers and even younger children, who cannot afford to purchase other psychoactive drugs. Unfortunately, they can be extremely toxic to the liver, kidneys, respiratory system, and the heart. "Sniffers" may develop hepatitis and arrhythmias (irregular heartbeat) and may suffer permanent brain damage.

11 SENSORY RECEPTION

Nosing Around

That charming protuberance called your nose knows more than you might think. In the upper reaches of the nasal cavity, patches of yellowish epithelium are crammed with some 10 million specialized neurons. The neurons sport hairlike cilia. And at the surfaces of the cilial cells are protein receptors for odor molecules. Molecules arriving in inhaled air are sensory calling cards of a rose's perfume, pungent garlic, and human body secretions, just to mention a few of the 10,000 different scents the average person can distinguish (Figure 11.1). Sensory receptors are gateways into the nervous system, as we will see in the following pages.

A recent exciting discovery indicates that there are several hundred *different* types of protein receptors in

Figure 11.1 The sense of smell—an ancient sensory capacity and just one of the ways humans gain information about the world.

the olfactory epithelium, each one constructed according to a genetic blueprint in a particular segment of DNA (a gene). There may even be as many as 1,000 different receptor types coded by a large family of "smell genes." Each receptor can latch onto odor molecules of a specific size and shape. Most likely, many "smells" result from stimulation of a combination of receptors.

By contrast, your taste buds distinguish only four basic types of flavors (sweet, sour, salty, and bitter), and your ability to visually distinguish colors relies on just three general types of photoreceptors (for red, green, or blue light wavelengths). In short, your ability to sniff the difference between pizza pie and apple pie may be the product of one of the more complex elements of the human nervous system.

In this chapter we turn to the means by which humans receive signals from the external and internal environments—and how we then decode those signals in ways that give rise to awareness of sounds, sights, odors, pain, and other sensations. Sensory neurons, nerve pathways, and brain regions are required for these tasks. Together they represent the portions of the nervous system that are called *sensory systems*.

KEY CONCEPTS

1. Sensory systems consist of specific types of sensory receptors, nerve pathways from those receptors to the brain, and brain regions that deal with sensory information.

2. Sensory receptors at or near the body surface detect specific stimuli that can give rise to sensations of vision, hearing, taste, smell, touch, externally applied pressure, temperature and pain.

3. Sensory receptors associated with skeletal muscles provide information about body position, balance, and movement. Receptors associated with internal organs other than skeletal muscles have roles in sensations of fatigue, hunger, thirst, nausea, and internal pressure and pain.

4. A sensation is an awareness of change in external or internal conditions. It results from four steps in which (a) sensory receptors detect a specific stimulus, (b) the stimulus energy is converted to signals that can give rise to action potentials, (c) information about the stimulus becomes encoded in the number and frequency of action potentials sent to the brain, and (d) specific brain regions translate the information into a sensation.

SENSORY SYSTEMS: AN OVERVIEW

Sensory systems are the front doors of the nervous system, the parts that let in information about changing conditions and allow the brain to become aware of pertinent changes occurring outside and inside the body. Each sensory system consists of (1) sensory receptors, (2) nerve pathways leading from the receptors to the brain, and (3) brain regions where sensory information is processed.

Sensory receptors are finely branched endings of sensory neurons or specialized cells associated with them, and most detect specific stimuli. Receptors for light stimuli are associated with the eyes, receptors for sound with the ears, receptors for pressure with the skin, and so forth. A *stimulus* such as light, heat, sound, or mechanical pressure is a form of energy that is capable of eliciting a response from a sensory receptor. Once detected, the stimulus energy is converted to a change in the membrane potential of the receiving cell. This change may lead to an action potential in this or an adjacent cell. By way of action potentials, information travels along nerve pathways to the brain.

When a stimulus registers with specific brain regions, the result is a **sensation**. We are consciously aware of sensations that are processed by the cerebral cortex. **Perception** results when the brain interprets a sensation and you understand what the sensation means. Splash hot tea on your hand, and you are conscious of the stimulus—and you understand that the intense heat is hurting you.

A sensory system consists of sensory receptors for specific stimuli, nerve pathways that conduct information from those receptors to the brain, and brain regions where the information is processed.

Information flowing through a sensory system may lead to conscious awareness of the stimulus.

Types of Sensory Receptors

Different sensory receptors are specialized to detect different kinds of stimuli. Many sensory receptors are positioned individually, like sentinels, in the skin and other body tissues. Others are part of specialized sensory organs, such as eyes and ears, which have structures that amplify or focus the stimulus energy.

There are five major types of sensory receptors, based on the type of stimulus they detect (see Table 11.1):

1. **Chemoreceptors** detect molecules of specific substances dissolved in the fluid surrounding the receptors.

Table 11.1 Receptors Associated with the Major Senses		
Category of Receptor	Examples	Stimulus
Chemoreceptors		
Internal chemical senses	Carotid bodies in blood vessel wall	Substances (CO_2, etc.) dissolved in extracellular fluid
Taste	Taste receptors of tongue	Substances dissolved in saliva, etc.
Smell	Olfactory receptors of nose	Odors in air, water
Mechanoreceptors		
Touch, pressure	Pacinian corpuscles in skin	Mechanical pressure against body surface
Stretch	Muscle spindle in skeletal muscle	Stretching
Auditory	Hair cells within ear	Vibrations (sound waves)
Balance	Hair cells within ear	Fluid movement
Photoreceptors		
Visual	Rods, cones of eye	Light
Thermoreceptors	Cold or warm receptors in skin; central thermoreceptors in hypothalamus, etc.	Heat or cold
Nociceptors	Free nerve endings in skin	Any stimulus that causes tissue damage and leads to sensation of pain

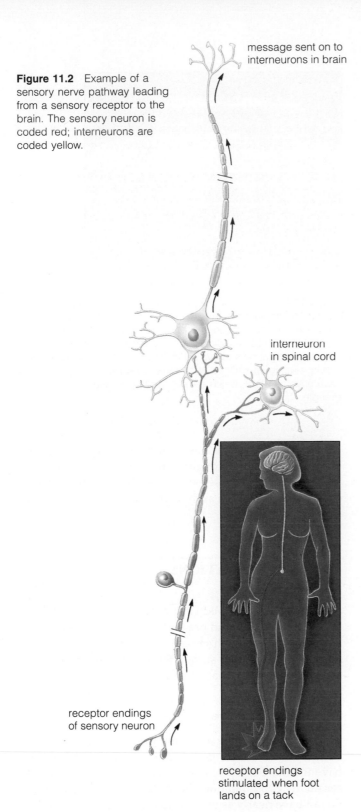

Figure 11.2 Example of a sensory nerve pathway leading from a sensory receptor to the brain. The sensory neuron is coded red; interneurons are coded yellow.

message sent on to interneurons in brain

interneuron in spinal cord

receptor endings of sensory neuron

receptor endings stimulated when foot lands on a tack

2. **Mechanoreceptors** detect mechanical energy associated with changes in pressure, changes in position, or acceleration.

3. **Photoreceptors** detect visible light.

4. **Thermoreceptors** detect heat energy.

5. **Nociceptors** (pain receptors) detect tissue damage.

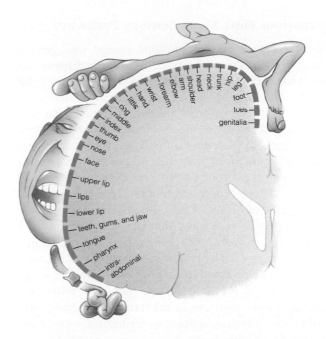

Figure 11.3 Body regions represented in the primary somatic sensory cortex. This region is a strip a little more than an inch (2.5 centimeters) wide, running from the top of the head to just above the ear on the surface of each cerebral hemisphere. The diagram is a cross-section through the right hemisphere of someone facing you. Compare this with Figure 10.21, which shows the motor cortex.

Sensory Pathways

Each sensory pathway starts at receptors or sensory neurons that are sensitive to a particular type of stimulus, such as light, cold, or pressure. As Figure 11.2 shows, the axons of sensory neurons lead into the central nervous system. There they might converge on a single interneuron or branch out and connect with several to many interneurons.

Sensory nerve pathways from different receptors lead indirectly or directly to different parts of the cerebral cortex, the outermost layer of the brain. For example, signals from receptors in the skin and joints travel by way of fiber tracts (formed from the axons of interneurons) in the spinal cord to the **somatic sensory cortex** (see Figure 10.20). Cells of this cortical region are laid out like a map corresponding to the body surface. Some map regions are larger than others; the body regions they represent are functionally more important and have more receptors, and thus are more sensitive to sensory stimulation. In humans, a large part of the primary somatic sensory cortex responds to receptors located in the fingers, thumbs, lips and tongue, as shown in Figure 11.3.

Similarly, the visual world is mapped onto the **visual cortex** (see Figures 10.20 and 11.16). The map region corresponding to the center of the visual field is larger than other regions.

Information Flow Along Sensory Pathways

Sensory receptors convert stimulus energy into graded changes in membrane potential. (Recall from Chapter 10 that a graded change is proportional to the strength of the stimulus that caused it.) If the changes reach the necessary threshold, they generate action potentials. Action potentials are propagated rapidly to the axon endings of the sensory neuron. There, the sensory neuron releases a neurotransmitter that affects the activity of neighboring interneurons (or motor neurons). The arrival of neurotransmitter molecules may trigger action potentials in the interneurons, which are part of nerve tracts leading to the brain.

What exactly do action potentials being propagated along sensory neurons "tell" the brain about the stimulus? First, the genetically determined network of neurons in the brain can interpret action potentials only in certain ways. That is why, for example, you "see stars" when you accidentally poke your eye in the dark. Even though the stimulus is mechanical, in this case the brain always interprets signals arriving from an optic nerve as "light."

Second, the stronger the stimulus, the more frequently receptors fire action potentials. Thus an individual receptor will detect a caress as well as a hammer blow. The brain senses the differences through frequency variations in the signals the receptor sends to it. Figure 11.4 shows frequency variations corresponding to differences in sustained pressure on human skin. Note that the signal amplitude is constant in all cases because action potentials are "all or none" events (page 231).

Third, strong stimulation recruits more receptors. You activate some receptors when you lightly touch skin on your arm, but you activate many more when you press hard on the same area. The increased disturbance sets off action potentials in many sensory axons at the same time. The brain interprets the combined activity as an increase in stimulus intensity.

The brain senses a stimulus based on which nerve pathways carry the incoming signals, the frequency of signals traveling along each axon of that pathway, and the number of axons that have been recruited.

Sometimes the frequency of action potentials decreases or stops even when a stimulus is being maintained at a constant strength. This is called **adaptation**. After you have put on clothing, for example, you may stop being aware of its pressure against your skin. Mechanoreceptors (*Pacinian corpuscles*) in your skin have fired only at the onset of the stimulus. These receptors are a rapidly adapting type. By contrast, when you pick up a warm muffin, thermoreceptors in your skin rapidly fire off action potentials. Then the frequency drops to a lower level that is maintained for as long as you hold the muffin. These receptors are a slowly adapting type.

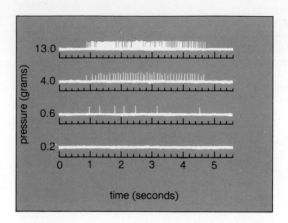

Figure 11.4 Action potentials recorded from a single pressure receptor of the human hand. The recordings correspond to variations in stimulus strength. A thin rod was pressed against the skin with the pressure indicated on the vertical axis of this diagram. Vertical bars above each thick horizontal line represent individual action potentials. Notice the increases in frequency (from bottom to top), which correspond to increases in stimulus strength.

Let's now focus on a few kinds of sensory receptors. Some kinds occur throughout the body; they contribute to general, *somatic sensations*. Other kinds are restricted to certain locations, such as the eyes or ears. They contribute to the *special senses*.

SOMATIC SENSATIONS

Somatic sensations, also called "general senses," include awareness of touch and pressure, heat and cold, pain, and limb movements. Such sensations start with receptor endings in skin and other surface tissues, in skeletal muscles, and in walls of internal organs.

Touch and Pressure

Mechanoreceptors are abundant in many places near the body surface. The fingertips and the tip of the tongue have the most receptors; these regions show the greatest sensitivity to stimulation. The back and much of the torso do not have nearly as many, and these regions are far less sensitive to stimulation.

Mechanoreceptors in the skin are the receiving ends (dendrites) of sensory neurons, with or without a surrounding capsule of epithelial or connective tissue. The free nerve endings (nociceptors) and Pacinian corpuscles shown in Figure 11.5 are examples. All skin mechanoreceptors respond easily to deformation of the skin resulting from pressure. But about half of them fire off action potentials only when the stimulus is first encountered and

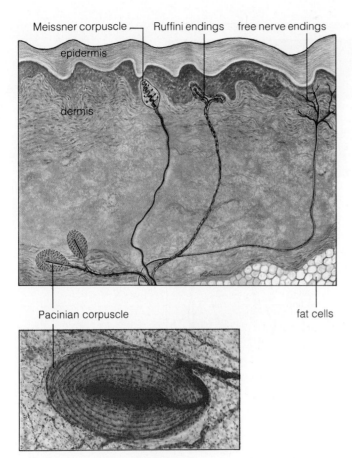

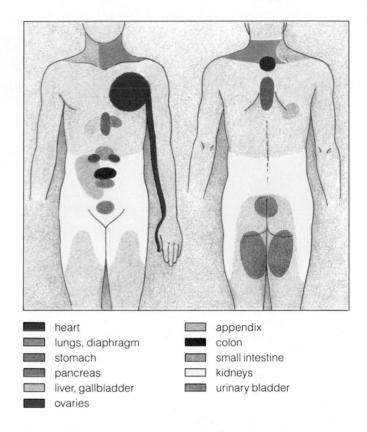

Figure 11.5 Tactile receptors in human skin. Free nerve endings contribute to sensations of temperature, light pressure, and pain. Pacinian corpuscles contribute to sensations of vibration. Meissner corpuscles are stimulated at the onset and end of sustained pressure; the Ruffini endings react continually to ongoing stimuli.

when it ends. They make you aware of touch, even light tickles and vibrations. The other half fire off action potentials as long as the stimulus is present. These are the mechanoreceptors that signal pressure.

Temperature Changes

When the temperature near the body surface remains fairly stable, thermoreceptors in skin quickly adapt and steadily fire off action potentials to the brain. When the temperature increases, so does the firing frequency. Free nerve endings in the skin, liver, and in the hypothalamus serve as "heat" receptors. We do not yet understand exactly how the body senses "cold."

Pain

Pain is the perception of injury to some body region. That perception begins with signals from nociceptors. These receptors occur in nearly all body tissues, but they are

Figure 11.6 Regions of referred pain. Sensations of certain internal disorders are localized to the skin areas indicated.

most abundant in the skin, joints, and around blood vessels. Nociceptors include free nerve endings that detect any stimulus causing tissue damage, such as a pinch or cut. Signals from these receptors are relayed to the part of the cerebrum called the parietal lobe, where they are processed (Figure 10.19).

Free nerve endings apparently detect any stimulus that is intense enough to damage the surrounding tissue. They respond to strong mechanical stimulation, intense heat or cold, and chemical irritation. When the stimulus intensity exceeds a certain threshold, the signals generated become translated into sensations of pain. *Anesthesia* literally means "no sensation"; medical anesthetics are chemicals that block transmission of pain-related action potentials to the central nervous system. *Analgesics* such as aspirin and ibuprofen are pain-reducing drugs.

Responses to pain often depend on the brain's ability to identify the affected tissue. For instance, if you get hit in the face with a snowball, you "feel" the contact on facial skin. For reasons that are not fully understood, however, the brain does not always associate internal injury with the tissue that is actually injured. Instead, it may associate the pain with a tissue in some other location. This phenomenon is called "referred pain" (Figure 11.6). A heart attack, for example, might be felt as pain in the skin above the heart and along the left shoulder and arm or neck.

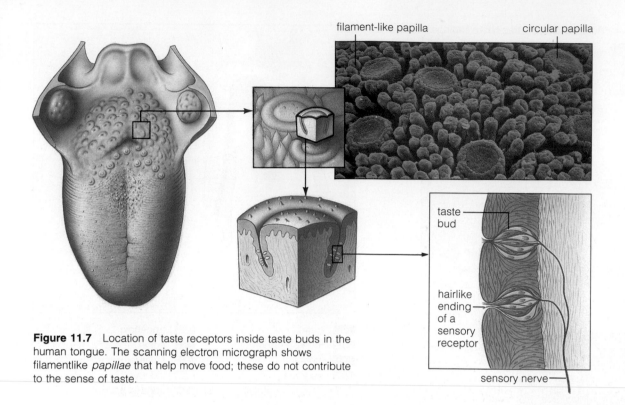

filament-like papilla   circular papilla

taste bud

hairlike ending of a sensory receptor

sensory nerve

Figure 11.7 Location of taste receptors inside taste buds in the human tongue. The scanning electron micrograph shows filamentlike *papillae* that help move food; these do not contribute to the sense of taste.

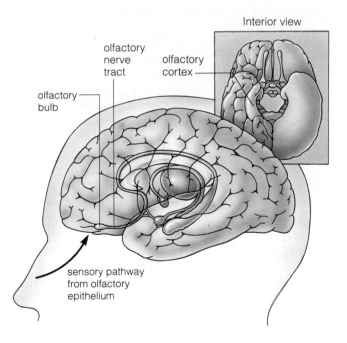

Interior view

olfactory nerve tract

olfactory cortex

olfactory bulb

sensory pathway from olfactory epithelium

Figure 11.8 Sensory nerve pathway leading from olfactory receptors in the nasal cavity to primary receiving centers in the brain.

Generally speaking, neurons leading to both the injured tissue and the tissue where pain is referred occur in the same region of the spinal cord.

Proprioception: "Muscle Sense"

Mechanoreceptors in skeletal muscle, joints, tendons, ligaments, and skin are responsible for awareness of limb movements. Stretch receptors are foremost among them. As described on page 236, stretch receptors are part of sensory organs called muscle spindles. Such organs are embedded in skeletal muscle tissue and run parallel to the muscle itself. They also occur in smooth muscle, as in respiratory bronchi and bronchioles. The response of stretch receptors to stimulation depends on how much and how fast the muscle is stretched.

SPECIAL SENSES

Taste and Smell

In the human body, *taste receptors* are part of sensory organs called taste buds (Figure 11.7). You have about 10,000 taste buds, which are scattered over the tongue, roof of the mouth, throat, and palate. Taste buds bear receptors for any (or all) of four general taste categories:

sweet, sour, salty, and bitter. The receptors bind molecules that are dissolved in saliva. Strictly speaking, the flavors of most foods are some combination of the four basic "tastes," plus sensory input from olfactory receptors in the nose. This olfactory element of taste is extremely important. It is the reason why anything that dulls your sense of smell—such as a head cold—also seems to diminish food's flavor.

Humans detect smells with the *olfactory receptors* described at the beginning of this chapter and depicted in Figure 11.8. From an evolutionary perspective, olfaction, the sense of smell, is one of the most ancient senses—and for good reason. Food and predators both give off chemical substances that can diffuse through water or air and so give clues or advance warning of their whereabouts. Even humans, with their relatively insensitive sense of smell, have about 10 million olfactory receptors in patches of olfactory epithelium in the upper nasal passages. (The nose of a bloodhound has more than 200 million.)

Recent studies suggest that when odor molecules bind to receptors on olfactory neurons in cells of the nose's olfactory epithelium, the resulting action potential travels directly to olfactory bulbs in the frontal area of the brain. There, other neurons forward the message to a center in the cerebral cortex, which interprets it as "fresh bread," "pine tree," and so forth.

Hearing

Hearing requires mechanoreceptors that can detect vibrations. A vibration is a wavelike form of mechanical energy. When sound waves travel into the ear, they reach a membrane (the tympanic membrane, or eardrum) that is stretched across their path and make it vibrate. In a series of steps we will trace shortly, the vibrations are relayed first through bone (where they are amplified) and then through fluids deep inside the ear. There, the moving fluid causes other membranes to vibrate, and those movements stimulate mechanoreceptors. The excited mechanoreceptors release a neurotransmitter that stimulates action potentials that travel along an auditory nerve leading from the receptors to the brain.

The Ear A human ear consists of three regions (Figure 11.9a). Sound waves enter the *outer ear*, and the vibrations are amplified in the *middle ear*. Within the *inner ear,* vibrations of different frequencies are "sorted out" as they stimulate different patches of receptors. The inner ear houses several structures, including *semicircular canals*, which are involved in balance, and the coiled **cochlea** involved in hearing. If we could uncoil the cochlea, we would see that a fluid-filled chamber folds around an inner *cochlear duct* (Figure 11.9b). In the coiled cochlea, each long "arm" of the outer chamber is considered a sepa-

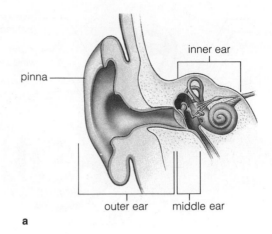

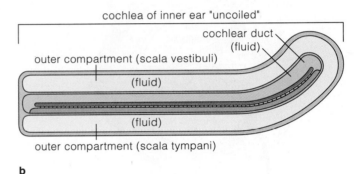

Figure 11.9 (**a**) Internal structure of the human ear. (**b**) The cochlea "uncoiled" to show its fluid-filled compartments.

rate compartment (the *scala vestibuli* and *scala tympani*, respectively).

Mechanism of Hearing Hearing begins when the external flaps of the outer ear collect sound waves and channel them inward through the auditory canal to the **tympanic membrane**—what most people call the eardrum. Sound waves cause the tympanic membrane to vibrate. This vibration in turn causes vibrations in a leverlike sequence of three tiny bones of the middle ear: the *malleus* ("hammer"), *incus* ("anvil"), and stirrup-shaped *stapes*. The vibrating bones transmit their motion to the *oval window*, an elastic membrane over the entrance to the cochlea. The oval window is much smaller than the tympanic membrane, and as the middle-ear bones vibrate against its small surface with the full energy that struck the tympanic membrane, the force of the original vibrations becomes amplified.

The amplified vibrations of the oval window are strong enough to produce pressure waves in the fluid

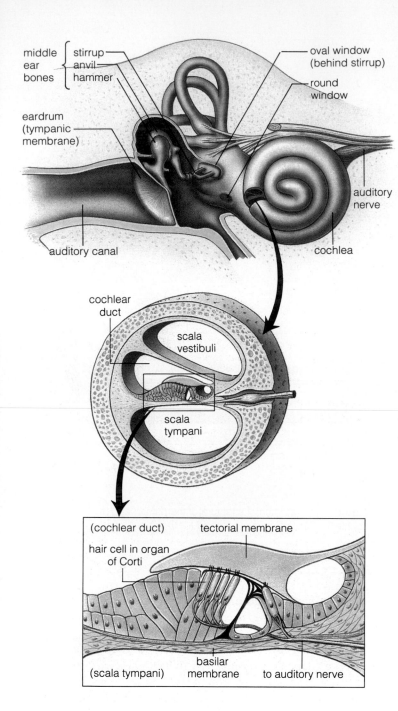

middle ear bones { stirrup / anvil / hammer }

oval window (behind stirrup)

round window

eardrum (tympanic membrane)

auditory nerve

auditory canal

cochlea

cochlear duct

scala vestibuli

scala tympani

(cochlear duct)   tectorial membrane

hair cell in organ of Corti

(scala tympani)   basilar membrane   to auditory nerve

Figure 11.10 The outer ear collects sound waves with external flaps, then channels them through the auditory canal to the eardrum (tympanic membrane). The waves cause the eardrum to vibrate. Tiny bones of the middle ear amplify the stimulus by transmitting the force of vibration to the smaller surface of the oval window. This is an elastic membrane over the entrance to the coiled inner ear. The oval window bows in and out, producing fluid pressure waves in two chambers of the inner ear (scala vestibuli and scala tympani).

Pressure waves become sorted out in the inner ear's third duct (cochlear duct). At different points, the cochlear duct's basilar membrane vibrates more strongly to sounds of different frequencies. Different patches of hair cells in the organ of Corti respond as vibrations of different regions of the membrane push them against the tectorial membrane. Signals from bent hair cells trigger action potentials that travel to the brain via an auditory nerve.

the basilar membrane to vibrate, and those movements bend hair cell projections embedded in the tectorial membrane. The bending stimulates affected hair cells to release a neurotransmitter, which triggers action potentials in auditory neurons. The auditory nerve carries the action potentials to the brain.

Pressure waves that vibrate at a higher frequency (more waves per second) set up vibrations in regions of the basilar membrane nearer the entrance to the coil, whereas waves of lower frequency cause vibrations nearer the tip of the coil. Vibrations of different regions of the basilar membrane bend the projections of different patches of hair cells. Studies suggest that the total number of hair cells that become stimulated in a given region determines the loudness of a sound. The tone or "pitch" of a perceived sound depends on the frequency of vibrations that excite different groups of hair cells—the higher the frequency, the higher the pitch.

Pressure waves traveling through the cochlea eventually push against the *round window*, a membrane at the far end of the cochlea. As the round window bulges outward toward the air-filled middle ear, it serves as a "release valve" for the force of the waves. Air also moves through an opening in the middle ear into the *Eustachian tube*. This tube runs from the middle ear to the throat (pharynx) and permits air pressure in the middle ear to be equalized with the pressure of outside air. When you change altitude (including riding in aircraft), this equalizing process makes your ears "pop."

Sounds such as amplified music and the thundering of jet engines are so intense that prolonged exposure to them can cause permanent damage to the inner ear (see *Focus on Environment*). Such recent technological developments exceed the functional range of the evolutionarily ancient hair cells in the ear.

within the chambers surrounding the cochlear duct (Figure 11.10). In turn, these waves are transmitted to the fluid in the cochlear duct. On the floor of the cochlear duct is a *basilar membrane*, and resting on the basilar membrane is a specialized **organ of Corti**, which includes sensory **hair cells**. Hairlike projections at the tips of the cells are embedded in an overhanging **tectorial** ("rooflike") **membrane**. (This noncellular "membrane" is the consistency of soft putty.) Pressure waves in the cochlear fluid cause

Noise Pollution: An Attack on the Ears

From the cacophony of city streets, "boom boxes," and rock concerts to roaring lawnmower engines, home appliances, and the screech of power tools, *noise pollution* has become an increasingly serious health problem for millions of people.

Consider some statistics. In the United States, roughly 10 million people, including a growing number of musicians, have lost some of their hearing due to damage caused by loud noise. A recent report from the National Institutes of Health estimates that high noise levels in the home, on the job, or in recreational pursuits puts another 20 million Americans at risk of hearing loss. Fully one-third of adults in the United States will suffer significant damage to their hearing by the time they are 65—and researchers believe that most cases are due to the long-term effects of living in a noisy world.

The loudness of a sound is measured in *decibels*. A conversation with a friend in a quiet room occurs at about 50 decibels, whereas rustling papers make noise at a mere 20 decibels. The delicate sensory hair cells in the inner ear (Figure *a*) begin to be damaged when a person is exposed to sounds louder than about 75-85 decibels over long periods. Unfortunately, our society is permeated with sounds above that threshold.

For example, a snowmobile or "boom box" stereo system cranks out sound at well over 100 decibels, as can a personal "Walkman"-type stereo if the volume is turned up too high. At 130 decibels—typical of a rock concert or shotgun blast—permanent damage can occur much more quickly. That is why targetshooters are well advised to wear ear plugs or hearing protectors. Many health professionals also recommend protective earwear for anyone who regularly operates a power mower or other noisy lawn-care equipment. Such protection is (or should be) mandatory for workers in some professions, such as airline service personnel or construction workers who routinely operate jackhammers or chain saws.

In developed societies today there are very few lifestyle choices that permit a person to avoid noise pollution completely. However, individuals *can* be aware of everyday hearing hazards, use ear protection when it seems prudent, and avoid frequent, prolonged exposure to very loud noise (such as stereo, rock concert, and aircraft noise). Following these simple guidelines can help ensure that your hearing lasts as long as you do.

Figure a Effect of intense sound on the inner ear. (**a**) Normal organ of Corti from a guinea pig, showing two rows of outer hair cells. (**b**) Organ of Corti after 24-hour exposure to noise levels approached by extremely loud music (2,000 cycles per second at 120 decibels).

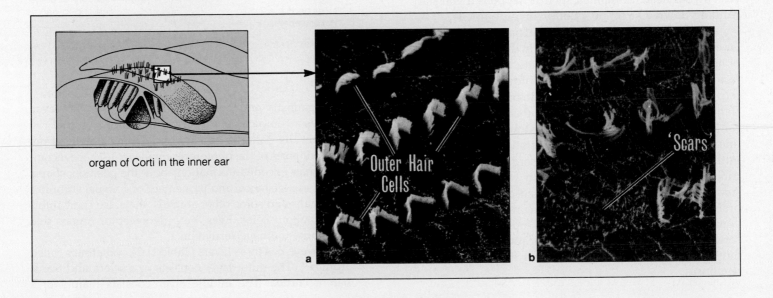

organ of Corti in the inner ear

Outer Hair Cells

'Scars'

a

b

Balance

Like most other animals, humans must have a sense of the "natural" position for the body (and its parts), given the predictable way they return to it after being tilted or turned upside-down. The baseline against which the brain assesses displacement from the natural position is called the "equilibrium position."

The sense of balance relies partly on input from receptors in the eyes, skin, and joints. It also relies on the **vestibular apparatus**, a closed system of fluid-filled canals and sacs in the inner ear (Figure 11.11a). The three **semicircular canals** are positioned at right angles to one another, corresponding to the three planes of space. Within them, sensory receptors detect rotational head movement in all directions, as well as acceleration and deceleration. At the base of each canal is a jellylike mass (the *cupula*); this mass rests atop a specialized region of epithelium that contains hair cells just like those in the cochlea. When fluid in the canals moves and presses on the soft mass, projections of the hair cells bend. As in the cochlea, this bending is the first step in a sequence of events that leads to action potentials that travel to the brain—in this case, along the vestibular nerve.

The vestibular apparatus also contains two fluid-filled sacs (the *utricle* and *saccule*), each of which contains an *otolith organ* containing hair cells embedded in a jellylike "membrane." This material also contains hard bits of calcium carbonate called *otoliths* ("ear stones"). Movements of the membrane and otoliths signal changes in the head's orientation relative to gravity, as well as straight-line acceleration and deceleration. For example, if you tilt your head to one side, the otoliths slide in that direction, the membrane mass shifts, and projections of the hair cells are bent (Figure 11.11b). The otoliths also press on hair cells if your head accelerates, such as happens if you start running down a street or set off in an automobile.

Action potentials from the different parts of the vestibular apparatus travel to reflex centers in the brainstem. The brain integrates the incoming signals with information from the eyes and muscles and then orders compensatory movements that help you keep your balance when you stand, walk, or move your body in other ways.

Motion sickness can occur when extreme or continuous linear, angular, or vertical motion overstimulates hair cells in the vestibular apparatus. Visual sensations often contribute to the sickness; fear and anxiety can also play roles. Action potentials triggered by the sensory input reach a brain center that governs the vomiting reflex. Nausea and vomiting are the chief symptoms in individuals who get carsick, airsick, or seasick. Medications that relieve motion sickness act on brain centers that govern these responses.

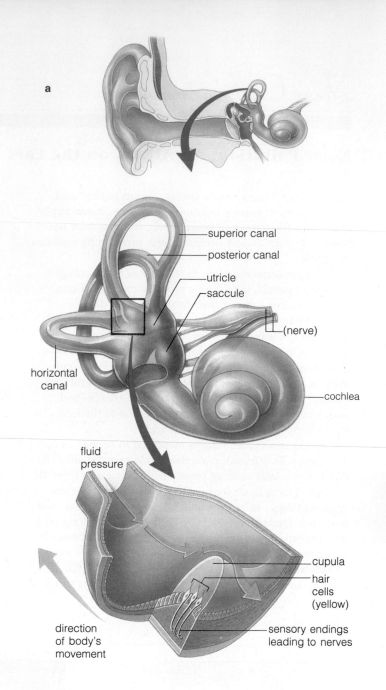

a

superior canal
posterior canal
utricle
saccule
(nerve)
horizontal canal
cochlea
fluid pressure
cupula
hair cells (yellow)
direction of body's movement
sensory endings leading to nerves

Vision

All organisms are sensitive to light. **Vision**, however, requires a complex neural program in the brain that can interpret the patterns of action potentials arriving from different parts of the photoreceptor system. Those incoming signals encode information about the position, shape, brightness, distance, and movement of a visual stimulus; in humans and some other animals, they also carry information about color. **Eyes** are photoreceptor organs that contribute to image formation.

The eye has three layers (Table 11.2), sometimes called "tunics." The outer layer consists of a sclera and transparent **cornea**. The middle layer consists mainly of a

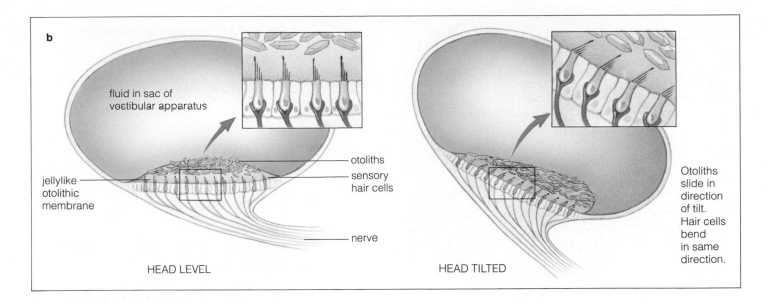

b

fluid in sac of
vestibular apparatus

jellylike
otolithic
membrane

otoliths

sensory
hair cells

nerve

HEAD LEVEL

HEAD TILTED

Otoliths
slide in
direction
of tilt.
Hair cells
bend
in same
direction.

Figure 11.11 (**a, left**) Vestibular apparatus, an organ of equilibrium that is given a workout when you ride a rollercoaster. Its three semicircular canals, positioned at angles corresponding to the three planes of space, detect changes in angular (rotational) movements. Its two sacs (utricle and saccule) each has an otolith organ that detects tilts in the head's position.

(**b, above**) Body movement in a given direction displaces fluid in the canal corresponding to that direction. The fluid pressure bends hair cells inside the canal, and these produce signals that travel to the brain. Moving in response to gravity, the otoliths bend projections of hair cells and so activate the cells.

Table 11.2 Components of the Human Eye	
Eye Region	Functions
Outer Layer	
Sclera	Protects eyeball
Cornea	Focuses light
Middle Layer	
Choroid (pigmented tissue)	Prevents light scattering
Iris	Controls amount of light entering via pupil opening
Ciliary body	Aids in focusing
Inner Layer	
Retina	Absorbs, converts light
Fovea	Increases visual acuity
Start of optic nerve	Transmits signals to brain
Other Components	
Lens	Finely focuses light on photoreceptors
Aqueous humor	Transmits light, maintains pressure
Vitreous humor	Transmits light, supports lens and eye

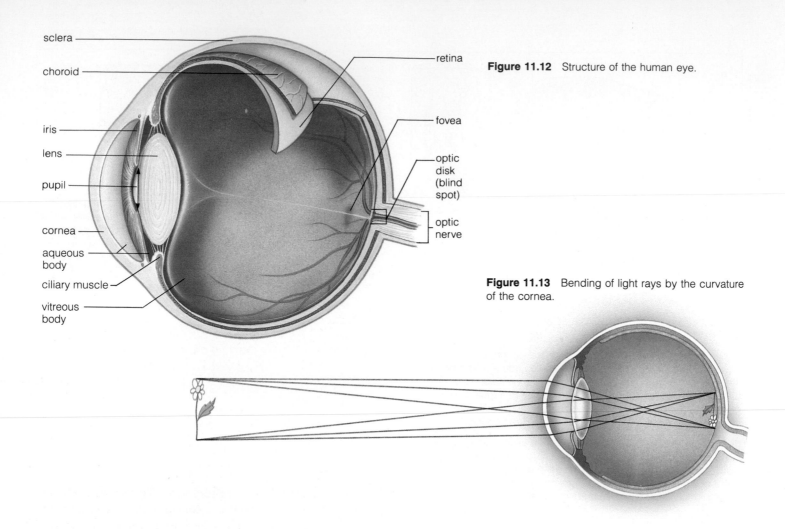

sclera

choroid

retina

Figure 11.12   Structure of the human eye.

iris

lens

fovea

pupil

optic
disk
(blind
spot)

cornea

optic
nerve

aqueous
body

ciliary muscle

Figure 11.13   Bending of light rays by the curvature
of the cornea.

vitreous
body

choroid, ciliary body, and iris. The key feature of the inner layer is the retina (Figure 11.12).

The *sclera*—the dense, fibrous "white" of the eye—protects most of the eyeball, except for a "front" region formed by the cornea. Moving inward, the thin, dark-pigmented *choroid* underlies the sclera. It prevents light from scattering inside the eyeball and contains most of the eye's blood vessels. Behind the transparent cornea is a circular, pigmented **iris** (after *irid,* meaning "colored circle") with a "hole" in its center. This *pupil* is the entrance for light. When bright light hits the eye, circular muscles in the iris contract and shrink the pupil. In dim light, radial muscles contract and enlarge the pupil. Behind the iris is a saucer-shaped **lens**, with onionlike layers of transparent proteins. Ligaments attach the lens to smooth muscle of the *ciliary body*; this muscle functions in focusing light, as we will see shortly. The lens focuses incoming light onto a dense layer of photoreceptor cells behind it, in the retina. A clear fluid (*aqueous humor*) bathes both sides of the lens, and a jellylike substance (*vitreous humor*) fills the chamber behind the lens.

The surface of the cornea is curved. Incoming light rays hit it at different angles, and as they pass through the cornea their trajectories (paths) bend. The rays con-

verge at the back of the eyeball. There, because of the way the rays were bent at the curved cornea, they form a pattern of stimulation that is upside-down and reversed left to right relative to the original light source (Figure 11.13).

Light rays from sources at different distances from the eye strike the cornea at different angles and will be focused at different distances behind it. Therefore, adjustments must be made so that the light will be focused precisely onto the retina. Normally, the lens can be adjusted so that the focal point coincides exactly with the retina. Ciliary muscle adjusts the shape of the lens. The muscle encircles the lens and attaches to it by fiberlike ligaments (Figure 11.12). When the muscle contracts, the lens bulges, so the focal point moves closer. When the muscle relaxes, the lens flattens, so the focal point moves farther back (Figure 11.14). Such adjustments are called **accommodation**. If they are not made, rays from very distant objects will be in focus at a point slightly in front of the retina, and rays from very close objects will be focused behind the retina.

Sometimes the lens cannot be adjusted enough to match the focal point with the retina. Sometimes also, the eyeball is not shaped quite right. The lens is too close or too far away from the retina, so accommodation alone

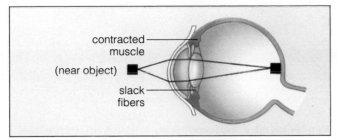

a Accommodation for near objects (lens bulges)

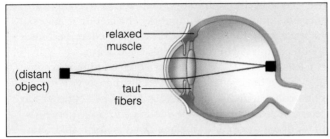

b Accommodation for distant objects (lens flattens)

Figure 11.14 Focusing light on the retina by adjusting the lens (visual accommodation). A muscle encircling the lens attaches to it by fiberlike ligaments. (**a**) Close objects are brought into focus when the muscles contract and makes the lens bulge, so the focal point moves closer. (**b**) Distant objects are brought into focus when the muscle relaxes and makes the lens flatten, so the focal point moves farther back.

cannot bring about a precise match. Eyeglasses can correct both problems, which are called farsightedness and nearsightedness, respectively (see *Focus on Wellness*).

Photoreception   Humans have a keen sense of vision and a well-developed retina. The **retina** is a thin, complex layer of neural tissue at the back of the eyeball. It has a basement layer—a pigmented epithelium that covers the choroid. Resting on the basement layer are densely packed photoreceptors that are functionally linked with a variety of neurons. Axons from some of these neurons converge to form the optic nerve at the back of the eyeball. The optic nerve is the trunk line to the thalamus, which sends information on to the visual cortex. The site where the optic nerve exits the eye is a "blind spot" because no photoreceptors are present there.

The photoreceptors of the retina are called **rod cells** and **cone cells** because of their shapes (Figure 11.15). Rods are sensitive to very low levels of light. They contribute to perception of movements by detecting changes in light intensity across the field of vision, but they do not detect variations in light wavelengths (colors). Typically they are abundant in the periphery of the retina, facilitating what we call "peripheral vision."

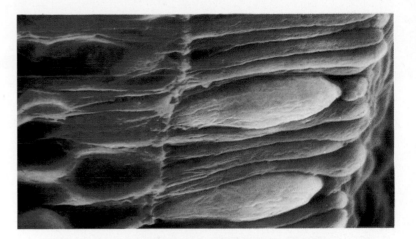

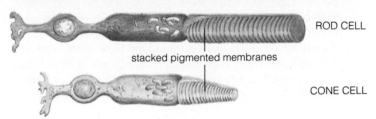

Figure 11.15   Structure of rods and cones, the photoreceptors of the human eye.

Cones respond to bright light. They contribute to sharp daytime vision and color perception. Pigments in different cone cells are selectively sensitive to wavelengths corresponding to red, green, or blue. Cones are densely packed in the *fovea*, a funnel-shaped depression near the center of the retina, where the overlying nerve tissue is thinner. This is the area of "central vision," where the light rays strike when you look directly at something. Cones at the fovea contribute most to visual acuity—that is, to precise discrimination between adjacent points in space. Loss of rods leads to poor night vision, but loss of cones—and the visual acuity they provide—can result in legal blindness.

Rods contain molecules of **rhodopsin** in their membranes. Each molecule consists of a protein (opsin) to which a side group (*cis*-retinal) is attached. The retinal is derived from vitamin A, which is one reason why lack of vitamin A in the diet can impair vision, especially at night. When the side group absorbs light energy, it is temporarily converted to a slightly different form, called *trans*-retinal:

cis-retinal   becomes   *trans*-retinal

The Vulnerable Eye

Your eyes are your single most important source of information about the outside world. Injury, disease, vitamin deficiencies, inherited malfunctions, or advancing age can disrupt eye functions, with various consequences.

Color Blindness   *Red-green color blindness* is a common genetic abnormality that shows up most often in males (Chapter 18). The retina lacks some or all of the cone cells with pigments that normally respond to light of red or green wavelengths. Most of the time, color-blind people merely have trouble distinguishing red from green. The rare few who are totally color blind have only one of three kinds of pigments that selectively respond to red, green, or blue wavelengths. They see the world only in shades of gray.

Focusing Problems   *Astigmatism* results from corneas with an uneven curvature; they cannot bend incoming light rays to the same focal point. *Nearsightedness* (myopia) commonly occurs when the vertical axis of the eyeball is shorter than the horizontal axis. It also occurs when the ciliary muscle responsible for adjustments in the lens contracts too strongly. As a result, images of distant objects are focused in front of the retina instead of on it (Figure *a*).

Farsightedness (hyperopia) is the opposite problem. The horizontal axis of the eyeball is shorter than the vertical axis (or the lens is "lazy"), so close images are focused behind the retina (Figure *b*).

Even a normal lens loses some of its natural flexibility as a person grows older. That is why people who once had perfect vision often must start wearing eyeglasses after about age 40.

Eye Diseases   The eye is vulnerable to infection and disease. For example, Herpes simplex, a virus that causes skin sores, also can infect the cornea and cause ulcers there. A common eye ailment is *conjunctivitis*, or "pinkeye," an inflammation of the transparent membrane (conjunctiva) that lines the eyelids and covers the sclera. Bacteria, viruses, and allergies are usual causes. *Trachoma* is a highly contagious disease that has blinded millions, mostly in North Africa and the Middle East. The culprit is a strain of the bacterium *Chlamydia trachomatis*, which also is responsible for several sexually transmitted diseases. The eyeball and the lining of the eyelids (conjunctiva) become damaged. The damaged tissue provides entry points for bacteria that can cause secondary infections. In time the cornea can become so scarred that blindness follows. In many less developed countries, blindness also results in children whose diets are severely lacking in vitamin A.

Age-Related Problems   *Cataracts*, a gradual clouding of the lens, is a problem associated with aging, although it also may arise through injury, diabetes, or overexposure to ultraviolet radiation (in sunlight). The clouding may skew the trajectory of incoming light rays. If the lens becomes totally opaque, light cannot enter the eye at all.

Glaucoma results when excess aqueous humor accumulates inside the eyeball. Blood vessels that service the retina collapse under the increased fluid pressure. Vision deteriorates as neurons of the retina and optic nerve die off. Although chronic glaucoma often is associated with advanced age, the problem actually starts in middle age. If detected early, the fluid pressure can be relieved by drugs or surgery before the damage becomes severe.

Eye Injuries   *Retinal detachment* may follow a physical blow to the head or an illness that tears the retina. As the semifluid vitreous humor oozes through the region, the retina becomes lifted from the underlying choroid. In time it may peel away entirely, leaving its blood supply behind. Early symptoms of injury include blurred vision, flashes of light that occur in the absence of outside stimulation, and loss of peripheral vision. Without medical intervention, the injured person may become totally blind in the injured eye.

New Technologies   Today a variety of tools are used to correct some eye disorders. In *corneal transplant surgery*, the defective cornea is removed, then an artificial cornea made of clear plastic or a natural cornea from a donor is stitched in place. Within a year, the patient is fitted with eyeglasses or contact lenses. Similarly, cataracts can sometimes be surgically corrected by removing the lens and replacing it with an artificial one. Severely nearsighted people may opt for *radial keratotomy*, a still-controversial surgical procedure in which tiny, spokelike incisions are made around the edge of the cornea to flatten it more. When all goes well, the adjustment eliminates the need for corrective lenses. Sometimes, however, the result is overcorrected or undercorrected vision. Serious scarring can also occur. A new version of this procedure, which uses a laser to do the "cutting," is under development and may help solve these problems.

Retinal detachment may be treatable with *laser coagulation*, a painless technique in which a laser beam seals off leaky blood vessels and "spot welds" the retina to the underlying choroid.

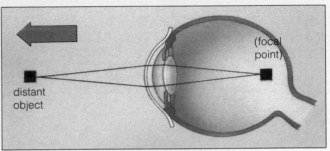

a Focal point in nearsighted vision. (The example shows flamingos in Tanzania, East Africa.)

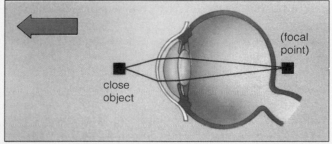

b Focal point in farsighted vision.

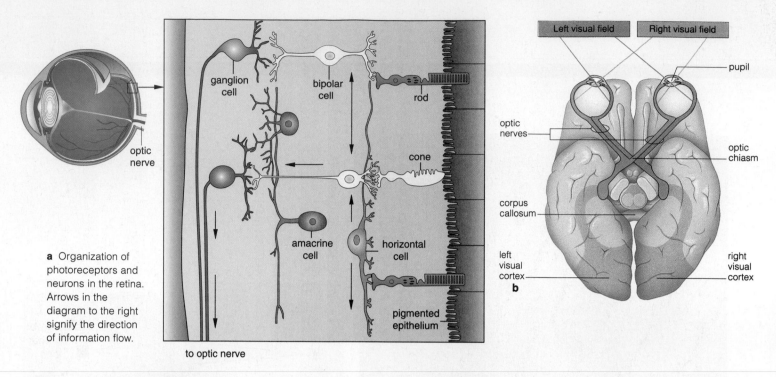

a Organization of photoreceptors and neurons in the retina. Arrows in the diagram to the right signify the direction of information flow.

to optic nerve

Figure 11.16 (**a**) Neurons and photoreceptors of the retina and (**b**) the sensory pathway to the visual cortex.

In this altered form, the side group initiates a series of chemical reactions within the photoreceptor. The reactions lead to a voltage change across the photoreceptor's plasma membrane. This local, graded potential causes the rod or cone to *reduce* its release of neurotransmitter. That chemical change then alters the activity of neighboring neurons.

Processing Visual Information How does the information from photoreceptors become translated into the sense of vision? Part of the answer is that the information moves in increasingly organized ways through *levels* of synapsing neurons—first in the retina, then in different parts of the brain. The information includes signals about form, movement, depth, color, and texture. Each type of signal seems to be processed along a separate communication channel. Many different channels run in parallel to the brain.

Only the rods and cones respond directly to light. When stimulated, they pass signals to *bipolar cells* (Figure 11.16). Groups of bipolar cells synapse with ganglion cells. Information traveling from photoreceptors to bipolar cells is modified by *horizontal cells*, whereas *amacrine cells* modify information traveling the next leg, from bipolar cells to ganglion cells, where action potentials are initiated. The axons of ganglion cells converge to form the *optic nerve* leading to the brain. The optic nerves from both eyes converge at the base of the brain. A portion of the axons of each nerve cross over here before continuing onward. This partial crossover, the *optic chiasm*, permits depth perception because it ensures that information from both eyes will reach both cerebral hemispheres. Most axons of the optic nerves lead into the thalamus, which passes on information to the visual cortex.

The visual cortex has several subdivisions, each with the whole visual field mapped onto it. The **visual field** is the portion of the outside world that is being detected by photoreceptors at any given time. A particular part of each cortex subdivision thus receives input from a particular part of the visual field. In each part of a subdivision, some neurons are sensitive to stimuli in the visual field that are oriented in one direction only—a line in the outside world that is horizontal, vertical, tilted left, or tilted right. (Figure 11.17 shows an example of this.) Nearby neurons may respond to stimuli oriented in a different direction.

Different bits of information that have reached the visual cortex are sent to different parts of the cerebral cortex. All the information is processed simultaneously in different cortical regions. Finally, signals are integrated to produce the organized electrical activity that gives rise to the sensation of sight.

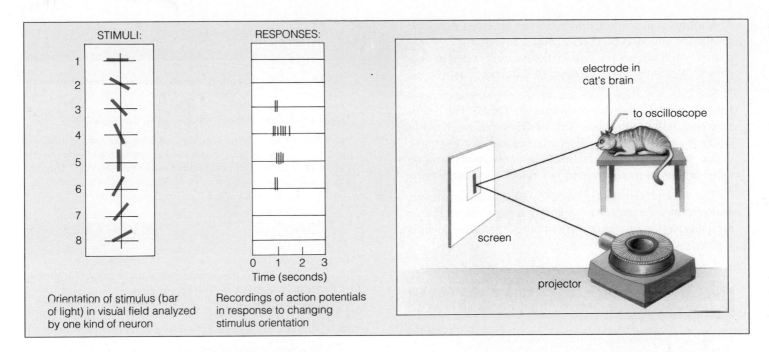

STIMULI: RESPONSES:

1
2
3
4
5
6
7
8

0   1   2   3
Time (seconds)

Orientation of stimulus (bar of light) in visual field analyzed by one kind of neuron

Recordings of action potentials in response to changing stimulus orientation

electrode in cat's brain

to oscilloscope

screen

projector

Figure 11.17 From signaling to visual perception. Neurons in the visual cortex are stacked in columns at right angles to the brain's surface. Connections run between neurons in each column and between different columns. Each column apparently deals with only one kind of stimulus, received from only one location. Visual perception seems to be based on the organization and synaptic connections between neurons in these columns. The neurons fall into a few categories. Those in each category seem to be tripped into action the same way. For instance, excitatory signals traveling up through the cortex activate certain neurons, which then send out inhibitory signals to other neurons. The excitatory and inhibitory signals between neurons form narrow bands of electrical activity.

Experiments show that the pattern of excitation through specific columns of neurons is highly focused. For example, David Hubel and Torsten Wiesel implanted electrodes in individual neurons in the brain of an anesthetized cat. Then they positioned the cat in front of a small screen. They projected images of different shapes (including a bar) onto the screen. When the bar was tilted at different angles, changes in electrical activity that corresponded to the different angles were recorded.

The strongest activity was recorded for one type of neuron when the bar image was nearly vertical (numbered 4 in the sketch). When the bar image was tilted slightly, the signals were less frequent. When the image was tilted past a certain angle, the signals stopped. In other experiments, a certain neuron fired only when an image of a block was moved from left to right across the screen; another fired when the image was moved from right to left.

SUMMARY

1. A stimulus is a specific form of energy or other event that the body is able to detect by means of sensory receptors. A sensation is an awareness (not always conscious) that stimulation has occurred. Perception is understanding that the sensation has been experienced.

2. Sensory receptors are endings of sensory neurons or specialized accessory structures associated with them. They respond to light, forms of mechanical energy, and other specific stimuli.

   a. Chemoreceptors, such as taste receptors, detect chemical substances dissolved in the body fluids that are bathing them.

   b. Mechanoreceptors, such as free nerve endings, detect mechanical energy associated with changes in pressure, changes in position, or acceleration.

   c. Photoreceptors, such as rods and cones of the retina, detect light.

   d. Thermoreceptors detect the presence of or changes in radiant energy from heat sources.

3. At receptors (or adjacent neurons, as in the ears and eyes), the stimulus triggers graded changes in the membrane potential. If threshold is reached, action potentials are generated. The action potentials travel on specific nerve pathways from the receptors to specific parts of the brain.

4. Variations in stimulus intensity are encoded in (a) the frequency of action potentials propagated along an information-carrying neuron and (b) the number of active neurons that convey signals related to that stimulus.

5. Somatic sensations include touch, pressure, temperature, pain, and proprioception (muscle sense). The receptors associated with these sensations are not localized in a single organ or tissue. Stretch receptors, for example, occur in skeletal muscles throughout the body and in smooth muscle in airways, blood vessels, and elsewhere.

6. The special senses are taste, smell, hearing, balance, and vision. The receptors associated with these senses typically reside in sensory organs or some other specific region.

Review Questions

1. Label the component parts of the human eye: *264*

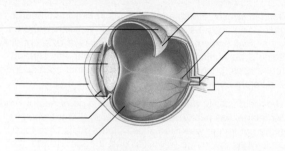

2. What is a stimulus? Receptor cells detect specific kinds of stimuli. When they do, what happens to the stimulus energy? 254

3. Give some examples of chemoreceptors and mechanoreceptors. 254

4. What is sound? How is vibration frequency related to sound? 259, 260

5. What is pain? Name one of the receptors associated with pain. 257

6. How does vision differ from photoreception? What sensory apparatus does vision require? 262

7. How does the eye focus the light rays of an image? What is meant by *nearsighted* and *farsighted*? 268

1. Astronauts living in a space vehicle eat much of their food directly from sealed pouches. They have often complained that such food tastes bland. Given what you know about the senses of taste and smell and the link between them, can you explain why?

2. While walking on the beach one evening you notice movement out of the corner of your eye, but when you turn and look directly at the spot you see only dim outlines of sand dunes. Why? (Assume there are no trees, grass, or other hiding places on the beach.)

1. A _____ is a specific form of energy that is capable of eliciting a response from a sensory receptor.

2. Awareness of a stimulus is called a _____.

3. _____ is understanding what particular sensations mean.

4. Each sensory system is composed of _____.
   a. nerve pathways from specific receptors to the brain
   b. sensory receptors
   c. brain regions that deal with sensory information
   d. all of the above are components of sensory systems

5. _____ detect mechanical energy associated with changes in pressure, in position, or acceleration.
   a. Chemoreceptors   c. Photoreceptors
   b. Mechanoreceptors   d. Thermoreceptors

6. Detecting chemical substances present in the body fluids that bathe them is the function of _____.
   a. thermoreceptors   c. mechanoreceptors
   b. photoreceptors   d. chemoreceptors

7. Which of the special senses is based on the following events: Membrane vibrations cause fluid movements, which lead to bending of mechanoreceptors and initiation of action potentials.
   a. taste   c. hearing
   b. smell   d. vision

8. The outer layer of the eye includes the _____.
   a. lens and choroid   c. retina
   b. sclera and cornea   d. both a and
                                   c are correct

9. The middle layer of the eye includes the _____.
   a. lens and choroid   c. retina
   b. sclera and cornea   d. start of optic nerve

10. Match each of the following terms with the appropriate description.

_____ somatic senses (general senses)
_____ special senses
_____ variations in stimulus intensity
_____ action potential
_____ sensory receptor
_____ stimulus

a. produced by strong stimulation
b. endings of sensory neurons or specialized cells next to them
c. taste, smell, hearing, balance, and vision
d. a specific form of energy that can elicit a response from a sensory receptor
e. frequency and number of action potentials
f. touch, pressure, temperature, pain, and proprioception

Key Terms

accommodation 264
adaptation 256
chemoreceptor 254
cochlea 259
cone cell 265
cornea 262
eye 264
hair cell 262
iris 266
lens 266
mechanoreceptor 257
nociceptor 257
organ of Corti 262
perception 256
photoreceptor 257
retina 267
rhodopsin 267
rod cell 267
semicircular canal 264
sensation 256
sensory receptor 256
sensory system 256
somatic sensation 258
somatic sensory cortex 257
tectorial membrane 262
thermoreceptor 257
tympanic membrane 261
vestibular apparatus 264
vision 264
visual cortex 257
visual field 270

Readings

Discover, June 1993. Entire issue devoted to articles on current understanding of the special senses.

Vander, A., J. Sherman, and D. Luciano. 1990. *Human Physiology,* 5th ed. New York: McGraw-Hill. Chapter 9 is a clear introduction to sensory systems.

Zeki, S. September 1992. "The Visual Image in Mind and Brain." *Scientific American.*

12 INTEGRATION AND CONTROL: ENDOCRINE SYSTEMS

Rhythms and Blues

The pineal gland, a lump of tissue in your brain, secretes the hormone melatonin. When the brain receives sensory signals from your eyes about the waning light at sunset, the pineal gland steps up its melatonin secretion, which is picked up by the bloodstream, which transports it to target cells—in this case, certain brain neurons. Those neurons are involved in sleep behavior, a lowering of body temperature, and possibly other physiological events. At sunrise, when the eye detects the light of a new day, melatonin production slows down. Your body temperature increases, and you wake up and become active.

The cycle of sleep and arousal is evidence of an internal *biological clock* that seems to tick in synchrony with daylength. The clock apparently is influenced by melatonin, and it can be disturbed by a change in circumstances. Jet lag is an example. A traveler from the United States to Paris starts off her vacation with four days of disorientation. Two hours past midnight she is sitting upright in bed, wondering where the coffee and croissants are. Two hours past noon she is ready for bed. Her body will gradually shift to a new routine as melatonin secretion becomes adjusted so that the hormone signals begin arriving at their target neurons on Paris time.

Figure 12.1 Too little light and too much melatonin, a hormone, may trigger seasonal affective disorder (SAD), or "winter blues," in some people.

Some people are affected by severe *winter blues*, or seasonal affective disorder (SAD). Symptoms typically include depression and an overwhelming desire to sleep (Figure 12.1). Their discomfort may result from a biological clock that is out of synchrony with the changes in daylength during winter (days are shorter and nights longer). Their symptoms worsen when they are given doses of melatonin. And they improve dramatically when they are exposed to intense light, which shuts down pineal activity.

Hormones, our main topic in this chapter, are part and parcel of life's tempos. Some are crucial to basic biological events, such as sexual development and reproduction. They and other signaling molecules are secreted by various organs and tissues and, as we will see, have a range of targets throughout the human body. Indeed, they ultimately control many body functions and influence many aspects of behavior.

KEY CONCEPTS

1. Hormones and other signaling molecules help integrate cell activities in ways that benefit the whole body. Some hormones help the body adjust to short-term changes in diet and activity levels and to changes in the external environment. Other hormones help bring about long-term adjustments involved in growth, development, and reproduction.

2. The hypothalamus and pituitary gland coordinate the activities of many endocrine glands. Their interactions control many of the body's functions.

3. Neural signals, hormonal signals, chemical changes in blood, and environmental cues trigger hormone secretions. Only cells with receptors for a specific hormone are its targets. Some hormones, such as steroid hormones, induce gene activation and protein synthesis in target cells. Protein hormones alter enzyme activity in target cells.

THE "ENDOCRINE SYSTEM"

The word *hormone* dates back to the early 1900s, when W. Bayliss and E. Starling were trying to figure out what triggers the secretion of pancreatic juices that act on food traveling through the gut of a dog. At the time, it was known that acids mix with food in the stomach, and that the pancreas secretes an alkaline solution when the acidic mixture has passed into the small intestine. Was the nervous system or something else stimulating the pancreatic response?

To find the answer, Bayliss and Starling blocked nerves—but not blood vessels—leading to a laboratory animal's small intestine. Later, when acidic food entered the intestine, the pancreas still responded to it. More telling, extracts of cells taken from the intestinal lining also induced its response. Glandular cells in the lining had to be the source of a pancreas-stimulating substance.

The substance came to be called secretin. Proof of its existence and mode of action confirmed a centuries-old idea: *Internal secretions released into the bloodstream influence the activities of tissues and organs.* Starling coined the word *hormone* for such internal glandular secretions (after the Greek *hormon*, meaning "to set in motion").

Later, researchers identified other hormones and their sources (Figure 12.2). In the human body the sources include but are not limited to the following:

Pituitary gland

Pineal gland

Thyroid gland

Parathyroid glands (four)

Adrenal glands (two)

Gonads (two)

Thymus gland

Pancreatic islets (many)

Endocrine cells of gut, liver, kidneys, placenta

The body's hormone sources came to be viewed as the "endocrine system." The name implies that there is a separate control system for the body, apart from the nervous system. (The Greek *endon* means "within"; *krinein* means "to separate.") However, biochemical studies and electron microscopy now show that the boundaries between the nervous and endocrine systems are not sharply defined. Glands thought to be free of neural control turned out to be well supplied with nerves. Some neurons actually secrete "hormones."

Given how intricately the endocrine system and nervous system overlap and interact, we begin here with the secretions that serve as chemical messengers, regardless of their origin.

Structures and functions of the endocrine system overlap with those of the nervous system, so that there is no clear division between them.

HORMONES AND OTHER SIGNALING MOLECULES

Cells take up and release chemical substances in response to changes in their surroundings. In an animal as complex as a human being, the responses of millions to billions of cells must be integrated in ways that benefit the whole body.

Integration is carried out by way of signaling molecules. These include hormones and other chemical secretions that alter the behavior of **target cells**. Any cell is a "target" if it has receptors for a specific signaling molecule and can alter its behavior in response to the molecule. A target may or may not be next to the secreting cell.

Different cells and glands in the human body produce at least four types of signaling molecules:

1. **Hormones** are secreted from endocrine glands, endocrine cells, and some neurons; then they are transported by the bloodstream to nonadjacent targets.

2. **Neurotransmitters** are released from neurons. Unlike hormones, neurotransmitters are not carried in the bloodstream. Instead, they act on immediately adjacent target cells, then are rapidly broken down or recycled. These were described in detail in Chapter 10.

3. **Local signaling molecules** are secreted from many cell types. They alter chemical conditions in the immediate vicinity, then are broken down swiftly.

4. **Pheromones**, which are secreted by some exocrine glands, diffuse through water or air and have targets outside the body. They act on cells of other animals of the same species. They are especially important in altering the behavior of the receiving individual.

Hormones are signaling molecules secreted by endocrine glands, endocrine cells, and some neurons. They are transported by the bloodstream, which distributes them to nonadjacent target cells.

Other types of signaling molecules include neurotransmitters, local signaling molecules, and pheromones.

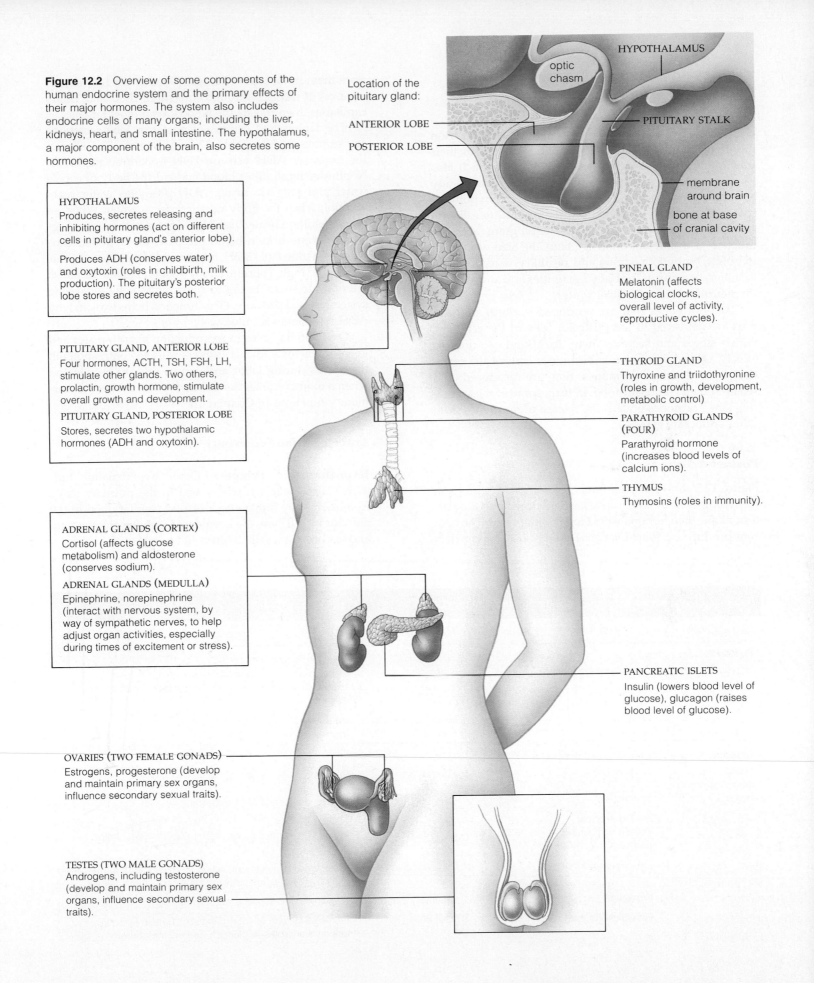

Figure 12.2 Overview of some components of the human endocrine system and the primary effects of their major hormones. The system also includes endocrine cells of many organs, including the liver, kidneys, heart, and small intestine. The hypothalamus, a major component of the brain, also secretes some hormones.

Location of the pituitary gland:

HYPOTHALAMUS

optic chasm

ANTERIOR LOBE

POSTERIOR LOBE

PITUITARY STALK

membrane around brain

bone at base of cranial cavity

HYPOTHALAMUS

Produces, secretes releasing and inhibiting hormones (act on different cells in pituitary gland's anterior lobe).

Produces ADH (conserves water) and oxytoxin (roles in childbirth, milk production). The pituitary's posterior lobe stores and secretes both.

PITUITARY GLAND, ANTERIOR LOBE

Four hormones, ACTH, TSH, FSH, LH, stimulate other glands. Two others, prolactin, growth hormone, stimulate overall growth and development.

PITUITARY GLAND, POSTERIOR LOBE

Stores, secretes two hypothalamic hormones (ADH and oxytoxin).

ADRENAL GLANDS (CORTEX)

Cortisol (affects glucose metabolism) and aldosterone (conserves sodium).

ADRENAL GLANDS (MEDULLA)

Epinephrine, norepinephrine (interact with nervous system, by way of sympathetic nerves, to help adjust organ activities, especially during times of excitement or stress).

OVARIES (TWO FEMALE GONADS)

Estrogens, progesterone (develop and maintain primary sex organs, influence secondary sexual traits).

TESTES (TWO MALE GONADS)

Androgens, including testosterone (develop and maintain primary sex organs, influence secondary sexual traits).

PINEAL GLAND

Melatonin (affects biological clocks, overall level of activity, reproductive cycles).

THYROID GLAND

Thyroxine and triidothyronine (roles in growth, development, metabolic control)

PARATHYROID GLANDS (FOUR)

Parathyroid hormone (increases blood levels of calcium ions).

THYMUS

Thymosins (roles in immunity).

PANCREATIC ISLETS

Insulin (lowers blood level of glucose), glucagon (raises blood level of glucose).

THE HYPOTHALAMUS AND PITUITARY GLAND

Deep in the brain is the **hypothalamus**. This brain region monitors internal organs and activities related to their functioning, such as eating and sexual behavior. It also monitors internal environmental conditions such as body temperature. A short stalk at the base of the hypothalamus connects it to the lobed **pituitary gland** (see Figure 12.2). The human pituitary weighs half a gram (about as much as a thumbtack) and is about the size of a small marble. The hypothalamus and pituitary work together to integrate many of the body's activities and so are called the **neuroendocrine control center**.

Two hormones that are produced in the hypothalamus are delivered to the **posterior lobe** of the pituitary, which stores and secretes them. In a kind of "domino effect," other hypothalamic hormones control the secretion of still other hormones that are produced and released by the **anterior lobe**. In turn, some of these anterior lobe hormones govern the release of hormones from other endocrine glands.

Posterior Lobe Secretions

Figure 12.3 shows the cell bodies of secretory neurons in the hypothalamus. Their axons extend down into the posterior lobe, then terminate next to a capillary bed. The neurons produce *antidiuretic hormone* (ADH) and *oxytocin*, then store them in the axon endings. When either hormone is released, it diffuses through interstitial fluid and enters capillaries, then travels the bloodstream to its targets.

The targets and actions of posterior lobe secretions are summarized in Table 12.1. As Figure 12.4 shows, for example, ADH acts on cells in kidney nephrons. Nephrons, recall, filter blood and rid the body of excess water and salts via urine. ADH promotes water reabsorption when the body must conserve water. When an injury results in heavy blood loss or some other event triggers a severe drop in blood pressure, the hypothalamus releases a flood of ADH into the bloodstream. At this unusually high concentration, ADH appears to cause arterioles in some tissues to constrict, thereby increasing systemic blood pressure. (This function helps explain why ADH is sometimes called *vasopressin*.)

Oxytocin has roles in reproduction in both males and females. For example, it triggers muscle contractions in the uterus during labor and causes milk to be released when a mother nurses her infant. We will revisit this hormone's functions in Chapter 13.

Anterior Lobe Secretions

Hypothalamic Triggers Other hypothalamic hormones also enter a capillary bed in the pituitary stalk; from there they move into a second capillary bed in the anterior lobe. There, the hormones leave the bloodstream and act on target cells (Figure 12.5).

Table 12.1	Hormones Released from the Human Pituitary Gland			
Pituitary Lobe	Secretions	Abbreviation	Main Targets	Primary Actions
Posterior				
Nervous tissue (extension of hypothalamus)	Antidiuretic hormone	ADH	Kidneys	Induces water conservation required in control of extracellular fluid volume (and, indirectly, solute concentrations)
	Oxytocin	—	Mammary glands Uterus	Induces milk movement into secretory ducts Induces uterine contractions
Anterior				
Mostly glandular tissue	Corticotropin	ACTH	Adrenal cortex	Stimulates release of adrenal steroid hormones
	Thyrotropin	TSH	Thyroid gland	Stimulates release of thyroid hormones
	Gonadotropins: Follicle-stimulating hormone	FSH	Ovaries, testes	In females, stimulates egg formation; in males, helps stimulate sperm formation
	Luteinizing hormone	LH	Ovaries, testes	In females, stimulates ovulation, corpus luteum formation in males, promotes testosterone secretion, sperm release
	Prolactin	PRL	Mammary glands	Stimulates and sustains milk production
	Growth hormone	GH	Most cells	Promotes growth in young; induces protein synthesis, cell division; roles in glucose, protein metabolism

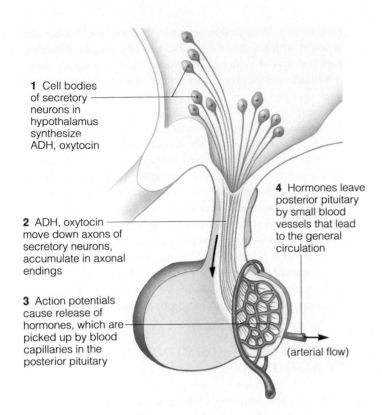

1 Cell bodies of secretory neurons in hypothalamus synthesize ADH, oxytocin

2 ADH, oxytocin move down axons of secretory neurons, accumulate in axonal endings

3 Action potentials cause release of hormones, which are picked up by blood capillaries in the posterior pituitary

4 Hormones leave posterior pituitary by small blood vessels that lead to the general circulation

(arterial flow)

Figure 12.3 Functional links between the hypothalamus and the posterior lobe of the pituitary.

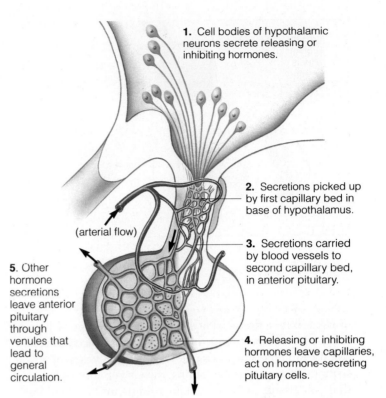

1. Cell bodies of hypothalamic neurons secrete releasing or inhibiting hormones.

2. Secretions picked up by first capillary bed in base of hypothalamus.

(arterial flow)

3. Secretions carried by blood vessels to second capillary bed, in anterior pituitary.

5. Other hormone secretions leave anterior pituitary through venules that lead to general circulation.

4. Releasing or inhibiting hormones leave capillaries, act on hormone-secreting pituitary cells.

Figure 12.5 Functional links between the hypothalamus and the anterior lobe of the pituitary.

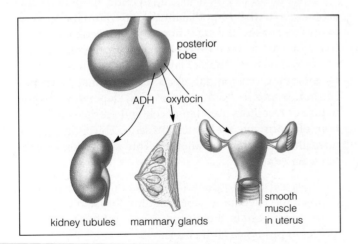

posterior lobe

ADH   oxytocin

kidney tubules   mammary glands   smooth muscle in uterus

Figure 12.4 Main targets of posterior lobe secretions.

Most hypothalamic hormones acting in the anterior lobe are **releasing hormones**; they cause target cells to secrete hormones of their own. Some are **inhibiting hormones**; they slow down hormone secretion from their targets (Table 12.2). In the table and elsewhere in this chapter, you may notice that the names of several stimulatory and releasing hormones include the suffixes -*tropin* or -*tropic*. These come from the Greek root *trope*, which means "to turn"—or as used here, "to turn on."

Table 12.2	Effects of Releasing and Inhibiting Hormones on Anterior Pituitary	
Hormone	Influences Secretion of:	Effect*
Corticotropin-releasing hormone (CRH)	Corticotropin (ACTH)	+
Thyrotropin-releasing hormone (TRH)	Thyrotropin (TSH)	+
Gonadotropin-releasing hormone (GnRH)	Follicle-stimulating hormone (FSH)	+
	Luteinizing hormone (LH)	+
Growth hormone-releasing hormone (GHRH)	Growth hormone (GH); also called somatotropin (STH)	+
Somatostatin	Growth hormone, TSH	−
Dopamine	Prolactin (PRL), LH, FSH	−

*Stimulatory (+) or inhibitory (−).

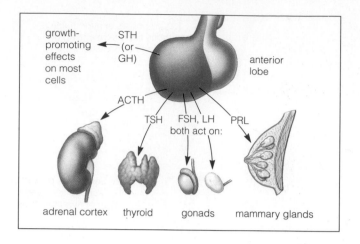

Figure 12.6 Main targets of anterior lobe secretions.

Research tells us that even though hormones have large effects on body functions, they are produced in very small amounts. For example, it took four years of dissecting 500 tons of sheep brains, then seven tons of hypothalamic tissue, to finally obtain a single milligram of the first known releasing hormone (thyrotropin-releasing hormone).

Anterior Pituitary Hormones The hypothalamic signals just described govern the secretion of six hormones from different cells of the anterior pituitary:

Corticotropin (adrenocorticotropic hormone):	ACTH
Thyrotropin (thyroid-stimulating hormone):	TSH
Follicle-stimulating hormone:	FSH
Luteinizing hormone:	LH
Prolactin:	PRL
Growth hormone (somatotropin):	GH (or STH)

ACTH acts on adrenal glands, and TSH acts on the thyroid gland, as described shortly. FSH and LH have important roles in reproduction, and we consider them in Chapter 13.

Prolactin influences a variety of activities (Figure 12.6), although it is best known for its role in stimulating and sustaining milk production in a woman's breasts (mammary glands) after other hormones have "primed" the tissues (page 324).

Growth hormone affects most body tissues. It stimulates protein synthesis and cell division in target cells, and it profoundly influences growth, especially of cartilage and bone. GH is equally important as a "metabolic hormone." Throughout life it stimulates cells to take up amino acids and promotes the breakdown and release of fats stored in adipose tissues, thereby increasing the amount of fatty acids available to cells. GH also moderates the rate at which cells take up glucose, which helps maintain satisfactory blood levels of that crucial cellular fuel.

Figure 12.7 shows what can happen with too little or too much GH. *Pituitary dwarfism* results when not enough GH is produced during childhood. Adults are similar in proportion to a normal person but much smaller. *Gigantism* results when excessive amounts of GH are produced during childhood. Adults are similar in proportion to a normal person but much larger.

During adulthood, when long bones can no longer lengthen, excessive GH secretion causes *acromegaly*. In this rare disorder, cartilage, bone, and other connective tissues of the hands, feet, and jaws thicken, as do epithelia of the skin, nose, eyelids, lips, and tongue (Figure 12.8). Skin thickens abnormally on the forehead and soles of the feet.

EXAMPLES OF HORMONES IN ACTION

Table 12.3 lists hormones from cells and glands other than the hypothalamus and pituitary. In the balance of this section we will focus on a few examples of endocrine activity to show how hormonal controls work.

There are three points to keep in mind about hormone action. First, the hypothalamus and pituitary govern many responses through homeostatic feedback loops (page 8). One or both detect a change in the blood level of a hormone from another endocrine gland and respond by inhibiting or stimulating secretion of that hormone. Second, responses to hormones vary, depending on the nature of receptors on a target cell and on the concentration of hormone molecules present. Third, hormones often interact with one another. Three kinds of interactions are common:

1. *Antagonistic interaction*. The effect of one hormone opposes the effect of another. Insulin, for example, promotes a decrease in the glucose level in the blood, and glucagon promotes an increase.

2. *Synergistic interaction*. The sum total of the actions of two or more hormones is necessary to produce the required effect on target cells. For instance, women cannot produce and secrete milk without the synergistic interaction of the hormones prolactin, oxytocin, and estrogen.

3. *Permissive interaction*. One hormone exerts its effect only when a target cell has become "primed" to respond in an enhanced way to that hormone. The priming is accomplished by previous exposure to another hormone. Pregnancy, for example, depends on

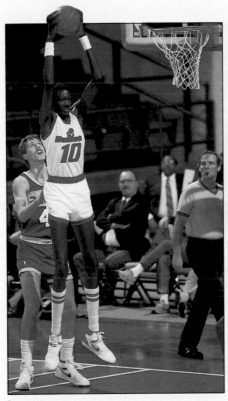

a

b

Figure 12.7 (**a**) Basketball player Manute Bol is 7 feet 6-3/4 inches tall owing to excessive growth hormone production during childhood. (**b**) The effect of GH on overall growth. The man in the center is affected by gigantism, which resulted from overproduction of the hormone during childhood. The man at right displays pituitary dwarfism, which resulted from underproduction of somatotropin during childhood. The man at the left is of average size.

age nine   sixteen

thirty-three   fifty-two

Figure 12.8 Acromegaly, which resulted from excess production of GH during adulthood. Before this female reached maturity, she was symptom-free.

the lining of the uterus being exposed first to estrogens, then to progesterone.

Responses to hormones may be influenced by hormone interactions, variations in hormone concentrations, and the nature of receptors on the target cell.

Synthesis and release of many hormones are governed by homeostatic feedback loops to the hypothalamus and pituitary.

In the next sections we will consider the actions of hormones from several sources, beginning with adrenal gland hormones.

Adrenal Gland Hormones

Hormones of the Adrenal Cortex Humans have a pair of adrenal glands, one perched above each kidney. The outer portion of each gland is the **adrenal cortex**. Here, two major types of hormones, glucocorticoids and mineralocorticoids, are secreted. Homeostatic feedback loops to the neuroendocrine control center govern those secretions.

Glucocorticoids influence metabolism in ways that help raise the blood level of glucose when it falls below a set point. In humans, *cortisol* is the primary glucocorticoid.

Among other effects, it promotes protein breakdown and stimulates the liver to take up amino acids, from which liver cells synthesize glucose in a process called *gluconeogenesis*. Cortisol also dampens the uptake of glucose from the blood and its use for cellular fuel by tissues such as skeletal muscle. This effect is sometimes called "glucose sparing." Lastly, cortisol promotes fat breakdown and use of the resulting fatty acids for energy. Glucose sparing is an extremely important mechanism of homeostasis: It helps ensure that the concentration of glucose in blood will be adequate to meet the needs of the brain, which generally cannot use other molecules for fuel.

Under routine conditions, a negative feedback mechanism operates to control cortisol secretion. When the blood level of the hormone rises above a set point, the hypothalamus begins to secrete less of the releasing hor-

Table 12.3   Hormone Sources Other Than the Hypothalamus and Pituitary

Source	Secretion(s)	Main Targets	Primary Actions
Adrenal cortex	Glucocorticoids (including cortisol)	Most cells	Promote protein breakdown and conversion to glucose in the liver. Cortisol inhibits glucose uptake, mainly in skeletal muscle
	Mineralocorticoids (including aldosterone)	Kidney	Promote sodium reabsorption; control salt, water balance
Adrenal medulla	Epinephrine (adrenalin)	Liver, muscle, adipose tissue	Raises blood level of sugar, fatty acids; increases heart rate, force of contraction
	Norepinephrine	Smooth muscle of blood vessels	Promotes constriction or dilation of blood vessel diameter
Thyroid	Triiodothyronine, thyroxine	Most cells	Regulates metabolism; has roles in growth, development
	Calcitonin	Bone	Lowers calcium levels in blood
Parathyroids	Parathyroid hormone	Bone, kidney, intestine	Elevates calcium levels in blood
Gonads:			
Testis (in males)	Androgens (including testosterone)	General	Required in sperm formation, development of genitals, maintenance of sexual traits; influences growth, development
Ovary (in females)	Estrogens	General	Required in egg maturation and release; prepares uterine lining for pregnancy; other actions same as above
	Progesterone	Uterus, breast	Prepares, maintains uterine lining for pregnancy; stimulates breast development
Pancreatic Islets	Insulin	Muscle, adipose tissue	Lowers blood sugar level
	Glucagon	Liver	Raises blood sugar level
	Somatostatin	Insulin-secreting cells of pancreas	Influences carbohydrate metabolism
Endocrine cells of stomach, gut	Gastrin, secretin	Stomach, pancreas, gallbladder	Stimulates activity of stomach, pancreas, liver, gallbladder
Liver	Growth factors	Most cells	Stimulate overall growth, development
Kidney	Erythropoietin*	Bone marrow	Stimulates red blood cell production
	Angiotensin*	Adrenal cortex, arterioles	Helps control blood pressure, aldosterone secretion
	Vitamin D_3	Bone, gut	Enhances calcium resorption and uptake
Heart	Atrial natriuretic peptide	Kidney, blood vessels	Increases sodium excretion; lowers blood pressure
Thymus	Thymosin, etc.	Lymphocytes	Has roles in immune responses
Pineal	Melatonin	Gonads (indirectly)	Influences daily biorhythms, seasonal sexual activity

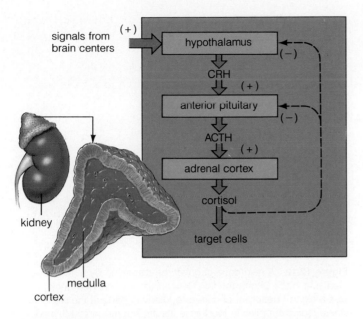

signals from brain centers (+)

Figure 12.9 Negative feedback controls over cortisol secretion.

Figure 12.10 President John F. Kennedy reportedly suffered from Addison's disease but kept the ailment a secret for political reasons.

mone CRH (Figure 12.9), the anterior pituitary responds by secreting less ACTH, and the adrenal cortex slows its secretion of cortisol. Daily cortisol secretion is highest in a healthy person when the blood glucose level is lowest, generally in the early morning. Chronic severe *hypoglycemia*, a persistent low concentration of glucose in the blood, can develop when a person has a cortisol deficiency. Then, the mechanisms that spare glucose and generate new supplies in the liver do not operate properly.

Glucocorticoids also suppress inflammatory responses to tissue injury or infection. When the body is abnormally stressed, the nervous system overrides the feedback control of cortisol secretion. A painful injury, severe illness, or allergic reaction may trigger shock, tissue inflammation, or both. Then, increased secretion of cortisol and other signaling molecules is essential to recovery. That is why cortisol-like drugs, such as cortisone, are administered following an asthma attack and episodes of serious inflammatory disorders. (Prolonged use of glucocorticoids has serious side effects, including suppression of the immune system.) Cortisone is the active ingredient in many over-the-counter creams and lotions for treating skin irritations.

Mineralocorticoids mainly regulate the concentrations of mineral salts, such as potassium and sodium, in extracellular fluid. The most abundantly produced mineralocorticoid is *aldosterone*. As described on page 213, aldosterone acts on the distal tubules of kidney nephrons, stimulating the reabsorption of sodium ions and the excretion of potassium ions. Sodium reabsorption in turn promotes reabsorption of water from the tubules as urine

is forming. Various circumstances, such as falling blood pressure or falling blood levels of sodium, can trigger aldosterone secretion.

A condition called *Addison's disease* arises when the adrenal cortex fails to secrete sufficient mineralocorticoids and glucocorticoids. Sufferers lose weight, and blood levels of glucose and sodium drop while blood potassium rises. The large sodium losses limit the kidney's ability to conserve water. One result can be dangerously low blood pressure and dehydration as the body loses large amounts of water. Figure 12.10 shows a famous Addison's disease patient. The condition can be easily treated with drugs that replace the missing hormones.

In a developing fetus and early in puberty, the adrenal cortex also secretes notable amounts of sex hormones. The main ones are male sex hormones (androgens), especially testosterone, although some female sex hormones (estrogens and progesterone) are also produced. As Chapter 14 describes, however, in adults most sex hormones are generated by the reproductive organs. Adrenal androgens may be responsible for the female sex drive. Some estrogens may continue to be produced even after menopause when a woman's primary estrogen sources, her ovaries, stop functioning.

Adrenal Medulla Hormones The **adrenal medulla** is the inner region of the adrenal gland (Figure 12.9). It contains modified neurons that secrete *epinephrine* and *norepinephrine* (once commonly called adrenalin and noradrenalin). These substances are considered hormones when they are released into the bloodstream, and neurotransmitters when they are released (in tiny amounts) by neurons. Sympathetic nerves carry stimulatory signals to

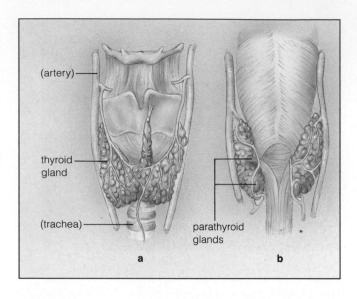

Figure 12.11 (a) Anterior view of the human thyroid gland. (b) Posterior view, showing the four parathyroid glands next to it.

Figure 12.12 A mild case of goiter, as displayed by Maria de Medici in 1625. During the late Renaissance, a rounded neck was a sign of beauty. It occurred regularly in parts of the world where iodine supplies in food were insufficient for normal thyroid function.

the adrenal medulla from the hypothalamus and other brain regions.

Both substances help regulate blood circulation and carbohydrate metabolism when the body is excited or stressed. For example, they increase heart rate, dilate arterioles in some body regions and constrict them in others, and dilate bronchioles. Thus the heart beats faster and harder, more blood volume is shunted to heart and muscle cells from other regions, and more oxygen flows to energy-demanding cells throughout the body. These are features of the "fight or flight" response described on page 237.

Thyroid Gland Hormones

The **thyroid gland** is located at the base of the neck, in front of the trachea (Figure 12.11). The main secretions of this gland, *thyroxine* (T_4) and *triiodothyronine* (T_3), influence overall metabolic rates, growth, and development. The thyroid also makes *calcitonin*, a hormone that lowers the level of calcium (and phosphate) in blood.

Thyroid hormones contain iodine and cannot be produced without it. In the absence of iodine, thyroid hormone levels in the blood decrease. The anterior pituitary responds by secreting thyroid-stimulating hormone (TSH). Excess TSH overstimulates the thyroid gland and causes it to enlarge. The resulting tissue enlargement is a form of *goiter* (Figure 12.12). Goiter caused by iodine deficiency is now uncommon in countries where iodized table salt is used. Although goiter itself is not usually a serious health threat, it can be a symptom of a serious disorder.

Insufficient thyroid output is called *hypothyroidism.* Hypothyroid adults have a slowed heart rate; they are sluggish, dry-skinned, intolerant of cold, and sometimes feel confused and depressed. Affected women often have menstrual disturbances. *Cretinism* is a severe hypothyroid condition in children, and it can arise from a genetic disorder that affects the thyroid gland in the fetus. If affected children do not receive treatment, their growth is stunted and they become mentally retarded. However, these effects can be prevented if the disorder is detected soon after birth. In the United States, newborns commonly are tested for hypothyroidism.

Excessive thyroid output can lead to *hyperthyroidism.* The most common hyperthyroid disorder is *Graves' disease.* Affected people show increases in metabolic rates, heart rate, and blood flow, and they lose weight even when they take in normal or increased amounts of food. They may be excessively nervous and agitated and typically have trouble sleeping. They are intolerant of heat and sweat profusely. Apparently, Graves' disease is an autoimmune disorder (page 185) in which a thyroid-stimulating antibody binds to thyroid cells and causes overproduction of thyroid hormones. Often individuals have a genetic predisposition to the disorder, which may be triggered by some environmental event. Although surgery and drug treatment are options, today treatment typically involves administering radioactive iodine, which destroys thyroid gland function. The individual then takes controlled doses of thyroid hormones to maintain normal functioning.

Figure 12.13 A child with rickets. Bowed legs are typical of the disorder.

Parathyroid Hormone

Some endocrine glands are not stimulated directly by hormones or nerves. Rather, they respond homeostatically to a chemical change in the immediate surroundings. The four **parathyroid glands**, which lie adjacent to the back of the thyroid (Figure 12.11b), are like this.

Previously we noted that calcitonin from the thyroid lowers the level of calcium in blood. This function is one side of a calcium balancing act that contributes to homeostasis by maintaining blood calcium levels within proper limits. When the concentration of calcium ions in blood plasma decreases, the parathyroid glands secrete *parathyroid hormone* (PTH). Their action affects how much calcium is available in blood for enzyme activation, muscle contraction, blood clotting, and many other tasks.

PTH stimulates bone cells to release calcium and phosphate, and the kidneys to reabsorb more calcium and excrete excess phosphate. PTH also helps activate vitamin D, which is a precursor to another hormone that functions along with PTH and calcitonin in calcium regulation. This hormone enhances calcium absorption from food in the GI tract. In children who have a vitamin D deficiency, developing bones do not mineralize properly because too little calcium and phosphate become incorporated into the forming bone tissue. This disorder, called *rickets*, is characterized by weak and deformed bones (Figure 12.13).

Hormones Produced by the Gonads

Homeostatic feedback loops also govern the function of **gonads**, the primary reproductive organs. Male gonads are testes (singular: testis) and female gonads, ovaries. Gonads produce reproductive cells or *gametes* —sperm in males, eggs in females. They also secrete estrogens, progesterone, and androgens (including testosterone). These sex hormones control reproductive function and the development of secondary sexual traits, such as breasts in women and a deeper voice in men. How they do this is an intricate story, and we postpone our discussion of them until Chapter 13.

Pancreatic Islet Hormones

In one respect, the pancreas is an exocrine gland associated with the digestive system: It secretes digestive enzymes and sodium bicarbonate. (Recall from Chapter 3 that exocrine gland products are not secreted into the bloodstream; they are secreted through a duct onto a free epithelial surface.) But the pancreas also has small clusters of endocrine cells scattered through it. There are about 2 million of these clusters; each one is a **pancreatic islet**. The islets are permeated with blood capillaries and contain three types of hormone-secreting cells:

1. *Alpha cells* secrete the hormone **glucagon**. Between meals, cells use the glucose delivered to them by the bloodstream. The blood glucose level decreases, at which time glucagon secretions cause glycogen (a storage polysaccharide) and amino acids to be converted to glucose in the liver. In such ways, *glucagon raises the glucose level in the blood.*

2. *Beta cells* secrete the hormone **insulin**. After meals, when the blood glucose level is high, insulin stimulates uptake of glucose by liver, muscle, and adipose cells especially. It also promotes synthesis of proteins, fats, and glycogen and inhibits protein conversion to glucose. Thus *insulin lowers the glucose level in the blood.*

3. *Delta cells* secrete *somatostatin*. This hormone acts on alpha and beta cells to inhibit secretion of glucagon and insulin, respectively. (Somatostatin is part of several hormonal control systems. For example, it is released from the hypothalamus to block secretions of growth hormone; it is also secreted by cells of the GI tract, where it acts locally to inhibit secretion of various substances involved in digestion.)

Figure 12.14 shows how the interplay among the pancreatic hormones helps keep blood glucose levels fairly constant despite great variation in when—and how much—we eat. When the body cannot produce enough insulin, or when insulin's target cells cannot respond to

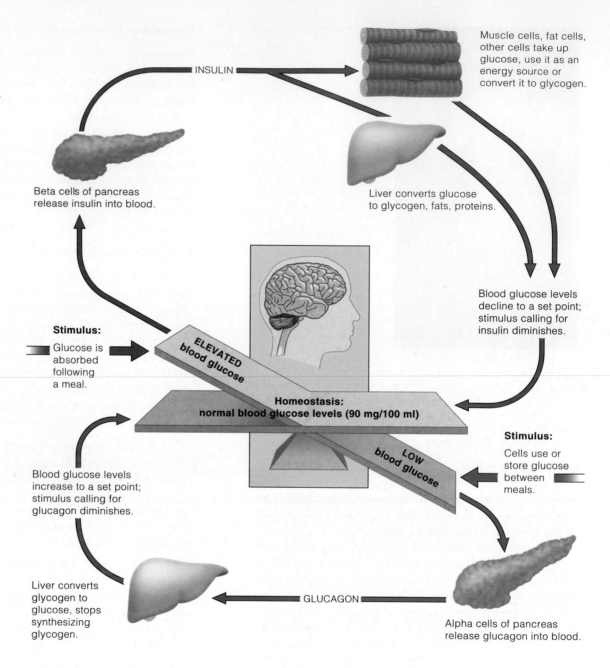

Figure 12.14 Some homeostatic controls over glucose metabolism. *Following* a meal, glucose enters the bloodstream faster than cells can use it. The blood glucose level rises, and pancreatic beta cells are stimulated to secrete insulin. Insulin's targets (mainly liver, fat, and muscle cells) use glucose or store it as glycogen or fat.

Between meals, the blood glucose level drops. Pancreatic alpha cells are stimulated to secrete glucagon. This hormone's target cells convert glycogen back to glucose, which enters the blood. Also, the hypothalamus prods the adrenal medulla to secrete other hormones that slow down conversion of glucose to glycogen in liver and muscle cells.

Muscle cells, fat cells, other cells take up glucose, use it as an energy source or convert it to glycogen.

INSULIN

Beta cells of pancreas release insulin into blood.

Liver converts glucose to glycogen, fats, proteins.

Blood glucose levels decline to a set point; stimulus calling for insulin diminishes.

Stimulus: Glucose is absorbed following a meal.

ELEVATED blood glucose

Homeostasis: normal blood glucose levels (90 mg/100 ml)

LOW blood glucose

Stimulus: Cells use or store glucose between meals.

Blood glucose levels increase to a set point; stimulus calling for glucagon diminishes.

Liver converts glycogen to glucose, stops synthesizing glycogen.

GLUCAGON

Alpha cells of pancreas release glucagon into blood.

it, the body does not store glucose in a normal fashion, and disorders in carbohydrate, protein, and fat metabolism occur.

Insulin deficiency can lead to *diabetes mellitus*. Blood glucose levels rise because cells cannot take up glucose from blood, and glucose accumulates in blood and is lost in the urine. (*Mellitus* means honey in Greek, and early physicians often tasted their patients' urine to confirm the diagnosis.) Excess glucose in urine promotes water loss with the glucose, and the body's water-solute balance is disrupted. Affected people become dehydrated and excessively thirsty. Their insulin-deprived (glucose-

starved) cells start breaking down proteins and fats for energy, and this leads to weight loss. Ketones—normal acidic products of fat breakdown—accumulate in the blood and urine. One result is excess water loss; another is *metabolic acidosis*, a lower than optimal pH in blood. These imbalances disrupt heart and brain functioning, sometimes lethally.

In type 1 or "insulin-dependent" diabetes, the body mounts an autoimmune response against its own insulin-secreting beta cells and destroys them. The disorder is the less common but more immediately dangerous of the two types of diabetes. The cause is not fully understood, although a combination of genetic susceptibility and viral infection may be involved. Because symptoms usually appear during childhood and adolescence, type 1 diabetes also is called "juvenile-onset diabetes." Patients survive with insulin injections.

In "type 2 diabetes," insulin levels are close to or above normal, but for various reasons target cells lack enough functional receptors and so cannot respond properly to the hormone. As affected people grow older, their beta cells deteriorate and they produce less and less insulin. Type 2 diabetes usually occurs in middle age and is sometimes called "maturity-onset diabetes." Obesity increases a person's risk of developing the disorder. Affected people can lead a normal life by controlling diet and weight, and sometimes by taking drugs that enhance insulin action or secretion.

Thymus Hormones

The lobed **thymus** is located behind the breastbone (sternum), between the lungs. As Chapter 7 described, T lymphocytes multiply, differentiate, and mature in this gland under the influence of hormones collectively called *thymosins*. You may recall that different categories of mature T cells, such as helper T cells and cytotoxic T cells, are key elements of immune system functioning (page 174). The thymus is large in children but shrinks to a relatively small size in adulthood.

Pineal Gland and Melatonin

So far, we have seen how endocrine glands and endocrine cells respond to other hormones, to signals from the nervous system, and to chemical changes in their surroundings. In humans and other animals, certain hormones are secreted or inhibited in response to cues from the external environment.

Until about 240 million years ago, vertebrates commonly had a third eye on top of the head. Lampreys (a type of jawless, parasitic fish) still have one, beneath the skin. A modified form of this photosensitive organ, the **pineal gland** described at the beginning of this chapter, persists in nearly all vertebrates. The pineal gland secretes the hormone melatonin into the blood and cerebrospinal fluid. Melatonin functions in the development of gonads and in reproductive cycles, and in sleep/wake cycles.

As the vignette explained, melatonin is secreted in the absence of light. This means that melatonin levels vary from day to night. The levels also change with the seasons, for winter days are shorter than summer days. In humans, decreased melatonin secretion might help trigger the onset of *puberty*, the age at which reproductive organs and structures start to mature. If disease destroys the pineal gland, puberty can begin prematurely.

Other Hormone Sources

Beyond the endocrine glands we have just considered, hormones are also manufactured and secreted by specialized cells in the heart, gastrointestinal tract, and elsewhere. For example, the heart atria secrete *atrial natriuretic peptide* or ANP, a hormone with various effects that include helping to regulate blood pressure. When blood pressure rises, ANP acts to inhibit the reabsorption of sodium ions—and hence water—in the kidneys. As a result, more water is excreted, blood volume decreases, and blood pressure falls.

Hormone secretion is influenced by neural signals, interactions between hormones, chemical changes in the surrounding tissues, and changes in a person's environment.

Secretion of many hormones is controlled by homeostatic feedback from the hypothalamus (neuroendocrine control center).

LOCAL SIGNALING MOLECULES

In humans (and other mammals), many cells can detect changes in the surrounding chemical environment and alter their activity, often in ways that either counteract or amplify the change. The cells secrete local signaling molecules, which act only on the secreting cell itself or in the immediate vicinity of change. Most of the signaling molecules are taken up so rapidly that not many are left to enter the general circulation. Prostaglandins and growth factors are examples of such secretions.

Prostaglandins

More than 16 different kinds of the lipids called **prostaglandins** have been identified in tissues throughout the body. In fact, the plasma membranes of most cells contain prostaglandins. They are released continually, but the rate of synthesis often increases in response to local chemical changes. The stepped-up secretion can influence neighboring cells as well as the prostaglandin-releasing cells themselves.

At least two prostaglandins help adjust blood flow through local tissues. When their secretion is stimulated by epinephrine and norepinephrine, they cause smooth muscle in the walls of blood vessels to constrict or dilate. Prostaglandins have similar effects on smooth muscle of airways in the lungs. Tissue inflammation and allergic responses to airborne dust and pollen may be aggravated by prostaglandins (page 184).

Prostaglandins have major effects on some reproductive events. Many women experience painful cramping and excessive bleeding when they menstruate, and both effects have been traced to prostaglandin action on smooth muscle of the uterus. (Aspirin and other antiprostaglandin drugs block synthesis of this local signaling molecule and alleviate the discomfort.) Prostaglandins also influence the *corpus luteum*, a glandular structure that develops from cells that earlier surrounded a developing ovum (egg) in the ovary. When pregnancy does not follow ovulation (the release of an ovum from the ovary), a corpus luteum self-destructs by producing copious amounts of prostaglandins. In response, capillaries that service the corpus luteum constrict, shutting off the blood supply. Along with oxytocin, prostaglandins also have roles in stimulating uterine contractions during labor. Prostaglandins in semen may stimulate uterine contractions that help move sperm deeper into the female reproductive tract.

Growth Factors

Signaling molecules called **growth factors** influence growth by regulating the rate at which certain cells divide. *Epidermal growth factor* (EGF), discovered by Stanley Cohen, influences the growth of many cell types. *Nerve growth factor* (NGF) is another example. NGF, discovered by Rita Levi-Montalcini, promotes survival and growth of neurons in the developing embryo. One experiment demonstrated that certain immature neurons survive indefinitely in tissue culture when NGF is present but die within a few days if it is not. NGF also may define the direction of growth for these embryonic neurons, laying down a chemical path that leads the elongating processes to target cells. *Focus on Science* on page 290 gives an inkling of how growth factors may one day be important tools for treating victims of heart attack, paralyzing spinal cord injuries, and other conditions.

SIGNALING MECHANISMS

Hormones and other signaling molecules have diverse effects. They induce target cells to take up substances and influence the transport of ions and other substances into and out of cells. They make cells alter the rate of protein synthesis, cause modification of proteins already present in the cell, and induce changes in the cell's shape and internal structure.

Two factors influence responses to hormonal signals. First, different hormones activate different cellular mechanisms. Second, not all cells *can* respond to a given signal, because not all cells have the same types of receptors. Many types of cells have receptors for cortisol, for exam-

ple. That is one reason why cortisol has such widespread effects. More specific hormones "home in" on only the few target cell types that carry the proper receptors.

Let's consider the effects on target cells of the two major types of hormone molecules: steroid hormones and nonsteroid hormones (Table 12.4).

Steroid Hormone Action

Steroid hormones are synthesized from cholesterol. They are lipid-soluble and diffuse directly across the lipid bilayer of a target cell's plasma membrane. Once inside, a steroid hormone molecule usually moves into the nucleus and binds to a receptor for it (Figure 12.15). In some cases, the molecule binds to a receptor in the cytoplasm, then the hormone-receptor complex moves into the nucleus. There, the shape of the complex allows it to interact with specific regions of DNA. Regions of DNA are the genes that contain instructions for making proteins. The complex stimulates (or inhibits) protein synthesis by switching certain genes on or off. It does not alter the activity of already existing proteins.

Testosterone is one of the steroid hormones that influence the development and adult function of male sexual organs. Development proceeds normally when target cells have functional receptors for testosterone. In *testicular feminization syndrome*, the receptors are defective. Genetically, the affected individual is male; he has functional testes that secrete testosterone. But none of the target cells can respond to the hormone, so the secondary sexual traits that develop are like those of females. Anabolic steroids used by athletes to promote muscle development (page 106) are synthetic variants of testosterone.

Nonsteroid Hormone Action

Nonsteroid hormones include amines, peptides, proteins, and glycoproteins. All are water-soluble and so cannot cross the lipid bilayer of the plasma membrane without assistance.

Table 12.4 Two Main Categories of Hormones	
Type of Hormone	Examples
Steroid	Estrogens, testosterone, aldosterone, cortisol
Nonsteroid:	
Amines	Norepinephrine, epinephrine
Peptides	ADH, oxytocin, TRH
Proteins	Insulin, GH, prolactin
Glycoproteins	FSH, LH, TSH

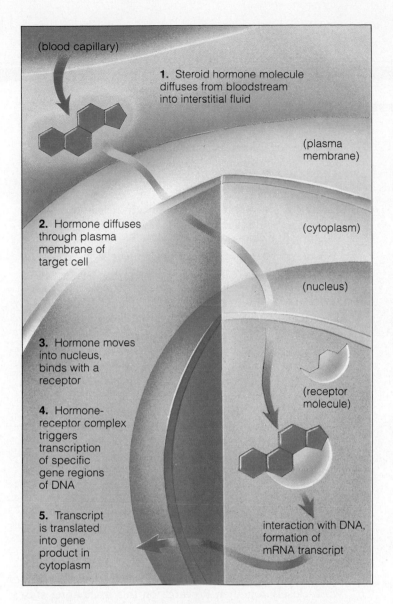

1. Steroid hormone molecule diffuses from bloodstream into interstitial fluid

(blood capillary)

(plasma membrane)

2. Hormone diffuses through plasma membrane of target cell

(cytoplasm)

(nucleus)

3. Hormone moves into nucleus, binds with a receptor

4. Hormone-receptor complex triggers transcription of specific gene regions of DNA

(receptor molecule)

5. Transcript is translated into gene product in cytoplasm

interaction with DNA, formation of mRNA transcript

Figure 12.15 Mechanism of steroid hormone action on a target cell. This same type of mechanism is also thought to occur for thyroid hormones, with one qualification: Transport proteins assist the movement of thyroid hormones across the plasma membrane.

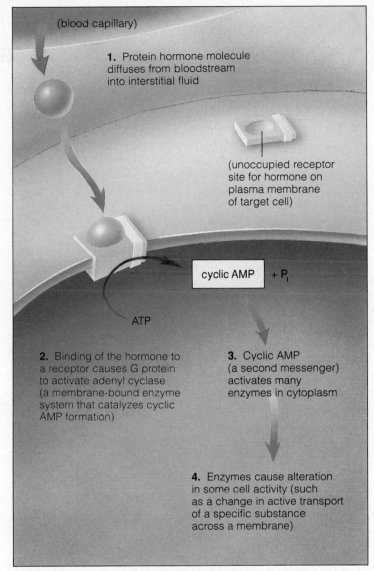

(blood capillary)

1. Protein hormone molecule diffuses from bloodstream into interstitial fluid

(unoccupied receptor site for hormone on plasma membrane of target cell)

cyclic AMP + P_i

ATP

2. Binding of the hormone to a receptor causes G protein to activate adenyl cyclase (a membrane-bound enzyme system that catalyzes cyclic AMP formation)

3. Cyclic AMP (a second messenger) activates many enzymes in cytoplasm

4. Enzymes cause alteration in some cell activity (such as a change in active transport of a specific substance across a membrane)

Figure 12.16 Mechanism of one protein hormone's action on a target cell. A second messenger inside the cell (in this case, cyclic AMP) amplifies the cellular response to the hormone.

Nonsteroid hormones bind to receptors at the target cell's plasma membrane. In some cases, the hormone-receptor complex moves into the cytoplasm by way of endocytosis (Figure 2.6 on page 44), then further action occurs inside the cell. In other cases, a hormone simply binds to receptors at the membrane surface. The hormone-receptor complex activates transport proteins or triggers the opening of channel proteins across the membrane. In either case, specific ions or other substances move inside and their cytoplasmic concentrations change. The changes influence specific cell activities.

Second Messengers Often, nonsteroid hormones stimulate the synthesis of **second messengers**. These are molecules inside the cell that mediate the response to a hormone. An example is **cyclic AMP** (cyclic adenosine monophosphate). First a hormone binds to a membrane receptor on a target cell. Binding of this "first messenger" causes a protein to alter the activity of a membrane-bound enzyme system (Figure 12.16). In response to the protein, the enzyme adenylate cyclase becomes activated. This enzyme speeds the conversion of ATP to cyclic AMP, the second messenger.

Communication Molecules Provide Hope for Healing

In the not too distant future, a doctor treating an automobile accident victim who has sustained severe injuries, including cuts, broken bones, and major damage to the spinal cord, could have a powerful new chemical arsenal at her disposal. Researchers are unraveling how an array of communication molecules, including growth factors, operate in body tissues. They are also seeking ways such knowledge can be transformed into treatments for specific kinds of tissue damage.

The list of known growth factors is growing rapidly. They include the epidermal growth factor (EGF) and nerve growth factor (NGF) described in the text, as well as fibroblast growth factor (FGF), platelet-derived growth factor (PDGF), transforming growth factors (TGFs), and tumor angiogenesis factors (TAFs).

Studies on rodents have shown that EGF and one type of TGF enhance the healing of some types of wounds. In one experiment, an antibiotic cream that contained TGF-*alpha* accelerated the rate at which burned skin regenerated. TGF-*beta* helps regulate the formation of bone and could possibly be marshalled to speed the healing of broken bones. TAFs that stimulate the development of new blood vessels in tumors could conceivably be harnessed to perform that function in cardiac muscle damaged by a heart attack. PDGF, which is released by platelets in blood plasma at the site of a wound (page 157), seems to promote the regeneration of smooth muscle cells in torn vessel walls—and could perhaps do the same for damaged smooth muscle elsewhere in the body.

Some of the most tragic injuries involve spinal cord damage that quickly results in the death of neurons—and in paralysis. Experiments have already demonstrated that NGF can help severed nerves regrow; there is also evidence that FGF serves to help maintain healthy neurons and in the repair of damaged tissues. (Fibroblasts are cells in loose connective tissue that are thought to give rise to various fiberlike molecules with structural roles.) In experimental trials, a substance called GM_1 appears to derail continued deterioration of damaged spinal cord neurons. GM_1 is a ganglioside, a normal component of nerve cell membranes that seems to serve as a communication link between cells. It is already being used experimentally on spinal injury patients, with encouraging results.

The hormone-receptor complex activates many molecules of adenylate cyclase, not just one. Each enzyme molecule increases the rate at which ATP molecules are converted to cyclic AMP. Each cyclic AMP molecule so formed then activates many enzyme molecules that catalyze another specific cellular function. Each of the activated enzyme molecules can convert a very large number of molecules into activated enzymes, and so forth. Soon the number of molecules representing the final cellular response to the initial signal is enormous. Thus, second messengers *amplify* a cell's response to a signaling molecule.

Pheromones

Bears, coyotes, dogs, rhesus monkeys, various insects, and many other animals produce pheromones that serve as sex attractants, territory markers, and other communication signals to members of their own species. Pheromones are released outside of the individual and then pass through the air or water and so reach another individual. Recent studies suggest that humans produce pheromones that can stimulate sexual responses during intimate encounters. Pheromone activity may also account for the common observation that menstrual cycles can become synchronized in women who live or work closely together. Research on this fascinating subject is continuing.

SUMMARY

1. The myriad cellular activities in the human body are integrated by way of signaling molecules: chemical secretions by one cell that adjust the behavior of other, target cells. A target cell has receptors to which specific signaling molecules can bind and elicit a cellular response. It may or may not be adjacent to the signaling cell.

2. Signaling molecules include hormones, neurotransmitters, local signaling molecules, and pheromones.

3. A neuroendocrine control center integrates many activities for the human body. This center consists of the hypothalamus and pituitary gland.

One goal of research on nerve growth factor (NGF) is to develop a treatment that could help repair spinal cord injuries that have paralyzed an estimated 200,000 Americans. Even with such nerve damage, people can live active, enjoyable lives—as this photograph attests.

4. Two hypothalamic hormones—ADH and oxytocin—are stored in and released from the posterior lobe of the pituitary. ADH regulates body water balance. Oxytocin influences contraction of the uterus and milk release from mammary glands.

5. Six other hypothalamic hormones, called releasing or inhibiting hormones, control the secretions by cells of the anterior lobe of the pituitary.

6. The anterior lobe of the pituitary produces and secretes six hormones. Two (prolactin and somatotropin, or growth hormone) have general effects on body tissues. The remainder (ACTH, TSH, FSH, and LH) act on specific endocrine glands.

7. Hormone secretion is influenced by neural signals, hormonal interactions, local chemical changes in the surrounding tissues, and changes in the external environment (outside the body).

8. Antagonistic, synergistic, and permissive interactions occur among hormones. The secretion of many hormones is controlled by homeostatic feedback to the neuroendocrine control center.

9. Some hormones such as parathyroid hormone (PTH) or insulin generally come into play when the extracellular concentration of a substance must be homeostatically controlled. Other hormones such as growth hormone have more prolonged, gradual, and often irreversible effects, such as those on development.

10. Cells respond to specific hormones or other signaling molecules only if they have receptors for them. Steroid hormones have receptors in the nucleus of target cells. Nonsteroid hormones (the amines, peptides, proteins, and glycoproteins) have receptors on the plasma membrane of target cells; responses to them are often mediated by a second messenger (such as cyclic AMP) inside the cell.

11. Steroid hormones trigger gene activation and protein synthesis. Most nonsteroid hormones alter the activity of proteins already present in target cells. These cellular responses contribute in some way to maintaining the internal environment or to the developmental or reproductive program.

13 REPRODUCTIVE SYSTEMS

Girls and Boys

An old joke says that you can tell the sex of a baby by the booties it wears—pink or blue. Actually, the determination of human sex is a bit more interesting than that. When you were conceived, a sperm cell from your father fertilized an egg cell in your mother's body. Among its total of 23 chromosomes, the egg carried an X chromosome, and among *its* 23 chromosomes the sperm carried either an X or a Y chromosome. Normally, from the time of conception the embryo that develops from the fertilized egg carries a combination of these parental sex chromosomes, either XX or XY. But for the first month or so of development, anatomically *you were neither male nor female.*

A gene on the Y chromosome governs the fork in the developmental road that determines a person's gender. In XY embryos, testes—the primary male reproductive organs—begin to develop about six weeks following conception (Figure 13.1). If the embryo is XX,

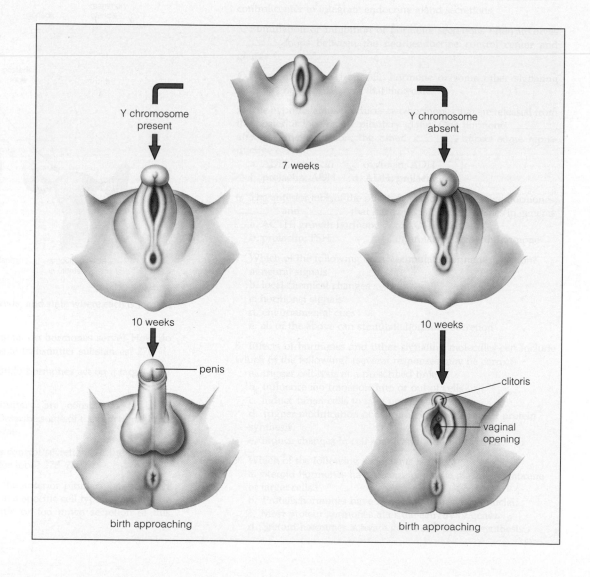

Y chromosome present

Y chromosome absent

7 weeks

10 weeks

10 weeks

penis

clitoris

vaginal opening

birth approaching

birth approaching

however, it lacks the gene and develops ovaries—the primary female reproductive organs. When no Y chromosome is present, female body structures begin to develop automatically. In males, the Y chromosome literally makes all the difference.

Once their development is under way, the testes start producing sex hormones, including testosterone. These hormones influence the development of the different organs that make up the male reproductive system. The newly forming ovaries also produce sex hormones, and these influence the development of the female reproductive system.

Not long ago researchers identified a region of the Y chromosome that appears to include the "master gene" for sex determination. It has been called Sry (for *sex-determining region of the* Y chromosome). So far, the same gene has been identified in the DNA of human males and in male chimpanzees, rabbits, pigs, horses, cattle, and tigers. In all females tested, the gene was absent because females lack the Y chromosome. Other tests with mice indicate that the gene region becomes active about the time that testes start developing. Biologists are currently exploring the hypothesis that Sry activity is part of a "gene cascade" that ultimately produces a baby boy.

In this chapter we focus on the male and female reproductive systems that come into being by way of these fascinating events. In the chapter that follows, the story continues as we trace the stages of human development from the fertilized egg to ripe old age.

KEY CONCEPTS

1. The human reproductive system consists of a pair of primary reproductive organs (testes in males, ovaries in females), accessory glands, and ducts. Testes produce sperm, ovaries produce eggs, and both release sex hormones in response to signals from the hypothalamus and pituitary gland.

2. Males produce sperm continuously from puberty onward. The hormones testosterone, GnRH, LH, and FSH control male reproductive functions.

3. Adult females are fertile on a cyclic basis. Each month during their reproductive years, eggs are released and the lining of the uterus is prepared for pregnancy. The hormones estrogen, progesterone, GnRH, FSH, and LH control this cyclic activity.

Figure 13.1 Inset photograph: The child this woman is expecting has inherited chromosomes from its parents that will determine its sex and the reproductive structures it will have. Diagram: External appearance of developing reproductive organs.

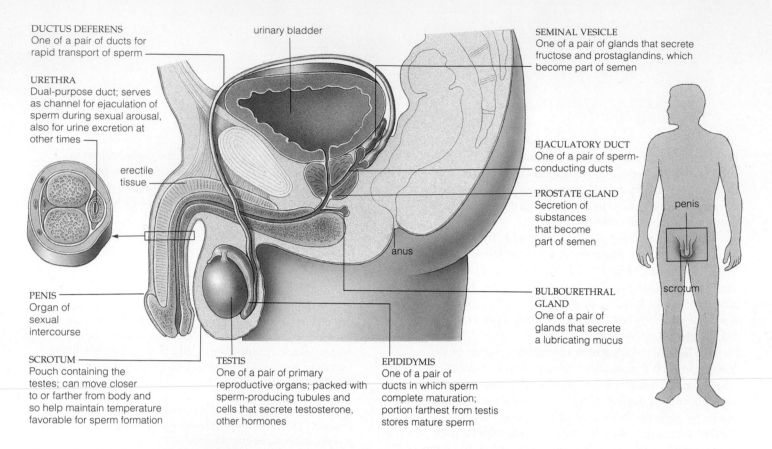

DUCTUS DEFERENS
One of a pair of ducts for rapid transport of sperm

urinary bladder

SEMINAL VESICLE
One of a pair of glands that secrete fructose and prostaglandins, which become part of semen

URETHRA
Dual-purpose duct; serves as channel for ejaculation of sperm during sexual arousal, also for urine excretion at other times

erectile tissue

EJACULATORY DUCT
One of a pair of sperm-conducting ducts

PROSTATE GLAND
Secretion of substances that become part of semen

anus

penis

PENIS
Organ of sexual intercourse

BULBOURETHRAL GLAND
One of a pair of glands that secrete a lubricating mucus

scrotum

SCROTUM
Pouch containing the testes; can move closer to or farther from body and so help maintain temperature favorable for sperm formation

TESTIS
One of a pair of primary reproductive organs; packed with sperm-producing tubules and cells that secrete testosterone, other hormones

EPIDIDYMIS
One of a pair of ducts in which sperm complete maturation; portion farthest from testis stores mature sperm

Figure 13.2 Components of the human male reproductive system and their functions.

REPRODUCTIVE SYSTEM

For both males and females, the reproductive system consists of a pair of primary reproductive organs (gonads), accessory glands, and ducts. Male gonads are **testes** (singular: testis), and female gonads are **ovaries.** Testes produce sperm; ovaries produce eggs. Both also secrete sex hormones. The hormones influence reproductive functions and the development of **secondary sexual traits** (Table 13.1), such as beard growth in males and increased fat deposits in the breasts of females. Such traits are distinctly associated with adult maleness and femaleness but play no direct role in reproduction.

Gonads look the same in all early human embryos. After seven weeks, the Sry gene on the Y chromosome triggers the development of embryonic gonads into testes. In embryos lacking the Sry gene, the rudimentary gonads develop into ovaries. At birth, gonads and accessory organs are already formed. They reach full size and become functional 12 to 16 years later.

Male Reproductive Organs

Figure 13.2 shows the main components of the male reproductive system and lists their functions. Besides testes and their associated ducts and glands, males have a shaftlike *penis* that contains three cylinders of spongy tissue. The tip of one cylinder expands into the mushroom-shaped *glans penis,* which is well endowed with sensory receptors. For religious or hygienic reasons, the folded *foreskin* that covers the glans penis sometimes is surgically removed in a procedure called *circumcision.*

Where Sperm Form The testes start forming as buds from the wall of the embryo's abdominal cavity. About two months before birth they descend into the *scrotum,* an outpouching of skin below the pelvic region. (Undescended testicles, or *cryptorchidism,* can be corrected by surgery, usually during childhood.) Sperm develop properly when the scrotum's interior is kept a few degrees cooler than normal body temperature. When it is cold outside, muscles that position the pouch draw it closer to the (warmer) body. When it is too warm, the muscles relax and lower the pouch.

As many as 300 wedge-shaped lobes divide the interior of each testis. Coiled inside each lobe are two or three

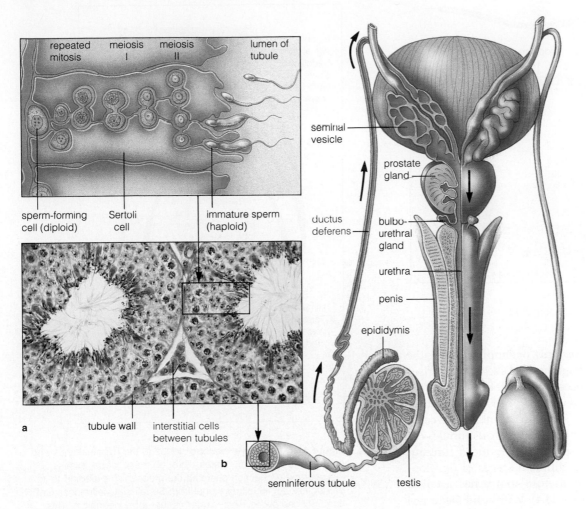

Figure 13.3 Where sperm form. (**a**) Inside a seminiferous tubule, a diploid cell divides. Repeated cell divisions, including some by meiosis, produce haploid cells that differentiate into sperm. In response to pituitary hormones, interstitial cells outside the tubule and Sertoli cells on the inside interact to produce testosterone, the hormone that promotes sperm production. (**b**) The route that mature sperm follow before being ejaculated from the penis of a sexually aroused male. This is a posterior view.

Table 13.1 Secondary Sexual Characteristics	
Female	Male
Growth of pubic hair and underarm (axillary) hair	Growth of pubic hair, underarm hair, and hair on face, chest, or other areas
Breast growth and other changes in body contours due to increased connective tissue and fat deposits in breasts, hips, buttocks	Increased size and mass of skeletal muscle
	Voice deepening due to enlargement of larynx

tubes, the **seminiferous tubules**, where sperm develop (Figure 13.3). Although a testis is only about 5 centimeters long, 125 meters of tubes are packed into it! Connective tissue between the tubes contains **interstitial cells** (Leydig cells) with endocrine functions. They secrete primarily the sex hormone testosterone.

Just inside the walls of these tubes are undifferentiated cells called *spermatogonia*. The cells undergo divisions, including a type called *mitosis* and a type called *meiosis*. Chapter 16 gives the details of mitosis and meiosis; here it is important to keep in mind that meiosis results in the specialized reproductive cells called *gametes* (sperm and eggs). As the introduction noted, human gametes have 23 chromosomes, including one sex chromosome. This is termed a "haploid" number of chromosomes because it is one-half the normal chromosome number (a "diploid" number of 46) of other human body cells. When two haploid gametes unite at fertilization, the diploid chromosome number is restored.

Spermatogonia develop into *primary spermatocytes*, which after a first round of meiotic division (meiosis I) are termed *secondary spermatocytes*. A second round of division (meiosis II) results in *spermatids*. The spermatids

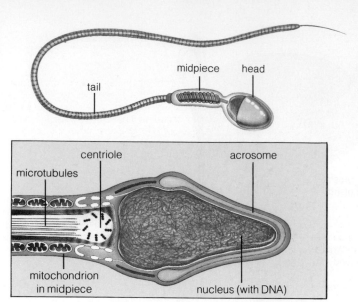

Figure 13.4 Structure of a mature human sperm.

gradually develop into *spermatozoa*, or simply **sperm**—the male gametes. The "tail" (flagellum) of each sperm arises at the very end of the process, which takes nine to ten weeks. All the while, the developing cells receive nourishment and chemical signals from adjacent **Sertoli cells**, which are the cells that line the seminiferous tubule.

From puberty onward, sperm are produced continuously. Many millions are in different stages of development on any given day. A mature sperm has a tail, a midpiece, and a head (Figure 13.4). Within the head is a nucleus containing DNA organized into chromosomes. An enzyme-containing cap, the **acrosome**, covers most of the head. Its enzymes help the sperm penetrate an egg at fertilization. In the midpiece, mitochondria supply energy for the tail's whiplike movements.

How Semen Forms When sperm move out of a testis, they enter a long coiled duct, the *epididymis*. At this time, the sperm are not fully developed. They are not capable of fertilization until they are exposed to secretions from the duct walls, which help them mature. Until sperm leave the body, they are stored in the last stretch of the epididymis. Just before ejaculation, they pass through a thick-walled tube, the **ductus deferens**, and on through a short *ejaculatory duct* into the urethra, the channel leading to the body surface.

Secretions from glands along the ductus deferens become mixed with sperm. The result is **semen**, a thick fluid that is expelled from the penis. *Seminal vesicles* secrete the sugar fructose, which nourishes sperm, and prostaglandins, which may trigger contractions in the female reproductive tract and assist sperm movement. Secretions from the **prostate gland**, which wraps around the urethra and ejaculatory ducts, probably help buffer acid conditions in the vagina. Vaginal pH is about 3.5–4, but sperm motility and fertility improve when it is

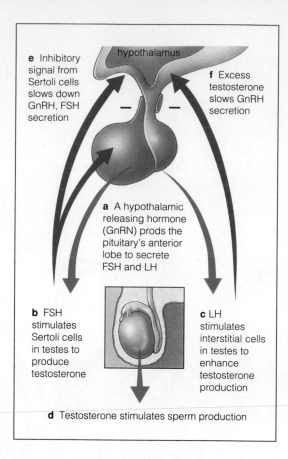

Figure 13.5 Negative feedback loops to the hypothalamus and pituitary gland from the testes. Through these loops, excess testosterone production shuts off the mechanisms leading to its production. An inhibitory signal from Sertoli cells influences GnRH and FSH secretions. The "signal" is the hormone inhibin. This helps maintain the testosterone level in amounts required for sperm formation.

about 6. *Bulbourethral glands* (also called Cowper's glands) secrete mucus-rich fluid into the urethra during sexual arousal. This fluid neutralizes acids in any traces of urine in the urethra, creating a more hospitable environment for the 150–350 million sperm that pass through the channel in a typical ejaculation.

Hormonal Control of Male Reproductive Functions

Three hormones directly control male reproductive functions. One is **testosterone**, produced by interstitial cells in the testes. The others, produced by the anterior lobe of the pituitary, are **LH** (luteinizing hormone) and **FSH** (follicle-stimulating hormone). They were named for their effects in females, but we now know that exactly the same hormones function in males.

Testosterone stimulates sperm production and controls the growth, form, and function of male reproductive structures. It stimulates sexual behavior and aggressive behavior. At puberty, rising testosterone production promotes the typical male secondary sexual traits, such as

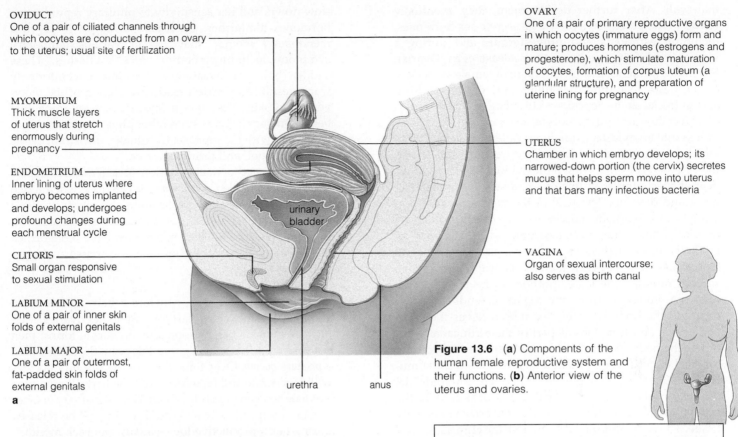

OVIDUCT
One of a pair of ciliated channels through which oocytes are conducted from an ovary to the uterus; usual site of fertilization

MYOMETRIUM
Thick muscle layers of uterus that stretch enormously during pregnancy

ENDOMETRIUM
Inner lining of uterus where embryo becomes implanted and develops; undergoes profound changes during each menstrual cycle

CLITORIS
Small organ responsive to sexual stimulation

LABIUM MINOR
One of a pair of inner skin folds of external genitals

LABIUM MAJOR
One of a pair of outermost, fat-padded skin folds of external genitals

urinary bladder

urethra   anus

a

OVARY
One of a pair of primary reproductive organs in which oocytes (immature eggs) form and mature; produces hormones (estrogens and progesterone), which stimulate maturation of oocytes, formation of corpus luteum (a glandular structure), and preparation of uterine lining for pregnancy

UTERUS
Chamber in which embryo develops; its narrowed-down portion (the cervix) secretes mucus that helps sperm move into uterus and that bars many infectious bacteria

VAGINA
Organ of sexual intercourse; also serves as birth canal

Figure 13.6 (**a**) Components of the human female reproductive system and their functions. (**b**) Anterior view of the uterus and ovaries.

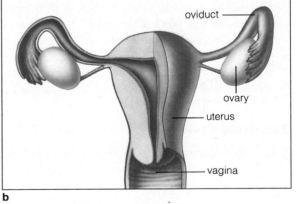

oviduct

ovary

uterus

vagina

b

growth of hair on the face and increased development of skeletal muscles.

The hypothalamus governs sperm production by controlling hormone interactions. When the blood level of testosterone falls, the hypothalamus secretes GnRH (gonadotropin-releasing hormone), which stimulates the anterior pituitary to secrete LH and FSH. These hormones move via the bloodstream to the testes. There, LH prompts interstitial cells to secrete testosterone. (LH is also called ICSH, for "interstitial cell-stimulating hormone.") The testosterone enters the seminiferous tubules, where spermatogonia are present. FSH also enters the tubules and diffuses into Sertoli cells, which are interspersed among the spermatogonia. They release a protein that causes spermatogonia to bind testosterone, which in turn stimulates spermatogonia to complete their development into sperm.

Over the course of a day, an adult male's blood level of testosterone normally does not fluctuate much. Whenever the level increases past a set point, negative feedback loops slow down GnRH secretion by the hypothalamus (Figure 13.5). A high blood level of testosterone also directly causes Sertoli cells to produce *inhibin*, a hormone that apparently inhibits the release of both FSH (from the anterior pituitary) and GnRH from the hypothalamus. Together these events trigger a sharp decline in testosterone secretion. With less testosterone available, sperm production drops.

Testes are the main reproductive organs in males. They produce sperm and the hormone testosterone, which influences sperm formation and male secondary sexual traits.

Hormone feedback loops operate among the testes, the hypothalamus, and the anterior pituitary.

Female Reproductive Organs

Figure 13.6 shows the main components of the female reproductive system and lists their functions. About eight weeks after a female embryo is conceived, the two ovaries (like the testes) begin to develop from buds in the abdominal wall. After further development, they eventually reside in the pelvic cavity. Ovaries secrete sex hormones, including **estrogens** and **progesterone**, and during a woman's reproductive years they release eggs. Ovarian hormones influence the development of female secondary sexual traits, including the "filling out" of breasts, hips, and buttocks as fat deposits accumulate in those areas.

An immature egg, or **oocyte**, released from an ovary moves into an **oviduct** (sometimes called a *Fallopian tube*), which is where fertilization may occur. Fertilized or not, the egg then continues down the oviduct into the hollow, pear-shaped **uterus**. In this organ, a new individual can grow and develop. The wall of the uterus consists of a thick layer of smooth muscle (the myometrium) and an interior lining, the **endometrium**, which consists of epithelial tissue. Layers of connective tissue, glands, and blood vessels lie between the endometrium and uterine smooth muscle. The lower portion of the uterus is the *cervix*. A muscular tube, the *vagina*, extends from the cervix to the body surface. This tube receives the penis and sperm and functions as part of the birth canal.

At the body surface are external genitals, collectively called the *vulva*, which include organs for sexual stimulation. Outermost are a pair of fat-padded skin folds, the *labia majora*. They enclose a smaller pair of skin folds, the *labia minora*, that are highly vascularized but have no fatty tissue. The labia minora partly enclose the **clitoris**, a small organ sensitive to stimulation that is developmentally analogous to the penis (see Figure 13.1). The urethra's opening is about midway between the clitoris and the vaginal opening. Notice that whereas in males the urethra carries both urine and sperm, in females the urethra is separate and is not involved in reproduction.

Menstrual Cycle

Like other female primates, female humans follow a **menstrual cycle**. As Table 13.2 indicates, it takes about 28 days to complete one cycle, although this can vary from month to month and from woman to woman. On the first day of the cycle, blood-rich fluid, about four to six tablespoons total, starts flowing out through the vaginal canal. While the cycle progresses, an oocyte matures and is released from an ovary. As you will see, hormones are also priming the endometrium to receive and nourish an embryo. If fertilization occurs and an embryo becomes implanted in the endometrium, menstrual cycling will stop for the duration of the pregnancy. If the oocyte is not fertilized, the endometrium will break down, and a new cycle will begin with the menstrual flow.

The first menstruation, or *menarche*, usually occurs between the ages of 10 and 16. Menstrual cycles continue until *menopause*, in the late forties or early fifties. By then, a woman's egg supply is dwindling, hormone secretions

slow down, and her sensitivity to pituitary reproductive hormones diminishes. Decreasing estrogen levels in a menopausal woman may trigger various temporary symptoms, including moodiness and "hot flashes." These sudden bouts of sweating and feeling uncomfortably warm result from widespread vasodilation of the blood vessels in skin. The *Human Impact* essay on aging that begins on page 331 describes other physiological changes associated with menopause. Eventually, menstrual cycles stop altogether, and fertility is over.

The absence of menstruation in a female of reproductive age is called *amenorrhea*. It can arise when a disorder of the ovaries, hypothalamus, or pituitary gland disrupts the cyclic release of hormones associated with the cycle. Female athletes sometimes have amenorrhea because regular, extremely strenuous exercise can trigger hormonal changes.

Ovarian Function   Even before a baby girl is born, about 2 million oocytes start forming in her ovaries. Meiotic cell division begins soon after an oocyte forms, then is arrested part way through. At this point the cell is called a *primary oocyte*. Over time, most of the primary oocytes will degenerate and be resorbed by the body. By the time a female is seven years old, about 300,000 primary oocytes remain. Of these, only about 400 to 500 will be released during her reproductive life—usually one each month.

Each primary oocyte becomes surrounded by a layer of **granulosa cells.** This surrounding cell layer nourishes the oocyte; together it and the primary oocyte make up a **follicle** (Figure 13.7).

Early in each menstrual cycle, the hypothalamus secretes GnRH, which stimulates the anterior pituitary to release FSH and LH. These hormones in turn stimulate several follicles to begin growing, but by a mechanism that is not well understood, one follicle usually becomes dom-

Table 13.2 Events of the Menstrual Cycle		
Phase	Events	Days of the Cycle*
Follicular phase	Menstruation; endometrium breaks down	1–5
	Follicle matures in ovary; endometrium rebuilds	6–13
Ovulation	Secondary oocyte released from ovary	14
Luteal phase	Corpus luteum forms; endometrium thickens and develops	15–28

*Assumes a 28-day cycle.

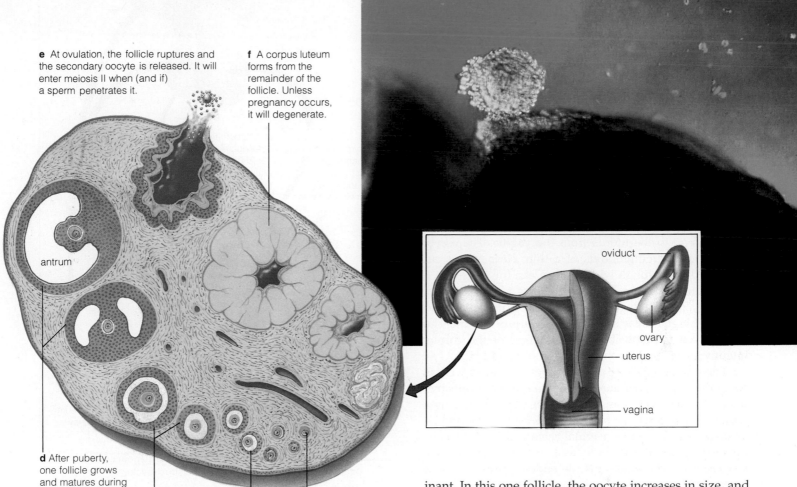

e At ovulation, the follicle ruptures and the secondary oocyte is released. It will enter meiosis II when (and if) a sperm penetrates it.

f A corpus luteum forms from the remainder of the follicle. Unless pregnancy occurs, it will degenerate.

antrum

oviduct

ovary

uterus

vagina

d After puberty, one follicle grows and matures during each menstrual cycle. The oocyte is now a *secondary* oocyte. Estrogen builds up in the space inside the follicle (antrum).

c Follicles form; primary oocyte is arrested in meiosis I

b Even before a female is born, meiosis I occurs in her primary oocytes

a One of many diploid reproductive cells (an oogonium) in a female embryo

Figure 13.7 Section through a human ovary. All the events shown occur in the same location in the ovary; the sketch merely shows the sequence of those events. Following menstruation, the menstrual cycle continues with the growth and maturation of a new follicle, followed by ovulation. At ovulation, a mature follicle ruptures and a secondary oocyte is released. Now a new phase begins, in which a corpus luteum forms from the remains of the follicle. It self-destructs if pregnancy does not occur. The photograph shows a secondary oocyte at the moment of ovulation.

inant. In this one follicle, the oocyte increases in size, and more cell layers form around it. Glycoprotein deposits build up between the oocyte and the cell layers, widening the space between them. In time they form a noncellular coating, the *zona pellucida*, around the oocyte.

In response to FSH and LH, cells outside the zona pellucida secrete several estrogens. Estrogen-containing fluid starts to accumulate inside the follicle (now called a secondary or Graafian follicle), and estrogen levels in the blood start to rise. About 8–10 hours before being released from the ovary, the oocyte completes the meiotic cell division (called meiosis I) that was arrested years before. Now, two cells are present: a large, **secondary oocyte** and a smaller *polar body*. The allocation of chromosomes to each cell assures that the secondary oocyte will be haploid—the chromosome number required for gametes.

About midway through the menstrual cycle the pituitary gland detects the rising estrogen levels and responds with a brief outpouring of LH. The LH triggers cellular contractions that make the fluid-filled follicle balloon outward, then rupture. The fluid escapes, carrying the secondary oocyte with it. This release of a secondary oocyte from an ovary is called **ovulation.**

A secondary oocyte released from an ovary enters an oviduct. Fingerlike projections from the oviduct (called *fimbriae*) extend over part of the ovary. Each "finger" bears beating cilia, and movements of the projections and their cilia sweep the oocyte into the channel. If fertilization takes place, it typically occurs while the oocyte is in the oviduct.

Uterine Function Estrogens released during the early phase of the menstrual cycle also contribute to changes that prepare the uterus for pregnancy. They stimulate the growth of the endometrium and its glands. Then, just before the midcycle surge of LH, cells of the zona pellucida also start secreting progesterone as well as estrogens. The progesterone causes blood vessels to grow rapidly in the thickened endometrium.

At ovulation, estrogens also act on the cervix, the narrow opening to the uterus from the vagina. The cervix now secretes large amounts of a thin, clear mucus, an ideal medium through which sperm can travel.

Another hormone dominates events after ovulation. Granulosa cells left behind in the follicle differentiate into a yellowish glandular structure, the **corpus luteum**. The name means "yellow body." Formation of the corpus luteum is triggered by the midcycle surge of LH.

The corpus luteum secretes progesterone as well as some estrogen. Progesterone prepares the reproductive tract for the arrival of an embryo. For example, it causes mucus in the cervix to become thick and sticky and so provide a barrier against vaginal bacteria that might enter the uterus through the cervix and endanger the embryo. Progesterone also maintains the endometrium during a pregnancy.

If a secondary oocyte becomes fertilized and implants in the uterine wall (page 314), cells associated with the developing early embryo will secrete a hormone called HCG (human chorionic gonadotropin) that maintains the corpus luteum. If an ovulated oocyte is not fertilized, the corpus luteum can persist for about 12 days. During that time, progesterone from the corpus luteum acts on the hypothalamus and the anterior pituitary, which decreases its FSH secretions (Figure 13.8). This prevents other follicles from developing until the menstrual cycle is over. During the last days of the cycle the corpus luteum degenerates. Apparently, it self-destructs by secreting prostaglandins, which interfere with its function.

After this, progesterone and estrogen levels fall rapidly, so the endometrium breaks down. Deprived of oxygen and nutrients, its blood vessels constrict, and its tissues die. Then, the menstrual cycle begins anew. Blood escapes from the ruptured walls of weakened capillaries. The blood and sloughed endometrial tissues make up the menstrual flow, which continues for three to six days. At the end of this period, rising estrogen levels stimulate the repair and growth of the endometrium.

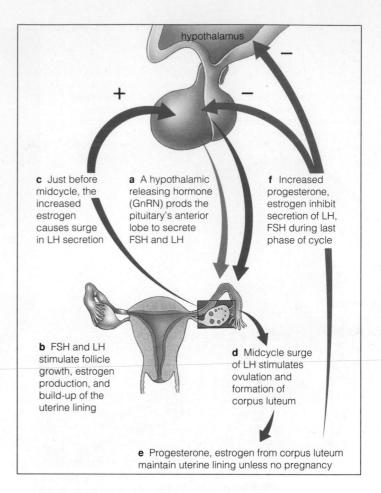

Figure 13.8 Feedback loops among the hypothalamus, anterior pituitary, and ovaries during the menstrual cycle.

Figure 13.9 summarizes the correlations between changing hormone levels and changes in the ovary and uterus during the menstrual cycle.

A pair of ovaries are the primary reproductive organs in females. Hormones, including estrogens, progesterone, LH, and FSH, control the cyclic maturation and release of eggs, and changes in the uterine lining.

Each year, 4–10 million American women are affected by *endometriosis*, the spread and growth of endometrial tissue outside the uterus. Estrogen acting on this tissue may lead to pain during menstruation, sexual relations, or urination. Endometrial scar tissue on the ovaries or oviducts can cause infertility. Endometriosis might arise when menstrual flow backs up through the oviducts and spills into the pelvic cavity. Or perhaps some embryonic cells were positioned in the wrong place before birth and are

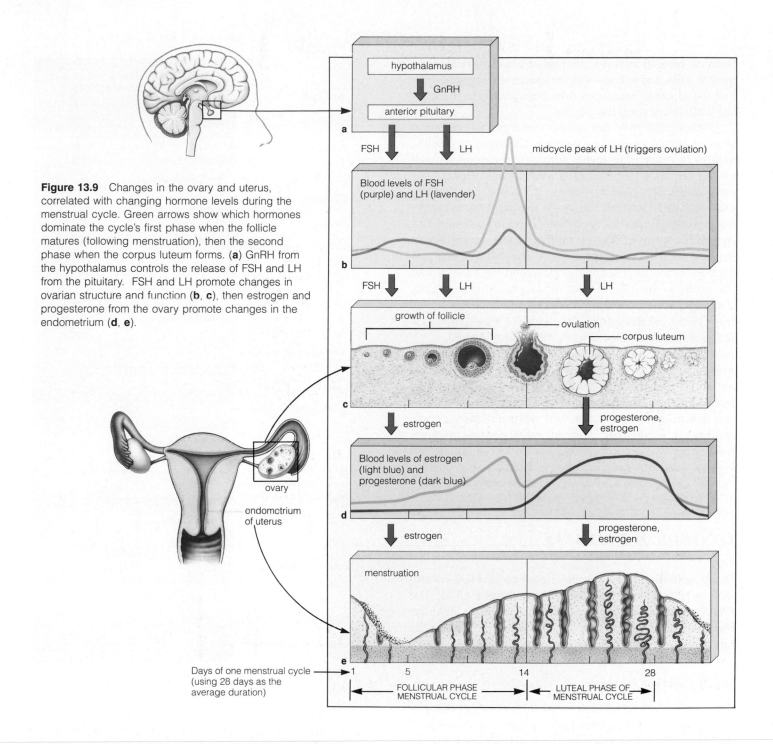

Figure 13.9 Changes in the ovary and uterus, correlated with changing hormone levels during the menstrual cycle. Green arrows show which hormones dominate the cycle's first phase when the follicle matures (following menstruation), then the second phase when the corpus luteum forms. (**a**) GnRH from the hypothalamus controls the release of FSH and LH from the pituitary. FSH and LH promote changes in ovarian structure and function (**b**, **c**), then estrogen and progesterone from the ovary promote changes in the endometrium (**d**, **e**).

hypothalamus

GnRH

anterior pituitary

FSH   LH   midcycle peak of LH (triggers ovulation)

Blood levels of FSH (purple) and LH (lavender)

FSH   LH   LH

growth of follicle   ovulation   corpus luteum

estrogen   progesterone, estrogen

Blood levels of estrogen (light blue) and progesterone (dark blue)

estrogen   progesterone, estrogen

menstruation

ovary

endometrium of uterus

Days of one menstrual cycle (using 28 days as the average duration)

1   5   14   28

FOLLICULAR PHASE MENSTRUAL CYCLE   LUTEAL PHASE OF MENSTRUAL CYCLE

stimulated to grow at puberty, when sex hormones become active.

Some women experience *premenstrual syndrome* (PMS), which can include a distressing combination of mood swings, fluid retention (edema), anxiety, backache and joint pain, food cravings, and other symptoms. PMS usually develops after ovulation and lasts until just before or just after menstruation begins. Although the precise cause of PMS is unknown, it seems clearly related to the cyclic production of ovarian hormones. A woman's doctor can recommend strategies for managing PMS, which often include diet changes, regular exercise, and use of diuretics or other drugs. Many women find that doses of vitamin B_6 and vitamin E help reduce pain and other symptoms.

Sexual Intercourse

Between times of sexual arousal, blood vessels leading into the three cylinders of the penis are constricted and the penis is limp. Upon arousal, blood flows into the cylinders faster than it flows out and collects in the spongy tissue, so the penis lengthens and stiffens. In females, arousal includes dilation of blood vessels in the genital area and causes vulvar tissues to engorge with blood and swell. Secretions flow from glands in the vaginal wall, lubricating the vagina.

Although it may not sound romantic, the technical term for sexual intercourse is **coitus**. During coitus, pelvic thrusts or other movements stimulate the penis as well as the female's vaginal walls and clitoral region. The mechanical stimulation causes rhythmic, involuntary contractions in smooth muscle in the male reproductive tract, especially the ductus deferens and the prostate. The contractions move sperm from their main storage site in the last portion of the epididymis. The contractions also force the contents of seminal vesicles and the prostate into the urethra, then ejaculation of semen into the vagina. (During ejaculation, a sphincter closes and prevents urine from being excreted from the bladder.) Ejaculation is a reflex response; once it begins, it cannot be halted.

Together, the muscular contractions, ejaculation, and associated sensations of release, warmth, and relaxation are called **orgasm**. Female orgasm involves similar events, including involuntary uterine and vaginal contractions and sensations of relaxation and warmth. Many people mistakenly believe that a woman cannot become pregnant if she does not experience orgasm. In fact, a female can become pregnant from intercourse regardless of whether orgasm occurs, and even if she is not sexually aroused.

CONTROL OF FERTILITY

Many sexually active people choose to exercise control over whether their activity will produce a child. The *Choices* feature explores some social and ethical dilemmas that control of fertility can entail. Here we consider the biological bases of different forms of birth control.

Birth Control Options

The most effective method of birth control is complete *abstinence*—no sexual intercourse whatsoever. However, the motivation to engage in sex has been evolving for more than 500 million years, and so it is probably unrealistic to expect many people to practice abstinence for long periods.

A modified form of abstinence is the *rhythm method*, also called the "fertility awareness" or *sympto-thermal method.* The idea is to avoid intercourse during the woman's fertile period, beginning a few days before ovulation and ending a few days after. Her fertile period is identified and tracked by keeping records of the length of her menstrual cycles, and sometimes by examining her cervical secretions and taking her temperature each morn-

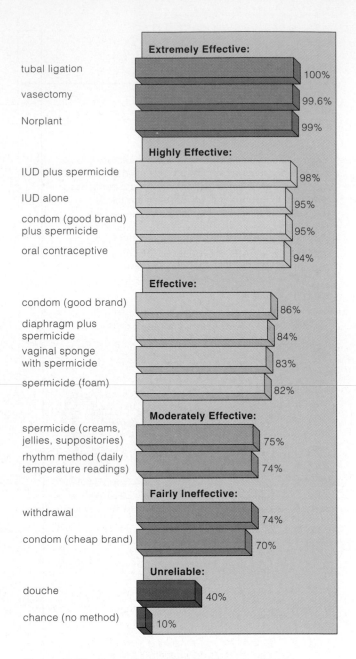

Figure 13.10 Comparison of the effectiveness of some contraceptive methods in the United States. Percentages shown are based on the number of unplanned pregnancies per 100 couples who used the method as the only form of birth control for one year. For example, "94% effectiveness" for oral contraceptives (the Pill) means that on average, 6 of every 100 women using them will become pregnant.

ing when she wakes up. (Core body temperature rises by one-half to one degree just after ovulation.) But ovulation can be irregular, and it can be easy to miscalculate. Also, sperm deposited in the vaginal tract a few days before ovulation may survive until ovulation. The method is inexpensive (a woman need only buy a thermometer) and

Abortion and Human Population Growth: Dilemmas of Fertility Control

About 3 percent of American women of childbearing age have had an abortion; more than 1,500,000 abortions are performed in the United States each year.

In recent years, legal and social conflict has raged in the United States over the issue of legalized abortion. Both sides have protested, marched, and lobbied lawmakers (Figure *a*). Fights have broken out between the factions, clinics have been bombed and burned, and clinic staff have been harassed. Several doctors have been shot.

There are few easy solutions to the ethical issues associated with fertility control. For some people, a crucial question is *When does life begin?* In formulating your own answer, consider that during her lifetime, a human female can produce as many as 400 eggs, all of which are alive. During one ejaculation, a human male can release up to half a billion sperm, which also are living cells. Does "life begin" only when a sperm and an egg fuse? Is there a difference between life in general and *a particular life?* Does "meaningful" life begin only when a fetus attains a certain age in the womb?

Globally, fertility control is equally controversial. Some cultures and religions strongly discourage efforts to limit births. Meanwhile, the human population growth rate spirals out of control (page 508). For example, in the time it takes you to read this chapter, about 10,700 babies were born. By the time you go to bed tonight, there will be 257,000 more people on earth than there were last night at that hour. Within a week, the number will reach 1,800,000—about the population of Massachusetts. World-wide population growth has so outstripped resources that each year millions face the horrors of starvation.

Is massive, global birth control a viable solution? Where do individual rights fit into the picture? Critics of the large-scale use of Norplant and the contraceptive injection Depo-Provera (page 306) fear that the drugs' side effects

have not been well-studied. Moreover, they worry that in some circumstances (poor or uneducated women, or those with a history of child abuse or drug addiction) authorities might coerce women into receiving injections or implants. Would coercion ever be justified? These are not easy questions, but most likely your generation will have to try to find answers for them.

---

does not require fittings and periodic checkups by a physician. But its practitioners do run a very high risk of pregnancy (Figure 13.10).

Withdrawal, or removing the penis from the vagina before ejaculation, dates back at least to biblical times. It is not very effective. Not only does withdrawal require strong willpower, but fluid released from the penis before ejaculation may contain sperm.

Douching, or rinsing out the vagina with a chemical right after intercourse, is next to useless. Sperm can move past the cervix and out of reach of the douche within 90 seconds after ejaculation.

Other methods involve physical or chemical barriers to prevent sperm from entering the uterus and moving to the oviducts. *Spermicidal foam* and *spermicidal jelly* are toxic to sperm. They are packaged in an applicator and placed in the vagina just before intercourse. These products are not reliable unless used with another device, such as a diaphragm or condom.

A *diaphragm* is a flexible, dome-shaped device that is inserted into the vagina and positioned over the cervix before intercourse. A diaphragm is fairly effective when fitted initially by a doctor, used with foam or jelly before each sexual contact, and inserted correctly with each use. A variation on the diaphragm, the *cervical cap*, is smaller and can be left in place for as long as three days with just a single dose of spermicide. It can be more difficult to fit properly, however. The *contraceptive sponge* is a soft, disposable disk that contains a spermicide and covers the cervix. It does not require a prescription or special fitting; it is simply wetted and inserted up to 24 hours before intercourse. Because the sponge can slip and become displaced, the method is about 84 percent reliable.

The *intrauterine device* or IUD is a small plastic or metal device that is placed into the uterus for up to two years at a time. It interferes with implantation of a fertilized egg into the uterine wall. Available by prescription, IUDs have been associated with severe menstrual cramping, increased risk of infection or perforation of the uterus, and other problems. "Improved" types now claim to address such concerns, but any woman considering an IUD should discuss the matter fully with her physician.

Condoms are thin, tight-fitting sheaths of latex or animal skin worn over the penis during intercourse. They are about 85 to 93 percent reliable, and latex condoms help prevent the spread of sexually transmitted diseases. However, condoms can tear and leak, which renders them useless. A pouchlike latex "female condom" that is inserted into the vagina has recently been approved.

The most widely used method of fertility control is *the Pill*, an oral contraceptive of synthetic estrogens and progesterones. It suppresses the normal release of anterior pituitary hormones (LH and FSH) required for eggs to mature and be released at ovulation. The Pill is a prescription drug. Formulations vary and are selected to match each patient's needs. That is why it is not wise for a woman to borrow oral contraceptives from someone else.

When a woman unfailingly remembers to take her daily dosage, the Pill is one of the most reliable methods of controlling fertility. It does not interrupt sexual intercourse, and the method is easy to follow. Often the Pill corrects erratic menstrual cycles and decreases cramping, and studies suggest that using the Pill reduces a woman's risk of ovarian cancer. However, the Pill has some side effects for a small number of users. In the first month or so of use it may cause nausea, weight gain, tissue swelling, and minor headaches. Continued use may lead to blood clotting in the veins of a few women (3 out of 10,000) predisposed to this disorder. Some cases of ele-

vated blood pressure and abnormalities in fat metabolism might be linked to gallbladder disorders in Pill users. Complications are much more likely to develop in women who smoke, and as a rule a physician will not prescribe the Pill for a smoker.

Norplant, an implant that is inserted under the skin of a woman's upper arm, is currently considered to be one of the most effective reversible methods of fertility control. It consists of six matchsticklike capsules that contain the hormone levonorgestrel, which prevents implantation of a fertilized egg. The capsules release the hormone slowly over five years. Over 500,000 women around the world have elected this option. There can be side effects, such as heavier menstrual periods. In addition, a woman who obtains Norplant from her private physician may pay hundreds of dollars for it.

In 1993, a progesterone drug called *Depo-Provera* was approved for contraceptive use in the United States. (The drug was previously approved for treating endometriosis and some other female reproductive disorders.) A single injection costs about $40, provides protection against pregnancy for three months by suppressing ovulation, and is more than 99 percent effective. This drug can have side effects, including weight gain, fatigue, abdominal pain, and gradual bone loss. Some studies suggest a link between Depo-Provera and development of breast cancer in younger women. For these and other reasons, its approval has been controversial.

In *vasectomy*, a tiny incision is made in a man's scrotum, and each ductus deferens is severed and tied off. The simple operation can be performed in 20 minutes in a physician's office, with only a local anesthetic. After vasectomy, sperm cannot leave the testes and so will not be present in semen. So far there is no firm evidence that vasectomy disrupts the male hormone system, and there seems to be no noticeable difference in sexual activity. Technically, vasectomies can be reversed, but residual scar tissue can block the tubes, preventing the passage of sperm.

For females, surgical intervention includes *tubal ligation*, in which the oviducts are cauterized or cut and tied off. Tubal ligation is usually performed in a hospital. A small number of women who have had the operation suffer recurring bouts of pain and inflammation of tissues in the pelvic region where the surgery was performed. The operation can be reversed, although major surgery is required and success is not always assured.

Once conception and implantation have occurred, the only way to terminate a pregnancy is *abortion*, the dislodging and removal of the embryo from the uterus. RU 486, a compound that blocks the action of progesterone (required for a fertilized egg to implant and begin development), can induce abortion during the first seven weeks of pregnancy. In Europe, it is administered under a doctor's supervision, then followed by a dose of prostaglandins, which stimulate uterine contractions. Although those opposed to abortion are currently fighting the use of RU 486 in the United States, other groups support its approval for various reasons. Aside from its potential use to end a pregnancy, there is evidence that RU 486 may be helpful in treating some types of breast and brain cancer. It holds promise for treating endometriosis as well.

At one time, abortions were generally forbidden by law in the United States unless the pregnancy endangered the mother's life. In 1973 the Supreme Court ruled (in *Roe v. Wade*) that the government does not have the right to forbid abortions during the early stages of pregnancy (typically up to five months). Before this ruling, there were dangerous, traumatic, and often fatal attempts to abort embryos, either by pregnant women themselves or by quacks. More recent court action has reaffirmed the legality of abortion while permitting states to impose some restrictions.

Abortions during the first trimester (12 weeks) are relatively rapid, painless, and free of complications. Abortions in the second and third trimesters will probably remain highly controversial unless the mother's life is threatened. For both medical and humanitarian reasons, however, it is generally agreed in this country that the preferred route to birth control is not through abortion but through control of conception in the first place.

Future Fertility Control Options Researchers are working to develop new and better methods of fertility control. Examples include biodegradable implants that don't require surgical removal, a two-year "pregnancy vaccine," a contraceptive for men that reduces sperm count, chemicals for nonsurgical sterilization, and male and female sterilization procedures that can be reversed more easily.

In Vitro Fertilization

Controls over fertility also extend in the other direction—to help childless couples who wish to conceive a child but have problems doing so. In the United States, about 15 percent of all couples cannot conceive a child because of sterility or infertility. For example, hormonal imbalances may prevent ovulation, or the oviducts may be blocked by effects of disease. Alternatively, the male's sperm count may be low, or his sperm may be defective in some way that reduces the likelihood of successful fertilization.

With *in vitro fertilization*—literally "fertilization in glass"—conception can occur externally, provided that sperm and oocytes obtained from the couple are normal.

First, FSH is administered to prepare the ovaries for ovulation. Then, in a technique called laparoscopy, a physician locates and removes preovulatory oocytes with a suction device. Before the oocytes are removed, a sample of the male's sperm is placed in a glass laboratory dish in a solution that simulates the fluid in oviducts. When suctioned oocytes are placed with the sperm, fertilization may occur. About 12 hours later, zygotes (fertilized eggs in their earliest stages of development) are transferred to a solution that will support further development. Two to four days after that, one embryo is transferred to the woman's uterus. The embryo implants in about 20 percent of cases, and each attempt costs several thousand dollars.

In vitro fertilization can produce several viable embryos at one time, and ones not used in a given procedure can be frozen and stored for long periods. The fate of unused embryos has prompted ethical debates and even bitter court battles between divorcing couples.

Recently, doctors in Belgium pioneered a variation on in vitro fertilization in which a single sperm is injected into an egg using a tiny glass needle. The method appears to have a higher success rate than traditional in vitro fertilization, and so far it has produced healthy babies for several hundred couples.

Intrafallopian Transfers

In a technique called GIFT, for *gamete intrafallopian transfer*, a couple's sperm and oocytes are collected and then placed into an oviduct (fallopian tube). About 20 percent of the time, the oocyte becomes fertilized and a normal pregnancy ensues. An alternative procedure is ZIFT (*zygote intrafallopian transfer*). Here, oocytes and sperm are brought together in a laboratory dish, where fertilization can give rise to a zygote. The zygote is then placed in one of the woman's oviducts. GIFT and ZIFT are about as successful as in vitro fertilization, at about half the cost. A variation on these methods is *embryo transfer*, in which a fertile female "donor" is inseminated with sperm from a male whose female partner is infertile. If the donor becomes pregnant, the developing embryo is transferred to the infertile woman's uterus. Embryo transfer is technically difficult and can have legal complications, and it is not yet a common solution to infertility.

Artificial Insemination

Artificial insemination is the placing of semen into the vagina or uterus by artificial means, usually a syringe, around the time of ovulation. The semen may come from a woman's partner, especially if the man has a low sperm count, because his sperm can be concentrated prior to the procedure. In *artificial insemination by donor* (AID), an anonymous donor can sell his sperm to a "sperm bank," which then charges a fee for insemination. AID produces about 20,000 babies in the United States every year.

SUMMARY

1. Humans have a pair of primary reproductive organs (sperm-producing testes in males, egg-producing ovaries in females), accessory ducts, and glands. Testes and ovaries also produce hormones that influence reproductive functions and secondary sexual traits.

2. The hormones testosterone, LH, and FSH directly control sperm formation. They are part of feedback loops among the hypothalamus, anterior pituitary, and testes.

3. The hormones estrogen, progesterone, FSH, and LH control egg maturation and release, as well as changes in the lining of the uterus (endometrium). They are part of feedback loops involving the hypothalamus, anterior pituitary, and ovaries.

4. In both males and females, GnRH from the hypothalamus stimulates the anterior pituitary to release LH and FSH.

5. The following events occur during a menstrual cycle:
   a. The cycle begins with the onset of menstrual flow, as the endometrial lining of the uterus is shed.
   b. A follicle, which is an oocyte surrounded by a cell layer, matures in an ovary. Under the influence of hormones produced by the follicle, the endometrium of the uterus starts to rebuild.
   c. A midcycle peak of LH triggers the release of a secondary oocyte from the ovary. This event is called ovulation.
   d. A corpus luteum forms from the remainder of the follicle. Its secretions prime the endometrium for fertilization. When fertilization occurs, the corpus luteum is maintained by the hormone HCG secreted by cells associated with the early embryo. Secretions from the corpus luteum help maintain the endometrium. When fertilization does not occur, the corpus luteum degenerates and the endometrial lining breaks down.

6. Control of human fertility raises important ethical questions. These questions extend to the physical, chemical, surgical, or behavioral interventions used in the control of unwanted pregnancies.

Review Questions

1. List the main organs of the human male reproductive tract and identify their functions. *296*

2. Which hormones influence male reproductive function? *296–297*

3. Label the component parts of the female reproductive tract:

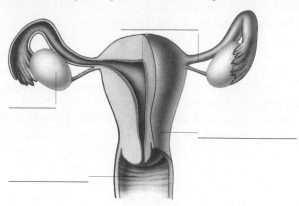

4. What is the menstrual cycle? Which hormones influence this cycle? *300–302*

5. List at least three events that are triggered by the surge of LH at the midpoint of the menstrual cycle. *302*

6. What changes occur in the endometrium during the menstrual cycle? *302*

Critical Thinking: You Decide *(Key in Appendix IV)*

1. Counselors sometimes advise women who wish to conceive a child to use an alkaline (basic) douche immediately before intercourse. What is their reasoning?

2. In the "fertility awareness" method of birth control, a woman gauges her fertile period each month by monitoring changes in the consistency of her vaginal mucus. What kind of specific information does such a method provide? How does it relate to the likelihood of getting pregnant?

Self-Quiz *(Answers in Appendix III)*

1. Besides producing gametes, human male and female gonads also produce sex hormones. The _____ and the pituitary gland control secretion of both.

2. _____ production is continuous from puberty onward in males; _____ production is cyclic and intermittent in females.
   a. egg; sperm   c. testosterone; sperm
   b. sperm; egg   d. estrogen; egg

3. Sperm formation is controlled through _____ secretions.
   a. testosterone   c. FSH
   b. LH   d. all of the above are correct

4. During the menstrual cycle, a midcycle surge of _____ triggers ovulation.
   a. estrogen   c. LH
   b. progesterone   d. FHS

5. Which is the correct order for one turn of the menstrual cycle?
   a. corpus luteum forms, ovulation, follicle forms
   b. follicle forms, ovulation, corpus luteum forms

Readings

"The Evolution of Sexes." July 17, 1992. *Science*. Fascinating discussion of why (perhaps) there are two sexes of human beings and many other organisms.

Franklin, D. July/August 1992. "The Birth Control Bind." *Health*.

Hales, D. 1992. *An Invitation to Health,* 5th ed. Redwood City, Calif.: Benjamin/Cummings. Chapters 10 and 11 contain a wealth of useful information on human sexuality and reproductive choices.

Smolowe, J. June 14, 1993. "New, Improved, and Ready for Battle." *Time*. Clear presentation of issues surrounding the controversy over the "abortion pill," RU 486.

14 DEVELOPMENT

The Journey Begins

At first, nothing spectacular seems to happen to the fertilized egg. Then, as the egg travels along the oviduct toward the uterus (Figure 14.1), a truly remarkable journey begins. The fertilized egg—a single cell—cleaves in one direction, then another and another, until it is a hollow ball of about 64 tiny cells. In one part of the ball's inner surface, the cells are organized into a clump. A space opens up inside the clump and fluid moves in; the space widens. The fluid is lifting the cell mass away from the inner surface, like tissue fluid does to the skin above a blister.

By this time the cells of the "blister" are spread out as a tiny oval disk with two layers. At less than a millimeter across, the disk could stretch out across the head of a straightpin with room to spare. But that speck is an embryo.

A third cell layer forms between the original two. Within those three layers, cells follow commands to

Figure 14.1 The billowing entrance to an oviduct, the tubelike road to the uterus. On that road, a sperm traveling from the opposite direction encountered an egg, and a remarkable developmental journey began.

change shape, to get moving. Part of one layer curves up and folds over. Other cells are elongating, making their component layer thicker. And still other cells are migrating to new destinations. These changes transform the pancakelike embryo into a pale, translucent crescent with surface bumps, tucks, and hollows.

By the third week after fertilization, the tubelike forerunners of the nervous system and circulatory system appear. By the fourth week, a heart starts beating rhythmically in the thickening wall of one tube. On the embryo's sides, budding regions of tissue mark the beginning of upper and lower limbs.

Five weeks into the journey, the limb buds have paddles—the start of hands and feet. Round dots appear under the transparent skin; the eyes are forming. Around and below them, cells migrate in concert, interacting in ways that sculpt out a nose, cheeks, and a mouth. By now, the embryo is recognizably a human in the making.

This was your beginning. Later, embryonic and fetal events filled in the details, rounded out contours, added flesh and fat and hair and nails to your peanut-sized body. That beginning, and all the subsequent events that unfolded inside your mother's body, is the story of this chapter.

KEY CONCEPTS

1. Human development begins when gametes form in the parents.

2. Development then proceeds through stages of fertilization, cleavage, gastrulation, organ formation, and growth and tissue specialization.

3. A key event in early embryonic development is the formation of three distinct tissue layers: endoderm, mesoderm, and ectoderm. Every tissue of the body arises from one of these layers.

4. Development does not end with the birth of a baby. It continues through childhood and adolescence and culminates as an adult undergoes aging. In a real sense, development ends only with death.

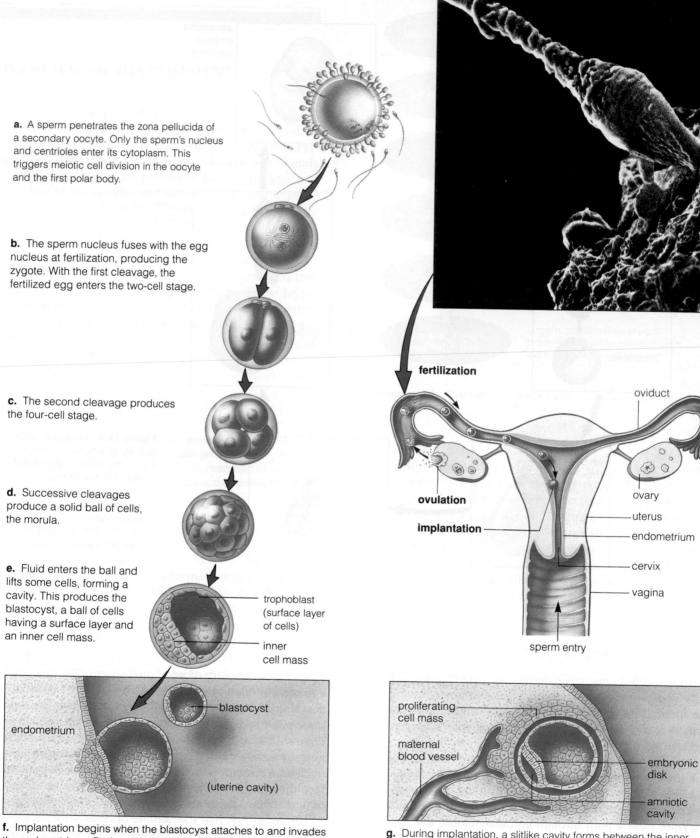

a. A sperm penetrates the zona pellucida of a secondary oocyte. Only the sperm's nucleus and centrioles enter its cytoplasm. This triggers meiotic cell division in the oocyte and the first polar body.

b. The sperm nucleus fuses with the egg nucleus at fertilization, producing the zygote. With the first cleavage, the fertilized egg enters the two-cell stage.

c. The second cleavage produces the four-cell stage.

d. Successive cleavages produce a solid ball of cells, the morula.

e. Fluid enters the ball and lifts some cells, forming a cavity. This produces the blastocyst, a ball of cells having a surface layer and an inner cell mass.

trophoblast (surface layer of cells)

inner cell mass

blastocyst

endometrium

(uterine cavity)

f. Implantation begins when the blastocyst attaches to and invades the endometrium. During the second week after fertilization, it slowly embeds itself in the endometrium.

fertilization

oviduct

ovulation

ovary

implantation

uterus

endometrium

cervix

vagina

sperm entry

proliferating cell mass

maternal blood vessel

embryonic disk

amniotic cavity

g. During implantation, a slitlike cavity forms between the inner cell mass and the surface layer of the embryo. The inner cell mass is transformed into a flattened, embryonic disk from which the embryo develops.

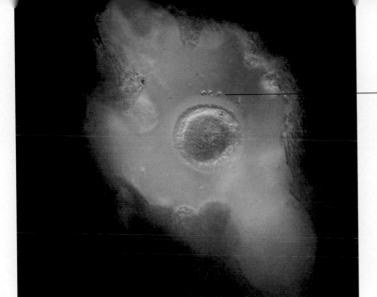

polar bodies

Figure 14.4 Fertilization and implantation in the uterus. The left photograph (page 314) shows a human sperm about to penetrate the membranous covering (zona pellucida) around a secondary oocyte. The right photograph shows the three polar bodies above a mature ovum; these will soon degenerate.

become a "whole"—a zygote with a full complement of 46 chromosomes. As the ensuing embryo develops, each cell will have all the DNA required to guide proper development and functioning of body parts.

Cleavage

Fertilization is followed by cell divisions that convert the zygote, a single cell, into a multicelled ball (Figure 14.4). (These cell divisions occur by mitosis; each new cell receives a full complement of 46 chromosomes.) Usually the embryo does not increase in size while the initial cleavages are proceeding. All the daughter cells collectively occupy the same volume as did the zygote, although they differ in size, shape, and activity.

Sometimes the two cells produced by the first cleavage division separate from each other and continue development into independent embryos. This is the origin of *identical twins,* or two complete, normal individuals having the same genetic makeup. Nonidentical or *fraternal twins* occur when two different eggs are fertilized at roughly the same time by two different sperm. Fraternal twins can have different sexes and do not necessarily resemble each other any more than do any other siblings.

Implantation

For the first three or four days after fertilization, the zygote travels down the oviduct. Sustained by nutrients stored in the ovum or present in maternal secretions, it undergoes the first cleavages. By the time the cluster of dividing cells reaches the uterus, it is a solid ball of cells, the *morula.* The ball becomes transformed into a **blastocyst,** a stage consisting of a surface layer of cells called

the **trophoblast** and a clump of cells to one side called the **inner cell mass** (Figure 14.4e).

Implantation occurs within a week after fertilization. It begins as the trophoblast adheres to the lining of the uterus, the endometrium. Trophoblast cells secrete enzymes that break down a tiny region of the lining, and projections of a subset of trophoblast cells invade and continue to digest the maternal tissues. This enzyme activity gives the blastocyst a foothold in the uterus, so to speak. It can now sink deep into the eroding endometrial lining, which closes over it.

Occasionally a fertilized egg implants not in the uterine wall, but in the oviduct (fallopian tube), or sometimes even in the external surface of the ovary or in the abdominal wall. This condition is called *ectopic pregnancy* (or "tubal" pregnancy). The developing embryo is contained within a fluid-filled sac (the amniotic sac), and this sac—or the oviduct itself—may eventually burst, causing pain, bleeding, dizziness, even death. For these reasons, ectopic pregnancy cannot go to full term and must be terminated by surgery. Although a woman who has had an ectopic pregnancy may later have a normal one, the condition sometimes results in permanent infertility. It occurs most commonly in women who use an IUD for birth control, whose oviducts have been scarred by an operation or infection (such as pelvic inflammatory disease; page 350), or who take progesterone-only birth control pills.

Implantation is completed 14 days after ovulation. The menstrual flow would begin now, if pregnancy had not occurred. The flow does not begin because the implanted blastocyst secretes HCG (human chorionic gonadotropin). This hormone, recall, prods the corpus luteum to keep on secreting estrogen and progesterone, which maintain the uterine lining. By the third week of

Figure 14.5 The two-layered embryonic disk. At this stage the developing amnion is a single layer of cells, and the yolk sac is roughly the same size as the embryonic disk. As development continues, the fluid-filled amniotic cavity will enlarge and ultimately will completely surround and protect the embryo (see Figure 14.6).

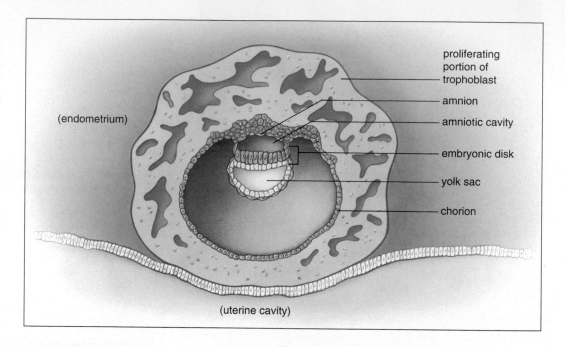

(endometrium)

proliferating portion of trophoblast

amnion

amniotic cavity

embryonic disk

yolk sac

chorion

(uterine cavity)

pregnancy HCG can be detected in the mother's blood or urine. At-home pregnancy tests use chemicals that change color when HCG is present in urine.

While implantation is proceeding, the inner cell mass of the blastocyst becomes transformed into an **embryonic disk** (Figure 14.5). This is the oval, flattened pancakelike structure described in the chapter introduction. Some cells of the disk will give rise to the embryo during the week following implantation, when all three germ layers will form. Others will give rise to vital membranes around it.

Membranes Around the Embryo

Several days after implantation, **extraembryonic membranes** form. One of these is a **yolk sac** that forms below the embryonic disk (Figures 14.6a and 14.7). It is the source of early blood cells and of germ cells that will become gametes; parts of the yolk sac give rise to the embryo's digestive tube. The other three extraembryonic membranes are the amnion, allantois, and chorion.

The innermost membrane is the **amnion** (Figure 14.6b). It develops into a fluid-filled sac that surrounds the embryo (later called the fetus). The amniotic fluid keeps the embryo from drying out, absorbs shocks, and acts as insulation. Moving outward, the next membrane is the **allantois**. As it develops, mesoderm on its surface gives rise to blood vessels that become housed within the **umbilical cord** (Figures 14.6c and 14.7). These vessels link the embryo and the placenta, which we consider shortly.

The **chorion** is a protective membrane around the embryo and the other membranes. It develops from the

trophoblast and continues the secretion of HCG that began when the blastocyst implanted. This maintains the uterine lining for the first three months of pregnancy. Thereafter, the placenta produces sufficient progesterone and estrogen to maintain the lining.

The Placenta

Three weeks after fertilization, almost a fourth of the inner surface of the uterus has become a spongy tissue, the developing **placenta**. By way of this tissue, the embryo receives nutrients and oxygen from the mother and sends out wastes to the mother's bloodstream in return. Such substances are transported by the blood vessels within the umbilical cord.

Although the fully developed placenta is considered an organ, it is helpful to think of it as an intimate association of the embryonic chorion and the superficial cells of the mother's endometrial lining where the embryo implanted. The "maternal side" of the placenta consists of a layer of endometrial tissue containing arterioles and venules. As the chorion develops (from the trophoblast), the tiny projections sent out from the blastocyst as it implants in the endometrium develop into many *chorionic villi*, each endowed with small blood vessels (Figure 14.7). While chorionic villi are developing, the erosion of endometrium that began with implantation continues. As capillaries in the endometrium are broken down, spaces in the eroding endometrial tissue fill with maternal blood. The chorionic villi extend into these spaces. As the embryo develops, its bloodstream remains distinct from

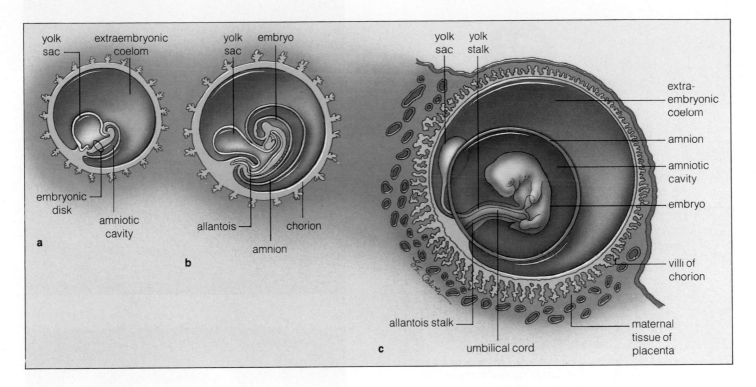

Figure 14.6 Stages in the development of extraembryonic membranes.

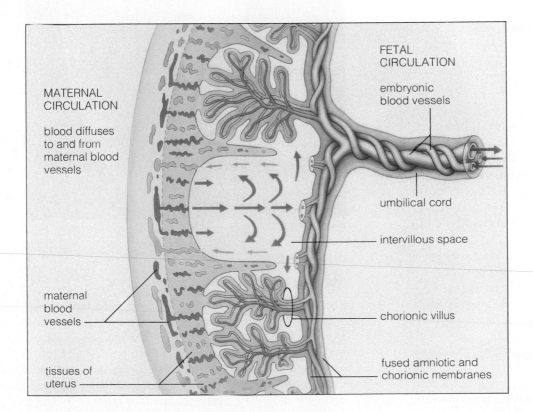

Figure 14.7 Relationship between fetal and maternal tissues in the placenta. The diagram shows how chorionic villi become progressively developed (from left to right across the illustration).

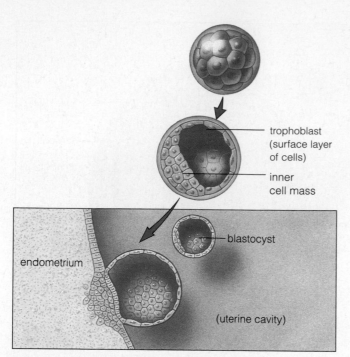

trophoblast
(surface layer
of cells)

inner
cell mass

blastocyst

endometrium

(uterine cavity)

Following fertilization, cell divisions in the zygote produce a ball of about 32 cells, the morula. The morula develops into a hollow blastocyst with an inner cell mass at one side and a surface layer that will give rise to the placenta and extraembryonic membranes. No larger than a pinpoint, at about day 7 the blastocyst reaches the uterus and implantation begins.

Figure 14.8 The remarkable developmental journey from fertilized egg to newborn human being.

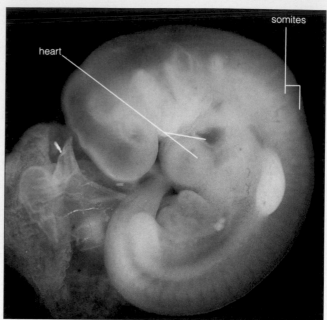

Week 3

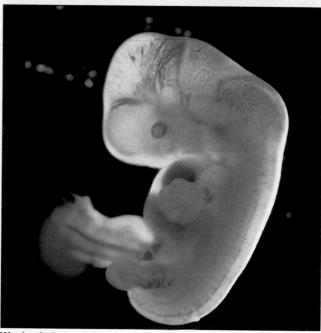

Weeks 4–5

the mother's. Oxygen and nutrients simply diffuse out of the mother's blood vessels, across the blood-filled spaces in the endometrium, then into the embryo's blood vessels. Carbon dioxide and other wastes diffuse in the opposite direction, leaving the embryo. *Focus on Science* describes several methods for screening for birth defects, including using cells withdrawn from chorionic villi in a procedure called *chorionic villus sampling.*

Besides nutrients and oxygen, many other substances taken in by the mother—including alcohol, caffeine, drugs, pesticide residues, and toxins in cigarette smoke—can cross the placenta, as can the virus that causes AIDS.

EMBRYONIC DEVELOPMENT

Figure 14.8 traces the truly remarkable developmental journey from fertilized egg to a newborn human being. The figure summaries provide additional details about the sequence of events we consider in this section and the next.

The First Eight Weeks

The embryonic period begins shortly after fertilization and lasts for eight weeks. It makes up most of the "first trimester," or three months, of the nine months of human gestation. This first trimester is a vulnerable period of embryonic development. By about seven days after fertilization the embryonic disk has formed. Now, gastrulation gets underway, leading to the formation of the three germ layers: ectoderm, mesoderm, and endoderm. Once the germ layers are established, the embryo's organs and organ systems begin to develop (*organogenesis*).

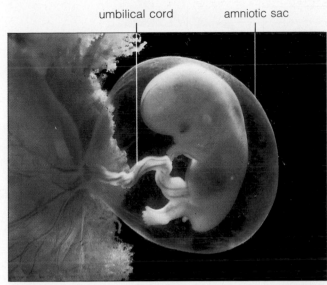

umbilical cord   amniotic sac

Weeks 6-8

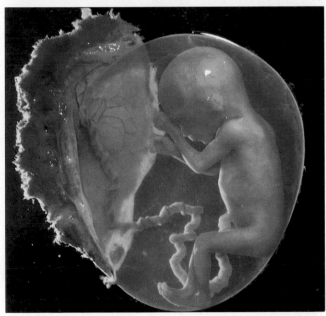

Months 3-4

Months 5-6

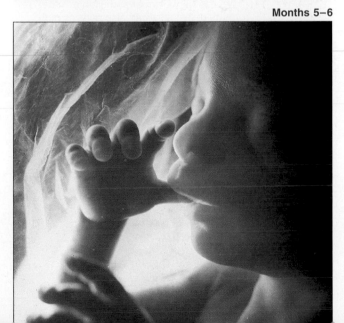

The mesoderm forms when cells migrate inward from the surface of the embryo, forming a thickened, raised band of cells called the *primitive streak* (Figure 14.9). This is the onset of gastrulation. Ectoderm remaining at the surface will give rise to the nervous system, as well as to certain glands and other structures (Table 14.1). Later, endoderm, the inner germ layer, will give rise to parts of the respiratory and digestive systems. Mesoderm will develop into the heart, muscles, bone, and many other internal organs. Although there is little (if any) increase in size during gastrulation, the cell rearrangements dramatically change the embryo's appearance. In particular, they establish a lengthwise axis for further development, once gastrulation is complete.

The end of gastrulation marks the beginning of *neurulation*, the first stage in the development of the nervous

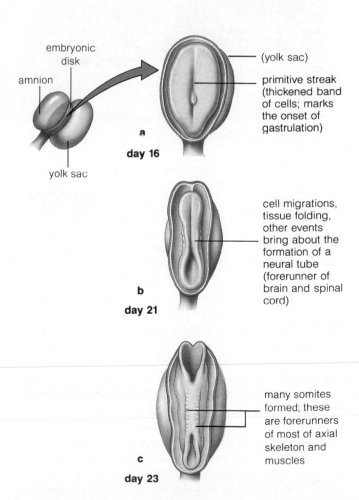

embryonic disk

amnion

yolk sac

(yolk sac)

primitive streak (thickened band of cells; marks the onset of gastrulation)

a
day 16

cell migrations, tissue folding, other events bring about the formation of a neural tube (forerunner of brain and spinal cord)

b
day 21

many somites formed; these are forerunners of most of axial skeleton and muscles

c
day 23

Figure 14.9 Transformation of the pancake-shaped embryonic disk into the early embryo. Shown are three dorsal views (the embryo's back).

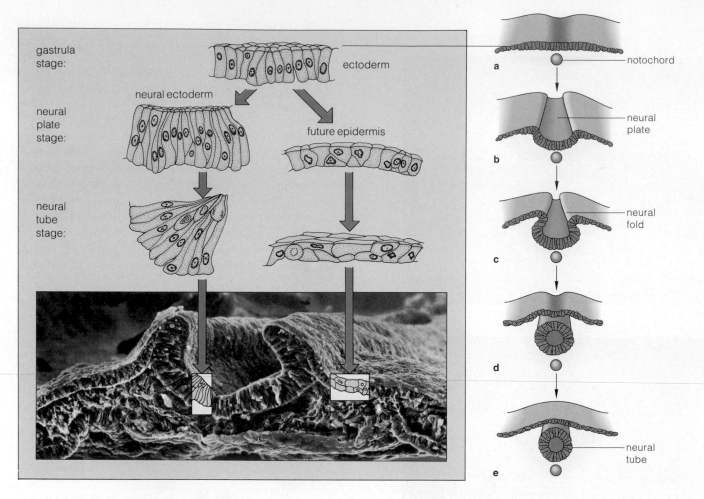

Figure 14.10 Example of morphogenesis: the changes in cell shape that underlie the formation of a neural tube (the forerunner of the brain and spinal cord). As gastrulation draws to a close, the ectoderm is a uniform sheet of cells (**a**). Some ectodermal cells elongate, forming a neural plate, then constrict at one end to become wedge-shaped. The changes in cell shape cause the ectodermal sheet to fold over the neural plate to form the neural tube (**b–e**). Other ectodermal cells flatten while these changes are occurring; they will become part of the epidermis. The scanning electron micrograph and diagrams show the neural tube and epidermis forming in a chick embryo.

system. Figure 14.10 shows how ectodermal cells at the midline of the embryo elongate and form a *neural plate,* the first sign that a region of ectoderm is on its way to developing into nervous tissue. Next, cells near the middle become wedge-shaped. Collectively, the changes in cell shape cause the neural plate to fold over and meet at the embryo's midline to form the *neural tube.* This tube is destined to become the brain and spinal cord.

One form of a birth defect called *spina bifida* ("split spine") occurs when the neural tube fails to develop properly and doesn't separate from ectoderm. The infant may be born with a portion of its spinal cord exposed within a cyst and have various kinds of neurological problems, including poor bowel and bladder control. Infection is a constant danger. Neural tube defects afflict two of every 1,000 babies born in the United States.

Table 14.1 Tissues and Organs Derived from the Three Germ Layers in Human Embryos

Germ Layer	Main Derivatives in the Adult
Endoderm	Various epithelia, as in the gut, respiratory tract, urinary bladder and urethra, and parts of the inner ear; also portions of the tonsils, thyroid and parathyroid glands, thymus, liver, and pancreas
Mesoderm	Cartilage, bone, muscle, and various connective tissues; gives rise to cardiovascular system (including blood), lymphatic system, spleen, and adrenal cortex
Ectoderm	Central and peripheral nervous systems; sensory epithelia of the eyes, ears, and nose; epidermis and its derivatives (including hair and nails), mammary glands, pituitary gland, subcutaneous glands, tooth enamel, and adrenal medulla

Data from Keith Moore, *The Developing Human.*

Focus on Science

Prenatal Diagnosis: Detecting Birth Defects

A growing number of options now enable more than 100 genetic disorders, such as cystic fibrosis, to be detected before birth.

Amniocentesis samples fluid from within the amnion, the sac that contains the fetus (Figure *a*). During the fourteenth to sixteenth week of pregnancy, the thin needle of a syringe is inserted through the mother's abdominal wall, into the amnion. The physician must take care that the needle doesn't puncture the fetus and that no infection occurs. Amniotic fluid contains sloughed fetal cells; as the syringe withdraws fluid, some of those cells are included. They are then cultured and tested for genetic abnormalities.

Chorionic villus sampling (CVS) uses tissue from the chorionic villi of the placenta. CVS is tricky. Using ultrasound, the physician must guide a tube through the vagina, past the cervix, and along the uterine wall, then remove a small sample of chorionic villus cells by suction. The method can be used by the eighth week of pregnancy, and results are available within days. However, CVS can damage the placenta—and hence the fetus. Both CVS and amniocentesis involve a small risk of triggering miscarriage.

An experimental method called FISH (for fluorescent in situ hybridization) analyzes the genetic material of fetal cells that have crossed the placenta and entered the mother's blood. The genetic material is exposed to segments of DNA, called probes, that are tagged with dyes that glow (fluoresce) when exposed to ultraviolet light. The DNA probes attach to chromosome regions that are associated with various genetic birth defects. In experiments, the method has correctly detected abnormalities in hundreds of fetal samples. FISH takes only about two days and does not disturb the protected fetal environment in the uterus.

Researchers have also developed methods of *embryo screening*. In preimplantation diagnosis, an embryo "conceived" by in vitro fertilization is analyzed for genetic defects using recombinant DNA technology. The testing occurs at the eight-cell stage (Figure *b*), which might be considered a "prepregnancy" stage because under normal circumstances the tiny ball of cells would not yet have implanted in the mother's uterus. A method called embryoscopy can be used to view older embryos that have already implanted. A physician threads a tiny viewing device through a hollow needle inserted into the mother's uterus. Embryo screening is designed to help parents who are at high risk of having children with a genetic birth defect. Even so, it is controversial, raising questions we will consider in the *Choices* essay in Chapter 18.

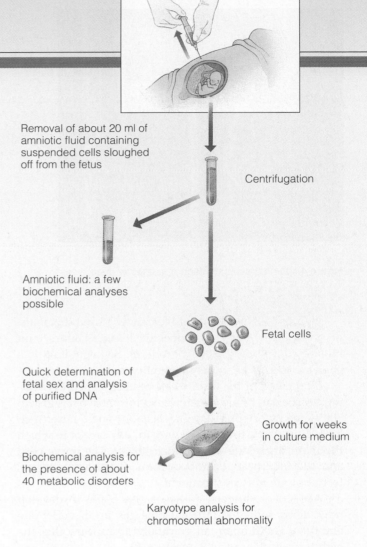

Removal of about 20 ml of amniotic fluid containing suspended cells sloughed off from the fetus

Centrifugation

Amniotic fluid: a few biochemical analyses possible

Fetal cells

Quick determination of fetal sex and analysis of purified DNA

Growth for weeks in culture medium

Biochemical analysis for the presence of about 40 metabolic disorders

Karyotype analysis for chromosomal abnormality

Figure *a* Procedure for amniocentesis.

Figure *b* An eight-cell stage human embryo.

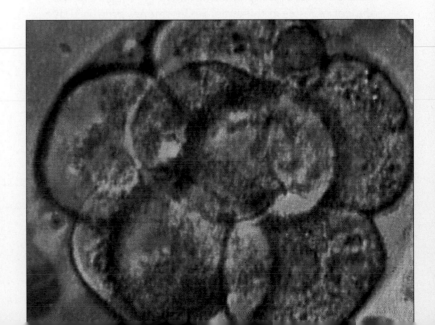

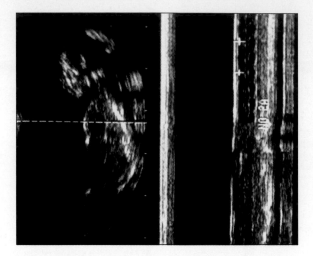

Figure 14.11 Ultrasound image of a fetus *in utero*.

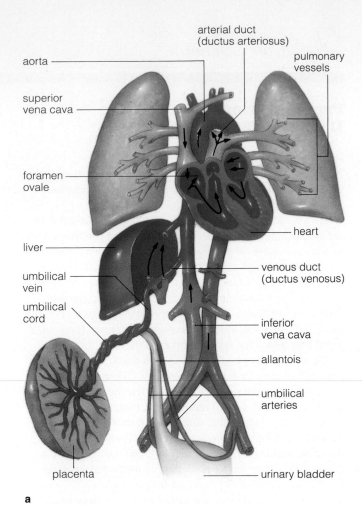

a

Figures 14.8 (week 3) and 14.9c show paired segments, or *somites*—the start of connective tissues, bones, and muscles developing from mesoderm. Segmentation is a characteristic of all vertebrate embryos.

By the end of the third week, a tubelike heart is beating. By the end of the fourth week, the embryo has grown 500 times its original size and is about one-quarter inch long. Its growth spurt gives way to four weeks in which the main organs develop rather slowly. The nerve cord and the four heart chambers form. Respiratory organs form but are not yet functional.

Arms, legs, fingers, and toes also begin to develop now, along with the tail that emerges in all vertebrate embryos. The human tail gradually disappears after the eighth week (Figure 14.8, weeks 6–8).

Gonad Development Male or female gonads begin to develop by the second half of the first trimester. As the introduction to Chapter 13 described, in an embryo that has inherited both X and Y sex chromosomes, a sex-determining region of the Y chromosome (Sry) triggers development of testes at this time. Sex hormones produced by the embryonic testes then influence the development of the entire reproductive system. An embryo with XX sex chromosomes will be female, and ovaries and other female reproductive structures begin to form in her body. Notice that no hormones are required to stimulate development as a female—all that is necessary is the *absence* of testosterone.

After eight weeks the embryo is just over one inch long, its organ systems are formed, and it is designated a **fetus**. As the first trimester ends, a heart monitor can detect the fetal heartbeat. The genitals are well formed, and a physician can often detect the baby's sex using ultrasound (Figure 14.11).

Fetal Circulation

Over time, maturation of organs and organ systems gradually readies the fetus for life outside its mother's body. In the case of the circulatory system, however, the path toward independence requires a detour. Several temporary bypass-vessels form and will function until birth. As Figure 14.12 shows, two *umbilical arteries* within the umbilical cord transport deoxygenated blood and metabolic wastes from the fetus to the placenta. There, the fetal blood gives up wastes and takes on nutrients, and exchanges gases with maternal blood. Fetal hemoglobin binds oxygen more readily than "normal" hemoglobin, which helps ensure that adequate oxygen will reach developing fetal tissues. The oxygenated blood, enriched with nutrients, returns from the placenta to the fetus in the *umbilical vein*.

Other temporary vessels divert blood past the lungs and liver. These organs do not develop as rapidly as some others, because (by way of the placenta) the mother's body can perform their functions. The lungs of a fetus are collapsed and will not become functional for gas exchange

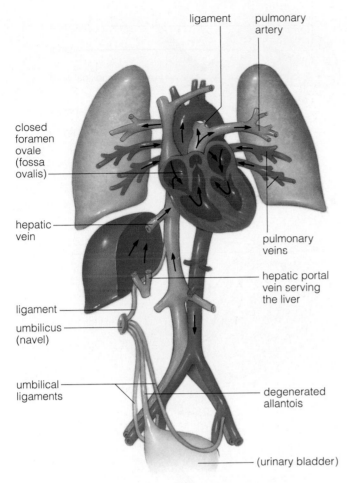

ligament   pulmonary artery

closed foramen ovale (fossa ovalis)

hepatic vein

pulmonary veins

hepatic portal vein serving the liver

ligament

umbilicus (navel)

umbilical ligaments

degenerated allantois

(urinary bladder)

b

Figure 14.12 The direction of blood flow in fetal circulation (arrows). (**a**, left page) Umbilical arteries carry deoxygenated blood from fetal tissues to the placenta. There blood picks up oxygen and nutrients from the mother's bloodstream and returns to the fetus via the umbilical vein. Blood bypasses the lungs, moving through the foramen ovale and the arterial duct. It bypasses the liver by moving through the venous duct. (**b**) At birth the foramen ovale closes, and the pulmonary and systemic circuits of blood flow become completely separate. The arterial duct, venous duct, umbilical vein, and portions of the umbilical arteries become ligaments, and the allantois degenerates.

until the newborn takes its first breaths outside the womb. Hence, lung tissues receive only enough blood to sustain their development. A little of the blood entering the heart's right atrium flows into the right ventricle and moves on to the lungs. Most of it, however, travels through a gap in the interior heart wall called the *foramen ovale* ("oval opening"), or into an arterial duct (*ductus arteriosus*) that bypasses the nonfunctioning lungs entirely.

Likewise, most blood bypasses the fetal liver because the mother's liver performs most liver functions (such as nutrient processing) until birth. Nutrient-laden blood from the placenta travels through a venous duct (*ductus*

venosus) past the liver and on to the heart, which pumps it to body tissues. At birth, a valvelike flap of tissue (*fossa ovalis*) normally closes over the foramen ovale, separating the pulmonary and systemic circuits of blood flow, and the arterial duct collapses (Figure 14.12b). The venous duct gradually closes during the first few weeks after birth.

Miscarriage, the spontaneous expulsion of the uterine contents, occurs in more than 20 percent of all conceptions and usually during the first trimester. Often a woman may not realize she was ever pregnant. Although many factors can trigger a miscarriage (also called spontaneous abortion), as many as half of all cases occur because the embryo (or the fetus) suffers from one or more genetic disorders that prevent normal development.

FETAL DEVELOPMENT

Months Four Through Six

The "second trimester" of development extends from the start of the fourth month to the end of the sixth. Figure 14.8 (months 3–4) shows what the fetus looks like at the sixteenth week of development. Movements of facial muscles produce frowns and squints. The sucking reflex also is evident. Before the second trimester draws to a close, the mother can easily sense movements of the fetal arms and legs.

When the fetus is five months old, it is about four and one-half inches long. Soft, fuzzy hair (the *lanugo*) covers its body. Its reddish skin is wrinkled and protected from abrasion by a thick, cheesy coating called the *vernix caseosa*. During the sixth month, eyelids and eyelashes form.

From the Seventh Month to Birth

The "third trimester" extends from the seventh month until birth. At seven months the fetus is five to six inches long, and during the seventh month its eyes open. Although the fetus is growing much larger and rapidly becoming "babylike," not until the middle of the third trimester will it be able to survive on its own if born prematurely or removed surgically from the uterus. However, with intensive medical support, fetuses as young as 23–25 weeks have survived early delivery. Although development appears to be relatively complete by the seventh month, few fetuses would be able to maintain a normal body temperature or breathe normally, even with the best medical care. Babies born before seven months' gestation often suffer from *respiratory distress syndrome*, as described on page 198, because their lungs lack surfactant and cannot expand adequately. The longer the baby can remain in its mother's uterus, the better its chances. By the ninth month, survival chances increase to about 95 percent.

Birth and Lactation

Birth, or **parturition**, takes place about 39 weeks after fertilization, usually within two weeks of the "due date" (280 days from the start of the last menstrual period). The birth process, "labor," begins when smooth muscle in the uterus starts to contract. Although the precise trigger for labor is still a bit of a mystery, there is evidence that rising levels of the hormone oxytocin (produced by the fetus) and of prostaglandins (produced by the placenta) stimulate contractions of the uterus. For the next 2–18 hours, the contractions will become stronger, more painful, and more frequent.

Three Stages of Labor   Labor is divided into three stages, somewhat akin to "before, during, and after." In the first stage, uterine contractions push the fetus against the mother's cervix. The contractions start out relatively mild and occur about every 15–30 minutes. As the cervix gradually dilates to a diameter of about 10 centimeters (four inches or "five fingers"), contractions become more frequent and intense. Usually, the amniotic sac ruptures during this stage, which can last up to 12 hours or more.

The second stage of labor, actual birth of the fetus, typically occurs less than an hour after full dilation. This stage is generally brief—under two hours. Strong contractions of the uterus and abdominal muscles occur every two or three minutes, and the mother feels an urge to push. Her efforts and the intense contractions move the soon-to-be newborn through the cervix and out through the vaginal canal, usually headfirst (Figure 14.13). Complications can develop if the baby begins to emerge in a "bottom-first" (*breech*) position, and the attending physician may use hands or forceps to aid the delivery.

After the baby is expelled, the third stage of labor gets underway. Uterine contractions force fluid, blood, and the placenta (now called *afterbirth*) from the mother's body. The umbilical cord—the lifeline to the mother—is now severed. Without the placenta to remove wastes, carbon dioxide builds up in the infant's bloodstream. Together with other factors, including handling by medical personnel, this stimulates control centers in the brain, which respond by triggering inhalation—the newborn's crucial first breath. As the lungs begin to function, the bypass vessels of the fetal circulation begin to close down and soon shut completely. The opening in the fetal heart, the foramen ovale, normally closes up slowly during the first year of life. A lasting reminder of this final stage is the scar we call the navel, the site where the umbilical cord was once attached.

The mother's cervix dilates during the first stage of labor, and the baby is born during the second stage. In the third stage, uterine contractions expel the placenta.

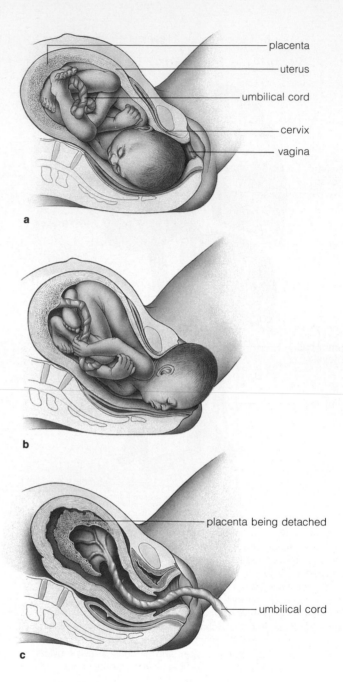

Figure 14.13   Expulsion of the fetus during birth. The placenta, fluid, and blood are expelled shortly afterward (this is the "afterbirth").

Lactation   During pregnancy, estrogen and progesterone were stimulating the growth of mammary glands and ducts in the mother's breasts (Figure 14.14). For the first few days after birth, those glands produce *colostrum*, a colorless fluid low in fat but rich in proteins, antibodies, minerals, and vitamin A. Then prolactin secreted by the pituitary stimulates milk production, or **lactation**.

Some Reflections on Infant Mortality

A useful indicator of overall health in a country or region is the *infant mortality rate,* the number of babies out of every 1,000 born each year that die before their first birthday. Because it reflects the general level of nutrition and health care, infant mortality is probably the single most important measure of a society's quality of life.

Interestingly, although the United States is one of the world's most technologically advanced countries, in 1993 it ranked only thirtieth in terms of infant mortality. That is, 29 countries had lower rates. Why the higher U.S. rate? There are several reasons, but studies show that two of the most important are drug addiction among pregnant women, and inadequate health care for poor women during pregnancy and for their babies after birth. In less developed countries, a high infant mortality rate usually indicates insufficient food (undernutrition), poor nutrition (malnutrition), and a high incidence of infectious disease, usually from contaminated drinking water.

Between 1965 and 1993, improved public health programs reduced infant mortality up to 35 percent in some countries. Even so, each year at least 12 million babies under the age of one year die of preventable causes.

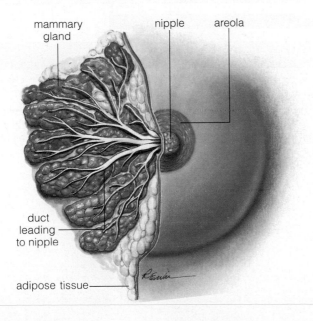

mammary gland

nipple

areola

duct leading to nipple

adipose tissue

Figure 14.14 Breast of a lactating female. This cutaway view shows the mammary glands and ducts. Chapter 20 describes how to examine breast tissues for cancer.

When the newborn suckles, the pituitary also releases oxytocin, which causes breast tissues to contract and so force milk into the ducts. Oxytocin also triggers uterine contractions that "shrink" the uterus back to its normal size.

Maternal Life-style and Early Development

The placenta is a highly selective filter that prevents many noxious substances in the mother's bloodstream from gaining access to the embryo or fetus. Even so, the developing individual is at the mercy of the mother's diet and health habits (see *Focus on Environment*).

Maternal Nutrition During pregnancy, a balanced diet that includes sufficient carbohydrates, amino acids, and fats or oils usually provides most necessary vitamins and minerals. The mother's vitamin needs are definitely increased, and most physicians recommend that pregnant women take vitamin supplements. The developing fetus is more resistant to vitamin and mineral deficiencies because the placenta preferentially absorbs vitamins and minerals from the mother's blood. Even so, recent evidence suggests that folic acid supplements may be important in preventing certain birth defects, including neural tube disorders.

In most cases, a woman should gain from 20 to 35 pounds during pregnancy. If she restricts her food intake too severely, especially during the last trimester, the newborn will be underweight. Significantly underweight infants face more postdelivery complications than do infants of normal weight and represent nearly half of all newborn deaths. As birth approaches, the growing fetus demands more and more nutrients from the mother's body. During this last phase of pregnancy, the mother's diet profoundly influences the course of development.

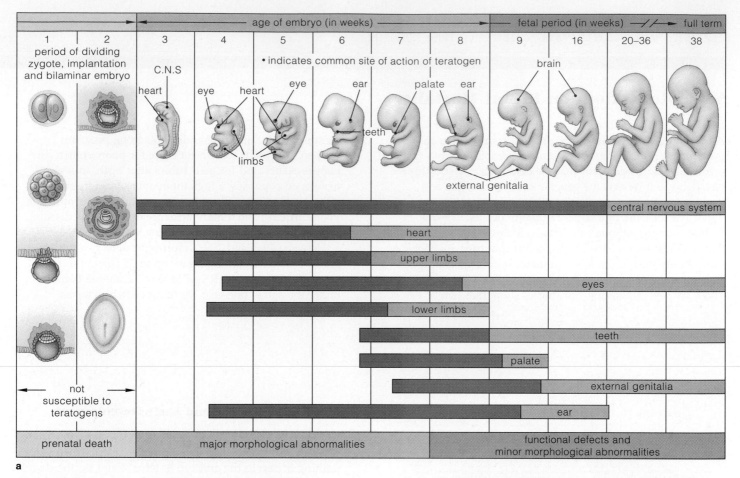

		age of embryo (in weeks)	fetal period (in weeks)	full term

1	2	3	4	5	6	7	8	9	16	20–36	38

period of dividing zygote, implantation and bilaminar embryo

• indicates common site of action of teratogen

central nervous system

heart

upper limbs

eyes

lower limbs

teeth

palate

external genitalia

ear

not susceptible to teratogens

prenatal death	major morphological abnormalities	functional defects and minor morphological abnormalities

a

Figure 14.15 **(a)** Critical periods of embryonic and fetal development. Light blue also indicates periods in which organs are most sensitive to damage from alcohol, viral infection, and so on. Numbers signify the week of development.

(b, below) An infant affected by fetal alcohol syndrome (FAS). Symptoms include a small head, low and prominent ears, poorly developed cheekbones, and a long, smooth upper lip. The child may be likely to encounter growth problems and abnormalities of the nervous system. About one newborn in 750 in the United States is affected by this disorder.

Poor nutrition damages most organs—particularly the brain, which undergoes its greatest growth in the weeks just before birth.

Risk of Infections Throughout pregnancy, some antibodies cross the placenta and help protect the developing individual from all but the most severe bacterial infections. However, certain viral diseases can have damaging effects (Figure 14.15a) if they are contracted during the first six weeks after fertilization, the critical time of organ formation. For example, if the woman contracts German measles during this period, there is a 50 percent chance that her embryo will become malformed. If she contracts the virus when the embryo's ears are forming, her newborn may be deaf. (German measles can be avoided by vaccination *before* pregnancy.) The likelihood of damage to the embryo diminishes after the first six

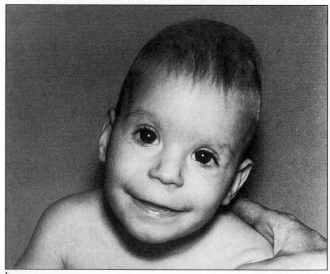

b

weeks. After the fourth month, the same infection has no apparent effect on the fetus.

Prescription Drugs, Illegal Drugs, and Alcohol

During the first trimester the embryo is highly sensitive to drugs. In the 1960s many women using the drug thalidomide during the first trimester gave birth to infants with missing or severely deformed arms and legs. Although it wasn't known at the time, thalidomide disrupts the events that induce development of limbs. Thalidomide was withdrawn from the market, but there is evidence that other tranquilizers (and sedatives and barbiturates) might cause similar, although less severe, damage. Use of certain anti-acne drugs, such as retinoic acid (Retin-A), increases the risk of facial and cranial deformities. The antibiotic tetracycline causes yellowed teeth. Streptomycin causes hearing problems and may affect the nervous system.

Cocaine, particularly crack cocaine, disrupts the function of the fetal nervous system as well as the mother's (page 250). Cocaine-addicted newborns are chronically irritable. They are also abnormally small because during prenatal development their body tissues were not provided with enough oxygen and nutrients. (Crack causes blood vessels to constrict.) Even though crack babies are abnormally fussy, they do not respond to rocking and other kinds of normal stimulation. It may be a year or more before they recognize their mother. Although authorities disagree about the long term effects of infant crack addiction, there is some evidence that even with treatment, severely affected babies may develop into children with moderate to severe problems in their emotional and social adjustment.

Like many other drugs, alcohol crosses the placenta and affects the fetus. *Fetal alcohol syndrome* (FAS) is a constellation of defects that can result from alcohol use by a pregnant woman. *FAS is the third most common cause of mental retardation in the United States.* It is characterized by facial deformities, poor coordination and, sometimes, heart defects (Figure 14.15b). Between 60 and 70 percent of alcoholic women give birth to infants with FAS. Some researchers suspect that drinking any alcohol at all during pregnancy may be dangerous for the fetus.

Effects of Cigarette Smoke

Cigarette smoking harms fetal growth and development. Research shows that women smokers are at greater risk of miscarriage, stillbirth, and premature delivery. Newborns of women who have smoked every day throughout pregnancy have a low birth weight. That is true even when the woman's weight, nutritional status, and all other relevant variables are identical to those of pregnant women who do not smoke. Studies also suggest that babies of smoking mothers may be at greater risk for heart abnormalities and be slower learning to read.

The critical period appears to be the last half of pregnancy. In one study, newborns of women who had stopped smoking by the middle of the second trimester were indistinguishable from those born to women who had never smoked.

POSTNATAL DEVELOPMENT AND AGING

Following birth, a prescribed course of further growth and development leads to adulthood. Table 14.2 summarizes all the prenatal (before birth) and postnatal (after birth) stages of life. A newborn is called a "neonate." Infancy lasts until about 15 months of age. Although a baby is born with all the brain neurons it will ever have,

Table 14.2 Stages of Human Development: A Summary	
Prenatal Period:	
1. Zygote	Single cell resulting from fusion of sperm nucleus and egg nucleus at fertilization
2. Morula	Solid ball of cells produced by cleavages
3. Blastocyst	Ball of cells with surface layer and inner cell mass
4. Embryo	All developmental stages from two weeks after fertilization until end of eighth week
5. Fetus	All developmental stages from the ninth week until birth (about 39 weeks after fertilization)
Postnatal Period:	
6. Newborn (neonate)	Individual during the first two weeks after birth
7. Infant	Individual from two weeks to about 15 months after birth
8. Child	Individual from infancy to about 12 or 13 years
9. Pubescent	Individual at puberty, when secondary sexual traits develop; girls between 10 and 16 years, boys between 13 and 16 years
10. Adolescent	Individual from puberty until about three or four years later; physical, mental, emotional maturation
11. Adult	Early adulthood (between 18 and 25 years), bone formation and growth completed. Changes proceed very slowly afterward.
12. Old age	Aging culminates in general body deterioration

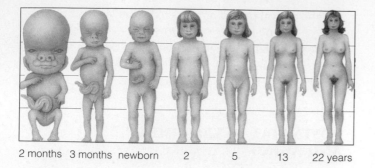

2 months 3 months newborn   2   5   13   22 years

Figure 14.16 Diagram of changes in the proportions of the human body during prenatal and postnatal growth.

during this period the nervous and sensory systems mature rapidly, and the body becomes longer through a series of "growth spurts." Figure 14.16 shows how body proportions change as development proceeds through childhood and adolescence. A key feature of adolescence is *puberty*, the arrival of sexual maturity as the reproductive organs begin to function. Sex hormones (page 296) trigger the appearance of pubic and underarm hair, other secondary sex characteristics, and behavior changes. A combination of hormones triggers another "growth spurt" at this time. Boys usually grow most rapidly between the ages of 12 and 15, whereas girls tend to grow most rapidly between the ages of 10 and 13. After several years, the influence of sex hormones causes the cartilaginous ends of long bones to harden into bone, and growth stops by the early twenties.

Adult Life and Aging

Following growth and differentiation, the cells of all complex animals gradually deteriorate. Humans are no different. Although in the United States the average life expectancy is 72 years for males and 79 years for females, we reach the peak of our physical potential in adolescence and early adulthood. A healthy diet, regular exercise, and other beneficial life-style habits can go far in keeping a person vigorous for decades of adult life. Even so, after about age 30, body parts begin to undergo structural changes, and there is a gradual loss of efficiency in bodily functions, as well as increased sensitivity to environmentally induced stress. This progressive cellular and bodily deterioration is built into the life cycle of all organisms in which differentiated cells become highly specialized. The process is called *senescence*, or simply **aging**.

Aging in humans leads to many structural changes in the body. Beginning around age 30, there is a gradual decline in bone and muscle mass, increased skin wrinkling, and more fat deposition. Less obvious are gradual physiological changes. For example, metabolic rates

decline in kidney cells, so the body cannot respond as effectively as it once did to changes in extracellular fluid volume and composition. Reflexes become slower, and eyesight, hearing, and taste perception diminish. A major change involves collagen, the fibrous protein that is present in the extracellular spaces of nearly all tissues. Collagen may represent as much as 40 percent of your body's proteins. With increasing age, new collagen fibers that are being produced are structurally altered—and such structural changes are bound to have widespread physical effects. For example, structurally altered collagen causes aging blood vessels to lose some of their elasticity, and that leads to less efficient blood circulation throughout the body.

No one knows what causes aging, although researchers have given us some interesting things to think about. For example, in laboratory culture, normal human embryonic cells divide about 50 times, then die. Such experiments suggest that normal cell types have a *limited division potential*. That is, the ability of a body cell to divide may be genetically programmed to decline at a particular stage of the life cycle. But does the change *cause* aging, or is it a *result* of the aging process? Consider that neurons, which do not divide at all after an early stage of embryonic development, still deteriorate gradually during the life of an animal. Their predictable deterioration might indicate that similar changes may be occurring in dividing cells throughout the body. The *Human Impact* section following this chapter probes our current understanding of aging in more detail and discusses some tantalizing hypotheses about the aging process. We close with a *Choices* essay on the controversial issues raised by life-prolonging technologies and changing attitudes toward both life and death.

SUMMARY

1. Human development proceeds through six stages:

   a. Gamete formation, during which the egg and sperm mature within the reproductive organs of the parents.

   b. Fertilization, which begins when a sperm penetrates an egg and is completed when the sperm and egg nuclei fuse.

   c. Cleavage, when the fertilized egg (zygote) undergoes cell divisions that form the early multicelled embryo (the blastocyst). The destiny of cell lineages is established in part by the sector of cytoplasm inherited at this time.

   d. Gastrulation, when the organizational framework of the whole body is laid out. Endoderm, ectoderm, and mesoderm form; all the tissues of the adult body will develop from these germ layers.

Life, Death, and Technology

Technology has begun to change the rules by which humans live and die. It has also raised serious social issues surrounding medical treatment for grave illness and medically assisted death.

One area of growing controversy is the cost of procedures such as organ and bone marrow transplants and complex heart surgeries. Increasingly, health insurers and government agencies are reluctant to pay for expensive or experimental therapies. In part, they argue that it is unwise to expend large sums when a treatment is a last-ditch attempt to save a desperately ill person's life—and when the odds against a cure are great. Another argument is that it is unfair to devote a large share of the "health care dollar" to only a few people, when the same money could benefit many individuals through prenatal care, vaccination programs, and similar efforts. On the other hand, critics of "rationed health care" fear that in such a system only the wealthy will have access to the most advanced procedures.

In various countries, controversy also rages around the "right to die." In recent years, Michigan physician Jack Kervorkian became an open practitioner of "assisted suicide," providing technical help to individuals who were ill and wished to end their lives in a painless manner. Kervorkian's clients have included people with incurable cancer, Alzheimer's disease, and other devastating conditions with no hope of recovery. Although in every case the ill person requested Kervorkian's help, many people vigorously oppose physician-assisted death on ethical, religious, or legal grounds. Others say adamantly that individuals have a "right to die." Several people have even sued (unsuccessfully) physicians and hospitals for "wrongful life" after being kept alive against their wishes.

Should we as a society espouse an unconditional "right to live" with the full spectrum of medical technology available to everyone on demand? Who will pay for it? Should such care be provided only in certain cases? What about a person's right to choose a "dignified death" when the alternative is a prolonged, agonizing decline? What if *your* life or health were in the balance?

e. Organ formation (organogenesis). The different organs start developing by a tightly orchestrated program of cell differentiation and morphogenesis. This is a critical time of early development.

f. Growth and tissue specialization, when organs enlarge overall and acquire their specialized chemical and physical properties. The maturation of tissues and organs continues into the fetal stage and beyond.

2. Cell differentiation and morphogenesis have their foundations in (a) the distribution of localized cytoplasmic determinants during cleavage and (b) cell interactions that begin during organogenesis.

3. Embryonic development depends on the formation of four extraembryonic membranes:

a. Yolk sac: parts give rise to the embryo's digestive tube and form blood cells and germ cells.

b. Allantois: its blood vessels become umbilical arteries and function in oxygen transport and waste excretion.

c. Amnion: a fluid-filled sac that surrounds and protects the embryo from mechanical shocks and keeps it from drying out.

d. Chorion: a protective membrane around the embryo and the other membranes; a primary component of the placenta.

4. The embryo and the mother exchange nutrients, gases, and wastes by way of the placenta (a spongy tissue of endometrium and extraembryonic membranes).

5. The placental barrier provides some protection for the embryo/fetus, but there may be harmful effects from the mother's nutritional deficiencies; infections; intake of prescription drugs, illegal drugs, and alcohol; and smoking.

6. Estrogen and progesterone stimulate growth of the mammary glands. At delivery, contractions of the uterus dilate the cervix and expel the fetus and afterbirth. After delivery, nursing causes the release of hormones that stimulate milk production and release.

7. As with all complex animals, the human body gradually shows changes in structure and a decline in efficiency (aging). Although aging is part of the life cycle of all animals having extensively specialized cell types, its precise cause is unknown.

Review Questions

1. Define and describe the main features of the following developmental stages: fertilization, cleavage, gastrulation, and organogenesis. *312, 318–322*

2. Define cell differentiation and morphogenesis, two processes that are critical for development. Which two mechanisms serve as the basic foundation for cell differentiation and morphogenesis? *312*

3. Summarize the development of an embryo and fetus. When are body parts such as the heart, nervous system, and skeleton largely formed? *318–323*

Critical Thinking: You Decide *(Key in Appendix IV)*

1. How accurate is the phrase "A pregnant woman must do everything for two"? Give some specifics to support your answer.

2. The renowned developmental biologist Lewis Wolpert once observed that birth, death, and marriage are not the most important events in human life—rather, gastrulation is. In what sense was he correct?

Self-Quiz *(Answers in Appendix III)*

1. Development cannot proceed properly unless each stage is successfully completed before the next begins, starting with _____.
   a. gamete formation   d. gastrulation
   b. fertilization   e. organ formation
   c. cleavage   f. growth, tissue specialization

2. During cleavage, the _____ becomes converted to a ball of cells, which in turn is transformed into the _____.
   a. zygote; blastocyst   c. ovum; embryonic disk
   b. trophoblast; embryonic disk   d. blastocyst; embryonic disk

3. In the week following implantation, cells of the _____ will give rise to the embryo.
   a. blastocyst   c. embryonic disk
   b. trophoblast   d. zygote

4. The developmental process called _____ produces the shape and structure of particular body regions.

5. _____ is the gene-guided process by which cells in different locations in the embryo become specialized.
   a. Implantation   c. Cell differentiation
   b. Neurulation   d. Morphogenesis

6. In a human zygote, the cell divisions of cleavage produce an embryonic stage known generally as a _____.
   a. zona pellucida   c. blastocyst
   b. gastrula   d. larva

7. Match the following developmental stages with its description.
   ____ cleavage   ____ cell differentiaton
   ____ gamete formation   ____ gastrulation
   ____ organ formation   ____ fertilization

   a. egg and sperm mature in parents
   b. sperm, egg nuclei fuse
   c. formation of germ layers
   d. zygote becomes a ball of cells called a morula
   e. cells come to have specific structures and functions
   f. starts when germ layers split into subpopulations of cells

8. Parts of the _____, an extraembryonic membrane, give rise to the embryo's digestive tube.
   a. yolk sac   c. amnion
   b. allantois   d. chorion

9. The _____, a fluid-filled sac, surrounds and protects the embryo from mechanical shocks and keeps it from drying out.
   a. yolk sac   c. amnion
   b. allantois   d. chorion

10. Blood vessels of the _____, an extraembryonic membrane, transport oxygen and nutrients to the embryo.
    a. yolk sac   c. amnion
    b. allantois   d. chorion

11. Substances are exchanged between the embryo and mother through the _____, which is composed of maternal and embryonic tissues.
    a. yolk sac   d. amnion
    b. allantois   e. chorion
    c. placenta

Key Terms

aging *328*
allantois *316*
amnion *316*
blastocyst *315*
cell differentiation *312*
chorion *316*
cleavage *315*
ectoderm *312*
embryonic disk *315*
endoderm *312*
extraembryonic membranes *316*
fertilization *312*
fetus *322*
gamete formation *312*
gastrulation *312*
germ layer *312*
implantation *315*
inner cell mass *315*
lactation *324*
mesoderm *312*
morphogenesis *312*
organogenesis *312*
ovum *313*
parturition *324*
placenta *316*
trophoblast *315*
umbilical cord *316*
yolk sac *316*
zygote *312*

Readings

Caldwell, M. November 1992. "How Does a Single Cell Become a Whole Body?" *Discover.*

Nilsson, L., et al. 1986. *A Child Is Born.* New York: Delacorte Press/Seymour Lawrence. Extraordinary photographs of embryonic development.

"The Science of Sex." June 1992. *Discover.*

Spence, A. 1989. *Biology of Human Aging.* Englewood Cliffs, N.J.: Prentice-Hall. Paperback.

Human Impact

THE BODY AS IT AGES

In humans, as in most other multicellular organisms, a predictable sequence of changes unfolds as we age, from youth to maturity and beyond. To a greater or lesser degree, these changes affect virtually all body tissues and organs.

Technically, aging or *senescence* begins the day we are born and continues until we die. However, most people equate aging with "getting old," a gradual loss of vitality as cells, tissues, and organs function less and less efficiently. Beginning at about age 40, our skin begins to noticeably wrinkle and sag, body fat tends to accumulate, and injuries to muscles and joints occur more readily and take longer to heal. Stamina declines, and we become increasingly susceptible to cardiovascular disorders, cancer, and degenerative ailments such as arthritis and Alzheimer's disease.

Biologists have a fairly good understanding of the physiological deterioration that occurs over time in various tissues and organ systems. For example, structural alterations in body proteins—including collagen, a key component of various connective tissues—may contribute to many changes we associate with aging. A collagen molecule is a polypeptide chain twisted into a spiral, and this shape is stabilized by molecular bonds that form "crosslinks" between segments of the chain. As a person grows older, new crosslinks develop, and the protein becomes more and more rigid. As collagen's structure changes, so may the structure and functioning of organs and blood vessels that contain it. Age-related cross-linking is also thought to affect many enzymes and possibly DNA.

As we will see, much less is known about the exact causes of such changes, though researchers currently are pursuing several promising hypotheses. Ultimately, however, aging is a steady decline in the finely tuned ebb and flow of substances and chemical reactions that maintain homeostasis.

Age-Related Changes in Selected Body Systems

Skin Skin changes often are early obvious signs of aging. The normal replacement of sloughed cells of the epidermis (through cell division) begins to slow. Cells called fibroblasts are a major component of connective tissue, and the number of fibroblasts in the dermis starts to decrease. Also, elastic fibers that give skin its flexibility become replaced with more rigid collagen. As a result of these changes, the skin becomes thinner and less elastic, and sags and wrinkles develop. Wrinkling increases as fat stores decline in the hypodermis, the subcutaneous layer beneath the dermis. The skin becomes dryer as sweat and

Figure *a* Moderate physical activity, consistently maintained as a person ages, slows age-related deterioration of the musculoskeletal system and promotes cardiovascular health. Throughout life, the benefits of regular exercise include aiding in weight control and promoting a positive outlook.

oil glands begin to break down and are not replaced. The loss of sweat glands and subcutaneous fat is one reason why older people tend to have difficulty regulating their body temperature. Hair follicles die or become less active, so there is a general loss of body hair. As pigment-producing cells die and are not replaced, remaining body hair begins to appear gray or white.

Muscles and Skeleton Fibers in skeletal muscle atrophy, in part because of a corresponding loss of motor neurons that synapse with muscle fibers. Muscles lose mass and strength, and lost muscle tends to be replaced by fat and later by collagen, although the extent of this replacement depends on a person's diet and level of exercise. In general, an 80-year old has about half the muscle strength he or she had at 30. The adage "use it or lose it" applies here; a program of regular physical activity can help retard all these changes (Figure *a*).

Bones become weaker, more porous, and brittle as a person ages. Over the years bones begin to lose some of their collagen and elastin, but the main cause of bone weakening is loss of calcium and other minerals. Some individuals may inherit a gene for faulty vitamin D receptors, which reduces the efficiency with which calcium is absorbed and used in bone formation. After about age 40, minerals begin to be removed from bone faster than they are replaced (see page 92). Women naturally have less

bone calcium than men do, and they begin to lose it earlier. As a result, older women are prone to develop osteoporosis. Without medical intervention, some women lose as much as 50 percent of their bone mass by age 70. In both women and men, the spinal vertebrae become closer together as the intervertebral disks deteriorate. This is why people often get shorter, losing about a centimeter every 10 years from middle age onward.

Nine out of ten people over 40 have some degree of joint breakdown. Cartilage in joints deteriorates over time, the surfaces of the joined bones begin to wear away, and the joints become more difficult to move. About 15 percent of adults develop *osteoarthritis*, a chronic inflammation of cartilage, in one or more joints. Although many factors probably contribute to osteoarthritis, it is most common in older people. This suggests that age-related changes in bone are often involved.

Cardiovascular and Respiratory Systems The heart and lungs function less efficiently with increasing age. In the lungs, walls of alveoli break down, so there is less total respiratory surface available for gas exchange. If a person does not develop an enlarged heart due to cardiovascular disease, the heart muscle becomes slightly smaller and its strength and ability to pump blood diminishes (Table 1). As a result, less blood and oxygen are delivered to muscles and other tissues. In fact, decreased blood supply may be a factor in age-related changes throughout the body. Blood transport is also affected by structural changes in aging blood vessels. Elastic fibers in blood vessel walls become replaced with connective tissue containing collagen or become hardened with calcium deposits, and so vessels become stiffer. Cholesterol plaques and fatty deposits often further narrow arteries and veins (page 160). Hence older people often have higher blood pressure. However, as with the musculoskeletal system, life-style choices such as not smoking, eating a low-fat diet, and regular exercise can help a person maintain a vigorous respiratory and cardiovascular system well past middle age.

Table 1	Some Physiological Changes in Aging			
Age	Maximum Heart Rate	Lung Capacity	Muscle Strength	Kidney Function
25	100%	100%	100%	100%
45	94%	82%	90%	88%
65	87%	62%	75%	78%
85	81%	50%	55%	69%

Note: Age 25 is the benchmark for maximal efficiency of physiological functions.
Source: "The Search for the Fountain of Youth," *Newsweek*, 5 March 1990.

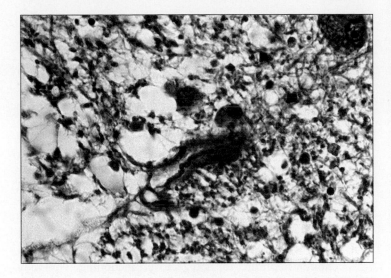
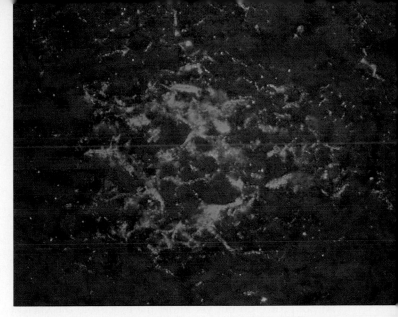

Figure *b* (Left) A neurofibrillary tangle in brain tissue from a patient with Alzheimer's disease. (Right) Three beta amyloid plaques. Older people who do not have Alzheimer's disease also develop neurofibrillary tangles and beta amyloid plaques, but not nearly as many.

Reproductive Systems and Sexuality Levels of most hormones often remain steady throughout life, but falling secretion of estrogens and progesterone trigger menopause in women, whereas falling testosterone causes reduced fertility in older men. Although menopause brings a woman's reproductive period to an end, men can and have fathered children into their eighties. Males and females both retain their capacity for sexual response well into old age.

Menopause usually begins in a woman's late forties or early fifties. Over a period of several years, her menstrual periods become irregular, then stop altogether as the ovaries become less and less sensitive to the hormones FSH and LH (page 302) and gradually stop secreting estrogen in response. Many women report relatively few unpleasant symptoms during menopause. However, declining estrogen levels may trigger "hot flashes" (intense sweating and uncomfortable warmth), thinning of the vaginal walls, and some loss of natural lubrication. Postmenopausal women are also at increased risk of osteoporosis (page 92) and heart disease. Small doses of estrogen administered as *hormone replacement therapy* (HRT) can counteract severe side effects of estrogen loss and may even help protect against uterine cancer. However, HRT is not without risks and must be carefully discussed with a knowledgeable physician.

After about age 50, men gradually begin to take longer to achieve an erection due to vascular changes that cause the penis to fill with blood more slowly. In the United States, more than half of men over 50 also experience bladder infections, urinary frequency, and other problems caused by age-related enlargement of the prostate gland.

Nervous System and Senses Neurons stop dividing once they form. Thus you are born with all the neurons you will ever have, and any that are lost will not be replaced. Brain neurons die steadily throughout life, and as they do the brain shrinks slightly, losing about 10 percent of its mass after 80 years. On the other hand, the brain has more neurons than are required for various functions, and there is evidence that when some types of neurons are lost or damaged, other neurons may produce new dendrites and synaptic connections and so take up the slack.

Over time, however, the death of some neurons and structural changes in others apparently do interfere with nervous system functions. For instance, in nearly anyone who lives to old age, tangled clumps of microtubules develop in the cytoplasm of many neuron cell bodies. These *neurofibrillary tangles* may disrupt normal cell metabolism, although their exact effect is not understood. Clotlike plaques containing protein fragments called *beta amyloid* also develop between neurons. The drug tacrine can temporarily help alleviate some symptoms of Alzheimer's disease, which include memory loss and disruptive personality changes.

In people who exhibit the dementia called *Alzheimer's disease* (AD), the brain tissue contains masses of neurofibrillary tangles. It also is riddled with large numbers of beta amyloid plaques (Figure *b*). For the time being, however, researchers disagree over whether the amyloid plaques are a cause of Alzheimer's or simply one effect of another, unknown disease process. For instance, the brains of Alzheimer's patients also have lower-than-normal amounts of the neurotransmitter acetylcholine, and some evidence suggests that this shortage may be related to beta amyloid buildup. Other possibilities under investigation include the hypothesis that AD results from chronic inflammation of brain tissue, similar to the inflammation that triggers arthritis.

We are beginning to understand the genetic foundations for at least some inherited forms of AD. In families that show a pattern of early-onset (before age 60) Alzheimer's, the disease has been linked to a mutated gene for the amyloid precursor protein on chromosome 21, or to an unknown gene on chromosome 14. In families in which late-onset AD is prevalent, chromosome 19 of affected individuals carries at least one copy of a gene calling for a variant form of a lipid-binding protein. Although we do not yet know the exact role the variant protein plays in AD, individuals who have two copies of the gene that encodes it (the maximum number of copies possible) have a 90 percent chance of developing Alzheimer's.

Even otherwise healthy people begin to have problems with short-term memory after about age 60 and may find it takes longer to process new information. Perhaps because aging CNS neurons tend to lose some of their insulating myelin sheath, older neurons do not conduct action potentials as efficiently. In addition, neurotransmitters such as acetylcholine may be released more slowly. This may be due to age-related changes in plasma membrane proteins, which cause the plasma membranes of synaptic vesicles to stiffen. As a result of such changes, movements and reflexes become slower, and some coordination is lost.

Sensory organs become less efficient at detecting or responding to stimuli: For example, people tend to become farsighted as they grow older because the lens of the eye loses its elasticity and is altered in other ways that prevent it from properly flexing to bend incoming light during focusing.

Other Systems Other organs and organ systems also change as the years go by. In the immune system, the number of T cells falls, B cells become less active, and the ability to recognize self markers on body cells declines. These and other events help account for the fact that older people are more prone to many illnesses, including autoimmune diseases such as rheumatoid arthritis. In the aging digestive tract, glands in mucous membranes lining the stomach and small and large intestines gradually degenerate, and fewer digestive enzymes are secreted.

Although it is vital for older people to maintain adequate nutrition, we require fewer food calories as we age. By age 50, the basal metabolic rate (page 124) is only 80–85 percent of what it was in childhood and will keep falling about 3 percent every decade. Hence people tend to gain weight as they enter middle age, unless they compensate

for a falling BMR by consuming fewer calories, increasing physical activity, or both.

The muscular walls of the large intestine, bladder, and urethra become weaker and less flexible. As the urinary sphincter is affected, many older people experience the urine leakage called *urinary incontinence*. Women who have borne children may have more trouble with urinary incontinence because their pelvic floor muscles are weak; in older men, prostate enlargement can cause difficulty in urination. The kidneys may continue to function well throughout life, despite the fact that nephrons gradually break down and lose some of their ability to maintain the balance of water and ions in body fluids (Table 1). At birth, a person's kidneys generally contain more than enough nephrons for satisfactory functioning. This may explain why the loss of nephrons in normal aging can often occur without major effects.

What Causes Aging?

No one yet knows for sure what factors cause the degenerative changes we associate with aging. Many hypotheses have been proposed, and there is tantalizing evidence for some of them.

Oxidative Damage Some researchers believe that many of the changes seen in aging tissues could result from damage caused by free radicals of oxygen. As Chapter 1 described, free radicals are by-products of normal cell metabolism. They carry an unpaired electron and can combine with (oxidize) and damage proteins, DNA, lipids, and other molecules. Aging cells in the heart, brain, skin, and other tissues typically contain clumps of lipofuscins, which are oxidized lipids. In tissue samples, the amount of oxidized proteins, including key cell enzymes, also increases with the age of the sample donor. Moreover, studies show that the DNA in cell mitochondria (page 49) is especially susceptible to free radical damage. This discovery suggests why, as the years pass, an organ might function less and less well. As more and more of its cells contain increasing numbers of damaged mitochondria, the cells would undergo a steady decline in the energy available to sustain metabolic activities.

Decline in DNA Repair A related hypothesis focuses on a possible decline in the ability of cells to repair damaged DNA, either in mitochondria or in the cell nucleus. DNA makes up genes, the instructions for building cell proteins, including many vital enzymes. Some genes regulate the activity of others and so indirectly control cell operations. If cells gradually lose the capacity for DNA self-repair, over time gene mutations would accumulate. Needed enzymes or other molecules might be in increasingly short supply, or cell operations might become deficient in other ways. Perhaps genetic changes cause cells to lose their ability to repair free radical damage. We already know that changes in various types of genes underlie most cancers (page 430), and it may not be a coincidence that the incidence of cancer rises sharply in older age groups.

Cell division is the mechanism by which the body grows, maintains certain tissues (such as skin and intestinal epithelium), and repairs tissue damage. Although the reasons are not clear, normal embryonic cells cultured in the laboratory tend to divide ever more slowly as they grow "older," and after about 50 rounds of division the process stops altogether. Researchers are now looking for evidence of how the gradual loss of the ability of body cells to proliferate may contribute to aging.

Readings

Ezzell, C. March 7, 1992. "Alzheimer's Alchemy." *Science News.*

Flieger, K. April 1992. "Prostate Problems Plague Older Men." *FDA Consumer.*

Olshansky, S. J., B. Carnes, and C. Cassel. April 1993. "The Aging of the Human Species." *Scientific American.*

Rusting, R. L. December 1992. "Why Do We Age?" *Scientific American.*

15 AIDS AND OTHER SEXUALLY TRANSMITTED DISEASES

Safer Sex

Every year about 12 million people in the United States, a quarter of them teenagers, confront a harsh reality of life: They discover that they have contracted a sexually transmitted disease (STD). STDs are more than socially embarrassing. Untreated, various ones can lead to excruciating pain, organ damage, prolonged illness, sterility, babies with birth defects, even death. The economics of this health problem are staggering. The cost of treatment far exceeds $2 billion a year—and this does not include the accelerating cost of treating AIDS patients.

AIDS, caused by the human immunodeficiency virus (HIV), has become the second leading cause of death among males ages 25–44, and its incidence is steadily increasing among women and teenagers of both genders. In many developing countries, this modern plague threatens to overwhelm health care delivery systems and to unravel decades of economic progress. In the United States, the rate of reported new HIV infections is greatest among women and teenagers.

Figure 15.1 There is goodness in togetherness, but for sexually active people there is also a very real threat of sexually transmitted diseases. As with other types of diseases, an ounce of prevention is worth a pound of cure—when a cure exists.

Reading the grim statistics—including 130,000 reported new cases each year of syphilis, 700,000 of gonorrhea, and half a million new genital herpes infections—you might conclude that there is no such thing as "safe sex." It's more accurate to say that sexual encounters (Figure 15.1) can sometimes be a source of infection by viruses, bacteria, fungi, and other disease agents. That is one reason why we include this chapter in a text on human biology. Another is that "safer sex" *is* possible. By learning more about the causes and symptoms of sexually transmitted diseases, you will be equipped not only to adopt behaviors that can reduce your personal risk, but you may be able to help educate others as well.

KEY CONCEPTS

1. Most of the agents of sexually transmitted diseases are viruses and bacteria. They are usually transmitted from infected to uninfected people during sexual intercourse.

2. Complications of untreated STDs can result in sterility and birth defects, irreversible damage to organs (including the brain), and death. There are currently no cures for genital herpes and HIV infections.

3. The chances of contracting STDs can be reduced by adopting behaviors that help prevent or significantly reduce the likelihood of exposure to disease agents.

AGENTS OF DISEASE

The human body is continually under siege from pathogens, including viruses, bacteria, fungi, protozoa, and parasitic worms. *Pathogen* means "disease-causing." Pathogens are capable of *infection*— a process in which a pathogen invades and then multiplies in the tissues of a host organism. An infection, in turn, can lead to *disease*—the serious malfunctioning of an organ, tissue, or organ system.

Reported cases of **sexually transmitted diseases** (STDs) have reached epidemic proportions, even in countries with high medical standards. The number of unreported cases may be up to 10 times higher. Most of the pathogens are bacteria and viruses. Usually they are transmitted from infected to uninfected people during sexual contacts. By and large, these pathogens are **parasites**. They must live on or within a host, from which they gain nutrients or other substances essential to their survival and reproduction—at the host's expense.

Viruses

Strictly speaking, a virus is not alive. A **virus** is a tiny, noncellular pathogen with two characteristics. It consists only of a nucleic acid core and a protein coat that sometimes is enclosed in a lipid envelope. And it can replicate only after its genetic material enters a host cell and directs the cellular machinery to synthesize the materials necessary to produce new virus particles. The genetic material of a given virus is either DNA or RNA, but not both. The viral protein coat, or *capsid*, comes in many different forms. Figure 15.2 diagrams the structure of the human immunodeficiency virus, HIV.

Viral Replication When a virus replicates, the viral DNA or RNA is copied repeatedly, viral proteins are synthesized, and many new virus particles are assembled inside a suitable host cell (Figure 15.3). A cell is a potential *host* if a virus can chemically recognize and lock onto certain molecular groups at the cell's surface. This gives each type of virus specificity; a flu virus can attack cells of repiratory epithelium but not liver cells, and a hepatitis virus attacks liver cells but not cells of the airways.

Viral replication proceeds through the following stages:

1.   The virus recognizes and attaches to molecules at the surface of a host cell.

2.   The whole virus or its genetic material alone (DNA or RNA) enters the cell's cytoplasm.

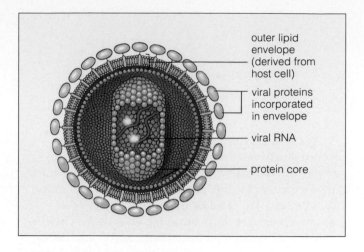

Figure 15.2   Lipid envelope around the capsid of HIV, the causative agent of AIDS.

3.   Information contained in the viral DNA or RNA directs the host cell to replicate viral nucleic acids and synthesize viral enzymes and other proteins.

4.   The viral nucleic acids, enzymes, and capsid proteins are assembled into new virus particles.

5.   The newly formed virus particles are released from the infected cell.

Viruses usually replicate by one of two pathways. In a *lytic pathway*, stages 1 through 4 proceed quickly, and virus particles are released as the cell ruptures (lyses), loses its contents, and dies. In *lysogenic pathways*, the virus does not kill the host cell outright. Instead, the infection enters a period of **latency**, in which viral genes remain inactive inside the host cell.

Retroviruses are RNA viruses that infect animal cells and follow lysogenic pathways of replication. After the viral RNA molecule enters the cytoplasm, a viral enzyme (reverse transcriptase) uses the RNA as a template and synthesizes a DNA "transcript." The transcript, not the RNA itself, integrates into the host DNA. This is what has happened in people infected with HIV.

Viruses and Disease Pathogenic viruses cause disease in several ways. A few viruses trigger cancer when they interact with the DNA in a host cell's chromosomes. More commonly, disease symptoms develop as a result of the death or malfunctioning of infected cells. Paradoxically, immune system responses to viral infection can also lead to symptoms we associate with disease. Fever and fatigue are just two examples. Another is inflammation. The inflammatory response can cause serious damage, especially if a major organ or organ system is affected. Table 15.1 lists a few types of animal viruses that are responsible for diseases as varied as warts, genital herpes, and the common cold, as well as AIDS.

As discussed on page 86, antibiotics are useless against viruses. Such drugs act by disrupting living *cells*—they cannot "kill" a virus particle, which has no metabolic pathways to disrupt.

Bacteria

Unlike viruses, bacteria are cellular. Bacteria differ from all other kinds of organisms in that they are *prokaryotes*. That is, they have no nucleus or other membrane-bound organelles. Their metabolic reactions take place at the plasma membrane, the bacterial equivalent of organelle membranes. Proteins are synthesized at ribosomes, which are distributed through the cytoplasm or attached to the plasma membrane. (All living organisms other than bacteria are *eukaryotes*—their cells have a nucleus and other membrane-bound organelles.)

Nearly all bacteria have a *cell wall* (Figure 15.4). In a large number of bacterial species the wall is composed of a tough mesh of *peptidoglycan*. This is a molecule in which complex polysaccharide strands are cross-linked to one another by peptides. Peptidoglycan makes the wall strong and semirigid, and it helps maintain the bacterial cell in one of three common shapes: a spherical cell called a

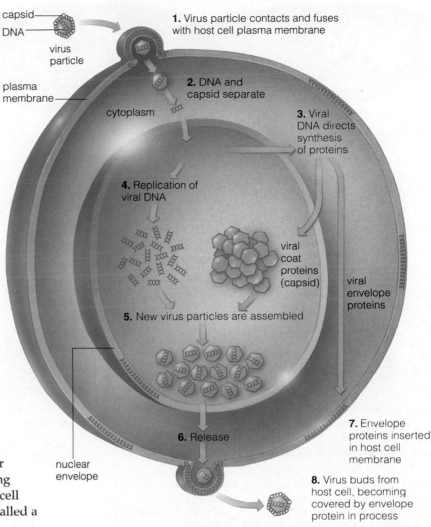

1. Virus particle contacts and fuses with host cell plasma membrane

2. DNA and capsid separate

3. Viral DNA directs synthesis of proteins

4. Replication of viral DNA

5. New virus particles are assembled

6. Release

7. Envelope proteins inserted in host cell membrane

8. Virus buds from host cell, becoming covered by envelope protein in process

Figure 15.3 Replication of an enveloped animal virus. In this case, replication occurs in the nucleus; in some viruses it takes place in the cytoplasm. Also, some other viruses acquire their envelope as they emerge from the nuclear envelope.

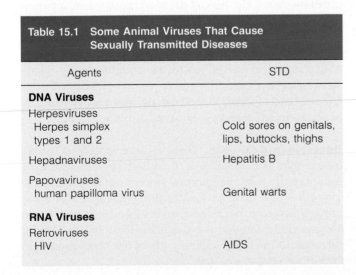

Table 15.1 Some Animal Viruses That Cause Sexually Transmitted Diseases

Agents	STD
DNA Viruses	
Herpesviruses Herpes simplex types 1 and 2	Cold sores on genitals, lips, buttocks, thighs
Hepadnaviruses	Hepatitis B
Papovaviruses human papilloma virus	Genital warts
RNA Viruses	
Retroviruses HIV	AIDS

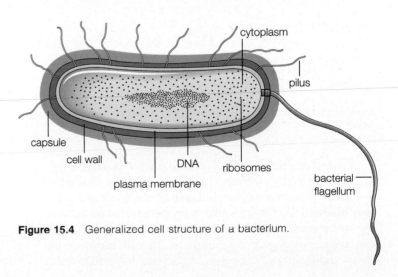

Figure 15.4 Generalized cell structure of a bacterium.

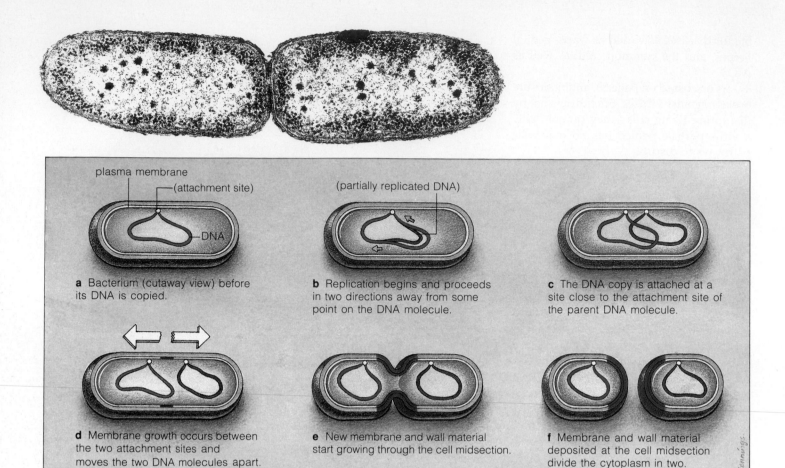

plasma membrane
(attachment site)
(partially replicated DNA)
DNA

a Bacterium (cutaway view) before its DNA is copied.

b Replication begins and proceeds in two directions away from some point on the DNA molecule.

c The DNA copy is attached at a site close to the attachment site of the parent DNA molecule.

d Membrane growth occurs between the two attachment sites and moves the two DNA molecules apart.

e New membrane and wall material start growing through the cell midsection.

f Membrane and wall material deposited at the cell midsection divide the cytoplasm in two.

Figure 15.5 Bacterial reproduction by binary fission, a cell division mechanism. The micrograph shows the bacterium *Pseudomonas* after new membrane and wall material divided the cytoplasm of the parent cell.

coccus, a rodlike *bacillus*, or a helical *spirillum* (or *spirochete*):

coccus (plural, cocci);
spherical

rod or bacillus (plural, bacilli);
cylindrical

spirillum (plural, spirilla);
helical

Some bacterial species have filamentous structures anchored to the cell wall and plasma membrane. One type is a *bacterial flagellum* (plural: flagella), a rather rigid protein filament that rotates like a propeller and so moves the cell through its fluid surroundings. Bacterial flagella can appear at one or both ends of the cell, alone or in small clusters. Some species of bacteria have filaments called *pili* that help the cell adhere to a variety of surfaces. For example, the bacterium that causes gonorrhea uses short pili to attach to mucous membranes.

Bacterial Reproduction Bacteria reproduce by way of *binary fission*. By this cell division mechanism, a parent cell divides into two genetically identical daughter cells following the replication of its DNA. Bacterial cells have only a single chromosome—a circular DNA molecule—to replicate and parcel out to the daughter cells. Bacterial replication generally occurs rapidly; under optimal circumstances, some bacteria can divide every 20 minutes or so. Figure 15.5 shows the steps of binary fission.

Bacteria and Disease Disease-causing bacteria secrete toxins that poison cells. If invading bacteria multiply more rapidly than the immune system can respond, large quantities of a toxin may enter the bloodstream or other tissues. Bacteria that release toxins into the bloodstream are especially dangerous because the toxin moves rapidly throughout the body and may affect the nervous system and other vital organ systems.

Viruses and bacteria cause most types of sexually transmitted diseases.

Viruses usurp the metabolic machinery of the host cell to produce new virus particles; some types of virus can become latent within the host. Bacteria are prokaryotes and replicate by binary fission. They too are capable of multiplying rapidly within a host's body.

With these general features of viruses and bacteria as background, we turn now to a survey of the major sexually transmitted diseases of humans. We will focus on a rogue's gallery of the most common STDs, most of which are caused by viruses or bacteria. We begin with the most troubling STD of all, AIDS.

HUMAN IMMUNODEFICIENCY VIRUS AND AIDS

AIDS is a constellation of disorders that follow infection by the **human immunodeficiency virus**, HIV (Figure 15.6). The virus cripples the immune system and leaves the body susceptible to infections and some otherwise rare forms of cancer. Currently there is no vaccine against this virus, and there is no cure for people already infected. There is a great deal about HIV and AIDS that we still do not understand. The word *hypothesis* crops up frequently in the following paragraphs. At this writing, after more than a decade of intensive research, researchers still seek answers to questions as basic as exactly how HIV breaks down the human immune response.

Physicians diagnose AIDS if a patient has a severely depressed immune system, tests positive for HIV, and has one or more of 26 "indicator diseases," including types of pneumonia, cancer, recurrent yeast infections, and drug-resistant tuberculosis.

A Global Perspective on HIV and AIDS

The World Health Organization (WHO) recently estimated that, worldwide, more than two million people, mostly adults and teenagers, have already died from AIDS. More than 13 million have become infected with HIV since the global epidemic started. Those numbers are increasing, and no one can say how many will have died or been infected by 2000. WHO estimates that 40 million people could be infected with HIV by the turn of the century. More pessimistic researchers place the figure at 110 million or more.

Roughly 70 percent of all HIV victims—around 9 million—live in Africa, where HIV is thought to have originated. Another 1.5 million are in the United States and

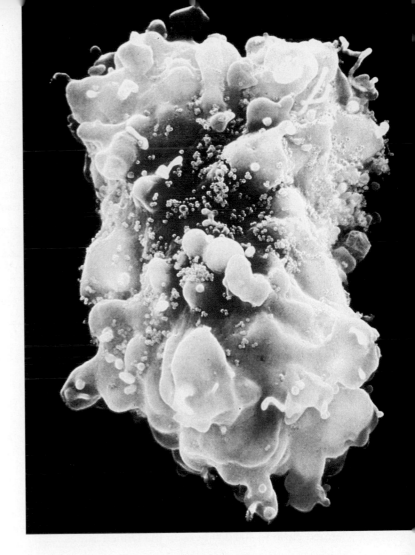

Figure 15.6 In this color-enhanced micrograph, the blue "balls" are spherical HIV particles spread over the surface of a helper T cell. The image is highly magnified. HIV particles are only one one-thousandth of a millimeter wide—a thousand of them could line up across the period at the end of this sentence.

Canada. At this writing, about 550,000 people are infected in Europe and Central Asia; 1.5 million in Latin America; at least 1.5 million in Asia, India, and Australia; and the remainder in other regions (Figure 15.7). The rate of new infections is growing especially rapidly in parts of South and Southeast Asia.

The devastating impacts of HIV infection in some African countries suggests what may be in store elsewhere unless prevention measures can significantly slow the spread of the virus. In hard-hit African cities, officials estimate that as much as *one-third* of the adult population is infected. These infected adults are parents, wage-earners, providers of services, or fill other vital roles in society. From what we know so far, at least 99 percent will eventually become seriously ill with AIDS and die—with horrible human, social, and economic costs. In the United

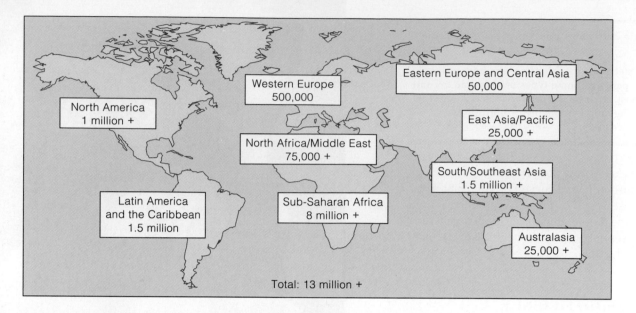

Figure 15.7 WHO estimates of the world distribution of cumulative HIV infections in adults, as of mid-1993. The rate of new infections is increasing most rapidly in Latin America and South/Southeast Asia (including India).

States, a total of 315,000 AIDS cases had been reported through mid-1993 (Figure 15.8). By 1995, AIDS will have claimed close to 400,000 lives in the United States.

A Brief History of HIV

Where did HIV come from? One prominent hypothesis is that HIV is a monkey or chimpanzee virus that "jumped species" at some point and began infecting humans. Genetic studies show that the two main classes of HIV, designated HIV-1 and HIV-2, are remarkably similar to various simian immunodeficiency viruses (SIV). SIVs infect nonhuman primates such as monkeys and chimpanzees; laboratory accidents have shown that they can also infect humans. In Africa, humans have long hunted and handled monkeys and other primates. In the process, they may have become infected with SIV, which then evolved into HIV.

Evidence of HIV infection has been uncovered in preserved blood samples that were taken in 1959 in the African nation of Zaire. Whatever its origin, some researchers believe that HIV has been around in human populations for some time. HIV readily mutates (changes genetically). One scenario is that the virus may have existed in a relatively harmless form in Central Africa for decades or even centuries, until more aggressive strains began to evolve. AIDS-type illnesses were first recorded in various countries in the 1970s and early 1980s. HIV

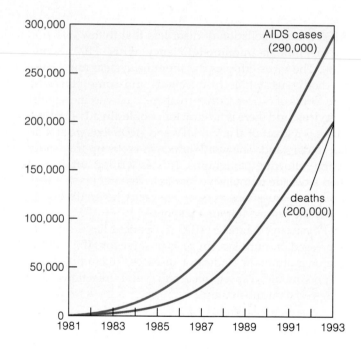

Figure 15.8 Cumulative number of AIDS cases and deaths in the United States from 1981 to the end of 1991. Even with massive public education programs, the curve of new infections has continued to rise, although not as steeply.

itself was finally identified in 1984. Worldwide, most AIDS patients are infected with HIV-1, although HIV-2 is more prevalent in parts of Africa and is also becoming common in India.

Within the two broad HIV classes (1 and 2), there are many genetic strains of the virus. Researchers at the Centers for Disease Control and Prevention in Atlanta have

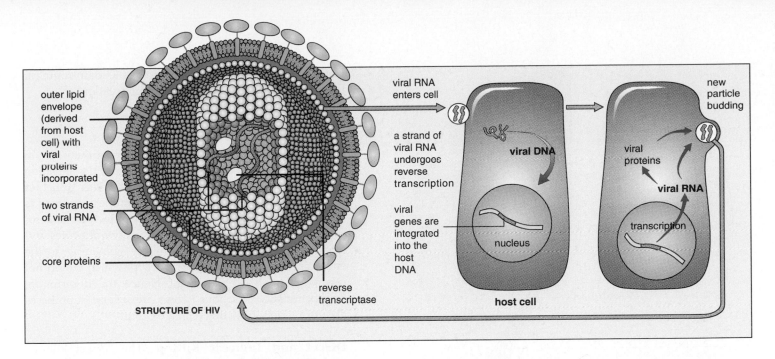

outer lipid envelope (derived from host cell) with viral proteins incorporated

two strands of viral RNA

core proteins

STRUCTURE OF HIV

reverse transcriptase

viral RNA enters cell

a strand of viral RNA undergoes reverse transcription

viral genes are integrated into the host DNA

viral DNA

nucleus

host cell

new particle budding

viral proteins

viral RNA

transcription

Figure 15.9 Infection cycle of HIV, a retrovirus. For clarity, the drawing shows only a single HIV budding from the host cell. In fact, many particles will be formed and will exit from each infected cell.

found that the different strains seem to fall into two distinct groups. Some seem best adapted to infecting cells of the immune system, whereas others may be more effective at entering cells in the mucous membranes lining reproductive structures such as the vagina. This second group is most common in Africa and parts of Asia, which could help explain why heterosexuals make up the majority of AIDS cases in those places. At present, in the United States and Europe most AIDS cases are caused by the first type.

HIV Structure and Replication

HIV is a retrovirus and so has genetic material of ribonucleic acid, RNA, rather than of deoxyribonucleic acid, DNA (page 410). A protein core surrounds the RNA and several copies of an enzyme called reverse transcriptase. The core itself is wrapped in a lipid envelope derived from the plasma membrane of a host helper T cell. Once inside a host cell, the enzyme uses the viral RNA as a template for making DNA. This DNA then becomes integrated into a host chromosome, the structure into which the host's genetic material is packaged (page 414). At this point, the integrated viral DNA is termed a *provirus*, and it typically remains latent in the host cell for a time. Eventually, however, the provirus's genetic instructions for making new virus particles are read out. A process called *transcription* rewrites the genetic message in DNA as RNA, and these RNA instructions then are "translated" into protein (Chapter 19). Figure 15.9 summarizes how

this process generates new HIV particles in an infected cell. We will return to this topic shortly to consider efforts by researchers to develop drugs that can chemically disrupt HIV replication—and so prevent HIV from reproducing once it enters a host.

How HIV Is Transmitted

HIV is transmitted when bodily fluids, especially blood and semen, of an infected person enter another person's tissues. The virus can enter the body through *any* kind of cut or abrasion, including on the penis, in the vagina or rectum, or in mucous membranes in the mouth. HIV-infected blood also can be present on toothbrushes and razors; on needles used to inject drugs intravenously, pierce ears, do acupuncture, or create tattoos; and on unsterilized dental or surgical equipment.

Before screening for HIV was implemented in 1985, contaminated blood supplies accounted for some cases among hemophiliacs and surgery patients. HIV also can be transmitted from infected mothers to their infants during pregnancy, birth, and breast-feeding. Cases of HIV transmission by way of donated tissues used for organ grafting have also been identified. In several countries, HIV has spread through reuse of unsterile needles by health care providers. There is no evidence that casual contact, insect bites, hugging, nonsexual touching, or sharing food can spread the virus.

The virus has been isolated from blood, semen, vaginal secretions, saliva, tears, breast milk, amniotic fluid, cerebrospinal fluid, and even urine. However, apparently only infected blood, semen, vaginal secretions, and breast milk contain HIV in concentrations that are high enough

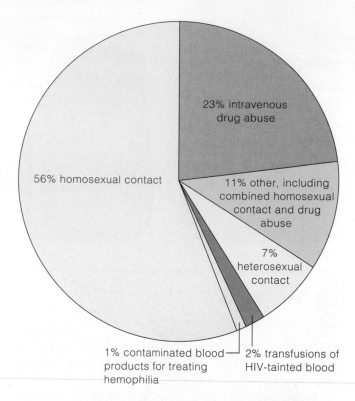

23% intravenous
drug abuse

56% homosexual contact

11% other, including
combined homosexual
contact and drug
abuse

7%
heterosexual
contact

1% contaminated blood
products for treating
hemophilia

2% transfusions of
HIV-tainted blood

Figure 15.10 Reported AIDS cases in the United States, by mode of transmission. These numbers are as of mid-1993. They cover individuals over the age of 13 but do not include the roughly 5,000 cases of young children with AIDS. Of those, 87 percent were infected when they were exposed to the mother's HIV-infected blood during gestation.

for successful transmission. People who have other sexually transmitted diseases—such as syphilis, herpes, and chancroid (page 348)—are at greater risk of HIV infection because they may have sores or other lesions that afford the virus easy entry into the body. HIV generally cannot survive for more than about one or two hours outside of an infected cell. Virus particles on needles and other objects are readily destroyed by disinfectants, including household bleach.

In the United States and other developed countries, HIV spread at first within three groups: male homosexuals, hemophiliacs and others who received HIV-tainted blood products, and intravenous drug abusers who shared needles that were contaminated with infected blood. As many as two-thirds of all drug abusers may now be infected with the virus. The incidence of HIV infection also is increasing among heterosexual males and females, including teenagers, in the general population (Figure 15.10). Many of these cases are due to intravenous drug abuse, but others are the result of heterosexual con-

tact. Currently, heterosexual contact is the main route for the spread of HIV through African populations.

HIV and the Immune System

HIV causes AIDS in humans by breaking down certain cells of the immune system. Over time, HIV infection depletes a person's supply of CD4 lymphocytes, or helper T cells, until the ability to mount immune responses is severely compromised. Other targets include macrophages and brain cells. The assault on lymphocytes and macrophages makes the body highly vulnerable to the entire spectrum of pathogens, many of which would not otherwise be life-threatening. (Hence the description, "opportunistic" infections.) There are several hypotheses about precisely how this damage occurs.

Direct and Indirect Killing The "direct killing" hypothesis holds that CD4 lymphocytes die after they are infected and the virus replicates within them. In laboratory tests, a viral protein (called gp160; gp stands for glycoprotein) binds to newly forming molecules of the cell's own CD4 receptor; this complex apparently fuses with various cell organelles and the cell dies.

Researchers are also intrigued by a viral protein called gp120, which is produced when viral enzymes within an infected cell cut gp160 into pieces. Molecules of gp120 make up the protein core of newly forming virus particles, and some also bud through the infected cell's plasma membrane. Like gp160, the gp120 molecules have a strong tendency to bind with CD4 receptors of helper T cells. Situated in an infected cell's plasma membrane, gp120 molecules apparently attach to the CD4 receptors of nearby *uninfected* T cells. In this way, many helper T cells become fused in large clumps, and none of the cells can take part in an immune response.

There is also evidence for "indirect killing" of helper T cells. Several experiments have found that apparently healthy helper T cells from HIV-infected people self-destruct (a phenomenon called *apoptosis*) when the cells come into contact with an antigen that activates them. Other research suggests that infection somehow marks uninfected helper T cells as "foreign," so that they are attacked and destroyed by the immune system's cytotoxic T cells (page 174).

HIV Infection

Soon after a person is infected, HIV begins to replicate and circulate in the bloodstream. At this stage, many people suffer flulike symptoms, which then disappear. This vanishing act is only superficial, however. Cells in the lymph nodes, spleen, and other organs of the lymphatic system remove HIV from the blood. In lymphoid tissues, HIV actively replicates, and more and more inactive helper T cells in those tissues become infected. Although HIV is far from "latent" during this period, it may take up to three years before blood tests can detect antibodies to various HIV proteins. The antibodies usually appear after about six months. When they do appear, they do not control the infection.

Eventually infected lymphocytes move into the bloodstream and may become activated when the body mounts an immune response. When that happens, the infected cell's genetic machinery—which now includes viral genes—is activated and new viruses are manufactured. They bud from the plasma membrane of the host helper T cell or are released when the membrane ruptures (Figure 15.11). With time, more and more T cells are destroyed or disabled.

As the population of functional helper T cells becomes seriously depleted, the person may begin to lose weight and experience joint and muscle pain, fatigue and malaise, nausea, bed-drenching night sweats, enlarged lymph nodes, various minor infections, and other symptoms. Full-blown AIDS often is heralded by specific indicator diseases, which begin to appear when a person has fewer than 500 CD4 T cells per milliliter of blood. These include opportunistic infections, such as a pneumonia caused by a protozoan (*Pneumocystis carinii*); a rapidly lethal form of tuberculosis; and persistent yeast infections of the mouth, throat, rectum, or vagina. Blue-violet or brown-colored spots on the legs and elsewhere are signs of *Kaposi's sarcoma*. This deadly form of cancer strikes males only. It affects blood vessels in the skin and some internal organs and is rare in individuals who are not HIV-infected.

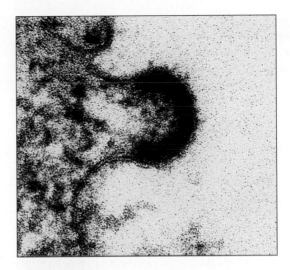

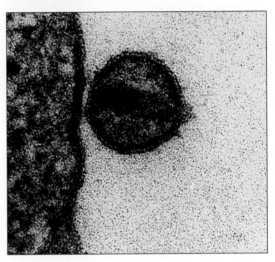

Figure 15.11   An HIV virus particle budding from the host cell's plasma membrane.

POSSIBILITIES FOR AIDS TREATMENT

Preventing HIV Replication

Realizing that HIV replication is a multistep process, researchers are seeking ways to disrupt one or more vital steps and so prevent HIV from reproducing. For example, it might be possible to chemically block the action of HIV protease, the enzyme that cuts gp160 into smaller fragments used in assembling new virus particles. HIV genes also include instructions for proteins such as *tat* and *rev*, which are synthesized early in replication and help regulate subsequent replication steps. Efforts are underway to develop anti-*rev* and anti-*tat* drugs that could knock the proteins out and so halt reproduction of the virus. The search also is proceeding for compounds that might disrupt the ability of HIV to bind to receptors by which it gains entry to cells.

Some AIDS patients are already receiving one or more of the drugs zidovudine (AZT), ddI (dideoxyinosine), and ddC (zalcitabine or dideoxycytidine). These compounds block the action of reverse transcriptase, so HIV particles that have not yet become integrated into a host chromosome cannot replicate. Other compounds that attack reverse transcriptase in different ways are also being tested. Anti-HIV drugs developed thus far appear to be only modestly helpful. For instance, AZT seems to benefit patients for only about one year.

Vaccines: Problems and Prospects

Modern medicine has developed vaccines to protect human populations from several viral diseases. Vaccination has eradicated smallpox, and vaccines against measles, mumps, and polio viruses are highly effective. Yet for a variety of reasons that we can only touch on here, it is much more difficult to develop a vaccine against HIV.

As Chapter 7 describes, a vaccine can confer long-term protection by stimulating the immune system to make antibodies to a viral antigen—a protein marker on the surface of the invading virus (page 174). However, a key problem in developing a vaccine against HIV is that when the virus integrates into a host cell's DNA, it is out of reach of antibodies, which do not enter cells. Even if researchers should discover a way to deliver antibodies into a cell, antibodies recognize protein markers—not the DNA of a provirus. Moreover, because HIV mutates so rapidly, it can exist in many different genetic forms, even in the same person. Because each form of the virus bears a slightly different marker, a vaccine would have to stimulate many different antibodies.

With these and other obstacles in mind, researchers are using various strategies in designing experimental vaccines. Many efforts rely on genetic engineering, which we explore in detail in Chapter 21. One approach is to base vaccines on HIV particles from which key genetic information has been removed. In theory, the altered virus would still be able to stimulate the immune system to produce protective antibodies but could not replicate or cause disease. Although this approach raises important safety questions, such vaccines may well become candidates for testing because of the rapid spread of HIV. Several dozen different vaccines of various types have already been tested on small groups of HIV-infected people to find out if vaccination can bolster their weakened immune systems. Larger-scale trials are being planned or implemented in countries around the globe.

Such trials are essential to the ultimate development of a safe, effective vaccine but entail many ethical problems. For example, if a vaccine does its job of stimulating anti-HIV antibodies, those who participate in a trial will thereafter test positive for the virus—and could become victims of discrimination. This and other ethical dilemmas will be unavoidable as the search for a vaccine moves forward.

AIDS Prevention

While the search for drug treatments and vaccines continues, it is crucial that every sexually active person be aware of the danger of HIV infection and adopt behaviors that minimize her or his risk of exposure. These include limiting the number of partners, using latex condoms, and avoiding any exchange of body fluids if there is a chance either person may be infected. Proper use of high-quality latex condoms, together with the spermicide nonoxynol-9, helps prevent infection. Caressing carries no risk—if there are no body fluids exchanged and or lesions or cuts through which the virus can enter the body.

To stop the spread of HIV and millions of deaths from AIDS, individuals must become convinced that any uninfected person who is sexually active and not in a mutually monogamous relationship with an uninfected partner must follow "safer sex" guidelines (see *Focus on Wellness*).

In addition, public health officials in hard-hit cities are promoting "clean needle" strategies, such as supplying IV-drug abusers with sterile needles. This is not support for drug abuse, but merely recognition that society has a huge stake in halting the spread of HIV. Free or low-cost, confidential testing for AIDS is available through public health facilities and many physicians' offices. There are now blood tests for HIV infection that take as little as 10 minutes, and as a society we may benefit most by ensuring that people who test positive for the virus have access to nonjudgmental counseling and appropriate treatments.

In the United States and some European countries, the increase in reported new HIV infections appears to be slowing. This positive trend seems to be evidence that people can and will change their behavior when it is clear that their lives may hang in the balance. At the same time, in parts of Asia and Latin America the incidence of new cases has begun to rise. Everywhere, a distressingly large percentage of people, especially younger people, say they do not practice safer sex, despite being well-educated about HIV and AIDS. The attitude seems to be, "It won't happen to me." As we are learning with HIV, however, ignoring the threat can be lethal. With the mounting medical, social, and economic consequences of AIDS, no one can afford to be complacent. Now and in the future, HIV and AIDS are everyone's problem.

OTHER SEXUALLY TRANSMITTED DISEASES

Gonorrhea

Unlike AIDS, **gonorrhea** can be cured by prompt diagnosis and treatment. Gonorrhea is caused by *Neisseria gonorrhoeae* (Figure 15.12). This bacterium (also called *gonococcus*) can infect epithelial cells of the genital tract, the rectum, eye membranes, and the throat. Since the mid-1980s its incidence has been decreasing somewhat, although among the urban poor gonorrhea still ranks near the top among communicable diseases in the United States. Each year there are about 700,000 new cases reported, and there may be anywhere from 3 million to 10 million unreported cases. Part of the problem is that the initial stages of the disease can be so uneventful that a carrier may be unaware of being infected.

In early stages of infection, males are more likely than females to detect the disease because the symptoms are more readily apparent. Within a week, yellow pus is dis-

Focus on Wellness

Prevention: Protect Yourself from STDs

The old saying "An ounce of prevention is worth a pound of cure" has never been truer than with sexually transmitted diseases. The only people who are not at risk of STDs are those who are celibate (never have sex) or who are in a long-term, mutually monogamous relationship in which no disease is present. Otherwise, health care professionals recommend the following guidelines to help you minimize your risk of acquiring or spreading an STD.

1. Use a latex condom, or insist that your partner use one. Doing so will significantly reduce your risk of exposure to the pathogens that cause HIV, gonorrhea, herpes, and other diseases. With the condom, use a spermicide that contains nonoxynol-9, which may help kill virus particles. Condoms are now available for males and females.

2. Limit yourself to one partner who has sex only with you. Promiscuous sex can endanger not only your health, but your life.

3. Get to know a prospective partner—before you have sex. Insist on a friendly but frank discussion of your sexual histories, including any previous exposure to an STD. The safest policy is to assume you are at risk and to take appropriate precautions.

4. If you decide to become sexually intimate, be alert to the presence of sores, scabs, a discharge, or any other sign of possible trouble in the genital area. Don't be embarrassed to look—much is at stake.

5. Keep yourself and your immune system healthy by getting sufficient rest, eating properly, and learning strategies for coping with stress.

6. If for no other reason, avoid alcohol and drug abuse. Studies show that both are correlated with unsafe sex practices.

Figure 15.12 *Neisseria gonorrhoeae*, or gonococcus, a bacterium that typically is seen as paired cells, as shown here. The threadlike pili evident in this electron micrograph help the bacterium attach to its host, upon which it bestows gonorrhea.

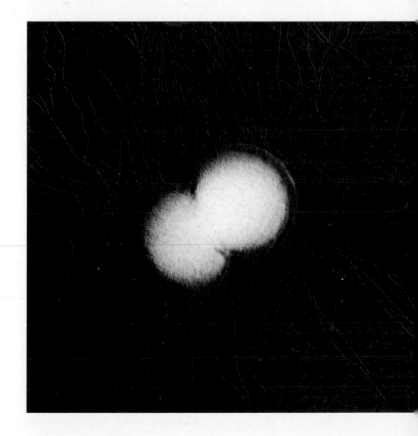

charged from the penis. Urination becomes more frequent and painful. Roughly 80 percent of females diagnosed with gonorrhea are described as "asymptomatic." Females may or may not experience a burning sensation while urinating. They may or may not have a slight vaginal discharge; even if they do, the discharge may not be perceived as abnormal. In the absence of worrisome symptoms, gonorrhea often goes untreated. The bacteria may spread into the oviducts. Eventually, there may be violent cramps, fever, vomiting, and sterility due to scarring and blocking of the oviducts (from pelvic inflammatory disease, page 350). A male can become sterile when untreated gonorrhea leads to inflammation of the testicles.

Treatment Antibiotics can kill the gonococcus and prevent complications of gonorrhea. Penicillin was once the most commonly used drug treatment. Unfortunately, since the 1970s a variety of antibiotic-resistant strains of

Choices: Biology and Society

Doing Unto Others: Personal Responsibility and STDs

Humans have probably been afflicted with sexually transmitted diseases since prehistoric times. We know for sure that syphilis and gonorrhea have been around for hundreds, if not thousands, of years. And for all those years people have probably felt worried, ashamed, guilty, embarrassed, or angry—or all of the above—about getting an STD. Admitting to a partner that you have an STD—and that *both* of you may require treatment—can be difficult. Yet most people recognize that they have an ethical obligation to inform partners when an STD enters the picture. In addition to simply being honest, dealing realistically with an infection involves the following:

1. Learn about and be alert for symptoms of STDs. If you have reason to think you have been exposed, abstain from sexual contact until a medical checkup rules out any problems. Self-treatment won't help. See a doctor or visit a clinic.

2. Take all prescribed medication. *Do not* share medication with a partner. Unless both of you take a full course of medication, chances of reinfection will be great. Your partner may need to be treated even if he or she does not show symptoms.

3. Avoid sexual activity until medical tests confirm that you are free from infection.

---

gonococcus have appeared. As a result, many physicians now order testing to determine the strain responsible for a particular patient's illness and then treat the infection with an appropriate antibiotic.

Many people wrongly believe that once cured of gonorrhea, they are "immune" to reinfection. Multiple reinfections can and do occur. People who have multiple sexual partners can use condoms to help avoid becoming infected. Condom use can also prevent a person who has gonorrhea from infecting others, one of the issues we explore in this chapter's *Choices* essay.

Syphilis

Syphilis is caused by a motile, spirochete bacterium, *Treponema pallidum*. Each year in the United States about 30,000 new cases are reported, although the actual incidence may be much higher. Since 1984, the incidence of syphilis has nearly doubled among females between ages 15 and 24. Infected women who become pregnant typically have miscarriages, stillbirths, or sickly and syphilitic infants.

The bacterium is transmitted by sexual contact. After it has penetrated exposed tissues, it produces a type of ulcer called a chancre ("shanker," Figure 15.13a) that teems with treponeme offspring. Usually the chancre is flat rather than bumpy, and it is not painful. It becomes visible between one and eight weeks following infection and is a symptom of the *primary stage* of syphilis. Using a technique called immunofluorescence, treponemes (Figure 15.13d) can be identified—and syphilis diag-

nosed—in a cell sample taken from a chancre. By then, however, bacterial cells have already moved into the lymph vascular system and bloodstream.

The *secondary stage* of syphilis begins about one to two months following the appearance of the chancre. Lesions can develop in mucous membranes, eyes, bones, and the central nervous system. A blotchy rash breaks out over much of the body (Figure 15.13b). Afterward, the infection enters a latent stage, which can last many years. Meanwhile, the disease produces no outward symptoms and can be detected only by laboratory tests.

Usually, the *tertiary stage* of syphilis begins from 5 to 20 years after infection. Lesions may develop in the skin (Figure 15.13c) and internal organs, including the liver, bones, and aorta. Scars form; the walls of the aorta can weaken. Treponemes also damage the brain and spinal cord in ways that lead to various forms of insanity and paralysis.

Treatment Penicillin effectively cures syphilis during the early stages. Even so, potential health damage is so serious that no one who is sexually active should take the disease lightly.

Chancroid

The bacterium *Haemophilis ducreyi* causes an STD called **chancroid**. Chancroid was once mainly a tropical disease, but due to increased international travel it is becoming more common in the United States and Europe. Chan-

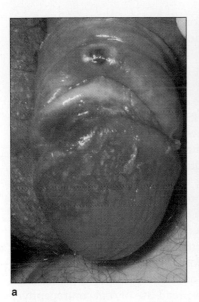
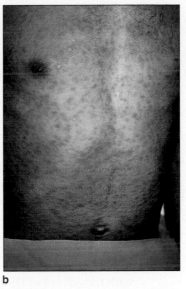
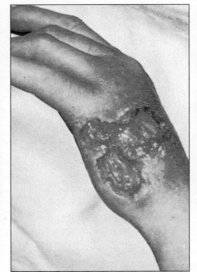
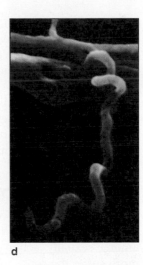

a b c

d

Figure 15.13 (**a**) Painless ulcer called a chancre ("shanker"), a sign of the first stage of syphilis. The sore appears anytime from about nine days to three months after infection, on the genitals or near the anus or the mouth. It literally teems with infective bacteria and remains for 6–10 weeks if the infection is not treated. (**b**) Rash typical of secondary syphilis. (**c**) Arm tissue showing the white lesions, called *gummas*, that occur during the tertiary stage of syphilis. (**d**) *Treponema pallidum*, the spirochete bacterium that causes syphilis.

croid produces soft, painful ulcers on the external genitals. Untreated, the infection can spread to lymph nodes in the pelvic area and destroy tissues there. Chancroid (as well as some other STDs) has been implicated in some cases of heterosexual transmission of HIV because the sores provide a site for the virus to enter the bloodstream. It can be treated by antibiotics.

Chlamydial Infections

The bacterium *Chlamydia trachomatis* is the culprit behind an infection commonly called **chlamydia**, and also is implicated in other sexually transmitted diseases (Figure 15.14). Each year, about 3 million to 10 million Americans are affected, a large proportion of them college students. More than 30 percent of newborns who are treated for eye infections and pneumonia developed those disorders after being infected with *C. trachomatis* during birth.

The bacterium infects cells of the genital and urinary tract. Infected men may have a discharge from the penis and a burning sensation when they urinate. Women may have a vaginal discharge as well as burning and itching sensations. Often, however, there is no outward evidence of infection—yet the bacterium can still be passed on to others.

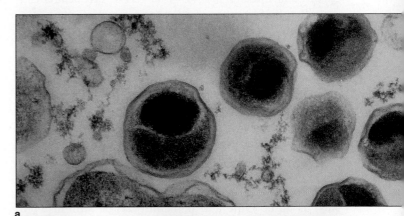

a

b

Figure 15.14 (**a**) The large spheres in this color-enhanced micrograph are *Chlamydia trachomatis*, magnified about 32,000 times. People under age 25 account for a large percentage of chlamydia infections in the United States. (**b**) Students at a workshop on STDs.

Following infection, the bacteria migrate to lymph nodes, which become enlarged and tender. The enlargement can impair lymph drainage and lead to pronounced swelling of the surrounding tissues.

Treatment   Chlamydial infections can be treated with antibiotics such as tetracycline. Although many infections have no long-term complications, *C. trachomatis* is a major cause of pelvic inflammatory disease.

Pelvic Inflammatory Disease

A condition called **pelvic inflammatory disease** (PID) strikes about 1 million women each year, most often among sexually active women under age 25. Millions more suffer the consequences of a past infection—PID is the leading cause of infertility among young women.

PID is one of the serious complications of gonorrhea, chlamydial infections, and other STDs. Physicians estimate that half or more of all cases are sexually transmitted. It also can arise when microorganisms that normally inhabit the vagina ascend into the pelvic region as a result of excessive douching. Most often, the uterus, oviducts, and ovaries are affected. Pain may be so severe that infected women often think they are having an attack of acute appendicitis. The oviducts may become scarred, which in turn can lead to ectopic pregnancy (page 315) as well as to sterility. Affected women can also develop chronic menstrual problems.

Treatment   As soon as PID is diagnosed, antibiotics such as tetracycline and penicillin can be administered. Severe cases require hospitalization and hysterectomy (removal of the uterus). A woman's partner should also be treated, even if the partner has no symptoms.

Nongonococcal Urethritis (NGU)

Males and females both can develop **nongonococcal** ("not caused by gonococcus") **urethritis**, an inflammation of the urethra that is usually simply called NGU. Doctors are now diagnosing more than a million cases each year. And although some medical experts lump NGU with chlamydia because a percentage of cases are caused by the *Chlamydia* bacterium, in other cases laboratory tests reveal the Herpes virus or fail to turn up any clear evidence of a pathogen. Symptoms in males are discomfort during urination and a puslike discharge from the urethra. As with chlamydia, females may have a slight vaginal discharge, some pelvic pain and urinary discomfort (similar to symptoms of *cystitis*), or no symptoms at all.

Although an antibiotic such as tetracycline can usually cure NGU, the disorder can be difficult to treat in cases in which no cause can be established. Relapses are common and complications can arise, including inflam-mation of the prostate or testicles in males and inflammation of the oviducts in women.

Genital Herpes

Infections with Herpes simplex viruses, or HSV, are extremely contagious. HSV is transmitted when any part of a person's body comes into direct contact with active viruses or sores that contain them (Figure 15.15). Mucous membranes of the mouth or genital area are especially susceptible to invasion, as is broken or damaged skin. Transmission seems to require direct contact; the virus does not survive for long outside the human body.

There are an estimated 30 million or more people with **genital herpes** in the United States alone. From 1965 to 1979, the number of reported cases increased by 830 percent. It probably is no accident that this rise coincided with the advent of widespread use of oral contraceptives and changing attitudes toward sexual activity. About 200,000 to 500,000 new cases of genital herpes are now reported annually.

The many strains of sexually transmitted HSV are classified as types 1 and 2. Type 1 strains infect mainly the lips, tongue, mouth, and eyes. Type 2 strains cause most (but not all) genital infections. Often, an infected person has both types. Symptoms most often occur 2–10 days after exposure to the virus, although sometimes they are mild or absent. Usually, small, painful blisters appear on the penis, vulva, cervix, urethra, or anal tissues. The sores can also occur on the buttocks, thighs, or back. During the initial flare-up a person may also have a fever and flulike symptoms for several days. Within three weeks the sores crust over and heal.

After the first sores disappear, sporadic reactivation of the virus can produce new, painful sores at or near the original site of infection. Recurrences may be triggered by sexual intercourse, emotional stress, menstruation, a rise in body temperature, or other infections.

If a pregnant woman has genital herpes, her infant can become infected at birth during a vaginal delivery. Contact with the mother's active lesions can lead to a form of herpes that is often fatal or can lead to mental retardation; lesions arising in the infant's eyes can cause blindness. If a woman has an active herpes infection near or at the time of delivery, her physician will recommend a cesarean section—surgical removal of the baby through an incision made in the mother's abdomen.

Treatment   At present there is no cure for HSV infection. Between flare-ups, the virus remains latent in nervous tissue in the spinal cord. Acyclovir (Zovirax), an antiviral drug that is most effective taken in pill form, inhibits the ability of the virus to synthesize DNA and so inhibits viral reproduction. Acyclovir has helped millions of people who are infected with HSV. It seems to reduce

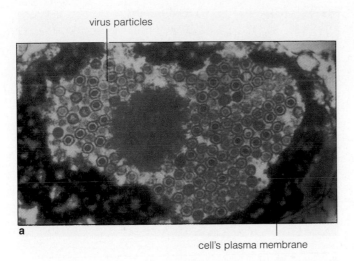

virus particles

cell's plasma membrane

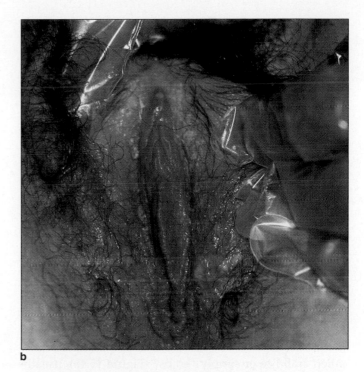

b

Figure 15.15 (**a**) Particles of *Herpes* virus in an infected cell. (**b**) A herpetic lesion on the vulva. Herpes infections below the waist may involve either type 1 or type 2 strains, but most genital herpes infections are type 2.

the shedding of virus particles from sores, and sores are often less painful and heal more rapidly. Over time, recurrences may also become both less frequent and less severe.

Human Papillomavirus

Genital warts are painless growths caused by infection of epithelium by the **human papillomavirus** (HPV). The warts can develop months or years after exposure to the virus and usually they occur in clusters on the penis, the cervix, or around the anus. There is strong evidence that females who have genital warts on the cervix, or who have had a partner who has genital warts, are at increased risk of developing cervical cancer. In fact, HPV is found in 90 percent of women with cervical cancer who are checked for the virus. Any woman who has a history of genital warts should inform her physician, who may want to schedule a *Pap smear*, a painless test for abnormal cell growth on the cervix, at least once a year.

Treatment   Genital warts are usually removed surgically, by freezing (with liquid nitrogen) or burning, or by application of the drug podophyllin, which destroys the wart tissue.

Type B Viral Hepatitis

The hepatitis B virus (HBV) is a DNA virus that, like HIV, is transmitted in blood or body fluids such as saliva, vaginal secretions, and semen. However, HBV is far more contagious than HIV. **Hepatitis B** is an increasingly common STD, striking about 300,000 people in the United States each

year. The virus attacks cells of the liver. A key symptom is jaundice, yellowing of the skin and whites of the eyes as the liver becomes unable to process bilirubin pigments produced by the breakdown of hemoglobin from red blood cells. In about 10 percent of cases the immune system cannot fully eliminate HBV, and the disease becomes chronic. Carriers may have no symptoms for years, yet can easily spread infection to their intimate contacts. Chronic hepatitis can lead to liver cirrhosis or cancer.

A blood test can reveal HBV or antibodies to it (which indicate that a person has been infected). The only treatment is rest. However, people at known risk for contracting the disease (such as health care workers and anyone who needs repeated blood transfusions) can receive a vaccination that provides immunity against the virus. Parents are advised to vaccinate infants for HBV, along with other routine vaccinations for childhood diseases.

The most prevalent sexually transmitted diseases in the United States are chlamydial infections, gonorrhea, genital warts, genital herpes, nongonococcal urethritis, type B hepatitis, chancroid, AIDS, and syphilis.

Pelvic inflammatory disease is also termed an STD because it appears most often as a complication of chlamydial infection and gonorrhea.

Pubic Lice and Scabies

Two tiny animal parasites can be transmitted by way of intimate body contact. Both are arthropods, "jointlegged" relatives of crabs and spiders. *Pubic lice*, also

Figure 15.16 Magnified 120 times, this crab louse looks rather vicious. Crab lice are tiny but are large enough to be visible on the skin, generally as mobile brownish dots. The mite uses claws at the end of its appendages to cling tenaciously to a hair shaft.

Figure 15.17 Flagella propel the protozoan parasite *Trichomonas vaginalis* through its fluid environment in the vagina. Structurally, flagella of animal cells such as this one differ from bacterial flagella, but all serve the function of aiding locomotion.

called crab lice or simply "crabs" (Figure 15.16), usually turn up in the pubic hair, although they can make their way to any hairy spot on the body. They cling tenaciously to individual hairs and attach their small, whitish eggs ("nits") to the base of the hair shaft. Intense itching and irritation result when the parasites bite into the skin and suck blood.

A mite causes *scabies*. As the parasite burrows in and lays eggs, it creates dark, undulating lines in the skin of the pubic area, armpits, around the nipples, or elsewhere. When the eggs hatch, they cause tremendous irritation and extreme itching. Both lice and scabies can be treated by applying antiparasitic drugs to the infested areas.

Vaginitis

The warm, moist environment of the vagina can provide a hospitable environment for a range of organisms, although the vagina's rather acidic pH usually keeps pathogens in check. When certain types of vaginal infections do occur, they can be transmitted to a partner during sexual intercourse. Any event that alters the usual chemical balance of the vagina (such as taking an antibiotic) can trigger overgrowth of *Candida albicans*, a type of yeast (a fungus) that is a normal inhabitant of the vagina. Symptoms of a vaginal yeast infection (candidiasis) include a white "cottage cheesy" discharge and itching and irritation of the vulva. An affected male may notice itching, redness, and flaking skin on the penis. Yeast infections are easily treated by over-the-counter and prescription antifungal medications. Doctors often recommend that both partners be treated to prevent reinfection.

Trichomonas vaginalis, a small protozoan parasite (Figure 15.17), can cause a severe inflammation of the vagi-

nal epithelium. Symptoms include a greenish, frothy, foul-smelling discharge and burning and itching of the vulva. A male who contracts *T. vaginalis* may experience painful urination and a discharge from the penis, the results of an inflamed urethra. When the infection is diagnosed, both partners are treated with the drug metronidazole.

SUMMARY

1. Viruses, bacteria, fungi, protozoa, and parasitic arthropods such as pubic lice and mites are all potential pathogens associated with sexually transmitted diseases. They live as parasites on or in their human host, from which they derive nutrients or other substances essential to their survival and reproduction.

2. HIV infection and the diminished immune response associated with AIDS is the most threatening of all STDs. The HIV virus cripples the immune system primarily by disabling helper T cells (CD4 lymphocytes) and macrophages, leaving the body defenseless against a wide range of pathogens. AIDS currently appears to be always fatal.

3. Viral STDs also include genital herpes, genital warts (HPV), and type B viral hepatitis. Bacteria are responsible for syphilis, gonorrhea, chlamydial infections, and chancroid.

4. Untreated STDs can lead to serious, even life-threatening complications. Scarring of the oviducts due to pelvic inflammatory disease is a major cause of female sterility. NGU can result in inflammation of reproductive

structures in both males and females; infants born to mothers infected with syphilis or genital herpes may suffer effects ranging from blindness to death. Chronic type B viral hepatitis can lead to liver cancer or cirrhosis.

5. Frequently, sexually transmitted infections are interconnected. Half of all people diagnosed with gonorrhea also have a chlamydial infection, and conditions such as genital herpes and syphilis increase the risk of HIV infection during risky sexual encounters.

Review Questions

1. Describe five ways the human immunodeficiency virus (HIV) can be transmitted. *343*

2. List at least four major STDs that are caused by bacteria. What are the major symptoms of each? In addition to AIDS, which STDs are viral in origin? (*throughout chapter*)

3. Pelvic inflammatory disease is a major cause of infertility in young women. Discuss the various STD sources of this disorder. *350*

4. Which STDs pose a danger to a fetus? To an infant during birth? What are those dangers? *343, 349, 350*

Critical Thinking: You Decide (*Key in Appendix IV*)

1. A cleanup worker in the cafe where you work on weekends has been diagnosed as HIV-positive, and a server has come down with type B hepatitis. Some other employees start a petition demanding that both people be required to wear a mask over the mouth and nose, not handle soiled dishes or food, and not use the employee restroom. Both infected employees strenuously object to the plan. You are asked to lead a discussion aimed at resolving the issue, and you decide to prepare a handout giving the scientific basis for making a decision in each case. What does the handout say?

2. You've been dating someone for a while, and now the relationship is getting serious. Everything seems fine, except that your potential partner refuses to consider using a condom if the two of you engage in sex. To allay your fears, the person shows you a lab report indicating a negative HIV blood test. You refuse to forego condom use. Why?

Self-Quiz (*Answers in Appendix III*)

1. The pathogens responsible for sexually transmitted diseases are mostly _____ and _____. A _____ is described as a noncellular infectious agent.

2. The majority of HIV infections occur through _____. In the vast majority of cases, the virus enters the body of a new host in infected _____, _____, _____, or _____.

3. True or false: Once a person is cured of a gonorrhea infection, that person has immunity to subsequent infection.

4. Of the following, which *cannot* be treated effectively with an antibiotic?
   a. gonorrhea   c. syphilis
   b. chancroid   d. type B hepatitis

5. All the following are true of chlamydial infections except:
   a. many victims are young adults
   b. often no outward evidence of infection
   c. often cause sterility in males
   d. are common culprits in PID

6. A woman with a history of STDs and chronic menstrual problems is having difficulty getting pregnant. Her infertility might well be traceable to _____.
   a. PID   c. human papilloma virus
   b. syphilis   d. ARC

7. Type B viral hepatitis resembles HIV in that _____.
   a. it is caused by a retrovirus
   b. the virus is transmitted in blood or body fluids such as semen and vaginal secretions
   c. it strikes about 300,000 people in the United States each year
   d. it causes jaundice

8. Match the following STD concepts:
   _____ NGU
   _____ HPV
   _____ *Trichomonas vaginalis*
   _____ HIV
   _____ genital herpes
   _____ PID
   _____ chancroid

   a. diminished immune response
   b. protozoan parasite
   c. may be caused by either of two types of the responsible virus
   d. may be caused by *Chlamydia* parasite or herpes virus
   e. causes genital warts
   f. implicated in some cases of HIV transmission
   g. a serious complication of gonorrhea, chlamydial infection, and other STDs

Key Terms

chancroid *348*
chlamydia *349*
genital herpes *350*
gonorrhea *346*
hepatitis B *351*
human immunodeficiency virus *341*
human papillomavirus (HPV) *351*

latency *338*
nongonococcal urethritis (NGU) *350*
parasite *338*
pelvic inflammatory disease (PID) *350*
sexually transmitted disease (STD) *338*
syphilis *348*
virus *338*

Readings

Alcamo, I. E. 1993. *AIDS: The Biological Basis*. Dubuque: Wm. C. Brown Publishers.

The Battle Against Infection. 1992. C. B. Clayman, ed. American Medical Association. Pleasantville, N.Y.: Reader's Digest Association, Inc. Clearly written summary of the basic science and medical aspects of sexually transmitted diseases.

Fuerst, M. May 1991. "An STD Primer." *American Health*.

Rathus, S. A. and S. Boughn. 1993. *AIDS: What Every Student Needs to Know*. Fort Worth: Harcourt Brace College Publishers. This 90-page booklet is an invaluable reference guide on HIV and AIDS.

Sperm on the surface of an egg. Only one sperm can successfully fertilize an egg. Their union into a single cell brings together genetic instructions from mother and father, a DNA blueprint that will guide the development and functioning of the new individual.

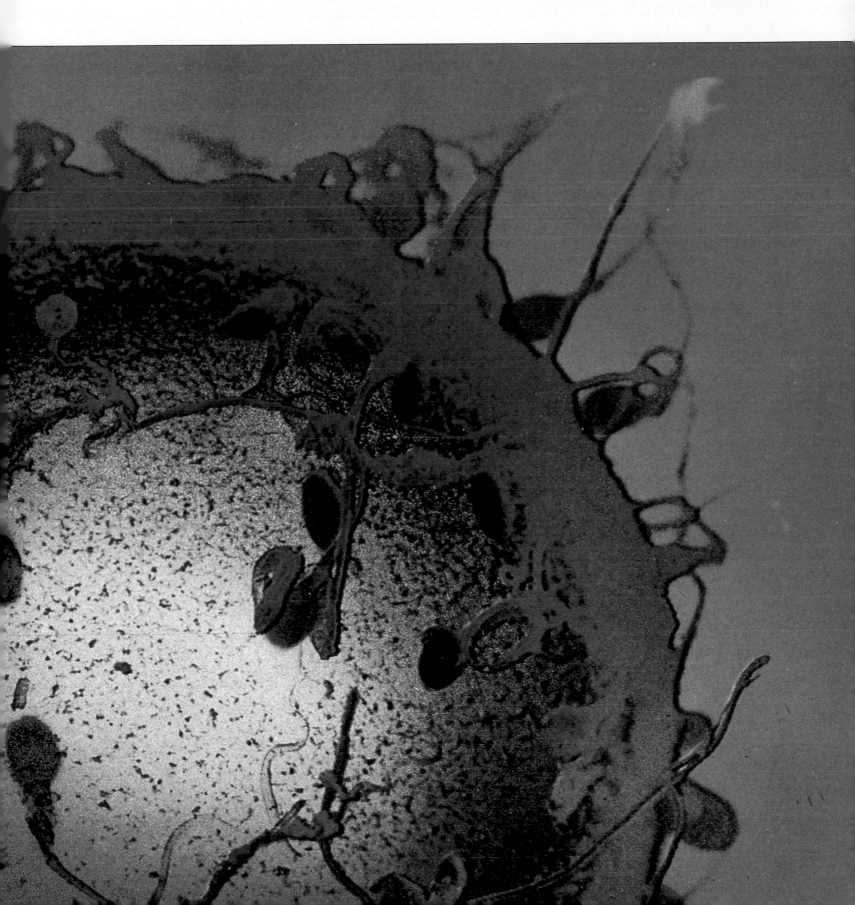

16 CELL REPRODUCTION

Trillions from One

Starting with the fertilized egg in your mother's body, a single cell divided in two, then the two into four, and so on, until billions of cells were growing, developing in specialized ways, and dividing at different times to produce your genetically prescribed body parts (Figure 16.1). Today your body is composed of trillions of cells. Cell divisions still occur within it. Every five days, for example, cell divisions replace the entire lining of your small intestine.

Depending on the end result of cell division, a cell's nucleus divides in one of two ways. One mechanism, mitosis, occurs in all dividing cells except those that give rise to gametes—sperm and eggs. The other mechanism, meiosis, occurs only in the reproductive organs, in gamete-producing stem cells. Mitosis and meiosis clearly resemble each other. But as we'll see, they are fundamentally different in the way they parcel out the bundles of DNA and protein called chromosomes.

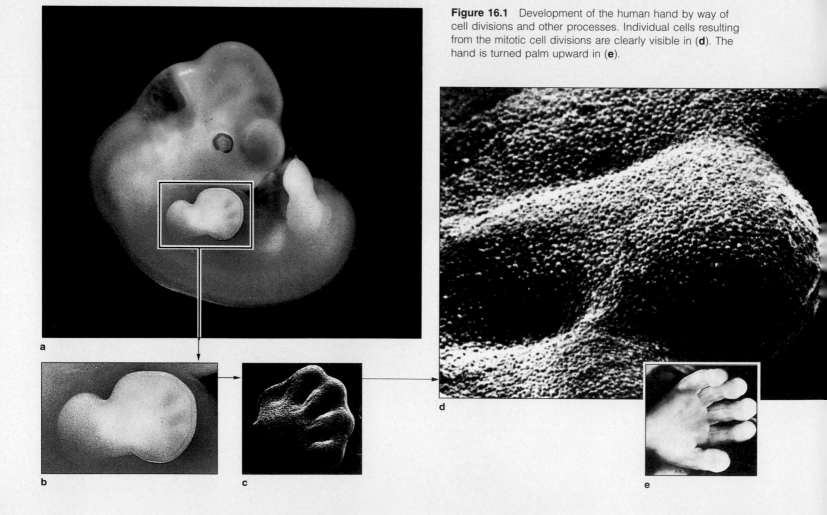

Figure 16.1 Development of the human hand by way of cell divisions and other processes. Individual cells resulting from the mitotic cell divisions are clearly visible in (**d**). The hand is turned palm upward in (**e**).

a

b

c

d

e

Thanks to mitosis, the body grows larger, wounds heal, and many tissues are renewed. It is meiosis, followed by the union of sperm and egg at fertilization, that makes sexual reproduction possible.

Extraordinary microscopic bodies called chromosomes carry the instructions—genes—that determine which traits a person inherits. How are chromosomes and their genes distributed into daughter cells? In this chapter and the next, we consider the answers (and best guesses) to questions about cell reproduction and other aspects of human heredity.

KEY CONCEPTS

1. When a cell divides, each of its two daughter cells must receive the same hereditary instructions (DNA) as the original cell had, and some of its cytoplasm. In humans and other eukaryotes, mitosis sorts out the DNA into two new nuclei. Another mechanism, called cytokinesis, divides the cytoplasm in two. Mitosis is the basis of body growth and tissue repair.

2. DNA molecules are usually packaged into chromosomes. Except for gametes (sperm and eggs), most cells in the human body contain two of each type of chromosome characteristic of our species. That is, they contain two full sets of hereditary information, one set from each parent. The total number of chromosomes in a body cell is called the *diploid* number, and body cells are thus diploid cells.

3. In a diploid cell, the two members of each pair of chromosomes (one from each parent) are *homologous*—they carry genetic information about the same traits.

4. Mitosis divides the nucleus of a cell in a way that ensures that daughter cells receive the same number of chromosomes as were in the parent cell.

5. Meiosis divides the nucleus of a germ cell in a way that *reduces* the diploid number of chromosomes by half. A resulting gamete will thus contain only one full set of chromosomes. This single set of chromosomes, half the diploid number, is called the *haploid* number. The union of two haploid gametes at fertilization restores the diploid number in the new individual.

6. During meiosis, each pair of chromosomes may swap segments and thereby exchange some hereditary instructions. Also, meiosis randomly assigns one chromosome of each pair to gametes. At fertilization, the particular egg and sperm that give rise to a new individual are also determined at random. All three events lead to variations in the traits of offspring.

DIVIDING CELLS: THE BRIDGE BETWEEN GENERATIONS

Overview of Division Mechanisms

In biology, reproduction means producing a new generation of individuals *or* a new generation of cells. Reproduction is part of a *life cycle*, a recurring series of events in which individuals grow, develop, maintain themselves, and reproduce according to instructions encoded in DNA, which they inherit from their parents. Reproduction typically begins with the division of single cells. It follows this basic rule: Each cell of a new generation must receive a duplicate of the parent cell's DNA and enough cytoplasmic machinery to start up its own operation.

DNA contains the genetic instructions for making proteins. Some proteins are structural materials. Many serve as enzymes during the synthesis of carbohydrates, lipids, and other building blocks of the cell, or in crucial reactions such as glycolysis and the Krebs cycle. Unless new cells receive the necessary DNA instructions, they will not grow or function properly.

Also, the cytoplasm of the parent cell already has operating machinery—enzymes, organelles, and so forth. When a daughter cell inherits what might look like a blob of cytoplasm, it really is getting "start-up" machinery for its operation until it has time to use its inherited DNA for growing and developing on its own.

The cells of humans and other eukaryotic organisms divide either by **mitosis** or **meiosis.** Strictly speaking, mitosis and meiosis divide only the genetic material, although we often speak of them as "cell division." Both mechanisms sort out and package DNA molecules into new nuclei for forthcoming daughter cells. The actual splitting of a parent cell into two daughter cells occurs by **cytokinesis**, the division of the cytoplasm. By way of cytokinesis, organelles and various substances in the cytoplasm are randomly distributed to daughter cells.

The body grows by mitotic division of body cells, which are called **somatic cells**. Tissues are repaired by the same mechanism. When you cut yourself with a knife, mitotic cell divisions produce new cells to replace the ones the knife damaged. Mitosis also is the basis for *asexual* reproduction in many plants, animals, and other eukaryotic organisms.

In contrast, meiosis occurs only in **germ cells** set aside for sexual reproduction—the oogonia in ovaries and spermatogonia in testes (Chapter 13). In these cells, recall, meiosis must precede the formation of gametes—that is, eggs and sperm. In order to understand the basic elements of mitosis and meiosis, you must first know more about chromosomes—the structures in the cell nucleus that carry hereditary information.

Some Key Points About Chromosomes

The DNA in your cells comes packaged in chromosomes. A **chromosome** is a single, very long molecule that consists of roughly equal amounts of DNA and protein. There are 46 of these chromosomes in the nucleus of a human body cell, and their physical organization differs at different times in a cell's life. When a cell is dividing, each chromosome is tightly coiled, or *condensed*, into a rodlike shape. Between divisions, the chromosomes are not as tightly coiled and have a threadlike configuration. When stained with a dye and viewed through almost any type of microscope, the uncondensed "threads" are visible as a somewhat jumbled-looking mass. This mass is called *chromatin*. Each chromosome is still threadlike when it is duplicated prior to cell division, and the two duplicate threads of the chromosome remain attached as **sister chromatids**:

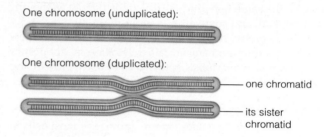

One chromosome (unduplicated):

One chromosome (duplicated):
— one chromatid
— its sister chromatid

Notice how the two chromatids of a duplicated chromosome become constricted at one small region. This region, the **centromere**, has sites where microtubules will attach and help move the chromosome when the nucleus divides:

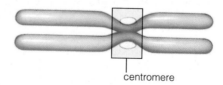

centromere

The preceding sketches are highly simplified. As Figure 16.2 suggests, the centromere's location is different on different chromosomes. Also, the two parallel strands that make up a DNA molecule don't look like a ladder; they are twisted together repeatedly like a spiral staircase (a "double helix") and are then condensed further together with proteins. They are also much longer than shown here. Chapter 19 describes DNA structure in detail.

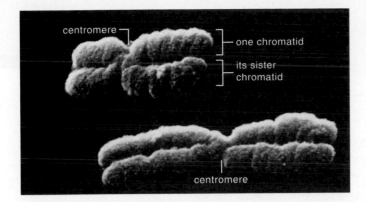

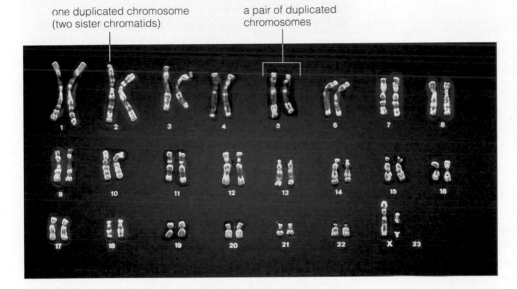

Figure 16.3 Forty-six chromosomes from a human male. Each chromosome is in the duplicated state. There are pairs of chromosomes (two of each type), which tells you that they came from a diploid cell. One member of each pair contains genetic instructions inherited from the father. The other member contains instructions from the mother.

Mitosis and the Chromosome Number

Human cells other than sperm or eggs normally each have two full sets of the chromosomes characteristic of our species—one set from each parent. Your parents each contributed a full set of 23 chromosomes to the fertilized egg that became you, so there are 46 chromosomes in each of your somatic cells. Any cell having two of each type of chromosome is said to be a **diploid** cell. Diploid cells, with two of each type, are referred to as 2*n*. The *n* stands for the number of chromosomes in one complete set. **Chromosome number** tells you how many of each type of chromosome are present in a cell.

With mitosis, a diploid parent cell produces two diploid daughter cells. *A diploid cell cannot function properly unless it receives one pair of each type of chromosome.* One member of each chromosome pair must come from the mother, and the other must come from the father. Shortly before a diploid cell divides, the genetic material is duplicated, so that the diploid number of chromosomes dou-

bles (yielding the sister chromatids just described). Mitosis allots half of this doubled genetic material to each new cell, so each daughter cell receives the full diploid number of chromosomes.

Mitosis produces daughter cells that have the same number of chromosomes as the parent cell. If a parent cell is diploid, its two daughter cells will be diploid also.

Figure 16.3 shows all the chromosomes in a human somatic cell. The chromosomes have been lined up as 23 pairs. Except for the sex chromosomes, X and Y, the two members of each chromosome pair are the same length and shape; both carry hereditary instructions for the same traits. Such corresponding chromosomes, one from each parent, are called **homologous chromosomes**, or simply *homologues* (from a Greek word meaning "to agree"). In general, homologues are the same length, have the same

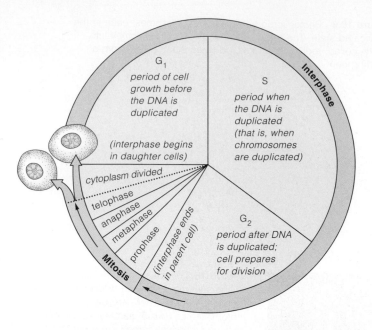

Figure 16.4 The cell cycle. This drawing has been generalized. The length of different stages varies greatly from one type of cell to the next.

shape, and carry genes for the same traits. The sex chromosomes are exceptions; although they are considered homologues, they differ in size and form and carry different genes. Certain traits, such as color-blindness, are X-linked (caused by genes on the X chromosome). Only a few traits are Y-linked. Chapter 18 describes such traits more fully.

The two chromosomes of each corresponding pair in a diploid cell are said to be homologous.

MITOSIS AND THE CELL CYCLE

Mitosis is a small part of the **cell cycle**, which begins when a cell forms and continues until that cell has divided. Normally, a cell destined to enter mitosis spends about 95 percent of the cell cycle in **interphase**. (Some cells, such as mature neurons of the central nervous system, become arrested in interphase and never divide again.)

During interphase, the cell becomes larger. It roughly doubles the number of its cytoplasmic components during a primary growth or G_1 phase, and during an S (synthesis) phase it duplicates its DNA through a process called DNA replication (Figure 16.4). During a second

Focus on Science

Henrietta's Immortal Cells

Henrietta Lacks was only 31 years old when cancer killed her in 1951. Runaway mitotic cell divisions are cancer's hallmark; within six months of her diagnosis, Henrietta's body was riddled with proliferating tumor cells. In another two months she was dead.

Chance can play a pivotal role in science. In 1951, George and Margaret Gey of Johns Hopkins University happened to be seeking a way to keep human cells dividing outside the body. Researchers could use such cell lines to probe basic life processes and to study diseases—including cancer—without having to experiment directly on humans. Among cell samples the Geys obtained from local physicians were some of Henrietta Lack's tumor cells.

Until then, no one had succeeded in getting human cells to live more than a few weeks in the laboratory. Neither the Geys nor their assistant Mary Kubicek expected the new batch of cancer cells, coded HeLa, to be different. But after only four days, the cells had undergone so many divisions that Kubicek had to subdivide the culture into more test tubes. Soon HeLa cells, the first successful human cell culture, were being shipped to scientists around the world.

In the decades since, researchers have published thousands of scientific papers based on work with HeLa cells. Much of that work applies directly to studies of cancer and other diseases. Henrietta's legacy—cells that are still alive and dividing, day after day—continues to be a legacy of hope for millions of people.

brief growth phase, G_2, the cell synthesizes proteins that will be used in mitosis. The duration of the cycle varies, depending on the type of cell involved. For instance, the cycle lasts 18 hours in human bone marrow cells and 25 hours in epithelial cells in your stomach. Chapter 19 examines the details of DNA replication.

Good health depends on the successful completion of cell cycles. Any error in DNA replication or misshipment of chromosomes may lead to problems. Even the timing and regulation of cell division must be precisely controlled. If the cycle proceeds uncontrollably in mature tissues, cancer follows (Chapter 20). A landmark case of unchecked cell divisions is described in *Focus on Science*.

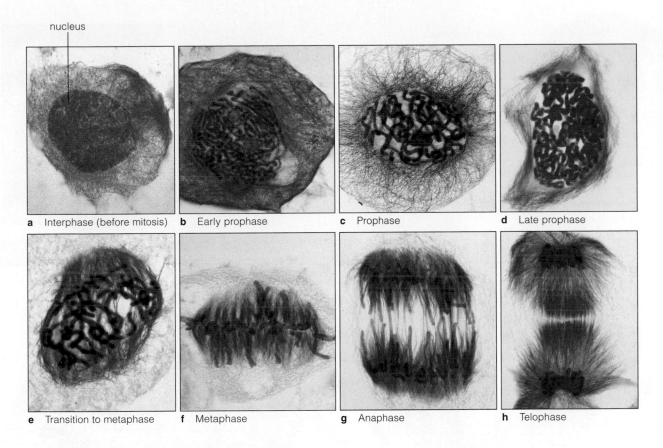

nucleus

a Interphase (before mitosis) **b** Early prophase **c** Prophase **d** Late prophase

e Transition to metaphase **f** Metaphase **g** Anaphase **h** Telophase

Figure 16.5 Mitosis in a cell from the African blood lily (*Haemanthus*). Chromosomes are stained blue, and the microtubules of the spindle are stained red.

THE FOUR STAGES OF MITOSIS

When a cell shifts from interphase to mitosis, it stops making new cell parts. A series of radical changes now occur through four continuous stages. In order, the stages of mitosis are **prophase**, **metaphase**, **anaphase**, and **telophase**.

Figure 16.5 will give you an idea of how dramatically chromosomes move about during the different stages. Chromosomes are pulled into specific positions by a **spindle apparatus**. A fully formed spindle consists of two sets of microtubules. The microtubules extend from the spindle's two end points (poles) and overlap at its equator, which lies midway between them. The spindle poles establish the destinations of chromosomes during mitosis (and meiosis).

The chromosome movements orchestrated by the spindle are essential to successful cell division. For instance, irradiation damages organisms by breaking cell chromosomes into fragments. If the cell then divides, the spindle

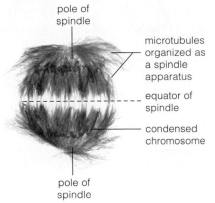

pole of spindle

microtubules organized as a spindle apparatus

equator of spindle

condensed chromosome

pole of spindle

cannot harness all the pieces during cell division, and the defective daughter cells may die (see *Focus on Environment*).

In many cells, including those of humans, a "microtubule organizing center" dictates where the spindle microtubules will arise. This center is sometimes called the *centrosome*, and in human cells it includes a pair of structures called centrioles. The position and orientation of centrioles in a parent cell seem to influence the organization of the cytoskeleton that will form in each daughter cell.

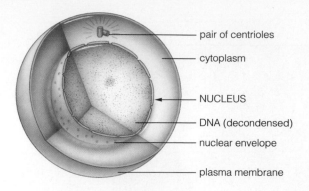

pair of centrioles

cytoplasm

NUCLEUS

DNA (decondensed)

nuclear envelope

plasma membrane

Cell at Interphase
The DNA is duplicated, then the cell prepares for division.

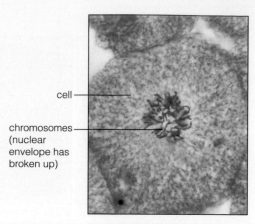

cell

chromosomes
(nuclear
envelope has
broken up)

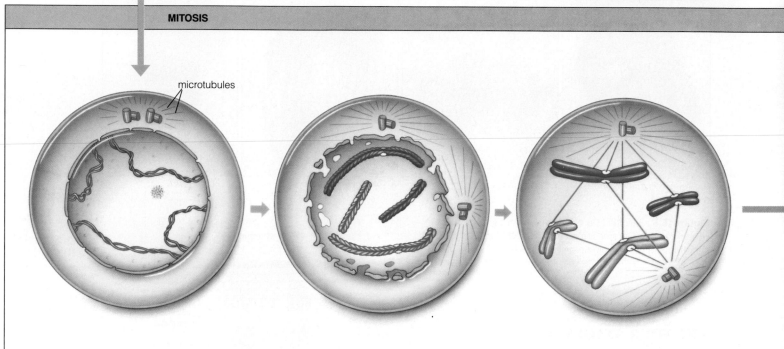

MITOSIS

microtubules

Early Prophase
The DNA and its associated proteins start to condense. The two chromosomes shaded purple were inherited from the male parent. The other two (blue) are their counterparts, inherited from the female parent.

Late Prophase
Chromosomes continue to condense. New microtubules are assembled, and they move the two centrioles toward opposite ends of the cell. The nuclear envelope starts to break up.

Transition to Metaphase
Microtubules penetrate the nuclear region. Together they form a spindle apparatus. They become attached to the sister chromatids of each chromosome.

Figure 16.6 Mitosis. This nuclear division mechanism assures that daughter cells will have the same chromosome number as the parent cell. For clarity, this diagram shows only two pairs of chromosomes from a diploid cell. With rare exceptions, the picture is more involved than this, as indicated by the micrographs of mitosis in a whitefish cell.

Prophase: Mitosis Begins

During prophase, the first and longest stage of mitosis, a dividing cell undergoes many changes (Figure 16.6). The cytoskeleton and nuclear envelope are dismantled, and

much of the cell's metabolic activity slows or stops. In the cytoplasm, the spindle begins to form. As prophase begins, chromosomes are in their threadlike, chromatin form. Each chromosome was duplicated earlier, during interphase, so it already consists of two sister chromatids joined at the centromere. In late prophase, the chromatids become condensed into thicker, rodlike forms. Somewhat like a knitter winds pieces of yarn into balls, the condensing of chromosomes packages them into "bundles" that can be easily moved.

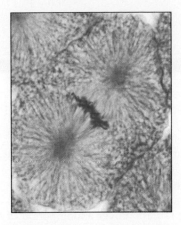
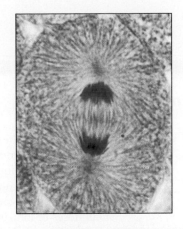
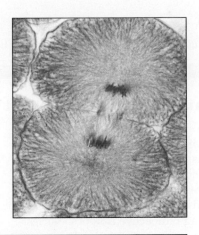

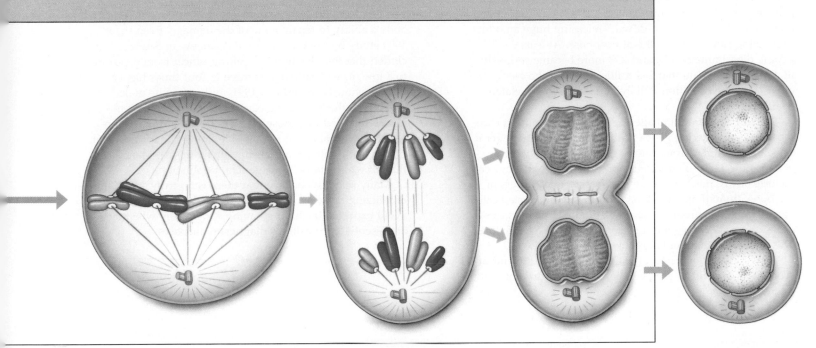

Metaphase

All chromosomes are lined up at the spindle equator. They are now in their most condensed form.

Anaphase

The attachment between the two sister chromatids of each chromosome breaks. The two are now independent chromosomes. They move to opposite spindle poles.

Telophase

Chromosomes decondense. New patches of membrane join to form nuclear envelopes around them. Most often, cytokinesis occurs before the end of telophase.

Interphase

Two daughter cells have formed. Each is diploid, with two of each type of chromosome—just like the parent cell.

Transition to Metaphase

Major interactions occur during the transition from prophase to metaphase. (Many researchers call this transitional period "prometaphase.")

The nuclear envelope breaks up completely into vesicles. Microtubules are now free to interact with the chromosomes. At first they attach randomly to them. The attachments become more organized, and the spindle takes on final form.

Now microtubules extending from each pole start pulling each chromosome in both directions (Figure 16.6). The two-way yanking orients sister chromatids toward opposite poles. Meanwhile, microtubules from both poles are ratcheting past each other, and the force pushes the poles farther apart. All the push-pull forces are balanced when the chromosomes reach the spindle equator.

When all of the duplicated chromosomes are aligned midway between the poles, the dividing cell is in metaphase (*meta-* means "midway between"). The alignment is crucial for the next stage of mitosis.

Focus on Environment   G. Tyler Miller, Jr.

Ionizing Radiation: Invisible Threat to Dividing Cells

Ionizing radiation can take many forms, and all of them can damage human cells. Natural sources include cosmic rays from outer space and radioactive radon gas in rocks and soil (page 433). Most nonnatural ionizing radiation comes from medical and dental X rays and from diagnostic tests and treatments using radioactive isotopes. Nuclear power plants are not major sources, *provided* they operate properly. In 1986 a nuclear reactor at Chernobyl in the former Soviet Union "melted down," releasing huge amounts of ionizing radiation into the atmosphere. At least 36 people died immediately, and 237 more became seriously ill. Officials estimate that 2–4 million others who were exposed to the radiation will eventually develop related health problems.

Ionizing radiation damages cells by breaking apart chromosomes and altering genes. If the damage occurs in germ cells, the resulting gametes may give rise to infants with genetic defects. If somatic cells are affected, the damage can include burns, miscarriages, eye cataracts, and cancers of the bone, thyroid, breast, skin, and lung. If an affected cell has fragmented chromosomes, the spindle apparatus cannot harness and move the fragments when the cell divides. The cell or its daughters may then die.

As we learned at Chernobyl and in Japan at the end of World War II, when a person receives a large dose of ioniz-ing radiation over a short time, the radiation inexorably destroys cells of the immune system, epithelial cells of the skin and intestinal lining, red blood cells, and other cell types. The results are raging infections, intestinal hemorrhages, anemia, and wounds that do not heal.

Small doses of ionizing radiation over a long period of time appear to cause less damage than the same total dosage given all at once. This may be due in part to the body's ability to repair some of the damage. Even so, a 1990 study by the U.S. National Academy of Sciences concluded that the likelihood of getting cancer from exposure to a low dose of radiation is three to four times higher than previously thought. A 1991 study of white male workers at the Oak Ridge National Laboratory in Tennessee revealed that those workers exposed to radiation well below established permissible limits had a 63 percent higher death rate from leukemia (bone marrow cancer) than of white males in the general population.

Many scientists believe that there is no truly "safe" dose of ionizing radiation. For individuals, the best policy is to avoid exposure, including having X rays only when they are essential for health or diagnostic purposes.

Anaphase: Two Sets of Independent Chromosomes

During **anaphase**, the two chromatids of each chromosome separate from each other. Spindle microtubules are still attached to each chromatid, and they move the separated chromatids to opposite poles. Now each former chromatid is an independent chromosome:

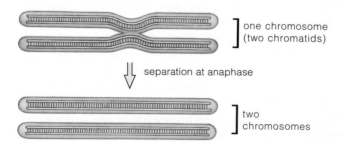

one chromosome (two chromatids)

separation at anaphase

two chromosomes

Telophase: Two Nuclei Form

Telophase begins once the separated chromosomes arrive at opposite spindle poles. The chromosomes are no longer harnessed to microtubules, and they decondense. Soon a nuclear envelope forms around each cluster of chromosomes. If the parent cell was diploid, each cluster will contain a pair of each type of chromosome. With mitosis, each new nucleus has the same chromosome number as the parent nucleus. Once the two nuclei form, telophase is over—and so is mitosis.

Cytokinesis

Division of the cytoplasm, or cytokinesis, usually coincides with the period from late anaphase through telophase. For most animal cells, deposits accumulate and form a ringlike layer around microtubules at the dividing

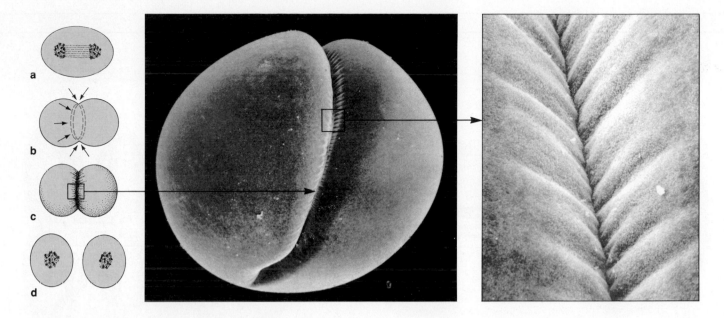

Figure 16.7 Cytokinesis in an animal cell. (**a**) Mitosis is complete and the spindle is disassembling. (**b**) Just beneath the plasma membrane, microfilament rings at the former spindle equator contract, like a purse string closing. (**c**, **d**) Continuing contractions divide the cell in two. The micrographs show how the plasma membrane sinks inward, defining the plane of cleavage.

cell's midsection. A shallow depression appears above the layer, at the cell surface (Figure 16.7). At this depression, called a **cleavage furrow**, microfilaments made of the contractile protein actin pull the plasma membrane inward and cut the cell in two. Organelles in the divided cytoplasm are randomly distributed to daughter cells.

Mitosis divides the nucleus of a somatic cell in a way that ensures that daughter cells receive the diploid number of chromosomes. It is the basis for growth and tissue repair.

Mitosis is a small part of the cell cycle. It consists of prophase, metaphase, anaphase, and telophase. Near the end of mitosis cytokinesis divides the cytoplasm.

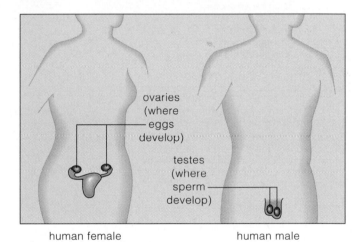

Figure 16.8 Germ cells in the testes and ovaries give rise to sperm and eggs, respectively. In the testes, primary spermatocytes undergo meiosis to give rise to spermatids that differentiate into sperm. In females, ovaries are the sites where primary oocytes undergo meiosis. Figures 16.14 and 16.15 show these events in more detail.

OVERVIEW OF MEIOSIS

Meiosis occurs in the nucleus of a dividing germ cell in ovaries or testes (Figure 16.8). Recall from Chapter 13 that human germ cells are *spermatogonia* in males and *oogonia* in females. Spermatogonia and oogonia are diploid, like other body cells. However, unlike other body cells, they can give rise to haploid gametes (sperm or eggs) by way of meiosis.

A mature human sperm or oocyte typically contains a single set of 23 chromosomes, rather than pairs of homologous chromosomes. Meiosis yields precisely this result. It is a **reductional division** that halves the diploid number of chromosomes (2n) to a **haploid** number (n). And not just any half: *Each haploid gamete ends up with one partner from each pair of homologous parent chromosomes.*

Recall that when a diploid cell is in interphase, the parental chromosomes are duplicated. Hence, to produce haploid gametes, meiosis divides the chromosomes in a

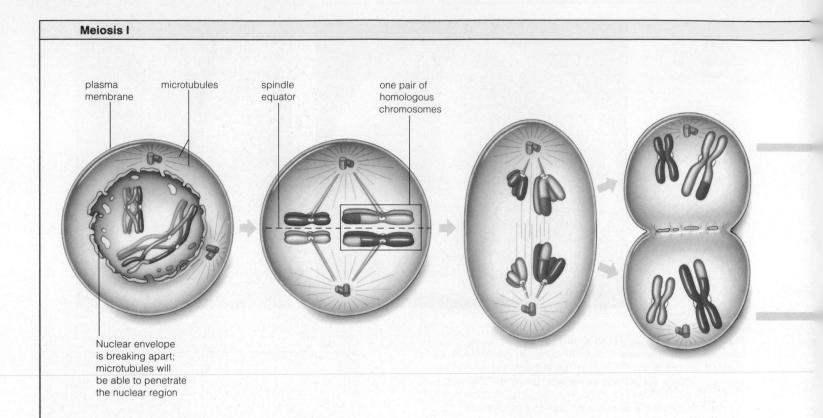

plasma membrane   microtubules   spindle equator   one pair of homologous chromosomes

Nuclear envelope is breaking apart; microtubules will be able to penetrate the nuclear region

Prophase I

By now, chromosomes are in the threadlike, duplicated form (each consists of two sister chromatids). Usually, all homologous pairs of chromosomes undergo crossing over. (For clarity, only one crossover is shown). Chromosomes start condensing to rodlike form. Each becomes attached to the microtubular spindle.

Metaphase I

All chromosomes are now positioned at the equator of the spindle.

Anaphase I

Each chromosome is separated from its homologue. The two are moved to opposite poles of the spindle.

Telophase I

When the cytoplasm divides, there are two haploid (*n*) cells; each has one chromosome of each type. Chromosomes are still in the duplicated state.

Figure 16.9 Meiosis: the nuclear division mechanism by which the parental number of chromosomes is reduced by half (to the haploid number) for forthcoming gametes. Only two pairs of homologous chromosomes are shown. Maternal chromosomes are shaded purple, and paternal ones blue.

cell nucleus not once but twice prior to cell division. Meiosis I produces two haploid cells with duplicated chromosomes, and meiosis II produces four haploid cells with single (unduplicated) chromosomes:

	MEIOSIS I		MEIOSIS II
DNA duplication during interphase	Prophase I Metaphase I Anaphase I Telophase I	*No DNA duplication between divisions*	Prophase II Metaphase II Anaphase II Telophase II

Overview of the Two Divisions

During meiosis I, each duplicated chromosome lines up with its partner, *homologue to homologue*. Next, the partners are separated from each other. Here we show just one pair of homologous chromosomes, but the same thing happens to all pairs in the nucleus, including sex chromosomes:

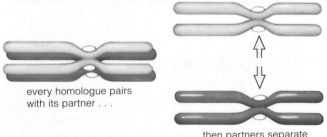

every homologue pairs with its partner . . .

. . . then partners separate

Cytokinesis follows. At this point, each daughter cell has a haploid number of chromosomes—but each chromosome is still duplicated and consists of joined sister chro-

There is no DNA replication between the two divisions

Prophase II

During the transition to prophase II, the two centrioles in each new cell were moved apart and a new spindle was assembled. Now, microtubules attach chromosomes to the spindle and start moving them toward the equator.

Metaphase II

All chromosomes are now positioned at the equator of the spindle.

Anaphase II

The attachment between the two chromatids of each chromosome breaks. Now the former "sister chromatids" are chromosomes in their own right and are moved to opposite poles of the spindle.

Telophase II

Four daughter nuclei form. When the cytoplasm divides, each new cell has a haploid number of chromosomes, all in the unduplicated state. One or all of the cells may develop into gametes.

matids. Now, in meiosis II, the sister chromatids of each chromosome are separated from each other:

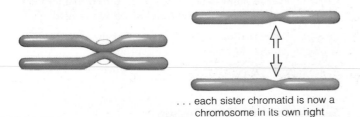

. . . each sister chromatid is now a chromosome in its own right

STAGES OF MEIOSIS

Prophase I: Genes Are Shuffled

Figure 16.9 traces the stages of meiosis. The stages are similar to the stages of mitosis and are called prophase, metaphase, anaphase, and telophase. They are designated I and II because there are two cell divisions in meiosis.

In the first stage, prophase I, a spindle forms and the nuclear envelope disappears. This is also a time of major gene shufflings between homologous chromosomes. As prophase I gets underway, homologues begin to pair up in a process called *synapsis* ("bringing together"). It is as if the homologues become stitched point by point along their entire length, with little space between them. (The X and Y chromosomes pair at one end only.) Keep in mind that each homologue is in the duplicated state and consists of two sister chromatids joined at a centromere. The four homologous chromatids make up a structure known as a **tetrad** (Greek, meaning "four"). This arrangement favors **crossing over** between *nonsister chromatids* of a tetrad. In a crossover, the nonsister chromatids of the tetrad exchange corresponding segments at the crossover points. Each segment contains a group of genes. As Figure 16.10 shows,

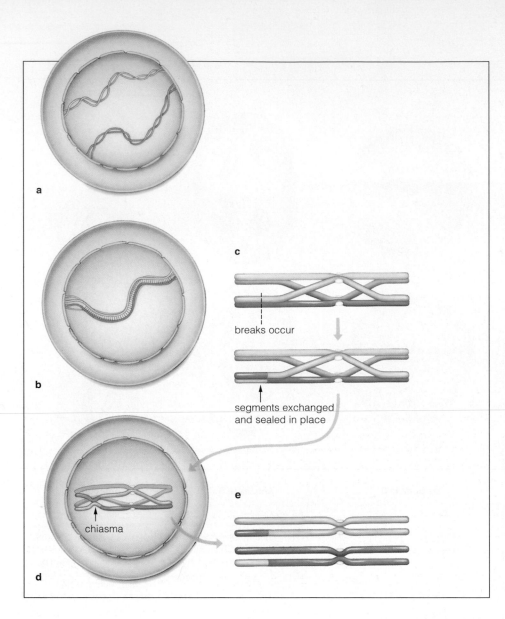

Figure 16.10 Prophase I of meiosis, when crossing over occurs between homologous chromosomes.

a Chromosomes become duplicated before meiosis begins. Early in prophase I, each duplicated chromosome is in threadlike form, attached at both ends to the nuclear envelope. Its two sister chromatids are so close together they look like a single thread.

b The two chromosomes become zippered together, so that all four chromatids are positioned close together.

c One or more crossovers occur at intervals along the chromosomes. In each crossover, two nonsister chromatids break at identical sites. They swap segments at the breaks, then enzymes seal the broken ends. (For clarity, the two chromosomes are shown in condensed form and pulled apart. Crossing over may seem more plausible when you realize that it occurs while chromosomes are extended like threads and tightly aligned.)

d As prophase I ends, the chromosomes continue to condense, becoming thicker, rodlike forms. They detach from the nuclear envelope and from each other—except at "chiasmata." Each chiasma is indirect evidence that a crossover occurred at some point in the chromosomes.

e Crossing over breaks up old combinations of alleles and puts new ones together in pairs of homologous chromosomes.

breaks occur

segments exchanged and sealed in place

chiasma

---

crossovers may occur within X-shaped configurations called *chiasmata* (singular: *chiasma*).

Genes can come in alternative forms called **alleles**. For example, the gene for earlobe shape has two alleles—one calls for attached earlobes, and the other calls for detached earlobes. Frequently, many of the alleles on one chromosome will not be exactly identical to the corresponding alleles on its homologue. With each crossover, then, truly unique combinations of genes may form.

This exchange of chromosome pieces is called **genetic recombination**, and it leads to variation in the traits of offspring. Such variation can be a major advantage for sexually reproducing organisms. In later chapters we will see that variations in traits can be the source of evolutionary modifications that enable a population of organisms, including humans, to adapt to a changing environment.

Crossing over in meiosis leads to genetic recombination as existing combinations of genes (alleles) are rearranged.

Genetic recombination is a major source of the variation that is present in the chromosomes of offspring.

Separating the Homologues

During metaphase I, whole chromosomes are shuffled. The spindle apparatus moves all the tetrads until they are lined up at the spindle equator. Figure 16.11 shows how just three pairs of homologues can be shuffled into any one of four possible lineups at metaphase I. In this case, 2^3 or 8 combinations of chromosomes are possible for the forthcoming gametes. A human germ cell has 23 pairs of homologous chromosomes, not just three. Thus 2^{23} or

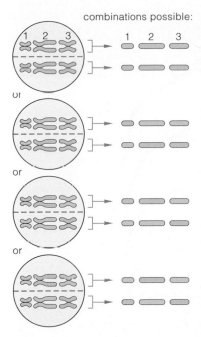

combinations possible:

Figure 16.11 Possible outcomes of the random alignment of three pairs of homologous chromosomes at metaphase I of meiosis. The three types of chromosomes are labeled 1, 2, and 3. Maternal chromosomes are purple; paternal ones are blue.

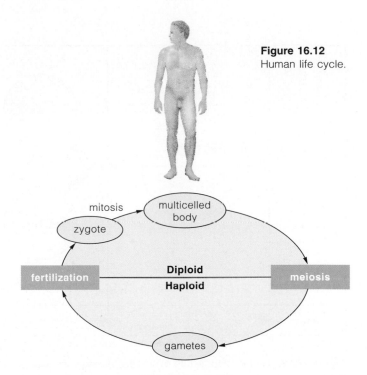

Figure 16.12 Human life cycle.

8,388,608 combinations of maternal and paternal chromosomes are possible every time a germ cell gives rise to sperm or eggs! Factor in crossing over, and the variation becomes even greater. Can you see why striking mixes of traits show up even in the same family?

During meiosis I, either one of a pair of homologous chromosomes can move to either pole of the spindle. As a result, different gametes end up with different mixes of maternal and paternal chromosomes. Like genetic recombination, this contributes to variations in the traits of offspring.

During anaphase I, each homologue is separated from its partner, an event called *disjunction*. Each set of sister chromatids is now called a **dyad**. The two dyads are moved to opposite poles of the spindle. So long as the separation takes place, it doesn't matter which dyad moves to which pole. Failure of homologues to separate during meiosis, called nondisjunction, can lead to birth defects (Chapter 18).

Telophase I is followed by cytokinesis. After a brief resting period (called interkinesis), the second phase of meiosis will get underway. DNA is not duplicated between the two meiotic divisions. But remember, each chromosome was duplicated earlier (during interphase) and is still in the duplicated form when meiosis II begins.

Separating the Sister Chromatids

Meiosis II is almost identical to mitosis. It occurs in the two daughter cells resulting from meiosis I, and its main function is to separate the two sister chromatids of each

chromosome in the daughter cells. Prophase II may be fleeting or may not occur at all. At metaphase II, each duplicated chromosome is moved to the spindle equator. Each duplicated chromosome is split at anaphase II; its former sister chromatids are now chromosomes in their own right.

As Figure 16.9 shows, at the close of anaphase II one of each type of chromosome moves to each spindle pole. During telophase II, new nuclear membranes form around the chromosomes after they have become clustered at the two poles. Meiosis is complete. Cytokinesis now pinches each cell in two. The end result of the two meiotic divisions is a set of four cells, each having a haploid number of unduplicated chromosomes.

Meiosis occurs in germ cells of the gonads (ovaries or testes). It reduces the diploid chromosome number by half, to the haploid number.

Meiosis results in gametes, the ova and sperm that function in sexual reproduction.

MEIOSIS AND THE LIFE CYCLE

How Gametes Form

As with other multicellular animals, the human life cycle proceeds from meiosis to gamete formation, fertilization, then growth of the new individual by way of mitosis (Figure 16.12). In males, meiosis and gamete formation are

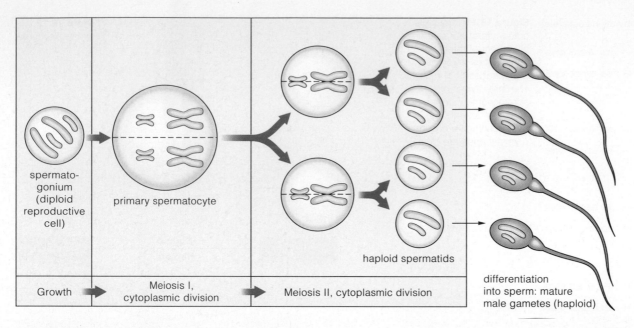

spermato-
gonium
(diploid
reproductive
cell)

primary spermatocyte

haploid spermatids

differentiation
into sperm: mature
male gametes (haploid)

Growth	Meiosis I, cytoplasmic division	Meiosis II, cytoplasmic division	

Figure 16.13 Generalized picture of spermatogenesis. For clarity, the nuclear envelopes are not shown.

called **spermatogenesis**; Figure 16.13 shows the steps. Typically a diploid germ cell increases in size. The resulting large, immature cell (a primary spermatocyte) undergoes meiosis. Eventually, four haploid spermatids are produced. The spermatids change in form, develop tails, and become sperm—the mature male gametes.

In females, meiosis and gamete formation are called **oogenesis** (Figure 16.14). Oogenesis differs from spermatogenesis in several important features. Compared to a primary spermatocyte, many more cytoplasmic components accumulate in a primary oocyte, the female germ cell that undergoes meiosis. Also, in females the cells formed after meiosis differ in their size and function.

The early stages of oogenesis take place in a developing female embryo. Until she reaches puberty, however, her primary oocytes are arrested in prophase I. Then, each month meiosis resumes in the (usually) one oocyte that is ovulated. Following meiosis I, this cell, the secondary oocyte, receives nearly all the cytoplasm; the other, much smaller cell is a polar body (page 299). Both cells enter meiosis II, but the process is arrested again at metaphase II. If fertilization occurs, meiosis II continues, and the outcome is one large cell and three extremely small polar bodies. The polar bodies serve as "dumping grounds" for three sets of chromosomes, so that the future egg ends up with the necessary haploid number. The large cell develops into the mature egg (ovum). Its ample supply of cytoplasm contains components that will help guide development of an embryo.

Biological Role of Meiosis and Fertilization

The diploid number of chromosomes is restored at fertilization, when the nuclei of two gametes fuse to form the zygote (Figure 16.15). Here we see why meiosis is so necessary. If gametes arose by mitosis and ended up as diploid cells, fertilization would double the number of chromosomes in cells of every new generation. The result in humans would be catastrophic. For instance, just a single extra chromosome 21 causes Down syndrome, which results in mental retardation and other birth defects (Chapter 18).

Like meiosis, fertilization contributes to variation in the traits of offspring. Think about the possibilities for humans. First, genes are shuffled during prophase I, when each chromosome takes part in two or three crossovers, on the average. Then, random alignments at metaphase I lead to one of 8,388,608 possible combinations of chromosomes in each gamete. Finally, of all the genetically diverse male and female gametes that are produced, *which* two will get together is a matter of chance. As you can see, the sheer number of new combinations brought together at fertilization is staggering!

The mix of different combinations of alleles from two different gametes at fertilization contributes to variation in the traits of offspring.

Meiosis and Mitosis Compared

Table 16.1 summarizes the differences between mitosis and meiosis. Mitosis produces genetically identical copies of the parent cell; by way of mitosis, the body grows and

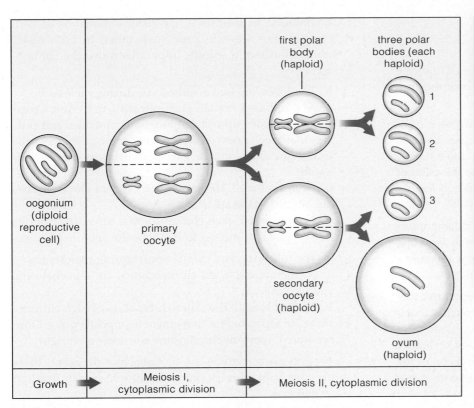

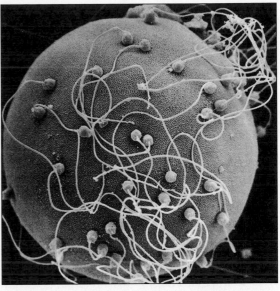

Figure 16.14 Generalized picture of oogenesis. This sketch is not drawn to the same scale as Figure 16.13. A primary oocyte is *much* larger than a primary spermatocyte, as indicated by the micrograph. Also, the polar bodies are extremely small compared to an ovum, as Figure 14.4 on page 314 shows.

Table 16.1 Basics of Mitosis and Meiosis

	Mitosis	Meiosis
Function	Growth, including repair and maintenance	Gamete production (sperm/eggs)
Occurs in	Somatic (body) cells	Germ cells in gonads (testes and ovaries)
Mechanism	One round of nuclear division plus cytokinesis	Two rounds of nuclear division plus cytokinesis
Outcome	Maintains diploid chromosome number ($2n \rightarrow 2n$)	Reduces diploid chromosome number ($2n \rightarrow n$)
Effect	Two diploid daughter cells	Four haploid daughter cells

replaces lost cells. Meiosis occurs only in germ cells used in sexual reproduction. Together with fertilization, it gives rise to new combinations of genes in offspring. Genes are the instructions for inherited traits, and in the next chapter we consider the "rules" of inheritance that apply to virtually all sexually reproducing organisms— and so to humans.

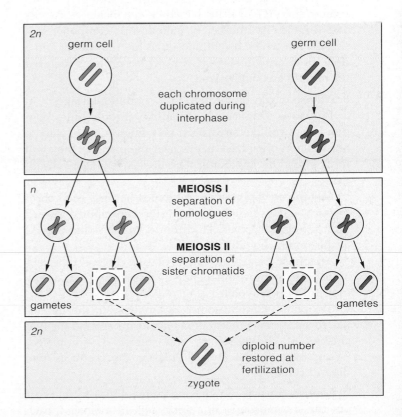

Figure 16.15 How fertilization restores a diploid ($2n$) number of chromosomes that meiosis had divided in half.

SUMMARY

1. Each cell of a new generation must receive a duplicate of all the parental DNA and enough cytoplasmic machinery to start up its own operation.

2. Somatic (body) cells of multicelled eukaryotes commonly have a diploid number of chromosomes. That is, they have two of each type of chromosome characteristic of the species. Except for sex chromosomes, the pairs of homologous chromosomes generally are alike in length, shape, and the traits they deal with.

3. Eukaryotic chromosomes are duplicated during interphase (between cell divisions). Whereas each was one DNA molecule (and associated proteins), now it consists of two, temporarily attached as sister chromatids.

4. In eukaryotes, mitosis maintains the parental number of chromosomes in each of two daughter nuclei. If the parental nucleus is diploid, the nucleus in each daughter cell will also be diploid. Mitosis is the basis of bodily growth for multicelled eukaryotes. Actual division of the cytoplasm, or cytokinesis, occurs toward the end of nuclear division or at some point afterward.

5. A cell destined to divide by mitosis spends a significant portion of the cell cycle in interphase, a period between mitotic divisions. Interphase includes a primary growth stage (G_1), during which the cell increases in mass and doubles its number of cytoplasmic components; a stage (S) in which its chromosomes are replicated; and a final, brief growth stage (G_2) in which proteins required for mitosis are synthesized.

6. Mitosis proceeds through four continuous stages:
a. Prophase. Duplicated, threadlike chromosomes condense into rodlike structures; new microtubules start to assemble in organized arrays near the nucleus; they will form a spindle apparatus. The nuclear envelope disappears.
b. Metaphase. Spindle microtubules harness each chromosome and orient its two sister chromatids toward opposite spindle poles. All chromosomes become aligned at the spindle equator.
c. Anaphase. Sister chromatids of each chromosome separate. Both are now independent chromosomes, and they move to opposite poles.
d. Telophase. Chromosomes decondense to the threadlike form. A new nuclear envelope forms around the two clusters of chromosomes, and mitosis is completed. Cytokinesis divides the cytoplasm; the result is two diploid cells.

7. Meiosis consists of two consecutive divisions that sort out the chromosomes in a germ cell. In meiosis I, each chromosome pairs with and then separates from its homologue. In meiosis II, the sister chromatids of each chromosome separate from each other. In both cases, microtubules of a spindle apparatus move the chromosomes.

8. The following key events occur during meiosis I:
a. At prophase I, homologues pair with each other. Crossing over breaks up old combinations of alleles and puts together new ones in the chromosomes. This genetic recombination leads to variation in traits among offspring.
b. At metaphase I, all pairs of homologous chromosomes are aligned at the spindle equator.
c. At anaphase I, each chromosome is separated from its homologue and moved to the opposite spindle pole.

9. The following key events occur during meiosis II:
a. At metaphase II, all chromosomes are moved to the spindle equator.
b. At anaphase II, the sister chromatids of each chromosome are separated for movement to opposite poles. Once separated, they are chromosomes in their own right.

10. Following cytokinesis (cytoplasmic division), there are four haploid cells, one or all of which may function as gametes.

Review Questions

1. Define the two types of nuclear division mechanisms that occur in eukaryotes. What is cytokinesis? *358*

2. Define somatic cell and germ cell. Which type of cell can undergo mitosis? *358*

3. What is a chromosome? What are chromosomes called during interphase, when they are not condensed? *358*

4. What are homologous chromosomes? Relate this concept to the diploid number. *359*

5. Describe the spindle apparatus and its general function in nuclear division processes. *361*

6. Name the four main stages of mitosis, and describe the main features of each stage. *361–364*

7. In a paragraph, summarize the similarities of and differences between mitosis and meiosis. *370*

8. Explain the significance of fertilization. *371*

Critical Thinking: You Decide (*Key in Appendix IV*)

1. Under normal circumstances you can't inherit both copies of a homologous chromosome from the same parent. Why? Assuming that no crossing over has occurred, what is the chance that one of your autosomes is an exact copy of the same chromosome possessed by your maternal grandmother?

2. The Mongol chief Genghis Khan died in 1227, at least 38 human generations ago. If one of your friends claims to be a direct descendant of Khan, is it likely that he or she actually carries some of the great warrior's chromosomal DNA?

Self-Quiz *(Answers in Appendix III)*

1. Eukaryotic DNA is distributed to daughter cells by _____ or _____, both of which are nuclear divisions.

2. Each kind of organism contains a characteristic number of _____ in each cell; each of those structures is composed of a _____ molecule with its associated proteins.

3. A pair of chromosomes that are similar in length, shape, and the traits they govern are called _____.
   a. diploid chromosomes
   b. mitotic chromosomes
   c. homologous chromosomes
   d. germ chromosomes

4. Somatic cells of multicelled organisms usually have a _____ number of chromosomes.

5. Interphase is the stage when _____.
   a. nothing occurs
   b. a germ cell forms its spindle apparatus
   c. a cell grows and duplicates its DNA
   d. cytokinesis occurs

6. After mitosis, each daughter cell contains genetic instructions that are _____ and _____ chromosome number of the parent cell.
   a. identical to the parent cell's; the same
   b. identical to the parent cell's; one half the
   c. rearranged; the same
   d. rearranged; one-half the

7. All of the following are stages of mitosis *except* _____.
   a. prophase
   b. interphase
   c. metaphase
   d. anaphase

8. A duplicated chromosome has _____.
   a. one chromatid
   b. two chromatids
   c. three chromatids
   d. four chromatids

9. Crossing over in meiosis _____.
   a. alters the chromosome alignments at metaphase
   b. occurs between sperm DNA and egg DNA at fertilization
   c. leads to genetic recombination
   d. occurs only rarely

10. Because of the _____ alignment of homologous chromosomes at metaphase I, gametes can end up with _____ mixes of maternal and paternal chromosomes.
   a. unvarying; different   c. random; duplicate
   b. unvarying; duplicate   d. random; different

11. Following meiosis and cytokinesis, there are _____ haploid cells, one or all of which may function as _____.
   a. two; body cells   c. four; gametes
   b. two; gametes   d. four; body cells

12. Match each stage of mitosis with the following key events.
   ____ metaphase   a. sister chromatids of each chromosome
   ____ prophase   separate and move to opposite poles
   ____ telophase   b. threadlike chromosomes condense and a
   ____ anaphase   microtubular spindle forms
                      c. chromosomes decondense, daughter
                         nuclei re-form
                      d. all chromosomes become aligned at
                         spindle equator

Key Terms

allele *368*	homologous chromosome *359*
anaphase *361*	interphase *360*
cell cycle *360*	meiosis *358*
centromere *358*	metaphase *361*
chromosome *358*	mitosis *358*
chromosome number *359*	oogenesis *370*
cleavage furrow *365*	prophase *361*
crossing over *367*	reductional division *365*
cytokinesis *358*	sister chromatid *358*
diploid *359*	somatic cell *358*
dyad *369*	spermatogenesis *370*
genetic recombination *368*	spindle apparatus *361*
germ cell *358*	telophase *361*
haploid *365*	tetrad *367*

Readings

Glover, D., C. Gonzalez, and J. Raff. June 1993. "The Centrosome." *Scientific American*. New insights about how the microtubule organizing center guides mitosis and affects cell movements and shape.

Gould, S. J. 1980. *The Panda's Thumb*. New York: Norton. The classic essay "Dr. Down's Syndrome" describes one outcome in humans when meiosis falters.

Murray, A. and M. Kirschner. March 1991. "What Controls the Cell Cycle?" *Scientific American*.

17 INHERITANCE

Ears and Other Parts

Actress Joan Chen has them. Actor Tom Cruise doesn't, and neither does basketball ace Charles Barkley. To see how *you* fit in with these folks, use a mirror to check your ears. Do they have fleshy lobes that can be flapped back and forth? If so, you and Chen have something in common. Or are the lobes attached to the side of your head and unflappable? If so, you are like Cruise and Barkley (Figure 17.1).

Whether a person is born with detached or attached earlobes depends on a single gene. That gene comes in slightly different molecular forms, called alleles. Only one form has information about detached lobes. The information is put to use while a human fetus is developing inside its mother. It calls for a death signal, which is sent to all the cells positioned between the newly forming lobes and the head. Without the signal, the cells don't die and earlobes don't detach.

We all have genes for thousands of traits, such as earlobes, cheeks, lashes, and chin configuration. Most of the traits vary in their details from one person to the next. Remember, humans inherit pairs of genes (that is, two alleles) on pairs of chromosomes. In some pairings, one allele has powerful effects and overwhelms the other's contribution to a trait. The less powerful allele is said to be recessive to the dominant one. If you have detached earlobes, dimpled cheeks, long lashes, or a fissure (indentation) in your chin, you carry at least one and possibly two dominant alleles that influence that trait in a particular way.

When both alleles of a pair are recessive, nothing masks their effect on a trait. You get attached earlobes with one pair of recessive alleles, a straight nose with another, and so forth.

How did we discover such properties of human genes? It all began with a monk named Gregor Mendel. By analyzing pea plants, generation after generation, Mendel found indirect but *observable* evidence of how parents transmit units of hereditary information—genes—to offspring. Mendel's work is a classic example of how a scientific approach can reveal important secrets about the natural world. To this day, it serves as the foundation for modern genetics.

a Tom Cruise

b Charles Barkley

c Joan Chen

d Gregor Mendel

1. Genes are units of information about heritable traits. Each kind of gene has a specific location on a specific chromosome. The molecular form of a gene may differ slightly from one individual to the next. The different molecular forms of a gene, called alleles, specify different versions of the same trait.

2. Human body cells are diploid—they have pairs of homologous chromosomes and so have pairs of each type of gene. A gene pair may consist of identical alleles, *or* it can consist of two different alleles of the gene.

3. The two genes of a pair are separated from each other during meiosis and end up in different gametes. Gregor Mendel found indirect evidence of this when he cross-bred pea plants having observable differences in a trait, such as purple or white flowers.

4. When a given gene pair on a set of homologous chromosomes sort into different gametes, this sorting is independent of how other gene pairs on other sets of homologous chromosomes sort into gametes.

5. When homologous chromosomes separate during meiosis, the genes on each one tend to stay together and move into the same gamete.

6. If the two genes of a pair specify different versions of a trait (they are nonidentical alleles), one may have more pronounced effects on the trait.

7. Two or more genes often influence the same trait, and some single genes influence many different traits. Gene interactions and environmental conditions may also alter gene expression.

Figure 17.1 Observable evidence of a trait governed by a single kind of gene. Neither actor Tom Cruise (**a**) nor basketball star Charles Barkley (**b**) carries the molecular form of this gene that specifies *detached* earlobes. Actress Joan Chen (**c**) carries it. So did Gregor Mendel (**d**), the founder of classical genetics.

PATTERNS OF INHERITANCE

The Origins of Genetics

Having read about meiosis in Chapter 16, you already have insight into the mechanisms of sexual reproduction. That is more than Gregor Mendel had. Mendel was born into a farm family in what later became Czechoslovakia. By the 1850s he had become a university-educated monk, specializing in studies of both theology and what was then called "natural history." Even so, Mendel did not know about chromosomes, so he could not have known that the chromosome number is reduced by half in gametes, then restored at fertilization. Yet Mendel was interested in discovering how traits were inherited in the offspring of sexually reproducing organisms, and to pursue this question he experimented with the garden pea plant, *Pisum sativum* (Figure 17.2). This plant is self-fertilizing: Its flowers produce sperm (pollen) *and* eggs, and fertilization occurs in the same flower. Based on his research, Mendel hypothesized that fertilization united "factors" from each parent that were the units of heredity. Today we would call such factors *genes*.

Some Terms Used in Genetics

The following list expresses some of Mendel's ideas in modern terms (see also Figure 17.3):

1. **Genes** are units of information about specific traits, and they are passed from parents to offspring. Each gene is a segment of DNA and has a specific location (locus) on a chromosome.

2. Diploid cells have a pair of genes for each trait, on pairs of homologous chromosomes.

3. Although both genes of a pair deal with the same trait, they may vary in their information about it. This happens when there are slight molecular differences in the DNA segments that make up each gene. Each version of the gene is called an **allele**. Contrasting alleles account for a great deal of the variation we see in traits. For example, one allele of a gene specifies a straight hairline, and another allele of the same gene specifies a widow's peak.

4. If a gene pair consists of two identical alleles, this is a **homozygous** condition (*homo*: same; *zygo*: joined together). If the gene pair consists of different alleles, this is a **heterozygous** condition (*hetero*: different).

5. An allele is **dominant** when its effect on a trait masks that of any **recessive** allele paired with it. A capi-

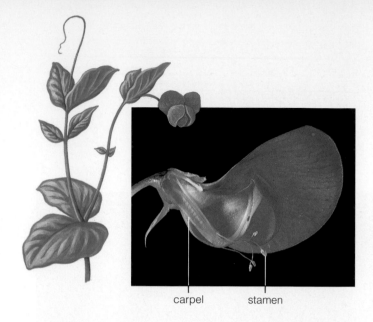

Figure 17.2 The garden pea (*Pisum sativum*), the focus of Mendel's experiments. A flower has been sectioned to show the location of its stamens and carpel. Sperm-producing pollen grains form in stamens. Eggs develop, fertilization takes place, and seeds mature inside the carpel.

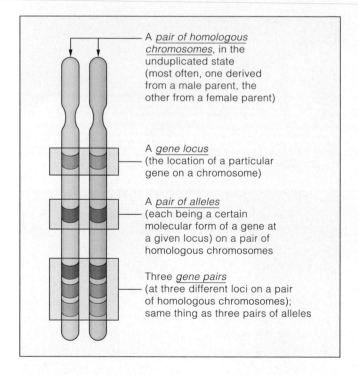

Figure 17.3 A few genetic terms illustrated. Diploid organisms such as humans have pairs of genes on pairs of homologous chromosomes. One chromosome of each pair is inherited from an individual's mother and the other from the father.

The genes themselves may have different molecular forms, called alleles. Different alleles specify slightly different versions of the same trait. An allele at one location on a chromosome may or may not be identical to its partner on the homologous chromosome.

tal letter represents a dominant allele, and a lowercase letter represents a recessive allele (for instance, *A* and *a* or *C* and *c*).

6. A *homozygous dominant* individual has a pair of dominant alleles (*AA*) for the trait being studied. A *homozygous recessive* individual has a pair of recessive alleles (*aa*). A *heterozygous* individual has a pair of non-identical alleles (*Aa*).

7. Two terms help keep the distinction clear between genes and the traits they specify. **Genotype** refers to the genes present in an individual. **Phenotype** refers to an individual's observable traits.

Table 17.1 summarizes the relationship between genotype and phenotype.

The Theory of Segregation

Mendel had an idea that a pea plant inherits separate units of information for a trait, one from each parent. Those units were Mendel's "factors," which we now call genes. In formulating his ideas, Mendel proposed three basic tenets: (1) There are discrete units of inheritance; (2) each diploid organism inherits two such units for each trait, one from each parent; and (3) in parents, different units assort independently into gametes.

To test his hypothesis, Mendel devised what is now called a **monohybrid cross**. *Mono-* means one, and a monohybrid cross addresses the question of how a single trait is passed to offspring. In a monohybrid cross, the parents have differing allele pairs for the trait in question. Although we do not carry out controlled crosses between individual humans, such crosses do occur naturally, and they show patterns of inheritance at work.

The following example illustrates a monohybrid cross involving the configuration of a person's chin. This trait is controlled by a gene that has two allelic forms. One allele, which calls for an indentation called a chin fissure (Figure 17.4), is dominant when it is present; we can represent this genotype as *C*. The recessive allele, which codes for a smooth chin, is *c*. In the example we consider here,

a   b

Figure 17.4 (**a**) The chin fissure, a heritable trait arising from a rather uncommon allele of a gene. Actor Kirk Douglas received a gene that influences this trait from each of his parents. At least one of those genes called for a chin fissure, and one is all it takes in this case. (**b**) This photograph shows what Mr. Douglas's chin might have looked like if he had inherited two ordinary forms of the gene instead.

one parent is homozygous for the *C* form of the gene and has a chin fissure, and the other is homozygous for the *c* form and has a smooth chin. The *CC* parent produces only *C* gametes, and the *cc* parent produces only *c* gametes:

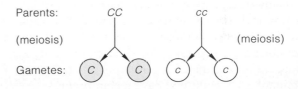

Each gamete receives one allele for the trait. This makes sense because in anaphase I in a germ cell, the homologous chromosomes separate from each other (Figure 16.9, page 366). We know already that the genes on homologues match up into equivalent pairs. So, when homologues separate into different gametes, the two genes of each pair separate also (Figure 17.5). This is known as the theory of **segregation**. Because meiosis also reduces the diploid chromosome number to the haploid number, each gamete contains one member of each chromosome pair.

The two copies of each gene present in a diploid organism segregate from each other during meiosis in germ cells. As a result, each gamete contains only one copy of each gene.

Table 17.1	Genotype and Phenotype Compared	
Genotype	Described as	Phenotype
CC	homozygous dominant	chin fissure
Cc	heterozygous (one of each allele; dominant form of trait observed)	chin fissure
cc	homozygous recessive	smooth chin

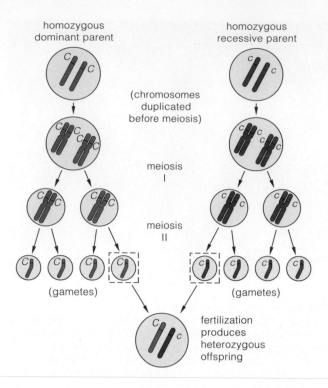

homozygous dominant parent

homozygous recessive parent

(chromosomes duplicated before meiosis)

meiosis I

meiosis II

(gametes)

(gametes)

fertilization produces heterozygous offspring

Figure 17.5 Example of a monohybrid cross, showing how one gene of a pair segregates from the other. Two parents that are each homozygous for a different version of a trait give rise only to heterozygous offspring.

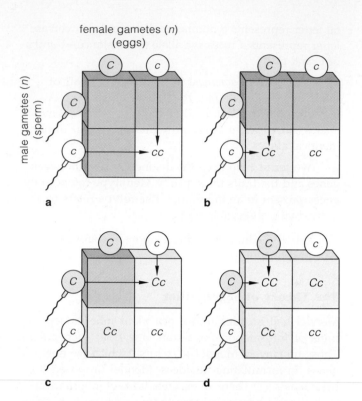

female gametes (*n*) (eggs)

male gametes (*n*) (sperm)

a

b

c

d

Figure 17.6 Punnett-square method of predicting the probable outcome of a genetic cross. In this example, the cross is between two heterozygous individuals. Circles represent gametes. Letters on gametes represent dominant or recessive alleles. The different squares depict the different genotypes possible among offspring.

The parent generation of a cross is designated by a **P**. Offspring of an initial cross are called the F_1 or *first filial* generation. In the example we are considering, each child will inherit a pair of differing alleles for the trait, one from each homozygous parent. The children will thus each be heterozygous for the chin genotype, or *Cc*. Because *C* is dominant, each child will have a chin fissure.

Suppose now that one of the *Cc* children grows up and marries another *Cc* person. From this second cross we derive the F_2 (second filial) generation. Figure 17.6 shows what genotypes will arise. Because half of each parent's gametes (sperm or eggs) are *C* and half are *c* (due to segregation at meiosis), four outcomes are possible every time a sperm fertilizes an egg:

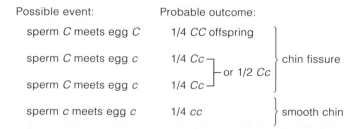

Possible event:	Probable outcome:	
sperm *C* meets egg *C*	1/4 *CC* offspring	
sperm *C* meets egg *c*	1/4 *Cc*	or 1/2 *Cc* } chin fissure
sperm *C* meets egg *c*	1/4 *Cc*	
sperm *c* meets egg *c*	1/4 *cc*	} smooth chin

The diagrams in Figure 17.6 show how to construct a **Punnett square**, a convenient tool for determining the probable outcome of genetic crosses. In this case, there is a 75 percent chance that a child from a cross between two *Cc* parents will have the dominant *C* allele and a chin fissure. When a large number of offspring are involved, a ratio of 3:1 is likely (Figure 17.7).

Fertilization is a chance event, which is why rules of probability apply to crosses. **Probability** simply means the number of times a certain outcome is likely to occur, divided by the total number of all possible outcomes (Figure 17.8).

Having a chin fissure doesn't affect a person's health. However, a fair number of human genetic disorders, including cystic fibrosis and sickle cell anemia, do involve single-gene defects and so follow a Mendelian inheritance pattern. We look in detail at human genetic disorders in Chapter 18. For the moment, it is important to realize two things:

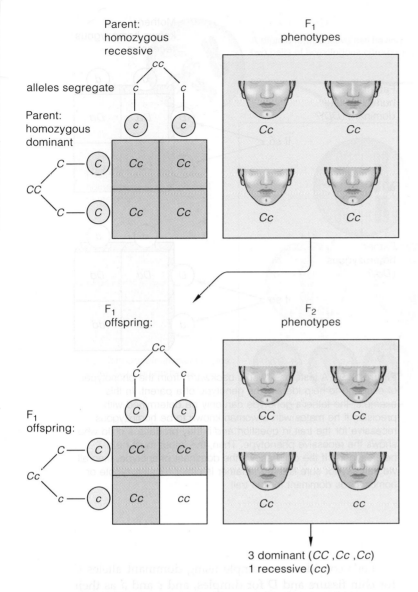

Figure 17.7 Results from a monohybrid cross, in which one parent is homozygous dominant for a trait and the other is homozygous recessive for the same trait. Notice that the dominant-to-recessive ratio is 3:1 for the second generation (F₂) offspring.

Parent: homozygous recessive

alleles segregate

Parent: homozygous dominant

F₁ phenotypes

F₁ offspring:

F₂ phenotypes

3 dominant (CC, Cc, Cc)
1 recessive (cc)

How to Calculate Probability

Step 1. Actual genotypes of parental gametes

In the cross $Cc \times Cc$, gametes have a 50–50 chance of receiving either allele (C or c) from each parent. Said another way, the probability that a particular sperm or egg will be C is 1/2, and the probability that it will be c is also 1/2:

probability of C: 1/2

probability of c: 1/2

Step 2. Probable genotypes of offspring

Offspring receive one allele from each parent. Three different combinations of alleles are possible in this cross. To figure the probability that a child will receive a particular allele combination, simply multiply the probabilities of the possible allele pairs:

probability of CC: $1/2 \times 1/2 = 1/4$

probability of Cc: $1/2 \times 1/2 = 1/4$ ⎤
probability of cC: $1/2 \times 1/2 = 1/4$ ⎦ 1/2

probability of cc: $1/2 \times 1/2 = 1/4$

Step 3. Probable phenotypes

Chin fissure: $1/4 + 1/4 + 1/4 = 3/4$
(CC, Cc, cC)

Smooth chin: 1/4
(cc)

Figure 17.8 Calculating probabilities. Simple multiplication lets you figure the probability that a child will inherit genes for a particular phenotype.

---

1.   The ratios predicted by probability don't necessarily turn up in a single family. They turn up consistently only when a large number of events is analyzed. You can test this for yourself by flipping a coin. Probability predicts that heads and tails should each come up about half the time, but you may have to flip the coin a hundred times to achieve a ratio close to 1:1. Likewise, probability predicts that parents with two or more children will have equal numbers of girls and boys, but this does not always happen.

2.   In a given situation, *probability remains constant.* The likelihood that a certain genotype will occur—say, a baby with the genotype for cystic fibrosis—is the same for every child no matter how many children a couple has. Based on the parents' genotypes, if the probability that a child will inherit a certain genotype is one in four, then each child of those parents has a one-in-four (25 percent) chance of inheriting the genotype. If the parents have three children without the trait, the fourth child still has only a one-in-four chance of inheriting it.

sibilities (Figure 17.11). Add them up, and you get nine combinations with a chin fissure and dimples, three with a chin fissure and no dimples, three with a smooth chin and dimples, and one with a smooth chin and no dimples. That is, you get a probable phenotypic ratio of 9:3:3:1. Whenever both parents are heterozygous for genes controlling two traits (i.e., *AaBb* x *AaBb*), this 9:3:3:1 ratio is the predicted outcome of the first generation offspring.

Figure 17.12 shows how to calculate the probability that a child will inherit genes for a particular set of two traits on different chromosomes.

VARIATIONS ON MENDEL'S THEMES

Mendel deliberately focused on traits that were either dominant or recessive, with no in-between. Yet many traits do not show up in populations in predicted Mendelian ratios. We now know that such deviations from Mendel's principles reflect interactions among genes. As you study the following section, keep in mind the basic function of genes. Genes are chemical instructions for building proteins. At the most fundamental level, a gene is "expressed" when its instructions are implemented and the protein in question is synthesized (page 418).

Dominance Relations in Humans

In some heterozygotes, a pair of contrasting alleles of a gene show **codominance**. They specify different phenotypes, and both are expressed. A classic example of codominance occurs in people who are heterozygotes for alleles of the gene that codes for hemoglobin, the oxygen-transporting pigment in red blood cells (page 190).

The allele that carries instructions for normal hemoglobin is Hb^A, and the allele for abnormal hemoglobin is Hb^S. Normally, red blood cells are disk-shaped. In the homozygous recessive condition ($Hb^S Hb^S$), the disease called *sickle cell anemia* results. Red blood cells become deformed into a sickle shape when the oxygen content of blood falls below a certain level (Figure 17.13). The sickled cells clump in blood capillaries, and the cells may even rupture. Blood flow can be so disrupted that oxygen-deprived tissues are catastrophically damaged, and

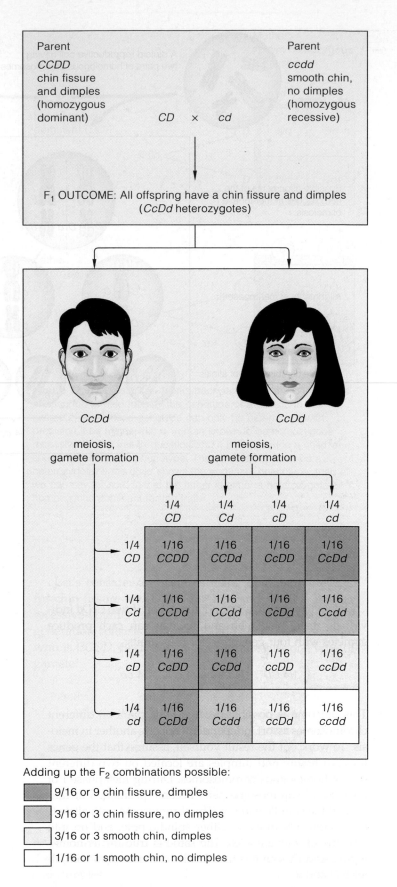

Figure 17.11 Results from a dihybrid cross between parents who are homozygous for different versions of two traits (dimples and chin fissure). *C* and *c* represent dominant and recessive alleles for chin fissure. *D* and *d* represent dominant and recessive alleles for dimples. The probabilities of certain combinations of phenotypes among F₂ offspring occur in a 9:3:3:1 ratio, on the average.

Figure 17.12 Probability also applies to dihybrid crosses.

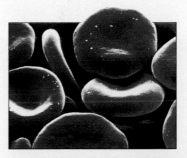

Figure 17.13 Normal and sickled red blood cells. Normal RBCs are biconcave discs. The sickle shape of the abnormal RBCs causes them to clump and rupture in capillaries.

homozygotes often die early in life. Heterozygotes ($Hb^A Hb^S$), on the other hand, have *sickle cell trait*. Their red blood cells contain both normal and abnormal hemoglobin. Under most environmental conditions they may show few symptoms because the one "good" allele provides enough normal hemoglobin in red blood cells to prevent the cells from sickling.

During a crisis, sickle cell anemia patients may receive blood transfusions, oxygen, antibiotics, and painkilling drugs. Scientists are also exploring the therapeutic potential of butyrate, a common food additive. There is evidence that butyrate can reactivate "dormant" genes responsible for fetal hemoglobin, which normally is produced only before birth and is an efficient oxygen carrier (page 322). Many states now require hospitals to screen newborn infants for sickle cell anemia so that appropriate drug therapy can be prescribed right away.

Genes and Blood Type Human blood types are another example of codominance. If you have type AB blood, for instance, you have a pair of codominant alleles that are both expressed in the stem cells that give rise to your red blood cells. Recall that a polysaccharide (a sugar) on the surface of red blood cells has different molecular forms that determine your blood type (page 157). A technique called ABO blood typing can reveal which form (or forms) of the polysaccharide a person has.

The gene specifying an enzyme that synthesizes this polysaccharide has three alleles, which influence the sugar molecule's form in different ways. Two alleles, I^A and I^B, are codominant when paired with each other. A third allele, i, is recessive. When paired with either I^A or I^B, its effect is masked. Whenever a gene has three or more alleles, we call this a **multiple allele system**. Table 17.2 summarizes the blood type alleles in humans.

Table 17.2	Human Blood Groups
Genotype	Blood Type (Phenotype)
$I^A I^A$ or $I^A i$	A
$I^B I^B$ or $I^B i$	B
$I^A I^B$	AB
ii	O

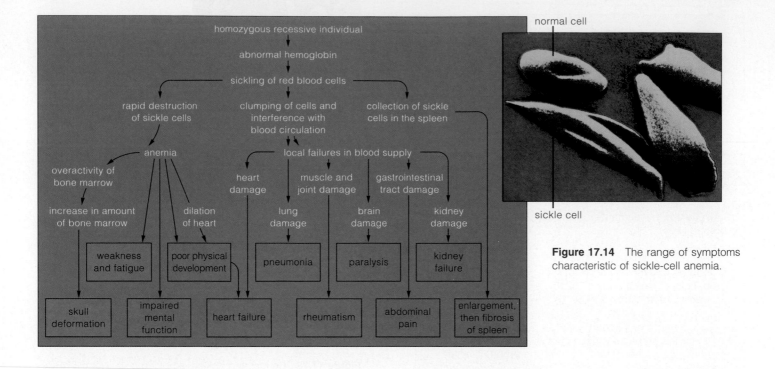

Figure 17.14 The range of symptoms characteristic of sickle-cell anemia.

Alleles I^A and I^B call for different versions of an enzyme that links a sugar chain to a lipid. (This happens in the Golgi body, where sugars are added to glycolipids.) The two enzymes attach different versions of the sugar chain, and this gives the glycolipid a special identity—either A, B, or O.

Which two alleles for this enzyme do you have? With either $I^A I^A$ or $I^A i$, your blood is type A. With $I^B I^B$ or $I^B i$, it is type B. With codominant alleles $I^A I^B$, both versions of the enzyme are produced, so the glycolipid has both forms of the sugar chain attached. In this case, your blood is type AB. If you are homozygous recessive (*ii*), a five-sugar (rather than six-sugar) polysaccharide is attached. Then, your blood type is neither A nor B, but type O.

Accurate blood typing is essential before a person receives a transfusion. When blood of two people mixes, all surface antigens must be compatible. Red blood cells from a donor who has an antigen incompatible with the recipient's blood will be recognized as foreign by the immune system and will trigger rejection reactions that can quickly lead to death.

Multiple Effects of Single Genes

In **pleiotropy**, one gene has several seemingly unrelated effects. For instance, mutation in a single gene results in a disorder called *osteogenesis imperfecta*. Infants with this disorder typically develop extremely fragile bones, weak tendons and ligaments, deafness, and blue coloration in the sclera of the eye. Likewise, the direct impact of the Hb^S allele is sickling of red blood cells, but that abnor-

mality can have many other harmful effects, including severe abdominal pain, impaired mental functioning, and pneumonia (Figure 17.14).

LESS PREDICTABLE VARIATIONS IN TRAITS

Penetrance and Expressivity

A gene can have an all-or-nothing effect on a trait—either you have dimples or you don't. In other cases, gene interactions as well as environmental factors may alter the *degree* to which a gene is actually expressed.

Penetrance refers to the percentage of individuals in which a particular genotype is expressed. The recessive gene that produces cystic fibrosis is completely penetrant; 100 percent of children born with it develop the disease (page 390). The gene for having extra fingers or toes (called *polydactyly*) is incompletely penetrant. Some people who inherit the dominant allele for this condition have the normal number of digits, and others have more (Figure 17.15). Some people with *campodactyly* have immobile, bent fingers on both hands. In others, the traits show up on one hand only. When expression of a gene can produce such a range of phenotypes, the gene is said to show "variable expressivity." The gene for campodactyly also is incompletely penetrant. In some people with the gene, the traits don't show up at all.

Studies of some rare genetic disorders have uncovered evidence that a trait may differ from person to per-

Mom and Pop Genes: Genomic Imprinting

No doubt Gregor Mendel would have been fascinated by the investigations of modern geneticists. In the case of a phenomenon called *genomic imprinting*, he might have been forced to consider adding a footnote to his basic principles. A basic tenet of Mendelian genetics is this: A given gene will give rise to the same trait, regardless of which parent it is inherited from. In the 1980s, however, researchers began to suspect that this is not always true. A "genome" is the sum total of genes a person (or any other organism) inherits from his or her parents. And in some cases, it seems that a given gene can result in *different* traits in offspring, depending on which parent the gene came from. Such genes must somehow be marked, or imprinted, as "from mother" or "from father."

Tantalizing evidence for genomic imprinting has come from odd inheritance patterns for certain disorders. For example, a rare inherited neck tumor (paraganglioma) is caused by a DNA segment on chromosome 11. However, genetic analysis shows that children develop the tumors only if they inherit the tumor-causing gene from their father. Girls who have the gene (and have tumors) do not pass it on to their children. The same connection between a paternal imprint and disease seems to hold true for several types of childhood cancer, including osteosarcoma, a bone cancer.

Two unusual genetic diseases that cause different types of mental retardation also have shed light on genomic imprinting. The diseases apparently can arise in two ways. In extremely rare cases, a chance glitch produces a zygote with two normal copies of chromosome 15, but *both copies are from one parent*. If the chromosomes come from the mother, the result is *Prader-Willi syndrome*, or PWS. If they come from the father, *Angelmann's syndrome* (AS) occurs. More often, an affected child inherits a homologous pair of chromosome 15, one from each parent, but one of the chromosomes has a region missing. As you might guess, if the defect is in the paternal chromosome PWS develops, and AS develops if the problem is in the maternal chromosome.

No one yet knows exactly how genetic material might become imprinted. Even so, our growing understanding of the phenomenon can be useful in genetic counseling. For example, parents who have a child with PWS or AS can be tested to determine whether the problem lies in an inherited chromosome 15 defect. If so, then the risk of having another retarded child may be relatively great. But if both parents have normal chromosomes, the probability that a second zygote would receive two copies of chromosome 15 from only one of them—and thus be affected—is extremely small.

son, depending on whether the gene(s) responsible for it are inherited from the mother or from the father. The phenomenon is called *genomic imprinting*, and it is the topic of this chapter's *Focus on Science*.

Differences in Gene Interactions   Some of the variations we observe in gene expression are likely the result of gene interactions. Bear in mind that the path from most precursors to their products (proteins) is actually a series of small metabolic steps, each controlled by enzymes, which are the products of genes. Genes or their products interact at many steps. If different people have varying combinations of alleles for those genes, their phenotypes may vary also.

Continuous Variation and Polygenic Traits

Many human traits are polygenic—they result from the combined expression of two or more genes. Skin and eye color are in this category. Both are the cumulative result

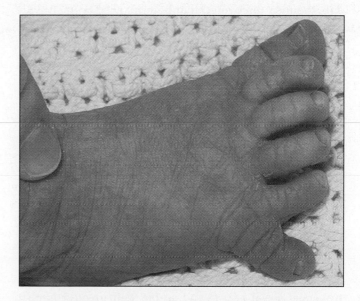

Figure 17.15   An example of polydactyly, a dominant trait.

of many genes involved in the stepwise production and distribution of melanin. Black eyes have abundant melanin deposits in the iris. Dark-brown eyes have less melanin, and light-brown or hazel eyes have still less (Figure 17.16). Green, gray, and blue eyes don't have green, gray, or blue pigments. Instead, they have so little melanin that we readily see blue wavelengths of light being reflected from the iris. Skin color probably results from the interactions of at least four genes. This explains our real-world observations of a tremendous amount of variation in human skin color.

For many polygenic traits a population may show **continuous variation**. That is, its members show a range of smaller, less pronounced differences in some trait. Continuous variation is especially evident in traits that are easily measurable, such as height (Figure 17.17). The more genes that affect a trait, the more its different versions will combine in a smooth continuum.

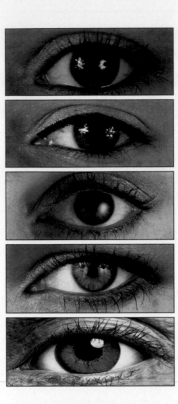

Figure 17.16 Samples from the range of continuous variation in human eye color. Different gene pairs interact to produce and deposit melanin. Among other things, this pigment helps color the eye's iris. Different combinations of alleles result in small differences in eye color. So the frequency distribution for the eye color trait appears to be continuous over the range from black to light blue.

Gene expression can vary, so that individuals of a population show varying phenotypes for each genotype.

Gene interactions and environmental factors can influence the degree to which a gene is expressed in different individuals.

Do Genes "Program" Behavior?

The identical twins shown in Figure 17.18 were separated at birth and raised in different communities. Yet in addition to having identical genes and looking alike, they have many behaviors in common—apparently down to the shape in which they trim their mustaches. When another set of separated identical twins was reunited in middle age, the men discovered that both married women with the same first name (twice!), both had given a son the same first and middle names, and both took vacations at the same beach in Florida.

Are these parallels coincidence, clever hoaxes, or evidence that aspects of behavior are inherited along with hair and eye color? Does "nature" have the major role in determining who we are, or is "nurture" responsible—or is the reality somewhere in between? Although the questions are intriguing, clear answers have proven very difficult to come by.

There is strong evidence that certain basic human behaviors, such as smiling to indicate pleasure and the crying of an infant when it is hungry, are indeed genetically programmed universals. Recently, scientists have also begun to look for connections between genes and alcoholism, some types of mental illness, violent behavior, and even sexual orientation. Such studies raise controversial social issues, and so far their greatest impact

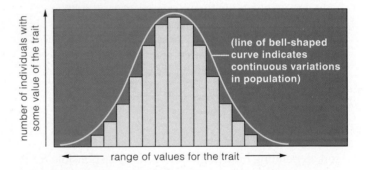

Figure 17.17 Continuous variation in traits. Most traits show continuous variation among the members of a population. Humans, for instance, are not all tall *or* short. They range from very short to very tall, with average heights being more common than either extreme. Often a bar graph is used to depict continuous variation. Here, the proportion of individuals in each category is plotted against the range of measured phenotypes.

has been to point up how little we really know about the biological basis of human behavior. In general, most human behavior is so complex that it is extraordinarily difficult to design scientifically rigorous experiments to test hypotheses about genetic links. For the time being, we can only marvel at this remarkable complexity and ponder the intriguing possibilities of human inheritance.

Figure 17.18 Identical twins who, although reared apart, still share many personality characteristics in addition to their physical likenesses. It is unclear how—or if—human personality traits and complex behaviors are related to genetic makeup.

SUMMARY

1. Mendel's hybridization studies with pea plants provided evidence that genes remain as discrete units when passed to offspring.

2. Monohybrid crosses (between parents showing different versions of a single trait) provide evidence that a gene can have different molecular forms (alleles), some of which are dominant over other, recessive forms.

3. Homozygous dominant individuals have two dominant alleles (*AA*) for the trait being studied. Homozygous recessives have two recessive alleles (*aa*). Heterozygotes have two contrasting alleles (*Aa*).

4. In the text example of a monohybrid cross (*CC* x *cc*), all F$_1$ offspring were *Cc*. Crosses between F$_1$ offspring resulted in the following combinations of alleles in F$_2$ offspring:

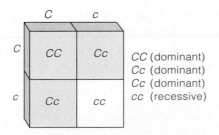

CC (dominant)
Cc (dominant)
Cc (dominant)
cc (recessive)

This produced an expected phenotypic ratio of 3:1.

5. The results of monohybrid crosses provide evidence in support of the principle of segregation. This principle states that (1) diploid organisms have pairs of genes, on pairs of homologous chromosomes, and (2) the two genes of each pair segregate from each other during meiosis, such that each gamete formed ends up with one or the other.

6. A dihybrid cross occurs between two homozygous parents showing different versions of two traits. The result in the F$_1$ generation is:

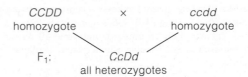

CCDD   ×   *ccdd*
homozygote   homozygote

F$_1$:   *CcDd*
all heterozygotes

After a second round of reproduction, between two heterozygous parents, the probable result is a 9:3:3:1 phenotypic ratio:

CcDd   ×   *CcDd*

9 dominant for both traits
3 dominant *A*, recessive for *b*
3 dominant *B*, recessive for *a*
1 recessive for both traits

7. Dihybrid crosses provide evidence of independent assortment during meiosis. In modern terms, pairs of homologous chromosomes tend to be sorted into one gamete or another independently of how other pairs of chromosomes are sorted out. Thus, the genes on the chromosomes are assorted independently.

8. Various factors can influence gene expression. Degrees of dominance exist between some gene pairs. Genes can also interact to produce some effect on a trait. In some cases a single gene can influence many seemingly unrelated traits.

1. Define the difference between: (a) gene and allele, (b) dominant allele and recessive allele, (c) homozygote and heterozygote, and (d) genotype and phenotype. *376*

2. State the theory of segregation. Does segregation occur during mitosis or meiosis? *377*

3. Distinguish between monohybrid and dihybrid crosses. What is a testcross, and why is it useful in genetic analysis? *377–380*

4. What is independent assortment? Does independent assortment occur during mitosis or meiosis? *380*

5. What is polygenic inheritance? What is the difference between penetrance and variable expressivity? *384–386*

9. Match each genetic term appropriately.
_____ dihybrid cross   a. *AA* x *aa*
_____ monohybrid cross,   b. *Aa*
       two heterozygotes   c. *AABB* x *aabb*
_____ homozygous condition   d. *Aa* x *Aa*
_____ heterozygous condition   e. *aa*
_____ homozygous parents

Critical Thinking: You Decide *(Key in Appendix IV)*

1. Identical twin sisters who were separated at birth and then reunited as adults discover that both are five feet four inches tall and weigh about 126 pounds. Both started walking at the age of 10 months, have chronic high blood pressure, and had chicken pox as children. Each twin also loves chocolate ice cream. Which of these similarities are *most* likely to be due to genetic influences (including development), and which are not? Give reasons for your conclusions.

2. About 300 people live on the island of Tristan da Cunha off the southwestern coast of Africa. For several centuries Tristan residents have intermarried. In the early 1960s, geneticists found that every island resident was homozygous for the same 10 recessive traits. Based on material in this chapter, offer a genetic explanation for this phenomenon.

Self-Quiz *(Answers in Appendix III)*

1. Alleles are _____.
   a. different molecular forms of a gene
   b. different molecular forms of a chromosome
   c. always homozygous
   d. always heterozygous

2. A heterozygote has _____.
   a. only one of the various forms of a gene
   b. a pair of identical alleles
   c. a pair of contrasting alleles
   d. a haploid condition, in genetic terms

3. The observable traits of an organism are its _____.
   a. phenotype   c. genotype
   b. sociobiology   d. pedigree

4. Offspring of a monohybrid cross *AA* x *aa* are _____.
   a. all *AA*   d. 1/2 *AA* and 1/2 *aa*
   b. all *aa*   e. none of the above
   c. all *Aa*

5. Second-generation offspring from a cross are the _____.
   a. F₁ generation
   b. F₂ generation
   c. hybrid generation
   d. none of the above

6. Assuming complete dominance, offspring of the cross *Aa* x *Aa* will show a phenotypic ratio of _____.
   a. 1:2:1   c. 9:1
   b. 1:1:1   d. 3:1

7. Which statement best fits the principle of segregation?
   a. Units of heredity are transmitted to offspring.
   b. Two genes of a pair separate from each other during meiosis.
   c. Members of a population become segregated.
   d. A segregating pair of genes is sorted out into gametes independently of how gene pairs located on other chromosomes are sorted out.

8. Dihybrid crosses of heterozygous individuals (*AaBb* x *AaBb*) lead to F₂ offspring with phenotypic ratios close to _____.

Genetics Problems *(Answers in Appendix II)*

1. One gene has alleles *A* and *a*. Another has alleles *B* and *b*. For each genotype listed, what type(s) of gametes can be produced? (Assume independent assortment occurs.)
   a. *AABB*   c. *Aabb*
   b. *AaBB*   d. *AaBb*

2. Still referring to Problem 1, what will be the possible genotypes of offspring from the following matings? With what frequency will each genotype show up?
   a. *AABB* x *aaBB*   c. *AaBb* x *aabb*
   b. *AaBB* x *AABb*   d. *AaBb* x *AaBb*

3. In one experiment, Mendel crossed a pea plant that produced only green pods with one that produced only yellow pods. All the F₁ plants had green pods. Which trait (green or yellow pods) is recessive? Explain how you arrived at your conclusion.

4. A gene on one chromosome controls a trait involving tongue movement. If you have a dominant allele of that gene, you can curl the sides of your tongue upward (see photo). If you have two recessive alleles of the gene, you cannot roll your tongue. A gene on a different chromosome controls whether your earlobes are attached or detached (Figure 17.1). People with detached earlobes have at least one dominant allele of the gene. These two genes are on different chromosomes and so they assort independently. Suppose a tongue-rolling woman with detached earlobes marries a man who has attached earlobes and can't roll his tongue. Their first child has attached earlobes and can't roll its tongue.
   a. What are the genotypes of the mother, the father, and the child?
   b. What is the probability that a second child will have detached earlobes and won't be a tongue-roller?

5. Go back to Problem 1, and assume you now study a third gene having alleles *C* and *c*. For each genotype listed, what type(s) of gametes can be produced:
   a. *AABBCC*   c. *AaBBCc*
   b. *AaBBcc*   d. *AaBbCc*

6. Your sister moves away and gives you her purebred Labrador retriever, a female named Dandelion. When you decide to breed Dandelion and sell the puppies to help pay your tuition, you discover that two of Dandelion's four siblings have developed a hip disorder that is traceable to the action of a single recessive gene. Dandelion herself shows no sign of the disorder. If you breed Dandelion to a male Labrador that does not carry the recessive allele, can you guarantee to a purchaser that the resulting puppies will also be free of the condition? Explain your answer.

7. The ABO blood system has been used to settle cases of disputed paternity. Suppose, as a geneticist, you must testify during a case in which the mother has type A blood, the child has type O blood, and the alleged father has type B blood. How would you respond to the following statements:

a. *Attorney of the alleged father:* "The mother has type A blood, so the child's type O blood must have come from the father. Because my client has type B blood, he could not have fathered this child."

b. *Mother's attorney:* "Further tests prove this man is heterozygous, so he must be the father."

8. Soon after a couple marries, tests show that both the man and the woman are heterozygotes for the recessive allele that causes sickling of red blood cells; they are both Hb^AHb^S. What is the probability that any of their children will have sickle cell trait? Sickle cell anemia?

9. A man is homozygous dominant for 10 different genes that assort independently. How many genotypically different types of sperm could he produce? A woman is homozygous recessive for eight of these ten genes and is heterozygous for the other two. How many genotypically different types of eggs could she produce? What can you conclude regarding the relationship between the number of different gametes possible and the number of heterozygous and homozygous gene pairs that are present?

10. As is the case with the mutated hemoglobin gene that causes sickle cell anemia, certain dominant alleles are crucial to normal functioning (or development). Some are so vital that when the mutant recessive alleles are homozygous, the combination is lethal and death results before birth or early in life. However, such recessive alleles can be passed on by heterozygotes (*Ll*). In many cases, these are not phenotypically different from homozygous normals (*LL*). If two heterozygotes mate (*Ll* x *Ll*), what is the probability that any surviving progeny will be heterozygous?

11. Bill and Marie each have flat feet, long eyelashes, and "achoo syndrome" (chronic sneezing). All are dominant traits. The genes for these traits each have two alleles, which we can designate as follows:

	Dominant	Recessive
Foot arch	A	a
Sneezing	S	s
Eyelash length	E	e

Bill is heterozygous for each trait. Marie is homozygous for all of them. What is Bill's genotype? What is Marie's genotype? If they have four children, what is the probability that each child will have the same phenotype as the parents? What is the probability that a child will have short lashes, high arches, and no "achoo syndrome"?

12. Aunt Tillie, your father's sister, has been diagnosed as allergic to a particular antibiotic. Rather than go into detail, her physician simply tells her that she has "inherited a bad protein." Tillie asks you for a fuller explanation. How would you explain the situation? (Hint: Remember that antibodies produced during a cell-mediated immune response are proteins, and proteins are encoded by genes.)

13. Referring to question 12, assume that the drug allergy trait results from a recessive allele. Neither of your parents exhibits the drug allergy, and you don't either. However, your mother is pregnant. Assuming that the problem is heritable in Mendelian fashion, what is the worst-case scenario for the likelihood that your new sibling will have the allergy?

Readings

Cummings, M. 1991. *Human Heredity*, 2nd ed. St. Paul: West Publishing Company.

Horgan, J. June 1993. "Eugenics Revisited." *Scientific American*. A skeptical look at studies aimed at finding connections between genes and human behavior.

Mendel, G. 1959. "Experiments in Plant Hybridization." Translation in J. Peters (editor), *Classic Papers in Genetics*. Englewood Cliffs, N.J.: Prentice-Hall.

"New Evidence Supports Genomic Imprinting." April 6, 1991. *Science News*.

18 CHROMOSOME VARIATIONS AND MEDICAL GENETICS

At the Mercy of Our Genes

Several years ago scientists identified the genetic cause of the disease cystic fibrosis. The breakthrough made news because CF is devastating. In the United States roughly one in every 2,000 babies is born with the disease, and many of those individuals have been likely to die from it before they reach 30.

Several different mutations in a single gene labeled CFTR can lead to the CF phenotype. The gene is carried on human chromosome number 7. When a mutated form is expressed, the outcome is a defective channel protein in the plasma membrane of cells in exocrine glands. Recall that such glands secrete a variety of substances, including digestive enzymes (in the pancreas), sweat, and mucus in the lungs and digestive tract (page 69). In CF, the genetic flaw affects the flow of certain ions into and out of cells. As a result, exocrine glands

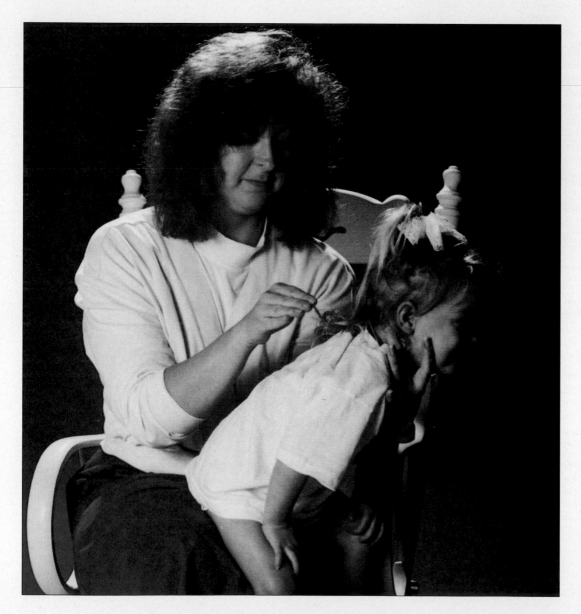

Figure 18.1 This little girl, born with a gene mutation that causes cystic fibrosis, is receiving percussion therapy as her mother pounds forcefully on her back. The procedure helps clear the child's lungs and bronchi of thick mucus, making breathing easier and reducing the chance of infection.

secrete an abnormally thick, gluey mucus. Ducts of the pancreas become so mucus-clogged that enzymes required for proper digestion of proteins, lipids, and carbohydrates never make it to the small intestine. Until their disease is diagnosed, affected children often suffer from severe malnutrition.

The most dangerous, frightening effect of CF occurs in the lungs. Cells lining respiratory passages crank out huge quantities of the sticky mucus, overwhelming the normal cleansing action of cilia. Breathing becomes extraordinarily difficult, and the bronchi and lungs become susceptible to infections such as pneumonia. Many CF children begin the day with a regimen of physical therapy designed to literally pound mucus from their lungs (Figure 18.1). Most deaths result from respiratory failure.

In this chapter we delve more deeply into the inheritance of chromosomes and the genes they carry. Human inheritance patterns run the gamut from totally unforeseeable changes in genetic instructions to the predictable appearance of a trait through many generations of a family. Cystic fibrosis is only one of several thousand disorders that can arise from errors in genes or chromosomes. Fortunately, as we will see in this chapter and the next, our growing understanding of the basis of genetic disorders holds the promise of more effective treatments—and improving lives.

KEY CONCEPTS

1. Genes are segments of DNA. One gene follows another along the length of a chromosome, and each one has its own position in that sequence.

2. Chromosomes are of two types: autosomes and sex chromosomes (X and Y). Diploid human cells contain 22 pairs of autosomes and one pair of sex chromosomes. Normal female sex chromosomes are XX and normal male sex chromosomes are XY.

3. The gene sequence does not necessarily remain intact through meiosis and gamete formation. Whenever crossovers take place, some genes in the sequence are removed and replaced with genes from the homologous chromosome. Crossing over occurs more often between genes that are farther apart in the sequence. It occurs less often between genes that are close together.

4. Inheritance of many traits, including genetic disorders, follows one of several common patterns. These include autosomal recessive inheritance, autosomal dominant inheritance, and X-linked recessive inheritance.

5. Genetic disorders can arise when the genome is altered. These genetic changes fall into three categories: (a) gene mutations, (b) abnormal changes in chromosome structure, and (c) changes in the number of chromosomes.

CHROMOSOMAL INHERITANCE

Chromosomes were first observed with a microscope in the 1880s. Since then, geneticists drawing on the work of Mendel and other early thinkers have learned a great deal about chromosomes, the genes they carry, and human inheritance. As we begin this chapter, here are some fundamental points to keep in mind about chromosomal inheritance:

1. Genes, the units of instruction for heritable traits, are segments of DNA. They are arranged along chromosomes in linear order.

2. From meiosis through fertilization, sexual reproduction assures that each new individual will have the same chromosome number as its parents.

3. Diploid ($2n$) cells have pairs of homologous chromosomes. Except for sex chromosomes, the homologues have the same length and shape and carry genes for the same traits. All of them, including sex chromosomes, line up with their partner at meiosis.

4. Genes located on nonhomologous chromosomes are inherited independently of each other, in accordance with the principle of independent assortment (page 380).

5. Genes on the same chromosome tend to stay together when chromosomes move about during meiosis. But crossing over (breakage and exchange of corresponding segments of homologues) can disrupt such linkages and lead to genetic recombination (Figure 18.2).

6. Chromosomal abnormalities sometimes occur. A chromosome segment may be deleted, duplicated, inverted, or moved to a new location. Chromosomes also may not separate properly at meiosis, and so gametes end up with an abnormal chromosome number.

Autosomes and Sex Chromosomes

In the early 1900s, microscopists discovered a chromosomal difference between the sexes. In 22 of the 23 pairs of chromosomes in a human diploid cell, the maternal and paternal chromosomes are roughly the same size and shape and carry genes for the same traits. These are **autosomes**. The remaining pair of chromosomes are **sex chromosomes**, and they differ between males and females. Sex chromosomes carry the basic hereditary instructions about *gender*—that is, whether a new individual will be male or female. Human females have two sex chromosomes, both designated "X." Males have one X chromosome and another, physically different chromosome designated "Y."

Today, special stains can be used to precisely characterize human autosomes and sex chromosomes at metaphase, when they are most condensed. Then, each type has a certain length, banding pattern, and so forth. (The bands appear because some regions condense more than others and absorb stains differently.) These features are used to create a **karyotype**, a cut-up, rearranged pho-

Figure 18.2 Crossing over and genetic recombination. Figure 16.11 on page 369 showed the stage in meiosis when crossing over occurs. Only a single crossover was illustrated, but in human cells every chromosome must undergo at least one crossover. Otherwise meiosis cannot be properly completed.

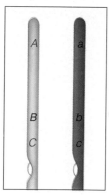 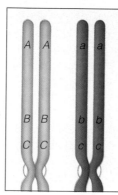   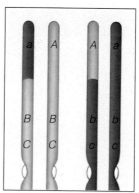

a Equivalent gene regions of a pair of homologous chromosomes at interphase, before DNA replication.

b The same regions after DNA replication at interphase. Both chromosomes are now in the duplicated state.

c At prophase I of meiosis, two of the nonsister chromatids break while aligned very tightly together.

d The nonsister chromatids swap segments, then enzymes seal the broken ends.

e The crossover led to genetic recombination between two of four chromatids. (Here they are shown after meiosis, as separate, unduplicated chromosomes.)

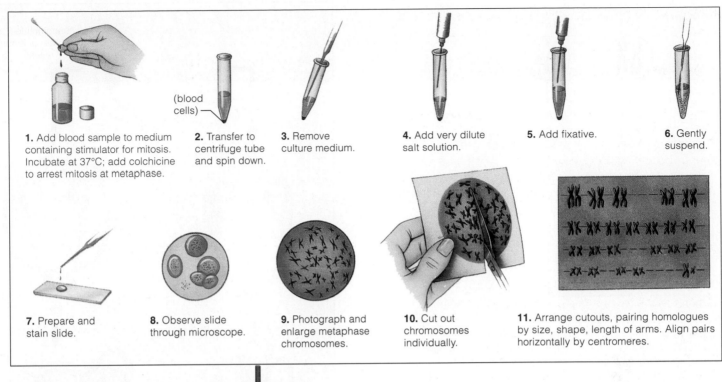

1. Add blood sample to medium containing stimulator for mitosis. Incubate at 37°C; add colchicine to arrest mitosis at metaphase.

2. Transfer to centrifuge tube and spin down.

(blood cells)

3. Remove culture medium.

4. Add very dilute salt solution.

5. Add fixative.

6. Gently suspend.

7. Prepare and stain slide.

8. Observe slide through microscope.

9. Photograph and enlarge metaphase chromosomes.

10. Cut out chromosomes individually.

11. Arrange cutouts, pairing homologues by size, shape, length of arms. Align pairs horizontally by centromeres.

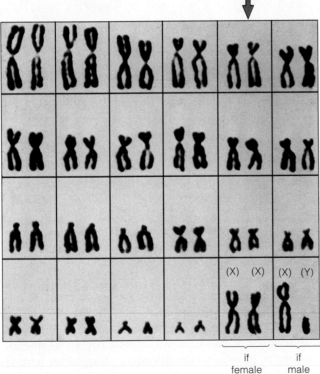

(X) (X)   (X) (Y)

if female   if male

Figure 18.3 Karyotype preparation. Human cells have a diploid chromosome number of 46. The nucleus contains 22 pairs of autosomes and 1 pair of sex chromosomes (X and Y). Each chromosome in this karyotype is duplicated and so consists of two sister chromatids.

tograph of the chromosomes of a cell at metaphase. The chromosomes are arranged in order, from largest to smallest, and the sex chromosomes are placed last (Figure 18.3).

Each gamete produced by a female (XX) carries an X chromosome. Half the gametes produced by a male (XY) carry an X and half carry a Y chromosome. A new individual's sex is determined by the father's sperm because the mother's oocytes can contain only an X chromosome, while the father's sperm can carry either an X or a Y chromosome. When a sperm and an egg both carry an X chromosome and combine at fertilization, the new individual will develop into a female. Conversely, when the sperm carries a Y chromosome, the new individual will be XY and develop into a male.

Recall from the introduction to Chapter 13 that among the very few genes on the Y chromosome is a "male-determining gene" labeled Sry. The presence of its product causes a new individual to develop testes. In the absence of that gene product, ovaries will develop automatically. The hormones secreted from testes and ovaries trigger the development of sexual characteristics.

The human X chromosome probably carries more than 300 genes. Like other chromosomes, it carries some genes associated with sexual traits, such as the distribution of body fat. But most of the genes on the X chromosome are concerned with *nonsexual* traits, such as blood-clotting functions.

A Y chromosome is much smaller than an X and carries far fewer genes. We know that these include genetic instructions for male sexual development, but geneticists

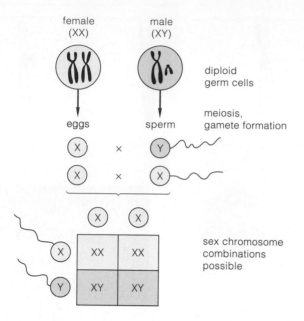

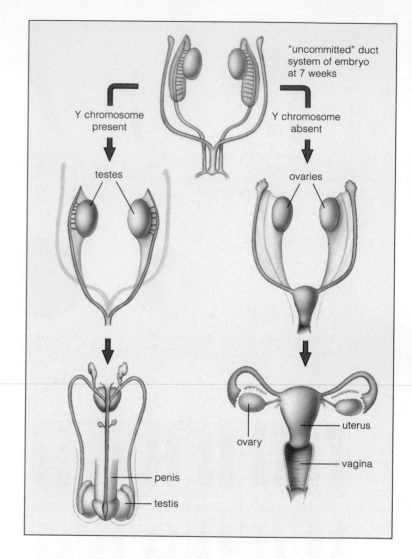

Figure 18.4 The pattern of sex determination in humans. (**above**) Only the sex chromosomes, not the autosomes, are shown. Males transmit their Y chromosome to their sons but not to their daughters. Males receive their X chromosome only from their mother. (**right**) Duct system in the early embryo that develops into a male or female reproductive system. Compare to Figure 13.1 on page 292.

disagree over the evidence for other roles of Y chromosome genes.

Any gene on the X or Y chromosome may be called a "sex-linked gene." However, researchers now use the more precise designations, **X-linked gene** or **Y-linked gene**. As we will see, X-linkage is a key factor in several genetic disorders. Figure 18.4 shows the pattern of sex determination in humans.

X Chromosome Inactivation The fact that females have two X chromosomes and males have only one raises an interesting question. Do females have twice as many X-linked genes as males do, and hence a double dose of their gene products? Studies show that for the most part the answer is no, because a compensating phenomenon called **X inactivation** occurs in females. Although the precise mechanism is not fully understood, it appears that most or all of the genes on one of a female's X chromosomes are "switched off" in early development, soon after the first cleavages of the zygote. The switch flips at random; in a given cell, inactivation can occur in either one of the X chromosomes. The inactivated X becomes condensed into a *Barr body* (Figure 18.5) that can be seen when a female's body cells are examined microscopically.

Once X inactivation takes place, the embryo continues to develop. Ultimately, every female's body is a mosaic of tissues in which patches of cells have one or the other of the original pair of X chromosomes she inherited from her parents.

Gene Mutations on Chromosomes Genes are segments of DNA on chromosomes. As you know, DNA consists of various types of nucleotides linked by chemical bonds (page 33). A *mutation* is a change in one or more of the nucleotides that make up a particular gene. Mutations can arise in various ways, which we consider in Chapter 19. A mutation may have no apparent effect on an individual, it can code for a new, beneficial phenotype—or, as in some of the disorders we consider here, it can call for a phenotype that seriously disrupts healthy functioning.

Linkage and Crossing Over

Research on the tiny fruit fly *Drosophila melanogaster* has contributed major insights to our modern understanding of basic genetic principles. For example, early researchers

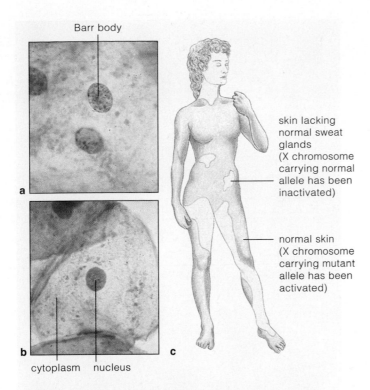

Barr body

a

b cytoplasm nucleus

c

skin lacking normal sweat glands (X chromosome carrying normal allele has been inactivated)

normal skin (X chromosome carrying mutant allele has been activated)

Figure 18.5 (**a**) Nucleus from a female cell, showing a Barr body. (**b**) Nucleus from male cell, which has no Barr body. (**c**) Pattern of gene expression in a woman affected by anhidrotic ectodermal dysplasia, a disorder in which patches of skin do not have normal sweat glands. Such "mosaic" tissue effects were first proposed by Mary Lyon. They arise through random inactivation of the X chromosome during embryonic development.

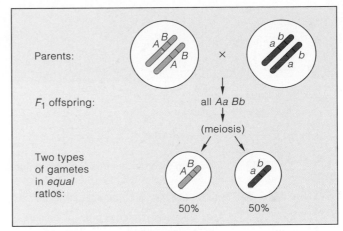

a Complete linkage (no crossing over)

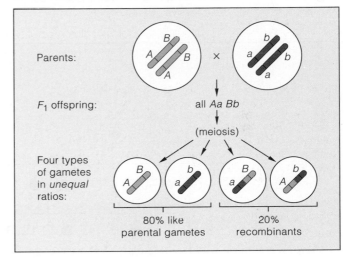

b Incomplete linkage owing to crossing over

Figure 18.6 How crossing over can affect gene linkage, using two gene loci as the example.

realized that many traits were being inherited *as a group* from one parent or the other. The term **linkage** is now used to describe the tendency of genes located on the same chromosome to end up together in the same gamete. We now know that linkage can be disrupted by crossing over, the breakage and exchange of segments between homologous chromosomes (Chapter 16 and Figure 18.2).

To grasp the relationship between linkage and crossing over, think about the location of any two genes on the same chromosome. The probability of a crossover occurring between those two genes is proportional to the distance separating them along the chromosome. Suppose that two genes A and B are twice as far apart as two other genes, C and D:

A B C D

We would expect crossing over to disrupt the linkages between A and B much more frequently than those between C and D.

Two genes are very closely linked when the distance between them is small; they nearly always end up in the same gamete (Figure 18.6a). Linkage is more vulnerable to crossing over when the distance between the two genes is greater (Figure 18.6b). Two genes very far apart on the same chromosome are affected by crossing over so often that they may appear to assort independently.

The patterns in which genes are distributed into gametes are so regular that they can be used to determine the positions of the various genes on a chromosome. Plotting the positions of genes on a chromosome is called **linkage mapping**.

There are an estimated 50,000 to 100,000 genes in the 23 types of human chromosomes. An ambitious, long-term international research effort called the Human Genome Project is now attempting to determine the pre-

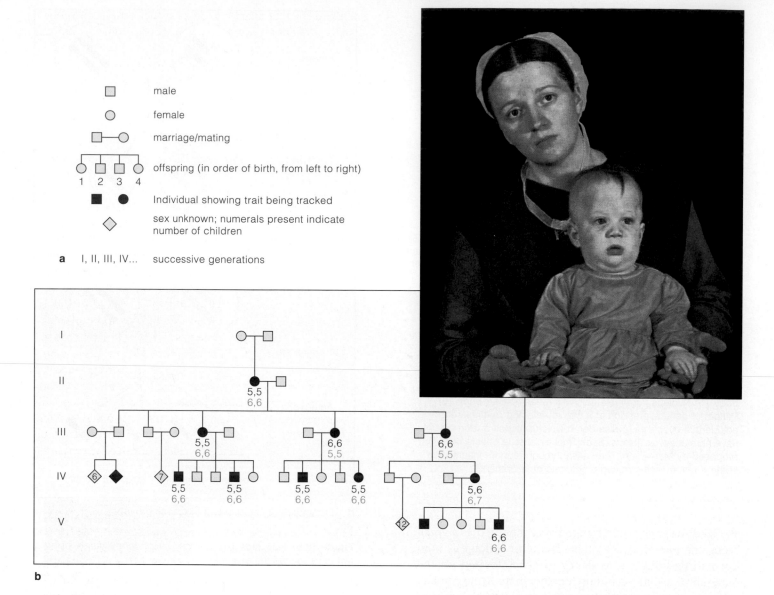

Figure 18.7 (**a**) Some symbols used in constructing pedigree diagrams. (**b**) Example of a pedigree for *polydactyly*, a condition in which an individual has extra fingers, extra toes, or both. Expression of the gene governing polydactyly can vary from one person to the next, an example of the variable expressivity described in Chapter 17. The number of fingers on each hand is shown in black numerals, and the number of toes on each foot is shown in blue. The phenotype of the female in generation I is uncertain.

Polydactyly is an example of how gene expression can vary. As a human embryo develops, a dominant allele *D* controls how many sets of bones will form within the body regions destined to become hands and feet. The *Dd* genotype varies in how it is expressed. The pedigree shown here indicates the kind of variation possible.

cise sequence of nucleotides on chromosomes, and then to identify the specific chromosome locations of all human genes. The project makes use of linkage mapping and also relies heavily on genetic engineering methods. We will consider it in more detail in Chapter 19.

Genes are arranged on chromosomes in a linear order.

The farther apart two genes are on a chromosome, the more often crossing over will occur between them.

Pairs of autosomes in a diploid cell carry genetic information about the same traits. Sex chromosomes (X and Y) carry information about gender. The X chromosome carries other information as well. Females are XX and males are XY.

Linkage mapping is a method for plotting the positions of genes on a chromosome.

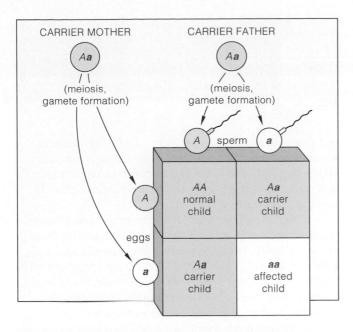

Figure 18.8 Possible phenotypic outcomes for autosomal recessive inheritance when both parents are heterozygous carriers of the recessive allele (shaded red here).

PATTERNS OF INHERITANCE

If you read the newspapers, you are probably aware that the field of human genetics is growing rapidly. **Pedigrees**, or charts of genetic connections among individuals, are important research tools. Figure 18.7 is an example. With pedigrees constructed by standardized methods, we can identify inheritance patterns and track genetic abnormalities through several generations. Pedigrees are clues to a trait's genetic basis. They may indicate whether the responsible gene is dominant or recessive and whether it occurs on an autosome or a sex chromosome. Gathering many family pedigrees increases the numerical base for analysis.

A person who is heterozygous for a recessive trait can be designated as a *carrier*. A carrier shows the dominant phenotype (few or no disease symptoms) but can produce gametes with the recessive allele and potentially pass on that allele to a child. If both parents are carriers for a genetic disorder, a child has a 25 percent chance of being homozygous recessive for the harmful allele (Figure 18.8).

Keep in mind that "abnormality" and "disorder" are not necessarily the same thing. *Abnormal* simply means deviation from the average, such as having six toes on each foot instead of five. A *genetic disorder* is an inherited condition that causes mild to severe medical problems.

It has been estimated that every human being carries an average of three to eight harmful recessive alleles. Why don't alleles that cause severe disorders simply disappear from a population? First, rare mutations put new copies of the alleles in the population. Second, in heterozygotes such a recessive allele is paired with a normal dominant one that may cover its functions, so it can be passed to offspring.

Autosomal Recessive Inheritance

When a trait results from a *recessive* allele on an autosome, two clues point to this condition. First, if both parents are heterozygous, there is a 50 percent chance that each child will be heterozygous and a 25 percent chance that it will be homozygous recessive (Figure 18.8). Second, if both parents are homozygous recessive, any child of theirs will be also.

Phenylketonuria (PKU) results from abnormal buildup of the amino acid phenylalanine. Affected people are homozygous recessive for a mutated form of a gene that normally codes for an enzyme that breaks down phenylalanine. If excess phenylalanine is diverted into other metabolic pathways, phenylpyruvic acid and other compounds may be produced. At high levels, phenylpyruvic acid can cause mental retardation. A diet that provides only the minimum required amount of phenylalanine will alleviate PKU symptoms. Diet soft drinks and many other products often are artificially sweetened with aspartame, which contains phenylalanine. Such products carry warning labels directed at phenylketonurics.

Cystic fibrosis (CF) is also an autosomal recessive condition that arises from a mutation in a gene on human chromosome 7. It is one of the common genetic disorders among Caucasians. Major symptoms are the buildup of massive amounts of sticky mucus in the lungs and pancreatic ducts. Recently, the National Institutes of Health approved experimental trials of a gene therapy for CF. Patients inhale a nasal spray containing genetically engineered "common cold" virus particles that carry normal genes for the affected protein. The trials are designed to determine whether the normal gene will begin to be expressed in lung cells "infected" by the virus. They may also reveal whether there are serious side effects of repeated "infections." The promise and potential perils of this and other types of gene therapy are the subject of the *Human Impact* section that begins on page 457.

There are at least half a dozen autosomal recessive disorders in which the defective gene product is an enzyme required in metabolic breakdown of lipids. Infants born with *Tay-Sachs* disease lack the enzyme hexosaminidase A, which is required to break down certain brain lipids (gangliosides). As a result, the material accumulates in brain and other nervous system tissue. Affected children seem normal until about six months of age. Then they progressively lose motor functions and become deaf,

Choices: Biology and Society

Promise—and Problems—of Genetic Screening

In Great Britain recently, a couple who had a child with cystic fibrosis underwent embryo screening (page 321) to ensure that their next baby would not have the disease. The woman's ovaries were hormonally stimulated to produce several eggs at once, then the eggs were fertilized in the laboratory with the father's sperm. When the resulting embryos reached the eight-cell stage, genetic engineering methods were used to test each embryo for the presence of CF alleles. Embryos that tested positive were discarded, while two normal embryos were placed in the mother's uterus. One of them implanted and developed into a healthy baby girl, named Chloe.

In theory, more than 4,000 human genetic disorders can be detected in the early embryo. Researchers already are screening embryos for sickle-cell anemia, Tay-Sachs disease, hemophilia, and both Duchenne and myotonic muscular dystrophy, among other disorders. The required procedures are costly. Yet there are couples who are carriers for such disorders and wish desperately to have a child—*if* the child can be guaranteed to be free of the genetic defect.

Procedures that can detect harmful alleles in carriers or in offspring come under the heading of *genetic screening*. Common prenatal screening options include amniocentesis and chorionic villus sampling (Chapter 13). The embryo screening method that produced baby Chloe is not yet widely available, but it has fueled the controversy over how to handle our growing ability to pinpoint defects in genes or chromosomes.

Is it morally wrong to abort a fetus or discard a developing embryo because it carries a genetic disorder—even one with highly damaging effects? Some people answer with a resounding "Yes!" Others believe on humanitarian grounds that parents faced with the certainty of a seriously affected child should have every option open to them—including *not* bringing such a child into the world.

Genetic screening of embryos could also permit parents to discard a normal embryo for personal or cultural reasons. The notion is not so far-fetched. In India and China, it is fairly common for a woman to request ultrasound testing to ascertain the sex of her fetus, and to abort the fetus if it has the "wrong" gender (usually female). Surveys in the United States have found that the majority of couples preferred their first child to be male. Should screening technology be a tool only for detecting potential serious disorders? Or can it legitimately be used for "custom conceiving" a "perfect" child?

---

blind, and mentally retarded. Most die before the age of three. The disorder is most common among children of eastern-European Jewish descent.

Biochemical tests before conception can determine whether either member of a couple carries the recessive allele. As this chapter's *Choices* describes, new and sometimes controversial methods of genetic screening can also detect the condition in an embryo or fetus.

Autosomal Dominant Inheritance

What clues indicate that a *dominant* allele on an autosome is responsible for a trait? First, because such alleles are usually expressed (even in heterozygotes), the trait typically appears in each generation. Second, if one parent is heterozygous and the other homozygous recessive, there is a 50 percent chance that any one child of theirs will be heterozygous. Figure 18.9 shows one pattern for autosomal dominant inheritance.

A few dominant alleles that cause severe genetic disorders persist in populations. Mutation may be the reason in some cases. In others, a dominant allele may not prevent reproduction, or its effect may not show up until after reproductive age.

One autosomal dominant allele causes *achondroplasia*, a type of dwarfism, in about 1 in 10,000 individuals. When limb bones develop in affected children, improper cartilage formation leads to unusually short arms and legs. Affected people are less than 4 feet, 4 inches tall. The dominant allele often has no other phenotypic effects in heterozygotes, which normally are fertile and so may reproduce. Homozygous dominant fetuses usually are stillborn.

Progeria, or early aging, is a rare autosomal dominant disorder that apparently arises through new mutations. Children affected by this disorder start to show signs of advanced aging when they are only five or six (Figure 18.10). Their skin wrinkles, their hair thins, they start to suffer arthritis, and their blood vessels show arterioscle-

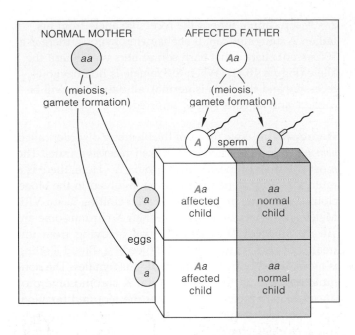

Figure 18.9 Possible phenotypic outcomes for autosomal dominant inheritance, assuming that the dominant allele (shown in red) is fully expressed in the carriers.

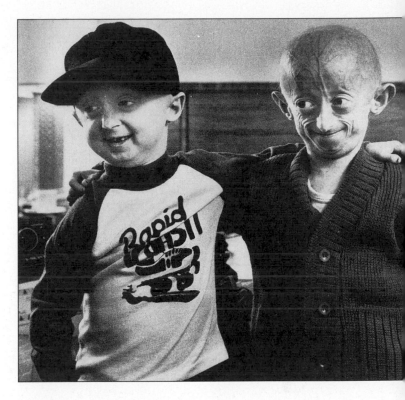

Figure 18.10 Two boys, both less than 10 years old, who met during a gathering of progeriacs at Disneyland in California. Progeria is a genetic disorder characterized by accelerated aging and extremely reduced life expectancy.

rosis. Frequently, affected youngsters die of heart disease before they are 10 years old.

Huntington's Disease *Huntington's disease*, a rare form of autosomal dominant inheritance, causes progressive degeneration of the basal nuclei of the brain (page 239). In about half the cases, symptoms emerge around age 40, after most people have already had children. In time, movements become convulsive, brain function deteriorates rapidly, and death follows. Using techniques of genetic engineering and recombinant DNA, researchers have found markers linked to the Huntington's gene at one tip of chromosome 4.

Homozygotes for the Huntington's allele die as embryos, so affected adults are always heterozygous. A person who has one parent with the disorder thus has a 50 percent chance of carrying the dominant allele. Testing can reveal whether an individual has the disease-causing allele. There is currently no cure for Huntington's disease. Some at-risk individuals choose not to have the test, and many elect to remain childless to avoid passing on the disorder.

Familial Hypercholesterolemia About 1 person in 500 is heterozygous for an allele that causes *familial hypercholesterolemia*. Individuals with this autosomal dominant condition have severely elevated blood cholesterol because they have fewer than the normal number of cell receptors for low-density lipoproteins (LDLs). Recall from Chapter 6 that LDLs bind cholesterol in the blood and thereby help remove it from the circulation. Rare individuals who are homozygous for the allele may develop severe cholesterol-related heart disease as children and usually die in early adulthood.

X-Linked Inheritance

Some genetic disorders fall in the category of **X-linked inheritance**: They are caused by genes on an X chromosome. Most X-linked disorders involve recessive alleles, and for a geneticist the telltale clues of an X-linked recessive condition are:

1. The recessive phenotype shows up far more often in males than in females. A recessive allele can be masked in females, who may inherit a dominant allele on their other X chromosome. Even though some of their cells may have the recessive allele (because X inactivation randomly inactivates some chromosomes with the "good" allele), other cells have the dominant form. Those cells apparently make enough of the proper gene product to prevent the disorder. A recessive allele cannot be masked in males, who have only one X chromosome.

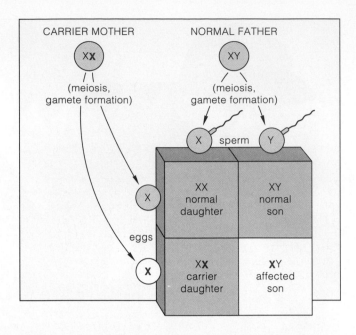

Figure 18.11 Possible phenotypic outcomes for X-linked inheritance when the mother carries a recessive allele on one of her X chromosomes (shown in red).

2. A son cannot inherit the recessive allele from his father. A daughter can. If she is heterozygous, there is a 50 percent chance that each son of hers will inherit the allele (Figure 18.11). When the female is homozygous recessive and the male is normal, all daughters will be carriers and all sons will be affected.

Hemophilia Two forms of the bleeding disorder called *hemophilia* are inherited as X-linked recessive traits. The most common of these is hemophilia A. Here, there is a mutation in the gene for a protein involved in the blood clotting mechanism (page 157), called clotting factor VIII. Males with a recessive allele on their X chromosome are always affected. They run the risk of dying from untreated bruises, cuts, or internal bleeding. Blood clotting is more or less normal in heterozygous females. The nonmutated gene on their one normal X chromosome produces enough factor VIII to cover the required function. Hemophilia B results in similar symptoms, but the clotting factor is factor IX.

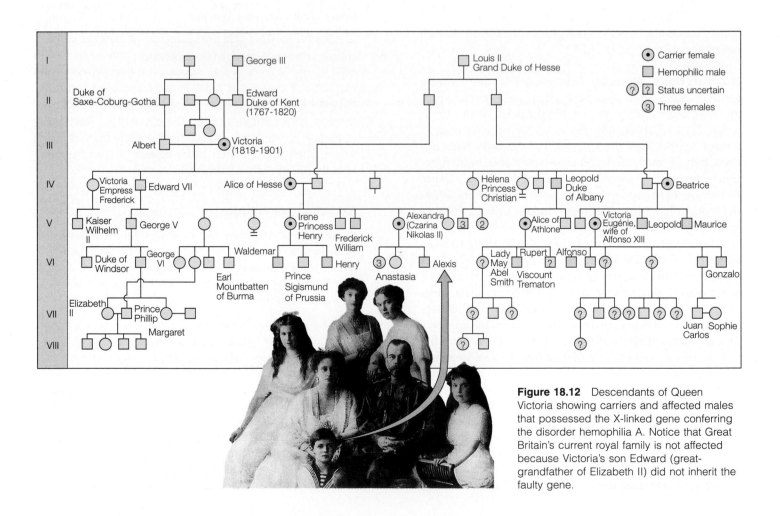

Figure 18.12 Descendants of Queen Victoria showing carriers and affected males that possessed the X-linked gene conferring the disorder hemophilia A. Notice that Great Britain's current royal family is not affected because Victoria's son Edward (great-grandfather of Elizabeth II) did not inherit the faulty gene.

Genotype	Male Phenotype	Female Phenotype
b^+b^+	Hair	Hair
b^+b	Bald	Hair
bb	Bald	Bald

Figure 18.13 Pattern baldness. The bald spot on the top of this man's head is a typical instance of pattern baldness. Geneticists usually denote the alleles for a sex-influenced trait with lowercase letters. Here, b^+ is "dominant" and b is recessive. Because of the influence of sex hormones, female heterozygotes have a full head of hair and male heterozygotes show pattern baldness. Both sexes show pattern baldness in the homozygous recessive condition.

Hemophilia A affects only about 1 in 7,000 human males. However, the frequency of the recessive allele was unusually high among 19th-century European royal families, whose members often intermarried. Queen Victoria of England was a carrier, as were two of her daughters (Figure 18.12). At one time, 18 of her 69 descendants were affected males or female carriers.

Duchenne Muscular Dystrophy The diseases lumped under "muscular dystrophy" all involve progressive wasting of muscle tissue. One X-linked type, *Duchenne muscular dystrophy* (DMD), affects 1 in 3,500 males, usually in early childhood. As muscles degenerate, affected boys become weak and unable to walk. Death due to cardiac or respiratory failure usually occurs by age 20. The normal gene that is mutated in DMD governs a protein called *dystrophin*, which gives structural support to the muscle cell plasma membrane. Lacking dystrophin, muscle fibers cannot withstand the physical stress of contraction and begin to break down. Eventually the whole muscle tissue is destroyed.

Color Blindness Red/green color blindness is an X-linked recessive trait. About 8 percent of males in the United States have this inconvenient disorder, which arises from mutation of an allele that codes for a protein (opsin) that binds visual pigments in cone cells of the retina (page 265). Females also can have red/green color blindness, but this occurs only rarely because a girl must inherit the recessive allele from both parents.

Sex-Influenced Inheritance

Some traits are *sex-influenced*—they appear more frequently in one sex than in the other, or the phenotype differs depending on whether the individual is male or female. The difference apparently reflects the varying influences of male and female sex hormones on gene expression. Genes for such traits are carried on autosomes, not on sex chromosomes. Pattern baldness (Figure 18.13) is an intriguing example. We can designate the "no baldness" form of the responsible gene b^+ and the "baldness" form b. A man will develop pattern baldness if he is homozygous bb, and *also* if he is heterozygous b^+b. Females develop pattern baldness only if they are bb, and then only after menopause, when their estrogen supply declines.

A pedigree is a tool for examining patterns of inheritance in a group of related individuals.

Major patterns of inheritance include autosomal dominant inheritance, autosomal recessive inheritance, X-linked inheritance, and sex-influenced inheritance.

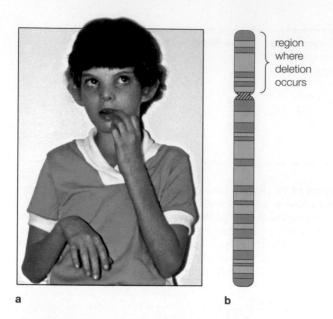

a

b

region
where
deletion
occurs

Figure 18.14 Deletion leading to cri-du-chat syndrome.
(**a**) Affected infants have an improperly developed larynx, and
their cries sound like a cat in distress. They show severe mental
retardation. A rounded face, small cranium, and misshapen ears
are outward symptoms. (**b**) Deletion occurs in a fragile region of
chromosome 5.

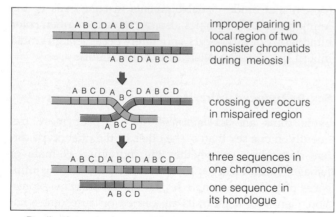

A B C D A B C D

improper pairing in
local region of two
nonsister chromatids
during meiosis I

A B C D A B C D

A B C D A B C D A B C D

crossing over occurs
in mispaired region

A B C D

A B C D A B C D A B C D

three sequences in
one chromosome

A B C D

one sequence in
its homologue

a Duplication

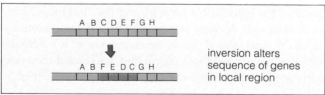

A B C D E F G H

A B F E D C G H

inversion alters
sequence of genes
in local region

b Inversion

Figure 18.15 Two kinds of changes in chromosome structure: a
duplication (**a**) and an inversion (**b**).

Figure 18.16 Normal (left) and fragile X (right) chromosomes.
Notice the dangling tip of the long arm of the chromosome.

CHROMOSOME VARIATIONS

Changes in Chromosome Structure

Occasionally a chromosome becomes abnormally rearranged. A **deletion** is a loss of a chromosome segment. The loss occurs when a segment of a chromosome breaks off. Viral attack, irradiation, or chemical action can cause such a break. The loss almost always means problems, for genes influencing one or more traits may be missing. For example, one deletion from human chromosome 5 leads to mental retardation and a malformed larynx. When affected infants cry, the sounds produced are more like meowing—hence the name of the disorder, *cri-du-chat* (meaning cat-cry). Figure 18.14 shows a child affected by the disorder.

A **duplication** is a gene sequence in excess of its normal amount in a chromosome. This happens, for example, when a deletion from one chromosome is inserted into its homologue (Figure 18.15a).

Fragile X syndrome, a leading cause of mental retardation, involves a variation on this theme. It arises when a sequence of three subunits (nucleotides) of a gene on the long arm of an X chromosome becomes repeated over and over (Figure 18.16). Like a fragile chemical thread, the long string of repeats easily breaks; apparently it also acts to shut down expression of the whole gene. The discovery that abnormally repeated sequences of a gene's subunits can cause disorders has already begun to shed light on several other diseases. One of these is schizophrenia; another is *myotonic dystrophy*, a form of muscular dystrophy that begins in adulthood and affects both men and women. Huntington's disease results when a particular nucleotide sequence is repeated from 40 to as many as 86 times.

An **inversion** is a chromosome segment that separated from the chromosome and then was inserted at the same place—but in reverse. The reversal alters the posi-

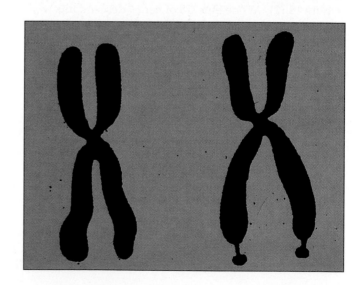

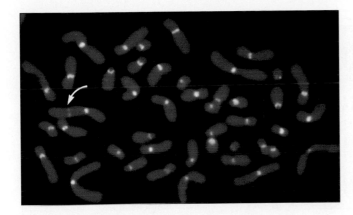

Figure 18.17 Translocation. These metaphase chromosomes were stained with substances that react with a protein structure near the centromere, called a kinetochore. It shows up as fluorescent yellow in micrographs. The arrow points to the inactivated kinetochores of chromosome 9, which fused with chromosome 11. (The normal kinetochores of chromosome 11 are visible to the right of the fusion point.)

Figure 18.18 Example of nondisjunction. Here, chromosomes fail to separate during anaphase I of meiosis and so change the chromosome number in resulting gametes.

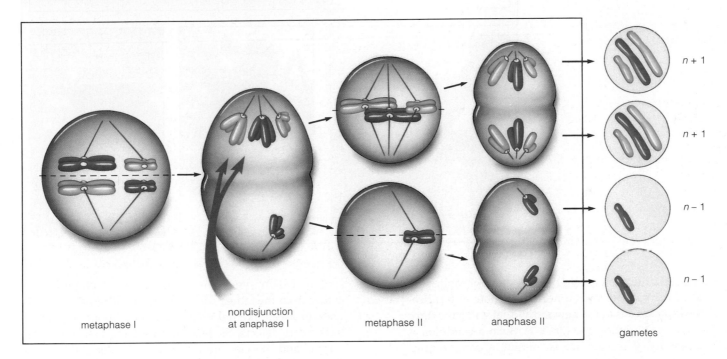

metaphase I   nondisjunction at anaphase I   metaphase II   anaphase II   gametes

$n + 1$

$n + 1$

$n - 1$

$n - 1$

tion and relative order of the genes on the chromosome (Figure 18.15b). We know relatively little about the effects of inversions on humans, possibly because those effects are so damaging that a developing embryo quickly dies.

Most often, a **translocation** is the transfer of part of one chromosome to a *non*homologous chromosome. In some types of human cancer, for example, a segment of chromosome 8 has been transferred to chromosome 14. Genes on that segment had been precisely regulated at their original location, but in the new chromosomal slot control over their expression seems to be lost.

Once in a while, translocation occurs in such a way that almost an entire chromosome becomes fused to a nonhomologous chromosome—and the two continue to function. This structural rearrangement changes the chromosome number. Figure 18.17 shows the result of such a fusion in a cell from a human female. Her diploid chromosome number is 45 instead of the normal 46.

Changes in Chromosome Number

Sometimes gametes or cells of a new individual end up with one more or one less than the parental number of chromosomes. This abnormal condition is called **aneuploidy**. It is a major cause of reproductive failure. Of the human embryos that have been miscarried and autopsied each year, about 70 percent are aneuploids.

An abnormal chromosome number may arise during mitotic or meiotic cell divisions or during the fertilization process. For example, one or more pairs of chromosomes may fail to separate during mitosis or meiosis, an event called **nondisjunction**. Perhaps a chromosome does not separate from its homologue at anaphase I of meiosis. Or perhaps sister chromatids of a chromosome do not separate at anaphase II. As Figure 18.18 shows, some or all of the gametes can end up either with one extra chromosome or with one less than the parental number.

Figure 18.19 (**a**) Karyotype of a girl with Down syndrome; brown arrows identify the trisomy of chromosome 21. (**b**) Relationship between the frequency of Down syndrome and the mother's age. Results are from a study of 1,119 children with the disorder who were born in Victoria, Australia, between 1942 and 1957.

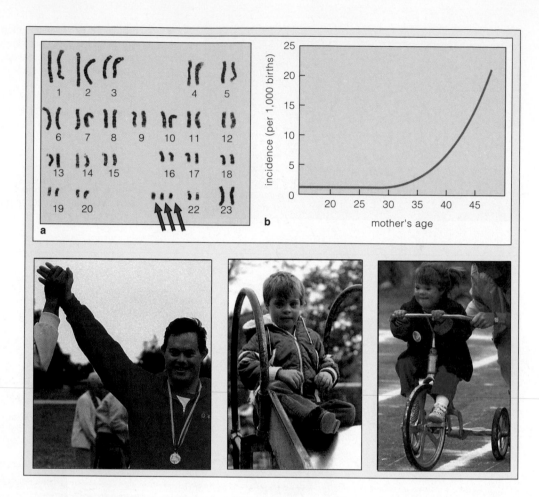

Figure 18.20 Individuals with Down syndrome.

Suppose a human gamete has one extra chromosome ($n + 1$). If it combines with a normal gamete at fertilization, the diploid cells of the new individual will have three of one type of chromosome ($2n + 1$). Such a condition is called **trisomy**. If the gamete is missing a chromosome, the new individual will have a chromosome number of $2n - 1$. One chromosome in its diploid cells will not have a homologue—a condition called **monosomy**.

Down Syndrome Sometimes the cells of an individual have three copies of chromosome 21, which is the smallest chromosome in human cells. That condition, called trisomy 21, leads to Down syndrome. (*Syndrome* simply means a set of symptoms characterizing a particular disorder; typically the symptoms occur together.) Figure 18.19a shows a karyotype of a girl with Down syndrome.

Trisomic 21 embryos are often miscarried, but about one child of every 1,000 live births in North America alone will have the disorder. Although the probability that a woman will give birth to a child with Down syndrome rises steeply after age 40 (Figure 18.19b), the father's age can also be a factor. Research suggests that increased age in the father contributes to increased nondisjunction for all chromosomes; geneticists estimate that in about 25 per-

cent of Down syndrome cases the extra chromosome comes from the father. Most affected children show moderate to severe mental retardation, and about 40 percent have heart defects. Skeletal development is slower than normal and muscles are rather slack. Older children are shorter than normal and have distinguishing facial features, including a small skin fold over the inner corner of the eyelid. With special training, affected individuals often participate in normal activities, and they enjoy life to the fullest extent allowed by their condition (Figure 18.20). Down syndrome is one of the genetic disorders that can be detected by prenatal diagnosis (page 321).

Turner Syndrome and Metafemales About 1 in every 5,000 newborns is destined to have Turner syndrome. Through a nondisjunction, affected individuals have a chromosome number of 45 instead of 46 (Figure 18.21). They are missing a sex chromosome (and a Barr body), so this is a type of sex chromosome abnormality. It is symbolized as XO. Turner syndrome occurs less often than other sex chromosome abnormalities, probably because most XO embryos are miscarried early in pregnancy. Affected people have a webbed neck and a distorted female phenotype. Their ovaries do not function,

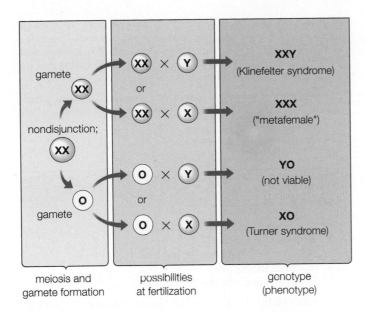

meiosis and gamete formation · possibilities at fertilization · genotype (phenotype)

Figure 18.21 Genetic disorders that result from nondisjunction of X chromosomes followed by fertilization involving normal sperm.

they are sterile, and secondary sexual traits fail to develop at puberty. Often they age prematurely and have shortened life expectancies.

Roughly 1 in 1,000 females is a "metafemale"—she has three X chromosomes. Two of these are condensed to Barr bodies, and most XXX females develop normally.

Klinefelter Syndrome Nondisjunction can give rise to males who show Klinefelter syndrome. In most cases the genotype is XXY (Figure 18.21). This sex chromosome abnormality occurs at a frequency of about 1 in 1,000 liveborn males. XXY males show low fertility and usually some mental retardation. Their testes are much smaller than normal, body hair is sparse, and there may be some breast enlargement. Injections of the hormone testosterone can reverse the feminized phenotype but not the sterility or mental retardation.

XYY Condition About 1 in every 1,000 males has one X and two Y chromosomes, an XYY condition. XYY males tend to be taller than average and some may show mild mental retardation, but most are phenotypically normal.

Testicular Feminizing Syndrome Infrequently, a person who is genetically of one gender develops sexual traits of the opposite gender. An example is *testicular feminizing syndrome* or "androgen insensitivity." In an affected genetic male (XY), a gene mutation on the X chromosome results in defective receptors for the male sex hormones (androgens), including testosterone and DHT

(dihydroxytestosterone). Normally, embryonic cells destined to become testes and other male reproductive structures bind one or more of the hormones and then proceed to develop. When their receptors are defective, however, they cannot bind the hormone and the embryo develops as an apparent female. At puberty, the individual develops female external characteristics, but has no internal female reproductive structures.

Genetic Counseling

Sometimes prospective parents suspect they are very likely to produce a severely afflicted child. Their first child or a close relative may show an abnormality, and they wonder if future children also will be affected. In such cases, clinical psychologists, geneticists, social workers, and other consultants may be brought in to give emotional support to parents at risk.

Counseling begins with accurate diagnosis of parental genotypes; this may reveal the potential for a specific disorder. Biochemical tests can be used to detect many metabolic disorders. Detailed family pedigrees can be constructed to aid the diagnosis. For disorders showing simple Mendelian inheritance patterns, it is possible to predict the chances of having an affected child—but not all disorders follow Mendelian patterns. Those that do can be influenced by other factors, some identifiable, others not. Even when the extent of risk has been determined with some confidence, prospective parents must know that the risk is the same for *each* pregnancy. If a pregnancy has one chance in four of producing a child with a genetic disorder, the same odds also apply to every subsequent pregnancy.

Chapter 14 described *prenatal tests*, such as amniocentesis and chorionic villus sampling, that can be used to detect genetic disorders in a fetus. This chapter's *Choices* discusses options for genetic screening. As with preconception counseling, in the event that a genetic defect is discovered, prospective parents may seek the advice and support of concerned social workers, mental health professionals, and even specialists in bioethics.

SUMMARY

1. Genes, the units of instruction for heritable traits, are arranged linearly on a chromosome.

2. Human diploid cells have two chromosomes of each type (one from the mother, one from the father). The two are homologues; they pair at meiosis. The sex chromosomes (X and Y) differ from each other in size, shape, and the genes they carry. All other pairs of chromosomes are

autosomes; they are roughly the same in size and shape and carry genes for the same traits.

3. Genes on the same chromosome tend to be linked—they tend to stay together during meiosis and end up in the same gamete. But crossing over can disrupt such linkages. The farther apart two genes are on a chromosome, the greater will be the frequency of crossing over and recombination between them.

4. A family pedigree is a chart of genetic relationships of individuals. It helps establish inheritance patterns and track genetic abnormalities through several generations.

5. In disorders involving autosomal recessive inheritance, a person who is homozygous for a recessive allele shows the recessive phenotype. Heterozygotes generally have no symptoms. Cystic fibrosis and Tay-Sachs disease are examples.

6. In autosomal dominant inheritance, a dominant allele usually is expressed to some extent. Disorders include Huntington's disease and achondroplasia.

7. A large number of genetic disorders are X-linked—the mutated gene occurs on the X chromosome. Males, who inherit only one X chromosome, typically are affected. Examples include hemophilia A and B and Duchenne muscular dystrophy.

8. Chromosome structure can be altered by deletions, duplications, inversions, or translocations. Chromosome number can be altered by nondisjunction (the failure of chromosomes to separate during meiosis or mitosis), so that gametes end up with one extra or one missing chromosome. Gene mutations, changes in chromosome structure, and changes in chromosome number tend to cause many genetic disorders.

Review Questions

1. What are the main differences between X and Y chromosomes? *393*

2. Why do harmful genes (alleles) persist in a population? *398*

3. What factors indicate that a trait is coded by a dominant allele on an autosome? *398*

4. What is the difference between an X-linked trait and a sex-influenced trait? *399*

5. Describe the general procedures of genetic counseling. *405*

Critical Thinking: You Decide *(Key in Appendix IV)*

1. A female athlete is disqualified from competition because chromosome testing shows that this individual is XY. Later, a medical exam reveals "androgen insensitivity," an abnormality in which an embryo's cells lack receptor proteins for the male hormone testosterone. As a result, female sex characteristics develop, and indeed the person clearly appears to be feminine. Knowing this, do you agree or disagree with the disqualification, and why?

2. Referring to question 1, would you consider androgen insensitivity a genetic disorder? Why or why not?

Self-Quiz *(Answers in Appendix III)*

1. _____ segregate during _____.
   a. Homologues; mitosis
   b. Genes on one chromosome; meiosis
   c. Homologues; meiosis
   d. Genes on one chromosome; mitosis

2. The two alleles of a gene on homologous chromosomes end up in separate _____.
   a. body cells
   b. gametes
   c. nonhomologous chromosomes
   d. offspring
   e. both b and d are possible

3. Genes on the same chromosome tend to remain together during _____ and end up in the same _____.
   a. mitosis; body cell
   b. mitosis; gamete
   c. meiosis; body cell
   d. meiosis; gamete
   e. both a and d

4. The probability of a crossover occurring between two genes on the same chromosome is _____.
   a. unrelated to the distance between them
   b. increased if they are closer together on the chromosome
   c. increased if they are farther apart on the chromosome
   d. impossible

5. Chromosome structure can be altered by _____.
   a. deletions
   b. duplications
   c. inversions
   d. translocations
   e. all of the above

6. Nondisjunction can be caused by _____.
   a. crossing over in meiosis
   b. segregation in meiosis
   c. failure of chromosomes to separate during meiosis
   d. multiple independent assortments

7. A gamete affected by nondisjunction would have _____.
   a. a change from the normal chromosome number
   b. one extra or one missing chromosome
   c. the potential for a genetic disorder
   d. all of the above

8. Genetic disorders can be caused by _____.
   a. gene mutations
   b. changes in chromosome structure
   c. changes in chromosome number
   d. all of the above

9. Match the following chromosome terms appropriately.

_____ crossing over
_____ deletion
_____ nondisjunction
_____ translocation
_____ gene mutation

a. a chemical change in DNA that may affect genotype and phenotype
b. movement of a chromosome segment to a nonhomologous chromosome
c. disrupts gene linkages during meiosis
d. causes gametes to have abnormal chromosome numbers
e. loss of a chromosome segment

Genetics Problems (*Answers in Appendix II*)

1. If a couple has a large family of six boys, what is the probability that a seventh child will be a girl?

2. Answer the following questions, given that human sex chromosomes are XX for females and XY for males.

   a. Does a male child inherit his X chromosome from his mother or father?

   b. With respect to an X-linked gene, how many different types of gametes can a male produce?

   c. If a female is homozygous for an X-linked gene, how many different types of gametes can she produce with respect to this gene?

   d. If a female is heterozygous for an X-linked gene, how many different types of gametes can she produce with respect to this gene?

3. Suppose that you have two linked genes with alleles *A,a* and *B,b*, respectively. An individual is heterozygous for both genes, as in the following:

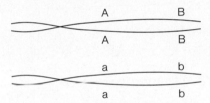

If the crossover frequency between these two genes is 0 percent, what genotypes would be expected among gametes from this individual?

4. Individuals affected by Down syndrome typically have an extra chromosome 21, so their cells have a total of 47 chromosomes. However, in a few cases of Down syndrome, 46 chromosomes are present. Included in this total are two normal-appearing chromosomes 21 and a longer-than-normal chromosome 14. Interpret this observation, and indicate how these few individuals can have a normal chromosome number.

5. The mugwump, a space alien, has a reversed sex-chromosome condition. The male is XX and the female is XY. However, perfectly good sex-linked genes are found to have the same effect as in humans. For example, a recessive, X-linked allele *c* produces red/green color blindness. If a normal female mugwump mates with a phenotypically normal male mugwump whose mother was color blind, what is the probability that a son from that mating will be color blind? A daughter?

6. A normal woman whose father had hemophilia A marries a man who also has hemophilia A. What is the chance their son will have the disorder?

7. Both parents in a couple are carriers for the recessive allele *p* that codes for the defective protein responsible for PKU. What is the symbol for the dominant allele? What is the probability that any of their four children will have PKU? Now, assume that this probable outcome actually occurs in the family, and mark the following pedigree chart accordingly:

How many of the children are carriers for PKU?

Readings

Cummings, M. 1991. *Human Heredity: Principles and Issues*. St. Paul, Minn.: West.

Elmer-Dewitt, P. October 5, 1992. "Catching a Bad Gene in the Tiniest of Embryos." *Time*.

"Huntington's Gene Finally Found." April 1993. *Science*.

19 DNA STRUCTURE AND FUNCTION

Cardboard Atoms and Bent-Wire Bonds

Throughout the early 20th century, biologists pondered one of the most fundamental questions in all science: What molecule serves as the book of genetic information in every living cell?

For decades researchers had focused on proteins, which could consist of virtually limitless combinations of amino acids. Heritable traits are awesomely diverse.

It made sense for the molecules that held the information for those traits to be structurally diverse. Yet, by the late 1940s, there was something about another cellular substance—DNA—that tugged at more than a few good minds. DNA consists of only four kinds of subunits. For DNA to be the "master molecule of life," the subunits would have to be arranged in a way that

Figure 19.1 James Watson and Francis Crick posing in 1953 by their newly unveiled model of DNA structure. Behind this photograph is a recent computer-generated model. It is more sophisticated in appearance yet basically the same as the prototype that was built nearly four decades before.

could carry a vast amount of information. And there would have to be a mechanism to copy that structure in a way that would faithfully transfer hereditary information to a new generation. Research published in 1950 left little doubt that DNA was indeed the genetic material, but still unknown were the crucial structural details that would explain how it functioned.

No doubt many brilliant researchers dreamed in 1950 of winning the race to elucidate the structure of DNA. Biochemist Linus Pauling—who had already discovered that a single gene could cause sickle-cell anemia—came very, very close. In the end, however, the glory went mostly to James Watson, a 25-year-old postdoctoral student from Indiana University, and to Francis Crick, a Cambridge University researcher who was still working on his Ph.D. Watson and Crick spent long hours arguing over everything they had read about the size, shape, and bonding requirements of DNA's subunits. They arranged and rearranged cardboard cutouts of the subunits and pestered chemists to identify potential bonds they might have overlooked. Crucial evidence came from the laboratory of Maurice Wilkins, where Rosalind Franklin was analyzing DNA samples using X-ray diffraction, a method that can reveal details of a molecule's structure. Aided by clues in the X-ray images, Watson and Crick painstakingly fashioned models of bits of metal held together with wire "bonds" bent at chemically correct angles. In 1953 they devised a model that fit what was known about DNA and the pertinent biochemical rules (Figure 19.1).

Watson and Crick had found the structure of DNA and ushered in the golden age of "molecular genetics." Together with Maurice Wilkins, they won the 1962 Nobel Prize for medicine. (Rosalind Franklin died before the prize was awarded.) The remarkable work of these scientists, and of others before and since, is the foundation for the concepts presented in this chapter.

KEY CONCEPTS

1. DNA stores hereditary instructions. The information in DNA is encoded in the linear sequence of nucleotides that make up DNA molecules.

2. In each DNA molecule, two strands of nucleotides are twisted together like a spiral staircase, forming a double helix.

3. Each chromosome consists of a single DNA molecule, together with a great number of proteins. Assemblies of DNA and certain proteins form nucleosomes, which are the basic structural units of chromosomes.

4. Genes are specific regions of DNA. They encode the information required to build proteins. The "code words" of a gene are sequences of nucleotide bases, read three at a time.

5. The path from genes to proteins has two steps. In the first, transcription, an enzyme "reads" a gene's sequence of bases and assembles a corresponding molecule of messenger RNA (ribonucleic acid). In the second step, translation, the messenger RNA codes directly for the assembly of a protein.

6. Occasionally one or more bases in a gene sequence are permanently deleted, added, or replaced. Such changes are gene mutations.

7. Gene activity is governed by regulatory proteins, hormones, and other molecules. Some of these regulatory elements operate on genes directly, turning a gene on or off or modifying its activity in some other way. Other regulators exert their effects on messenger RNA or on the protein products of genes.

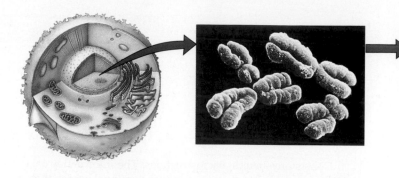

All chromosomes contain DNA. What does DNA contain? Only four kinds of nucleotides. Each nucleotide has a five-carbon sugar (shaded red). That sugar has a phosphate group attached to the fifth carbon atom of its ring structure. It also has one of four kinds of nitrogen-containing bases (shaded blue) attached to its first carbon atom. The nucleotides differ only in which base is attached to that atom:

OVERVIEW OF DNA FUNCTION AND STRUCTURE

DNA has two main functions. First, it stores the vast number of hereditary instructions required for the building and functioning of each and every body cell, tissue, and organ. Second, by way of *self-replication*, it copies those hereditary instructions in a way that faithfully conveys them to a new generation. The key to both these remarkable feats lies in DNA's structure.

A DNA molecule is built from four kinds of **nucleotides**, the building blocks of nucleic acids (Figure 19.2). A DNA nucleotide consists of a five-carbon sugar (deoxyribose), a phosphate group, and one of the following nitrogen-containing bases:

adenine (A)	guanine (G)	thymine (T)	cytosine (C)

Although many researchers contributed knowledge in the race to discover DNA's structure, Watson and Crick were the first to recognize that DNA consists of two strands of nucleotides twisted together into a *double* helix. Nucleotides in a strand are linked together, like box cars in a train, by strong covalent bonds. Weaker hydrogen bonds link the bases of one strand with bases of the other

Figure 19.2 The nucleotide subunits of DNA. The notation "5'" in the structural formulas identifies the carbon atoms to which other parts of the molecule are attached.

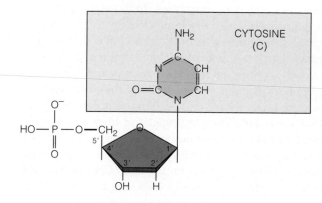

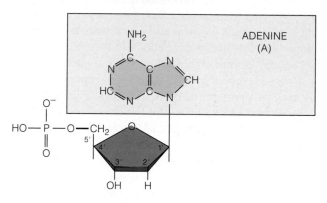

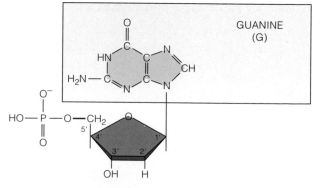

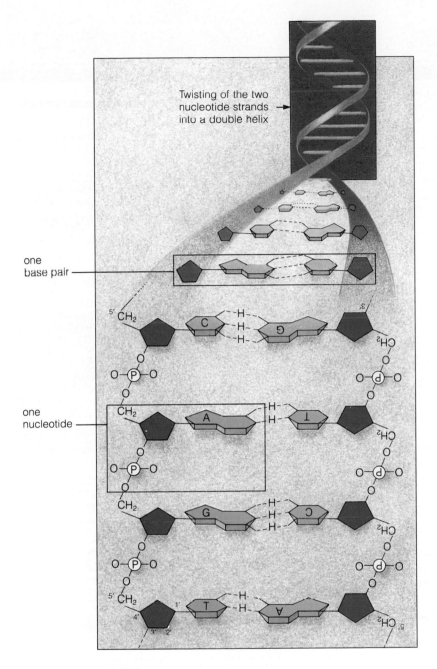

Twisting of the two
nucleotide strands
into a double helix

one
base pair

one
nucleotide

C — G

A — T

G — C

T — A

Figure 19.3 Arrangement of bases (blue) in the DNA double helix.

At various times in a cell's life, regions of the double helix unwind so that a cell gains access to specific genes. In general, a **gene** is a region of DNA that codes for a single polypeptide chain. Such chains are the basic structural units of proteins, and a later section describes the precise steps by which genes give rise to polypeptide chains. Some genes code only for certain types of RNA (the tRNAs and rRNAs described shortly). *Focus on Science* describes some of the classic research that uncovered the basic function of genes.

The two long nucleotide strands of the double helix also unwind and separate from each other when DNA replicates. As described next, the exposed bases of each strand serve as a pattern, or *template*, upon which a new strand is built.

DNA REPLICATION

Assembly of Nucleotide Strands

Before a cell divides, DNA is duplicated in a process called **replication**. Replication gives rise to the sister chromatids described in Chapter 16, the duplicated chromosomes that are parceled out to daughter cells in mitosis and meiosis.

Hydrogen bonds hold together the two nucleotide strands making up a DNA double helix. Enzymes can readily break those bonds. Through enzyme action, one strand starts to unwind from the other at distinct sites, leaving the bases exposed. Free nucleotides pair with the exposed bases, following the base-pairing rule (A with T and G with C). They are linked by hydrogen bonds. Each parent strand remains intact while a new companion strand is assembled on it, nucleotide by nucleotide.

As replication proceeds, the newly formed double-stranded molecule twists back into a double helix. Because one strand is from the old molecule—that strand is said to be *conserved*—

(Figure 19.3). The shapes of the bases and their sites available for hydrogen bonding restrict which bases can pair up. Adenine always pairs with thymine, and cytosine always pairs with guanine. Thus, only two kinds of **base pairs** occur in DNA: A—T and G—C. In a DNA molecule, the amount of adenine always equals the amount of thymine, and the amount of guanine always equals the amount of cytosine.

DNA Detectives: Discovering the Connection Between Genes and Proteins

In the early 1900s a physician, Archibald Garrod, was tracking metabolic disorders that seemed to be heritable (they kept recurring in the same families). Blood or urine samples from affected people contained abnormally high levels of a substance known to be produced at a certain step in one metabolic pathway. Most likely, the enzyme at the *next* step in the pathway was defective and could not use that substance. Because the pathway was blocked from that step onward, unused molecules of the substance accumulated in the body:

A ⟶ B ⟶ C ⟶ C ✗ D
pathway is blocked

Only one thing distinguished affected people from normal ones: They had inherited one metabolic defect. Thus, Garrod concluded, specific "units" of inheritance (genes) function through the synthesis of specific enzymes. As later researchers pursued this idea, it became known as the "one gene, one enzyme" hypothesis.

The hypothesis was refined through studies of sickle-cell anemia (page 382). This heritable disorder arises from the presence of abnormal hemoglobin in red blood cells. The abnormal molecule is designated HbS instead of HbA. In 1949 Linus Pauling and Harvey Itano subjected HbS and HbA molecules to *electrophoresis*. This is a way to measure how fast and in what direction an organic molecule will move in response to an electric field.

As shown in Figure *a*, if you place a mixture of different proteins in a slab of gel, each type will move toward one end of the slab or the other when voltage is applied to it. The rate and direction of movement depend partly on a molecule's net surface charge.

Electrophoresis studies showed that HbS and HbA molecules move toward the positive pole of the field—but HbS does so more slowly. HbS, it seemed, has fewer negatively charged amino acids.

Later, Vernon Ingram pinpointed the difference. Hemoglobin, recall, consists of four polypeptide chains (page 142). Two are designated alpha and the other two, beta. As Figure *b* shows, in each beta chain of HbS, one amino acid (valine) has replaced another (glutamate). Glutamate car-

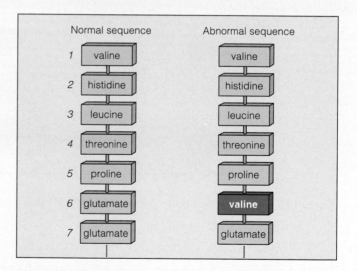

ries a negative charge; valine has no net charge. Thus HbS behaved differently in the electrophoresis studies.

This discovery suggested that two genes code for hemoglobin—one for each kind of polypeptide chain—and that genes code for proteins in general, not just for enzymes.

And so a more precise hypothesis emerged: *One gene codes for the amino acid sequence of one polypeptide chain—the structural unit of proteins.*

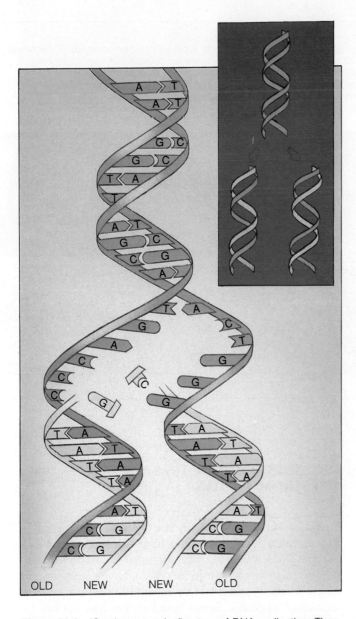

OLD   NEW   NEW   OLD

Figure 19.4 "Semiconservative" nature of DNA replication. The original two-stranded DNA molecule is shown in blue. A new strand (yellow) is assembled on each of the two original strands.

each "new" DNA molecule is really half-old, half-new. Hence this replication mechanism is called *semiconservative replication* (Figure 19.4).

Prior to cell division, the double-stranded DNA molecule unwinds and is replicated. Each parent strand remains intact—it is said to be conserved—and a new, complementary strand is assembled on each one.

Focus on Environment

When DNA Can't Be Fixed

A child born with a genetic condition called xeroderma pigmentosum literally faces a dim future. She or he cannot be exposed to sunlight, even briefly, without risking disfiguring skin tumors and early death from cancer. The disease is a tragic result of faulty DNA repair—there is a defect in one or more of the genes that are thought to govern repair processes in body cells.

The ultraviolet wavelengths in light from the sun, tanning lamps, and other environmental sources can cause molecular changes in DNA. Among other things, UV light can promote covalent bonding between neighboring thymine bases in a nucleotide strand. Thus the two nucleotides to which the bases belong are combined into an abnormal, bulky region of the DNA molecule called a "thymine dimer." Normally, a DNA repair mechanism eliminates thymine dimers. The mechanism requires at least seven gene products, and mutation in one or more of the required genes can prevent the repair machinery from functioning. Thymine dimers can accumulate in the skin cells of people whose DNA has such mutations. The accumulation triggers development of serious lesions, including skin cancer.

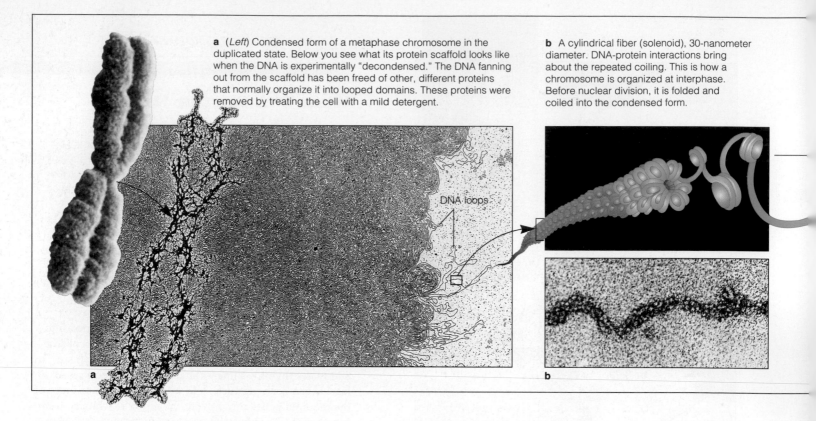

a (*Left*) Condensed form of a metaphase chromosome in the duplicated state. Below you see what its protein scaffold looks like when the DNA is experimentally "decondensed." The DNA fanning out from the scaffold has been freed of other, different proteins that normally organize it into looped domains. These proteins were removed by treating the cell with a mild detergent.

b A cylindrical fiber (solenoid), 30-nanometer diameter. DNA-protein interactions bring about the repeated coiling. This is how a chromosome is organized at interphase. Before nuclear division, it is folded and coiled into the condensed form.

DNA loops

Figure 19.5 Levels of organization of DNA in a chromosome.

During replication, enzymes and other proteins unwind the DNA molecule, keep the two strands separated at unwound regions, and assemble a new strand on each one. For example, **DNA polymerases** govern the assembly of nucleotides on a parent strand.

DNA Repair

DNA polymerases, DNA ligases, and other enzymes also perform DNA repair. If the sequences of bases in one strand of a double helix becomes altered, DNA polymerases "read" the complementary sequence on the other strand. With the aid of other repair enzymes, they restore the original sequence. *Focus on Environment* (previous page) gives an example of what can happen when the excision/repair function is impaired.

ORGANIZATION OF DNA IN CHROMOSOMES

Stretched out, the DNA in the 46 chromosomes of a human cell is about two meters long—a little over six feet. All that DNA would become a tangled mess if it were not organized in some way. In fact, the DNA of humans and all other eukaryotes is tightly bound with many proteins, including **histones**. Some histones are like spools for winding up small stretches of DNA. Each histone-DNA spool is a **nucleosome** (Figure 19.5a and b).

The nucleosome string along the DNA molecule becomes coiled repeatedly through interactions between the histones and DNA; this process increases its diameter, as Figure 19.5 shows. Further folding produces a series of loops of various sizes. Proteins other than histones seem to serve as a structural "scaffold" for the loops.

PROTEIN SYNTHESIS

Role of RNA

The path from genes to proteins has two steps, called transcription and translation. Both involve molecules of ribonucleic acid, or **RNA**. Most often, RNA is single-

c A chromosome immersed in a salt solution loosens up to a beads-on-a-string organization. The "string" is DNA. Each "bead" is a nucleosome (**d**). Short stretches of the chromosome may look like this when DNA is being replicated or when a gene's instructions are being read. Each nucleosome consists of a double loop of DNA around a core of proteins (eight histone molecules). Another histone (H1) stabilizes the arrangement.

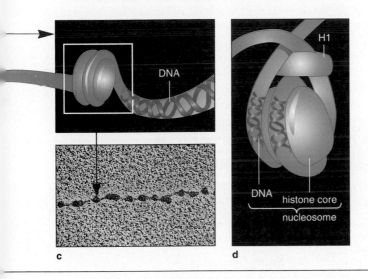

c

d

stranded. Structurally, it is much like a strand of DNA. Its nucleotides consist of a sugar (ribose), a phosphate group, and a nitrogen-containing base. However, its bases are adenine, cytosine, guanine, and **uracil**. We can summarize these differences as follows:

	DNA	RNA
Sugar:	deoxyribose	ribose
Bases:	adenine, cytosine, guanine, thymine	adenine, cytosine, guanine, uracil

Like the thymine in DNA, the uracil in RNA base-pairs with adenine.

In **transcription**, molecules of RNA are produced on DNA templates in the nucleus. In **translation**, RNA molecules move from the nucleus into the cytoplasm, where they are used as templates for assembling polypeptide chains. After translation, one or more polypeptide chains become folded into protein molecules. As you know, those proteins have structural and functional roles in cells. For example, they serve as enzymes in biosynthesis, as membrane receptors and channels, as transport proteins, and so forth. They also include most of the enzymes and other proteins that take part in DNA replication, RNA synthesis, and protein synthesis.

The *central dogma of molecular biology* summarizes this flow of information in cells. It states that DNA is transcribed into RNA, and then RNA is translated into protein:

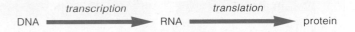

Genes are transcribed into three types of RNA molecules:

ribosomal RNA (rRNA) — a nucleic acid chain that combines with certain proteins to form a *ribosome*, a "workbench" on which a polypeptide chain is assembled

messenger RNA (mRNA) — a linear sequence of nucleotides that carries protein-building instructions; this "code" is delivered to the ribosome for translation into a polypeptide chain

transfer RNA (tRNA) — another nucleic acid chain that serves as an adaptor molecule; it can pick up a specific amino acid *and* pair with an mRNA code word for that amino acid

Only mRNA eventually becomes translated into a protein product. The other two types of RNAs have specific roles during translation.

The flow of information in cells starts when DNA is transcribed into RNA.

Different RNAs are translated into cell proteins.

Transcription: DNA into RNA

In transcription, an RNA strand is assembled on a DNA template according to the base-pairing rules:

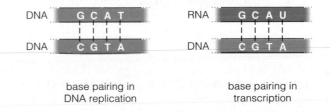

base pairing in DNA replication

base pairing in transcription

Transcription takes place in the cell nucleus, but it differs from DNA replication in key respects. First, only the gene segment serves as the template—not the whole DNA strand. Second, different enzymes, called RNA polymerases, are involved. Third, transcription results in only a single-stranded molecule, not one with two strands.

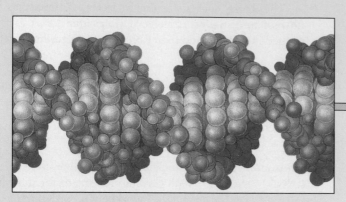

A gene region of a DNA double helix. One of the two nucleotide strands is about to be transcribed into an RNA molecule.

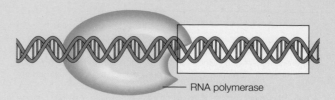

RNA polymerase

a An RNA polymerase molecule binds to a "start" site in the DNA. It will use the base sequence positioned downstream from that site as a template for linking nucleotides into a strand of RNA.

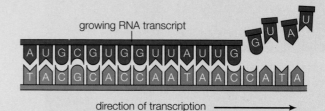

growing RNA transcript

direction of transcription →

b During transcription, RNA nucleotides are based-paired, one after another, with the exposed bases on the DNA template.

Figure 19.6 Transcription: the synthesis of an mRNA molecule on a DNA template.

Transcription starts at a **promoter**, a base sequence that signals the start of a gene. Proteins help position an RNA polymerase on the DNA so that it binds with the promoter. The enzyme moves along the DNA, joining nucleotides together (Figure 19.6). When it reaches a base sequence that serves as a signal to cut the RNA loose, the RNA is released as a transcript.

Newly formed pre-mRNA is an unfinished molecule that must be modified before its protein-building instructions can be used. Just as a film editor might cut some frames and splice together others to make the final version of a movie, a cell tailors its pre-mRNA. For example, newly formed transcripts contain portions called introns and exons. **Introns** are base sequences that do *not* get translated into an amino acid sequence. **Exons** are the regions that get translated into proteins. As Figure 19.7 shows, before the mRNA leaves the nucleus the introns are snipped out and the exons are spliced together.

Translation: RNA into Protein

The Genetic Code The sequence of nucleotides in DNA, which, in turn, is transcribed into mRNA, is a gene that specifies the amino acid sequence of a polypeptide chain. Which amino acids are placed in the polypeptide

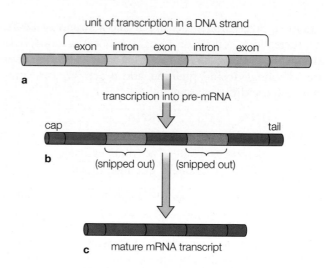

Figure 19.7 Transcription and modification of newly formed mRNA. The cap simply is a nucleotide with functional groups attached. The tail is a string of adenine nucleotides.

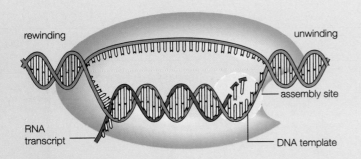

c Throughout transcription, the DNA double helix unwinds just in front of the RNA polymerase. Short lengths of the newly forming RNA strand temporarily wind up with the DNA template strand. Then they unwind from it—and the two strands of DNA wind up together again.

A U G C G U G G U U A U U G G U A U

new RNA transcript

d At the end of the gene region, the last stretch of the RNA is unwound from the DNA template and so is released.

First Letter	Second Letter				Third Letter
	U	C	A	G	
U	phenylalanine	serine	tyrosine	cysteine	U
	phenylalanine	serine	tyrosine	cysteine	C
	leucine	serine	stop	stop	A
	leucine	serine	stop	tryptophan	G
C	leucine	proline	histidine	arginine	U
	leucine	proline	histidine	arginine	C
	leucine	proline	glutamine	arginine	A
	leucine	proline	glutamine	arginine	G
A	isoleucine	threonine	asparagine	serine	U
	isoleucine	threonine	asparagine	serine	C
	isoleucine	threonine	lysine	arginine	A
	(start) methionine	threonine	lysine	arginine	G
G	valine	alanine	aspartate	glycine	U
	valine	alanine	aspartate	glycine	C
	valine	alanine	glutamate	glycine	A
	valine	alanine	glutamate	glycine	G

chain is determined by the **genetic code**. The code is actually quite simple. The bases of the nucleotides in mRNA are read in groups of three, or "triplets." Each triplet is called a **codon** and specifies an amino acid (Figure 19.8). There are 64 different codons. Sixty-one specify amino acids. For example, AUG specifies methionine, and CUG specifies leucine. Three (UGA, UAA, and UAG) are stop signals. They prevent ribosomes from adding more amino acids to a polypeptide chain. As Figure 19.8 illustrates, most amino acids can be specified by more than one codon. (Glutamate, for example, is coded by GAA and GAG.) The AUG codon also establishes the reading frame for translation; that is, ribosomes usually start their "three-bases-at-a-time" selections beginning with the first AUG in an mRNA strand.

Hereditary instructions for building proteins are encoded in the nucleotide sequence of DNA, which is transcribed into a corresponding nucleotide sequence in RNA.

A codon (base triplet) specifies a given amino acid.

Roles of tRNA and rRNA

Cells have pools of free amino acids and free tRNA molecules in the cytoplasm. Each tRNA has an **anticodon**, a nucleotide triplet that can base-pair with codons. It also has a molecular "hook," an attachment site for amino acids:

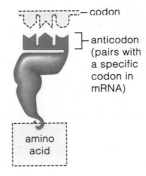

When different types of tRNAs bind to different codons, they automatically position their attached amino acids in the order specified by the mRNA sequence. Thus tRNAs are the "translators" of mRNA.

Figure 19.8 The genetic code. The codons in an mRNA molecule are nucleotide bases read in blocks of three. Sixty-one of these base triplets correspond to specific amino acids. Three others serve as signals that stop translation. In this diagram, the left column shows the first of the three nucleotides in each codon in mRNA. The middle columns show the second nucleotide; the right column shows the third nucleotide. Reading from left to right, for instance, the triplet U G G corresponds to tryptophan. Both U U U and U U C correspond to phenylalanine.

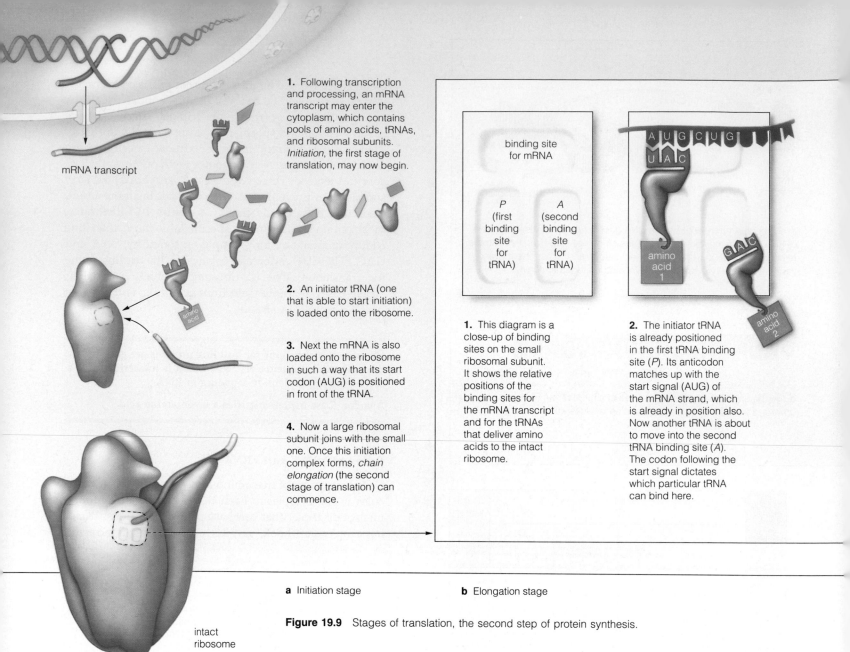

1. Following transcription and processing, an mRNA transcript may enter the cytoplasm, which contains pools of amino acids, tRNAs, and ribosomal subunits. *Initiation*, the first stage of translation, may now begin.

mRNA transcript

2. An initiator tRNA (one that is able to start initiation) is loaded onto the ribosome.

3. Next the mRNA is also loaded onto the ribosome in such a way that its start codon (AUG) is positioned in front of the tRNA.

4. Now a large ribosomal subunit joins with the small one. Once this initiation complex forms, *chain elongation* (the second stage of translation) can commence.

binding site for mRNA

P
(first binding site for tRNA)

A
(second binding site for tRNA)

amino acid 1

amino acid 2

1. This diagram is a close-up of binding sites on the small ribosomal subunit. It shows the relative positions of the binding sites for the mRNA transcript and for the tRNAs that deliver amino acids to the intact ribosome.

2. The initiator tRNA is already positioned in the first tRNA binding site (*P*). Its anticodon matches up with the start signal (AUG) of the mRNA strand, which is already in position also. Now another tRNA is about to move into the second tRNA binding site (*A*). The codon following the start signal dictates which particular tRNA can bind here.

intact ribosome

a Initiation stage

b Elongation stage

Figure 19.9 Stages of translation, the second step of protein synthesis.

The tRNA molecules meet up with an mRNA strand at binding sites within ribosomes. Each ribosome has two subunits, which are assembled in the nucleus from rRNA and proteins:

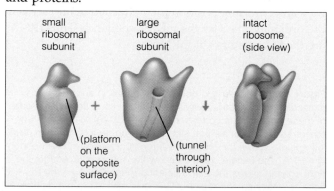

small ribosomal subunit

large ribosomal subunit

intact ribosome (side view)

(platform on the opposite surface)

(tunnel through interior)

The large and small ribosome subunits are shipped separately into the cytoplasm. They join together during translation.

Stages of Translation

Translation proceeds through three stages called initiation, elongation, and termination (Figure 19.9).

In *initiation,* an "initiator" tRNA that can start translation *and* a small ribosomal subunit become loaded onto the end of an mRNA. (The site on the ribosome where the RNAs are located is called the "platform.") This complex moves along the mRNA until it encounters an AUG codon. Then a large ribosomal subunit joins with the small one. We now have an *initiation complex* that includes an intact ribosome, an mRNA, and an initiator tRNA. The next stage can begin.

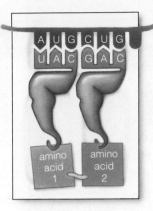

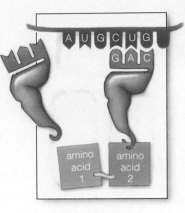

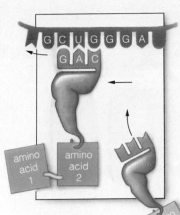

3. An enzyme breaks the bond between the initiator tRNA and the amino acid hooked to it. At the same time, a peptide bond forms between the two amino acids. After these events have occurred, the initiator tRNA will be released from the ribosome.

4. Two amino acids are now hooked onto the second tRNA. The tRNA will move into the *P* site, sliding the mRNA strand with it by one codon. With this movement, the third codon of the mRNA will become aligned above the *A* site.

5. A third tRNA is about to move into the *A* site. Its anticodon will base-pair with the third codon of the mRNA transcript. A peptide bond will form between amino acids 2 and 3.

6. Steps 3 through 6 are repeated again and again. The polypeptide chain continues to grow this way until a stop codon is reached. Then the chain and the mRNA will be released. Also, the two ribosomal subunits will separate. This marks the last stage of translation, *chain termination* (**c**).

In *elongation* (Figure 19.9b), the mRNA strand passes between the ribosome subunits, like a thread being moved through the eye of a needle. One after another, tRNA molecules arrive with the various amino acids specified by the mRNA's codons. Enzymes built into the ribosome join the amino acids together (they catalyze peptide bonds between the amino acids), and the polypeptide chain grows.

In *termination,* a stop codon is reached and there is no tRNA with a corresponding anticodon. The ribosome interacts with a protein called a release factor. This causes the ribosome and polypeptide chain to detach from the mRNA (Figure 19.9c). The detached chain joins the pool of polypeptides in the cytoplasm or enters the rough ER of the cytomembrane system (page 419) for further processing.

Commonly, the same mRNA transcript is translated repeatedly in a given period. It threads through many ribosomes arranged one after another in assembly-line fashion (an arrangement called a *polysome*). In this way, cells can produce many copies of a polypeptide chain from

1. Once a stop codon is reached, the mRNA transcript is released from the ribosome:

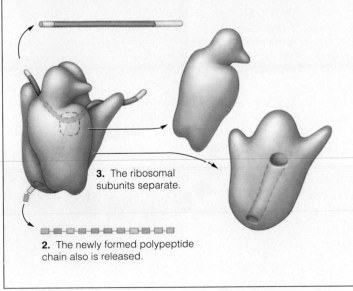

3. The ribosomal subunits separate.

2. The newly formed polypeptide chain also is released.

c Termination stage

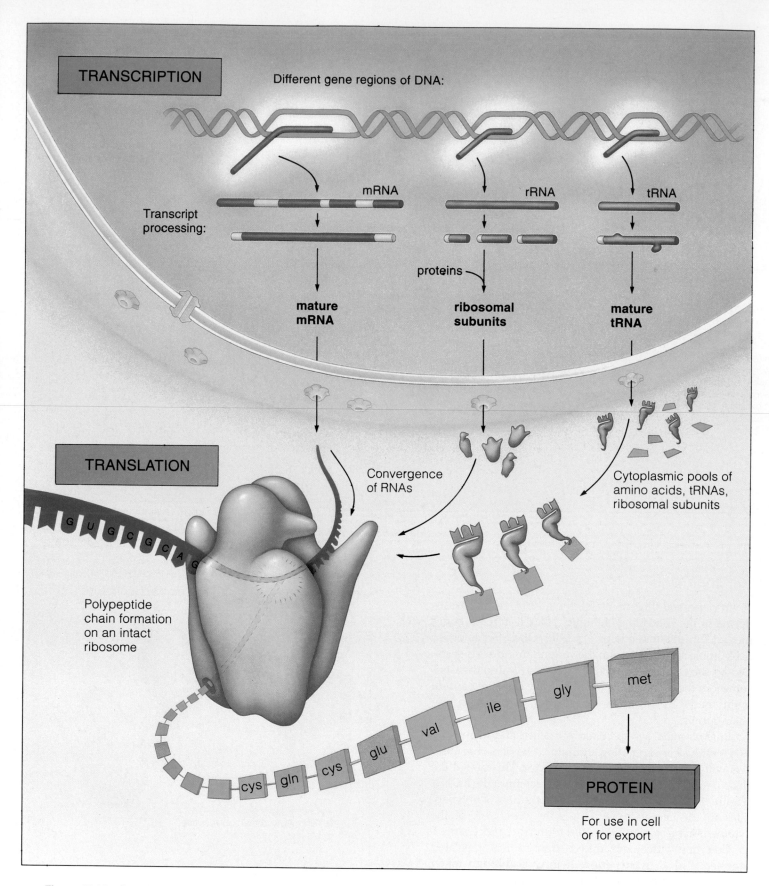

Figure 19.10 Summary of transcription and translation, the two steps leading to protein synthesis.

the same transcript. Figure 19.10 summarizes the flow of information along the path leading from genes to proteins.

Genetic instructions transcribed from DNA into mRNA are then translated into a polypeptide chain in stages called initiation, elongation, and termination.

Translation takes place in the cell cytoplasm, on the surface of ribosomes.

MUTATION

Types of Mutations

Every so often, genes change. One base is substituted for another in the DNA sequence, extra bases are inserted, or a few bases are lost. These small-scale changes in the nucleotide sequence of DNA are **point mutations** (Table 19.1). When a mutation occurs in a germ cell, it can be passed on to the next generation. When it occurs in a somatic cell, the mutation will not be heritable.

Some gene mutations are induced by **mutagens**, agents that attack a DNA molecule and modify its structure. Examples of mutagens include some viruses, ultraviolet radiation, many chemicals, and free radicals. Often, mutagens are also *carcinogens,* agents that can cause cancer (page 432). Mutagens can increase the *rate* at which mutation occurs by causing more changes in the DNA than would occur if no mutagen were present.

Other mutations occur spontaneously. These include replication errors, as when adenine wrongly pairs with a cytosine on a DNA template strand.

Proofreading enzymes might detect errors in replication or nucleotides altered by mutagens, then "fix" the wrong base:

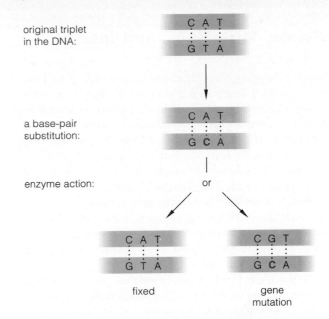

original triplet in the DNA:

a base-pair substitution:

enzyme action: or

fixed   gene mutation

If the enzymes remove the wrong base from a mismatched pair, the result is a type of spontaneous mutation called a *base-pair substitution*. Sickle-cell anemia results from a substitution mutation in the DNA strand coding for the beta chain of hemoglobin. In a chain 150 amino acids long, only one amino acid is substituted for another—yet, as we now know, the consequences of the substitution can be severe. (Several other human diseases, including some instances of heart disease, are now thought to arise from mutations in DNA that resides not in cell nuclei, but in mitochondria; see *Focus on Wellness*.)

Table 19.1	**Types of Point Mutations**	
Change	Effect on Base Sequence	Outcome
substitution	AAG CAG GGA GGA TTC GTC CCC CCT ↑ C instead of T correct reading frames	altered amino acid in a protein
frameshift	AAGCAGGGAGGA AAGGCAGGGAGGA ↑ nucleotide added (or deleted), so reading frame shifts	wrong amino acid strung together in protein
transposable element*	AAGCAGGGAGGA AAGCCACAGCAGGGAGGA movable DNA inserted	gene inactivated or otherwise altered, so phenotype changes

*This illustration of a transposable element is simplified for clarity. Usually, such elements consist of thousands of nucleotide pairs. They are also double-stranded.

Disease, Mutations, and mtDNA

Disorders ranging from blindness to an irregular heartbeat may have their origin in mutated genes—but not in genes carried on chromosomes in the cell nucleus. Instead, the real culprits are the cell "powerhouses," mitochondria.

Mitochondria, remember, lie outside the nucleus in the cell cytoplasm. By way of cellular respiration, they produce the ATP that fuels cell metabolism (page 49). Curiously, they have their own cache of genetic material, called mtDNA. The mtDNA in human cells is maternal—it comes only from the mother's egg that was fertilized at conception. This is because only the head of a sperm cell enters an egg at fertilization, while all of a sperm's mitochondria remain behind in its tail.

As in nuclear DNA, genes in mtDNA can mutate. When mutation occurs in all of a cell's mitochondria, the usual result is less energy available to the cell. The outcome is variable, however, because different cell types have different numbers of mitochondria, ranging from a few hundred to thousands. Also, some of those mitochondria may be normal. The difference comes about during early development.

Mitochondria are located in the cell cytoplasm; a mutagen that acts on a germ cell in a woman's ovaries may trigger a mutation in some mitochondria but not others. During meiosis, portions of the germ cell cytoplasm go to the different daughter cells, the future oocytes. By chance, one daughter cell may get many mutant mitochondria while another may get only a few. Mutant and normal mitochondria are also sorted randomly into different parts of the early embryo during cleavage.

As a result of these factors, a child may be born with defective mitochondria in some tissues or organs but not others. The problem is most serious in cardiac muscle, skeletal muscle, and the brain, which require large amounts of ATP for proper functioning. Some cases of heart disease and epilepsy are suspected to arise in this way, as does one type of diabetes. Researchers also have located the mutated gene responsible for a form of adult blindness—again, in mtDNA.

Another type of spontaneous change is a *frameshift mutation.* Here, one or more nucleotide pairs are added to or deleted from a DNA molecule. This puts the nucleotide sequence out of phase, so that the reading frame shifts during protein synthesis. Because genetic instructions are read out of frame, an abnormal protein is synthesized.

Spontaneous mutations occur quite frequently—on average, every 1 per 1,000 nucleotide pairs. Repair mechanisms fix most of these errors, so that the actual rate (after repair) is about 1 per billion nucleotides.

Transposable Elements and Pseudogenes

An intriguing type of gene rearrangement involves **transposable elements**, originally described in corn by Barbara McClintock. There are two types of these "jumping genes." A *transposon* is a DNA region that can "jump" to new locations in the same DNA molecule or in a different one. In humans and other mammals, the most common transposable elements are *retroposons*. This kind of element moves to a new location by way of an intermediate step involving RNA. It is transcribed into RNA, which is then transcribed *back* into DNA by reverse transcriptase. (Recall from Chapter 15 that this is how retroviruses such as HIV become transcribed into DNA and inserted into a host's DNA.) The DNA copy is then inserted into the host's genome in a new location. This sequence of steps can be repeated over and over, so that the host's DNA comes to include many copies of the retroposon. For instance, the DNA in a diploid human cell typically contains thousands of copies of a transposable element called *Alu* (from the name of an enzyme that can cleave the element's nucleotide sequence).

Transposable elements may inactivate the genes into which they become inserted or alter their functioning in other ways. For example, researchers have found that insertion of an *Alu* element into the gene coding for a certain mRNA can result in a mutation that triggers the genetic disorder neurofibromatosis. In a number of hemophilia patients, the disorder has been traced to insertion of a probable retroposon called LINE (for Long INterspersed Elements) into a gene coding a normal blood-clotting factor.

Many retroposons may be derived from infectious retroviruses that became integrated into a host organism's DNA. Others, such as *Alu* elements, may be **pseudogenes** (*pseudo:* false). Pseudogenes are faulty (mutated) copies of normal genes. Many are never transcribed; others may be transcribed and translated, but often their protein products do not function. Together, pseudogenes and other transposable elements may make up as much as 20 percent of human DNA.

A point mutation is a deletion, addition, or substitution of one to several bases in the nucleotide sequence of a gene. Mutations are sometimes spontaneous but are also caused by exposure to certain viruses, chemicals, and radiation.

REGULATING GENE ACTION

Most of the different cells of your body carry the same genes. Many of those genes carry instructions for synthesizing proteins that are essential to any cell's structure and functioning. Yet each type of cell also uses a small subset of genes in specialized ways. For example, every cell carries the genes for hemoglobin, but only the precursors of red blood cells activate those genes. Only the white blood cells known as lymphocytes activate genes for antibodies. *Each cell controls which genes are active and which gene products appear, when, and in what amounts.* Some genes might be switched on and off throughout the individual's life. Other genes might be turned on only in certain cells and only at certain times.

Controls over genes are exerted by way of proteins, hormones, and other molecules that interact with DNA, RNA, or gene products. For example, **regulatory proteins** enhance or suppress the rate of transcription.

A gene may be associated with one or more DNA sequences that do *not* code for a protein but instead interact with regulatory proteins. When a regulatory protein binds with one of these noncoding sequences, it may trigger transcription of the gene into mRNA (or shut it down altogether). For instance, like many other organisms, humans have "heat shock" genes that are activated when body temperature rises above 37°C (98.6°F), as when you have a fever. Rising cell temperature first causes a shape change in an inactive regulatory protein called heat shock factor (HSF). The change activates the protein, which binds to a nucleotide sequence "upstream" from a group of heat shock genes. The genes are then transcribed to mRNA, and the mRNA is translated into proteins that help repair heat-induced damage to other cell proteins.

Hormones as Control Agents Hormones have widespread effects on gene activity in many cell types. In humans, for instance, the pituitary gland secretes growth hormone (page 278). Most cells have receptors for growth hormone, which helps control synthesis of proteins required for cell division and, ultimately, body growth. On the other hand, some hormones affect only certain cells at certain times. The hormone prolactin, which activates genes in mammary gland cells that have exclusive responsibility for milk production, is a good example. Liver cells and heart cells have the same genes, but they have no means of responding to signals from prolactin.

Although most cells have the same genes, some of those genes are activated or suppressed in different ways to produce differences in cell structure or function.

SUMMARY

1. Deoxyribonucleic acid, or DNA, is the master blueprint of hereditary instructions in cells. It is assembled from small organic molecules called nucleotides.

2. All DNA nucleotides have a five-carbon sugar (deoxyribose) and a phosphate group. They also have one of four nitrogen-containing bases: adenine, thymine, guanine, or cytosine.

3. In a DNA molecule, two nucleotide strands are twisted together into a double helix. The bases of one strand pair with bases of the other strand (by hydrogen bonding).

4. There is constancy in base pairing in a DNA molecule. Adenine pairs with thymine (A—T), and guanine pairs with cytosine (G—C).

5. In DNA replication, the two strands of the double helix unwind from each other. Then, a new strand consisting of a complementary sequence of nucleotides is assembled on each existing strand. Two double-stranded molecules result. In each one, one strand is "old" (it is conserved) and the other is "new." Replication requires many enzymes and other proteins.

6. A eukaryotic chromosome consists of one DNA molecule bound with many proteins. Interactions between the DNA and proteins give rise to a coiled and looped structural organization.

7. The path from DNA to proteins has two steps, transcription and translation. In *transcription*, an exposed region of one strand of the DNA double helix serves as the template for assembling an RNA strand. In *translation*, three classes of RNA molecules (mRNA, rRNA, and tRNA) interact to convert the gene's message into a lin-

ear sequence of amino acids—a polypeptide chain. Such chains are the structural units of proteins.

8. The 64 triplet nucleotide codons that correspond to amino acids are collectively called the genetic code. The code words are a sequence of nucleotides that are read in blocks of three (base triplets). Each base triplet in mRNA is a codon; the complementary base triplet in tRNA is an anticodon.

9. Overall, the protein-building instructions in DNA are preserved through the generations. But crossing over and recombination, changes in chromosome structure or number, and gene mutations can change parts of those instructions.

10. Gene expression is controlled by many interacting elements, including control sites built into DNA molecules, regulatory proteins, and hormones. Their interactions govern which gene products appear, at what times, and in what amounts.

Review Questions

1. DNA is composed of four different kinds of nucleotides. Name the three molecular parts of a nucleotide. Name the four different kinds of nitrogen-containing bases that may occur in the nucleotides of DNA. *410*

2. Are the products specified by DNA assembled *on* the DNA molecule? If so, state how. If not, tell where they are assembled and on which molecules. *411*

3. Figure 19.6 shows the steps by which hereditary instructions are transcribed from DNA into RNA, which is then translated into proteins. Study this figure and then, on your own, write a description of this sequence, taking care to define the terms *transcription* and *translation*. *416*

4. Define *genetic code*. Is the same basic genetic code used for protein synthesis in all living organisms? *417*

5. Define the three types of RNA. What is a codon? An anticodon? *415–417*

Critical Thinking: You Decide *(Key in Appendix IV)*

1. Which mutation would be more harmful: a mutation in DNA or one in mRNA? Explain your answer.

Self-Quiz *(Answers in Appendix III)*

1. Which of the following is *not* a nitrogen-containing base of DNA?
   a. adenine d. cytosine
   b. thymine e. uracil
   c. guanine

2. Base pairing in the DNA molecule follows which configuration?
   a. A—G, T—C
   b. A—C, T—G
   c. A—U, C—G
   d. A—T, C—G

3. A single strand of DNA with the base sequence C-G-A-T-T-G would be complementary to the sequence _____.
   a. C-G-A-T-T-G
   b. G-C-T-A-A-G
   c. T-A-G-C-C-T
   d. G-C-T-A-A-C

4. DNA replication produces _____.
   a. two half-old, half-new double-stranded molecules
   b. two double-stranded molecules, one with the old strands and one with newly assembled strands
   c. three new double-stranded molecules, one with both strands completely new and two that are discarded
   d. none of the above

5. Genetic information in DNA is transferred to RNA strands during _____, the first step in protein synthesis.
   a. replication   c. multiplication
   b. duplication   d. transcription

6. During transcription, base pairing is similar to that of DNA replication except that _____.
   a. cytosine in DNA pairs with guanine in RNA
   b. adenine in DNA pairs with uracil in RNA
   c. thymine in DNA pairs with adenine in RNA
   d. guanine in DNA pairs with cytosine in RNA

7. _____ starts when two ribosomal subunits, an initiator tRNA, and a mRNA transcript come together.
   a. Transcription
   b. Replication
   c. Subduction
   d. Translation

8. The coded genetic instructions for forming polypeptide chains are carried to the ribosome by _____.
   a. DNA   c. mRNA
   b. rRNA   d. tRNA

9. The function of tRNA is to _____.
   a. deliver amino acids to the ribosome
   b. pick up genetic messages from rRNA
   c. synthesize mRNA
   d. all of the above

10. How many amino acids are coded for in this mRNA sequence: CGUUUACACCGUCAC?
   a. three   d. seven
   b. five   e. more than seven
   c. six

11. Match these DNA concepts appropriately.
   ____base pairs
   ____chromosome (metaphase)
   ____semiconservative replication
   ____double helix
   ____disrupts genetic instructions
   ____transcription
   ____translation

   a. two nucleotide strands twisted together
   b. A—T G—C
   c. one DNA strand old, the other new
   d. structure results from interactions between DNA and proteins
   e. ribosomes, tRNAs, and mRNAs convert genetic messages into polypeptides
   f. one DNA strand serves as the template for RNA synthesis
   g. mutation

Key Terms

adenine *410*	nucleotide *410*
anticodon *417*	point mutation *421*
base pair *411*	promoter *416*
codon *417*	pseudogene *423*
cytosine *410*	regulatory protein *423*
DNA polymerase *414*	replication *411*
exon *416*	ribosomal RNA *415*
gene *411*	RNA *414*
genetic code *417*	thymine *410*
guanine *410*	transcription *415*
histone *414*	transfer RNA *415*
intron *416*	translation *415*
messenger RNA *415*	transposable element *422*
mutagen *421*	uracil *415*
nucleosome *414*	

Readings

Beardsley, T. August 1991. "Smart Genes." *Scientific American*. A fascinating discussion of how organisms from humans to fruit flies control their genes by regulating transcription.

Watson, J. 1978. *The Double Helix*. New York: Atheneum. Highly personal view of scientists and their methods, interwoven into an account of how DNA structure was discovered.

Wolfe, S. L. 1992. *Molecular and Cellular Biology*. Belmont, Calif.: Wadsworth. Clear descriptions of basic genetic events in cells.

20 CANCER: A CASE STUDY OF GENES AND DISEASE

The Body Betrayed

The tiny mass of tissue, only a few million cells strong, may go unnoticed for 10, 20, even 30 years. In fact, not until the mass becomes a clump of perhaps 10 billion cells will it be felt, worried about, and finally—reluctantly—discussed with a doctor.

Cancer is a grim fact of human life. It strikes one in three people in the United States (Figure 20.1) and kills one in five—about 1,400 cancer deaths every day, just over half a million in a year. On average, cancer strikes more males than females, but the pattern varies

Figure 20.1 One out of three people in the United States will be diagnosed with cancer during their lifetime. These young women are taking steps to limit their risk of lung cancer, the chief cancer killer of both men and women.

depending on the type of cancer involved. Thus the number of new lung cancer cases is increasing faster among women due to a corresponding increase in the number of women who began smoking cigarettes in the last several decades. Adult leukemias and bladder cancers occur more often in men, but males and females are about equally susceptible to colon cancer. Cancer strikes more often in older age groups, although it is the leading cause of death among children ages 3–14. Fortunately, we are rapidly increasing our understanding of and ability to treat many cancers, including cancers of the breast, ovary, and skin.

Every cancer begins with a normal cell in which the controls over cell division are lost. In this sense cancer is a betrayal, a bit of the body that turns *against* the body. More than 100 specific cancer types are known. They have names like retinoblastoma and fibrosarcoma that identify the type of tissue in which the first cancer cell develops. Authorities estimate that about 15 percent of cancers arise from genetic changes caused by viral infection; 5 percent come from an inherited mutation; the rest, about 80 percent, develop due to mutations triggered by environmental factors, including certain chemicals and radiation.

Statistics hint at the impact of cancer on human beings, but they don't tell us much about the biological causes and effects of cancers, and they don't reveal recent advances in our understanding of the genetic events that underlie this group of diseases. These are the topics we consider in this case study, along with cancer treatments and life-style choices that can affect personal cancer risk.

KEY CONCEPTS

1. Cancer arises when genetic controls over cell division are lost.

2. Cancer develops through a multistep process. It requires mutations in a series of genes, including tumor suppressor genes that normally operate in cells.

3. Up to 80 percent of cancers may be triggered by "environmental" factors. These include mutagens such as ultraviolet radiation and chemical carcinogens that switch on oncogenes (cancer-causing genes), turn off tumor suppressor genes, or both.

4. Life-style choices can limit a person's risk of developing many types of cancer. The most important choices appear to involve diet and tobacco use.

TUMORS AND CANCER

As genes switch on and off, they determine what a cell's specialized function will be, when and how rapidly the cell will grow and divide, even when it will *stop* dividing and when it will die. If cells overgrow, the result is a defined mass of tissue called a **tumor**. This mass is a *neoplasm*, which literally means "new growth."

A tumor is not necessarily "cancer." As Figure 20.2a shows, the cells of a *benign* tumor are often enclosed by a capsule of connective tissue, and within the capsule they are organized in an orderly array. They also tend to grow slowly and to be well differentiated (structurally specialized), much like normal cells of the same tissue. Benign tumors typically stay put in the body, push aside but don't invade surrounding tissue, and usually can be easily removed by surgery. Benign tumors *can* threaten health, as when they occur in the brain. Nearly everyone has at least several of the benign tumors we call moles. Most of us also have or have had some other type of benign neoplasm, which is usually destroyed by the immune system.

Dysplasia ("bad form") is an abnormal change in the sizes, shapes, and organization of cells in a tissue. It is often a precursor to cancer. Under the microscope, the borders of a cancerous tumor often appear ragged (Figure 20.2b), and its cells form a disorganized clump. Cancer cells usually also have characteristics that enable them to behave very differently from normal body cells.

Characteristics of Cancer Cells

Structural Abnormalities   A cancer cell generally has an abnormally large nucleus, less cytoplasm than usual, and is *poorly differentiated*. That is, cancer cells often do not have the clear structural specializations of cells in mature body tissues. The extent of differentiation of cancer cells can be medically important. In general, the less differentiated cancer cells are, the more readily they break away from the primary tumor and spread the disease.

When a normal cell becomes transformed into a cancer cell, additional changes occur. The cytoskeleton shrinks, becomes disorganized, or both. Proteins of the plasma membrane are lost or altered, and new, different ones appear. These changes are passed on to the cell's descendants: When a transformed cell divides, its daughter cells are cancer cells too.

Uncontrolled Growth   Cancer cells lack normal controls over cell division. Contrary to popular belief, cancer cells do not necessarily divide more rapidly than normal cells do, but they do increase in number faster. Normally, the death of cells closely balances the production of new ones through mitosis, so that the cells are arranged in an orderly tissue. In a developing cancerous tumor, however, at any given moment more cells are dividing than are

Figure 20.2   (**a**) Sketch of a benign tumor. Cells appear nearly normal, and the tumor mass is encapsulated within connective tissue. (**b**) A cancerous neoplasm. Due to the abnormal growth of cancer cells, the tumor is a disorganized heap of cells, some of which break off and invade surrounding tissues (metastasis).

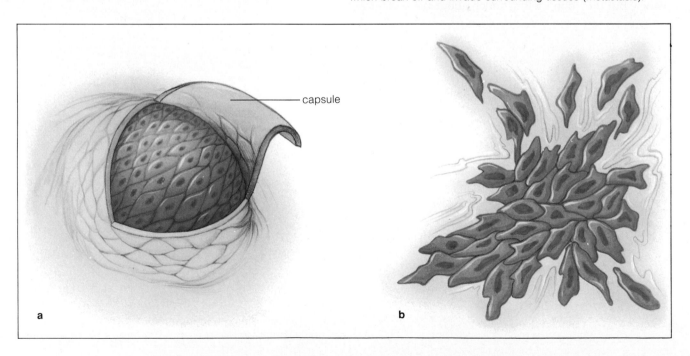

capsule

a

b

Figure 20.3 Scanning electron micrograph of a cancer cell surrounded by some of the body's white blood cells that may or may not be able to destroy it.

dying. As this unregulated cell division continues, the cancer cells do not respond to crowding. Whereas a normal cell stops dividing once it comes into contact with another cell, a cancerous cell continues to divide. As a result, cancer cells pile up in a disorganized heap—which is why tumors are lumpy.

Cancer cells also lack strong cell-to-cell junctions with neighboring cells, and as Figure 20.3 shows, they form extensions (pseudopodia) that enable them to move about. This property gives cancer cells the dangerous ability to break away from the parent tumor and invade surrounding tissue (metastasis), including the lymphatic system and circulatory system. It is this property that makes cancer *malignant*.

Some cancer cells produce interleukin-2, which stimulates cell division. They also display receptors for it. Cancer cells also secrete a growth factor called *angiogenin* that encourages new blood vessels to grow around the tumor. This is not surprising because tumor growth requires a large supply of nutrients and oxygen supplied via the bloodstream. In fact, researchers are attempting to determine whether it is possible to "starve" some tumors to death by blocking angiogenin's effects.

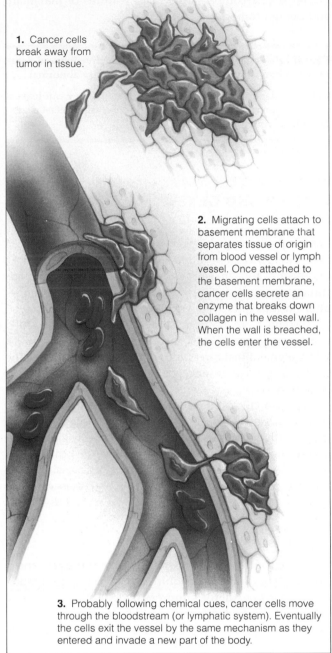

1. Cancer cells break away from tumor in tissue.

2. Migrating cells attach to basement membrane that separates tissue of origin from blood vessel or lymph vessel. Once attached to the basement membrane, cancer cells secrete an enzyme that breaks down collagen in the vessel wall. When the wall is breached, the cells enter the vessel.

3. Probably following chemical cues, cancer cells move through the bloodstream (or lymphatic system). Eventually the cells exit the vessel by the same mechanism as they entered and invade a new part of the body.

Figure 20.4 Stages in metastasis.

Metastasis Cancer cells that break away from a primary tumor can move through layers of tissue, enter blood or lymphatic vessels, and migrate to other locations where they establish new cancer sites (Figure 20.4). For example, cells from a primary tumor in the breast or prostate gland may move to the bone marrow, lungs, brain, and elsewhere. The process is called **metastasis**.

Table 20.1 compares the main features of malignant and benign tumors.

Cancer cells lack normal controls over cell division and organization into tissues. They are also structurally abnormal.

Cancer cells are often poorly differentiated. They can leave a primary tumor, invade surrounding tissue, and metastasize to other parts of the body.

CANCER AND GENES

Mammals, many fish species, and plants all get cancer. Fossil evidence shows that dinosaurs did also. In those organisms and in our own bodies, cancer develops through a multistep process in which genetic changes alter normal controls over cell division.

Certain DNA sequences on chromosomes are called **proto-oncogenes**. "Proto" means before; an **oncogene** is a DNA segment that can induce cancer in a normal cell. The distinction between oncogenes and proto-oncogenes is important. Proto-oncogenes are normal genes that regulate cell growth and development. They code for proteins that include growth factors (signals sent by one cell to trigger growth in other cells), regulatory proteins involved in cell adhesion, and the protein signals for cell division. If some event changes their structure or expression, they may become altered into oncogenes and then code for defective proteins that remove controls on cell division.

Tumor Suppressor Genes

An oncogene acting alone probably cannot trigger cancer. The onset of cancer also requires the absence or mutation of at least one **tumor suppressor gene**. Some of these genes encode protein products that operate to keep cell growth and division within normal bounds. Others may act to keep cells anchored in place.

Several lines of evidence have begun to shed light on the roles of specific tumor suppressor genes. Studies of the childhood eye cancer *retinoblastoma* have revealed that the disease develops when a child inherits only one normal copy of a tumor suppressor gene called Rb. If that allele mutates, the disease results. Inherited forms of lung, breast, and skin cancer have been linked to the absence of other tumor suppressor genes.

The p53 Gene   A tumor suppressor gene called p53 is especially interesting because it appears to help prevent cancerous changes in many types of tissue. This gene codes for a regulatory protein that seems to turn on proto-oncogenes that stop cell division at specified times. When p53 is absent, the controls don't operate and cell division continues unchecked. The plot changes when p53 mutates. When a thymine is inserted where a guanine should be in the base sequence of one nucleotide, the resulting faulty protein seems to *promote* the development of cancer, possibly by activating an oncogene. Researchers have identified the mutated p53 genes in cells from cancers of the breast, bone, colon, skin, lung, bladder, cervix, and brain.

Studies of p53 show how basic scientific research can be of major practical benefit. The fact that an absent or defective p53 gene correlates with so many cancers has become the basis for diagnostic tests that screen for the gene.

Oncogene Activation

An oncogene may be present in a cell and never cause trouble because some condition in the cell, such as the presence of a suppressor gene, prevents the oncogene from being expressed. Several kinds of events can change this state of affairs. The oncogene or a related suppressor gene may mutate in a way that triggers expression. A break in a chromosome followed by a translocation (page 403) may move an oncogene away from a regulatory nucleotide sequence that would normally prevent it from being expressed. Or, new genetic material may be introduced into a cell (as by viral infection) and disrupt controls.

Table 20.1 Comparison of Benign and Malignant Tumors		
	Malignant Tumor	Benign Tumor
Rate of growth	Rapid	Slow
Nature of growth	Invades surrounding tissue	Expands within tissue
Spread	Metastasis by way of bloodstream and lymphatic system	Stays localized
Cell differentiation	Usually poor	Nearly normal

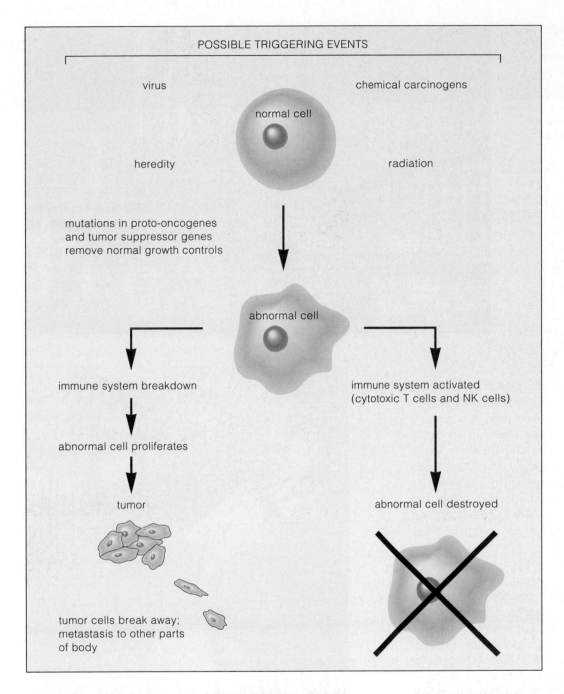

POSSIBLE TRIGGERING EVENTS

virus

chemical carcinogens

normal cell

heredity

radiation

mutations in proto-oncogenes
and tumor suppressor genes
remove normal growth controls

abnormal cell

immune system breakdown

immune system activated
(cytotoxic T cells and NK cells)

abnormal cell proliferates

tumor

abnormal cell destroyed

tumor cells break away;
metastasis to other parts
of body

Figure 20.5 Overview of steps in carcinogenesis.

Cancer develops through a multistage process in which gene changes remove normal controls over cell division.

Oncogenes have a major role in inducing cancer in a normal cell. Proto-oncogenes may become oncogenes if some event changes their structure or the way they are expressed.

The development of cancer also requires the absence or mutation of at least one tumor suppressor gene.

Carcinogenesis: The Onset of Cancer

The transformation of a normal cell into a cancerous one is called **carcinogenesis**. In this section we trace the steps that can be involved (Figure 20.5).

"Inherited" Cancer We have all heard of cancer "running in a family." In fact, heredity plays a major role in a small percentage of cancer cases. If a mutation exists in a germ cell (sperm or egg) and it removes controls over a proto-oncogene on a chromosome, the defect can be

Figure 20.6 (a) Environmental activists in Louisiana protesting the unusually high rates of cancers along a 137-mile stretch known as "Cancer Alley," in which oil refineries and petrochemical plants have discharged large amounts of carcinogenic chemicals. (b) Home garden chemicals are another avenue by which mutagenic or carcinogenic substances can come into contact with the human body.

a

b

passed on to offspring. A person may be more susceptible to cancer if a mutation or a chromosomal abnormality leaves them with only one "good" copy of a gene. If a later mutation alters the lone good allele, an oncogene may be left free of controls. This is what happens in familial retinoblastoma, described earlier. However, most instances of inherited cancer susceptibility are probably more complicated. Patterns of familial breast, colon, and lung cancer suggest that several genes may be involved, including genes that control aspects of cell metabolism and responses to hormones.

Viruses Viruses are thought to cause about 15 percent of cancers. Sometimes a viral infection can alter a proto-oncogene when the viral DNA becomes inserted at a certain position in the host cell DNA. For example, the viral gene could take the place of a regulatory sequence that normally prevents a proto-oncogene from switching on (or off) at the wrong time. Other viruses simply carry oncogenes as part of their genetic material and insert them into the host's DNA. Most viruses linked to human cancer (such as the Epstein-Barr virus associated with Burkitt's lymphoma) are DNA viruses, although new research is focusing on cancer-triggering effects of retroviruses.

Chemical Carcinogens There are thousands of known chemical **carcinogens**, cancer-causing substances that can cause a mutation in DNA (Figure 20.6). A carcinogen that directly causes a mutation is sometimes called an "initiator." The list includes many compounds that are by-products of the industrialization of human societies, such as asbestos, coal tar, vinyl chloride, and benzene (Table 20.2). It includes hydrocarbons in cigarette

smoke and on the charred surfaces of barbequed meats, and carcinogenic substances in dyes and pesticides (see *Focus on Environment*). One of the first carcinogens to be recognized was chimney soot (more accurately, substances soot contains), which frequently caused cancer of the scrotum in chimney sweeps. Plants and fungi can produce "natural" carcinogens; aflatoxin, a metabolic

Cancer and Environmental Chemicals

According to the Food and Drug Administration, about 40 percent of the food bought in supermarkets contains detectable residues of one or more of the active ingredients used in pesticides in the United States. Approximately 3 percent of this food has levels of one or more pesticides above the legal limit. The residues are especially likely to be found in tomatoes, grapes, apples, lettuce, oranges, potatoes, beef, and dairy products. Imported produce, including fruits, vegetables, and coffee beans, can also carry significant pesticide residues—sometimes including pesticides deemed so dangerous that they are banned in the United States. The results of this long-term worldwide experiment, in which the subjects are human consumers, may never be known because it is almost impossible to determine that a certain level of a specific chemical caused a particular cancer or some other harmful effect.

In 1987 the National Academy of Sciences reported that the active ingredients in 90 percent of all fungicides, 60 percent of all herbicides, and 30 percent of all insecticides in use in the United States may cause cancer in humans. According to the *worst-case estimate* in this study, exposure to pesticides in food causes 4,000 to 20,000 cases of cancer a year in the United States.

Cancer is only one possible harmful effect of long-term exposure to low levels of pesticides, which collectively contain more than 600 different chemicals. Some scientists are becoming increasingly concerned about possible genetic mutations, birth defects, nervous system disorders, and effects on the immune and endocrine systems. In 1993, new questions were raised about the health effects in children of low-level pesticide residues. Levels established as "safe" have traditionally been based on adult body weights and patterns of food consumption, whereas those parameters differ for children. Findings of one study suggest that by the age of one year, an average American child's exposure to some cancer-causing pesticides in and on food will be greater than federal guidelines currently consider safe over an *entire lifetime*.

Although residues of some pesticides can be (and should be) removed from the surfaces of fruits and vegetables by washing before eating, in our modern world avoiding pesticide exposure is not so easy. Many people are exposed to pesticides from community spraying programs to control mosquitoes and other pests; pesticide-sprayed lawns, gardens, golf courses, and roadsides; and living near sprayed croplands, rangelands, and forests.

by-product of a fungus that attacks peanuts and corn, causes liver cancer. For this reason, some authorities advise against eating "raw" peanut butter. Commercial peanut butters are safe because processing kills the aflatoxin fungus.

Some chemicals may be "precarcinogens" that cause gene changes only *after* they have been altered by metabolic activity in the cell. Others are cancer "promotors" —they are not carcinogens per se but make carcinogens more potent if both the carcinogen and the promoter act on a cell at the same time.

Radiation   Radiation can damage DNA, causing mutations that lead to cancer. Sources include ultraviolet radiation from sunlight and sunlamps, medical and dental X rays and some radioactive materials used to diagnose diseases, "background" radiation from cosmic rays and radon gas in soil and water, and gamma rays emitted from nuclear reactors and radioactive wastes.

Sun exposure is probably the greatest radiation risk factor for most people. (Skin cancer, described later, is the most frequently diagnosed cancer in the Northern Hemi-

Table 20.2 Selected Industrial Chemicals Linked to Cancer	
Chemical/Substance	**Type of Cancer**
Benzene	Leukemias
Vinyl chloride	Liver, various connective tissues
Various solvents	Bladder, nasal epithelium
Ether	Lung
Asbestos	Lung, epithelial linings of body cavities
Arsenic	Lung, skin
Radioisotopes	Leukemias
Nickel	Lung, nasal epithelium
Chromium	Lung
Hydrocarbons in soot, tar smoke	Skin, lung

sphere.) It is also wise to avoid unnecessary X rays. In addition, there is concern about the accumulation of radioactive radon gas in homes, schools, and office buildings. Buildings and well water can be tested for radon, and measures (sometimes quite costly) can be taken to reduce radon levels.

Breakdowns in Immunity

Typically, when a normal cell becomes cancerous, certain proteins at the cell surface become altered and function like foreign antigens—the "nonself" tags that mark a cell for destruction by cytotoxic T cells and natural killer cells (see Figure 7.7 on page 177). There is evidence that a healthy immune system regularly detects and destroys some types of cancer cells. This protective function can break down if the immune system becomes suppressed by any of a range of factors, including drugs (such as those used to prevent rejection of transplanted organs) and mental states such as anxiety and severe depression.

In some cases, the presence of a growing cancer may also suppress the immune system. In addition, cancerous cells that bear tumor antigens may lack other "costimulator" molecules that are required to trigger an immune response. In some cases tumor antigens may be chemically disguised or masked. For whatever reason, the transformed cells are free to divide uncontrolled. As we see in the next section, efforts are underway to mobilize immune system weapons, especially monoclonal antibodies, as powerful cancer treatments.

DIAGNOSIS AND TREATMENT

Detection: Heeding the Body's Warnings

Table 20.3 lists seven common cancer warning signs. These can help people spot cancer in its early stages, when treatment is most effective. Routine *cancer screening* becomes important as a person ages; some recommended cancer screening tests are listed in Table 20.4.

To confirm (or help rule out) cancer, various types of tests can refine the diagnosis. Blood tests can detect **tumor markers**, substances produced by specific types of cancer cells or by normal cells in response to the cancer. For example, the hormone HCG (human chorionic gonadotropin) is a highly specific marker for certain cancers of reproductive organs in both men and women. (In males, cancerous germ cells in the testicles produce HCG.) Radioactively labeled monoclonal antibodies, which home in on tumor antigens, have become extremely useful for pinpointing the location and sizes of tumors of the colon, brain, bone, and some other tissues. *Medical imag-*

ing of tumors includes methods such as magnetic resonance imaging (MRI; shown in Figure 20.7), X rays, ultrasound, and computerized tomography (CT). The definitive cancer detection tool is **biopsy**. A small piece of suspect tissue is removed from the body through a hollow needle or exploratory surgery. A pathologist then microscopically examines cells of the tissue sample to look for the characteristic features of cancer cells.

Cancer Treatments

Surgery, drug treatment, and irradiation of tumors have long been the major weapons against cancer. Surgery works when a tumor is fully accessible and has not spread.

Chemotherapy is the use of drugs to kill cancer cells. Most anticancer drugs are designed to kill dividing cells. They disrupt DNA replication during the S phase of the cell cycle or disrupt mitosis by interfering with formation of the mitotic spindle. The drugs fluorouracil and vincristine (derived from the rosy periwinkle) are among the compounds that work this way. Unfortunately, such drugs are also toxic to rapidly dividing healthy cells such as hair cells, stem cells in bone marrow, lymphocytes of the immune system, and epithelial cells of the gut lining. Hence chemotherapy patients can suffer side effects such as nausea and vomiting, hair loss, anemia, and reduced immune responses. Radiation likewise kills cancer cells *and* healthy cells in the irradiated area. *Adjuvant therapy* (*adjuvant* means help) combines surgery and a less toxic dose of chemotherapy. For example, a patient might receive enough chemotherapy to shrink a tumor, then have surgery to remove what remains.

Researchers are also developing monoclonal antibodies as "magic bullets" to deliver lethal doses of radiation or anticancer drugs to tumor cells while sparing healthy cells. Experimental trials with radioactive monoclonal antibodies have resulted in partial or complete remission

Table 20.3   The Seven Warning Signs of Cancer*
Change in bowel or bladder habits and function
A sore that does not heal
Unusual bleeding or bloody discharge
Thickening or lump
Indigestion or difficulty swallowing
Obvious change in a wart or mole
Nagging cough or hoarseness

* Notice that the first letters of the signs spell the advice CAUTION.
Source: American Cancer Society

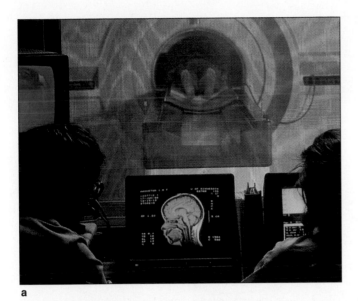

a

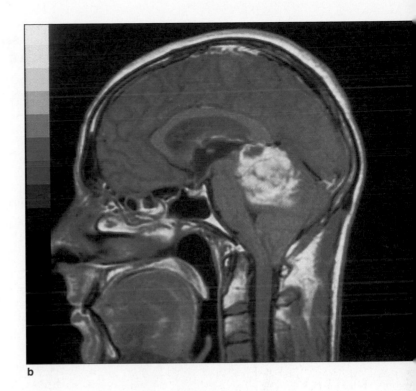

b

Figure 20.7 (**a**) MRI scanning—just one way to obtain images of organs and tissues that are not readily accessible by other means. MRI does not use X rays. Instead, the patient is placed inside a chamber that is surrounded by a magnet. When the machine is turned on, it generates a magnetic field. As a result, nuclei of hydrogen and certain other atoms in the body line up and absorb energy. A computer analyzes information about the responses of the nuclei and uses it to generate an image of soft tissues. (**b**) MRI image of a cancerous tumor, here in the brain.

Table 20.4 Recommended Cancer Screening Tests

Test or Procedure	Cancer	Sex	Age	Frequency
Breast self-examination	Breast	Female	20+	Monthly
Mammogram	Breast	Female	40–49 50+	Every 1–2 years Yearly
Testicle self-examination	Testicle	Male	18+	Monthly
Sigmoidoscopy	Colon	Male, Female	50+	Every 3–5 years
Fecal occult blood test	Colon	Male, Female	50+	Yearly
Digital rectal examination	Prostate, colorectal	Male, Female	40+	Yearly
Pap test	Uterus, cervix	Female	18+ and all sexually active women	Every other year until age 35; yearly thereafter
Pelvic examination	Uterus, ovaries, cervix	Female	18–39 40+	Every 1–3 years w/Pap Yearly
Endometrial tissue sample	Endometrial	Female	At menopause for women at high risk	Once
General checkup		Male, Female	20 39 40+	Every 3 years Yearly

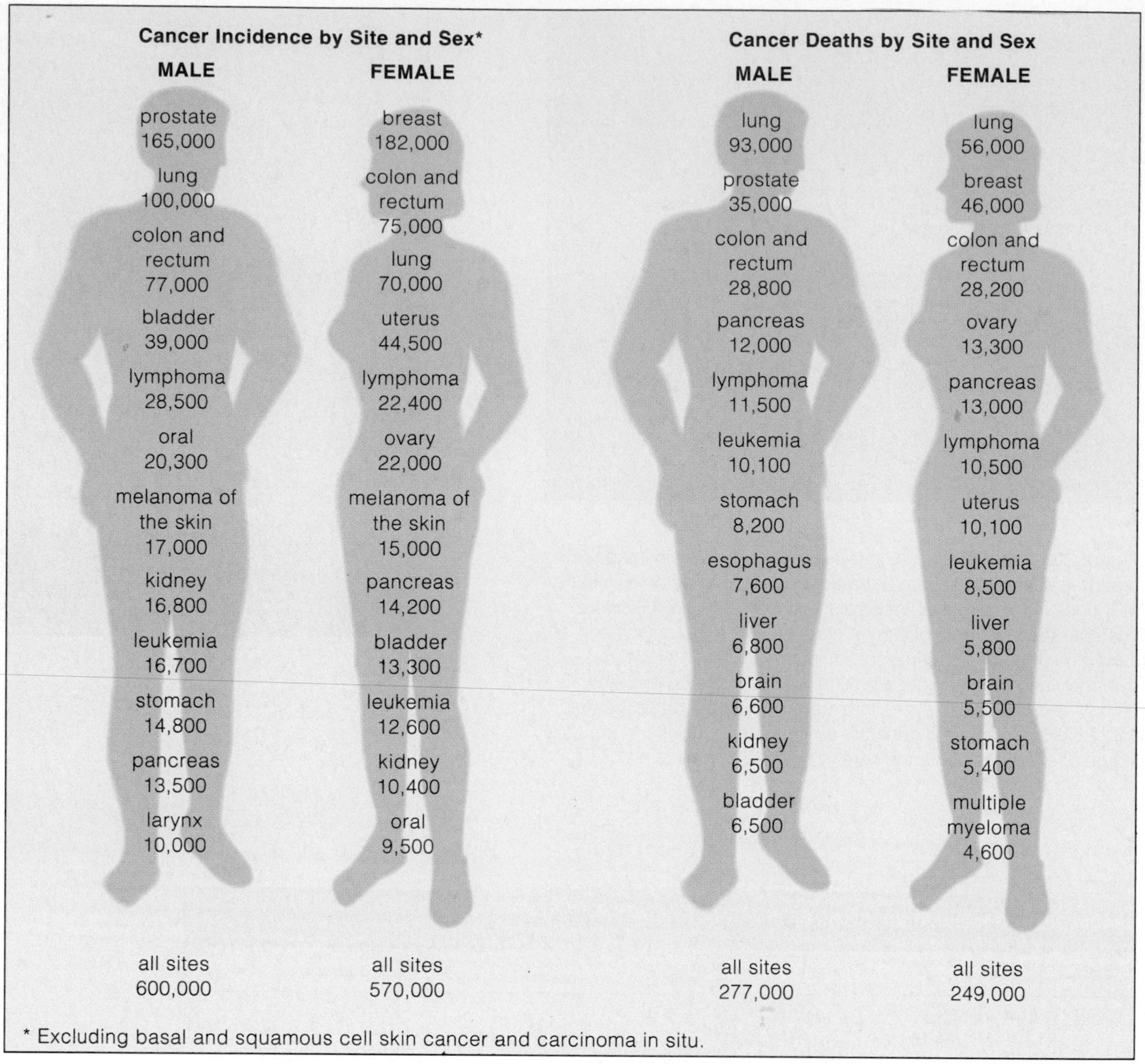

Cancer Incidence by Site and Sex*

MALE	FEMALE
prostate 165,000	breast 182,000
lung 100,000	colon and rectum 75,000
colon and rectum 77,000	lung 70,000
bladder 39,000	uterus 44,500
lymphoma 28,500	lymphoma 22,400
oral 20,300	ovary 22,000
melanoma of the skin 17,000	melanoma of the skin 15,000
kidney 16,800	pancreas 14,200
leukemia 16,700	bladder 13,300
stomach 14,800	leukemia 12,600
pancreas 13,500	kidney 10,400
larynx 10,000	oral 9,500
all sites 600,000	all sites 570,000

Cancer Deaths by Site and Sex

MALE	FEMALE
lung 93,000	lung 56,000
prostate 35,000	breast 46,000
colon and rectum 28,800	colon and rectum 28,200
pancreas 12,000	ovary 13,300
lymphoma 11,500	pancreas 13,000
leukemia 10,100	lymphoma 10,500
stomach 8,200	uterus 10,100
esophagus 7,600	leukemia 8,500
liver 6,800	liver 5,800
brain 6,600	brain 5,500
kidney 6,500	stomach 5,400
bladder 6,500	multiple myeloma 4,600
all sites 277,000	all sites 249,000

* Excluding basal and squamous cell skin cancer and carcinoma in situ.

Figure 20.8 Summary of incidence of and deaths from common cancers, by site and sex. Data are estimates for the United States, 1993. Courtesy American Cancer Society.

in a few patients with a type of adult leukemia. Another goal is to link tumor-specific monoclonal antibodies with lethal doses of cytotoxic drugs. So far, results have been mixed, but this is a very active area of cancer research.

Immune Therapy Another promising prospect for cancer treatment is **immune therapy**. The idea is to give patients substances that will trigger a strong immune response against cancer cells. Recall from Chapter 7 that many cells produce and release interferon following a viral attack. Interferon can activate cytotoxic T cells and natural killer cells, both of which can recognize and kill various types of cancer cells. So far, however, giving large doses of interferon has been useful only against some rare forms of cancer. Another approach is to develop cancer "vaccines" that stimulate cytotoxic T cells to recognize and destroy body cells bearing an abnormal surface protein. In some patients with advanced malignant melanoma (page 441), doses of an experimental vaccine have caused tumors to regress.

Interleukins—signaling molecules (lymphokines) produced by the immune system's lymphocytes—are also potential anticancer weapons. Cytotoxic T cells grown in culture with IL-2 become "lymphokine-activated killer cells" or LAKs. In clinical trials, several hundred patients with advanced kidney cancer have received LAKs com-

Life-style Choices and Cancer Prevention

None of us can control factors in our heredity or biology that might lead one day to cancer, but each of us *can* make life-style decisions that promote health. The American Cancer Society recommends the following strategies for limiting your cancer risk:

1. Avoid tobacco in any form, including "secondary smoke" from others.

2. Maintain a desirable weight. Being more than 40 percent overweight increases the risk of cancers of the colon, breast, prostate, gallbladder, ovary, and uterus.

3. Eat a low-fat diet that includes plenty of vegetables, fruits, and fiber (Figure *a*). Cut down on total fat intake. A low-fiber, high-fat diet is associated with cancer of the colon, prostate, and possibly other tissues. Cruciferous vegetables (members of the mustard family) such as cabbage, kale, broccoli, and cauliflower appear to contain anticancer compounds. As noted in Chapter 5 (page 131), antioxidants such as beta carotene and certain vitamins may have roles in cancer prevention.

4. Drink alcohol moderately, if at all. Heavy alcohol use, especially in combination with smoking, increases risk for cancers of the mouth, larynx, throat, esophagus, and liver.

5. If you are a woman entering menopause, consult carefully with your doctor about using estrogen, which can increase the risk of endometrial cancer.

6. Consider carefully whether your job or residence exposes you to such industrial agents as nickel, chromate, vinyl chloride, benzene, asbestos, and agricultural pesti-

Figure a A high-fiber, low-fat diet that includes cruciferous vegetables may be just what the doctor ordered to help limit personal cancer risk.

cides, which are associated with various cancers. If you are not sure of your exposure at work, ask. Federal regulations give you the legal right to know.

7. Avoid unnecessary X rays and protect your skin from excessive sunlight.

---

bined with large amounts of more IL-2. Although the treatment can have serious side effects, enough patients have responded favorably that regulators have approved it for limited use.

SOME MAJOR TYPES OF CANCER

In general, a cancer is named according to the type of tissue in which it first forms. For instance, cancers of connective tissues such as muscle and bone are *sarcomas*.

Various types of *carcinomas* arise from epithelium, including cells of the skin and epithelial linings of internal organs. When a cancer begins in the body or ducts of a gland, it is called an *adenocarcinoma*. *Lymphomas* are cancers of lymphoid tissues in organs such as lymph nodes, and cancers arising in blood-forming regions—mainly stem cells in bone marrow—are *leukemias*.

This section surveys some common cancers in the United States. Figure 20.8 summarizes the most recent data for males and females. *Focus on Wellness* lists some strategies for reducing your cancer risk through intelligent life-style choices.

Female Breast and Reproductive System

Breast Cancer About one woman in nine develops breast cancer—more than 180,000 new cases each year. Of all cancers in women, breast cancer currently ranks second only to lung cancer as a cause of death. Obesity, late childbearing, early puberty, late menopause, and excessive levels of estrogen (and perhaps other hormones) seem to play roles in the development of breast cancer. Some researchers have proposed that environmental toxins concentrated in fatty breast tissue have a role. A family history of breast cancer increases the risk. Although 80 percent of breast lumps are *not* cancer, a woman should seek medical advice about any breast lump, thickening, dimpling, breast pain, or discharge.

Chances for cure are excellent if breast cancer is detected early and treated promptly. That is why a woman should examine her breasts every month, about a week after her menstrual period. Figure 20.9 shows the steps of a self-exam, as recommended by the American Cancer Society. Low-dose mammography (breast X ray) is the most effective method for detecting small cancers in the breast; it is 80 percent reliable. The American Cancer Society recommends an annual mammogram for women over 50 and for younger women at high risk.

Treatment depends mainly on the extent of the disease. In *modified radical mastectomy,* the affected breast tissue, overlying skin, and lymph nodes in adjacent tissues are removed, but muscles of the chest wall are left intact. If a breast tumor is small, the preferred treatment may be *lumpectomy.* This procedure is less disfiguring than mastectomy because it leaves some breast tissue in place. In both cases, removed lymph nodes are examined to determine the need for further treatment and to predict the prospects of a cure.

Cancers of the uterus most often affect the endometrium (uterine lining) and the cervix. Various types are treated by surgery, radiation, or both. The incidence of uterine cancers is falling, in part because precancerous phases of cervical cancer can be easily detected by the *Pap smear* that is part of a routine gynecological examination. Risk factors for cervical cancer include having multiple sex partners, early age of first intercourse, cigarette smoking, and genital warts (human papillomavirus, page 351). Endometrial cancer is more common during and after menopause; women who take supplemental estrogen during menopause are at higher risk.

Ovarian cancer is often lethal because symptoms, mainly an enlarged abdomen, do not appear until the cancer is advanced and has already metastasized. In a few women the first sign is abnormal vaginal bleeding or vague abdominal discomfort. Risk factors include family history of the disease, childlessness, and a history of breast cancer. New hope for victims of ovarian (and breast) cancers has come from the recent discovery of taxol, a compound derived from the Pacific yew tree. In 20–30 percent of advanced cases, taxol shrinks tumors significantly (by killing dividing tumor cells). To meet the high demand for taxol while minimizing the need to cut down large numbers of Pacific yews, researchers are using biotechnology to engineer yeasts (a type of fungus) that can produce the compound in quantity.

Male Reproductive System

About 5,000 cases of cancer of the testis are diagnosed annually in the United States. In its early stages, testicular cancer is painless. However, it can spread to lymph nodes in the abdomen, chest, neck, and, eventually, the lungs. Once a month from high school onward, men should examine each testis separately after a warm bath or shower (when the scrotum is relaxed). The testis should be rolled gently between the thumb and forefinger to check for any type of lump, enlargement, or hardening. Such changes may or may not cause discomfort, but only a physician can rule out the possibility of disease. Surgery is the usual treatment, and the success rate is high when the cancer is caught before it can spread.

Prostate cancer is the second leading cause of cancer deaths in men (after lung cancer). There are no confirmed risk factors. Symptoms include various kinds of urinary problems, although these can also signal nothing more than a noncancerous enlarged prostate. The American Cancer Society recommends that men over 40 have an annual digital rectal examination, which enables a physician to feel the prostate and detect any unusual lumps. A blood test called PSA can also screen for larger-than-normal amounts of a substance called prostate-specific antigen. If a physician still suspects cancer after these two tests have been performed, the next step is a tissue biopsy. The cure rate for prostate tumors detected early is over 90 percent.

Oral and Lung Cancers

Cancers of the mouth, tongue, salivary glands, and throat are most common among smokers and people who use smokeless tobacco (snuff), especially if they are also heavy alcohol drinkers. In 1993, the American Cancer Society reported over 29,000 new cases of oral cancer and nearly 8,000 deaths, mostly among men. Of people who are diagnosed with oral cancer, only about half survive for five years, even with treatment.

Lung cancer kills more people than any other cancer. Tobacco smoking is the overwhelming risk factor, followed by exposure to asbestos, industrial chemicals such as arsenic, and radiation (including radon contamination in homes). For a smoker, a combination of these factors

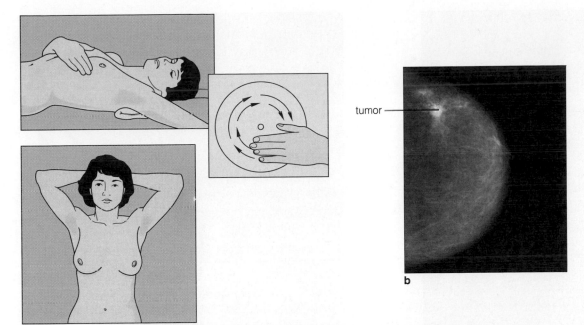
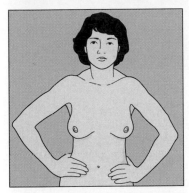
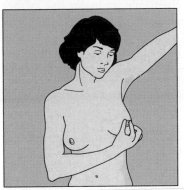

tumor

1. Lie down and put a folded towel or pillow under your left shoulder, then put your left hand behind your head. With the right hand (fingers flat), begin the examination of your left breast by following the outer circle of arrows shown. Gently press the fingers in small, circular motions to check for any lump, hard knot, or thickening. Next, follow the inner circle of arrows. Continue doing this for at least three more circles, one of which should include the nipple. Then repeat the procedure for the right breast.

2. For a complete examination, repeat the procedure of step 1 while standing in a shower or tub (hands glide more easily over wet skin).

3. Stand before a mirror, lift your arms over your head, and look for any unusual changes in the contour of your breasts, such as a swelling, dimpling, or retraction (inward sinking) of the nipple. Also check for any unusual discharge from the nipple.

If you discover a lump or any other change during a breast self-examination, it's important to see a physician at once. Most changes are not cancerous, but let the doctor make the diagnosis.

Figure 20.9 (a) How to perform a breast self-examination. (**b**) A mammogram showing a tumor, which a biopsy indicated to be cancer. (The white patches at the front of the breast are milk ducts and fibrous tissue.)

a

b

significantly increases the odds of developing cancer. For nonsmokers, especially children and spouses of smokers, inhaling "secondhand" tobacco smoke also poses a significant cancer risk. The Environmental Protection Agency estimates that lung cancer resulting from long-term breathing of secondhand smoke kills 3,000 people in the United States each year.

On average, a lung cancer victim smokes for 20 years before serious symptoms occur. The incidence has decreased in men, but a rise in the relative number of female smokers in the last several decades is now reflected in the fact that lung cancer has surpassed breast cancer as the leading cancer killer of women. Warning signals include a nagging cough, shortness of breath,

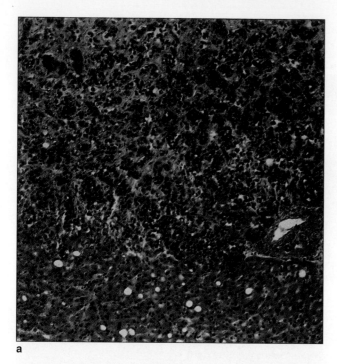

a

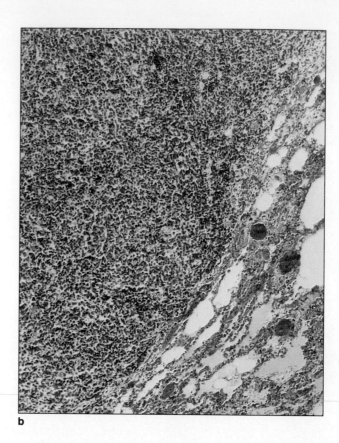

b

Figure 20.10 Photomicrographs of (**a**) adenocarcinoma and (**b**) large cell carcinoma of the lung.

chest pain, bloody sputum (coughed-up phlegm), unexplained weight loss, and frequent respiratory infections or pneumonia.

Four types of lung cancer account for 90 percent of cases. About one-third of lung cancers are *squamous cell carcinomas,* affecting squamous epithelium in the bronchi. Another 48 percent are either *adenocarcinomas* or *large-cell carcinomas* (Figure 20.10). A fourth type, *small-cell carcinoma,* spreads rapidly and kills 99 percent of its victims within five years of diagnosis.

Cancers of the Digestive System and Related Organs

Cancers of the stomach and pancreas are usually adenocarcinomas of duct cells, and often they are not detected until they have spread to other organs. Cigarette smoking is a risk factor for pancreatic cancer, while stomach cancer may be associated with heavy alcohol consumption and a diet rich in smoked, pickled, and salted foods. Liver cancer is uncommon in the United States, but growing evidence suggests that hepatitis B infection can trigger it.

Most cases of colon cancer are adenocarcinomas. Warning signs include a change in bowel habits, rectal bleeding, and blood in the feces. Family history of colon cancer or inflammatory bowel disease is a major risk factor, as is a lack of fiber in the diet.

Urinary System Cancers

Carcinomas of the bladder and kidney account for about 70,000 new cancer cases each year. The incidence of both types of cancer is higher in males, and smoking and exposure to certain industrial chemicals are major risk factors for both. Evidence suggests that consuming chlorinated water also increases risk. Kidney cancer easily metastasizes via the bloodstream to the lungs, bone, and liver. An inherited type, *Wilm's tumor,* is one of the most common of all childhood cancers.

Cancers of the Blood and Lymphatic System

Lymphomas develop in lymphoid tissues in organs such as the lymph nodes, spleen, and thymus. They include the diseases known as *Hodgkin's disease, non-Hodgkin's lymphoma,* and *Burkitt's lymphoma.* Risk seems to increase along with infections—such as HIV—that impair immune system functioning. Burkitt's lymphoma is most common in parts of Africa, where it seems to develop especially in children who become infected with both the Epstein-Barr virus (which also causes mononucleosis) and with malaria. Veterans of the Vietnam conflict who were exposed to the defoliant "Agent Orange" also have higher

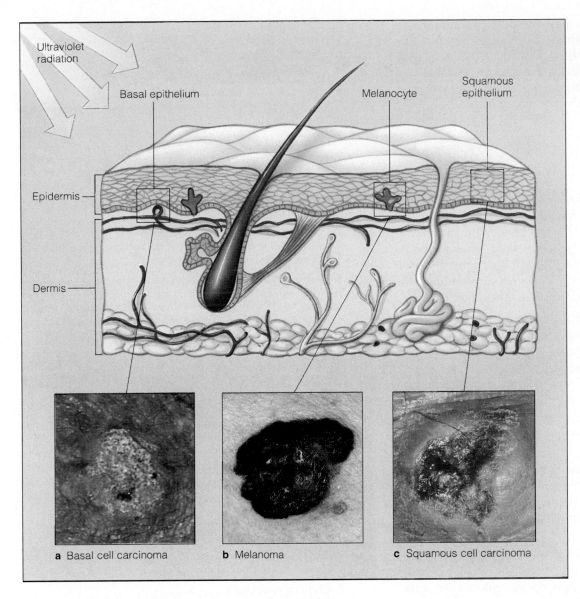

Figure 20.11 (a) Basal cell carcinoma, (b) malignant melanoma, and (c) squamous cell carcinoma. Each type of skin cancer has a distinctive appearance. Malignant melanoma can be deadly because it metastasizes aggressively.

Ultraviolet radiation

Basal epithelium

Melanocyte

Squamous epithelium

Epidermis

Dermis

a Basal cell carcinoma

b Melanoma

c Squamous cell carcinoma

rates of lymphoma. Symptoms include enlarged lymph nodes, rashes, weight loss, and fever. Intense itching and night sweats also are typical of Hodgkin's disease. Chemotherapy and radiation are standard treatments, and new treatments using targeted monoclonal antibodies are being developed against lymphoma.

Leukemias are cancers in which stem cells in bone marrow overproduce white blood cells. Some types are the most common childhood cancers, but other types are diagnosed in more than 70,000 adults each year. Risk factors include Down syndrome, exposure to chemicals such as benzene, and radiation exposure. Many leukemias can be effectively treated with chemotherapy using two compounds—vincristine and vinblastine—derived from species of periwinkle plants that grow only on the African island of Madagascar. Environmentalists often cite these

drugs and other plant-derived compounds such as taxol as reasons why humans should attempt to preserve natural areas where as-yet-undiscovered medicinal plants may be present.

Skin Cancer

Skin cancers collectively are the most common of all cancers. For all types, fair skin and exposure to ultraviolet radiation in sunlight (or tanning salons) are risk factors. The most dangerous kind, malignant melanoma, is actually a cancer of melanin-producing cells called melanocytes. It is highly dangerous because in its later stages it metastasizes aggressively. Squamous cell (epidermal) carcinomas start out as scaly, reddened bumps (Figure 20.11). They grow rapidly and can spread to adjacent lymph nodes unless they are surgically removed. Basal cell carcinomas begin as small, shiny bumps and slowly grow into ulcers with beaded margins. Basal cell carcinoma and squamous cell carcinoma are much more common than malignant melanoma and usually are easily treated by minor surgery in a doctor's office.

SUMMARY

1. Cancer arises when genetic controls over cell division are lost. Cancer cells have specific characteristics that set them apart from normal cells. These include structural abnormalities, including (typically) poor differentiation and altered surface proteins; uncontrolled growth with absence of contact inhibition; invasion of surrounding tissues; and the ability to metastasize to other body regions.

2. Cancer develops through a multistep process involving several genetic changes. Initially, mutation or some other event may alter a proto-oncogene into a cancer-causing oncogene. Infection by a virus can also insert an oncogene into a cell's DNA or disrupt normal controls over a proto-oncogene. In addition, one or more tumor suppressor genes probably must be missing or become mutated before a normal cell can be transformed into a cancerous one.

3. A predisposition to some cancers is inherited. Other causes of carcinogenesis are viral infection, chemical carcinogens, radiation, faulty immune system functioning, and possibly breakdown in normal DNA repair mechanisms.

4. Standard cancer treatments include surgery, chemotherapy, and tumor irradiation. Targeted monoclonal antibodies and immune therapy using interferons and interleukins are other treatment options currently under development.

5. Life-style choices such as the decision not to use tobacco, to maintain a low-fat, high-fiber diet, and to avoid unnecessary exposure to X rays and chemical carcinogens can help limit personal cancer risk.

1. How are cancer cells structurally different from normal cells of the same tissue? What is the relevance of altered surface proteins to uncontrolled growth? *428–429*

2. Write a short paragraph that summarizes the roles of proto-oncogenes, oncogenes, and tumor suppressor genes in carcinogenesis. *430*

3. List the four main categories of cancer tumors. *437*

Critical Thinking: You Decide *(Key in Appendix IV)*

1. Some people are concerned that exposure to electromagnetic fields (EMFs) emanating from power lines and small appliances may trigger cancerous changes in cells. What are some specific kinds of biological events you would expect researchers to probe in order to establish an EMF-cancer link?

2. A textbook on cancer contains the following statement: "Fundamentally, cancer is a failure of the immune system." Why do you think the author wrote this?

3. Ultimately, cancer kills because it spreads and disturbs homeostasis. Consider, for example, a kidney cancer that metastasizes to the lungs and liver. What are some specific homeostatic mechanisms that the spreading disease could disrupt?

Self-Quiz *(Answers in Appendix III)*

1. A tumor is _____.
   a. malignant by definition
   b. always enclosed by connective tissue
   c. a mass of tissue that may be benign *or* malignant
   d. usually slow-growing

2. Cancer cells _____.
   a. lack normal controls over cell division
   b. secrete the growth factor angiogenin
   c. display altered surface proteins
   d. do not respond to contact inhibition
   e. all of the above

3. The onset of cancer seems to require the activity of an oncogene plus the absence or mutation of at least one _____.

4. Chemical carcinogens _____.
   a. include viral oncogenes
   b. can bind DNA and cause a mutation
   c. must be ingested in food
   d. include some rare vitamins

5. So far as we know, carcinogenesis is *not* triggered by _____.
   a. breakdowns in DNA repair
   b. a breakdown in immunity
   c. radiation
   d. hormone deficiency
   e. inherited gene defects

6. Tumor suppressor genes _____.
   a. occur normally in cells
   b. promote metastasis
   c. enter cells by way of viral infection
   d. probably affect the development of cancer only in rare cases

7. _____ is the definitive method for detecting cancer.

8. The goal of immune therapy is to _____.
   a. cause defective T cells in the thymus to disintegrate
   b. activate cytotoxic T cells
   c. dramatically increase the numbers of circulating macrophages
   d. promote the secretion of monoconal antibodies

Key Terms

biopsy *434*
carcinogen *432*
carcinogenesis *431*
chemotherapy *434*
dysplasia *428*
immune therapy *436*

metastasis *429*
oncogene *430*
proto-oncogene *430*
tumor *428*
tumor marker *434*
tumor suppressor gene *430*

Readings

Boon, T. March 1993. "Teaching the Immune System to Fight Cancer." *Scientific American*.

Cancer Facts and Figures. American Cancer Society. Free pamphlet published annually; an excellent summary of the latest information on cancer, risk factors, and treatments.

Kupchella, C. 1987. *Dimensions of Cancer*. Belmont, Calif.: Wadsworth. Clear, accurate discussion of cancer causes, types, treatments, and prevention.

Radetsky, P. March 1993. "Magic Missiles." *Discover*. Fascinating nontechnical description of current research on monoclonal antibodies as anticancer weapons.

21 BIOTECHNOLOGY

Ingenious Genes

In the 1960s a world-shaking revolution began quietly in test tubes and laboratory culture dishes. The revolutionaries were scientists who discovered how enzymes can snip DNA molecules into specific pieces and then rejoin the pieces in new arrangements. They learned, too, how to use bacteria and viruses as messengers that could cross the boundaries between cells of one species and cells of another, carrying a cargo of genetic information.

As you read this, investigators are using such methods to insert human genes into mouse embryos as part of the search for a treatment of Alzheimer's disease. Others are attempting to alter genes in the AIDS virus (HIV), aiming to develop effective vaccines. "Bioengineered" sheep, pigs, and cows already produce milk that contains human proteins, such as hemoglobin and blood clotting factors. Most likely, your corner drug-

a

b

Figure 21.1 (**a**) Diabetic injecting insulin from genetically engineered bacteria. (**b**) Commercial production of interferon.

store carries bioengineered products, including the hormone insulin required by people with type I diabetes—insulin produced by vats of genetically altered bacteria (Figure 21.1).

The revolution of biotechnology has staggering potential for medicine, agriculture, and industry. In 1993 nearly 200 applications for patents on bioengineered animals were pending government approval, and no doubt more have been submitted since then. For example, you may soon find yourself munching french fries from potatoes that absorb less oil during frying because they are unusually high in starch, a quality conferred by starch-coding genes from the intestinal bacterium *E. coli*. Other plant biotechnology projects aim to develop crops that are resistant to drought or insects or contain higher levels of essential amino acids.

With so many actual and potential applications to human health and well-being, we are well advised to understand the basics of biotechnology. In this chapter we will also consider some of the ecological, social, and ethical questions such methods raise.

KEY CONCEPTS

1. Genetic experiments have been occurring in nature for billions of years as a result of gene mutations, crossing over and recombination, and other genetic events. Humans are now engineering genetic changes by way of recombinant DNA technology.

2. Recombinant DNA technology involves three basic activities. First, DNA molecules are cut into fragments. Second, the fragments are inserted into carriers, or cloning vectors. Third, the fragments are screened to determine which ones contain genes of interest. Those that do are copied. The genes and, in some cases, their protein products are produced in large amounts that can be used for research and practical applications.

3. Like all new technologies, biotechnology raises social, legal, ecological, and ethical questions regarding its benefits and risks.

RECOMBINANT DNA TECHNOLOGY

For more than 3 billion years, nature has been conducting genetic experiments through events such as mutation and chromosomal crossing over. Genetic messages have changed countless times; this is the source of life's diversity. For thousands of years humans have been changing the genetic character of species by artificially selecting individuals for controlled matings. The results have included beef cattle, the nutritious grain we call corn, seedless watermelons, the tangelo (tangerine x grapefruit), and the mule (donkey x horse).

Today we also analyze and even engineer genetic changes through **recombinant DNA technology**. DNA from one source is cut and then spliced together, or *recombined*, with DNA from another source, using enzymes as "scissors and glue." The recombinant DNA can then be reinserted into one of the source organisms or transplanted into a different one. Commonly, recombinant DNA is inserted into bacteria or other types of cells that rapidly replicate their DNA and divide. The cells replicate the foreign DNA right along with their own. In short order a population of bacteria can produce useful quantities of recombinant DNA molecules. If the genes are transcribed and translated, the engineered bacteria (or cells of some other recombinant organism) can produce proteins that function normally. Many engineered genes are already altering specific traits of organisms.

Recombinant DNA technology grew out of experiments with bacteria. All bacteria have a single nucleoid, a circular DNA molecule. Many types also have **plasmids**—small, circular molecules of "extra" DNA with only a few genes (Figure 21.2). They can be copied independently of the nucleoid's replication. Portions of a plasmid can even be copied in the laboratory. Plasmids are commonly used as **cloning vectors**, elements that ferry foreign DNA into a cell. Other common vectors are bacteria-infecting viruses called *bacteriophages*. (Often, such viruses are referred to simply as "phages.")

Sometimes one bacterium may transfer plasmid genes to another bacterium, and occasionally the transferred plasmid becomes integrated into the recipient's chromosome. The result is a recombinant DNA molecule. In nature, bacteria use enzymes called *restriction enzymes* to cut apart foreign DNA that has been injected into a bacterial cell, often by bacteriophages. As we see next, researchers use these enzymes to accomplish the DNA cutting and pasting of recombinant DNA technology.

Figure 21.2 A ruptured bacterial cell (*Escherichia coli*). Notice the larger bacterial chromosome and the smaller plasmids (blue arrows).

In recombinant DNA technology, DNA from different sources is spliced together, then reinserted into a source organism or transplanted into a different one.

If engineered genes become integrated into a host cell's DNA, they can be more easily studied. They can also be expressed and give rise to functional proteins.

Producing Restriction Fragments

Bacteria are equipped with many different types of restriction enzymes, called *restriction endonucleases*. Each type makes a cut only wherever a specific and very short nucleotide sequence occurs in DNA. Cuts at two such sequences in the same molecule produce a fragment. In many cases, each cut is staggered, so the fragment is single-stranded at each end:

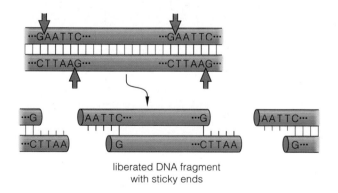

liberated DNA fragment
with sticky ends

The short, single-stranded ends of a DNA fragment are described as "sticky ends." This is because they can base-

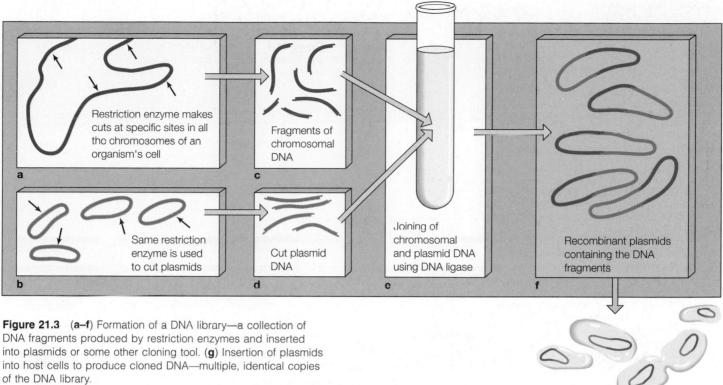

Restriction enzyme makes cuts at specific sites in all the chromosomes of an organism's cell

a

Fragments of chromosomal DNA

c

Same restriction enzyme is used to cut plasmids

b

Cut plasmid DNA

d

Joining of chromosomal and plasmid DNA using DNA ligase

e

Recombinant plasmids containing the DNA fragments

f

Plasmids inserted into host cells for amplification

g

Figure 21.3 (**a–f**) Formation of a DNA library—a collection of DNA fragments produced by restriction enzymes and inserted into plasmids or some other cloning tool. (**g**) Insertion of plasmids into host cells to produce cloned DNA—multiple, identical copies of the DNA library.

pair with any other DNA molecule cut by the same restriction enzyme.

Knowing the sites where different restriction enzymes will make cuts, a researcher can select those enzymes that will cut out a piece of DNA containing a particular gene. The pieces are called **restriction fragments**. When such fragments and cut plasmids are mixed together, they join up (base-pair) at the cut sites. Another enzyme, **DNA ligase**, seals the base pairings. A collection of DNA fragments produced by restriction enzymes and incorporated into plasmids make up a **DNA library** (Figure 21.3). Because this library consists of individual molecules, physically it is almost vanishingly small. To obtain adequate quantities for research or medical uses, the DNA library must be *amplified*—that is, copied over and over—into useful amounts.

Plasmids (or bacteriophages) carrying foreign DNA can readily enter bacteria, yeasts, or other cells that reproduce rapidly. A growing population of such cells can quickly amplify recombinant DNA they contain. For example, a laboratory culture that begins with just a single bacterium can contain more than a million identical bacteria in less than 10 hours. Repeated DNA replications and cell divisions of the host cells yield **cloned DNA**—multiple, identical copies of DNA fragments, inserted into plasmids.

In cases in which the DNA sequence to be cloned is very long (as is the case with most human genes), a

researcher may create a YAC, or "yeast artificial chromosome." YACs consist of several different yeast genes combined with other DNA of a researcher's choosing. For various reasons, it is easier to make very large recombinant molecules if YACs are used. YACs are cloned inside yeast cells.

Recombination is made possible by restriction enzymes that can make cuts in DNA molecules. Restriction enzymes make it possible to recombine DNA segments from different sources.

PCR Amplification of Specific Fragments

The **polymerase chain reaction**, or PCR, is a common method of amplifying specific fragments of DNA. For instance, researchers may suspect that a specific region of DNA on a human chromosome contains a gene for a particular trait. They can use PCR to obtain copies of the DNA region for further study. First, they synthesize short nucleotide sequences that will base-pair with complementary sequences known to be located just before and just after the region. The short sequences are mixed with enzyme molecules, free nucleotides, and all the DNA from a human cell. Then the mixture is subjected to repeated, precise cycles of high and low temperatures.

region of interest (yellow)
in a DNA double helix

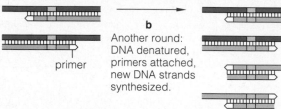

a
DNA denatured.
Primers (white)
attach to the now-
single DNA strands.
New DNA strand
synthesized behind
primers on each
template strand.

Figure 21.4 Polymerase chain reaction used in gene amplification, as described in the text.

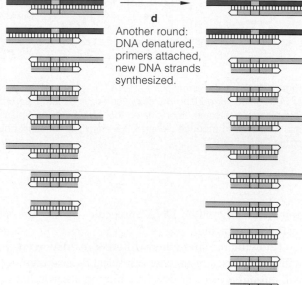

primer

b
Another round:
DNA denatured,
primers attached,
new DNA strands
synthesized.

c
Another round:
DNA denatured,
primers attached,
new DNA strands
synthesized.

d
Another round:
DNA denatured,
primers attached,
new DNA strands
synthesized.

e
Continued rounds of amplification quickly yield great numbers of identical fragments. Those fragments contain the DNA region of interest.

During the cycles, the two strands of all the DNA molecules unwind from each other. Then, as Figure 21.4 illustrates, the short sequences base-pair with any exposed complementary bases, making a short stretch of double-stranded DNA. An enzyme, **DNA polymerase**, uses the short double-stranded sites as "primers" to start replication. (A primer is necessary because the polymerase cannot initiate replication on a single-stranded nucleotide sequence.) DNA polymerase is derived from a bacterium that thrives in hot springs, even in hot-water heaters. It is not destroyed at the elevated temperatures required to unwind the two strands of a DNA fragment, and also at lower temperatures required for base-pairing.

The reactions quickly amplify a single DNA molecule to many billions of molecules. Thus PCR can amplify samples with little DNA, as might be found in a single hair left at the scene of a crime. It also can be used to amplify segments of fragmented DNA, such as that found in ancient human remains such as mummies and the remains of some fossilized organisms.

Locating Genetically Altered Host Cells

We study genes because we want to learn about or use their protein products. It takes living cells to transcribe and translate genes into proteins. When DNA is mixed with living cells such as bacteria, only some of them will successfully take up the DNA. Those that do are said to be *transformed*. To identify transformed colonies of cells, researchers use **DNA probes** (Figure 21.5)—short DNA sequences that usually are assembled from radioactively labeled nucleotides (page 410). The probe can base-pair

with part of the gene being studied. (Such base-pairing between nucleotide sequences from different sources is called *nucleic acid hybridization*.)

The first step is to select those cells that have taken up the recombinant plasmids. Most plasmids contain genes that make the bacteria resistant to antibiotics. The prospective host cells are placed on a culture medium that contains the antibiotic. The antibiotic prevents growth of all cells *except* the ones housing plasmids (because the plasmids carry the antibiotic-resistance genes).

The next step is to identify the particular bacterial cells having recombinant plasmids with the gene of interest.

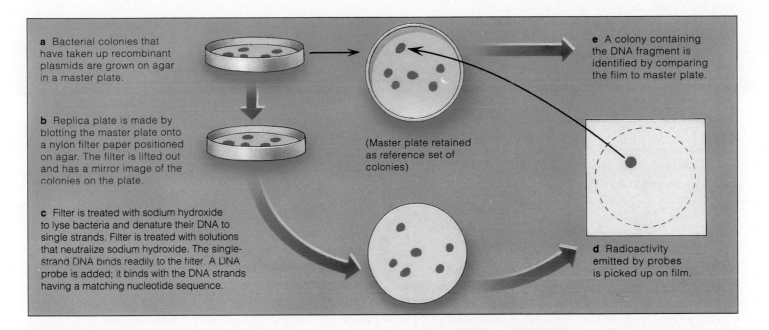

a Bacterial colonies that have taken up recombinant plasmids are grown on agar in a master plate.

b Replica plate is made by blotting the master plate onto a nylon filter paper positioned on agar. The filter is lifted out and has a mirror image of the colonies on the plate.

c Filter is treated with sodium hydroxide to lyse bacteria and denature their DNA to single strands. Filter is treated with solutions that neutralize sodium hydroxide. The single-strand DNA binds readily to the filter. A DNA probe is added; it binds with the DNA strands having a matching nucleotide sequence.

(Master plate retained as reference set of colonies)

e A colony containing the DNA fragment is identified by comparing the film to master plate.

d Radioactivity emitted by probes is picked up on film.

Figure 21.5 Use of a DNA probe to identify the colony of bacterial cells that have taken up genes and successfully integrated them into their own DNA.

As the bacteria divide, they form colonies. In *replica plating*, the colonies grow on agar (a solid gel) in a petri dish. When the agar is blotted against a nylon filter, some cells stick to it in locations that mirror the locations of the original colonies (Figure 21.5). Next, a series of solutions are used to rupture the cells, fix the released DNA onto the filter, and to make the double-stranded DNA molecules unwind. Then DNA probes are added. The probes hybridize only with gene regions having the proper base sequence. The probe-hybridized DNA emits radioactivity and allows a researcher to tag the colonies harboring the gene of interest.

Like molecular bloodhounds, DNA probes can help diagnose genetic or infectious diseases. For instance, a probe can be constructed so that it will base-pair with the nucleotide sequence of a specific disease gene—such as the gene for cystic fibrosis or Huntington's disease—or with a bit of viral DNA or RNA. Just such a method now enables laboratories to detect HIV infection much earlier than was once possible.

Expressing a Cloned Gene

Even if a host cell takes up a gene, it may not be able to translate it into protein. Human genes, for example, have noncoding regions (introns). Recall from Chapter 19 that transcripts of those genes cannot be translated until introns are snipped out and coding regions (exons) are spliced together into a mature mRNA transcript. Bacte-

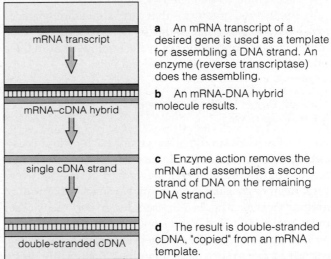

a An mRNA transcript of a desired gene is used as a template for assembling a DNA strand. An enzyme (reverse transcriptase) does the assembling.

b An mRNA-DNA hybrid molecule results.

c Enzyme action removes the mRNA and assembles a second strand of DNA on the remaining DNA strand.

d The result is double-stranded cDNA, "copied" from an mRNA template.

mRNA transcript

mRNA–cDNA hybrid

single cDNA strand

double-stranded cDNA

Figure 21.6 Making cDNA from an mRNA transcript.

rial enzymes don't recognize the splice signals, so bacterial host cells can't directly translate human DNA into functional proteins.

However, bacteria can translate **cDNA**, which has been "copied" from a mature mRNA transcript for the desired gene. By a process called **reverse transcription**, an mRNA transcript of a desired gene is used as a template for assembling a DNA strand. The result is a "hybrid" DNA/RNA molecule (Figure 21.6b). Enzymes remove the RNA from the double-stranded hybrid molecule, then assemble a complementary DNA strand on the strand remaining. This double-stranded cDNA, copied from mRNA, can now be inserted into a plasmid and amplified.

The enzyme *reverse transcriptase* accomplishes reverse transcription. Reverse transcriptase is obtained from retroviruses, which are RNA viruses that enter a host cell and use the enzyme to convert their viral RNA into double-stranded DNA. This DNA then becomes integrated into the host cell's DNA. As Chapter 15 noted, some important human pathogens are retroviruses (including the AIDS virus). Yet because their DNA products may naturally recombine with host DNA, modified versions of such particles are also being explored as vectors for gene therapy (page 457).

APPLICATIONS OF GENETIC ENGINEERING

Causing genetic change by deliberately manipulating DNA has come to be called *genetic engineering*. Genetically-engineered organisms that carry some foreign genes are said to be *transgenic*. In this section we consider some of the current applications of biotechnology that involve transgenic microbes, plants, and animals.

Uses of Genetically Engineered Microorganisms

For centuries humans have been using bacteria and yeasts as biochemical workhorses in the manufacture of cheeses, bread, yogurt, and alcoholic beverages. In the 20th century microorganisms have been harnessed to produce antibiotics (such as penicillin from fungi) and many other medicinal drugs. Today, plasmids in bioengineered bacteria carry a range of human genes, and those genes are expressed to produce large quantities of clinically useful human proteins. Many of these proteins, such as growth hormone, once were available only in tiny amounts and were extremely costly because they had to be chemically extracted from glandular tissues. In addition to growth hormone, human proteins currently produced by bacteria include insulin and interferons (Table 21.1). Insulin, once available only from the pancreases of slaughtered pigs and cattle, helps regulate glucose metabolism (page 285) and is essential to the survival of type I diabetics. Interferons have roles in immunity (page 183).

Vaccines   Bacteria may also be used as factories to manufacture *vaccines*, substances that trigger the development of antibodies to a particular antigen (such as an influenza virus) and so provide immunity to that antigen (page 182). Vaccines have traditionally been made using weakened or killed microbes, which carry proteins on their surface that trigger antibody production by the infected individual. However, it is possible to make a vaccine by inserting the gene for the antigen protein into a plasmid, then growing the transgenic bacteria to produce large quantities of the protein. The protein is added to the vaccine, which can then mobilize a person's immune system against that particular antigen. So far, bioengineered vaccines have been developed for hepatitis B and rabies. Various laboratories are working to genetically engineer an AIDS vaccine.

Bioremediation   Genetically engineered bacteria may also become major weapons in efforts to clean up environmental pollution, a process called *bioremediation*. The *Focus on Environment* essay gives some examples of this "environmental biotechnology." It involves strategies for exploiting genes that permit some types of bacteria to metabolically break down toxic substances to harmless compounds.

Safety Concerns   Some bacteria used in genetic engineering experiments are harmless to begin with, and in many cases they are also modified (by mutation) so that they cannot survive outside the laboratory. Even so, there is legitimate concern about possible risks of introducing genetically engineered bacteria into humans or the environment (see *Choices*). Government regulations now govern the release of genetically engineered organisms into the environment, and they are regularly reevaluated in the light of new research.

Table 21.1 Some Cloned Human Gene Products Approved for Use or Under Development	
Protein	Condition Treated
Insulin	Diabetes
Growth hormone	Pituitary dwarfism
Erythropoietin	Anemia
Factor VIII	Hemophilia A
Factor IX	Hemophilia B
Interleukin-2	Cancer
Tumor necrosis factor	Cancer
Interferons	Some cancers, viral infections
Monoclonal antibodies	Infectious diseases
Atrial natriuretic factor	High blood pressure
Tissue plasminogen factor (tPA)	Heart attack, stroke

a   b

c   d

Figure 21.7 Examples of genetically engineered plants having commercial importance. (**a**) Some worms attack buds of cotton plants but not the modified plant shown in (**b**). A normal potato plant susceptible to viral attack is shown in (**c**); a genetically modified strain is shown in (**d**).

Genetic Modification of Plants

By 1995 more than 600 field tests of genetically engineered plants will be in progress in some 20 countries. Some are geared toward creating improved varieties of crop plants. Others are exploiting plants that, like bacteria, can serve as factories for producing medically and commercially useful proteins.

For example, researchers can insert DNA fragments into a Ti plasmid, which is extracted from a bacterium (*Agrobacterium tumefaciens*) that infects many flowering plants. Some genes on the plasmid normally cause tumors to form. When the Ti plasmid is used to ferry genes, however, the harmful genes are first removed and desired ones substituted. Inserted into plant cells, the foreign genes can be integrated into the plant's chromosomes and are sometimes expressed normally in the plant tissues.

Similar methods have already produced some practical successes, including tomatoes that resist spoiling and cotton plants genetically engineered to resist worm attacks (Figure 21.7)—a potential alternative to heavy applications of insecticides. Efforts are also underway to

Microbes That Clean Up Pollutants

As a group, bacteria have a rather astonishing quality. They are so diverse metabolically that they can gain energy for life processes from breaking down substances that are highly toxic, even carcinogenic. In the process, many noxious substances—including crude oil, toxic heavy metals such as mercury, and pesticides—are chemically altered to a safer form.

"Oil-eating" bacteria, for example, oxidize the hydrocarbons in petroleum to carbon dioxide. Mercury-resistant bacteria metabolically process molecules of the metal (which causes severe neurological damage in humans) into a nontoxic compound.

Often, the genes that code for these metabolic feats are carried on plasmids—and so are ready candidates for bioengineering. A number of "de-tox" genes have been isolated, cloned, and transferred to other bacteria. Recently, for example, plasmids with genes that code for enzymes that break down hydrocarbons were successfully transferred to marine bacteria. The bacteria have since been used to help clean up oil spills off Texas and elsewhere.

Recombinant plasmids also have been put together using genes obtained from bacteria that normally break down various contaminants in toxic waste dumps. The goal is to engineer microbes that can be used to treat industrial wastes and possibly help clean up polluted soils. The U.S. Environmental Protection Agency has identified more than 2,000 seriously contaminated toxic-dump sites in the United States, many of them endangering public groundwater supplies. Biotechnology may well become an invaluable tool for restoring such places to environmental health.

engineer corn, soybeans, and other food plants so that they will be more tolerant to drought and pesticides, or to improve the balance of essential amino acids. Studies show that gene modifications can also stimulate plants to produce larger-than-normal quantities of commercially or nutritionally valuable starches, oils, and other substances.

For some species of plants that cannot be grown easily from seeds, methods of *tissue culture* can be used to multiply the number of plants for field transplanting. Basically, somatic cells of the parent plant are removed and placed on a nutrient medium that promotes tissue growth; small lumps of the tissue are then divided and grown in a new chemical solution that promotes development of a fully differentiated plant. The strawberries

Choices: Biology and Society

Issues for a Biotechnological Society

Some people say that no matter what the species of organism, DNA should never be altered. But as we noted earlier, nature alters DNA all the time. The real issue is how to bring about beneficial changes without harming ourselves or the environment.

In the 1980s activists entered a test plot and ripped out strawberry plants that had been sprayed with a bacterium that had been genetically altered to help prevent ice crystals from forming on the plants (Figure *a*). Although previous studies had shown that the "ice-minus" bacteria posed no threat to the environment or human health, many people have voiced similar concerns about releasing recombinant organisms into the environment. Researchers themselves have moved into the vanguard of efforts to develop regulations and guidelines that protect the public interest while permitting beneficial advances. Current concerns include the following:

• Transgenic bacteria or viruses could mutate, possibly becoming new pathogens in the process. However, as described on page 450, regulations require recombinant microbes to be altered in ways that will prevent them from reproducing.

• Bioengineered plants could escape from test plots and pose an ecological risk by becoming "superweeds" resistant to herbicides and other control measures.

• Crop plants with added insect resistance could set in motion an evolutionary see-saw in which new, even more formidable insect pests arise.

• Transgenic fish that feed voraciously and grow rapidly could displace natural species, wrecking the delicate ecological balance in streams and lakes.

• Major ethical problems could arise if people undertake "eugenic engineering"—using biotechnology to insert genes into normal individuals (or sperm or eggs) simply for the purpose of creating individuals that are "more desirable."

• Genetic screening could provide a means of discrimination by insurance companies and others against people

Figure a Spraying an experimental strawberry patch in California with "ice-minus" bacteria. (Government regulations required that the sprayer use elaborate protective gear.)

who carry a gene predisposing them to a disorder, even if such a person shows no sign of ill health.

These examples only touch on some of the profound social and ethical issues raised by recombinant DNA technology and genetic engineering. With new advances coming every day, it may be time for everyone to think about the benefits *and* risks of living in a biotechnological society.

a

b

Figure 21.8 (**a**) Ten-week-old mouse littermates, the one on the left weighing 29 grams, and the one on the right, 44 grams. The larger mouse grew from a fertilized egg into which the gene for human somatotropin (growth hormone) had been inserted. (**b**) This transgenic goat carries human genes that code for the production of tissue plasminogen activator (tPA), an enzyme that helps break up blood clots.

you buy in the grocery store are often the fruit of plants propagated through tissue culture.

Recently, researchers have begun to insert human genes for certain proteins into plants. One such protein, serum albumin, is widely used in preparations given to burn patients and other ill people to replace lost body fluids. Researchers have already extracted useful amounts of the protein from genetically altered potato and tobacco plants.

Genetic Modification of Animals

Animal cells do not accept plasmids, but they can be "micro-injected" with foreign DNA. In 1982, researchers introduced the gene for growth hormone from rats into fertilized mouse eggs. When the mice grew much larger than their normal littermates and had dramatically higher blood concentrations of the hormone, it was clear that the rat gene had become integrated into the mouse DNA and was being expressed. When the gene for human growth hormone was successfully introduced and expressed in mice, the result was the "super rodent" shown in Figure 21.8a.

Similar experiments with large domesticated animals have begun to show some success. As the chapter opening vignette noted, transgenic sheep, pigs, and cows may become sources of important pharmaceutical drugs. The goat in Figure 21.8b developed from a fertilized egg that was injected with recombinant DNA consisting of goat gene sequences spliced to human genes for tissue plas-

minogen activator, or tPA. This protein dissolves blood clots in heart attack and stroke patients, and the goat's milk contains it. Although tPA and other human proteins are currently made by recombinant bacteria, so-called "barnyard biotechnology" is exciting because in some cases transgenic livestock have the potential to produce large quantities of drugs faster and more cheaply than bacteria cultured in huge industrial vats.

People who have hemophilia A now can obtain the needed blood clotting factor VIII from a product called Recombinate. This genetically engineered drug is produced by Chinese hamster ovary cells in which genes for human factor VIII have been inserted. Factor VIII produced in this way eliminates the need to obtain it from human blood—and so eliminates the risk of transmitting blood-borne pathogens such as HIV.

The Human Genome Project

Scientists in more than 1,000 laboratories around the world are collaborating on a massive Human Genome Project. One goal is to create a physical map of all human chromosomes, identifying the locations of specific genes. Another, even more daunting goal is to determine the sequence of nucleotides in the entire complement of human DNA. The YACs described earlier are used to clone copies of human genes for analysis. The two smallest human chromosomes, the Y chromosome and chro-

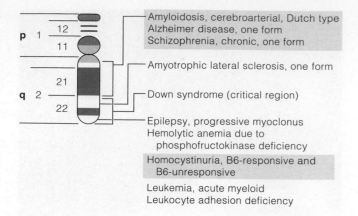

Figure 21.9 Map of human chromosome 21 showing genes correlated with specific diseases. In all, researchers have mapped 13 genes to this chromosome. The upper arm of the chromosome is marked p and the lower arm, q. In this case, each arm consists of a single region, labeled 1 and 2, respectively. In other chromosomes, each arm may have two or more regions. Various stains used in chromosome analysis produce a series of bands within each region. The small numbers just to the left of each arm indicate specific bands. A combination of letters and numbers indicates the chromosome region where a given gene is found; for instance, the gene for one form of amyotrophic lateral sclerosis (ALS) is 21q2.22.

mosome 21 (Figure 21.9), were extensively mapped in 1992, and scores of markers have been mapped to other chromosomes. For example, scrutiny of chromosome 21 has revealed that this chromosome carries genes for a form of Alzheimer's disease (page 333), some forms of epilepsy, and for one type of amyotrophic lateral sclerosis (ALS), a rare disease that progressively destroys motor neurons. In all, of 2,000 markers mapped on chromosomes, 400 are already linked to genetic disorders.

"Gene machines" can currently sequence a whopping 10,000 DNA nucleotides a day, and more efficient machines are becoming available. However, given that there are an estimated 3 billion nucleotides in the 23 pairs of human chromosomes, it will take 10 to 15 years to obtain a detailed picture of the human genome. Simply managing the resulting data will be mind-boggling—the list of base sequences in human DNA might fill 480,000 single-spaced typed pages!

Although mapping and sequencing the human genome will undoubtedly increase our understanding of the genetic basis of human traits and diseases, it also has raised thorny issues. The project will cost many millions of dollars, much of it public tax money. Some researchers question whether those funds might be better spent on other programs, such AIDS or cancer research (to name just two possibilities). There is criticism that U.S. laboratories are attempting to patent the information they uncover—pre-

sumably for future commercial gain—while in France the information is being made freely available to all. Some individuals have expressed concern that if the sequenced DNA comes only from a limited number of people, it may not represent the true diversity of the human genome. As the project goes forward, these all are issues that the public and policymakers will have to address.

RFLPs An essential tool for sequencing the human genome is the **RFLP** ("riff-lip"), an acronym for *restriction fragment length polymorphism*. Polymorphism means "many forms," and it refers to the fact that each person's DNA has a unique pattern of sites where restriction enzymes make their cuts. Hence the lengths of restriction fragments from each person's DNA will be slightly different from those of any other person. (Only identical twins have identical DNA.) The pattern variations can be detected by gel electrophoresis (Figure 21.10).

RFLPs can be used to locate mutant alleles responsible for genetic disorders. For example, the mutant allele coding for the defective hemoglobin chain in sickle-cell anemia has a unique restriction site that is *not* present in normal DNA.

Another kind of RFLP is being used in legal proceedings. Each person has a *DNA fingerprint*, a unique array of RFLPs inherited from each parent in a Mendelian pattern. In cases in which a child's parentage is in dispute, a comparison of RFLPs can help resolve the issue. Attorneys are also using RFLPs to help convict or clear individuals charged with murder or rape. Even a few drops of blood or semen contain enough DNA for its DNA fingerprint to be compared with a suspect's DNA.

Gene Therapy

Generally speaking, **gene therapy** involves inserting one or more normal genes into an organism's body cells to correct a genetic defect. The first attempts at human gene therapy are now underway, involving desperately ill victims of skin cancer, cystic fibrosis, and other disorders. The *Human Impact* essay following this chapter describes some of these attempts.

Altering Gene Expression Several approaches aim to develop drugs that act on DNA or RNA in specific ways. For example, genetic engineering methods are being used to create drugs that can block the synthesis of defective proteins by shutting down transcription of DNA to messenger RNA (page 415). One target is the virus that causes herpes infection: Scientists are working to develop a drug that can enter an infected cell and shut down transcription of viral DNA without harming expression of the cell's own genes. Research on such *transcription factors* could also lead to drugs that can turn off oncogenes responsible for various cancers (page 430) or that shut down mutated alleles for diseases such as sickle-cell anemia.

Another approach is to block protein synthesis by way of molecules that can prevent mRNA from being translated in the cell cytoplasm. Engineered "antisense" RNA molecules are one option. These are short stretches of nucleic acid that can bind to specific mRNAs and prevent translation. Alternatively, **ribozymes** are bits of RNA that act like enzymes and, when inserted into a cell, cut up specific mRNA sequences before they can be translated. Ribozymes will probably first be used in antiviral drugs.

SUMMARY

1. Genetic "experiments" have been occurring in nature for billions of years. Mutation, crossing over and recombination at meiosis, and other natural events have all contributed to the current diversity that we see among organisms.

2. Humans have been manipulating the genetic character of different species for thousands of years. The emergence of recombinant DNA technology in the past few decades has enormously expanded our capacity to cause genetic change. Recombinant DNA technology is founded on procedures by which DNA molecules can be cut into fragments, inserted into plasmids or some other cloning tool, then multiplied in a population of rapidly dividing cells.

3. A DNA clone is any DNA sequence that has been amplified in dividing cells. DNA sequences also can be amplified in test tubes by the polymerase chain reaction.

4. Recombinant DNA technology and genetic engineering have enormous potential for research and applications in medicine, agriculture, and industry. As with any new technology, potential benefits must be weighed against potential risks, including ecological and social disruptions.

5. Although the new technology has not developed to the extent that human genes can readily be modified, the social, legal, ecological, and ethical questions it raises should be explored in detail before such an application is possible.

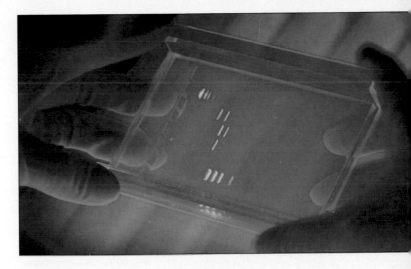

Figure 21.10 Columns, running from left to right, in which DNA fragments (labeled with a fluorescent dye) have been separated from one another. The fragments form a luminous banding pattern that reflects differences in their length.

modified host cells. Briefly describe one of the methods used in each of these categories. *446–447, 453*

4. Genetic researchers use plasmids as _____.

5. Rejoined cut DNA fragments from different organisms is best known as _____.
   a. cloning genes
   b. mapping genes
   c. recombinant DNA
   d. conjugating DNA

6. Repeated DNA replications and cell divisions of the host cells yield _____.

7. Using the metabolic machinery of a bacterial cell to produce multiple copies of genes carried on recombinant plasmids is _____.
   a. a way to create a DNA library
   b. bacterial conjugation
   c. mapping a genome
   d. DNA amplification

8. A collection of DNA fragments produced by restriction enzymes and incorporated into plasmids is a _____.
   a. DNA clone   c. hybridized sequence
   b. DNA library   d. gene map

9. The polymerase chain reaction _____.
   a. is a natural reaction in bacterial DNA
   b. cuts DNA into fragments
   c. amplifies DNA sequences in test tubes
   d. inserts foreign DNA into bacterial DNA

10. Recombination is made possible by _____ that can make cuts in DNA molecules.

Key Terms

cDNA 449
cloned DNA 447
cloning vector 446
DNA library 447
DNA ligase 447
DNA polymerase 448
DNA probe 448
gene therapy 454

plasmid 446
polymerase chain reaction 447
recombinant DNA technology 446
restriction fragment 447
reverse transcription 449
RFLP 454
ribozyme 455

Readings

"Biotech's Second Generation." May 8, 1992. *Science.*

Drlica, K. 1992. *Understanding DNA and Gene Cloning. A Guide for the Curious,* 2nd ed. New York: Wiley.

Gasser, C. and R. Fraley. June 1992. "Transgenic Crops." *Scientific American.*

Thompson, L. 1994. *Correcting the Code.* New York: Simon & Schuster. Up-to-date nontechnical book on human gene therapy.

GENE THERAPY: DRUGSTORE DNA

White blood cells inside a blood vessel. Some people who are candidates for gene therapy have drastically reduced numbers of these infection-fighting cells.

Genes in a bottle? In the not-so-distant future, your local pharmacist may be filling prescriptions for gene therapy drugs, as now happens with genetically engineered insulin used by diabetics.

Diseases such as cystic fibrosis and hemophilia result when a single gene is mutated, producing a malfunctioning protein. Many cancers develop when a chemical carcinogen or some other factor upsets normal gene regulation, and one or more genes in a cell turn on or off at the wrong time (page 430). **Gene therapy** attempts to replace mutated genes with normal ones, which will encode normal proteins, or to insert genes that restore normal controls over gene activity. An inserted gene might also code for a protein that blocks a cell surface receptor associated with a disorder. For example, it might be possible to treat arthritis by inserting genes coding for proteins that block the binding of substances that cause inflammation. Researchers are actively exploring all these possibilities (see Table I).

At this writing more than 20 clinical gene therapy trials have been approved in the United States. Dozens more are underway in other countries.

Strategies for Transferring Genes

Different genes are built of different numbers of base pairs, and the size of a gene helps determine what sort of mechanism can be used to insert it into a host cell. Animal cells do not take up plasmids, so plasmids cannot be

Table I Examples of Gene Therapy Trials*	
Disease	Inserted Gene Codes for:
ADA deficiency	Adenosine deaminase
Cystic fibrosis	CFTR
Malignant melanoma	Interleukin-2
Hemophilia B	Factor VIII
Advanced cancers	Tumor necrosis factor
Kidney cancer	Interleukin-2
Lung cancer	Antisense *ras*/p53
Ovarian cancer, brain cancer	Thymidine kinase (enzyme that promotes attack by the cell-destroying drug gancyclovir)
AIDS	Thymidine kinase

*List compiled as of 1993

directly used as vectors in gene therapy. Smaller genes can be carried into animal cells by viruses; larger ones must enter a host cell some other way.

Transfection—Direct Gene Transfer In *transfection*, cells cultured in the laboratory are directly exposed to DNA segments that contain a gene of interest. Although the exact mechanisms are not well understood, some of the foreign DNA becomes integrated into the host cell's genome. Exposing the host cells to a weak electric current, called *electroporation*, seems to help. Even so, this strategy is inefficient: Sometimes only one cell in 10 million incorporates a new gene.

Injection DNA can be injected into fertilized eggs. This method can be successful about 10 percent of the time and has been used to create hybrid mouse-human cells that are then used in other procedures. Thus far it has *not* been tried with human eggs—an experiment that would probably be extremely controversial.

Viruses A small gene can be inserted into a virus, which can then "infect" a host cell and carry the desirable gene with it. Several gene therapy trials are using retroviruses in this way because such RNA viruses easily infect animal cells. Once inside the host cell, the viral RNA is transcribed into DNA (by way of reverse transcriptase), and usually the foreign DNA is integrated into the host cell's DNA. The first step in gene therapy is to remove certain segments of genetic instructions from the virus, so that it cannot reproduce and cause disease. In their place, a researcher substitutes the gene to be transferred. Next the virus is permitted to infect target cells. Retroviruses are extremely efficient at integrating into the host DNA, increasing the odds that the new gene will become active. Some gene therapy experiments use other viruses that typically infect the kinds of host cells being manipulated.

Pitfalls of Using Viruses There are pitfalls involved with using viruses to introduce genes. Suppose, for example, that a virus is engineered to infect cells with a "good" copy of a gene that is mutated in the host DNA. When an engineered virus integrates into a host cell's DNA, there is no guarantee that the new gene will be in its normal position—and if it is not, the gene may not be transcribed into protein. Some viruses often undergo deletions or other rearrangements that alter any genes they carry and so make the genes ineffective. There is also the chance that a virus containing an introduced gene might "escape" and infect other individuals who do not need the new gene. Finally, there is a slight possibility that an altered virus may revert to a pathogenic form.

Obstacles At present, only a few types of human cells—notably skin and bone marrow stem cells—can successfully be removed, induced to take up cloned DNA, grown in the laboratory, and then reinserted into the body. Thus, this method can be used only with genes that are expressed in those tissues. There are no reliable procedures for making large quantities of certain vectors that are commonly used to carry new DNA into host cells. In addition, many introduced genes turn off within a few days or weeks. In spite of these obstacles, researchers are conducting important clinical trials of various gene therapy methods.

Gene Therapy in Action

Various approaches to gene therapy are being tested in the United States and elsewhere. This list of experimental trials is only the tip of the gene therapy iceberg—researchers are becoming remarkably clever at devising ways to test and improve their rapidly developing technology.

ADA Recently, 11-year-old Cynthia Cutshall became the subject of the first federally approved gene therapy test for humans. Born with a defective gene that codes for the enzyme adenosine deaminase (ADA), the girl suffers from *severe combined immune deficiency,* or SCID. This set of disorders is brought on by a drastic reduction or complete absence of two subpopulations of white blood cells: the B and T cells of the immune system.

Using recombinant DNA methods, researchers introduced copies of the ADA gene into some of the child's T cells, which were then stimulated to divide. Later, about a billion copies of the genetically engineered cells were delivered through a plastic tube into the young patient's bloodstream. Each month since, the girl has received another T cell infusion. In 1993 doctors extracted stem cells from her bone marrow, used a recombinant retrovirus to introduce functional ADA genes into them, and put the stem cells back in place. It may take several years to determine whether the method works. If it does, the child could be permanently cured of her disorder because bone marrow stem cells give rise to white blood cells throughout life.

Cystic Fibrosis Cystic fibrosis is another major target of gene therapy. A study now underway involves 10 adult patients, and early results have been promising. Here, a normal copy of the deficient CFTR gene (page 390) is inserted into a disabled adenovirus, a particle that infects epithelial cells of the human respiratory tract that malfunction in CF. Patients then inhale a mist that carries the engineered virus particles containing normal CFTR genes. Because epithelial cells regularly die and are replaced, any that take up the virus and begin to produce normal CFTR proteins will be lost after several months. As a result, this potential therapy offers a treatment for CF, but not a cure.

Hypercholesterolemia An inherited form of this disease causes extremely high cholesterol levels and is a leading cause of heart disease worldwide. The cause is a defective gene for LDLs—the low-density lipoprotein receptors produced in the liver that normally bind with cholesterol in blood and transport it into cells. In gene therapy trials, surgeons have removed portions of several patients' livers and have grown the liver cells in the laboratory. The cultured cells are then exposed to a retrovirus carrying a normal LDL receptor gene. Finally, some of each patient's genetically "corrected" liver cells are reinjected into a vein leading to the liver in hopes that the cells will proliferate and function normally.

Cancers Gene therapy may offer a potent weapon against various types of cancers. Trials have targeted the deadly skin cancer malignant melanoma, a virulent form of lung cancer, leukemia, and cancers of the brain, ovaries, kidneys, and other organs.

In some approaches, tumor cells are first removed from a patient and grown in the laboratory. Genes for an interleukin (the substance that helps activate white blood-cells of the immune system) are then introduced into the cells, and the cells are reinserted into the body. In theory, interleukins produced by the tumor cells may act as "suicide tags" that stimulate T cells of the immune system to recognize cancerous cells and attack them.

Other trials are exploring the benefits of adding the gene for a powerful antitumor agent called tumor necrosis factor (TNF) to tumor-infiltrating lymphocytes, or TILs. Tumor necrosis factor helps kill tumors by shutting off their blood supply, but it also can have the same effect on healthy tissues. As their name suggests, TILs home in specifically on tumors. One goal of such experiments is to determine whether TILs carrying the TNF gene can deliver enough TNF to kill a targeted tumor without harming surrounding tissues.

Researchers may also develop ways to alter the expression of genes that apparently play major roles in cancer development. For example, "non-small-cell" lung cancer tumors have been directly injected with retrovirus particles carrying two types of foreign genetic information. One is a stretch of "antisense" DNA—a DNA segment that can bind to a complementary nucleotide sequence, which happens to be a gene. The antisense DNA is designed to block the expression of a mutant form of a gene called *ras*; the mutated form causes cells to grow out of control. The second type of genetic information is a working copy of the tumor suppressor gene, p53. This gene normally acts to prevent tumor growth, but in many types of cancerous cells it has stopped functioning. Chapter 20 discusses tumor suppressor genes in more detail.

Gene Therapy and Your Future

All the methods of gene therapy described here have two things in common: They are expensive, and they require a great deal of specialized equipment and expertise. At the moment, most can benefit only a few of the millions of people who suffer from cancers, not to mention inherited disorders such as ADA deficiency and cystic fibrosis. Authorities agree that the key to widespread use of gene therapy is the development of vectors, such as specially engineered viruses, that can deliver genes to target cells with the same safety and reliability that we now take for granted with antibiotics, insulin, and other medicinal drugs. Many experts predict that as gene therapy becomes more commonplace, its costs will drop significantly. Will the day come when you will be able to get a "gene shot" at your doctor's office or DNA at the drugstore? The answer is probably yes—and it may be sooner than you think.

A smoke cloud over South America's Amazon River basin photographed by astronauts in the Space Shuttle Discovery in 1988. The cloud was caused by the clearing and burning of tropical forests, pasture, and croplands, activities that continue today at a rapid pace. If the cloud were placed over North America, it would cover more than one-third of the area of the 48 contiguous United States. Human impacts on earth's ecosystem are a central topic in Unit IV.

22 PRINCIPLES OF EVOLUTION

Floods and Fossils

Nearly 500 years ago, Michelangelo was finishing his painting of the Great Deluge on the ceiling of the Sistine Chapel in Rome (Figure 22.1). Another artist, Leonardo da Vinci, was brooding about seashells that lay embedded in rock layers in mountains of northern Italy. The traditional view was that the shells were strewn about during the Deluge. Yet da Vinci wondered how even a great flood could wash them inland and up the mountainsides, which were hundreds of kilometers from the sea. Da Vinci also perceived that the rock layers were arranged in strata, one above the other like cake layers. His journals tell us that he suspected the distinct groups of shells in those layers might be the fossilized remains of ancient marine communities, buried not in a great flood, but gradually over vast stretches of time. By the early 1700s fossil shells, bones, plant parts, and impressions of footprints, animal tracks, and burrows were being accepted as evidence of life in the past.

By the mid-18th century extensive canal building, quarrying, and mining operations were revealing regular patterns in the color and composition of stratified rock layers. Soon it became apparent that certain fossils occur in distinctive types of layers, sometimes over considerable distances. For example, cliffs on the English side of the English Channel reveal the same fossil sequences as cliffs on the French side.

Some scientists used the correlation between fossils and strata as a geologic mapping tool. Others began using them to analyze the possible connections between past geologic events and the history of life. As you will see in this chapter, fossils and other sources of information afford convincing evidence of change in populations of organisms through time—in other words, evolution.

1. Evolution is defined as change through time. The lines of evidence that species evolve include sequences of changing fossils in distinct layers of the earth and comparisons of body structure, patterning, and biochemistry.

2. As Charles Darwin perceived, members of the same population share the same heritable traits. Yet the members don't look or act exactly the same. As we now know, variation in any heritable trait results when there are different alleles for the trait among members of a population.

3. Within a population, one allele for a trait may become more or less common relative to other alleles for that trait. An allele may even disappear. Allele frequencies change through mutation, genetic drift, gene flow, and natural selection. Only mutation produces *new* alleles.

4. Natural selection is not an "agent," purposefully searching for the "best" individuals in a population. It simply is the *difference in survival and reproduction* that has occurred among individuals that differ in one or more traits.

5. Genetic differences may build up between isolated populations of the same species. When one population diverges genetically from the others, it may evolve into a new species. For sexually reproducing organisms such as humans, speciation has occurred when members of the separated populations no longer can interbreed and produce fertile offspring under natural conditions.

6. Events that lead to speciation are examples of *microevolution*. Groupings more inclusive than species (such as the genus, a collection of closely related species) arise by *macroevolution*.

7. Charles Darwin proposed that species evolve by natural selection. The central point of this theory is that a population evolves when some traits (phenotypes) become more common and others decrease or disappear over time.

8. The evolution of life on earth has been linked, from the time of origin to the present, to the physical and chemical evolution of the earth.

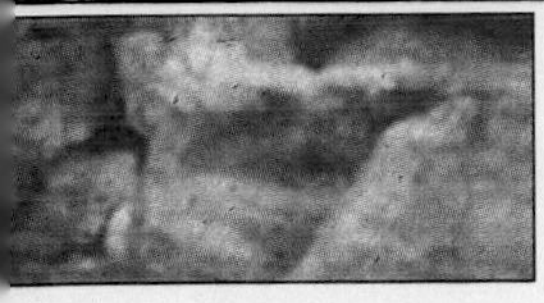

Figure 22.1 (*Left*) From the Sistine Chapel, Michelangelo's painting of the onset of a catastrophic flood, traditionally called the Great Deluge. (*Right*) A modern-day photographer captures the intricate structural pattern of fossilized ammonite shells. About 65 million years ago the last ammonites perished, along with many other groups of organisms. That mass extinction is but one piece of the evolutionary puzzle.

OVERVIEW OF EVOLUTION

The history of life on earth spans nearly 4 billion years. It is a story of how species originated, persisted or went extinct, and stayed put or radiated into new environments. The overall "plot" of the story is **evolution**, or change in lines of descent over time. The term **microevolution** refers to cumulative genetic changes that give rise to new species. **Macroevolution** applies to the large-scale patterns, trends, and rates of change among *groups* of species. In this chapter we will examine the basic principles of evolutionary change. Chapter 23 surveys what we currently know of the evolutionary events leading to modern humans.

Variation in Populations

Evolution results from changes in the genetic makeup of populations of organisms. By definition, a **population** is a group of individuals of the same species occupying a given area. There is a great deal of genetic variation within and among populations of the same species.

Overall, the members of a population have similar traits (phenotypes). They have the same general form and appearance (*morphological* traits), their body structures function the same way (*physiological* traits), and they respond the same way to certain basic stimuli (*behavioral* traits). However, the details of traits vary greatly from one individual to another. For instance, human hair varies in color, texture, amount, and distribution over the body. As we know from previous chapters, this example only hints at the immense genetic variation in human populations. Populations of most other species of organisms show the same kind of variation, as Figure 22.2 shows.

Sources of Variation   The members of a population generally have inherited the same number and kinds of genes. But remember, a given gene may have slightly different molecular forms, called alleles. Different combinations of alleles are inherited, and this leads to variations in traits. Whether your hair is black, brown, red, or blond depends on *which* alleles of certain genes you inherited from your mother and father. As described in earlier chapters, five events contribute to that mix of alleles:

1. Gene mutation (produces new alleles)

2. Crossing over at meiosis (leads to new combinations of alleles in chromosomes)

3. Independent assortment at meiosis (leads to mixes of maternal and paternal chromosomes in gametes)

Figure 22.2   Variation in shell color and banding patterns among populations of a snail species found on islands of the Caribbean. Variation in traits arises from different combinations of alleles carried by different members of the populations.

4. Fertilization (puts together combinations of alleles from two parents)

5. Changes in chromosome structure or number (leads to the loss, duplication, or alteration of alleles)

The abundance of each kind of allele in a population is called that allele's *frequency*. *Microevolution* refers to changes in allele frequencies brought about by mutation, natural selection, and other processes (Table 22.1) we consider in the next section.

Table 22.1	Major Microevolutionary Processes
Mutation	A heritable change in DNA
Genetic drift	Random fluctuation in allele frequencies over time, due to chance occurrences alone
Gene flow	Change in allele frequencies as individuals leave or enter a population
Natural selection	Change or stabilization of allele frequencies due to differences in survival and reproduction among variant members of a population

A population is a group of individuals of the same species occupying a given area.

Although individuals in a population share many general traits, in nearly all populations there is great underlying genetic variation.

Evolution is change in lines of descent over time. It results from changes in the genetic makeup of populations of organisms.

PROCESSES OF MICROEVOLUTION

Mutation: Sole Source of New Alleles

Recall that a mutation is a heritable change in DNA. In evolutionary terms, it is a significant event. *Mutations are the source of alleles that have been accumulating in different lineages for about 4 billion years.*

Mutations occur relatively rarely. Whether a mutation proves to be harmful, neutral, or beneficial depends on how its product interacts with other genes *and* with the environment. The majority of mutations are probably harmful, altering traits in such a way that an individual cannot survive or reproduce. For example, humans have typically lived in environments in which small cuts, bruises, or other minor injuries are relatively common. Before the advent of effective medical treatments, hemophiliacs, whose blood does not clot properly, could die (and often did) at a young age from such minor injuries. That is why the frequencies of the mutated alleles for various hemophilias historically have been low. As Chapter 19 noted, beneficial traits typically *improve* some aspect of an individual's functioning in the environment and so improve the chances to survive and reproduce. Neutral traits—such as attached earlobes in humans—may be of no great help to an individual's survival, but under prevailing conditions they do not seriously damage it either.

Mutations are the source of new alleles, hence of heritable variation in traits.

A mutated gene may code for a trait that is harmful, neutral, or beneficial, depending on how the trait affects an individual's ability to survive and reproduce under prevailing conditions.

Natural Selection

Natural selection probably accounts for more changes in allele frequencies than any other microevolutionary process. In the 19th century Charles Darwin discovered this major process by correlating his understanding of inheritance with certain features of populations and the environment. He published his ideas in a classic book, *On the Origin of Species.* Here we express the main points of Darwin's correlation in light of modern genetics:

1. Individuals of a population vary in form, function, and behavior.

2. Much of the variation in a population is *heritable.* This means that more than one kind of allele exists for genes that give rise to the traits.

3. Some forms of a trait are more *adaptive* than others. This means that they improve chances of surviving and reproducing.

4. **Natural selection** is the *difference in survival and reproduction* that has occurred among individuals that differ in one or more traits. (This differential in survival and reproduction is sometimes informally referred to as "survival of the fittest.")

5. A population is *evolving* when some forms of a trait are becoming more or less common relative to the other forms. The shifts are evidence of changes in the relative abundances of alleles for that trait.

6. Life's diversity is the outcome of changes in allele frequencies in different lines of descent over time. Such changes result in the splitting of a single species into two related ones.

As natural selection proceeds over time, organisms come to have characteristics that suit them to conditions in a particular environment—a trend we call **adaptation**.

People often speak of the "theory" of evolution. Yet decades of rigorous scientific observations have yielded so much evidence of evolution that it is more properly considered a fundamental principle of the living world. We consider just a few examples of this evidence in later sections.

Genetic Drift and Gene Flow

Allele frequencies in a population can change randomly through the generations because of chance events alone. This is called **genetic drift**. Often the change is most rapid in small populations. In one type of genetic drift, called the *founder effect*, a few individuals leave a population and manage to establish a new one. Compared to the population left behind, the founders happen to carry fewer (or more) alleles for certain traits. Simply by chance, the allele frequencies in the new population differ from the old. The founder effect probably explains why there are virtually no Native Americans with type B blood. Genes for that blood group must not have been present in the few individuals who originally crossed the Bering Strait from Asia into what is now Alaska at least 10,000 years ago.

Allele frequencies also can change as individuals leave a population (emigration) or new individuals enter it (immigration). This physical flow of alleles, or **gene flow**, helps keep neighboring populations genetically similar. Over time it tends to counter the differences between populations that are brought about through mutation, genetic drift, and natural selection. In this age of international travel, the pace of gene flow among human populations has increased dramatically.

Reproductive Isolation and Speciation

For sexually reproducing organisms, a **species** is a unit consisting of one or more populations of organisms that usually closely resemble each other physically and physiologically. Individuals of a species are able to interbreed and produce fertile offspring under natural conditions. No matter how diverse those individuals become, they remain members of the same species so long as they can continue to interbreed successfully and share a common pool of alleles. A woman lawyer in India may never meet up with an Icelandic fisherman, but there is no biological reason why the two could not mate and produce children. Neither individual, however, could mate with a chimpanzee, even though chimps and humans are closely related species and have a large number of genes in common. In evolutionary terminology, we would say that humans and chimpanzees are *reproductively isolated*. There are enough genetic differences between them that they cannot interbreed successfully.

A species is a unit of one or more populations of individuals that can interbreed under natural conditions and produce fertile offspring.

Populations of different species are reproductively isolated; that is, under normal conditions they cannot interbreed.

Reproductive isolation develops when gene flow between two populations stops. This can happen in various ways; it often occurs when two populations become separated geographically. For example, geological changes in the eastern United States have frequently split small rivers into separate streams. The newly separated fish populations may then encounter slightly different environments. In those different environments, mutation, natural selection, and genetic drift begin to operate independently in each population. And these processes can change the allele frequencies of each in different ways. Differences among separated populations may come to include **reproductive isolating mechanisms**—aspects of structure, function, or behavior that reduce the likelihood of successful interbreeding. The two populations may breed in different seasons or have different mating rituals, or there

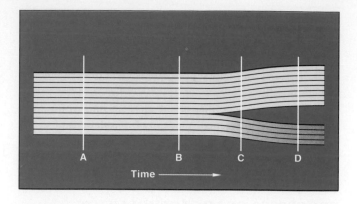

Figure 22.3 Divergence, the first step on the path to speciation. Each horizontal line represents a different population. Because evolution is gradual, we cannot say at any one point in time that there are now two species rather than one. At time A there is only one species. At D there are two. At B and C the split has begun but is far from complete.

may be changes in body structures that physically impede mating. Other types of isolating mechanisms prevent zygotes or hybrid offspring from developing properly.

An accumulation of differences in allele frequencies among isolated populations is called *divergence* (Figure 22.3). Divergence may become the first step on the road to **speciation**, the evolutionary process by which species originate. When the genetic differences between isolated populations become great enough, their members will not be able to interbreed successfully under natural conditions. At this point, the populations have become separate species.

Rates of Evolutionary Change

Although we know that factors such as gene mutation and natural selection ultimately trigger speciation, biologists disagree about the pace and timing of microevolution.

According to the traditional model, called *gradualism*, new species emerge through many small changes in form over long spans of time. In other words, microevolution is constantly going on in imperceptibly tiny increments to produce new species. By contrast, according to the model called *punctuated equilibrium*, most evolutionary changes occur in bursts. In this scenario, each species undergoes a spurt of changes in form when it first branches from the parental lineage, then changes little for the rest of its time on earth.

There is evidence that the driving force behind these rapid changes in species may be dramatic changes in climate or some other aspect of the physical environment. This type of change suddenly (in evolutionary terms) alters the physical conditions to which populations of organisms have become adapted. For example, the onset

Figure 22.4 (**a**) Plate tectonics, the arrangement of the earth's crust in rigid plates that split apart, move about, and collide with one another. (**b**) About 240 million years ago earth's land masses were joined in a massive supercontinent, Pangaea. By about 40 million years ago plate movements had split Pangaea into isolated land masses, including Africa, South America, Australia, and Eurasia.

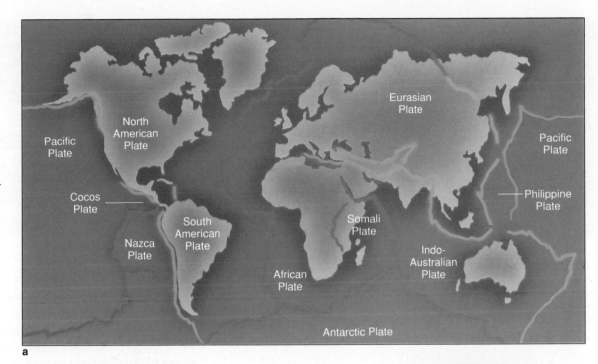

a

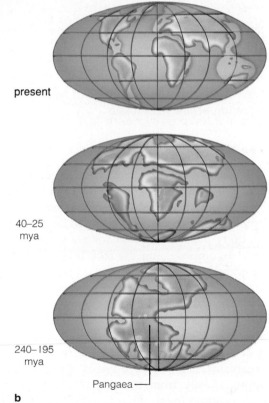

present

40–25 mya

240–195 mya

Pangaea

b

of an ice age changes the living conditions for many land-dwelling species in affected areas. A major shift in ocean currents (which help determine water temperature) might have the same effect on marine life forms. Punctuated equilibrium could help explain why the fossil record provides little evidence of a continuum of microevolution—there are few transitional forms between closely related species. It seems likely that both models have a place in explaining the whole history of life.

EVIDENCE FOR MACROEVOLUTION

Biogeography

Even before Charles Darwin, naturalists were finding indirect support for the idea of evolution in what today is called **biogeography**, the study of the world distribution of plants and animals. Biogeography addresses the basic question of why certain species (and higher groupings) occur where they do. For example, why do Australia, Tasmania, and New Guinea have species of monotremes (egg-laying mammals such as the duckbilled platypus), while such animals are absent from other regions of the world where the living conditions are similar? And why do we find weasels, skunks, and otters on virtually every continent except Australia? The simplest explanation for such biogeographical patterns is that species occur where they do either because they evolved there from ancestral species, or because they dispersed there from someplace else.

Modern studies of *plate tectonics*, the movement of plates of the earth's crust, show that earlier in earth's history all present-day continents, including Africa and South America, were parts of a massive "supercontinent" called Pangaea (Figure 22.4). By determining the locations of plates at different times in earth's history, researchers

a

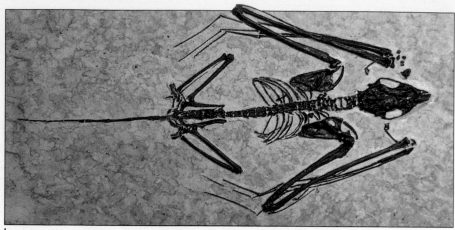

b

Figure 22.5 Some representative fossils. (**a**) Fossilized leaves from *Archeopteri*, which probably was on the evolutionary road that led to gymnosperms (such as pine trees) and angiosperms (flowering plants). (**b**) The complete skeleton of a bat that lived 50 million years ago. Fossils of this quality are extremely rare.

taxa. Through **comparative morphology**, researchers reconstruct evolutionary history on the basis of information contained in the observed patterns of body form. This work involves detailed studies of embryos at different stages of development as well as the adult form.

can shed light on possible dispersal routes for some groups of organisms and when (in geological history) such movements occurred. Key support for hypotheses about the evolutionary history of different groups of organisms comes from a range of sources, none of them more important than fossils.

The Fossil Record

A **fossil** is recognizable, physical evidence of an organism that lived long ago. Physical evidence for the long history of life comes from fossilized skeletons, shells, leaves, seeds, tracks, and such. To be preserved as a fossil, body parts or impressions are usually buried before they decompose, and the rock layers in which they are entombed must not be disturbed much over time. Fossil insects are sometimes found preserved in fossilized tree sap, which we call amber. A few fossils, such as the early plant and bat skeleton shown in Figure 22.5, are spectacularly complete and well preserved. They tell us a great deal about the structure, function, and behavior of extinct organisms. In all we have fossils of about 250,000 species, dating mostly from the past 600 million years. Using radioisotopes (page 18), researchers have established the ages of fossils that are millions, even billions, of years old.

Comparative Morphology

More evidence for evolution comes from detailed comparisons of body form and structural patterns in major

Comparative Embryology At certain stages of development, the embryos of different organisms within a phylum or some other major group bear striking resemblance to one another. Often the similarities indicate evolutionary relationships. For example, gill arches, a tail, and other structures of the sort shown in Figure 22.6 appear in all vertebrate embryos at corresponding stages of development.

Such structural similarities are one of the reasons fishes, amphibians, reptiles, birds, and mammals are said to belong to the same subphylum (Vertebrata) despite the large variation among adult forms. During vertebrate evolution, any mutations that disrupted an early stage of development would have had devastating effects on the organized interactions required for later stages. Embryos of different groups remained similar because mutations that altered early steps in development were selected against.

How, then, did the *adults* of different vertebrate groups get to be so different? At least some differences resulted from mutations that altered *rates of growth* of body parts formed during later stages. Notice, in Figure 22.7, how the proportions of chimpanzee and human skull bones are alike at birth. Thereafter, they change dramatically for chimps but only slightly for humans. Chimps and humans arose from the same ancestral stock, and their genes are nearly identical. However, at some point on the separate evolutionary road leading to humans, a regulatory gene probably mutated. Since then, instead of promoting the rapid growth required for dramatic changes in skull bones, the mutated gene has blocked it.

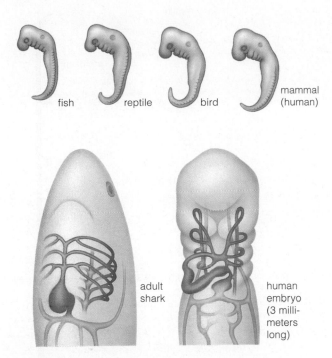

fish   reptile   bird   mammal (human)

adult shark   human embryo (3 millimeters long)

Figure 22.6 Evidence of evolution from comparative embryology. Aortic arches, a two-chambered heart, and other structures develop in a fish embryo. These structures, which persist in adult fishes, function in respiration in aquatic environments. They also develop in the early embryos of amphibians, reptiles, birds, and mammals. Shown here, the aortic arches (*red*), a two-chambered heart (*orange*), and certain veins (*blue*) in an early human embryo. Notice how these structures resemble those of an adult shark. They provide evidence that basic developmental processes have been retained during the evolution of different groups of vertebrates.

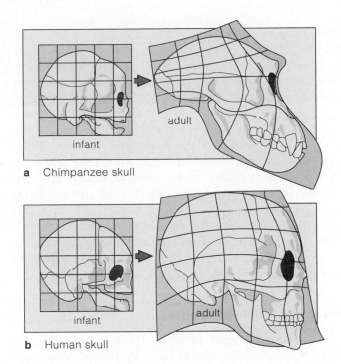

a Chimpanzee skull

b Human skull

Figure 22.7 Comparison of proportional changes in a chimpanzee skull (**a**) and a human skull (**b**). Both skulls are quite similar in infants. Imagine that these representations of infant skulls are paintings on a blue rubber sheet divided into a grid. Stretching the sheet deforms the grid's squares. For the adult skulls, differences in size and shape within corresponding grid sections reflect differences in growth patterns.

Vestigial Structures As Figure 22.8 indicates, the bodies of humans, pythons, and other organisms can have apparently useless "vestigial" structures. Like ear-wiggling muscles in humans, such body parts are left over from a time when more functional versions were important for an ancestor.

Homologous and Analogous Structures When populations of a species branch out in different evolutionary directions, they diverge in appearance, functions, or both. Yet the related species remain alike in many ways, for their evolution proceeds by modification of a shared body plan. In such species we see **homologous structures**. These are the *same* body parts modified in different ways in different lines of descent from a common ancestor (*homo-* means the same).

For example, the same ancestral organism probably gave rise to most land-dwelling vertebrates, which have homologous structures. Apparently their common ancestor

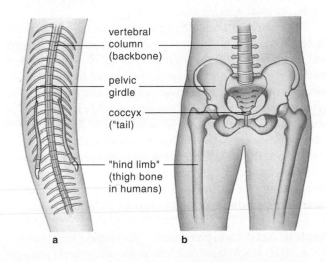

vertebral column (backbone)

pelvic girdle

coccyx ("tail)

"hind limb" (thigh bone in humans)

a   b

Figure 22.8 Bony parts of a python (**a**) correspond to the pelvic girdle of other vertebrates, including humans (**b**). In the snake, small vestigial "hind limbs" protrude through the skin on the underside of the body—remnants from a limbed ancestor. The human coccyx is a similar vestige remaining from an ancestral species that had a bony tail, although certain muscles still attach to it.

had five-toed limbs. The limbs diverged in form and became wings in pterosaurs, birds, and bats (Figure 22.9). All these wings are homologous—they have the same parts. The five-toed limb also evolved into the flippers of porpoises and the anatomy of your own forearms and fingers.

Sometimes body parts in organisms that do *not* have a recent common ancestor come to resemble one another in form and function. These **analogous structures** (from *analogos*, meaning similar to one another) arise when different lineages evolve in the same or similar environments. Different body parts, which were put to similar uses, became modified through natural selection and ended up resembling one another in structure and function. This pattern of change is called **morphological convergence**. Figure 22.10 illustrates a clear case of convergence in sharks, penguins, and porpoises.

Comparative Biochemistry

The genes and gene products (proteins) of different species contain information about evolutionary relationships. Figure 21.10 on page 455 showed how gel electrophoresis of proteins can be used to identify even a single amino acid substitution in the polypeptide chain of a protein. Genes dictate the amino acid sequence, so if the sequence is the same (or nearly so) in comparable proteins from two different species, the genes coding for those proteins point to a shared ancestor. The closer the match, the more recently the two species shared that common ancestor. The more remote the match, the more distantly the two species are related.

For example, organisms ranging from aerobic bacteria to corn plants to humans produce cytochrome *c*, a protein of electron transport chains. From various kinds of studies we know that the gene that codes for the protein has changed very little over huge spans of time. The cytochrome *c* sequence of 104 amino acids is identical in humans and chimps. It is the same except for one amino acid in rhesus monkeys. (On the basis of such data, would you assume that humans are more closely related to a chimpanzee or a rhesus monkey?) By contrast, the cytochrome *c* of chickens differs from that of humans by 18 amino acids, and that of yeasts by 56 amino acids.

Nucleic Acid Comparisons Nucleotide sequences of DNA and RNA are riddled with structural changes caused by mutation. This suggests that the extent to which a DNA or RNA strand from one species will base-pair with a comparable strand from another species is a rough measure of evolutionary distance between them.

Some nucleic acid comparisons are based on **DNA-DNA hybridization**. In this method, the two strands of the double-stranded DNA from two species are induced to unwind from each other (page 448). Then the single

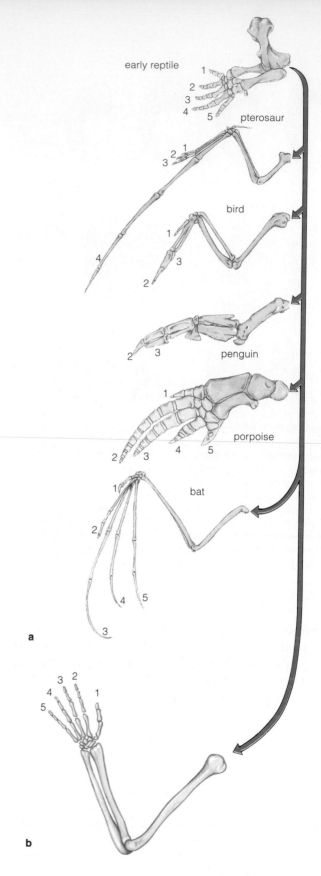

Figure 22.9 (**a**) Divergence in the vertebrate forelimb, starting with the generalized form of ancestral early reptiles. Diverse forms have emerged even while many similarities in the number and position of bones have been preserved. (**b**) Bones of the human forearm and hand.

shark   pectoral fin

penguin   flipper derived from a wing

porpoise   flipper derived from a foreleg

Figure 22.10 Morphological convergence among sharks, penguins, and porpoises. Selection for adaptations that permitted rapid swimming resulted in superficially similar shapes among these three kinds of vertebrates, even though they are only remotely related.

strands are allowed to recombine into "hybrid" molecules. Heating breaks the hydrogen bonds between the two hybridized strands. The heat energy required to pull them apart is a measure of the bonding between the strands and therefore of the similarity between them. Among other discoveries related to DNA-DNA hybridization, we now know that the endangered giant Panda is truly a member of the bear family, whereas the red panda (despite its name) belongs to the raccoon family.

Other methods used to compare DNA (and sometimes RNA) include PCR, gene sequencing, and restriction mapping—using the restriction enzymes described in Chapter 21 to obtain DNA fragments from different species. Gene sequencing and restriction mapping both enable researchers to analyze genetic similarities and differences among species. As a *Focus* box in Chapter 23 will describe more fully, studies of the DNA in mitochondria (mtDNA) are beginning to shed light on the controversy surrounding the geographical origin of the human species.

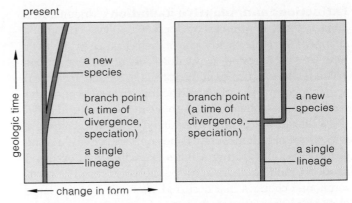

present

← change in form →

a Softly angled branching means speciation occurred through gradual changes in traits over geologic time.

b Horizontal branching means traits changed rapidly around the time of speciation. Vertical continuation of a branch means traits of the new species did not change much thereafter.

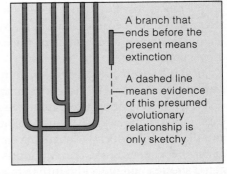

c Many branchings of the same lineage at or near the same point in geologic time means that an adaptive radiation occurred.

A branch that ends before the present means extinction

A dashed line means evidence of this presumed evolutionary relationship is only sketchy

Figure 22.11 How to read evolutionary tree diagrams. The branch points represent times of divergence and speciation as brought about by microevolutionary processes.

Do changes from the original species form proceed slowly or rapidly? Many small changes may accumulate gradually within a lineage over long spans of time. (This is the premise of the *gradual* model of rates of change in an evolving lineage.) Conversely, species may originate through rapid spurts of change over hundreds or thousands of years, then change little for the next 2–6 million years. (This is the premise of the *punctuation* model, which has most changes occurring about the time of speciation.)

Branchings, Extinctions, and Adaptive Radiations

Evolutionary Trees The fossil record, comparative morphology, and comparative biochemistry each provide insight into "who came from whom." Typically, information about this continuity of relationship among species is shown as **evolutionary trees**. In these treelike diagrams, branches represent separate lines of descent from a common ancestor. As Figure 22.11 indicates, each branch point in a tree represents a time of divergence and speciation.

Extinctions and Adaptive Radiations   Inevitably, a lineage loses a number of species as local conditions change. The rather steady rate of their disappearance over time is "background extinction." Regional extinctions occur when the background rate is exceeded in a local area. By contrast, a **mass extinction** is an abrupt, widespread rise in extinction rates above the background level. It is a catastrophic, global event in which major groups are wiped out simultaneously. Dinosaurs and many marine groups died out during a mass extinction that occurred about 65 milllion years ago—possibly the result of environmental changes that occurred after one or more large meteorites struck the earth during a short period of time.

In an **adaptive radiation**, new species of a lineage fill a wide range of habitats during bursts of microevolutionary activity. Many adaptive radiations have occurred during the first few million years after a mass extinction. For instance, fossil evidence reveals that this happened after dinosaurs became extinct. Many new species of mammals arose and radiated into the habitats the dinosaurs vacated.

In the last few hundred years human activities have caused the extinction of thousands of species. The extinction rate is accelerating as we cut down forests, fill in wetlands, and otherwise destroy the habitats of other animals and plants with which we share the earth. *Focus on Environment* considers some current efforts to rescue endangered species, and we will return to this topic in Chapters 24 and 25.

ORGANIZING THE EVIDENCE: CLASSIFICATION

How can we organize the hundreds of thousands of known species of organisms in a meaningful way? For centuries, naturalists (and later, biologists) used morphological similarities (similar body structures) to group organisms. More recently, many biologists prefer to classify life forms using a system based on phylogeny. **Phylogeny** is the evolutionary history of a group of organisms, starting with the most ancestral forms and including the branches leading to all of their descendants.

Classification requires the identification of each new species, which is assigned a unique name. (This work is called *taxonomy*.) In a *binomal system* devised centuries ago by Carolus Linnaeus, a two-part name is used. As Chapter 1 indicated, the first part is the genus name (plural: *genera*). All species that are similar to one another and distinct from others in certain traits are grouped in the same genus. The second part of the name designates the particular species within the genus.

Focus on Environment

Genes of Endangered Species

Barbara Durrant and many other biologists at zoos, private conservation centers, and wildlife refuges are scrambling to buy time for the world's most *endangered species*. Populations of such species are so small that they are poised at the brink of extinction. Although the chief threat to most endangered species is destruction of their habitat, for some a major problem is the absence of genetic diversity.

Consider cheetahs, which must have gone through a severe *bottleneck*—a type of genetic drift in which some event nearly wipes out a population. Whether it was disease or some other catastrophe that befell cheetahs, it drastically reduced their numbers and resulted in extreme genetic uniformity. These sleek, swift cats are so genetically uniform that a patch of skin from one can be successfully grafted onto another, even when they are not closely related. (In other animal species, grafts rarely take hold, even among littermates.)

Today only 20,000 or so cheetahs exist in the wild and in zoos. In the absence of genetic variation, their populations are extremely susceptible to abnormalities, diseases, and environmental changes. For example, up to 70 percent of a typical male cheetah's sperm may be abnormal, perhaps a result of inbreeding following a bottleneck.

Intense hunting and other human activities have endangered another big cat, the Florida panther. With fewer than 50 individuals, the last remaining population has become highly inbred. Parents may mate with their offspring if no other mates are available. Researchers at the National Zoo retrieve DNA from unfertilized eggs of "road-killed" female panthers. With genetic engineering, the DNA may find use in captive breeding programs.

Current **classification systems** organize species into a series of ever more inclusive groupings, called *taxa*, beginning with the species. Typically these groupings include the following taxa:

kingdom
  phylum (or division, in the case of plants)
    class
      order
        family
          genus
            species

Figure 22.12 Representation of the primordial earth, about 4 billion years ago. Within another 500 million years, diverse types of living cells would be present on the surface.

Table 22.2 shows how these groupings are used in the classification of humans.

Table 22.2	Classification of Humans
Kingdom	Animalia
Phylum	Chordata
Class	Mammalia
Order	Primates
Family	Hominidae
Genus	*Homo*
Species	*sapiens* (only living species of this genus)

EVOLUTION AND EARTH HISTORY

When all of the diverse kinds of evidence for evolution are carefully pieced together, an important fact emerges: *The evolution of life has been linked, from its origin to the present, to the physical and chemical evolution of the earth.*

Early Earth and Its Atmosphere

Billions of years ago, explosions of dying stars ripped through our galaxy and left behind a dense cloud of dust and gas that extended trillions of kilometers in space. The cloud flattened out into a slowly rotating disk; at the extremely dense, hot center of the disk our sun was born. Farther out from the center, planets were taking shape. Radioactive dating with ^{238}uranium indicates that the earth formed 4.6 billion years ago. An early atmosphere developed as gases formed in the earth's molten interior or trapped below the crust were forced to the outside. This atmosphere had very little free oxygen—a condition that favored the origin of life. (Free oxygen disrupts the structure of amino acids, nucleotides, and other biological molecules exposed to it.)

Early on, water vapor must have been released from the breakdown of rocks during volcanic eruptions, but it would have evaporated in the intense heat blanketing the crust (Figure 22.12). In time the crust cooled, and rains

Figure 22.13 One of the oldest known fossils, dated at 3.5 billion years old. It is a string of walled cells unearthed in the Warrawoona rocks of Western Australia.

started stripping mineral salts from the parched rocks. Salt-laden waters collected in depressions in the crust and formed the early seas.

Origin of Life

Figure 22.13 shows fossil cells discovered in the Warrawoona rock formation in Western Australia—a fossil that has been dated at 3.5 billion years old! The first living cells probably emerged between 4 billion and 3.8 billion years ago. Before something as complex as a cell was possible, however, biological molecules must have come about through *chemical evolution*. In fact, researchers have been able to put together a reasonable scenario for the emergence of life.

Synthesis of Biological Molecules Rocks collected from Mars, meteorites, and the earth's moon—which all formed at the same time as the earth, from the same cosmic cloud—contain precursors of biological molecules. Possibly, sunlight, lightning, or heat escaping from the earth's crust supplied enough energy to drive condensation of precursors into more complex organic molecules. In the early 1950s a chemistry graduate student named Stanley Miller took the likely constituents of the early atmosphere—hydrogen, methane, ammonia, and water—and mixed them in a reaction chamber (Figure 22.14). He recirculated the mixture and for several days bombarded it with a spark discharge to simulate lightning. Within a week, many amino acids and other organic compounds had formed.

In other experiments that simulated conditions on the early earth, glucose, ribose, deoxyribose, and other sugars were produced from formaldehyde. Adenine was pro-

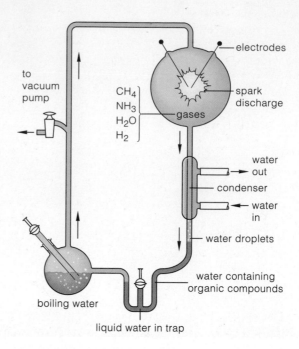

Figure 22.14 Stanley Miller's apparatus used in studying the synthesis of organic compounds under conditions believed to have been present on the early earth. (The condenser cools circulating steam and causes water to condense into droplets.)

duced from hydrogen cyanide. Adenine plus ribose occur in ATP, NAD, and other nucleotides.

Many experiments have focused on how proteins first arose. By one scenario, clay in the muck of tidal flats and estuaries served as templates (structural patterns) for the spontaneous assembly of proteins and other organic compounds. Clay consists of thin, stacked layers of aluminosilicates with metal ions at its surface. Clay and metal ions attract amino acids. Experiments show that clay that is warmed by sunlight and alternately dried out and moistened enhances condensation reactions that yield complex organic compounds.

Proteins that formed on some clay templates may have had the shape and chemical behavior needed to function as weak enzymes in hastening bonds between amino acids. If certain templates promoted such bonds, they would have selective advantage over other templates in the chemical competition for available amino acids. Perhaps selection was at work before the origin of cells, favoring the chemical evolution of enzymes.

Metabolic Pathways A defining characteristic of life is metabolism—the reactions by which cells harness energy and use it to drive their activities, including biosynthesis. During the first 600 million years of earth history, enzymes, ATP, and other molecules could have assembled spontaneously in places where they were in close physical proximity. If so, their close association would have promoted chemical interactions—and the beginning of metabolic pathways.

Self-Replicating Systems During the 300 million years after the first rains began, organic compounds accumulated in the shallow waters of the earth. The first self-replicating systems of interacting molecules must have emerged in this organic "soup."

Simple, self-replicating systems of RNA, enzymes, and coenzymes have been created in the laboratory. Experiments with these systems suggest that RNA might have formed on clay templates. Did RNA later replace clay as information-storing templates for protein synthesis? Perhaps, but existing RNA is too chemically inept to serve this role. And we still don't know how DNA entered the picture. Until the likely chemical ancestors of RNA and DNA are discovered, the story of life's origin will be incomplete.

The First Cells Before cells, there must have been "proto-cells." They may have been little more than membrane-bound sacs, protecting information-storing templates and various metabolic agents from the environment, such as amino acids. Experiments show that such sacs can form spontaneously. In one classic study, Sidney Fox heated amino acids until they formed protein chains, which he placed in hot water. After the chains cooled, they self-assembled into small, stable spheres called *microspheres*. Like cell membranes, the spheres were selectively permeable to different substances. They also picked up lipids from the water, and a lipid-protein film formed at their surface. In other experiments, lipids self-assembled into small, water-filled sacs that had many properties of cell membranes.

In short, there are major gaps in the story of life's origins. However, there is also good experimental evidence that chemical evolution could have given rise to the molecules and structures characteristic of life. Figure 22.15 summarizes some key events in that chemical evolution, which apparently led to the first cells.

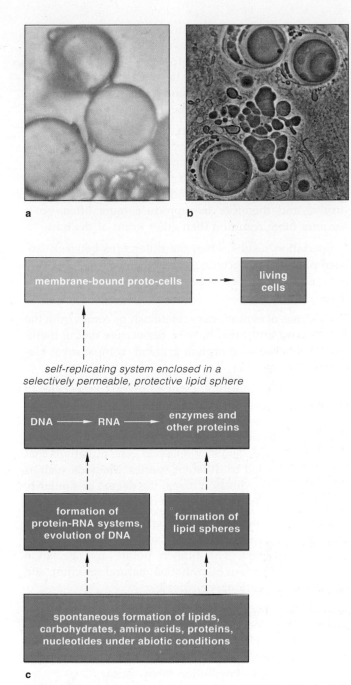

Figure 22.15 Microspheres of (**a**) proteins and (**b**) lipids that self-assembled under abiotic conditions. (**c**) One proposed sequence of events that led to the first self-replicating systems and later to the first living cells.

SUMMARY

1. Evolution proceeds by modifications of already existing species. Hence, there is a continuity of relationship among all species, past and present.

2. In a population, genetic variability gives rise to variations in traits. The frequencies of different genes (alleles) change as a result of four processes of microevolution: mutation, genetic drift, gene flow, and natural selection. The large-scale patterns, trends, and rates of change among groups of species (higher taxa) over time are called macroevolution.

3. Natural selection is a difference in survival and reproduction among members of a population that vary in one or more traits. That is, under prevailing conditions, one form of a trait may be more adaptive. Its bearers tend to survive and therefore to reproduce more often, so it becomes more common than other forms of the trait.

4. Speciation results when the differences between isolated populations become so great that their members are not able to interbreed successfully under natural conditions.

5. Evidence of evolutionary relationships comes from the fossil record and earth history, radioactive dating methods, comparative morphology, and comparative biochemistry.

6. Comparative morphology often reveals similarities in embryonic development that indicate an evolutionary relationship. They may reveal homologous structures, shared as a result of descent from a common ancestor. Comparative biochemistry relies on gene mutations that have accumulated in different species. Methods such as DNA-DNA hybridization reveal the degree of similarity among genes and gene products of different species.

7. Family trees may be constructed to show evolutionary relationships. The branches of such trees are lines of descent (lineages). The branch points are speciation events brought about by mutation, natural selection, and other microevolutionary processes.

8. A mass extinction is a catastrophic event in which major lineages perish abruptly. An adaptive radiation is a burst of evolutionary activity; a lineage rapidly produces many new species. Both events have changed the course of biological evolution many times.

Review Questions

1. What is the difference between microevolution and macroevolution? *464*

2. Explain the difference between:
a. homologous and analogous structures *469–470*
b. divergence and convergence *466*

3. Describe the chemical and physical characteristics of the earth 4 billion years ago. How do we know what it was like? *473–475*

Critical Thinking: You Decide *(Key in Appendix IV)*

1. Humans can inherit various alleles for the liver enzyme ADH (alcohol dehydrogenase), which breaks down ingested alcohol. People of Italian and Jewish descent commonly have a form of ADH that detoxifies alcohol very rapidly. People of northern European descent have forms of ADH that are moderately effective in alcohol breakdown, while people of Asian descent typically have ADH that is less efficient at processing alcohol. Explain why researchers have been able to use this information to help trace the origin of human use of alcoholic beverages.

2. The Atlantic Ocean is widening, and the Pacific and Indian oceans are closing. What are some possible biological consequences of the forthcoming formation of a second Pangaea?

Self-Quiz *(Answers in Appendix III)*

1. Large-scale patterns, trends, and rates of changes in *groups of species* over time are called _____.

2. The rate of evolution has been _____ in some lineages, but in others it has been much more rapid, with _____ of speciation events followed by lengthy periods with little evidence of change.

3. The fossil record of evolution correlates with evidence from _____.
   a. the geologic record
   b. radioactive dating
   c. comparative morphology
   d. comparative biochemistry
   e. all of the above

4. Comparative biochemistry _____.
   a. is based mainly on the fossil record
   b. often reveals similarities in embryonic development stages that indicate evolutionary relationships
   c. is based on mutations that have accumulated in the DNA of different species
   d. compares the proteins and the DNA from different species to reveal relationships
   e. both c and d are correct

5. _____ reveals mutations that have accumulated in the DNA of different species.
   a. Fossil evidence   c. DNA hybridization
   b. Embryonic development   d. Comparative morphology

6. Comparative morphology _____.
   a. is based mainly on the fossil record
   b. often reveals similarities in embryonic development stages that indicate an evolutionary relationship
   c. shows evidence of divergences and convergences in body parts among certain major groups
   d. compares the proteins and the DNA from different species to reveal relationships
   e. both b and c are correct

7. The frequencies of different genes (alleles) change as a result of four processes of microevolution: _____, _____, _____, and _____.

8. A difference in survival and reproduction among members of a population that vary in one or more traits is called _____.

Key Terms

Key Terms

adaptation *465*

adaptive radiation *472*

analogous structures *470*

biogeography *467*

classification system *472*

comparative morphology *468*

DNA-DNA hybridization *470*

evolution *464*

evolutionary tree *471*

fossil *468*

gene flow *466*

genetic drift *465*

homologous structure *469*

macroevolution *464*

mass extinction *472*

microevolution *464*

morphological convergence *470*

natural selection *465*

phylogeny *472*

population *464*

reproductive isolating
    mechanism *466*

speciation *466*

species *466*

Readings

"All the Way with RNA." January 1993. *Discover*. Intriguing non-technical discussion of RNA as the probable enzyme catalyst for earth's first proteins.

Horgan, J. February 1991. "Trends in Evolution: In the Beginning . . ." *Scientific American*.

Udall, J. R. September/October, 1991. "Launching the Natural Ark." *Sierra*. Thoughtful article on ways humans can help save other species from extinction.

23 HUMAN EVOLUTION

The Cave at Lascaux and the Hands of Gargas

Half a century ago, on a warm autumn day, four boys out for a romp stumbled into an intricately tunneled cave near Lascaux in southwest France. What they discovered stunned the world. Magnificent images swept out across the cave walls (Figure 23.1). Between 17,000 and 20,000 years ago, by the flickering light of crude oil lamps, artists captured the graceful, dynamic lines of bison, stags, lions, stallions, ibexes, a rhinoceros, and a heifer now named the Great Black Cow.

About 5,000 years earlier, artists put more than 150 outlines and imprints of hands on walls in the cave of Gargas, in the Pyrenees Mountains. Prehistoric art also exists in other caves of France, Spain, and Africa.

Who were the people who did this? From fossils we know they were anatomically like us. From the way they planned and executed their art, we sense a level of abstract thinking that is unique to humans.

Figure 23.1 Part of the human cultural heritage—prehistoric cave paintings, a unique outcome of a long history of biological evolution.

The quality of "humanness" did not materialize out of thin air. Our story began around 65 million years ago, with the origin of primates in tropical forests. In turn, the primate story began more than 200 million years ago, with the origin of mammals. The first animals emerged between 2.5 billion and 750 million years ago—and so on back in time, to the origin of the first cells.

As we probe the branches of our family tree, keep this greater story in mind. *Our "uniquely" human traits emerged through modification of gene-based traits that had already evolved in ancestral forms.* At each branch in the family tree, mutations led to workable changes in traits that proved useful in prevailing environments. From this perspective, "ancient" cave paintings are the legacy of individuals who departed only yesterday, so to speak. The artists of Lascaux and Gargas are not remote from us. They *are* us.

KEY CONCEPTS

1. Humans are primates, and the fossil record gives evidence of this group's evolutionary history.

2. Trends in primate evolution included modifications of the skeleton that began when four-legged primates were adapting to life in trees. The evolution of humanlike forms that could walk upright, on two legs, built upon those changes.

3. Other trends were modifications in the hands that led to increased dexterity and the ability to manipulate objects, and increased reliance on the sense of vision. Also, the brain became larger and more complex.

4. These trends in primate evolution were the foundation for a key characteristic in the human lineage, termed "plasticity"—the capacity to remain flexible and adapt to a wide range of challenges imposed by unpredictable, complex environments.

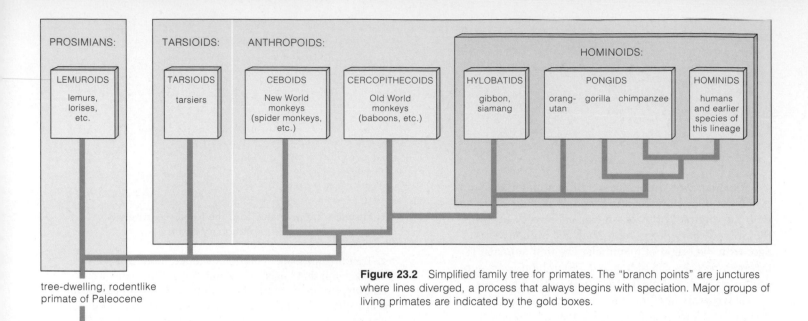

Figure 23.2 Simplified family tree for primates. The "branch points" are junctures where lines diverged, a process that always begins with speciation. Major groups of living primates are indicated by the gold boxes.

THE MAMMALIAN HERITAGE

Humans are members of the class Mammalia. Like other vertebrates, mammals have an internal skeleton with two key features: a nerve cord within a column of bones (backbone) and a skull that houses sense organs and a three-part brain (hindbrain, midbrain, and forebrain).

Mammals typically have hair. And of all vertebrates, only female mammals nourish their young with milk produced by mammary glands. Adult mammals feed and protect their young for an extended period and serve as models for their behavior. The young have an inborn capacity to learn and repeat a set of behaviors that have survival value. Some species even show behavioral flexibility. They can expand on the basics with novel forms of behavior.

Mammals also are known by their **dentition**—the type, number, and size of teeth. As we will see, fossil remnants of teeth and jaws from human ancestors serve as a window on their diet and life-styles.

THE PRIMATES

From Primate to Human: Key Evolutionary Trends

Figure 23.2 shows a family tree for **primates**. This order includes prosimians (including lemurs) and the anthropoids, including monkeys, apes, and humans (Figure 23.3). Apes, humans, and their recent common ancestors

are classified as **hominoids**. Only the members of the human lineage are classified further as **hominids**.

Most primates live in tropical or subtropical forests, woodlands, or savannas (open grasslands with a few stands of trees). Like their ancient ancestors, the vast majority are tree dwellers. Yet no one major feature sets "the primates" apart from other mammals. Each primate lineage evolved in a distinct way and has its own defining traits. Five trends define the lineage we are concerned with here. They were set in motion by selection pressures that operated when primates first started adapting to life in the trees, and they contributed to the emergence of modern humans. The five trends are:

1. Skeletal changes leading to upright walking, which freed the hands for new functions.

2. Changes in bones and muscles leading to refined hand movements.

3. Less reliance on a sense of smell and more on daytime vision.

4. Changes leading to fewer, less specialized teeth.

5. Brain elaboration and changes in the skull that led to speech. These developments became interlocked with each other and with cultural evolution.

Upright Walking Of all primates, only humans can stride freely on two legs for long periods of time. This habitual two-legged gait, called **bipedalism**, emerged through skeletal reorganizations in our primate ancestors.

Compared with monkeys and apes, humans have a shorter, S-shaped, and somewhat flexible backbone. The position and shape of the human backbone, shoulder blades, and pelvic girdle are the basis of bipedalism (Figure 23.4). These skeletal traits may have emerged around the time of the divergence that led to the hominids.

a

b

c

d

Figure 23.3 Representative primates. Gibbons (**a**) have limbs and a body adapted for brachiation (swinging arm over arm through the trees). Monkeys are quadrupedal (four-legged) climbers, leapers, and runners, as the spider monkey in (**b**) demonstrates. Tarsiers (**c**) are vertical clingers and leapers. (**d**) The most familiar primate of all.

monkey   gorilla   human

Figure 23.4 Comparison of the skeletal organization and stance of modern monkeys, apes (the gorilla is shown here), and humans. Modifications of the basic mammalian plan have allowed three distinct modes of locomotion. The quadrupedal monkeys climb and leap, and apes climb and swing by their forelimbs. Both modes of locomotion are well suited for life in the trees. On the ground, apes engage in "knuckle-walking." Humans are habitual two-legged walkers.

Precision Grips and Power Grips The first mammals were four-legged and spread apart the toes on their four feet to help support body weight as they walked or ran. Primates still can spread their toes or fingers. Many also make cupping motions, as when monkeys bring food to the mouth. Among ancestral tree-dwelling primates, modifications in the handbones allowed fingers to be wrapped around objects (*prehensile* movements) and the thumb and tip of each finger to touch each other (*opposable* movements). Hands also began to be freed from load-bearing functions during feeding because forelimbs were only partially used for locomotion. Much later, on the evolutionary road leading to humans, refinements in hand movements led to the precision grip and power grip:

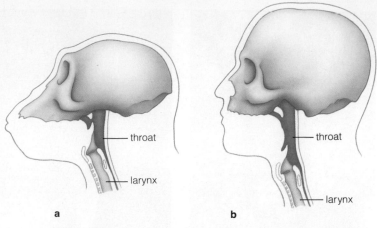

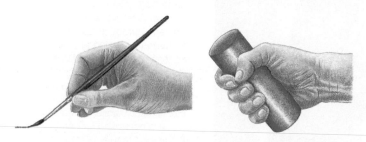

These hand positions gave early humans a tremendous evolutionary advantage: the capacity to make and use tools. Toolmaking, in turn, was the foundation for uniquely human technologies and cultural development.

Enhanced Stereoscopic Vision Many animals, including primates, have stereoscopic vision: They can detect images in three dimensions (including depth perception). Early primates had an eye on each side of the head. Later ones had forward-directed eyes. For animals living in trees, the latter arrangement would be a selective advantage because it is better for sampling shapes and movements in three dimensions. Through other modifications, primate visual systems also became capable of responding to variations in color and in light intensity (dim to bright). Such visual stimuli are also typical of life in the trees.

Changes in Dentition Monkeys and apes have long canines and rectangular jaws. Humans have smaller, squarish teeth of about equal length and a bow-shaped jaw. Teeth and jaws became modified on the road from early primates to humans. The changes were adaptations that accompanied a shift from eating insects, then fruits and leaves, and on to a mixed (omnivorous) diet that included relatively hard foodstuffs.

Brain Expansion and Reorganization Life on tree branches favored shifts in reproductive and social behavior. For example, for parents moving from tree to tree, it

Figure 23.5 The structural basis of speech. (**a**) Early humans had a skull with a flattened base. (The same configuration exists in modern chimpanzees.) Their larynx was not that far below the skull, so the throat volume was small. (**b**) In modern humans, the base of the skull angles down sharply during development, and this moves the larynx down also. Notice the resulting increase in the volume of the throat—the area in which most speech sounds are produced.

would have been much easier to carry, feed, and protect only a single offspring instead of a litter (which would be more suited to a burrow or nest). With fewer young, however, there is selective pressure for an increased parental investment—the nurturing of each offspring in ways that help ensure its survival. In many primate lineages there was a trend toward the forming of strong parental bonds with their young, intense maternal care, and a longer learning period during which offspring could become self-sufficient.

Before this, the brain had been increasing in size and complexity. Now, however, brain regions concerned with encoding and processing information (the cerebral cortex) started to expand dramatically. The adaptive benefits of new behaviors stimulated development of those regions, which served as the biological foundation for the evolution of more new behaviors. *Brain modifications and behavioral complexity became highly interlocked.*

The interlocking is most evident in the parallel evolution of human culture and the human brain. Here we define **culture** as the sum total of behavior patterns of a social group, passed between generations by learning and by symbolic behavior, especially language. The capacity for language arose among ancestral humans. And it arose through changes in the skull and expansion of parts of the brain (Figure 23.5).

Primate Origins

Primates evolved from ancestral mammals more than 60 million years ago. The first ones resembled small rodents or tree shrews (Figure 23.3c). Like tree shrews, they may

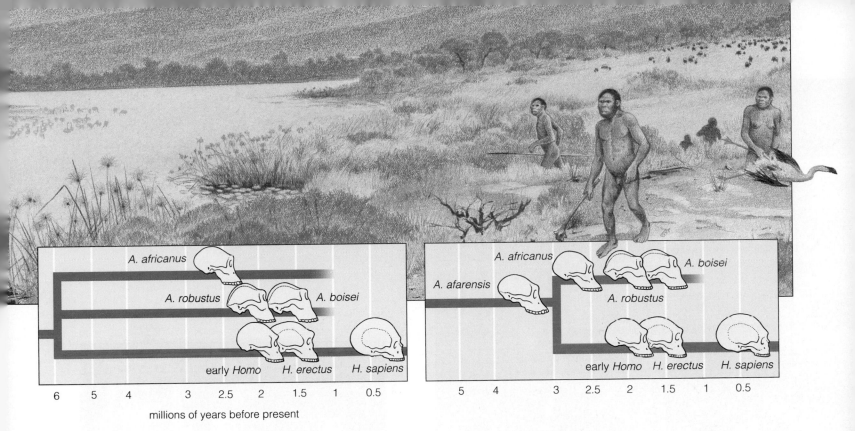

millions of years before present

have had huge appetites and foraged at night for insects, seeds, buds, and eggs near the forest floor. They had a long snout and a good sense of smell, useful for detecting predators or food. They could claw their way up through the shrubbery, although not with much speed.

Between 54 and 38 million years ago (the Eocene), most primates were staying in the trees. Fossils give evidence of increased brain size, a shorter snout, enhanced daytime vision, and refined grasping movements. How did these traits evolve?

Consider the trees. They offered food and safety from ground-dwelling predators. *They also were a place of uncompromising selection.* Imagine dappled sunlight, boughs swaying in the wind, colorful fruit tucked among the leaves, perhaps predatory birds. A long snout would not have been very adaptive—air currents disperse odors. A brain that assessed depth, shape, movement, and color would have been a definite plus. So would a brain that worked fast when its owner was running, swinging, and leaping (especially!) from branch to branch. Distance, body weight, winds, and suitability of the destination had to be estimated, and adjustments for miscalculations had to be quick. By 35 million years ago (the dawn of the Oligocene), tree-dwelling anthropoids had evolved in tropical forests. They included ancestors of monkeys, apes, and humans.

During the Miocene (22.5 million to 5 million years ago), continents began to assume their current positions. Climates were becoming cooler and drier. An adaptive radiation of apelike forms—the first hominoids—occurred

Figure 23.6 Reconstruction of one of the environments in which the first humanlike forms (early hominids) emerged. Their evolutionary connections with one another and with later hominids are not understood. Several phylogenetic trees have been proposed, including the two shown here for australopiths and for species on the road to modern humans.

during this epoch. By 13 million years ago, ape populations were scattered throughout Africa, Europe, and southern Asia. Among them were chimpanzee-sized **dryopiths**. Most Miocene apes became extinct around this time. However, a few fossils, together with molecular comparisons, point to three divergences that occurred between 10 million and 5 million years ago. Two gave rise to the gorilla and chimpanzee lineages. The third gave rise to the early hominids, including the ancestors of humans.

THE HOMINIDS

The first hominids evolved sometime around the Miocene-Pliocene boundary (5–10 million years ago). This was a "bushy" period of evolution—rapid branchings and radiations produced a variety of humanlike forms. We have only a limited number of fossil fragments of each, so we don't really know how they were related. In Africa, the climate was becoming cooler and dryer, and rain forests were giving way to mixed woodlands and grasslands (Figure 23.6).

This changing environment apparently exerted strong selective pressure on early hominids. Although we have only a limited number of fossil fragments and so have lit-

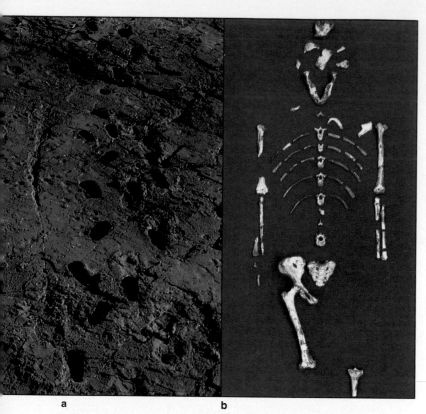

Figure 23.7 (**a**) Footprints made in soft, damp volcanic ash 3.7 million years ago at Laetoli, Tanzania, as discovered by Mary Leakey. The arch, big toe, and heel marks are those of upright hominids. (**b**) Fossil remains of Lucy, one of the earliest known australopiths (*A. afarensis*). Lucy was only 1.1 meters (3 feet 8 inches) tall. The density of her limb bones is indicative of very strong muscles.

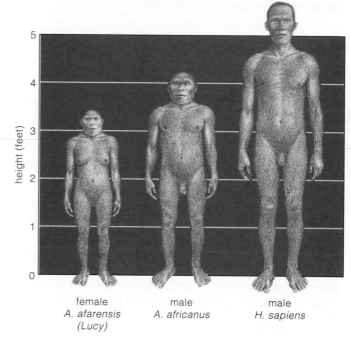

Figure 23.8 Comparison of size and stature of australopiths with modern humans. The drawing of *A. africanus* is the most speculative; until more fossil evidence becomes available for this group, we can make only educated guesses about typical body size and many other features.

Figure 23.9 Comparison of skull shapes of the early hominids relative to modern humans *(Homo sapiens)*. The drawings are not all to the same scale. White areas of skulls are reconstructions.

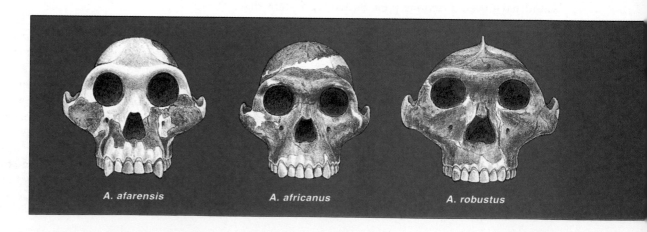

tle firm evidence of relationships between groups, we do know that all had three features in common:

Bipedalism

Teeth adapted for consuming a varied (omnivorous) diet

Further brain expansion and elaboration

So, early hominids had use of hands, they could survive on new kinds of food, and they could *learn* to adapt. These three features gave them **plasticity**—an ability to remain flexible and adapt to a wide range of demands.

All three features emerged through gene-based modifications of traits that can be observed among the other primates. *They were based on the primate heritage.*

Australopiths

The earliest known hominids are called **australopiths** ("southern ape"). They can be grouped into two broad categories:

1. Gracile forms (slightly built), currently named *Australopithecus afarensis* and *A. africanus.*

2. Robust forms (muscular, heavily built), including *A. boisei* and *A. robustus.*

The australopiths were transitional forms. Like large apes, they were small-brained (about 500 cubic centimeters). Like later hominids, they were bipedal. Legbone fragments, 4 million years old, have muscle attachment sites like those of two-legged walkers. Or look at Figures 23.7 and 23.8, which show the reconstructed skeleton and the form of a female dubbed Lucy. Her thighbones angled inward so that her body weight was centered directly beneath the pelvis—a sure sign of bipedalism. Thighbones angle straight downward in apes, which have a waddling, four-legged gait. Most telling, the australopiths left footprints (Figure 23.7a).

Like later hominids, australopiths had slightly bowed jaws. The robust forms apparently ate a mostly vegetarian diet, because their teeth show adaptations for grinding seeds, nuts, and other tough plant material. The teeth of others suggest they were more omnivorous.

Stone Tools and Early *Homo*

By about 2.4 million years ago, hominids had started making stone tools. (Before then, they could have used sticks and other perishable materials as tools, much as chimpanzees do today.) Compared to australopiths, **early *Homo*** species had smaller faces, more generalized teeth, and larger brains with a more developed cerebral cortex (Figure 23.9). This hominid apparently was a scavenger and gatherer of plant material, small animals, and insects. At some point, early *Homo* began using rocks to crack open animal bones and expose the soft, edible marrow. Individuals started using sharp-edged flakes of rock to scrape flesh from animal bones. Eventually, this hominid started to work stone implements into specific shapes.

The earliest known "manufactured" tools were crudely chipped pebbles discovered by Mary Leakey at Olduvai Gorge. This African gorge cuts through a sequence of sedimentary deposits, with the more recently deposited layers containing ever more sophisticated tools. The sequence gives insight into the increasingly refined exploitation of a major food source—game herds of the open grasslands.

Homo erectus

Our direct ancestors emerged between 1.8 million and 400,000 years ago, during the Pleistocene. *Homo erectus* had archaic traits, such as a heavy browridge (Figure 23.9). Yet in many other traits, members of this species were like modern humans. Later on, transitional forms evolved larger brain sizes that could just as easily be grouped with our own species.

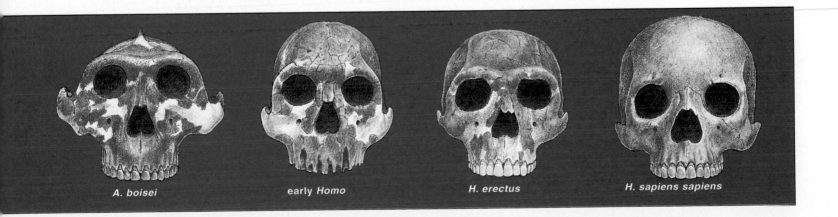

A. boisei   early Homo   H. erectus   H. sapiens sapiens

H. erectus faced huge environmental challenges. Global temperatures were declining. More than once, glaciers advanced and retreated over vast areas of north-central Europe, Asia, and North America. Between glacial episodes, when temperatures rose, *H. erectus* migrated out of Africa and into Southeast Asia, China, and possibly Europe—just about every place that could be reached on foot. Its brain was put to the test during the treks into new environments. Increases in brain complexity were probably interlocked with rapid cultural evolution. Over time, *H. erectus* became better at toolmaking and learned how to control fire.

Homo sapiens

Somewhere between 300,000 and 400,000 years ago, early modern humans *(H. sapiens)* evolved in Europe, the Near East, and China. Compared to their ancestors, early **Homo sapiens** had thinner facial bones, a larger brain, and a rounder, higher skull. About 150,000 years ago, the skull's base began to change in some groups. It gradually angled downward, forcing the larynx down (Figure 23.5). This change permitted complex human language.

There is no scientific consensus about the role of spoken language among the Neandertals. (Their name is sometimes also spelled Neanderthal.) Fossils of this group of early humans were first discovered in Germany's Neander Valley. The group arose 130,000 years ago. Their skull base was flatter than that of *H. erectus*. Although large-brained, they had heavy facial bones and large browridges.

Neandertal populations existed in France, central Europe, Israel, and Iraq. They lived at the edges of forests and in caves, rock shelters, and open-air camps, and they learned to exploit the abundant game herds of the world's vast grasslands.

Neandertals disappeared about 35,000 or 40,000 years ago. That is about the time that they came into contact with anatomically modern humans, who arose in Africa approximately 100,000 years ago. These modern groups they learned more diverse and specialized ways of obtaining food. In northern regions, for example, they learned to anticipate the seasonal migrations of grazing animals in the grasslands and to plan community hunts. They developed rich cultures and were the creators of exquisite tools and artistic treasures at Lascaux and elsewhere.

By one theory, *H. sapiens* evolved from different hominid stocks in each major geographic region. By another theory, *H. sapiens* evolved in Africa from a single ancestral stock, then radiated into different parts of the world over the past 100,000 years (see *Focus on Science*). If so, did they displace native hominid populations entirely, or were there instances of gene flow? At this writing we simply don't have the answers. We can only look forward to new discoveries that may help clarify the picture.

Focus on Science

Where Is Humanity's Homeland?

Exactly where did the first modern humans arise? The search for an answer to this question has sparked a heated controversy. Everyone agrees that anatomically modern humans arose from *Homo erectus*. But one view holds that this happened only in Africa, while *H. erectus* populations in other regions were replaced when the "new" group, early *Homo sapiens*, arrived. Another view holds that the origin occurred in Asia. A third,"multiregional" hypothesis proposes that *H. erectus* migrated out of Africa to Asia and Europe and then gave rise to anatomically modern humans in each of these regions.

The argument bears on present-day human diversity. Modern humans are all very much alike. Regardless of superficial differences, each of us shares more than 99 percent of our genes with everyone else on earth. However, the populations that have traditionally inhabited "Old World" geographical regions (mainly Africa, Asia, and Europe) do tend to differ with respect to a few physical traits, especially the shapes of the skull and the incisors (front teeth in the upper and lower jaw). For example, in a typically Asian skull the face is somewhat flattened, while in an African or European skull the facial bones project more.

The African and Asian origin models both hold that such differences are simply variations that arose in isolated early populations of *Homo sapiens*. Each cites evidence from the fossil record, and also from a more high-tech source: analysis of mitochondrial DNA. Mutations occur in mtDNA 10 times more frequently than they do in nuclear DNA. Using gene-sequencing methods, it is possible to closely compare mtDNA samples and the mutations they contain. Many researchers believe this information sheds light on the timing of evolutionary branchings. Some scientists believe that mtDNA analysis points clearly to an African origin for *Homo sapiens*. At least one laboratory interprets the data in favor of Asia as the single point of origin.

In the multiregional hypothesis, the slight anatomical differences among modern human racial groups could reflect the evolution of *Homo sapiens* from different groups of *H. erectus* in places outside Africa. As evidence, some scholars point to the fact that fossil skull fragments that seem intermediate between *H. erectus* and *H. sapiens* have turned up in China, Indonesia, and southeastern Europe—and they share features with modern human skulls from the same areas. For the moment, there are problems with these finds, however. Few researchers have been able to inspect them, and the dates of the discovery sites have not yet been firmly established. Until we have more definitive evidence for one view or another, the controversy seems sure to go on.

From 40,000 years ago to the present, human evolution has been almost entirely cultural rather than biological, and so we leave the story. From the biological perspective, however, we can make these concluding remarks: Humans spread throughout the world by rapidly devising the cultural means to deal with a broad range of environments. Compared with their predecessors, modern humans developed rich and varied cultures, moving from "stone-age" technology to the age of "high tech."

SUMMARY

1. Primates include prosimians (lemurs and related forms) and anthropoids (including monkeys, apes, and humans). Apes and humans are hominoids; only the early and modern human forms are further classified as hominids.

2. The following evolutionary trends for primates are related largely to their tree-dwelling ancestry:

   a. From a four-legged gait to bipedalism in the hominids (habitual free-striding, two-legged gait). This trend involved changes in the shoulders, backbone, pelvic girdle, legs, and feet.

   b. Increased manipulative skills owing to modification of the hands, which began to be freed from their locomotion function among tree-dwelling primates.

   c. Less reliance on the sense of smell and more reliance on enhanced daytime vision, including color vision and depth perception.

   d. From specialized to omnivorous eating habits.

   e. Brain expansion and reorganization. Larger, more complex brains are correlated with increasingly sophisticated technology (from simple to refined tools), and with social development in the human lineage. All of these trends were the foundation for the remarkable *plasticity* of the hominids (their capacity to remain flexible and to adapt to a wide range of demands imposed by a complex environment).

3. The first hominids (australopiths) emerged between 10 and 5 million years ago. All were bipedal, with a larger brain than their predecessors. Some were omnivores; others were largely vegetarians.

4. Fossils of the first known representative of the human genus date from 2.5 million years ago. Early *Homo* was omnivorous, larger brained and taller than its predecessors, and used simple tools.

5. *Homo erectus* fossils associated with abundant cultural artifacts date from about 1.8 million to 400,000 years ago. This form was adapted to a wide range of habitats. Fully human forms (*H. sapiens*) emerged between 300,000 and 400,000 years ago, possibly from *H. erectus* stock. By 40,000 years ago, modern forms had evolved. From that point on, cultural evolution has outstripped biological evolution of the human form.

Review Questions

1. What are the general evolutionary trends that occurred among the primates as a group? What way of life apparently was the foundation for these trends? *480*

2. What is the difference between "hominoid" and "hominid"? Are we hominoids, hominids, or both? *480*

3. Describe the key characteristics of hominid evolution. How do they relate to the concept of plasticity? *485–486*

Critical Thinking: You Decide *(Key in Appendix IV)*

1. In 1992 the frozen body of a Stone Age man was discovered in the Austrian Alps. Although this "Iceman" died about 5,300 years ago, his body is amazingly intact. Indeed, researchers have begun to analyze DNA extracted from bits of his tissue. Can these studies tell us something about early human evolution? Explain your reasoning.

Self-Quiz *(Answers in Appendix III)*

1. Primates include _____.
   a. lemurs   c. apes   e. all of the above
   b. monkeys   d. humans

2. _____ are hominoids; only _____ are hominids.
   a. Lemurs; monkeys and their immediate ancestors
   b. Apes and humans; humans and their recent ancestors
   c. Monkeys; apes and their recent ancestors
   d. Monkeys, apes, and humans; apes and their recent ancestors

3. The key trends in primate evolution began in the _____.
   a. savanna   c. trees
   b. water   d. forest floor

4. Which of the following did *not* occur on the evolutionary road leading to humans?
   a. bipedalism
   b. hand modification that increased manipulative skills
   c. shift from omnivorous to specialized eating habits
   d. less reliance on smell, more on vision
   e. brain expansion and reorganization
   f. all were evolutionary trends

5. Early hominids displayed great plasticity. This means that _____.
   a. they were adapted to a wide range of demands in complex environments
   b. they had flexible bones that cracked easily
   c. they were limber enough to swing through the trees
   d. they were adapted for a narrow range of demands in complex environments

6. The oldest known primates date from _____.
   a. 25 million to 13 million years ago
   b. 65 million to 54 million years ago
   c. 38 million to 25 million years ago
   d. 13 million to 2 million years ago

7. The first known hominids are generally classified as _____.
   a. *Homo*   c. cercopiths
   b. dryopiths   d. australopiths

8. The first known hominids included types that were _____.
   a. bipedal and with larger brains than their predecessors
   b. omnivores
   c. vegetarians
   d. all of the above

9. Fossils of early *Homo*, the first known representative of the human line, date from _____ million years ago.
   a. 8   c. 4
   b. 6   d. 2.5

10. Match the following early primates with their descriptions.
   _____ australopiths   a. first known representatives of the
   _____ *Homo*   human line
   _____ anthropoids   b. only humans and their recent ancestors
   _____ *Homo sapiens*   c. fully modern humans
   _____ hominids   d. the first hominids
                          e. monkeys, apes, humans

Key Terms

australopith 485	hominid 480
bipedalism 486	hominoid 480
culture 482	*Homo erectus* 485
dentition 480	*Homo sapiens* 486
dryopith 483	plasticity 485
early *Homo* 485	primate 480

Readings

Milton, K. August 1993. "Diet and Primate Evolution." *Scientific American*.

"Reading the Bones for Modern Human Origins." August 14, 1992. *Science*. Clear summary of some current controversies and fascinating new evidence in the search for human origins.

"Shaking the Tree." December 1992. *Scientific American*. How DNA analysis may help pinpoint the locale—or locales—where modern humans arose.

24 ECOSYSTEMS

Lessons from Antarctica

In 1961, an international treaty set aside Antarctica as a reserve for scientific research. In short order, that pristine continent began to be trashed. At first only a few oil drums and discarded tires piled up. What harm could that do in such a desolate place? Then scientific "villages" expanded. Researchers tossed junked equipment into the nearshore waters, which simultaneously were being used as dumps for raw sewage, chemical wastes, and other garbage.

Antarctica was also becoming a new destination of cruise ships. Thousands of pampered tourists began to arrive, ferried from ship to land across spectacular channels glistening with pancake ice (Figure 24.1). Cameras clicking, they disturbed penguin colonies and trampled fragile vegetation. In Antarctica's climate, their footprints will take decades to erase.

Antarctica may have rich deposits of oil, gold, uranium, and other valuable minerals. By 1988 the treaty nations were poised to begin exploring for—then extracting—those substances. Nations concerned about feeding their human populations also began to eye Antarctica's krill, tiny shrimplike creatures that are a major food source for baleen whales and other members of the ecosystem. Those animals, in turn, are food for others.

Figure 24.1 Nearing the shore of an Antarctic island, tourists get close—maybe too close—to nature. Humans have many impacts on ecosystems, many of which we are only beginning to understand.

Because Antarctica seems so remote from the rest of the world, perhaps it is easier to grasp how harm to one species or one part of the environment there might lead to collapse of the whole. In 1991, in the face of mounting concerns (and publicity) about serious damage to Antarctica's fragile ecosystem, the treaty nations imposed a 50-year ban on mineral exploration. Research stations began to properly dispose of wastes and to treat their sewage. Tour operators began to supervise their clients more strictly. In spite of these gains, huge Russian trawlers now harvest Antarctic krill; the long-term consequences of this activity remain to be seen.

Beyond Antarctica, are there profound interconnections between organisms and the physical environment in other ecosystems on earth? Of course, the answer is yes. The nature of such relationships is our main focus in this chapter.

KEY CONCEPTS

1. An ecosystem is a complex of organisms and their physical environment, linked by the flow of energy and the cycling of materials. It is an open system, with inputs and outputs of both energy and nutrients. In nearly all cases, energy enters an ecosystem from the sun.

2. There is a one-way flow of energy through an ecosystem. Producer organisms directly or indirectly nourish consumers, decomposers, and detritivores, which interact as part of food webs.

3. Water and nutrients move from the physical environment, through organisms, then back to the environment. These so-called biogeochemical cycles occur on a global scale.

4. Human activities are disrupting the cycles and so are reducing the stability of ecosystems.

a

b

Figure 24.2 Major types of ecosystems on earth are called "biomes." They include land biomes such as the warm desert near Tucson, Arizona, shown in (**a**), and aquatic realms such as the mountain lake in (**b**). This particular lake is in the Canadian Rockies.

INTRODUCTION TO ECOLOGY

The living world encompasses what biologists call the **biosphere**—those regions of the earth's crust, waters, and atmosphere in which organisms live. The earth's surface is remarkably diverse. In climate, soils, vegetation, and animal life, its deserts differ from hardwood forests, which differ from tropical rain forests, prairies, and arctic tundra. Oceans, lakes, and rivers differ in their physical properties and arrays of organisms (Figure 24.2).

Ecology is the study of the interactions of organisms with one another and with the physical environment. The general type of place in which a species normally lives is its **habitat**. For example, muskrats live in a streambank habitat, damselfish in a coral reef habitat. The habitat of any organism characteristically has certain physical and chemical features, as well as certain other species that typically live in the same place. Humans live in "disturbed" habitats, which have been deliberately altered for agriculture, urban development, and other purposes. In any given habitat the populations of all species directly or indirectly associate with one another as a **community**.

Figure 24.3 Some of the producers, consumers, and decomposers of the arctic tundra: sedges, mosses, and other plants, along with the lemming (eater of plant parts); the snowy owl (eater of lemmings); and a fungal decomposer. As in all ecosystems, these organisms interact with one another and with the physical environment through a one-way flow of energy and a cycling of materials.

You may have heard someone speak of an organism's ecological niche (pronounced nitch). The **niche** of a species is a "package deal." It consists of all the physical, chemical, and biological conditions a species needs to live and reproduce in an ecosystem. Examples of those conditions include the amount of water, oxygen, and other nutrients a species needs, the ranges of temperature it can tolerate, the places it finds food, the predators that feed on it and parasites that attack it, and the competitors that vie for the same limited resources it needs. *Specialist* species have narrow niches. They may be able to use only one or a few types of food or live only in one type of habitat. For example, the red-cockaded woodpecker builds its nest cavity mainly in longleaf pines that must be at least 75 years old. Humans and houseflies are examples of *generalist* species with a broad niche. Both can live in a range of habitats, eat many types of food, and so forth.

An **ecosystem** is a community of organisms interacting with one another *and* with the physical environment through a flow of energy and a cycling of materials. Figure 24.3 shows some of the interacting organisms in one type of ecosystem, arctic tundra.

a

b

c

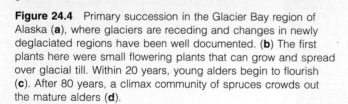

d

Figure 24.4 Primary succession in the Glacier Bay region of Alaska (**a**), where glaciers are receding and changes in newly deglaciated regions have been well documented. (**b**) The first plants here were small flowering plants that can grow and spread over glacial till. Within 20 years, young alders begin to flourish (**c**). After 80 years, a climax community of spruces crowds out the mature alders (**d**).

Communities

Communities of organisms make up the *biotic*, or living, portions of an ecosystem. New communities may arise in habitats initially devoid of life or in disturbed but previously inhabited areas such as abandoned pastures. Immature or disturbed areas are ecologically unstable. Through a process called **succession**, the first species in the habitat thrive, then are replaced by other species, which are replaced by others in orderly progression until the composition of species becomes steady as long as other conditions remain the same. This more or less stable array of species is the "climax community."

In *primary succession*, changes begin when pioneer species colonize a barren habitat, such as a recently deglaciated region (Figure 24.4). In *secondary succession*, a community progresses toward the climax state after parts of the habitat have been disturbed. For example, this pattern occurs in abandoned fields, where wild grasses and other plants quickly take hold when cultivation stops. Natural disasters, such as forest fires triggered by lightning, can also return a once-stable climax community to an earlier successional stage.

With these concepts as background, we can look more closely at how ecosystems are organized and how they function.

ORGANIZATION OF ECOSYSTEMS

Although there are many different types of ecosystems, they are all alike in many aspects of their structure and function. Nearly every ecosystem runs on energy from

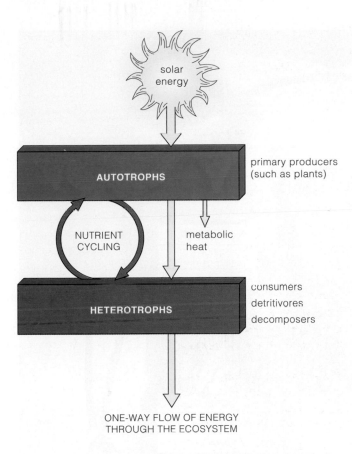

solar
energy

AUTOTROPHS

primary producers
(such as plants)

NUTRIENT
CYCLING

metabolic
heat

HETEROTROPHS

consumers

detritivores

decomposers

ONE-WAY FLOW OF ENERGY
THROUGH THE ECOSYSTEM

Figure 24.5 Generalized model of the one-way flow of energy
and the cycling of materials through ecosystems.

Autotrophs secure nutrients as well as energy for the entire system. As they grow, they take up water and carbon dioxide (as sources of oxygen, carbon, and hydrogen) along with dissolved minerals, including nitrogen and phosphorus. Such materials are building blocks for carbohydrates, lipids, proteins, and nucleic acids. When decomposers and detritivores get their turn at this organic matter, they can break it down completely to inorganic bits. If those bits are not washed away or otherwise removed from the system, autotrophs can use them again as nutrients.

It is important to keep in mind that ecosystems are *open* systems. They are not self-sustaining. Ecosystems require a continual *energy input* (as from the sun) and often *nutrient inputs* (as from minerals carried by erosion into a lake). Ecosystems also have *energy output* and *nutrient outputs*. Energy cannot be recycled; in time, most of the energy originally fixed by autotrophs is lost to the environment in the form of metabolic heat. Nutrients typically are cycled, but some are still lost (as through soil leaching). The balance of this section deals with the inputs, internal transfers, and outputs of ecosystems.

An ecosystem is a complex of producers, consumers, decomposers, and detritivores *and* the physical environment, all interacting through energy flow and materials cycling.

Feeding Relationships in Ecosystems

Each species in an ecosystem fits somewhere in a hierarchy of feeding relationships called **trophic levels** (from *troph*, meaning nourishment). A key aspect of ecosystem functioning is the transfer of energy from one trophic level to another.

Primary producers, which gain energy directly from sunlight, make up the first trophic level. Photosynthetic autotrophs in a lake (including cyanobacteria and aquatic plants) are examples. Snails and other herbivores feeding directly on the producers are at the next trophic level. Birds and other primary carnivores that prey directly on the herbivores form a third level. A hawk that eats a mouse is a secondary carnivore. Decomposers, humans, and many other organisms can obtain energy from more than one source. They cannot be assigned to a single trophic level and are more like "trophic groups."

Food Webs

A straight-line sequence of who eats whom in an ecosystem is sometimes called a **food chain**. However, you will have a hard time finding such a simple, isolated sequence. Typically, the same food resource is part of more than one

the sun. Plants and other photosynthetic organisms are **autotrophs** (self-feeders). They capture sunlight energy and convert it to forms they can use to build organic compounds from simple inorganic substances. By securing energy from the physical environment, autotrophs serve as the **producers** for the entire system (Figure 24.5).

All other organisms in the system are **heterotrophs**, not self-feeders. They depend directly or indirectly on energy stored in the tissues of producers. Some heterotrophs are **consumers**, which feed on the tissues of other organisms. Within this category, *herbivores* such as grazing animals and insects eat plants, *carnivores* such as lions and snakes eat animals, and *parasites* reside in or on living hosts and extract energy from them. Most humans are *omnivores*—they feed on a variety of food types. Other heterotrophs, the **detritivores**, get energy from partly decomposed particles of organic matter. They include crabs, nematodes, and earthworms. Still other heterotrophs, the **decomposers**, include fungi and bacteria that extract energy from the remains or products of organisms.

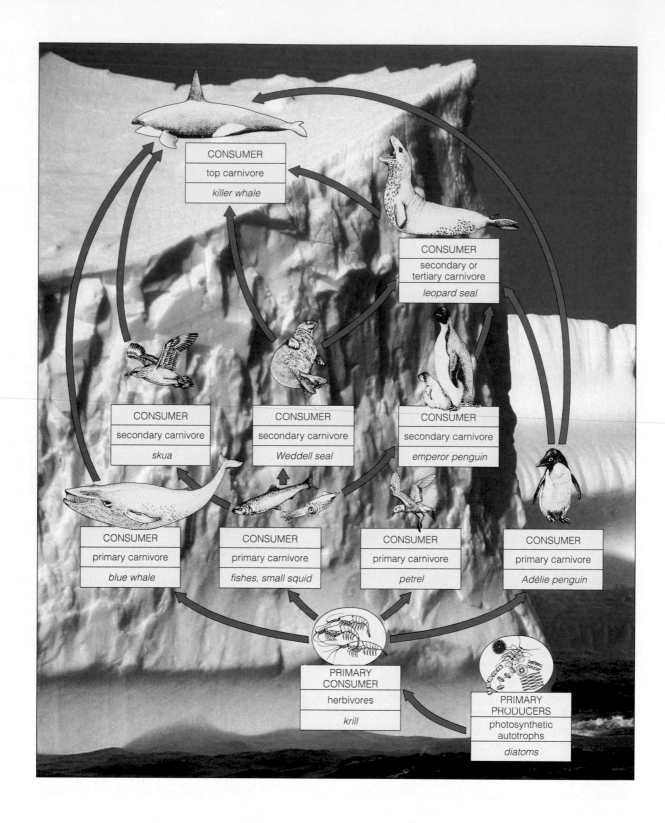

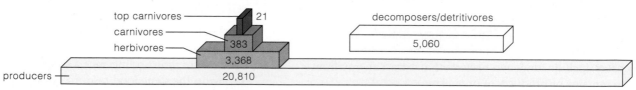

chain. This is especially true of plants and other resources at low trophic levels. It is more accurate to view food chains as cross-connecting with one another as **food webs**. Figure 24.6 shows an Antarctic food web.

A food web is a network of crossing, interlinked food chains encompassing primary producers and an array of consumers and decomposers.

By way of food webs, different species in an ecosystem are profoundly interconnected.

ENERGY FLOW THROUGH ECOSYSTEMS

Primary Productivity

By way of *photosynthesis*, plants trap light energy and convert it into the chemical energy of organic compounds. Thus we and all other consumers owe plants a great debt, for without this fundamental biological process there would be no energy-rich compounds to serve as fuel for our cells—and no energy flow in ecosystems.

To get an idea of how energy flow is studied, consider a land ecosystem in which trees, shrubs, and grasses are the primary producers. The rate at which the ecosystem's producers capture and store a given amount of energy is its **primary productivity**. How much energy actually gets stored depends on (1) how many plants are present and (2) the balance between energy gained from photosynthesis and energy lost by way of aerobic respiration in the plants. This energy "bottom line" is *net primary production*. It can be influenced by various factors, such as the number and size of the producers in the ecosystem, the amount of rainfall, and the range of temperature.

Gross primary productivity is the total rate of photosynthesis for an ecosystem during a specified interval.

Net primary productivity is the rate of energy storage in plant tissues in excess of the rate of aerobic respiration by the plants themselves.

Consumption of producers by heterotrophs affects the rate of energy storage.

Figure 24.6 (*top figure at left*) Simplified picture of a food web in the Antarctic. There are many more participants, including an array of decomposer organisms.

Figure 24.7 (*bottom figure at left*) Pyramid of energy flow during one year at an aquatic ecosystem, Silver Springs, Florida.

Major Pathways of Energy Flow

On land, only a small part of the energy from sunlight becomes fixed in plants. The plants themselves use as much as half of what they store (they lose metabolically generated heat). Other organisms tap into the fixed energy that is conserved in plant tissues, remains, or wastes. They, too, lose heat to the environment.

The heat losses represent a one-way flow of energy out of the ecosystem. In **grazing food webs**, energy flows from plants to herbivores, then through some array of carnivores. In **detrital food webs**, it flows mainly from plants through decomposers and detritivores (Figure 24.7).

In most cases, the greatest portion of net primary production passes through detrital food webs. In a marsh, for instance, most of the stored energy is not even used until plant parts die and become available for detrital food webs. When cattle graze on pasture plants, large amounts of undigested plant parts become available for decomposers and detritivores.

Here are the points to remember about major pathways of energy flow through ecosystems on land:

1. Energy flows into ecosystems from an outside source, which in most cases is the sun.

2. Energy flows through the food webs of ecosystems. Living tissues of photosynthesizers are the basis of grazing food webs. The remains of photosynthesizers and consumers are the basis of detrital food webs.

3. Energy leaves ecosystems mainly by loss of metabolic heat, which each organism generates.

Ecological Pyramids

The trophic structure of an ecosystem is often represented as an "ecological pyramid" in which producers form a base for successive tiers of consumers above them. Some pyramids are based on "biomass," which is the weight of all the members at each trophic level. Here is a pyramid of biomass (grams per square meter) for an aquatic ecosystem at Silver Springs, Florida:

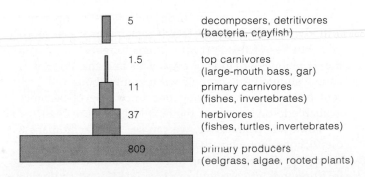

5	decomposers, detritivores (bacteria, crayfish)
1.5	top carnivores (large-mouth bass, gar)
11	primary carnivores (fishes, invertebrates)
37	herbivores (fishes, turtles, invertebrates)
809	primary producers (eelgrass, algae, rooted plants)

A more accurate way to represent an ecosystem's trophic structure is an **energy pyramid.** This kind of pyra-

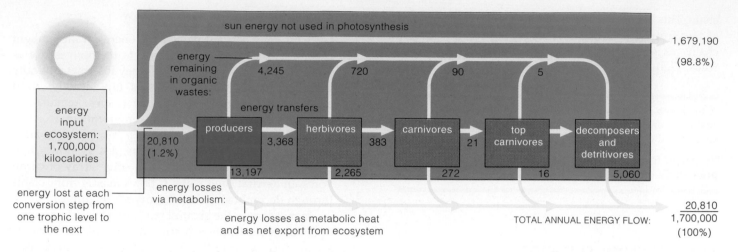

Figure 24.8 Annual energy flow, measured in kilocalories per square meter per year, for Silver Springs, Florida. The producers are mostly aquatic plants. The carnivores are insects and small fishes; top carnivores are larger fishes. The energy source (sunlight) is available all year long.

Photosynthetic autotrophs trap only 1.2 percent of incoming solar energy and use it to assemble new plant biomass. They themselves use more than 63 percent of what they trap. Herbivores get only 16 percent. Most of the stored energy in herbivores is used in metabolism or ends up with decomposers; only 11.4 percent enters carnivores, which burn up most of it. Only 5.5 percent reaches top carnivores. Decomposers recycle all the biomass from other trophic levels. In time, all of the 5,060 kilocalories transferred through the system will appear as metabolically generated heat.

This diagram is oversimplified, for no community is isolated from others. Organisms and materials constantly drop into the springs. Organisms and materials are slowly lost by way of a stream that leaves the springs.

mid is based on the energy lost as organisms at one trophic level become food for the next. As Figure 24.7 shows, energy pyramids always have a large energy base at the bottom. They accurately reflect the ever-diminishing amount of energy flowing through ecosystems.

To construct an energy pyramid, a researcher measures how much energy each type of individual in the system takes in, burns up during metabolism, and stores in body tissues, and how much remains in waste products. Studies of this sort are extremely revealing. Given the metabolic demands of organisms and the amount of energy shunted into organic wastes, only about 6–16 percent of the energy entering one trophic level becomes available to organisms at the next level (Figure 24.8).

From this perspective, we can see that for humans a diet rich in plant foods removes much less energy from an ecosystem than does eating meat. For example, the energy in 10 pounds of corn will produce one pound of human tissue, if you eat the corn. However, if a steer eats the corn and then you eat steaks and hamburger, the same pound

of human tissue "costs" nearly 100 pounds of corn. Some ecologists advocate "eating low on the food chain" because, among other reasons, such a diet can help conserve energy resources that go into producing human food.

The amount of useful energy flowing through consumer trophic levels declines at each energy transfer. It also declines as metabolically generated heat is lost and as food energy becomes shunted into organic wastes.

Transfer of Harmful Compounds Through Ecosystems

Human activities have major and minor effects on the functioning of ecosystems. We will consider some of these effects in Chapter 25, but for now a classic example will underscore the point.

During World War II we began using DDT, the first of the synthetic organic pesticides, to control mosquitoes and many other pests. The method worked, but it entailed unsuspected hazards. DDT is a chlorinated hydrocarbon compound that is almost insoluble in water, and it breaks down very slowly. However, it is soluble in fat and tends to accumulate in fatty tissues. Hence (as we now know), DDT can show **biological magnification**—an increase in concentration of a nondegradable substance as it is passed along food chains. Most of the DDT in the organisms a consumer eats during its lifetime will become concentrated in the consumer's tissues.

After the war, DDT began to infiltrate food webs. In streams flowing through forests where DDT was sprayed to control spruce budworms, salmon started dying. In sprayed croplands DDT killed off not only pests, but beneficial insects as well. In time, the effects of biological magnification started showing up. Most devastated were species at the top of food chains—peregrine falcons,

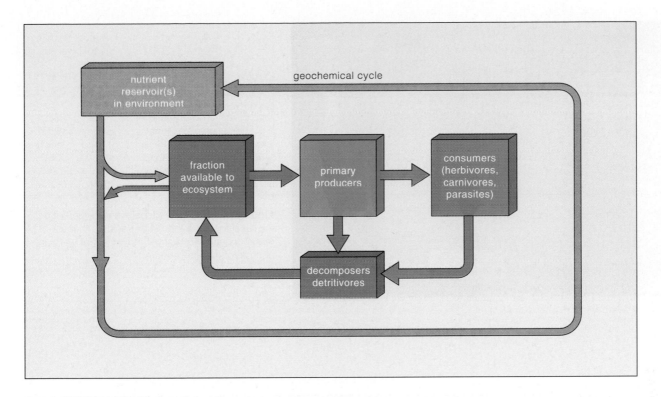

Figure 24.9 Generalized model of nutrient flow through a land ecosystem. The overall movement of nutrients from the physical environment, through organisms, and back to the environment constitutes a biogeochemical cycle.

ospreys, and bald eagles among them. DDT interferes with the deposition of calcium in egg shells, and the birds' eggs were breaking open, killing the developing embryos.

Since the 1970s, most applications of DDT have been banned in the United States. Some endangered bird species such as ospreys are making a comeback. However, imported fruits and vegetables may still carry DDT residues. Health officials recently recommended that a fishery off the coast of Los Angeles be closed because the fish contain DDT from industrial waste discharges that ended 25 years ago. Sampling has shown that many (if not most) people have DDT "stored" in their body fat. It also is detectable in human breast milk.

BIOGEOCHEMICAL CYCLES

The availability of nutrients as well as energy profoundly affects the trophic structure of ecosystems. Besides the carbon, oxygen, and hydrogen plants get from water and air, they require about 13 mineral elements, including nitrogen and phosphorus. With mineral deficiencies, plant growth suffers—and accordingly, so does the primary productivity of the whole ecosystem.

The elements essential for life tend to move in **biogeochemical cycles**. They are transferred from the environment to organisms, then back to the environment. The physical environment serves as a large reservoir through which elements move rather slowly, compared to how rapidly they are exchanged between organisms and the environment. Figure 24.9 shows a model of the relationship between geochemical cycles and most ecosystems.

There are three types of biogeochemical cycles. In the **hydrologic cycle**, oxygen and hydrogen move in the form of water molecules. In **atmospheric cycles**, a large portion of a given element occurs in a gaseous phase in the atmosphere. Carbon and nitrogen cycle in this way. In **sedimentary cycles**, an element does not have a gaseous phase. It moves from land to the seafloor and only "returns" to land through long-term geological uplifting. Phosphorus is such an element; it moves from land to sediments in the seas, then back to the land. The earth's crust is the main storehouse for this and other minerals.

Hydrologic Cycle: Water

Ocean currents, clouds, winds, and rainfall are all part of the global hydrologic cycle. Driven by energy from the sun, the waters of the earth move slowly and on a vast

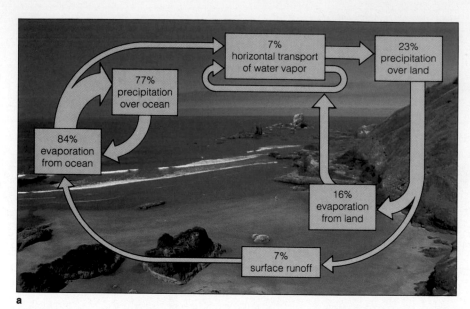

	Annual Volume (10^18 grams)
ocean	1,380,000
sedimentary layers	210,000
evaporation from ocean	319
precipitation over ocean	283
precipitation over land	95
evaporation from land	59
runoff and groundwater	36
atmospheric water vapor	13

b

Figure 24.10 (**a**) The global water cycle, listing the reservoirs and the major transfers that are made annually. (**b**) Annual movement of water into and out of the atmosphere.

Figure 24.11 The global carbon cycle. To the left, the movement of carbon through marine ecosystems; to the right, its movement through terrestrial ecosystems.

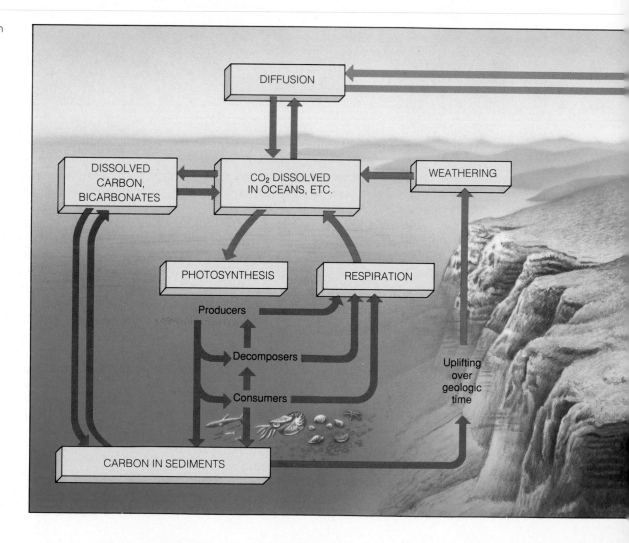

scale through the atmosphere, on or through the uppermost layers of land masses, to the oceans, and back again (Figure 24.10). Water moves into the atmosphere by evaporation from land and water surfaces. Plant leaves also release water molecules (transpiration). Water remains in the atmosphere as vapor, clouds, and ice crystals. It falls back as precipitation—mostly rain or snow.

Water released as rain or snow remains on land for about 10 to 120 days, depending on the season and where it falls. Some evaporates. Rivers and streams carry the rest to the seas. With large-scale evaporation from the seas, the cycle begins again.

A *watershed* is a region where precipitation becomes funneled into a single stream or river. Studies of watersheds have revealed fundamental links among the vegetation in an ecosystem, water movement, and soil nutrients. Water transports nutrients into and out of ecosystems, but plants greatly influence how fast many nutrients move through the system. For example, through their roots trees and other plants can "mine" soil for minerals and other nutrients and then store those essential

substances in their tissues. Roots also tend to hold soils in place. When land is stripped of its vegetation, such as when a forest is clear-cut and the cut-over area burned, both soil and nutrients tend to be washed away, damaging the entire ecosystem.

Carbon Cycle

In the **carbon cycle**, carbon moves from reservoirs in the atmosphere and oceans, through organisms, then back to the reservoirs (Figure 24.11). Carbon enters the atmosphere mainly by way of aerobic respiration, fossil fuel burning, and volcanic eruptions, which release carbon from rocks deep in the earth's crust. Carbon exists as a gas in the atmosphere (mostly as carbon dioxide, CO_2). About half of all the carbon entering the atmosphere each year moves into two large "holding stations"—that is, into accumulated plant biomass and the ocean.

Each year photosynthesizers capture airborne or dissolved carbon dioxide and incorporate billions of metric tons of its carbon atoms into organic compounds. How-

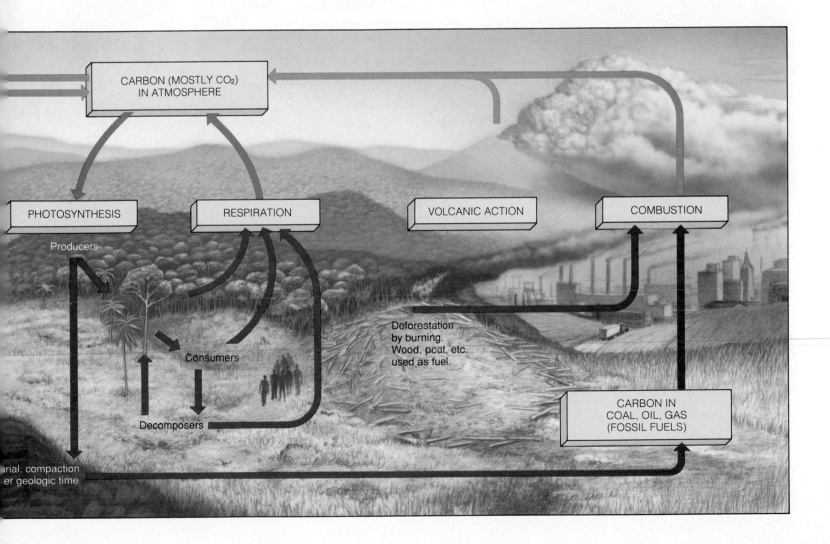

1. Sunlight penetrating the atmosphere warms the earth's surface.

2. The earth's surface radiates heat (infrared wavelengths) to the atmosphere, and some escapes into space. Greenhouse gases and water vapor absorb some infrared wavelengths and reradiate a portion of them toward the earth.

3. When greenhouse gases build up in the atmosphere, more heat is trapped near the earth's surface. Ocean surface temperatures rise, more water vapor enters the atmosphere, and the earth's surface temperature increases.

a

ever, several human activities are disrupting this natural cycling. These actions may be triggering global warming and intensifying what has come to be called the "greenhouse effect."

The Greenhouse Effect and Global Climate
Atmospheric concentrations of carbon dioxide, water, ozone, methane, nitrous oxide, and chlorofluorocarbons (CFCs) profoundly influence the average temperature near the earth's surface, and that temperature influences global climates. (As Chapter 25 describes, CFCs are also destroying the ozone layer.) Collectively, molecules of these gases act somewhat like a pane of glass in a greenhouse—hence the name "greenhouse gases." They allow wavelengths of visible light to reach the earth's surface, but they absorb longer, infrared wavelengths—that is, heat—much of which is reradiated back toward the earth.

Heat builds up in the lower atmosphere, a natural warming called the **greenhouse effect** (Figure 24.12a).

Greenhouse gases help keep the earth from being a cold and lifeless planet. However, largely as a result of human activities, the levels of greenhouse gases other than water vapor have been increasing (Figure 24.12b). Most of the increase is attributed to the burning of fossil fuels. Deforestation adds to it; when wood burns, carbon is released. Today, little remains of the once-great forests of North America and Europe, and vast tracts of tropical forests are being cleared and burned at a rapid rate (see Figure 25.15). Thus the world's plant biomass is shrinking—and so is photosynthetic activity, which withdraws carbon dioxide from the atmosphere.

Without doubt, the greenhouse effect does exist and the concentrations of many greenhouse gases are rising. However, there is some question about how much global

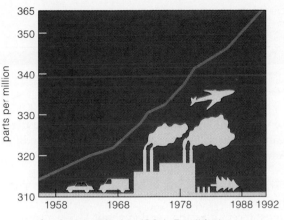

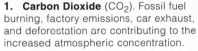

1. Carbon Dioxide (CO_2). Fossil fuel burning, factory emissions, car exhaust, and deforestation are contributing to the increased atmospheric concentration.

b Greenhouse gases.

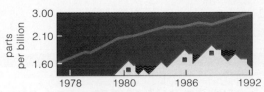

2. Chlorofluorocarbons (CFCs). These are used in plastic foams, air conditioners, refrigerators, and industrial solvents.

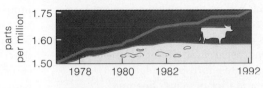

3. Methane (CH_4). This is produced by anaerobic bacteria in swamps, landfills, and termite activities. It is produced also by bacteria in the digestive tract of cattle and other ruminants.

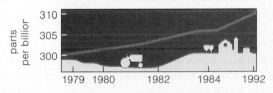

4. Nitrous Oxide (N_2O). This is a natural by-product of denitrifying bacteria. It also is released in great amounts from fertilizers and animal wastes, as in livestock feedlots.

Figure 24.12 (**a**, left) The greenhouse effect. (**b**) Recently documented increases in atmospheric concentrations of four greenhouse gases.

warming is occurring, and how fast. The uncertainty arises because projections are based on computer models of the earth's climate system, which is staggeringly complex and poorly understood.

Before 1993, researchers generally agreed that global climate only changes slowly, over hundreds or thousands of years. Then ice core drilling projects in Greenland provided evidence to the contrary. According to the record of carbon dioxide and other gases trapped in layers of ice that formed over the past 250,000 years, global climate has shifted drastically and often. It did so abruptly at least three times during a warm interglacial period that began about 125,000 years ago. Possibly in less than a decade, average world temperature plummeted 20°F.

The question now is: Will increases or decreases in average world temperature trigger *sudden* climate shifts? For the time being we can only speculate about the answer to this question. *Focus on Environment* describes some likely consequences of sudden global warming.

Nitrogen Cycle

Of all nutrients influencing the growth of land plants, nitrogen—a component of all proteins and nucleic acids—is often the one in shortest supply. For this reason, it is a key *limiting resource* for many organisms. The atmosphere is the largest nitrogen reservoir. About 80 percent of it is composed of gaseous nitrogen (N_2), a form most organ-

isms cannot use. Some bacteria, volcanic action, and lightning can "fix" N_2, converting it into forms that can be used in ecosystems. Nitrogen is lost from ecosystems through metabolic activities of bacteria that release the fixed nitrogen. On land, nitrogen is also lost by way of soil leaching, but leaching adds nitrogen to aquatic ecosystems such as streams, lakes, and oceans.

Today, most of the nitrogen in soils has been put there by nitrogen-fixing organisms. In **nitrogen fixation**, a few kinds of bacteria convert N_2 to ammonia. Certain cyanobacteria are the nitrogen fixers of aquatic ecosystems. *Rhizobium* and *Azotobacter* bacteria are nitrogen fixers in many land ecosystems. Collectively they fix about 200 million metric tons of nitrogen each year.

Once nitrogen is fixed, it becomes available to other organisms in the ecosystem. It moves into the tissues of plants by way of interactions with free-living or symbiotic nitrogen fixers. (A *symbiosis* is a relationship of continuous, intimate contact between interacting species.) For instance, species of *Rhizobium* "infect" the roots of soybeans and other legumes, forming root nodules where the bacteria live and fix nitrogen from the soil. Root nodules on clover in your front lawn can do the same. Ammonium and other nitrogen-containing substances also become available to plant roots when the nitrogen fixers (and other organisms) die and decompose. Plants are the only nitrogen source for animals, which feed directly or indirectly on them.

Focus on Environment

From Greenhouse Gases to a Warmer—or Colder—Planet?

Sophisticated computer models tell us that global warming might well trigger a reorganization of warm and cold ocean currents. Such currents are part of complex interactions that determine the global distribution of heat. If their patterns suddenly changed, climates in tropical regions could switch to frigid cold in a matter of years. Imagine people in San Francisco stomping through deep winter snow, as Alaskans do now. The cold would extend to millions of acres of California's most productive farmland. No matter where you live, your food supply would be affected. Imagine seawater freezing into massive glaciers that rapidly expand through Canada and northern Europe. Among other effects, we could see massive migrations of people away from those regions and into other areas where they would require housing, jobs, health care, and other resources.

Now, imagine that climate conditions again reverse after several decades. As described elsewhere in this chapter, recent geological studies show that rapid climate "flip-flops," from warmer to cooler and back again, have occurred previously in earth's history (Figure *a*). Melting of glacial ice would create vast lakes. Draining of the

immense volume of lake water into the oceans would disrupt ocean currents all over again. The prolonged flooding of the American Midwest during the summer of 1993, described as the "flood of the century," would pale by comparison to the destructive effects on towns, cities, and agriculture.

The most recent glaciation ended about 11,500 years ago. Since then, world temperatures have been rather stable. This stability may have lulled us into complacency about the potential effects on climate of increasing levels of greenhouse gases.

What if, 50 or 60 years from now, all the predictions have turned out to be wrong and human activities have caused no significant global warming trend? We would still have reaped the immense ecological and societal benefits associated with preserving forest ecosystems, developing renewable, nonpolluting energy sources, and other steps that many scientists and policymakers are urging nations to undertake *now*. More to the point, what if, 50 or 60 years from now, today's predictions turn out to have been right—and we did nothing when we had the chance?

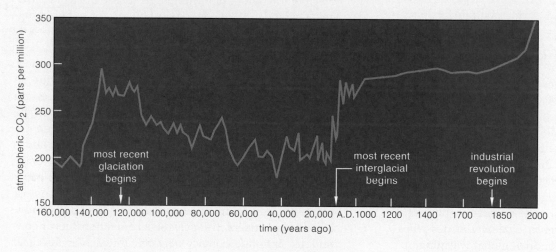

Figure *a* Shifts in atmospheric concentrations of carbon dioxide, correlated with the most recent glaciation and interglacial period during the past 160,000 years.

Nitrogen fixation is only one aspect of the nitrogen cycle illustrated in Figure 24.13. As you can see from the diagram, nitrogen enters and leaves the system by several interconnecting pathways. In *ammonification*, bacteria and fungi break down the nitrogen-containing wastes and remains of plants and animals. The decomposers use the

released amino acids and proteins for growth and give up the excess as ammonia or ammonium, some of which is picked up by plants. In **nitrification**, nitrifying bacteria produce nitrite (NO_2^-) from ammonia or ammonium. Still other nitrifying bacteria use nitrite in metabolism and produce another nitrogen compound, nitrate (NO_3^-),

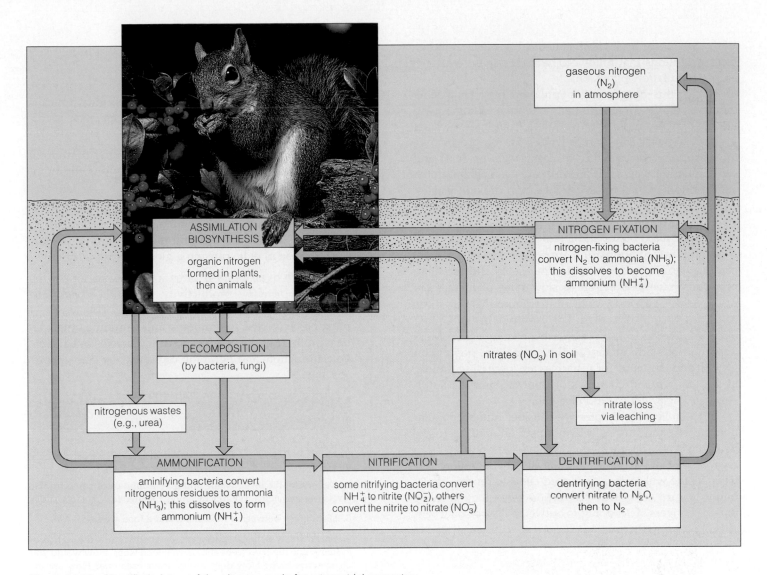

Figure 24.13 Simplified picture of the nitrogen cycle for a terrestrial ecosystem.

which may be utilized by plants or become raw material for **denitrification**. In this process, bacteria convert nitrate or nitrite to N_2 and a small amount of nitrous oxide (N_2O). Some of the N_2 becomes raw material for nitrogen fixation, but some also is lost to the atmosphere.

In agricultural regions a great deal of nitrogen that might otherwise return to the nitrogen cycle departs in harvested plant tissues. Erosion and leaching remove more. European and North American farmers traditionally have rotated crops, as when they alternate wheat with nitrogen-fixing legumes such as clover, alfalfa, or soybeans. Such crop rotation has helped maintain soils in stable and productive condition, sometimes for thousands of years.

Modern agriculture, by contrast, depends on nitrogen-rich fertilizers. With plant breeding, fertilizers, and pest control, the crop yields per acre have doubled and even quadrupled over the past 40 years. Unfortunately, enormous amounts of fossil fuel energy go into fertilizer production. In fact, we still commonly pour more energy into the soil (in the form of fuels, fertilizers, and other chemicals) than we are getting out of it in the form of food. In addition, nitrogen in the runoff from irrigated, fertilized fields has become a major pollutant in waterways (page 519).

Humans and the Biosphere

When the first humans emerged some 2.4 million years ago, their population sizes were small, and their interactions with earth's ecosystems were not significant. About 10,000 years ago, however, agriculture began in earnest, and it laid the foundation for rapid population growth.

With agriculture, and with the medical, social, and industrial revolutions that followed, humans began to experience a population explosion.

Today, our burgeoning population may be placing demands on the biosphere that cannot be sustained. We are taking vast amounts of resources from it, and we are giving back vast amounts of wastes. In the process, we are destroying the stability of ecosystems on land and contaminating water supplies. We are even changing the composition of the atmosphere. Carbon dioxide wastes, which are contributing to the greenhouse effect and possibly global warming, are just one example.

In the United States and a few of the wealthier countries of Europe, population growth is slowing or becoming stabilized. Conservation measures and recycling programs have also begun to slow rates of increase in resource use per individual. But resource use in those countries was already high. For example, in the United States the average person uses as much energy as six Mexicans and 500 Ethiopians. At the same time, population growth and demands for resources are increasing rapidly in the less affluent countries of Central America, South America, Asia, Africa, and elsewhere.

In the following chapter we consider human population growth and our collective impact on earth's ecosystems. It will take decades—and in some cases, centuries—to reverse some of the harmful trends already in motion. Perhaps people will make a concerted effort to reverse global trends toward ecosystem destruction when they perceive that the dangers of *not* doing so outweigh the personal benefits of ignoring them.

SUMMARY

1. An ecosystem is a whole complex of producers, consumers, detritivores, and decomposers *and* their physical environment, all interacting through a flow of energy and a cycling of materials.

2. Ecosystems are open systems, with inputs and outputs of energy as well as nutrients. With few exceptions, photosynthetic autotrophs are the primary producers. They secure energy from sunlight and take up nutrients. These producers then supply the energy-rich compounds used by other members of the systems. Energy moves through ecosystems in a one-way flow. Nutrients such as nitrogen and phosphorus are mostly recycled within the ecosystem. In general, water and carbon move quite freely into and out of ecosystems.

3. Energy fixed by photosynthesis passes through grazing food webs and detrital food webs, which typically are interconnected in the same ecosystem. In both cases,

energy is lost (as heat) through aerobic respiration and other metabolic activities.

4. Biogeochemical cycles include the movement of water, nutrients, and other elements and compounds from the physical environment, to organisms, then back to the environment.

5. Ecosystems on land have predictable rates of nutrient losses that generally increase when the land is cleared or otherwise disturbed.

6. Fossil fuel burning and conversion of natural ecosystems to cropland or grazing land are contributing to increased atmospheric concentrations of carbon dioxide. The increase may be causing a global warming trend.

7. Nitrogen availability is often a limiting factor for the total net primary productivity of land ecosystems. Gaseous nitrogen is abundant in the atmosphere, but it must be converted to ammonia and to nitrates that can be used by primary producers. A few species of bacteria, volcanic action, and lightning can cause the conversion.

Review Questions

1. Define ecosystem. Why do autotrophs play such a central role in an ecosystem? *493, 495*

2. Define trophic level. Name and give examples of some trophic levels in ecosystems. What is the energy source for each level? *495*

3. Distinguish between a food chain and a food web. *495–496*

4. There are two major pathways of energy flow through ecosystems: grazing food webs and detrital food webs. How would you characterize each? How does energy leave each one? How are the two types of food webs interconnected? *497*

5. What is a biogeochemical cycle? Give an example of one and describe the reservoirs and organisms involved. *499–505*

Critical Thinking: You Decide *(Key in Appendix IV)*

1. If you were growing a vegetable garden, what variables might affect its net primary production—that is, the amount of energy stored in the organic compounds of plant tissues?

2. Does your own life-style affect the possible buildup of greenhouse gases? If so, how?

Self-Quiz *(Answers in Appendix III)*

1. An _____ is a complex of organisms and their physical environment, linked by a flow of _____ and a cycling of _____.

2. Energy flow through organisms in nearly all ecosystems begins with the process of _____, and it flows in _____ direction.

3. Ecosystems are open systems that are not self-sustaining; they require inputs of _____ and _____, and they have outputs of _____ and _____.

4. Global movements of water or nutrients between the physical environment and organisms and then back to the environment are _____ cycles.

5. Trophic levels can be described as _____ in an ecosystem.
   a. structured feeding relationships
   b. who eats whom
   c. a hierarchy of energy transfers
   d. any one of the above descriptions is appropriate

6. An apparent global warming trend may be due to increases in greenhouse gases that are being brought about by _____.
   a. fossil fuel burning
   b. gases released from fertilizers and livestock wastes
   c. deforestation, especially of vast tropical forests
   d. CFCs in refrigerators and other machinery
   e. all of the above contribute to greenhouse gases

7. Match the following ecosystem terms.

   _____ primary producers
   _____ consumers
   _____ decomposers
   _____ detritivores
   _____ nitrogen availability
   _____ ecosystem components
   _____ global warming
   _____ ammonium
   _____ biogeochemical cycles

   a. feed on partly decomposed particles of organic matter
   b. break down remains or products of other organisms
   c. composed of herbivores, carnivores, omnivores, parasites
   d. composed of photosynthetic autotrophs
   e. movement of water or nutrients from environment to organisms and back to environment
   f. form of nitrogen suitable for primary producers
   g. limiting factor for total net primary productivity of land ecosystems
   h. producers, consumers, detritivores, decomposers, and physical environment
   i. attributed to increases in greenhouse gases

Readings

Chiles, J. July/August 1992. "High-Tech Conservation." *Nature Conservancy.* How computers and other new technologies have revolutionized the work of ecologists.

Miller, G. T. 1994. *Living in the Environment*, 8th ed. Belmont, Calif.: Wadsworth.

Smith, R. L. 1992. *Elements of Ecology*, 3rd ed. New York: Harper Collins.

25 IMPACTS OF HUMAN POPULATIONS

Nightmare Numbers

Suppose that this year the U.S. Congress passes legislation to control population growth. By law, there can be no more than three children per family. After the third child is born, the father must be sterilized, with or without his consent. *It would never happen here*, you may be thinking. Such an invasion of privacy would never be tolerated in our society. Besides, family size is not much of an issue in North America, where rates of food production and standards of hygiene and medical care are among the world's highest.

Elsewhere in the world, especially where living conditions are already marginal, many populations are growing at alarming rates. Consider India (Figure 25.1). Its population, which already surpasses that of North and South America combined, grows by about 2 percent annually. Most people there do not have adequate

food, shelter, or medical care. Each *week*, 100,000 enter the job market, with little hope for gainful employment. Each *day*, 100 acres of croplands that provide food for the population are removed from agriculture. Why? Too many salts have built up in the intensively irrigated soil; India does not get enough rain to flush them out.

Birth control programs sponsored by the government of India have not worked well. Administering the programs has been difficult, for many people live in remote villages. Information must be conveyed by word of mouth because of widespread illiteracy. Many children die of disease and starvation, so villagers often resist limiting family size. Without a large family, they ask, who would survive and help a father tend fields? Who will go to the cities and earn money to send back home? Who will care for parents when they are too old

Figure 25.1 This crowded street scene in India shows a sampling of the more than 5.5 billion humans on earth. In this chapter we turn to the principles governing the growth and sustainability of populations, including our own.

to work? How can a father otherwise know he will be survived by a son? By Hindu tradition, a son must conduct the last rites so that the soul of his dead father will rest in peace.

In 1976, out of desperation, government officials subjected some men to compulsory vasectomies. Public outrage over the policy contributed to the eventual downfall of Indira Gandhi's government, and the law was rescinded. By the end of 1993 there were more than 5 billion people on earth. Of those, 901 million were living in India.

Is there a way out of such dilemmas? Should the wealthier, less densely populated nations that now use most of the world's resources learn to get by more efficiently, on less? Should they donate surplus food to less fortunate nations? Would donations help, or would

they encourage dependency and further population growth? What would happen if the benefactor nations suffered severe droughts year after year and had trouble meeting their own resource requirements?

Whether we consider humans or any other kind of organism, *certain ecological principles govern the growth and sustainability of populations over time*. This chapter describes these principles, then shows how they apply to the past, present, and future growth of the human population.

KEY CONCEPTS

1. Ecological principles govern the growth and sustainability of all populations over time. For example, all populations face limits to growth, for no environment can indefinitely sustain a continuously increasing number of individuals.

2. Population growth generally follows certain patterns. When a population grows exponentially, it increases in size by ever larger amounts per unit of time. The human population has grown exponentially since the mid-18th century. Today we have the population size, the technology, and the cultural inclination to use energy and modify the environment at astonishing rates.

3. The accumulation of human-generated pollutants is disrupting complex interactions among the atmosphere, oceans, and land in ways that may have the most serious kinds of consequences for human populations in the near future.

POPULATION DYNAMICS

Characteristics of Human Populations

Populations as a whole display certain characteristics, including density, distribution, and age structure. Population *density* is the number of individuals per unit area or volume, such as the number of minnows in each liter of water in a small stream or the number of people living within the city limits of Des Moines. *Distribution* refers to the general pattern in which the population's members are dispersed through its habitat. Humans are social animals, and our populations tend to be clumped in villages, towns, and cities (Figure 25.2), where we interact with one another and have access to jobs and other resources.

The **age structure** of a population is the relative proportion of individuals of each age. These are often divided into prereproductive, reproductive, and postreproductive age categories. The middle category represents the **reproductive base** for the population. Figures 25.3 and 25.4 show age structure diagrams for populations growing at different rates. In these diagrams, ages 15 to 44 are used as the average range of childbearing years (for both men and women). The one for Kenya has a broad reproductive base, for example, and the population of that country is growing rapidly.

Population Size and Patterns of Growth

Over a given time span, the numerical size of a population depends on two factors: how many individuals enter it by birth or immigration, and how many leave it by death or emigration:

$$\begin{pmatrix} \text{population} \\ \text{growth} \\ \text{rate} \end{pmatrix} = \begin{pmatrix} \text{births} \\ + \\ \text{immigration} \end{pmatrix} - \begin{pmatrix} \text{deaths} \\ + \\ \text{emigration} \end{pmatrix}$$

If immigration and emigration are balanced, then population size is stable when the birth rate is balanced over the long term by the death rate. When this balance between births and deaths exists, there is **zero population growth**.

It is worth noting that in reality immigration may be a major contributor to a nation's population growth. In the United States, for example, immigration (legal and otherwise) accounts for more than 30 percent of annual population growth.

Figure 25.2 Roughly 8,300 cities in the world have a population of 25,000 or more. Most humans live in or near such places, like the city shown in this photograph.

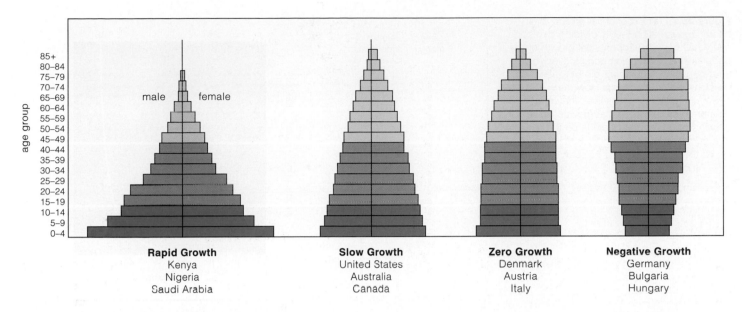

Figure 25.3 Age structure diagrams for several countries with rapid, slow, zero, and negative population growth rates. Dark green indicates the prereproductive base. Purple indicates reproductive years; light blue, the postreproductive years. The portion of the population to the left of the vertical axis in each diagram represents males; the portion to the right represents females.

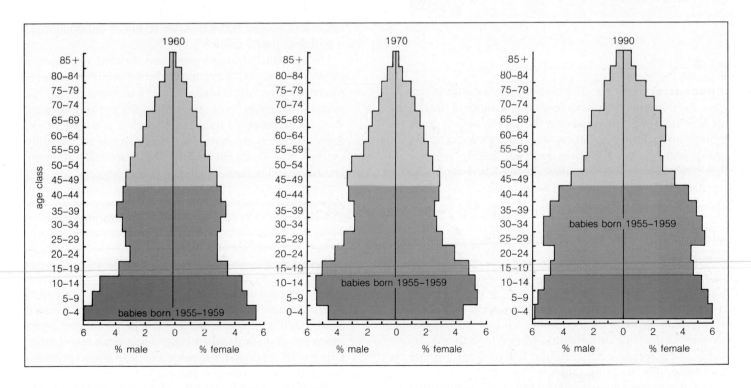

Figure 25.4 Age structure of the U.S. population in 1960, 1970, and 1990. The population bulge of babies who were born between 1955 and 1959 will slowly move up.

Figure 25.5 (**a**) Exponential growth for a bacterial population that is dividing by fission every half hour. (**b**) Exponential growth of the population when division occurs every half hour but when 25 percent die between divisions. Although deaths slow the rate of increase, in themselves they are not enough to stop exponential growth. (**c**) A growing population of bacteria on the surface of a human tooth.

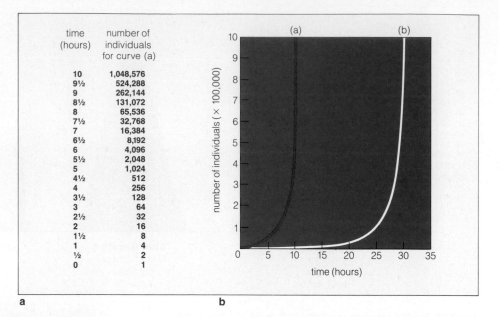

time (hours)	number of individuals for curve (a)
10	1,048,576
9½	524,288
9	262,144
8½	131,072
8	65,536
7½	32,768
7	16,384
6½	8,192
6	4,096
5½	2,048
5	1,024
4½	512
4	256
3½	128
3	64
2½	32
2	16
1½	8
1	4
½	2
0	1

a   b

c

Exponential Growth

Populations can increase in size when the number of births exceeds the number of deaths. But how fast and how much can they increase in a given period? Imagine a population that in a given year has 1,000 members ($N = 1,000$). In each succeeding year, for each 1,000 members, 10 members die but 50 are born. The population's rate of increase (r) would be:

$$r = \frac{\text{births} - \text{deaths}}{N} = \frac{50 - 10}{1,000} = 0.04 \text{ or } 4\% \text{ (here, 40) per year}$$

As long as r holds constant, any population will show **exponential growth**. The number of its individuals increases in *doubling* increments—from 2 to 4, then 8, 16, 32, 64, and so on. For a population with a 4 percent rate of increase, this doubling rate is 17.5 years.

We can observe a limited period of exponential growth in the laboratory by putting a single bacterium in a culture flask with a supply of nutrients. After 30 minutes the bacterium divides, and 30 minutes later the two divide into four. If no cells die between divisions, the population will double every 30 minutes. The larger the population base becomes, the more bacteria there are to divide. After only 9-1/2 hours (19 doublings), the population will exceed 500,000. After 10 hours (20 doublings) it will soar past 1 million!

In a population undergoing unrestricted, exponential growth, a J-shaped curve results when population size is plotted against time (Figure 25.5). Populations of this type increase in size by ever larger amounts per unit of time. When nothing stops its growth, such a population will grow exponentially, even when the birth rate only slightly exceeds the death rate. Only the time scale changes. Doubling still occurs—it simply takes longer.

As long as the birth rate remains even slightly above the death rate, a population will grow. If the rates remain constant, it will grow exponentially.

Biotic Potential The **biotic potential** of a population is its *maximum* rate of increase under ideal conditions. It is the rate that might be achieved when food and living space are abundant, when other organisms aren't interfering with access to those resources, and when no predators or disease agents are present.

Biotic potential is not the same for every species. For many bacteria, it is 100 percent every half hour; for humans and other large mammals, it is between 2 and 5 percent per year. The differences arise through variations in (1) how soon individuals start reproducing, (2) how often reproduction occurs, and (3) how many offspring are born each time.

A population may grow exponentially even when it is not expressing its full biotic potential. It is biologically possible for human females to bear 20 children or more, although few have done so. Yet the human population has been growing exponentially since the mid-18th century. At any given time, the *actual* rate of increase is influenced by environmental circumstances affecting human society.

Limiting Factors and Carrying Capacity Most often, environmental circumstances prevent a population from reaching its full biotic potential. For instance, when an essential resource such as food or water is in short supply, it becomes a **limiting factor** on population growth. Predation (as by disease organisms), competition for living space, and buildup of toxins are other examples of limiting factors. The number of such factors can be enormous, and their effects can vary.

The concept of limiting factors is important because it defines the **carrying capacity**—the number of individuals of a given species that can be sustained indefinitely by the resources in a given area. Some experts believe that earth has the resources to support from 7 to 12 billion humans, with a reasonable standard of living for many. A few say that 20–48 billion humans could inhabit the planet, provided everyone ate a diet only of grains (among other restrictions). Others believe that the current human population of 5.6 billion is already exceeding its carrying capacity through soil erosion, deforestation, pollution of groundwater supplies, and changes in global climate. All these viewpoints share the premise that *overpopulation is the root of many, if not most, of the environmental problems the world now faces.*

A low-density population starts growing slowly and goes through a rapid growth phase, and then growth levels off once the carrying capacity is reached. This pattern is called **logistic growth**. A plot of logistic growth gives us an S-shaped curve (Figure 25.6). This curve is only a simple approximation of what goes on in nature, however. Because environmental conditions vary, carrying capacity also can vary over time.

Carrying capacity is the number of individuals of a species that can be sustained indefinitely by resources in a given area.

Factors that limit human population growth include availability of various types of resources, predation by disease organisms, and effects of pollution.

Because these factors vary in their effects over time, both the carrying capacity and the population size can fluctuate.

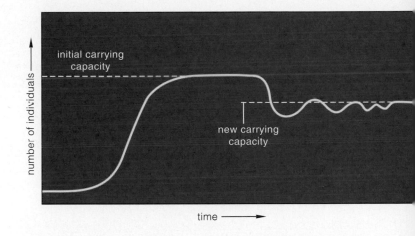

Figure 25.6 Idealized S-shaped curve characteristic of logistic growth. Following a rapid growth phase, growth slows and the curve flattens out as the carrying capacity is reached. Variations can occur in S-shaped growth curves, as when changed environmental conditions bring about a decrease in the carrying capacity. This happened to the human population in Ireland in the late 19th century, when a fungus wiped out the potatoes that were the mainstay of the diet.

Checks on Population Growth

When a growing population's density increases, high density and overcrowding result in competition for resources. They also put individuals at increased risk of being killed by infectious diseases and parasites. These are **density-dependent controls** on population growth. Once density decreases, the pressures ease and the population may grow once more.

A classic example is the *bubonic plague* that killed 25 million Europeans—roughly one-third of the population—during the 14th century. *Yersinia pestis*, the bacterium responsible, normally lives in wild rodents; fleas transmit it to new hosts. It spread like wildfire through the cities of medieval Europe because human dwellings were crowded together, sanitary conditions were poor, and the rats were abundant.

Density-independent controls can also operate. These are events such as floods, earthquakes, or other natural disasters that cause more deaths or fewer births regardless of whether the members of a population are crowded or not.

Life History Patterns

Like all species, humans have a characteristic life span, although few people reach the maximum age possible. Death is more probable at some ages and less so at others. Also, individuals are more likely to reproduce at some ages than at others, and these ages vary from one species to the next. The study of such age-specific patterns is the subject of *demography*.

A **life table** such as the one in Table 25.1 is a summary of the age-specific patterns of birth and death for a given population in a given area. Such tables were originally developed by insurance companies and are used to help set the price of life or health insurance for people of different ages.

HUMAN POPULATION GROWTH

In 1994, the human population totaled over 5.6 billion (Figure 25.7). In that one year almost 100 million more individuals were born—an average of 1.9 million more per week, 273,000 per day. This exponential growth is the consequence of advances in agriculture, industrialization, sanitation, and health care. It took *2 million years* for the human population to reach the first billion. It took only 130 years to reach the second billion, 30 years to reach the third, 15 years to reach the fourth, *and only 12 years to reach the fifth.* By 2000, some 6.2 billion humans will inhabit the earth.

The **demographic transition model** (Figure 25.8) links changes in population growth with changes that unfold during four stages of economic development. In the *preindustrial stage,* there is little population growth because living conditions are harsh and birth and death rates are both high. In the *transitional stage,* industrialization begins, food production rises, and sanitation and health care improve. Death rates drop, but birth rates remain high, so the population grows rapidly over a long period. Then growth starts to level off as living conditions improve. In the *industrial stage*—when industrialization is in full swing—population growth slows, mostly because

Table 25.1 Life Table for the U.S. Human Population, 1989*				
Age Interval (category for individuals between the two ages listed)	Survivorship (number alive at start of age interval, per 100,000 individuals)	Mortality (number dying during the age interval)	Life Expectancy (average lifetime remaining at start of age interval)	Number of Reported Live Births for Total U.S. Population
0–1	100,000	986	75.3	
1–5	99,104	192	75.0	
5–10	98,912	117	71.1	
10–15	98,975	132	66.2	11,486
15–20	98,663	429	61.3	506,503
20–25	98,234	551	56.6	1,077,598
25–30	97,683	606	51.9	1,263,098
30–35	97,077	737	47.2	842,395
35–40	96,340	936	42.5	253,878
40–45	95,404	1,220	37.9	44,401
45–50	94,184	1,766	33.4	1,599
50–55	92,148	2,727	28.9	
55–60	89,691	4,234	24.7	
60–65	85,357	6,211	20.8	
65–70	79,146	8,477	17.2	
70–75	70,669	11,470	13.9	
75–80	59,199	14,598	10.9	
80–85	44,601	17,448	8.3	
85 +	27,153	27,153	6.2	
				Total: 4,040,958

*Compiled by Marion Hansen, based on data from U.S. Bureau of the Census, Statistical Abstract of the United States, 1992 (edition 112).

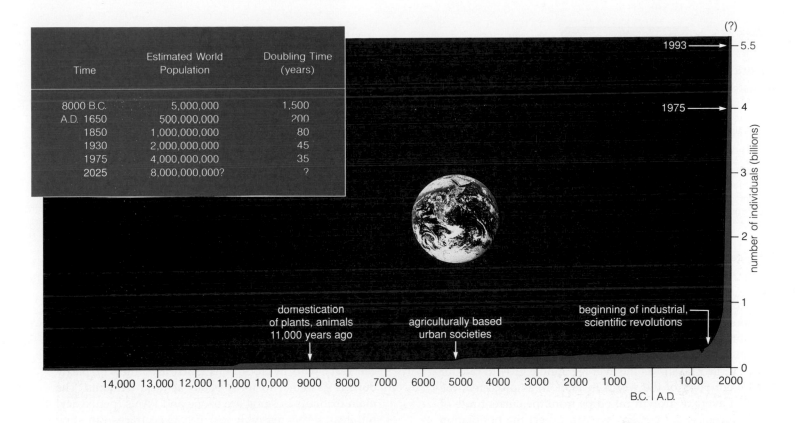

Time	Estimated World Population	Doubling Time (years)
8000 B.C.	5,000,000	1,500
A.D. 1650	500,000,000	200
1850	1,000,000,000	80
1930	2,000,000,000	45
1975	4,000,000,000	35
2025	8,000,000,000?	?

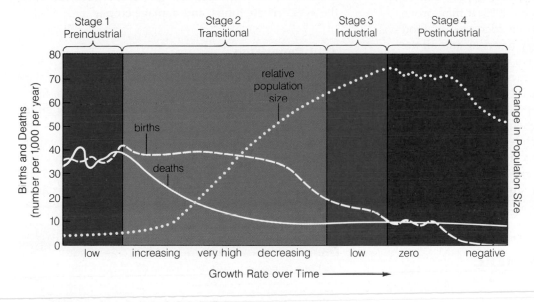

Figure 25.7 The curve of global human population growth. The vertical axis of the graph represents world population, in billions. (The slight dip between the years 1347 and 1351 shows the time when 25 million people died in Europe as a result of bubonic plague.) The growth pattern over the past two centuries has been exponential, sustained by revolutions in agriculture industrialization and improvements in health care.

Figure 25.8 The demographic transition model of changes in population size as correlated with changes in economic development.

urban couples regulate family size. Many decide that raising children is expensive and that having too many puts them at an economic disadvantage. In the *postindustrial stage*, zero population growth is reached. Then the birth rate falls below the death rate, and the population slowly decreases in size.

Today, the United States, Canada, Australia, Japan, countries of the former Soviet Union, New Zealand, and most countries of western Europe are in the industrial

stage, and their growth rate is slowly decreasing. Eighteen countries, including Sweden, the United Kingdom, Germany, and Hungary, are close to, at, or slightly below zero population growth.

Mexico and other developing countries are in the transition stage. They may not stay there, however. In many countries in this stage, population growth exceeds economic growth, and it is difficult if not impossible for such nations to afford the costs of fossil fuels and other

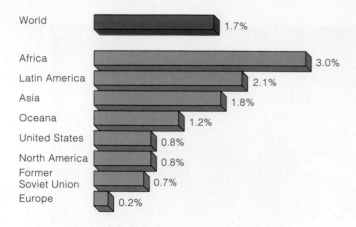

World 1.7%

Africa 3.0%

Latin America 2.1%

Asia 1.8%

Oceana 1.2%

United States 0.8%

North America 0.8%

Former Soviet Union 0.7%

Europe 0.2%

Figure 25.9 Average annual growth rate for different parts of the world in 1990.

resources that drive industrialization. Some transitional countries and their populations may return to the harsh conditions of the preceding stage.

Figure 25.9 shows the average annual growth rate for different parts of the world in 1990. If, as projected, the world average growth rate dips from the current rate of 1.8 percent to 1.7 percent, we can expect the population to double and soar past 11 billion by 2025. It may prove impossible to achieve corresponding increases in food production, drinkable water, energy reserves, and all the wood, steel, and other materials we use to meet everyone's basic needs—something we are not even doing now. As we see in the next section, there is evidence that harmful by-products of our activities—pollutants—are changing the land, seas, and the atmosphere in ominous ways. From what we know of the principles governing population growth, unless there are spectacular technological breakthroughs it is realistic to expect an increase in human death rates. Although our stupendously accelerated growth continues, it cannot be sustained indefinitely.

CHANGES IN THE ATMOSPHERE

Air Pollutants and Smog

Each day earth's atmosphere receives more than 700,000 metric tons of pollutants from the United States alone. **Pollutants** are substances that adversely affect the health, activities, or survival of a population.

Table 25.2 lists the major classes of air pollutants. They include carbon dioxide, sulfur oxides, nitrogen oxides, and chlorofluorocarbons (CFCs). They also include photochemical oxidants formed by interactions between sunlight and many of the chemicals we release into the atmosphere.

Two types of smog (gray air and brown air) can occur in major cities. Where winters are cold and wet, **industrial smog** forms as a gray haze over industrialized cities that burn coal and other fossil fuels for heating, manufacturing, and producing electric power. The burning releases airborne dust, smoke, ashes, soot, asbestos, oil, bits of lead and other heavy metals, and sulfur oxides. Industrial smog was the cause of London's 1952 air pollution disaster in which 4,000 people died. Like London, New York, Pittsburgh, and Chicago once were gray-air cities until they restricted coal burning. Now, most industrial smog forms in cities of developing countries, including China and India, as well as in countries of eastern Europe.

In warm climates where there is relatively little precipitation to cleanse the air, **photochemical smog** forms as a brown, smelly haze over large cities. Mexico City has become a symbol of smog-afflicted cities (Figure 25.10). The main culprit is nitric oxide, produced mostly by cars and other vehicles with internal combustion engines. Nitric oxide reacts with oxygen in the air to form nitrogen dioxide. When exposed to sunlight, nitrogen dioxide can react with hydrocarbons (such as spilled or partly burned gasoline) to form photochemical oxidants. The main oxidants in smog are ozone and PANs (peroxyacyl nitrates). PANs resemble tear gas; even traces can sting the eyes, irritate lungs, and damage crops.

Acid Deposition

Oxides of sulfur and nitrogen are among the most dangerous air pollutants. Coal-burning power plants, factories, and metal smelters are the main sources of sulfur dioxides. Vehicles, power plants that burn fossil fuels, and nitrogen fertilizers are the main sources of nitrogen oxides.

Most sulfur and nitrogen dioxides dissolve in atmospheric water to form a weak solution of sulfuric acid and nitric acid. Winds can distribute them over great distances

Table 25.2 Major Classes of Air Pollutants	
Carbon oxides	Carbon monoxide (CO), carbon dioxide (CO_2)
Sulfur oxides	Sulfer dioxide (SO_2), sulfer trioxide(SO_3)
Nitrogen oxides	Nitric oxide (NO), nitrogen dioxide (NO_2), nitrous oxide (N_2O)
Volatile organic compounds	Methane (CH_4), benzene (C_6H_6), chlorofluorocarbons (CFCs)
Photochemical oxidants	Ozone (O_3), peroxyacyl nitrates (PANs), hydrogen peroxide (H_2O_2)
Suspended particles	Solid particles (dust,soot,asbestos, lead, etc.), liquid droplets (sulfuric acid, oils, dioxins, pesticides)

before they fall to earth in rain and snow. This is *wet* **acid deposition**. Acid rain can be much more acidic than normal rainwater, sometimes becoming as acidic as lemon juice (pH 2.3). The acids chemically attack marble, metals, mortar, plastic, even nylon stockings. They also disrupt the physiology of organisms and the chemistry of ecosystems. (Tiny particles of these substances may be airborne for a while and then fall to earth as *dry* acid deposition.)

The precipitation in much of eastern North America is 30–40 times more acidic than it was several decades ago, and croplands and forests are suffering (Figure 25.11). All fish populations have vanished from more than 200 lakes of the Adirondack Mountains of New York. Acidic pollutants originating in industrial regions of England and Germany are key factors in the destruction of large tracts of forests in northern Europe. Such pollutants also are emerging as a serious problem in heavily industrialized parts of Asia, Latin America, and Africa.

Damage to the Ozone Layer

Ozone is a molecule of three oxygen atoms (O_3). Ozone occurs in two regions of earth's atmosphere, and it is somewhat of a Jekyll-and-Hyde molecule. In the troposphere, the region closest to the earth's surface, ozone is a component of smog. As such it can damage the human respiratory system as well as other animal life and plants. However, ozone in the next atmospheric layer—the stratosphere (17–48 kilometers, or 11–30 miles above the earth)—intercepts harmful ultraviolet radiation that can

cause skin cancer and eye cataracts in humans (page 433), and can severely damage other organisms. This protective screen has been thinning. Each year, from September through mid-October, an ozone "hole" appears over the Antarctic. It extends over an area about the size of the continental United States. Satellites and high-altitude planes have been monitoring the ozone hole for more than

Figure 25.10 Mexico City under its self-generated blanket of photochemical smog on a bright sunny morning.

Figure 25.11 Dead spruce trees in the Whiteface Mountains of New York (**a**) and in Germany (**b**). Acid deposition, perhaps in combination with prolonged exposure to other pollutants, seems to be contributing to the rapid destruction of trees in these and other forests. In some cases, prolonged exposure to multiple air pollutants directly damages the trees, especially conifers. In many cases, the pollutants weaken the trees and make them more susceptible to drought, disease, and insect attacks.

a

b

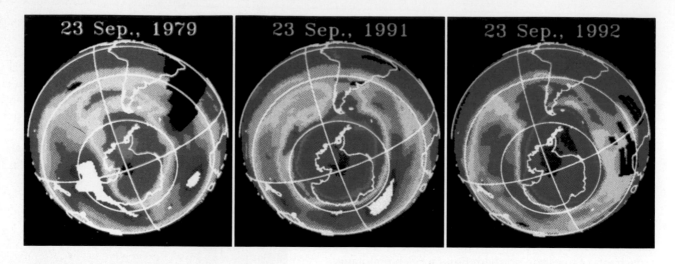

Figure 25.12 Expansion of the seasonal ozone hole over Antarctica, as recorded by special high-altitude planes. The lowest ozone values (the "hole") are indicated by magenta and purple. Droplets of sulfuric acid, released during the 1991 volcanic eruption of Mount Pinatubo, contributed to the thinning.

a decade (Figure 25.12). Since 1987 the ozone layer over the Antarctic has been thinning by about half every year. Since 1969 it has decreased by 3 percent over heavily populated regions of North America, Europe, and Asia. By 2050 the ozone layer over those regions may decrease by 10–25 percent.

As the ozone layer thins, more harmful ultraviolet radiation reaches the earth's surface. Already, skin cancers are increasing dramatically, and cataracts may become more common. Ultraviolet radiation can weaken the immune system, making humans more vulnerable to some viral and parasitic infections. Reducing the ozone layer also may damage phytoplankton, microscopic photosynthetic organisms that are the basis of food webs in freshwater and marine ecosystems. Phytoplankton are also a factor in maintaining the composition of the atmosphere because they act to remove carbon dioxide from surface waters and release oxygen.

Chlorofluorocarbons More than any other factor, chlorofluorocarbons (CFCs) are responsible for ozone depletion. These compounds of chlorine, fluorine, and carbon are widely used as propellants in aerosol sprays, as coolants in refrigerators and air conditioners, and for other industrial and commercial uses. They are also released during volcanic activity.

When a CFC in the stratosphere absorbs ultraviolet light, it gives up a chlorine atom. The chlorine can react with ozone (O_3), yielding oxygen (O_2) and chlorine monoxide. Reactions between oxygen and chlorine monoxide release more chlorine atoms, and each one can effectively destroy as many as 10,000 molecules of ozone!

In 1992, officials from 93 countries participating in a United Nations Environment Program revised an earlier treaty and agreed to a new timetable for phasing out ozone-depleting substances. By this agreement, CFCs and several other substances will be banned after 1996. Phase-out schedules for other offending chemicals were also moved up because of new evidence that in some regions ozone depletion is occurring more rapidly than suspected. At the same time, efforts to develop substitutes for CFCs are being speeded up. The agreement is a step in the right direction, although some fear that it is too little and too late. Up to 95 percent of the CFCs released since 1955 are still making their way up to the stratosphere. You, your children, and your grandchildren will be living with their destructive effects.

CHANGES IN THE HYDROSPHERE

Three of every four humans do not have enough clean water to meet basic needs. Most of earth's water is salty (in oceans). Of every million liters of water on our planet, only six liters are readily usable for human activities. As our population grows exponentially, so do demands and impacts on this limited water supply.

Impact of Large-Scale Irrigation

About a third of the food produced today grows on irrigated land (Figure 25.13). Water is piped into agricultural fields from groundwater or diverted from lakes or rivers.

Irrigation water often contains large amounts of mineral salts, and where soil drainage is poor, evaporation may cause salt buildup or **salinization**. Soil salinization can decrease yields and eventually kill crops. Worldwide, salinization is estimated to have reduced yields on 25 percent of all irrigated cropland. Improperly drained irrigated lands also can become waterlogged, so that soil around plant roots becomes saturated with toxic saline water.

Large-scale irrigation is also a major factor in depletion of groundwater supplies. For instance, U.S. farmers from South Dakota to Texas withdraw so much water from the Ogallala aquifer (Figure 25.14) that the annual overdraft (the amount of water not replenished) is nearly equal to the annual flow of the Colorado River! As a result, the already-low water tables in much of the region are dropping rapidly, and stream and underground spring flows are dwindling. *Subsidence*—the sinking of land when groundwater is withdrawn—is increasing. In coastal areas, overuse of groundwater can cause saltwater intrusion into human water supplies.

Problems with Water Quality

In many regions, agricultural runoff pollutes public water sources with sediments, pesticides, and plant nutrients. Power plants and factories pollute water with chemicals (including many known carcinogens), radioactive materials, and excess heat (thermal pollution). Such pollutants may accumulate in lakes, rivers, and bays before reaching their ultimate destination, the oceans. Contaminants from human activities have even begun to turn up in supposedly "pure" water in underground aquifers.

Many people view the oceans as "out of sight, out of mind" dumps for human refuse. Cities throughout the world dump untreated sewage, garbage containing potentially dangerous medical wastes, and other noxious debris into coastal waters. Cities along rivers and harbors maintain shipping channels by dredging the polluted muck and barging it out to sea. We do not yet know the full impact of such practices on fisheries from which humans obtain food. We do know that researchers have found toxins—including PCBs, industrial pollutants that are carcinogenic to humans—in the tissues of fish caught in the deep ocean.

The United States has facilities for treating the liquid wastes from about 70 percent of the population and 87,000 industries. The remaining wastes, mostly from suburban and rural populations, are treated in lagoons or septic tanks or discharged—untreated—directly into waterways. Studies show that water pollution intensifies as rivers flow toward the oceans and that water treatment often does not remove toxic wastes from industries upstream from a city. Do you know where your own water supply has been?

Figure 25.13 Irrigation-dependent crops growing in the Sahara Desert of Algeria.

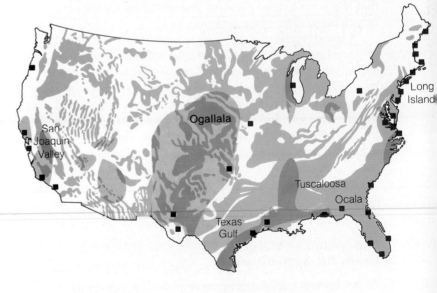

Figure 25.14 Major underground aquifers containing 95 percent of all fresh water in the United States. These aquifers are being depleted in many areas and contaminated elsewhere through pollution and saltwater intrusion. Blue areas indicate major aquifers; gold, the areas of groundwater depletion. The black boxes indicate areas of saltwater intrusion.

CHANGES IN THE LAND

Solid Wastes

Billions of metric tons of solid wastes are dumped, burned, and buried annually in the United States alone. This includes 50 billion nonreturnable cans and bottles. Fully one-half of the total volume of solid wastes are paper products.

Paper is relatively easily recycled. In the Netherlands, 53 percent of wastepaper is recycled, and in Japan the figure is 50 percent. In 1992, almost 40 percent of the wastepaper in the United States was recycled. This figure could be improved with a small amount of effort. Simply recycling the nation's Sunday newspapers would save 500,000 trees per week. Using recycled paper also reduces air pollution that results from paper manufacturing by 95 percent, lowers water pollution by 35 percent, and takes 30–50 percent less energy than making new paper.

Instead of recycling solid wastes, as occurs in natural ecosystems, we bury them in landfills or burn them in incinerators. Incinerators can add heavy metals and other toxic pollutants to the air and leave a highly toxic ash that must be disposed of safely. In any event, land available and acceptable for landfills is scarce and becoming scarcer. Because all landfills eventually "leak," they pose a threat to groundwater supplies. That is one reason why communities increasingly take the "NIMBY" approach to landfills—Not In My Back Yard. Although more and more people now participate in recycling programs, it seems that few want to take personal responsibility for the non-recyclable garbage they generate.

A transition from a throwaway mentality to one based on recycling and reuse is affordable, technically feasible, and environmentally essential. Consumers can play a role by refusing to buy goods that are lavishly wrapped and boxed, packaged in indestructible containers, or designed for one-time use. Individuals can ask the local post office to turn off their daily flow of junk mail. They can also urge local governments to develop resource recovery centers by which existing dumps and landfills become urban "mines" from which usable materials can be sorted out and recovered. Several such centers now operate successfully in different parts of the United States.

Conversion of Marginal Lands for Agriculture

Today, human population growth is forcing expansion of agriculture onto marginally productive land. Almost 21 percent of the earth's land is now being used for agriculture. Another 28 percent may be suitable for cropland or grazing land, but its potential productivity is so low that conversion may not be worth the cost.

Valiant efforts have been made to improve crop production on existing land. Under the banner of the **green revolution**, research has been directed toward improving the varieties of crop plants for higher yields and exporting modern agricultural practices and equipment to developing countries. Unfortunately, the green revolution involves intensive, mechanized agriculture. It is based on massive inputs of fertilizers and pesticides and ample irrigation to sustain high-yield crops. It is based also on fossil fuel energy to drive farm machines. Crop yields *are* four times as high as from traditional methods. But the modern practices use up a hundred times more energy and mineral resources, which makes them and the food they produce too expensive for both farmers and consumers in developing countries.

Overgrazing of livestock on marginal lands is a prime cause of **desertification**—the conversion of grasslands, rain-fed cropland, or irrigated cropland to a less productive, desertlike state. It occurs as vegetation is removed, leaving few or no plant roots to prevent wind and water from eroding the soil away. The subsoil then bakes rock-hard, and plants can no longer grow there. Worldwide, about 9 million square kilometers have become desertified over the past 50 years, and the trend is continuing.

Deforestation

Pressures for increased food production are most intense in areas of Central and South America, Asia, the Philippines, the Middle East, and Africa. There, human populations are rapidly expanding into marginal lands, often by cutting down forests. **Deforestation**, the removal of all trees from large tracts of land, leads to loss of fragile surface layers of soil and disrupts watersheds.

In the tropics, deforestation has become a critical problem. Through **shifting cultivation** (once called "slash-and-burn agriculture"), forests are cut and burned, then the ashes tilled into the soil. For a year or two the land produces fair crops. But rainwater quickly leaches nutrients from the exposed soils, and cleared plots soon become infertile and are abandoned.

Deforestation also can change regional patterns of rainfall and have global repercussions. Between 50 and 80 percent of the water vapor above tropical forests is released from the trees. Without trees, annual precipitation declines and the region gets hotter and drier. Moreover, the trees of these vast forested regions help maintain the global cycling of carbon and oxygen through their photosynthetic activities (page 501). When they are harvested or burned, the carbon stored in their biomass is released to the atmosphere as carbon dioxide. This may play a role in increasing the greenhouse effect.

Almost half of the world's expanses of tropical forests have already been cleared for cropland, grazing land, and other uses. Deforestation is greatest in Brazil, Indonesia, Colombia, and Mexico (Figure 25.15). At present rates of clearing, only Brazil and Zaire will have large tracts of

a

Figure 25.15 (**a**) The diverse occupants of various tropical rain forests include jaguars. (**b**) El Yunque rain forest, Puerto Rico. (**c**) Countries in which the largest destruction of tropical forests is occurring. Red shading signifies where 2,000–14,800 square kilometers are deforested annually; orange signifies regions of more moderate deforestation (100–1,900 square kilometers).

b

c

A Planetary Emergency: Loss of Biodiversity

On March 24, 1900, in Ohio, a young boy shot the last known wild passenger pigeon. More recently, fisheries biologist John Musick of the Virginia Institute of Marine Science reported his latest findings on the decline in populations of the dusky shark. Over the past 25 years, Musick has tracked the numbers of the dusky and other shark species that spend much of their lives off the mid-Atlantic coast of the United States and have become popular as human food. Musick's numbers show that, like codfish in parts of the North Atlantic, today the dusky shark has all but disappeared. Based on similar evidence from many scientists, the National Marine Fisheries Service in 1993 imposed catch limits on commercial shark fishing operations. Fleet operators responded with a lawsuit.

Sooner or later all species become extinct, but humans have become a primary factor in the premature extinction of more and more species. Already our actions are leading to the premature extinction of at least six species per hour.

Many biologists consider this epidemic of extinction an even more serious problem than depletion of stratospheric ozone or global warming because it is happening faster and is irreversible. Such rapid extinction cannot be balanced by speciation because it takes many thousands of generations for new species to evolve.

Globally, tropical deforestation (Figure 25.15) is the greatest killer of species, followed by destruction of coral reefs. However, plant extinctions can be more important ecologically than animal extinctions since most animal species depend directly or indirectly on plants for food, and often for shelter as well. Humans also have traditionally depended on plants as sources of medicines. Indeed, 40 percent of prescription drug sales in the United States involve natural plant products. One striking example is the rosy periwinkle of Madagascar (Figure *a*), from which we derive the anticancer drugs vincristine and vinblastine (page 434). As it happens, humans have destroyed 90 percent of the vegetation on Madagascar, so we may never know if it was home to plant sources of other life-saving drugs as well. Meanwhile, in the United States we are going about the business of habitat destruction at a dizzy-

Figure *a*   The rosy periwinkle found in the threatened tropical forests of Madagascar.

ing pace. Among other things, we have cut down 90–95 percent of old growth forests and drained half the wetlands, which filter human water supplies and provide homes for waterfowl and juvenile fishes. At least 500 species native to these areas have been driven to extinction, and dozens more are endangered.

The underlying causes of wildlife extinction are human population growth and economic policies that fail to value the environment. Instead, they promote unsustainable exploitation. As our population grows we clear, occupy, and damage more land to supply food, fuel, timber, and other resources. In wealthy countries, affluence leads to greater-than-average resource use per person. In less-affluent nations, the combination of rapid population growth and poverty push the poor to cut forests, grow crops on marginal land, overgraze grasslands, and poach endangered animals.

How can we help stem the tide of extinctions? Among other strategies, individuals can support efforts to reduce deforestation, projected global warming, ozone depletion, and poverty—the greatest threats to earth's wildlife *and* to the human species.

tropical forest in 2010. By 2035, most of those forests also will be gone. In the Pacific Northwest of the United States and in British Columbia, Canada, at least 90 percent of old-growth fir forests have been cut, and there is strong pressure from logging interests to cut what remains. As earth's forests disappear, so do the species of plants and animals that inhabit them. This critical loss of biodiversity is the topic of *Focus on Environment*.

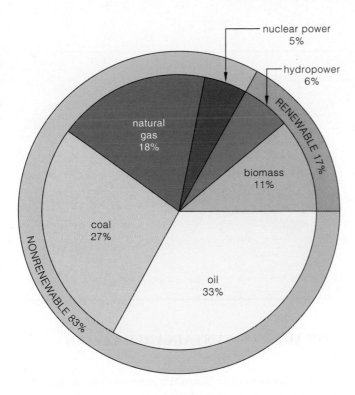

Figure 25.16 World consumption of nonrenewable and renewable energy sources in 1986.

Figure 25.17 This solar power plant (foreground) in Sacramento, California, may be the wave of the future, at least in some regions of the world. It uses solar-powered photovoltaic cells to produce electricity. The nuclear power plant in the background is no longer operating.

CONCERNS ABOUT ENERGY USE

Paralleling the J-shaped curve of human population growth is a steep rise in total and per capita energy consumption. The rise is due not only to increased numbers of energy users, but also to extravagant consumption and waste. Some sources of energy, such as hydropower and direct solar energy, are renewable. Others, such as coal and petroleum, are not. Currently, 79 percent of the energy stores being tapped are not renewable (Figure 25.16).

Fossil Fuels

Fossil fuels such as coal, petroleum, and natural gas are the carbon-containing remains of plants and other organisms that lived hundreds of millions of years ago. They are nonrenewable resources.

Even with stringent conservation efforts, known petroleum and natural gas reserves may be depleted during the 21st century. One "worst-case" scenario predicts they will be gone by 2035. As petroleum and natural gas deposits become depleted in easily accessible areas, we begin to seek new sources. Our explorations have been taking us to wilderness areas in Alaska and to other fragile environments, such as the continental shelves. After subtracting the energy costs of finding and extracting oil and gas from these remote areas, the *net* energy yield may not be worth the environmental costs. For example, we

still do not know the long-term impacts of the 11-million-gallon spill from the tanker *Exxon Valdez* in Alaska's coastal waters in 1989, nor of the more recent, much larger spill off the coast of Great Britain.

In theory, world coal reserves can meet the energy needs of the entire human population for at least several centuries. But coal burning has been the largest single source of air pollution worldwide. Fossil fuel burning also puts carbon dioxide into the atmosphere and so adds to the greenhouse effect.

Nuclear Energy

Today, nuclear power plants dot the landscape. At one time, proponents of nuclear power predicted that by 2000 it would meet the energy needs of 70 percent of the world's people. Although industrialized nations that are poor in energy resources now depend heavily on nuclear power, in most countries plans to develop more nuclear energy have been delayed or canceled. The cost, efficiency, environmental impact, and safety of nuclear energy are being seriously questioned. At present, nearly all electricity-generating methods have average costs below those of new nuclear power plants. Solar energy (Figure 25.17) is the exception, but by 2000 solar energy with natural gas backup is expected to be lower in cost also.

What about safety? Although most of us are aware of the devastating effects of nuclear weapons blasts, radioactivity escaping from a nuclear plant during normal oper-

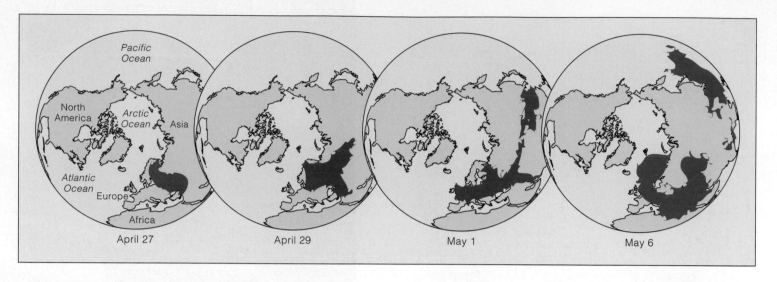

Figure 25.18 Global distribution of radioactive fallout after explosions ripped through the Chernobyl nuclear plant on April 26, 1986.

Pacific Ocean

North America

Arctic Ocean

Asia

Atlantic Ocean

Europe

Africa

April 27   April 29   May 1   May 6

ation is actually less than the amount released from a coal-burning plant of the same capacity. However, there is the potential danger of a **meltdown** if the reactor core becomes overheated. As happened at Chernobyl in Ukraine in 1986, the reactor core could melt through its thick concrete containment slab and into the earth, contaminating groundwater and releasing lethal amounts of radiation. Thirty-one people died immediately, and several hundred others died later of radiation sickness. Inhabitants of entire villages were relocated; their former homes were bulldozed under. The radioactive fallout put an estimated 300–400 million people throughout Europe at increased risk of leukemia and other radiation-induced disorders (Figure 25.18). In 1979 a nuclear accident occurred in a commercial nuclear facility at Three Mile Island near Harrisburg, Pennsylvania. Through mechanical failures and human error, one reactor came within 30–60 minutes of a total meltdown.

The wastes produced by a nuclear reactor are an enormously radioactive, extremely dangerous collection of materials. So-called "high-level" wastes must be isolated for at least 10,000 years. If one kind of plutonium isotope (^{239}Pu) is not removed, the wastes must be kept isolated for a quarter of a million years! Recently the U.S. government approved a permanent underground storage program for radioactive wastes. The wastes are to be sealed in ceramic material, placed in steel cylinders, then buried underground in supposedly stable rock formations (in the Yucca Mountains of Utah) that are free from exposure to corrosive saltwater. The first depository is not expected to open until 2010, however, and scientific and political difficulties may postpone the opening until much later.

THE IDEA OF SUSTAINABLE LIVING

You may have read or heard the expressions "sustainable agriculture" or "sustainable development." A **sustainable society** is one that manages its population growth and economic activities in ways that prevent serious damage to the environment. Every life form must obtain energy and other resources from its environment in order to grow and reproduce. The key to sustainable living is simple: We must not take or use resources in a manner or amount that irreparably harms ecosystems—or causes them to collapse altogether.

Elsewhere in this chapter you have read of human activities that in the long run are unsustainable. Our heavy dependence on fossil fuels is an obvious one, because we know that the earth has a limited supply of fossil fuels and eventually the supply will run out. Most likely, we also cannot long continue widespread deforestation, overfishing of the oceans, and pollution of the atmosphere with greenhouse gases and ozone-destroying chemicals (among other activities). Environmental experts disagree about the precise level of degradation the earth's land and water ecosystems and its atmosphere can sustain. Most agree, however, that without significant changes in human behavior and technologies, our planet will become unfit for human habitation.

Environmental scientist and author G. Tyler Miller, Jr., who has contributed many *Focus on Environment* essays to this text, cites many other strategies to promote sustainability. Here is a sampling of them:

1. Significantly slow human population growth, our greatest global environmental crisis.

2. Implement agricultural and forestry practices that minimize soil erosion and water pollution from runoff of sediment, fertilizers, and pesticides.

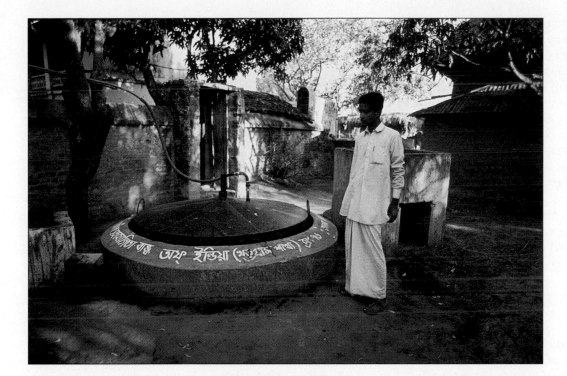

Figure 25.19 The raw material input for this biogas digester is animal dung, and the output is methane gas that can be used as a household fuel.

3. Develop appropriate technologies to address needs in different places. An **appropriate technology** uses resources efficiently, is adapted to local (or regional) needs and conditions, and can fairly readily be expanded, reduced, or repaired. The biogas digester shown in Figure 25.19 is an example. More than 750,000 of these devices operate in villages in India (and more than 6 million in China). The droppings of village animals are the raw material for the digester, which contains anaerobic bacteria that decompose the waste and convert it into biogas—a mixture that is about 60 percent methane, the principal component of natural gas. The methane serves as fuel for cooking and heating, and farmers use solid residues of the process as fertilizer on food crops or trees.

4. Make environmental considerations a part of the decision-making processes of all government agencies and all industries.

5. Strengthen and rigorously enforce laws and regulations that limit the discharge of industrial pollutants. Require industries to take full financial responsibility for their polluting activities. This will help make pollution control, not pollution, the most profitable option.

6. Develop efficient public transit systems, and automobiles that do not run on fossil fuels.

7. To encourage conservation and development of new, nonpolluting technologies, end government subsidies that enable consumers or businesses to pay far less than the actual cost of certain resources, such as fossil fuels.

8. Subsidize products and projects, such as solar power and cleanup of toxic wastes, that serve public needs in earth-friendly ways.

9. Focus personal consumption on basic needs: a balanced diet, health care, decent housing, clean air and water, and sanitation. Repair, reuse, and recycle.

10. On a personal, local, or national level, become active in organized efforts aimed at conserving environmental resources—parks, clean air, a river, earth's ozone shield—whatever matters most to you. Individual action counts.

Epilogue

The large and growing numbers of humans and our disregard for the limits of ecosystems have put our species at grave risk. The final straw could be a prolonged global food shortage, irrevocable poisoning of water supplies, passing a critical threshold for ozone depletion, or some other nightmarish disaster. However, one reason we include a chapter such as this in a human biology text is that we firmly believe that *it is not too late*. Humans have the capacity to anticipate events *before* they happen. We are not locked into responding only after irreversible change has begun. We all have the capacity for adapting to a future that we can partly shape. We can, for example, learn to live with less. Far from being a return to primitive simplicity, this could be one of the most intelligent behaviors our species ever engaged in.

It is probably not too extreme to say that our species's survival depends on coming up with some alternatives for the future. It depends on designing and constructing human ecosystems that are in harmony not only with what we define as basic human values, but also with biological reality. Our values and expectations must be adapted to the altered world in which we now live, because the principles of energy flow and resource use that govern the survival of all systems of life do not change. In the final analysis, we must come to terms with these principles and ask ourselves what our long-term contribution will be to the world of life.

SUMMARY

1. The growth rate of a population depends on the birth rate, death rate, and the rates of immigration and emigration. If the birth rate exceeds the death rate by a constant amount, the population will grow exponentially (assuming immigration and emigration remain zero).

2. Carrying capacity is the number of individuals of a species that can be sustained indefinitely by the available resources in a given area.

3. Population size is determined by carrying capacity, competition, and other factors that limit population growth. Limiting factors vary in their relative effects and over time, so population size also can change over time.

4. Some limiting factors, such as competition for resources, disease, and predation, are density-dependent.

5. Currently, human population growth varies from zero in some of the more-developed countries to more than 4 percent per year in some of the less-developed countries.

6. Rapid growth of the human population during the past two centuries was possible largely because of our capacity to expand into new environments, and because of agricultural and technological developments that increased the carrying capacity.

Review Questions

1. Why do populations that are not restricted in some way tend to grow exponentially? *512*

2. If the birth rate equals the death rate, what happens to the growth rate of a population? If the birth rate remains slightly higher than the death rate, what happens? *510*

3. At current growth rates, how many years will elapse before another billion people are added to the human population? *514*

4. Write a short essay about a hypothetical population that shows either one of the following age structures. Describe what might happen to younger and older age groups when members move into new categories. *511*

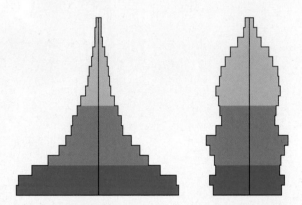

5. If a third of the world population is now below age 15, what effect will this age distribution have on the growth rate of the human population? What sorts of humane recommendations would you make that would encourage this age group to limit the number of children they plan to have? *515*

Critical Thinking: You Decide *(Key in Appendix IV)*

1. How have humans increased the carrying capacity of their environments? Have they avoided some of the limiting factors on population growth, or is the avoidance an illusion?

2. The poet T. S. Eliot once wrote that the world would end "not with a bang, but a whimper." Applying this grim thought to the topics of this chapter, design a 10-point plan for proving Eliot wrong.

Self-Quiz *(Answers in Appendix III)*

1. _____ is the study of how organisms interact with one another as well as with their physical and chemical environment.

2. The rate at which a population grows or declines depends on the _____ .
   a. birth rate
   b. death rate
   c. immigration rate
   d. emigration rate
   e. all of the above

3. Populations grow exponentially when _____ .
   a. birth rate remains above death rate and neither changes
   b. death rate remains above birth rate
   c. immigration and emigration rates are equal (a zero value)
   d. emigration rates exceed immigration rates
   e. both a and c combined are correct

4. The number of individuals of a species that can be sustained indefinitely by the resources in a given region is the _____.
   a. biotic potential
   b. carrying capacity
   c. environmental resistance
   d. density control

5. Which of the following factors does *not* affect sustainable population size?
   a. predation
   b. competition
   c. available resources
   d. pollution
   e. each of the above can affect population size

6. Population growth controls such as resource competition and disease are said to be _____.
   a. density independent
   b. population sustaining
   c. population dynamics
   d. density dependent

7. During the past two centuries, rapid growth of the human population has occurred largely because of _____.
   a. increased birth rate worldwide
   b. increased death rate worldwide
   c. carrying capacity reduction
   d. carrying capacity expansion

8. Match the following population ecology terms.
   ____ carrying capacity
   ____ exponential growth
   ____ population growth rate
   ____ density-dependent controls

   a. examples are disease and predation
   b. depends on birth, death, immigration, and emigration rates

   c. number of individuals of a given species that can be sustained indefinitely by the resources in a given area
   d. increases in population size by ever larger amounts per unit of time

Key Terms

acid deposition *517*
age structure *510*
appropriate technology *525*
biotic potential *512*
carrying capacity *513*
deforestation *520*
demographic transition model *514*
density-dependent control *513*
density-independent control *513*
desertification *520*
exponential growth *512*
fossil fuel *523*
green revolution *520*

industrial smog *516*
life table *514*
limiting factor *513*
logistic growth *513*
meltdown *524*
photochemical smog *516*
pollutant *516*
reproductive base *510*
salinization *519*
shifting cultivation *520*
sustainable society *524*
zero population growth *510*

Readings

Cohen, J. November 1992. "How Many People Can the Earth Hold?" *Discover*.

Dixon, T. F., J. Boutwell, and G. Rathjens. February 1993. "Environmental Change and Violent Conflict." *Scientific American*.

"MegaCities." January 11, 1993. *Time*. A thought-provoking discussion of the global impact of a burgeoning human population in urban centers.

Miller, G. T. 1994. *Living in the Environment*, 8th ed. Belmont, Calif.: Wadsworth. This author consistently pulls together information on human population growth into a coherent picture.

APPENDIX I
Periodic Table of the Elements

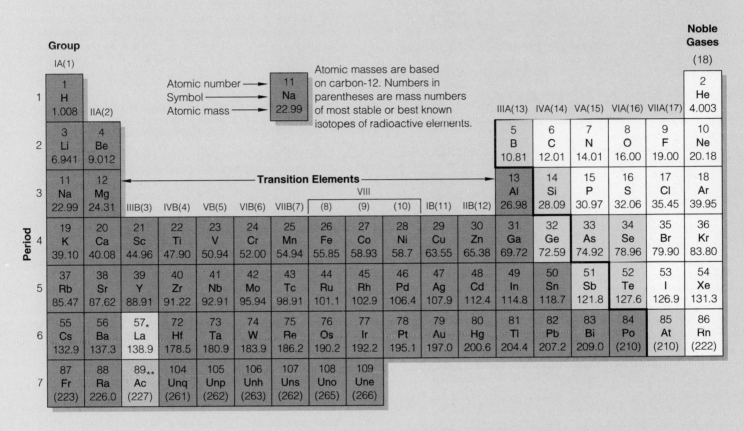

Group

IA(1)

Noble Gases (18)

Atomic number → 11
Symbol → Na
Atomic mass → 22.99

Atomic masses are based on carbon-12. Numbers in parentheses are mass numbers of most stable or best known isotopes of radioactive elements.

1 H 1.008																	2 He 4.003
3 Li 6.941	4 Be 9.012											5 B 10.81	6 C 12.01	7 N 14.01	8 O 16.00	9 F 19.00	10 Ne 20.18
11 Na 22.99	12 Mg 24.31											13 Al 26.98	14 Si 28.09	15 P 30.97	16 S 32.06	17 Cl 35.45	18 Ar 39.95
19 K 39.10	20 Ca 40.08	21 Sc 44.96	22 Ti 47.90	23 V 50.94	24 Cr 52.00	25 Mn 54.94	26 Fe 55.85	27 Co 58.93	28 Ni 58.7	29 Cu 63.55	30 Zn 65.38	31 Ga 69.72	32 Ge 72.59	33 As 74.92	34 Se 78.96	35 Br 79.90	36 Kr 83.80
37 Rb 85.47	38 Sr 87.62	39 Y 88.91	40 Zr 91.22	41 Nb 92.91	42 Mo 95.94	43 Tc 98.91	44 Ru 101.1	45 Rh 102.9	46 Pd 106.4	47 Ag 107.9	48 Cd 112.4	49 In 114.8	50 Sn 118.7	51 Sb 121.8	52 Te 127.6	53 I 126.9	54 Xe 131.3
55 Cs 132.9	56 Ba 137.3	57* La 138.9	72 Hf 178.5	73 Ta 180.9	74 W 183.9	75 Re 186.2	76 Os 190.2	77 Ir 192.2	78 Pt 195.1	79 Au 197.0	80 Hg 200.6	81 Tl 204.4	82 Pb 207.2	83 Bi 209.0	84 Po (210)	85 At (210)	86 Rn (222)
87 Fr (223)	88 Ra 226.0	89** Ac (227)	104 Unq (261)	105 Unp (262)	106 Unh (263)	107 Uns (262)	108 Uno (265)	109 Une (266)									

Period

Transition Elements

VIII

IIIB(3) IVB(4) VB(5) VIB(6) VIIB(7) (8) (9) (10) IB(11) IIB(12)

IIIA(13) IVA(14) VA(15) VIA(16) VIIA(17)

IIA(2)

Inner Transition Elements

	58 Ce 140.1	59 Pr 140.9	60 Nd 144.2	61 Pm (145)	62 Sm 150.4	63 Eu 152.0	64 Gd 157.3	65 Tb 158.9	66 Dy 162.5	67 Ho 164.9	68 Er 167.3	69 Tm 168.9	70 Yb 173.0	71 Lu 175.0
Lanthanide Series 6 *														
Actinide Series 7 **	90 Th 232.0	91 Pa 231.0	92 U 238.0	93 Np 237.0	94 Pu (244)	95 Am (243)	96 Cm (247)	97 Bk (247)	98 Cf (251)	99 Es (252)	100 Fm (257)	101 Md (258)	102 No (259)	103 Lr (260)

APPENDIX II
Answers to Genetics Problems

Chapter 17

1. a: *AB*
   b: *Ab* and *aB*
   c: *Ab* and *ab*
   d: *AB, aB, Ab,* and *ab*

2. a: *AaBB* will occur in all the offspring.
   b: 25% *AABB*; 25% *AaBB*; 25% *AABb*; 25% *AaBb*
   c: 25% *AaBb*; 25% *Aabb*; 25% *aaBb*; 25% *aabb*
   d: 1/16 *AABB* (6.25%)
      1/8 *AaBB* (12.5%)
      1/16 *aaBB* (6.25%)
      1/8 *AABb* (12.5%)
      1/4 *AaBb* (25%)
      1/8 *aaBb* (12.5%)
      1/16 *AAbb* (6.25%)
      1/8 *Aabb* (12.5%)
      1/16 *aabb* (6.25%)

3. Yellow is recessive. Because the first-generation plants must be heterozygous and had a green phenotype, green must be dominant over the recessive yellow.

4. a: Mother must be heterozygous for both genes; father is homozygous recessive for both genes. The first child is also homozygous recessive for both genes.
   b: The probability that the second child will not be able to roll the tongue and will have detached earlobes is 1/4 (25%).

5. a: *ABC*
   b: *aBc*
   c: *ABC, ABc, aBc, aBC*
   d: *ABC, abc, AbC, AbC,*
      Abc, ABc, abC, aBc

6. Because Dandelion does not exhibit the recessive hip disorder, she must be either homozygous dominant (HH) for this trait, or heterozygous (Hh). If the father is homozygous dominant (HH), then he and Dandelion cannot produce gametes that are homozygous recessive (hh), and so none of their offspring will have the undesirable phenotype. However, if Dandelion is a heterozygote for the trait, notice that the probability is 1/2 (50%) that a puppy will be heterozygous (Hh) and so carry the trait.

7. a: The mother must be heterozygous ($I^A i$). The man having type B blood could have fathered the child if he were also heterozygous ($I^B i$).
   b: If the man is heterozygous, then he *could be* the father. However, because any other type B heterozygous male could also be the father, one cannot say that this particular man absolutely must be. Actually, any male who could contribute an O allele (*i*) could have fathered the child. This would include males with type O blood (*ii*) or type A blood who are heterozygous.

8. The probability is 1/2 (50%) that a child of this couple will be a heterozygote and have sickle cell trait. The probability is 1/4 (25%) that a child will be homozygous for the sickling allele and so will have sickle cell anemia.

9. For these ten traits, all the man's sperm will carry identical genes. He cannot produce genotypically different sperm. The woman can produce eggs with four genotypes. This example underscores the fact that the more heterozygous gene pairs that are present, the more genetically different gametes are possible.

10. The mating is $Ll \times Ll$.
    Progeny genotypes: 1/4 *LL* + 1/2 *Ll* + 1/4 *ll*.
    Phenotypes: 1/4 homozygous survivors (*LL*)
    1/2 heterozygous survivors (*Ll*)
    1/4 lethal (*ll*) nonsurvivors

11. Bill's genotype: *Aa, Ss, Ee*
    Marie's genotype: *AA, SS, EE*
    No matter how many children Bill and Marie have, the probability is 100% that each child will have the parents' phenotype. Because Marie can produce only dominant alleles, there is no way that a child could inherit a *pair* of recessive alleles for any of these three traits—and that is what would be required in order for the child to show the recessive phenotype. Thus the probability is zero that a child will have short lashes, high arches, and no "achoo syndrome."

12. Every protein (polypeptide chain) in the body, including antibodies produced in immune responses (by B cells), is coded by a single gene. Aunt Tillie apparently has B cells that make antibodies that recognize a certain antibiotic as "foreign," triggering an immune response. The most important point here is that, strictly speaking, it is not the protein per se that is inherited, but the gene (allele) that encodes the protein. Chapter 19 describes the route from a gene to a protein in detail.

13. Given that the allergy trait results from a recessive allele, we know two things: Tillie must be homozygous recessive, *and* your father may be either homozygous for the dominant allele, or heterozygous for this trait. Because your mother does not have the allergy, she is also either homozygous dominant or heterozygous. At worst, if both parents are heterozygotes, each offspring has a 1/4 chance (25%) of being homozygous recessive and having the allergy.

Chapter 18

1. The probability is the same for each child: 1/2, or 50%.

2. a: From his mother.
   b: A male can produce two types of gametes with respect to an X-linked gene. One type will possess only a Y chromosome and so lack this gene; the other type will have an X chromosome and will have the X-linked gene.
   c: A female homozygous for an X-linked gene will produce just one type of gamete containing an X chromosome with the gene.
   d: A female heterozygous for an X-linked gene will produce two types of gametes. One will contain an X chromosome with the dominant alleles, and the other type will contain an X chromosome with the recessive allele.

3. A 0% crossover frequency means that 50% of the gametes will be *AB* and 50% will be *ab*.

4. If the longer-than-normal chromosome 14 represented the translocation of most of chromosome 21 to the end of a normal chromosome 14, then this individual would be afflicted with Down syndrome due to the presence of this attached chromosome 21 as well as two normal chromosomes 21. The total chromosome number, however, would be 46.

5. Using c as the symbol for color blindness and C for normal vision, then the cross can be diagrammed this way:

$$C(Y) \text{ female} \times Cc \text{ male}$$

In mugwumps, a son receives one sex-linked allele from each of his parents, but a daughter inherits her unpaired sex-linked allele solely from her father. In this cross, half of the sons will be CC and half wil be Cc, but none will be color blind. Of the daughters, half will be C(Y) and half will be c(Y). There is a 50% chance that a daughter will be color blind. Note: This answer is backward from the way it would be in humans.

6. The probability is 1/2, or 50%.

7. The dominant allele symbol is P. The probability that a child will have PKU is 25%. The pedigree chart should show one offspring, either male (box) or female (circle) colored in to show the presence of the PKU trait. All or none of the children may be carriers; the probability that a child will be heterozygous (Pp) is 50%.

APPENDIX III
Answers to Self-Quizzes

Introduction

1. DNA
2. Metabolism
3. Homeostasis
4. adaptive
5. mutations
6. d
7. c
8. c

Chapter 1

1. carbon
2. a
3. carbohydrates, lipids, proteins, nucleic acids
4. c
5. c
6. d
7. b
8. b
9. c
10. c, e, b, d, a

Chapter 2

1. e
2. cytoskeleton
3. c
4. a
5. c
6. b, g, f, d, c, a, e
7. d
8. d
9. c, e, d, a, b
10. b, a, c

Chapter 3

1. epithelial, connective, muscle, nervous
2. All answers are activities for which the human body is structurally and functionally adapted.
3. d
4. c
5. d
6. b
7. a
8. a, b, c
9. c
10. a
11. Receptors, integrator, effectors
12. d, e, c, b, a

Chapter 4

1. Skeletal and muscular
2. skeletal, smooth, cardiac
3. d
4. d
5. b
6. c
7. d, e, b, a, c

Chapter 5

1. digestive, circulatory, respiratory, urinary
2. digesting, absorbing, eliminating
3. caloric, energy
4. carbohydrates
5. essential amino acids, essential fatty acids
6. b
7. c
8. d
9. c
10. e, d, a, c, b

Chapter 6

1. circulatory system, lymphatic system
2. Arteries, veins; capillaries and venules; arterioles
3. d
4. d
5. c
6. a
7. d
8. b
9. d, b, a, c
10. c, d, a, b

Chapter 7

1. a
2. d
3. c
4. d
5. a
6. d
7. d
8. b
9. d
10. c, b, a, e, d, f

Chapter 8

1. Oxygen, carbon dioxide
2. gas, pressure gradient
3. epithelium, diffuse
4. air flow, blood flow
5. d
6. b
7. d
8. d
9. d
10. e, a, d, c, b

Chapter 9

1. filtration, reabsorption, secretion
2. c
3. c
4. d
5. d
6. a
7. d
8. c
9. d
10. b, d, e, c, a

Chapter 10

1. stimuli, neurons
2. action potentials
3. neurotransmitter
4. integration
5. d
6. c
7. b
8. a
9. c
10. e, d, b, c, a

Chapter 11

1. stimulus
2. sensation
3. Perception
4. d
5. b
6. d
7. c
8. b
9. a
10. f, c, e, a, b, d

Chapter 12

1. Hormones, neurotransmitters, local signaling molecules, pheromones
2. hypothalamus, pituitary gland
3. negative feedback
4. receptors
5. a
6. d
7. e
8. b, c, d, e
9. d
10. d, b, e, c, a, f

Chapter 13

1. hypothalamus
2. b
3. d
4. c
5. b

Chapter 14

1. a
2. a
3. c
4. morphogenesis
5. c
6. b
7. d, a, f, e, c, b
8. a
9. c
10. b
11. c

Chapter 15

1. bacteria, viruses, virus
2. sexual contact; blood, semen, vaginal secretions, breast milk
3. false
4. d
5. c
6. a
7. b
8. d, e, b, a, c, g, f

Chapter 16

1. mitosis; meiosis
2. chromosomes; DNA
3. c
4. diploid
5. c
6. a

7. b
8. b
9. c
10. d
11. c
12. d, b, c, a

Chapter 17

1. a
2. c
3. a
4. c
5. b
6. d
7. b
8. d
9. c, d, e, b, a

Chapter 18

1. c
2. e
3. e
4. c
5. e
6. c
7. d
8. d
9. c, e, d, b, a

Chapter 19

1. e
2. d
3. d
4. a
5. d
6. b
7. d
8. c
9. a
10. b
11. b, d, c, a, h, g, f

Chapter 20

1. c
2. e
3. tumor suppressor gene
4. b
5. d
6. a
7. Biopsy
8. b

Chapter 21

1. variation
2. genetic engineering
3. Plasmids
4. vectors

5. c
6. cloned DNA
7. d
8. b
9. c
10. restriction enzymes

Chapter 22

1. macroevolution
2. gradual; bursts
3. e
4. e
5. c
6. e
7. mutation, genetic drift, gene flow, natural selection
8. natural selection

Chapter 23

1. e
2. b
3. c
4. c
5. a
6. b
7. d
8. d
9. d
10. a, d, e, c, b

Chapter 24

1. ecosystem; energy; materials
2. photosynthesis; one
3. energy; nutrients; energy (including heat), nutrients
4. biogeochemical
5. b
6. e
7. d, c, b, a, g, h, i, f, e

Chapter 25

1. Ecology
2. e
3. e
4. b
5. e
6. d
7. a
8. c, d, b, a

APPENDIX IV
Key to Critical Thinking Questions

Introduction

1. Key: the different approaches to "truth" inherent in scientific inquiry and in subjective spiritual, moral, and ethical approaches to issues. Another problem you might consider: Based on discussions in this chapter, can anyone—including scientists—ever claim to know the "whole" truth about a particular question?

Chapter 1

1. Key: understanding the chemical nature of carbohydrates. Such basic knowledge is very useful when it comes to evaluating health claims of diet books and other commercial products.

2. Key: the mathematical relationship between values on the pH scale.

3. Key: knowing that cream is mainly animal fat, and that syrup is mainly water and a sugar such as fructose.

Chapter 2

1. See *Focus on Science*, page 39.

2. Key: the biochemical limits on mechanisms by which cells, especially muscle cells, can obtain energy to do work. In explaining the second observation, think of a "nonaerobic" exercise as one that, unlike aerobic exercise, does not require the body to make adjustments (as in breathing rate) in order to supply cells with the oxygen needed to keep aerobic cellular respiration going.

Chapter 3

1. Key: knowing the basic functions of epidermis.

2. Key: knowing the roles collagen plays in the body.

Chapter 4

1. Key: review the various types of muscle fibers, and relate their characteristics to the athletic requirements of sprinting versus the prolonged activity required of a marathoner's leg muscles.

2. Key: production of GH and the maximum length of an individual's long bones are both genetically determined.

Chapter 5

1. Key: consider the health impacts of an active life-style and a low-fat diet.

2 and 3. Key: consider the different caloric demands of an older, less active person compared to those of a normally active child. Also, keep in mind that developing youngsters require ample protein.

Chapter 6

1. Key: the relationship among blood pressure and blood volume, cross-sectional area of systemic vessels, and cardiac output.

2. Key: review the function(s) of each type of formed element in blood.

Chapter 7

1. Key: the function of macrophages, compared to the function of T cells.

2. Key: review mechanism by which memory cells recognize invaders.

Chapter 8

1. Key: review health effects of second-hand smoke; effects of smoking on developing fetus.

2. Key: hemoglobin binds much more readily with carbon monoxide than it does with oxygen. Also: consider the partial pressure of oxygen in air versus that in a tank of pure oxygen.

Chapter 9

1. Key: the function of ADH in regulating water/solute balance.

2. Key: review chapter section on filtration, reabsorption.

3. Key: review the mechanism of urine concentration.

Chapter 10

1. Key: review the various ways neurotransmitters can affect post-synaptic neurons.

2. Key: consider the blood-brain barrier and the functions of cerebrospinal fluid.

Chapter 11

1. Key: the role of olfaction in taste perception.

2. Key: review the differing functions of rods and cones.

Chapter 12

1. Key: the biological functions of melatonin.

2. Key: review the roles of anterior lobe secretions, including effects on target cells in other endocrine glands.

Chapter 13

1. Key: what chemical conditions in the vagina are most conducive to fertilization?

2. Key: review the role of estrogen at ovulation.

Chapter 14

1. Key: take into account respiration, circulation, nutrient intake, waste excretion, immune functions, and functions of other basic body systems.

2. Key: What is the immediate outcome of gastrulation? Your answer to this question should also consider the long-term developmental consequences if gastrulation goes awry.

Chapter 15

1. Key: compare modes of transmission and relative" contagiousness" of HIV, hepatitis B.

2. Key: review discussions of HIV transmission and prevention of infection.

Chapter 16

1 and 2. Key: review stages of meiosis and discussions of sources of genetic variation in offspring (such as crossing over and random alignment of chromosomes at metaphase. Also: keep in mind that in each new generation, offspring receive half their chromosomes from each parent.

Chapter 17

1. Key: review discussion of twin studies.

2. Key: consider the genetic makeup of possible gametes resulting from matings between (a) two homozygotes for recessive and dominant alleles for a trait, between (b) heterozygotes for the same trait, and between (c) homozygotes for the recessive form of the trait. Also: keep in mind that a recessive trait need not be harmful (blue eyes being a case in point).

Chapter 18

1. Key: among other complicated issues, here you must consider whether (in your view) a person's gender is determined by "genetic sex" or by reproductive structures. Also: review the discussion of sex determination at the beginning of Chapter 14. In addition, note that the secondary sex characteristics of normal males, which are due to the sensitivity of cells to testosterone, include greater muscle mass—which is why males typically are physically stronger than females.

2. Key: consider the link between genes and proteins.

Chapter 19

1. Key: consider the links among DNA, mutation, and inheritance.

Chapter 20

1. Key: review the types of cellular changes currently known to be associated with the onset of cancer.

2. Key: consider current hypotheses about the roles of immune system cells in recognizing and eliminating cancerous cells.

3. Key: review kidney functions that help maintain homeostasis (Chapter 9), as well as lung functions (Chapter 8) and liver functions (Chapter 5).

Chapter 21

1. Key: Genetic engineering and recombinant DNA techology are most applicable to plant and animal breeding, commercial and industrial microbiology (including use of bacteria and yeast in food processing), and to medical fields such as gene therapy and cancer treatments. You may also wish to relate potential new benefits or problems of genetic engineering to methods that have been or are being developed for manipulating human gametes and embryos.

2. Key: review genetic engineering and recombinant DNA methods described in this chapter. Also: keep in mind that pigs are mammals and so their growth and development entails many of the same gene-guided events—including production of specific hormones and other proteins—that occur in humans.

3. Key: what does "transgenic" mean?

Chapter 22

1. Key: consider that (a) proteins, including enzymes, are encoded by genes; (b) different populations of a species may show differing gene (allele) frequencies, and (c) various lines of evidence shed light on the geographical movements of early human populations.

2. Key: review discussions of biogeography and evolutionary processes such as gene flow.

Chapter 23

1. Key: review the time frame within which human evolution has occurred.

Chapter 24

1. Key: have in mind the environmental inputs—energy and nutrients—that affect plant growth.

2. Key: review discussion of the greenhouse effect on page 502.

Chapter 25

1. Key: consider changes in health care, agriculture, and public sanitation in the past few decades and centuries; for example, before penicillin was discovered in the 1940s, people often died of infections we would today consider minor. Also: review the discussion of limits on population growth.

2. Key: to survive, our species must grapple with environmental problems with energy and imagination. Although different people will focus their efforts on different aspects of our current predicament, it is generally agreed that a key element of sustaining a livable earth is controlling human population growth.

GLOSSARY OF BIOLOGICAL TERMS

ABO blood typing Method of characterizing an individual's blood according to whether one or both of two protein markers, A and B, are present at the surface of red blood cells. The O signifies that neither marker is present.

abortion Spontaneous or induced expulsion of the embryo or fetus from the uterus.

absorption The movement of nutrients, fluid, and ions across the gastrointestinal tract lining and into the internal environment.

accommodation In the eye, adjustments of the lens position that move the focal point forward or back so that incoming light rays are properly focused on the retina.

acid A substance that releases hydrogen ions in water.

acid deposition The falling to earth of sulfur and nitrogen oxides; called *wet acid deposition* when it occurs in rain or snow, and *dry acid deposition* when it occurs as airborne particles of sulfur and nitrogen oxides.

acrosome An enzyme-containing cap that covers most of the head of a sperm and helps the sperm penetrate an egg at fertilization.

actin (AK-tin) A globular contractile protein. In muscle cells, actin interacts with another protein, myosin, to bring about contraction.

action potential An abrupt, brief reversal in the steady voltage difference (resting membrane potential) across the plasma membrane of a neuron.

active site A crevice on the surface of an enzyme molecule where a specific reaction is catalyzed.

active transport The pumping of one or more specific solutes through a transport protein that spans the lipid bilayer of a cell membrane. Most often, the solute is transported against its concentration gradient. The protein is activated by an energy boost, as from ATP.

adaptation [L. *adaptare*, to fit] In evolutionary biology, the process of becoming adapted (or more adapted) to a given set of environmental conditions. Of sensory neurons, a decrease in the frequency of action potentials (or their cessation) even when a stimulus is maintained at constant strength.

adaptive behavior A behavior that promotes the propagation of an individual's genes and that tends to increase in frequency in a population over time.

adaptive radiation A burst of speciation events, with lineages branching away from one another as they partition the existing environment or invade new ones.

adaptive trait Any aspect of form, function, or behavior that helps an organism survive and reproduce under a given set of environmental conditions.

ADH Antidiuretic hormone. Produced by the hypothalamus and released by the posterior pituitary, it stimulates reabsorption in the kidneys and so reduces urine volume.

adipose tissue A type of connective tissue having an abundance of fat-storing cells and blood vessels for transporting fats.

ADP Adenosine diphosphate. A nucleotide coenzyme that accepts unbound phosphate or a phosphate group to become ATP.

ADP/ATP cycle In cells, a mechanism of ATP renewal. When ATP donates a phosphate group to other molecules (and so energizes them), it reverts to ADP, then forms again by phosphorylation of ADP.

adrenal cortex (ah-DREE-nul) Outer portion of the adrenal gland; its hormones have roles in metabolism, inflammation, maintaining extracellular fluid volume, and other functions.

adrenal medulla Inner region of the adrenal gland; its hormones help control blood circulation and carbohydrate metabolism.

aerobic respiration (air-OH-bik) [Gk. *aer*, air, and *bios*, life] The main energy-releasing metabolic pathway of ATP formation, in which oxygen is the final acceptor of electrons stripped from glucose or some other organic compound. The pathway proceeds from glycolysis through the Krebs cycle and electron transport phosphorylation. A typical net yield is 36 ATP for each glucose molecule.

age structure Of a population, the number of individuals in each of several or many age categories.

agglutination (ah-glue-tin-AY-shun) The clumping together of foreign cells that have invaded the body (as pathogens or in tissue grafts or transplants). Clumping is induced by cross-linking between antibody molecules that have already bound antigen at the surface of the foreign cells.

aging A range of processes, including the breakdown of cell structure and function, by which the body gradually deteriorates. All organisms showing extensive cell differentiation undergo aging.

agranulocyte Class of white blood cells that lack granular material in the cytoplasm; includes the precursors of macrophages (monocytes) and lymphocytes.

AIDS Acquired immunodeficiency syndrome. A set of chronic disorders following infection by the human immunodeficiency virus (HIV), which destroys key cells of the immune system.

alcoholic fermentation Anaerobic pathway of ATP formation in which pyruvate from glycolysis is broken down to acetaldehyde, which accepts electrons from NADH to become ethanol, and NAD is regenerated. Its net yield is two ATP.

aldosterone (al-DOSS-tuh-roan) Hormone secreted by the adrenal cortex that helps regulate sodium reabsorption.

allantois (ah-LAN-twahz) [Gk. *allas*, sausage] Of vertebrates, one of four extraembryonic membranes that form during embryonic development. In humans, it functions in early blood formation and development of the urinary bladder.

allele (uh-LEEL) For a given location on a chromosome, one of two or more slightly different molecular forms of a gene that code for different versions of the same trait.

allele frequency Of a given gene locus, the relative abundances of each kind of allele carried by the individuals of a population.

allergy An immune response made against a normally harmless substance.

all or none principle Principle that states that individual cells in a muscle's motor units always contract fully in response to proper stimulation. If the stimulus is below a certain threshold, the cells do not respond at all.

alveolar sac (al-VEE-uh-lar) Any of the pouchlike clusters of alveoli in the lungs; the major sites of gas exchange.

alveolus (ahl-VEE-uh-lus), plural alveoli [L. *alveus*, small cavity] Any of the many cup-shaped, thin-walled outpouchings of respiratory bronchioles. A site where oxygen diffuses from air in the lungs to the blood, and carbon dioxide diffuses from blood to the lungs.

amino acid (uh-MEE-no) A small organic molecule having a hydrogen atom, an amino group, an acid group, and an R group covalently bonded to a central carbon atom. The subunit of polypeptide chains, which represent the primary structure of proteins.

ammonification (uh-MOAN-ih-fih-KAY-shun) Together with decomposition, a process by which certain bacteria and fungi break down nitrogen-containing wastes and remains of other organisms.

amnion (am-NEE-on) Of land vertebrates, one of four extraembryonic membranes. It becomes a fluid-filled sac in which the embryo (and fetus) can grow, move freely, and be protected from sudden temperature shifts and impacts.

anaerobic pathway (AN-uh-row-bik) [Gk. *an*, without, and *aer*, air] Metabolic pathway in which a substance other than oxygen serves as the final acceptor of electrons that have been stripped from substrates.

analogous structures Body parts, once different in separate lineages, that were put to comparable uses in similar environments and that came to resemble one another in form and function. They are evidence of morphological convergence.

anaphase (AN-uh-faze) The stage at which microtubules of a spindle apparatus separate sister chromatids of each chromosome and move them to opposite spindle poles. During anaphase I of meiosis, the two members of each pair of homologous chromosomes separate. During anaphase II, sister chromatids of each chromosome separate.

aneuploidy (AN-yoo-ploy-dee) A change in the chromosome number following inheritance of one extra or one less chromosome.

antibiotic [Gk. *anti*, against] A normal metabolic product of certain microorganisms that kills or inhibits the growth of other microorganisms.

antibody Any of a variety of Y-shaped receptor molecules with binding sites for specific antigens. Only B cells produce antibodies, then position them at their surface or secrete them.

anticodon In a tRNA molecule, a sequence of three nucleotide bases that can pair with an mRNA codon.

antigen (AN-tih-jen) [Gk. *anti*, against, and *genos*, race, kind] Any molecular configuration that is recognized as foreign to the body and that triggers an immune response. Most antibodies are protein molecules at the surface of infectious agents or tumor cells.

antigen-MHC complex Unit consisting of fragments of an antigen molecule, bound to MHC proteins. MHC complexes displayed at the surface of an antigen-presenting cell such as a macrophage promote an immune response by lymphocytes.

antigen-presenting cell A macrophage or other cell that displays antigen-MHC complexes at its surface and so promotes an immune response by lymphocytes.

anus Terminal opening of the gastrointestinal tract.

aorta (ay-OR-tah) [Gk. *airein*, to lift, heave] Main artery of systemic circulation; carries oxygenated blood away from the heart to all body regions except the lungs.

aortic body Any of several receptors in artery walls near the heart that respond to changes in levels of carbon dioxide (and to some degree, oxygen) in arterial blood.

appendicular skeleton (ap-en-DIK-yoo-lahr) Bones of the limbs, hips, and shoulders.

appendix A slender projection from the cup-shaped pouch (cecum) at the start of the colon.

arrythmia Irregular or abnormal heart rhythm, sometimes caused by stress, drug effects, or coronary artery disease.

arteriole (ar-TEER-ee-ole) Any of the blood vessels between arteries and capillaries. They are control points where the volume of blood delivered to different body regions can be adjusted.

artery Any of the large-diameter blood vessels that conduct deoxygenated blood to the lungs and oxygenated blood to all body tissues. The thick, muscular artery wall allows arteries to smooth out pulsations in blood pressure caused by heart contractions.

asexual reproduction Mode of reproduction by which offspring arise from a single parent and inherit the genes of that parent only.

atherosclerosis Condition in which an artery's wall thickens and loses its elasticity, and the vessel becomes clogged with lipid deposits. In the artery wall, abnormal smooth muscle cells accumulate, and there is an increase in connective tissue. Plaques consisting of lipids, calcium salts, and fibrous material extend into the artery lumen, disrupting blood flow.

atmosphere A region of gases, airborne particles, and water vapor enveloping the earth; 80 percent of its mass is distributed within 17 miles of the earth's surface.

atmospheric cycle A biogeochemical cycle in which the atmosphere is the largest reservoir of an element. The carbon and nitrogen cycles are examples.

atom The smallest unit of matter that is unique to a particular element.

atomic number The number of protons in the nucleus of each atom of an element; it differs for each element.

ATP Adenosine triphosphate (ah-DEN-uh-seen try-FOSS-fate) A nucleotide composed of adenine, ribose, and three phosphate groups. As the main energy carrier in cells, it directly or indirectly delivers energy to or picks up energy from nearly all metabolic pathways.

atrioventricular node (AV node) In the septum dividing the heart atria, a site that contains bundles of conducting fibers. Stimuli arriving at the AV node from the cardiac pacemaker (sinoatrial node) pass along the bundles and continue on via Purkinje fibers to contractile muscle cells in the ventricles.

atrioventricular valve One-way flow valve between the atrium and ventricle in each half of the heart.

atrium (AY-tree-um) Upper chamber in each half of the heart; the right atrium receives deoxygenated blood (from tissues) entering the pulmonary circuit of blood flow, and the left atrium receives oxygenated blood from pulmonary veins.

australopith (oss-TRAH-low-pith) [L. *australis*, southern, and Gk. *pithekos*, ape] Any of the earliest known species of hominids; that is, the first species of the evolutionary branch leading to humans.

autoimmune response Misdirected immune response in which lymphocytes mount an attack against normal body cells.

autonomic nervous system (ah-toe-NOM-ik) Those nerves leading from the central nervous system to the smooth muscle, cardiac muscle, and glands of internal organs and structures—that is, to the visceral portion of the body.

autosomal dominant inheritance Condition arising from the presence of a dominant allele on an autosome (not a sex chromosome). The allele is always expressed to some extent, even in heterozygotes.

autosomal recessive inheritance Condition arising from a recessive allele on an autosome (not a sex chromosome). Only recessive homozygotes show the resulting phenotype.

autosome Any of the chromosomes that are of the same number and kind in both males and females of the species.

autotroph (AH-toe-trofe) [Gk. *autos*, self, and *trophos*, feeder] An organism able to build its own large organic molecules by using carbon dioxide and energy from the physical environment. Compare heterotroph.

axial skeleton (AX-ee-uhl) In vertebrates, the skull, backbone, ribs, and breastbone (sternum).

axon Of a neuron, a long, cylindrical extension from the cell body, with finely branched endings. Action potentials move rapidly, without alteration, along an axon; their arrival at axon endings may trigger the release of neurotransmitter molecules that influence an adjacent cell.

B lymphocyte, or B cell The only white blood cell that produces antibodies, then positions them at the cell surface or secretes them as weapons in immune responses.

bacterial conjugation The transfer of plasmid DNA from one bacterial cell to another.

bacterial flagellum Of many bacterial cells, a whiplike motile structure that does not contain a core of microtubules.

bacteriophage (bak-TEER-ee-oh-fahj) [Gk. *bakte–rion*, small staff, rod, and *phagein*, to eat] Category of viruses that infect bacterial cells.

Barr body In the cells of female mammals, a condensed X chromosome that was inactivated during early embryonic development.

basal body A centriole that, after having given rise to the microtubules of a flagellum or cilium, remains attached to its base in the cytoplasm.

basal metabolic rate Amount of energy required to sustain body functions when a person is resting, awake, and has not eaten for 12–18 hours.

base A substance that releases OH⁻ in water.

base pair A pair of hydrogen-bonded nucleotide bases in two strands of nucleic acids. In a DNA double helix, adenine pairs with thymine, and guanine with cytosine. When an mRNA strand forms on a DNA strand during transcription, uracil (U) pairs with the DNA's adenine.

base sequence The particular order in which one nucleotide base follows the next in a strand of DNA or RNA.

basement membrane Noncellular layer of mostly proteins and polysaccharides that is sandwiched in between an epithelium and underlying connective tissue.

basophil Fast-acting white blood cells that secrete histamine and other substances during inflammation.

biogeochemical cycle The movement of an element such as carbon or nitrogen from the environment to organisms, then back to the environment.

biogeographic realm [Gk. *bios*, life, and *geographein*, to describe the surface of the earth] Any of six major land regions, each having distinguishing types of plants and animals and generally retaining its identity because of climate and geographic barriers to gene flow.

biological clock Internal time-measuring mechanism that has a role in adjusting an organism's daily activities, seasonal activities, or both in response to environmental cues.

biological magnification The increasing concentration of a nondegradable or slowly degradable substance in body tissues as it is passed along food chains.

biomass The combined weight of all the organisms at a particular trophic (feeding) level in an ecosystem.

biome A broad, vegetational subdivision of a biogeographic realm shaped by climate, topography, and composition of regional soils.

biopsy Diagnostic procedure in which a small piece of tissue is removed from the body through a hollow needle or exploratory surgery, and then examined for signs of a particular disease (often cancer).

biosphere [Gk. *bios*, life, and *sphaira*, globe] All regions of the earth's waters, crust, and atmosphere in which organisms live.

biosynthetic pathway A metabolic pathway in which small molecules are assembled into lipids, proteins, and other large organic molecules.

biotic potential Of a population, the maximum rate of increase per individual under ideal conditions.

bipedalism A habitual standing and walking on two feet, as by ostriches and humans.

blastocyst (BLASS-tuh-sist) [Gk. *blastos*, sprout, and *kystis*, pouch] In mammalian development, a blastula stage consisting of a hollow ball of surface cells and an inner cell mass.

blastula (BLASS-chew-lah) An embryonic stage consisting of a ball of cells produced by cleavage.

blood A fluid connective tissue composed of water, solutes, and formed elements (blood cells and platelets); it carries substances to and from cells and helps maintain an internal environment that is favorable for cell activities.

blood pressure Fluid pressure, generated by heart contractions, that keeps blood circulating.

blood-brain barrier Set of mechanisms that helps control which blood-borne substances reach neurons in the brain.

bone The mineral-hardened connective tissue of bones. *Bones* are organs that function in movement and locomotion, protection of other organs, mineral storage, and (in some bones) blood cell production.

brain Organ that receives, integrates, stores, and retrieves information, and coordinates appropriate responses by stimulating and inhibiting the activities of different body parts.

brain case The eight bones that together surround and protect the brain.

brainstem The vertebrate midbrain, pons, and medulla oblongata, the core of which contains the reticular formation that helps govern activity of the nervous system as a whole.

bronchiole A component of the finely branched bronchial tree inside each lung.

bronchus, plural bronchi (BRONG-cuss, BRONG-kee) [Gk. *bronchos*, windpipe] Tubelike branchings of the trachea that lead into the lungs.

buffer A substance that can combine with hydrogen ions, release them, or both. Buffers help stabilize the pH of blood and other fluids.

bulk A volume of fiber and other undigested material that absorption processes in the colon cannot decrease.

cancer A type of malignant tumor, the cells of which show profound abnormalities in the plasma membrane and cytoplasm, abnormal growth and division, and weakened capacity for adhesion within the parent tissue (leading to metastasis). Unless eradicated, cancer is lethal.

capillary [L. *capillus*, hair] A thin-walled blood vessel that functions in the exchange of gases and other substances between blood and interstitial fluid.

capillary bed Dense capillary networks containing true capillaries where exchanges occur between blood and tissues, and also thoroughfare channels that link arterioles and venules.

carbaminohemoglobin A hemoglobin molecule that has carbon dioxide bound to it; $HbCO_2$.

carbohydrate [L. *carbo*, charcoal, and *hydro*, water] A simple sugar or large molecule composed of sugar units. All cells use carbohydrates as structural materials, energy stores, and trans-

portable forms of energy. The three classes of carbohydrates include monosaccharides, oligosaccharides, and polysaccharides.

carbon cycle A biogeochemical cycle in which carbon moves from its largest reservoir in the atmosphere, through oceans and organisms, then back to the atmosphere.

carbonic anhydrase Enzyme in red blood cells that catalyzes the conversion of unbound carbon dioxide to carbonic acid and its dissociation products, thereby helping maintain the gradient that keeps carbon dioxide diffusing from interstitial fluid into the blood.

carcinogen (kar-SIN-uh-jen) An environmental agent or substance, such as ultraviolet radiation, that can trigger cancer.

carcinogenesis The transformation of a normal cell into a cancerous one.

cardiac conduction system Set of noncontractile cells in heart muscle that spontaneously produce and conduct the electrical events that stimulate heart muscle contractions.

cardiac cycle (KAR-dee-ak) [Gk. *kardia*, heart, and *kyklos*, circle] The sequence of muscle contraction and relaxation constituting one heartbeat.

cardiac pacemaker Sinoatrial (SA) node; the basis of the normal rate of heartbeat. The self-excitatory cardiac muscle cells that spontaneously generate rhythmic waves of excitation over the heart chambers.

cardiovascular system Organ system that is composed of blood, the heart, and blood vessels and that functions in the rapid transport of substances to and from cells.

carotid body Any of several sensory receptors that monitor oxygen (and to some extent, carbon dioxide) levels in blood; located at the point where carotid arteries branch to the brain.

carrier protein Type of transport protein that binds specific substances and changes shape in ways that shunt the substances across a plasma membrane. Some carrier proteins function passively, others require an energy input.

carrying capacity The maximum number of individuals in a population (or species) that can be sustained indefinitely by a given environment.

cartilage A type of connective tissue with solid yet pliable intercellular material that resists compression.

cartilaginous joint Type of joint in which cartilage fills the space between adjoining bones; only slight movement is possible.

cDNA Any DNA molecule copied from a mature mRNA transcript by way of reverse transcription.

cell [L. *cella*, small room] The smallest living unit; an organized unit that can survive and reproduce on its own, given DNA instructions and suitable environmental conditions, including appropriate sources of energy and raw materials.

cell count The number of cells of a given type in a microliter of blood.

cell cycle Events during which a cell increases in mass, roughly doubles its number of cytoplasmic components, duplicates its DNA, then undergoes nuclear and cytoplasmic division. It extends from the time a new cell is produced until it completes its own division.

cell differentiation The gene-guided process by which cells in different locations in the embryo become specialized.

cell-to-cell junction Of multicelled organisms, a point of contact that physically links two cells or that provides functional links between their cytoplasm.

cell theory A theory in biology, the key points of which are that (1) all organisms are composed of one or more cells, (2) the cell is the smallest unit that still retains a capacity for independent life, and (3) all cells arise from preexisting cells.

cell wall A rigid or semirigid wall outside the plasma membrane that supports a cell and imparts shape to it; a cellular feature of plants, fungi, protista, and most bacteria.

central nervous system The brain and spinal cord.

centriole (SEN-tree-ohl) A cylinder of triplet microtubules that gives rise to the microtubules of cilia and flagella.

centromere (SEN-troh-meer) [Gk. *kentron*, center, and *meros*, a part] A small, constricted region of a chromosome having attachment sites for microtubules that help move the chromosome during nuclear division.

cerebellum (ser-ah-BELL-um) [L. diminutive of cerebrum, brain] Hindbrain region with reflex centers for maintaining posture and refining limb movements.

cerebral cortex Thin surface layer of the cerebral hemispheres. Some regions of the cortex receive sensory input, others integrate information and coordinate appropriate motor responses.

cerebrospinal fluid Clear extracellular fluid that surrounds and cushions the brain and spinal cord.

cerebrum (suh-REE-bruhm) Part of the forebrain; the most complex integrating center.

chancroid Clinical term for the sexually transmitted disease caused by infection with the bacterium *Haemophilis ducreyi*. Infection results in soft, painful ulcers on the external genitals and can spread to pelvic tissues.

channel protein Type of transport protein that serves as a pore through which ions or other water-soluble substances move across the plasma membrane. Some channels remain open, while others are gated and open and close in controlled ways.

chemical bond A union between the electron structures of two or more atoms or ions.

chemical synapse (SIN-aps) [Gk. *synapsis*, union] A small gap, the synaptic cleft, that separates two neurons (or a neuron and a muscle cell or gland cell) and that is bridged by neurotransmitter molecules released from the presynaptic neuron.

chemiosmotic theory (keem-ee-oz-MOT-ik) Theory that an electrochemical gradient across a cell membrane drives ATP formation. Metabolic reactions cause hydrogen ions (H^+) to accumulate in a compartment formed by the membrane. The combined force of the resulting concentration and electric gradients propels hydrogen ions down the gradient, through channel proteins. Through enzyme action at these proteins, ADP and inorganic phosphate combine to form ATP.

chemoreceptor (KEE-moe-ree-SEP-tur) Sensory receptor that detects chemical energy (ions or molecules) dissolved in the surrounding fluid.

chemotherapy The use of therapeutic drugs to kill cancer cells.

chlamydia Sexually transmitted disease caused by infection by the bacteria *Chlamydia trachomatis*. The bacterium infects cells of the genital organs and urinary tract.

chlorofluorocarbon (klore-oh-FLOOR-oh-car-bun), or CFC One of a variety of odorless, invisible compounds of chlorine, fluorine, and carbon, widely used in commercial products, that are contributing to the destruction of the ozone layer above the earth's surface.

chorion (CORE-ee-on) Of placental mammals, one of four extraembryonic membranes; it becomes a major component of the placenta. Absorptive structures (villi) that develop at its surface are crucial for the transfer of substances between the embryo and mother.

chromatid Of a duplicated eukaryotic chromosome, one of two DNA molecules and its associated proteins. One chromatid remains attached to its "sister" chromatid at the centromere until they are separated from each other during a nuclear division; then each is a separate chromosome.

chromosome (CROW-moe-soam) [Gk. *chroma*, color, and *soma*, body] Of eukaryotes, a DNA molecule with many associated proteins. A bacterial chromosome does not have a comparable profusion of proteins associated with the DNA.

chromosome number Of eukaryotic species, the number of each type of chromosome in all cells except dividing germ cells or gametes.

cilium (SILL-ee-um), plural cilia [L. *cilium*, eyelid] Of eukaryotic cells, a short, hairlike projection that contains a regular array of microtubules. Cilia serve as motile structures, help create currents of fluids, or are part of sensory structures. They typically are more profuse than flagella.

circadian rhythm (ser-KAYD-ee-un) [L. *circa*, about, and *dies*, day] Of many organisms, a cycle of physiological events that is completed every 24 hours or so, even when environmental conditions remain constant.

circulatory system An organ system consisting of the heart, blood vessels, and blood; the system transports materials to and from cells and helps stabilize body temperature and pH.

cladistics An approach to biological systematics in which organisms are grouped according to similarities that are derived from a common ancestor.

cladogram Branching diagram that represents patterns of relative relationships among organisms based on discrete morphological, physiological, and behavioral traits that vary among taxa being studied.

clavicle Long, slender "collarbone" that connects the pectoral girdle with the sternum (breastbone).

cleavage Stage of development when mitotic cell divisions convert a zygote to a ball of cells, the blastula.

cleavage furrow Of a cell undergoing cytoplasmic division, a shallow, ringlike depression that forms at the cell surface as contractile microfilaments pull the plasma membrane inward. It defines where the cytoplasm will be cut in two.

climate Prevailing weather conditions for an ecosystem, including temperature, humidity, wind speed, cloud cover, and rainfall.

climax community Following primary or secondary succession, the array of species that remains more or less steady under prevailing conditions.

clonal selection theory Theory that lymphocytes activated by a specific antigen rapidly multiply and differentiate into huge subpopulations of cells, all having the parent cell's specificity against that antigen.

cloned DNA Multiple, identical copies of DNA fragments that have been inserted into plasmids or some other cloning vector.

cochlea Coiled, fluid-filled chamber of the inner ear. Sound waves striking the ear drum become converted to pressure waves in the cochlear fluid, and the pressure waves ultimately cause a membrane to vibrate and bend sensory hair cells. Signals from bent hair cells travel to the brain, where they may be interpreted as sound.

codominance Condition in which a pair of nonidentical alleles are both expressed, even though they specify two different phenotypes.

codon One of a series of base triplets in an mRNA molecule, most of which code for a sequence of amino acids of a specific polypeptide chain. (Of 64 codons, 61 specify different amino acids and three of these also serve as start signals for translation; one other serves only as a stop signal for translation.)

coenzyme A type of nucleotide that transfers hydrogen atoms and electrons from one reaction site to another. NAD^+ is an example.

coevolution The joint evolution of two or more closely interacting species; when one species evolves, the change affects selection pressures operating between the two species, so the other also evolves.

cofactor A metal ion or coenzyme; it helps catalyze a reaction or serves briefly as an agent that transfers electrons, atoms, or functional groups from one substrate to another.

colon (CO-lun) The large intestine.

community The populations of all species occupying a habitat; also applied to groups of organisms with similar lifestyles in a habitat.

compact bone Type of dense bone tissue that makes up the shafts of long bones and outer regions of all bones. Narrow channels in compact bone contain blood vessels and nerves.

comparative morphology [Gk. *morph*, form] Anatomical comparisons of major lineages.

complement system A set of about 20 proteins circulating in blood plasma with roles in nonspecific defenses and in immune responses. Some induce lysis of pathogens, others promote inflammation, and others stimulate phagocytes to engulf pathogens.

compound A substance in which the relative proportions of two or more elements never vary. Organic compounds have a backbone of carbon atoms arranged as a chain or ring structure. The simpler, inorganic compounds do not have comparable backbones.

concentration gradient A difference in the number of molecules (or ions) of a substance between two adjacent regions, as in a volume of fluid.

condensation Enzyme-mediated reaction leading to the covalent linkage of small molecules and, often, the formation of water as a by-product.

cone cell In the retina, a type of photoreceptor that responds to intense light and contributes to sharp daytime vision and color perception.

connective tissue A category of animal tissues, all having mostly the same components but in different proportions. These tissues contain fibroblasts and other cells, the secretions of which form fibers (of collagen and elastin) and a ground substance (of modified polysaccharides).

consumers [L. *consumere*, to take completely] Of ecosystems, heterotrophic organisms that obtain energy and raw materials by feeding on the tissues of other organisms. Herbivores, carnivores, omnivores, and parasites are examples.

continuous variation A more or less continuous range of small differences in a given trait among all the individuals of a population.

control group In a scientific experiment, a group used to evaluate possible side effects of a test involving an experimental group. Ideally, the control group should differ from the experimental group only with respect to the variable being studied.

core temperature The body's internal temperature, as opposed to temperatures of the tissues near its surface. Normal human core temperature is about 37 degrees C (98.6 degrees F).

cornea Transparent tissue in the outer layer of the eye, which causes incoming light rays to bend.

coronary artery Either of two arteries leading to capillaries that service cardiac muscle.

corpus callosum (CORE-pus ka-LOW-sum) A band of 200 million axons that functionally link the two cerebral hemispheres.

corpus luteum (CORE-pus LOO-tee-um) A glandular structure; it develops from cells of a ruptured ovarian follicle and secretes progesterone and some estrogen, both of which maintain the lining of the uterus (endometrium).

cortex [L. *cortex*, bark] In general, a rind-like layer; the kidney cortex is an example.

covalent bond (koe-VAY-lunt) [L. *con*, together, and *valere*, to be strong] A sharing of one or more electrons between atoms or groups of atoms. When electrons are shared equally, the bond is nonpolar. When electrons are shared unequally, the bond is polar—slightly positive at one end and slightly negative at the other.

cross-bridge formation Of muscle cells, the interaction between actin and myosin filaments that is the basis of contraction.

crossing over During prophase I of meiosis, an interaction between a pair of homologous chromosomes. Their nonsister chromatids break at the same place along their length and exchange corresponding segments at the break points. Crossing over breaks up old combinations of alleles and puts new ones together in chromosomes.

culture The sum total of behavior patterns of a social group, passed between generations by learning and by symbolic behavior, especially language.

cyclic AMP Cyclic adenosine monophosphate. A nucleotide that has roles in intercellular communication, as when it serves as a second messenger (a cytoplasmic mediator of a cell's response to signaling molecules).

cytokinesis (sigh-toe-kih-NEE-sis) [Gk. *kinesis*, motion] Cytoplasmic division; the splitting of a parental cell into daughter cells.

cytomembrane system [Gk. *kytos*, hollow vessel] Organelles, functioning as a system to modify, package, and distribute newly formed proteins and lipids. Endoplasmic reticulum, Golgi bodies, lysosomes, and a variety of vesicles are its components.

cytoplasm (SIGH-toe-plaz-um) [Gk. *plassein*, to mold] All cellular parts, particles, and semifluid substances enclosed by the plasma membrane except for the nucleus.

cytosine (SIGH-toe-seen) A pyrimidine; one of the nitrogen-containing bases in nucleotides.

cytoskeleton A cell's internal "skeleton." Its microtubules and other components structurally support the cell and organize and move its internal components.

cytotoxic T cell Type of T lymphocyte that directly kills infected body cells and tumor cells by lysis.

decomposers [L. *de-*, down, away, and *companere*, to put together] Of ecosystems, heterotrophs that obtain energy by chemically breaking down the remains, products, or wastes of other organisms. Their activities help cycle nutrients back to producers. Certain fungi and bacteria are examples.

deforestation The removal of all trees from a large tract of land, such as the Amazon Basin or the Pacific Northwest.

degradative pathway A metabolic pathway by which molecules are broken down in stepwise reactions that lead to products of lower energy.

deletion A change in a chromosome's structure after one of its regions is lost as a result of irradiation, viral attack, chemical action, or some other factor.

demographic transition model Model of human population growth in which changes in the growth pattern correspond to different stages of economic development. These are a preindustrial stage, when birth and death rates are both high; a transitional stage; an industrial stage; and a postindustrial stage, when the death rate exceeds the birth rate.

denaturation (deh-nay-chur-AY-shun) Of any molecule, the loss of three-dimensional shape following disruption of hydrogen bonds and other weak bonds.

dendrite (DEN-drite) [Gk. *dendron*, tree] A short, slender extension from the cell body of a neuron.

denitrification (dee-nite-rih-fih-KAY-shun) The conversion of nitrate or nitrite by certain bacteria to gaseous nitrogen (N_2) and a small amount of nitrous oxide (N_2O).

density-dependent controls Factors, such as predation, parasitism, disease, and competition for resources, that limit population growth by reducing the birth rate, increasing the rates of death and dispersal, or all of these.

density-independent controls Factors such as storms or floods that increase a population's death rate more or less independently of its density.

dermis The layer of skin underlying the epidermis, consisting mostly of dense connective tissue.

desertification (dez-urt-ih-fih-KAY-shun) The conversion of grasslands, rainfed cropland, or irrigated cropland to desertlike conditions, with a drop in agricultural productivity of 10 percent or more.

detrital food web Of most ecosystems, the flow of energy mainly from plants through detritivores and decomposers.

detritivores (dih-TRY-tih-vorz) [L. *detritus*; after *deterere*, to wear down] Heterotrophs that consume dead or decomposing particles of organic matter. Earthworms, crabs, and nematodes are examples.

diaphragm (DIE-uh-fram) [Gk. *diaphragma*, to partition] Muscular partition between the thoracic and abdominal cavities, the contraction and relaxation of which contributes to breathing. Also, a contraceptive device used temporarily to prevent sperm from entering the uterus during sexual intercourse.

diastole Relaxation phase of the cardiac cycle.

diffusion Net movement of like molecules (or ions) down their concentration gradient. In the absence of other forces, molecular motion and random collisions cause their net outward movement from one region into a neighboring region where they are less concentrated (because collisions are more frequent where the molecules are most crowded together).

digestive system An internal tube from which ingested food is absorbed into the internal environment; divided into regions specialized for food transport, processing, and storage.

dihybrid cross An experimental cross in which offspring inherit two gene pairs, each consisting of two nonidentical alleles.

diploid number (DIP-loyd) For many sexually reproducing species, the chromosome number of somatic cells and of germ cells prior to meiosis. Such cells have two chromosomes of each type (that is, pairs of homologous chromosomes). Compare haploid number.

directional selection Of a population, a shift in allele frequencies in a steady, consistent direction in response to a new environment or to a directional change in the old one. The outcome is that forms of traits at one end of the range of phenotypic variation become more common than the intermediate forms.

disaccharide (die-SAK-uh-ride) [Gk. *di*, two, and *sakcharon*, sugar] A type of simple carbohydrate, of the class called oligosaccharides; two monosaccharides covalently bonded.

disruptive selection Of a population, a shift in allele frequencies to forms of traits at both ends of a range of phenotypic variation and away from intermediate forms.

distal tubule The tubular section of a nephron most distant from the glomerulus; a major site of water and sodium reabsorption.

divergence Accumulation of differences in allele frequencies between populations that have become reproductively isolated from one another.

DNA Deoxyribonucleic acid (dee-OX-ee-rye-bow-new-CLAY-ik). For all cells (and many viruses), the molecule of inheritance. A category of nucleic acids, each usually consisting of two nucleotide strands twisted together helically and held together by hydrogen bonds. The nucleotide sequence encodes the instructions for assembling proteins, and, ultimately, a new individual.

DNA-DNA hybridization See nucleic acid hybridization.

DNA fingerprint Of each individual, a unique array of RFLPs, resulting from the DNA sequences inherited (in a Mendelian pattern) from each parent.

DNA library A collection of DNA fragments produced by restriction enzymes and incorporated into plasmids.

DNA ligase (LYE-gase) Enzyme that seals together the new base-pairings during DNA replication; also used by recombinant DNA technologists to seal base-pairings between DNA fragments and cut plasmid DNA.

DNA polymerase (poe-LIM-uh-rase) Enzyme that assembles a new strand on a parent DNA strand during replication; also takes part in DNA repair.

DNA probe A short DNA sequence that has been assembled from radioactively labeled nucleotides and that can base-pair with part of a gene under investigation.

DNA repair Following an alteration in the base sequence of a DNA strand, a process that restores the original sequence, as carried out by DNA polymerases, DNA ligases, and other enzymes.

DNA replication Of cells, the process by which the hereditary material is duplicated for distribution to daughter nuclei. An example is the duplication of eukaryotic chromosomes during interphase, prior to mitosis.

dominant allele In a diploid cell, an allele that masks the expression of its partner on the homologous chromosome.

dryopith A type of hominoid, one of the first to appear during the Miocene about the time of the divergences that led to gorillas, chimpanzees, and humans.

ductus deferens Tube leading to the ejaculatory duct; one of several tubes through which sperm move after they leave the testes just prior to ejaculation.

duplication A change in a chromosome's structure resulting in the repeated appearance of the same gene sequence.

dysplasia An abnormal change in the sizes, shapes, and organization of cells in a tissue.

early *Homo* A type of early hominid that may have been the maker of stone tools that date from about 2.5 million years ago.

ecology [Gk. *oikos*, home, and *logos*, reason] Study of the interactions of organisms with one another and with their physical and chemical environment.

ecosystem [Gk. *oikos*, home] An array of organisms and their physical environment, all of which interact through a flow of energy and a cycling of materials.

ecosystem analysis Method of predicting unforeseen effects of disturbances to an ecosystem, based on computer programs and models.

ectoderm [Gk. *ecto*, outside, and *derma*, skin] The outermost primary tissue layer (germ layer) of an embryo, which gives rise to the outer layer of the integument and to tissues of the nervous system.

effector A muscle (or gland) that responds to signals from an integrator (such as the brain) by producing movement (or chemical change) that helps adjust the body to changing conditions.

effector cell Of the differentiated subpopulations of lymphocytes that form during an immune response, the type of cell that engages and destroys the antigen-bearing agent that triggered the response.

egg A mature female gamete; also called an ovum.

electron Negatively charged unit of matter, with both particulate and wavelike properties, that occupies one of the orbitals around the atomic nucleus. Atoms can gain, lose, or share electrons with other atoms.

electron transport system An organized array of enzymes and cofactors, bound in a cell membrane, that accept and donate electrons in sequence. When such systems operate, hydrogen ions (H^+) flow across the membrane, and the flow drives ATP formation and other reactions.

element Any substance that cannot be decomposed into substances with different properties.

embryo (EM-bree-oh) [Gk. *en*, in, and probably *bryein*, to swell] Of animals generally, the stage formed by way of cleavage, gastrulation, and other early developmental events.

embryonic disk In early development, the oval, flattened cell mass that gives rise to the embryo shortly after implantation.

endocrine system System of cells, tissues, and organs that is functionally linked to the nervous system and that exerts control by way of its hormones and other chemical secretions.

endocytosis (EN-doe-sigh-TOE-sis) Movement of a substance into cells; the substance becomes enclosed by a patch of plasma membrane that sinks into the cytoplasm, then forms a vesicle around it. Phagocytic cells also engulf pathogens or prey in this manner.

endoderm [Gk. *endon*, within, and *derma*, skin] The inner primary tissue layer, or germ layer, of an embryo, which gives rise to the inner lining of the gut and organs derived from it.

endometrium (en-doh-MEET-ree-um) [Gk. *metrios*, of the womb] Inner lining of the uterus consisting of connective tissues, glands, and blood vessels.

endoplasmic reticulum or ER (en-doe-PLAZ-mik reh-TIK-yoo-lum) An organelle that begins at the nucleus and curves through the cytoplasm. In rough ER (which has many ribosomes on its cytoplasmic side), many new polypeptide chains acquire specialized side chains. In many cells, smooth ER (with no attached ribosomes) is the main site of lipid synthesis.

endotherm Animal such as a human that maintains body temperature from within, generally by metabolic activity and controls over heat conservation and dissipation.

energy The capacity to do work.

energy carrier A molecule that delivers energy from one metabolic reaction site to another. ATP is the most widely traveled of these; it readily donates energy to nearly all metabolic reactions.

energy pyramid A pyramid-shaped representation of an ecosystem's trophic structure, illustrating the energy losses at each transfer to a different trophic level.

enzyme (EN-zime) One of a class of proteins that greatly speed up (catalyze) reactions between specific substances. The substances that each type of enzyme acts upon are called its substrates.

eosinophil Fast-acting, phagocytic white blood cell that takes part in inflammation but not in immune responses.

epidermis The outermost tissue layer of a multicelled animal.

epiglottis A flaplike structure at the start of the larynx, the position of which directs the movement of air into the trachea or food into the esophagus.

epiphyseal plate Region of cartilage that covers either end of a growing long bone, permitting the bone to lengthen. The epiphyseal plate is replaced by bone when growth stops in late adolescence.

epithelium (ep-ih-THEE-lee-um) A tissue consisting of one or more layers of adhering cells that covers the body's external surfaces and lines its internal cavities and tubes. Epithelium has one free surface; the opposite surface rests on a basement membrane between it and an underlying connective tissue. Epidermis or skin is an example.

erythrocyte (eh-RITH-row-site) [Gk. *erythros*, red, and *kytos*, vessel] Red blood cell.

esophagus (ee-SOF-uh-gus) Tubular portion of the digestive system that receives swallowed food and leads to the stomach.

essential amino acid Any of eight amino acids that the body cannot synthesize and must be obtained from food.

essential fatty acid Any of the fatty acids that the body cannot synthesize and must be obtained from food.

estrogen (ESS-tro-jen) A sex hormone that helps oocytes mature, induces changes in the uterine lining during the menstrual cycle and pregnancy, and maintains secondary sexual traits; also influences body growth and development.

eukaryotic cell (yoo-carry-AH-tic) [Gk. *eu*, good, and *karyon*, kernel] A type of cell that has a "true nucleus" and other distinguishing membrane-bound organelles. Compare prokaryotic cell.

evolution, biological [L. *evolutio*, act of unrolling] Change within a line of descent over time. A population is evolving when some forms of a trait are becoming more or less common relative to the other kinds of traits. The shifts are evidence of changes in the relative abundances of alleles for that trait, as brought about by mutation, natural selection, genetic drift, and gene flow.

evolutionary tree A treelike diagram in which branches represent separate lines of descent from a common ancestor.

excitatory postsynaptic potential or EPSP One of two competing signals at an input zone of a neuron; a graded potential that brings the neuron's plasma membrane closer to threshold.

excretion Any of several processes by which excess water, excess or harmful solutes, or waste materials leave the body by way of the urinary system or certain glands.

exocrine gland (EK-suh-krin) [Gk. *es*, out of, and *krinein*, to separate] Glandular structure that secretes products, usually through ducts or tubes, to a free epithelial surface.

exocytosis (ek-so-sigh-TOE-sis) Movement of a substance out of a cell by means of a transport vesicle, the membrane of which fuses with the plasma membrane, so that the vesicle's contents are released outside.

exon Of eukaryotic cells, any of the nucleotide sequences of a pre-mRNA molecule that are spliced together to form the mature mRNA transcript and are ultimately translated into protein.

experiment A test in which some phenomenon in the natural world is manipulated in controlled ways to gain insight into its function, structure, operation, or behavior.

expiration Expelling air from the lungs; exhaling.

exponential growth (ex-po-NEN-shul) Pattern of population growth in which greater and greater numbers of individuals are produced during the successive doubling times; the pattern that emerges when the per capita birth rate remains even slightly above the per capita death rate, putting aside the effects of immigration and emigration.

external respiration Movement of oxygen from alveoli into the blood, and of carbon dioxide from the blood into alveoli.

extinction, background A steady rate of species turnover that characterizes lineages through most of their histories. *Mass extinction* is an abrupt increase in the rate at which major taxa disappear, with several taxa being affected simultaneously.

extracellular fluid In animals generally, all the fluid not inside cells; includes plasma (the liquid portion of blood) and interstitial fluid (which occupies the spaces between cells and tissues).

extracellular matrix A material, largely secreted, that helps hold many animal tissues together in certain shapes; it consists of fibrous proteins and other components in a ground substance.

extraembryonic membranes Membranes that form along with a developing embryo, including the yolk sac, amnion, allantois, and chorion.

F_1 (first filial generation) The offspring of an initial genetic cross.

F_2 (second filial generation) The offspring of parents who are the first filial generation from a genetic cross.

FAD Flavin adenine dinucleotide, a nucleotide coenzyme. When delivering electrons and unbound protons (H^+) from one reaction to another, it is abbreviated $FADH_2$.

family pedigree A chart of genetic relationships of the individuals in a family through successive generations.

fat A lipid with a glycerol head and one, two, or three fatty acid tails. The tails of saturated fats have only single bonds between carbon atoms and hydrogen atoms attached to all other bonding sites. Tails of unsaturated fats additionally have one or more double bonds between certain carbon atoms.

fatty acid A long, flexible hydrocarbon chain with a —COOH group at one end.

feedback inhibition Of cells, a control mechanism by which the production (or secretion) of a substance triggers a change in some activity that in turn shuts down further production of the substance.

femur Thigh bone; longest and strongest bone of the body.

fermentation [L. *fermentum*, yeast] A type of anaerobic pathway of ATP formation; it starts with glycolysis, ends when electrons are transferred back to one of the breakdown products or intermediates, and regenerates the NAD^+ required for the reaction. Its net yield is two ATP per glucose molecule broken down.

fertilization [L. *fertilis*, to carry, to bear] Fusion of a sperm nucleus with the nucleus of an egg, which thereupon becomes a zygote.

fetus Term applied to an embryo after it reaches the age of eight weeks.

fever Body temperature that has climbed above the normal set point, usually in response to infection. Mild fever promotes an increase in body defense activities.

fibrous joint Type of joint in which fibrous connective tissue unites the adjoining bones and no cavity is present.

filtration In urine formation, the process by which blood pressure forces water and solutes out of glomerular capillaries

and into the cupped portion of a nephron wall (glomerular capsule).

flagellum (fluh-JELL-um), plural flagella [L., whip] Tail-like motile structure of many free-living eukaryotic cells; it has a distinctive 9 + 2 array of microtubules.

fluid mosaic model Model of membrane structure in which proteins are embedded in a lipid bilayer or attached to one of its surfaces. The lipid molecules give the membrane its basic structure, impermeability to water-soluble molecules, and (through packing variations and movements) fluidity. Proteins carry out most membrane functions, such as transport, enzyme action, and reception of signals or substances.

follicle (FOLL-ih-kul) In an ovary, a primary oocyte (immature egg) together with the surrounding layer of cells.

food chain A straight-line sequence of who eats whom in an ecosystem.

food web A network of cross-connecting, interlinked food chains encompassing primary producers and an array of consumers, detritivores, and decomposers.

forebrain Brain region that includes the cerebrum and cerebral cortex, the olfactory lobes, and the hypothalamus.

fossil Recognizable evidence of an organism that lived in the distant past. Most fossils are skeletons, shells, leaves, seeds, and tracks that were buried in rock layers before they decomposed.

FSH Follicle-stimulating hormone. The name comes from its function in females, in whom FSH helps stimulate follicle development in ovaries. In males, it acts in the testes as part of a sequence of events that trigger sperm production.

functional group An atom or group of atoms that is covalently bonded to the carbon backbone of an organic compound and that influences its behavior.

gall bladder Organ of the digestive system that stores bile secreted from the liver.

gamete (GAM-eet) [Gk. *gamete–s*, husband, and *gamete–*, wife] A haploid cell that functions in sexual reproduction. Sperm and eggs are examples.

ganglion (GANG-lee-un), plural ganglia [Gk. *ganglion*, a swelling] A distinct clustering of cell bodies of neurons in regions other than the brain or spinal cord.

gastrointestinal (GI) tract The digestive tube, extending from the mouth to the anus and including the stomach, small and large intestines, and other specialized regions with roles in food transport and digestion.

gastrulation (gas-tru-LAY-shun) The stage of embryonic development in which cells become arranged into two or three primary tissue layers (germ layers); in humans, the layers are an inner endoderm, an intermediate mesoderm, and a surface ectoderm.

gene A unit of information about a heritable trait that is passed on from parents to offspring. Each gene has a specific location on a chromosome.

gene flow A microevolutionary process; a physical movement of alleles out of a population as individuals leave (emigrate) or enter (immigrate); allele frequencies change as a result.

gene frequency More precisely, allele frequency: the relative abundances of all the different alleles for a trait that are carried by the individuals of a population.

gene locus A given gene's particular location on a chromosome.

gene pair In diploid cells, the two alleles at a given locus on a pair of homologous chromosomes.

gene pool Sum total of all genotypes in a population. More accurately, allele pool.

gene therapy Generally, the transfer of one or more normal genes into the body cells of an organism in order to correct a genetic defect.

genetic code [After L. *genesis*, to be born] The correspondence between nucleotide triplets in DNA (then in mRNA) and specific sequences of amino acids in the resulting polypeptide chains; the basic language of protein synthesis.

genetic disorder An inherited condition that results in mild to severe medical problems.

genetic drift A microevolutionary process; a change in allele frequencies over the generations due to chance events alone.

genetic engineering Altering the information content of DNA through use of recombinant DNA technology.

genetic recombination Presence of a new combination of alleles in a DNA molecule compared to the parental genotype; the result of processes such as crossing over at meiosis, chromosome rearrangements, gene mutation, and recombinant DNA technology.

genital herpes Infection of tissues in the genital area by a herpes simplex virus; an extremely contagious sexually transmitted disease.

genome All the DNA in a haploid number of chromosomes of a given species.

genotype (JEEN-oh-type) Genetic constitution of an individual. Can mean a single gene pair or the sum total of the individual's genes. Compare phenotype.

genus, plural genera (JEEN-us, JEN-erah) [L. *genus*, race, origin] A taxon into which all species exhibiting certain phenotypic similarities and evolutionary relationship are grouped.

germ cell One of a cell lineage set aside for sexual reproduction; germ cells give rise to gametes. Compare somatic cell.

germ layer One of two or three primary tissue layers that form during gastrulation and that gives rise to certain tissues of the adult body. Compare ectoderm; endoderm; mesoderm.

gland A secretory cell or multicelled structure derived from epithelium and often connected to it.

glomerular capillaries The set of blood capillaries inside the glomerular capsule of the nephron.

glomerular (Bowman's) capsule The ballooned region of a nephron wall that forms a cup for water and other solutes filtered from the clustered blood capillaries of the glomerulus.

glomerulus (glow-MARE-you-luss) [L. *glomus*, ball] The first portion of the nephron, where water and solutes are filtered from blood.

glucagon (GLUE-kuh-gone) Hormone that stimulates conversion of glycogen and amino acids to glucose; secreted by alpha cells of the pancreas when the flow of glucose decreases.

glucocorticoid Hormone secreted by the adrenal cortex that influences metabolic reactions that help maintain the blood glucose level.

glyceride (GLISS-er-eyed) One of the molecules, commonly called fats and oils, that has one, two, or three fatty acid tails attached to a glycerol backbone. They are the body's most abundant lipids and its richest source of energy.

glycerol (GLISS-er-all) [Gk. *glykys*, sweet, and L. *oleum*, oil] A three-carbon molecule with three hydroxyl groups attached; together with fatty acids, a component of fats and oils.

glycogen (GLY-kuh-jen) In animals, a storage polysaccharide that is a main food reserve; can be readily broken down into glucose subunits.

glycolysis (gly-CALL-ih-sis) [Gk. *glykys*, sweet, and *lysis*, loosening or breaking apart] Initial reactions of both aerobic and anaerobic pathways by which glucose (or some other organic compound) is partially broken down to pyruvate with a net yield of two ATP. Glycolysis proceeds in the cytoplasm of all cells, and oxygen has no role in it.

Golgi body (GOHL-gee) Organelle in which newly synthesized polypeptide chains as well as lipids are modified and packaged in vesicles for export or for transport to specific locations within the cytoplasm.

gonad (GO-nad) Primary reproductive organ in which gametes are produced.

gonorrhea Clinical term for the sexually transmitted disease caused by the bacterium *Neisseria gonorrhoeae*. This bacterium can infect epithelial cells of the genital tract, rectum, eye membranes, and the throat.

graded potential Of neurons, a local signal that slightly changes the voltage difference across a small patch of the plasma membrane. Such signals vary in magnitude, depending on the stimulus. With prolonged or intense stimulation, they may spread to a trigger zone of the membrane and initiate an action potential.

granulocyte Class of white blood cells that have a lobed nucleus and various types of granules in the cytoplasm; includes neutrophils, eosinophils and basophils.

gray matter The dendrites, neuron cell bodies, and neuroglial cells of the spinal cord and cerebral cortex.

grazing food web Of most ecosystems, the flow of energy from plants to herbivores, then through an array of carnivores.

green revolution In developing countries, the use of improved crop varieties, modern agricultural practices (including massive inputs of fertilizers and pesticides), and equipment to increase crop yields.

greenhouse effect Warming of the lower atmosphere due to the presence of greenhouse gases: carbon dioxide, methane, nitrous oxide, ozone, water vapor, and chlorofluorocarbons.

ground substance The intercellular material made up of cell secretions and other noncellular components.

growth factor A type of signaling molecule that can influence growth by regulating the rate at which target cells divide.

guanine A nitrogen-containing base; present in one of the four nucleotide building blocks of DNA and RNA.

habitat [L. *habitare*, to live in] The type of place where an organism normally lives, characterized by physical features, chemical features, and the presence of certain other species.

hair cell Type of mechanoreceptor that may give rise to action potentials when bent or tilted.

haploid number (HAP-loyd) The chromosome number of a gamete that, as an outcome of meiosis, is only half that of the parent germ cell (it has only one of each pair of homologous chromosomes). Compare diploid number.

HCG Human chorionic gonadotropin. A hormone that helps maintain the lining of the uterus during the menstrual cycle and during the first trimester of pregnancy.

heart Muscular pump that keeps blood circulating through the animal body.

helper T cell Type of T lymphocyte that produces and secretes chemicals that promote formation of large effector and memory cell populations.

hemoglobin (HEEM-oh-glow-bin) [Gk. *haima*, blood, and L. *globus*, ball] Iron-containing, oxygen-transporting protein that gives red blood cells their color.

hemostasis (hee-mow-STAY-sis) [Gk. *haima*, blood, and *stasis*, standing] Stopping of blood loss from a damaged blood vessel through coagulation, blood vessel spasm, platelet plug formation, and other mechanisms.

hepatic portal vein Vessel that receives nutrient-laden blood from villi of the small intestine and transports it to the liver, where excess glucose is removed. The blood then returns to the general circulation via a hepatic vein.

hepatitis B An extremely contagious, sexually transmitted disease caused by infection by the hepatitis B virus. Chronic hepatitis can lead to liver cirrhosis or cancer.

herbivore [L. *herba*, grass, and *vovare*, to devour] Plant-eating animal.

heterotroph (HET-er-oh-trofe) [Gk. *heteros*, other, and *trophos*, feeder] Organism that cannot synthesize its own organic compounds and must obtain nourishment by feeding on autotrophs, each other, or organic wastes. Animals, fungi, many protista, and most bacteria are heterotrophs. Compare autotroph.

heterozygous (het-er-oh-ZYE-guss) [Gk. *zygoun*, join together] For a given trait, having nonidentical alleles at a particular locus on a pair of homologous chromosomes.

hindbrain One of the three divisions of the brain; the medulla oblongata, cerebellum, and pons; includes reflex centers for respiration, blood circulation, and other basic functions; also coordinates motor responses and many complex reflexes.

histone Any of a class of proteins that are intimately associated with DNA and that are largely responsible for its structural (and possibly functional) organization in eukaryotic chromosomes.

homeostasis (hoe-me-oh-STAY-sis) [Gk. *homo*, same, and *stasis*, standing] A physiological state in which the physical and chemical conditions of the internal environment are being maintained within tolerable ranges.

homeostatic feedback loop An interaction in which an organ (or structure) stimulates or inhibits the output of another organ, then shuts down or increases this activity when it detects that the output has exceeded or fallen below a set point.

hominid [L. *homo*, man] All species on the evolutionary branch leading to modern humans. *Homo sapiens* is the only living representative.

hominoid Apes, humans, and their recent ancestors.

Homo erectus A hominid lineage that emerged between 1.5 million and 300,000 years ago and that may include the direct ancestors of modern humans.

Homo sapiens The hominid lineage of modern humans that emerged between 300,000 and 200,000 years ago.

homologous chromosome (huh-MOLL-uh-gus) [Gk. *homologia*, correspondence] One of a pair of chromosomes that resemble each other in size, shape, and the genes they carry, and that line up with each other at meiosis I. The X and Y chromosomes differ in these respects but still function as homologues.

homologous structures The same body parts, modified in different ways, in different lines of descent from a common ancestor.

homozygous (hoe-moe-ZYE-guss) Having two identical alleles at a given locus (on a pair of homologous chromosomes).

homozygous dominant condition Having two dominant alleles at a given locus (on a pair of homologous chromosomes).

homozygous recessive condition Having two recessive alleles at a given gene locus (on a pair of homologous chromosomes).

hormone [Gk. *hormon*, to stir up, set in motion] Any of the signaling molecules secreted from endocrine glands, endocrine cells, and some neurons that the bloodstream distributes to nonadjacent target cells (any cell having receptors for that hormone).

human genome project A basic research project in which researchers throughout the world are working together to sequence the estimated 3 billion nucleotides present in the DNA of human chromosomes.

human immunodeficiency virus (HIV) A retrovirus; the pathogen that causes AIDS (acquired immune deficiency syndrome).

human papillomavirus (HPV) Virus that causes genital warts; HPV infection is suspected of having a role in the development of some cases of cervical cancer.

humerus The long bone of the upper arm.

hydrogen bond Type of chemical bond in which an atom of a molecule interacts weakly with a neighboring atom that is already taking part in a polar covalent bond.

hydrogen ion A free (unbound) proton; a hydrogen atom that has lost its electron and so bears a positive charge (H^+).

hydrologic cycle A biogeochemical cycle, driven by solar energy, in which water moves slowly through the atmosphere, on or through surface layers of land masses, to the ocean and back again.

hydrolysis (high-DRAWL-ih-sis) [L. *hydro*, water, and Gk. *lysis*, loosening or breaking apart] Enzyme-mediated reaction in which covalent bonds break, splitting a molecule into two or more parts, and H^+ and OH^- (derived from a water molecule) become attached to the exposed bonding sites.

hydrophilic [Gk. *philos*, loving] A polar substance that is attracted to the polar water molecule and so dissolves easily in water. Sugars are examples.

hydrophobic [Gk. *phobos*, dreading] A nonpolar substance that is repelled by the polar water molecule and so does not readily dissolve in water. Oil is an example.

hydrosphere All liquid or frozen water on or near the earth's surface.

hypodermis A subcutaneous layer having stored fat that helps insulate the body; although not part of skin, it anchors skin while allowing it some freedom of movement.

hypothalamus [Gk. *hypo*, under, and *thalamos*, inner chamber, or possibly *tholos*, rotunda] A brain center that monitors visceral activities (such as salt-water balance, temperature control, and reproduction) and that influences related forms of behavior (as in hunger, thirst, and sex).

hypothesis A possible explanation of a specific phenomenon.

immune response A series of events by which B and T lymphocytes recognize a specific antigen, undergo repeated cell divisions that form huge lymphocyte populations, and differentiate into subpopulations of effector and memory cells. Effector cells engage and destroy antigen-bearing agents. Memory cells enter a resting phase and are activated during subsequent encounters with the same antigen.

immune therapy In cancer treatment, the use of specific chemicals that trigger a strong immune response in the patient's body against cancer cells. In general, immune therapies are still experimental.

immunization Various processes, including vaccination, that promote increased immunity against specific diseases.

immunoglobulin Any of the five classes of antibodies. Different immunoglobulins participate in specific ways in defense and immune responses. Examples are IgM antibodies (first to be secreted during immune responses) and IgG antibodies (which activate complement proteins and neutralize many toxins).

implantation Series of events in which a trophoblast (pre-embryo) invades the endometrium (lining of the uterus) and becomes embedded there.

incomplete dominance Of heterozygotes, the appearance of a version of a trait that is somewhere between the homozygous dominant and recessive conditions.

independent assortment Mendelian principle that each gene pair tends to assort into gametes independently of other gene pairs located on nonhomologous chromosomes.

induced-fit model Model of enzyme action whereby a bound substrate induces changes in the shape of the enzyme's active site, resulting in a more precise molecular fit between the enzyme and its substrate.

industrial smog A type of gray-air smog that develops in industrialized regions when winters are cold and wet.

inflammation, acute In response to tissue damage or irritation, phagocytes and plasma proteins, including complement proteins, leave the bloodstream, then defend and help repair the tissue. Proceeds during both nonspecific and specific (immune) defense responses.

inheritance The transmission, from parents to offspring, of structural and functional patterns that have a genetic basis and are characteristic of each species.

inhibiting hormone A signaling molecule produced and secreted by the hypothalamus that controls secretions by the anterior lobe of the pituitary gland.

inhibitor A substance that can bind with an enzyme and interfere with its functioning.

inhibitory postsynaptic potential, or IPSP Of neurons, one of two competing types of graded potentials at an input zone; tends to drive the resting membrane potential away from threshold.

inner cell mass In early development, a clump of cells in the blastocyst that will give rise to the embryonic disk.

insertion The end of a muscle attached to the bone that moves most when the muscle contracts.

inspiration The drawing of air into the lungs; inhaling.

integration, neural [L. *integrare*, to coordinate] Moment-by-moment summation of all excitatory and inhibitory synapses acting on a neuron; occurs at each level of synapsing in a nervous system.

integrator Of homeostatic systems, a control point where different bits of information are pulled together in the selection of a response. The brain is an example.

integument A protective body covering such as skin.

interferon Protein produced by T cells and that interferes with viral replication. Some interferons also stimulate the tumor-killing activity of macrophages.

interleukin One of a variety of communication signals, secreted by macrophages and by helper T cells, that drive immune responses.

intermediate filament A cytoskeletal component that consists of different proteins in different types of animal cells.

internal respiration Movement of oxygen into tissues from the blood, and of carbon dioxide from tissues into the blood.

interneuron Any of the neurons in the brain and spinal cord that integrate information arriving from sensory neurons and that influence other neurons in turn.

interphase Of cell cycles, the time interval between nuclear divisions in which a cell increases its mass, roughly doubles the number of its cytoplasmic components, and finally duplicates its chromosomes (replicates its DNA).

interstitial cell In testes, cells in connective tissue around the seminiferous tubules that secrete testosterone and other signaling molecules.

interstitial fluid (in-ter-STISH-ul) [L. *interstitus*, to stand in the middle of something] That portion of the extracellular fluid occupying spaces between cells and tissues.

intervertebral disk One of a number of disk-shaped structures containing cartilage that serve as shock absorbers and flex points between bony segments of the vertebral column.

intron A noncoding portion of a newly formed mRNA molecule.

inversion A change in a chromosome's structure after a segment separated from it was then inserted at the same place, but in reverse. The reversal alters the position and order of the chromosome's genes.

ion (EYE-on) An atom or a compound that has gained or lost one or more electrons and hence has acquired an overall negative or positive charge.

ionic bond An association between ions of opposite charge.

iris Of the eye, a circular pigmented region behind the cornea with a "hole" in its center (the pupil) through which incoming light enters.

isotonic Equality in the relative concentrations of solutes in two fluids; for two fluids separated by a cell membrane, there is no net osmotic (water) movement across the membrane.

isotope (EYE-so-tope) For a given element, an atom with the same number of protons as the other atoms but with a different number of neutrons.

J-shaped curve A curve, obtained when population size is plotted against time, that is characteristic of unrestricted, exponential growth.

joint An area of contact or near-contact between bones.

juxtaglomerular apparatus In kidney nephrons, a region of contact between the arterioles of the glomerulus and the distal tubule. Cells in this region secrete renin, which triggers hormonal events that stimulate increased reabsorption of sodium.

karyotype (CARRY-oh-type) Of eukaryotic individuals (or species), the number of metaphase chromosomes in somatic cells and their defining characteristics.

keratin A tough, water-insoluble protein manufactured by most epidermal cells.

keratinization (care-at-in-iz-AY-shun) Process by which keratin-producing epidermal cells of skin die and collect at the skin surface as keratinized "bags" that form a barrier against dehydration, bacteria, and many toxic substances.

kidney One of a pair of organs that filter mineral ions, organic wastes, and other substances from the blood and help regulate the volume and solute concentrations of extracellular fluid.

kilocalorie 1,000 calories of heat energy, or the amount of energy needed to raise the temperature of 1 kilogram of water by 1°C; the unit of measure for the caloric value of foods.

kinetochore A specialized group of proteins and DNA at the centromere of a chromosome that serves as an attachment point for several spindle microtubules during mitosis or meiosis. Each chromatid of a duplicated chromosome has its own kinetochore.

Krebs cycle Together with a few conversion steps that precede it, the stage of aerobic respiration in which pyruvate is completely broken down to carbon dioxide and water. Coenzymes accept the

unbound protons (H^+) and electrons stripped from intermediates during the reactions and deliver them to the next stage.

lactate fermentation Anaerobic pathway of ATP formation in which pyruvate from glycolysis is converted to the three-carbon compound lactate, and NAD^+ (a coenzyme used in the reactions) is regenerated. Its net yield is two ATP.

lactation The production of milk by hormone-primed mammary glands.

lacteal Small lymph vessel in villi of the small intestine that receives absorbed triglycerides. Triglycerides move from the lymphatic system to the general circulation.

large intestine The colon; a region of the GI tract that receives unabsorbed food residues from the small intestine and concentrates and stores feces until they are expelled from the body.

larynx (LARE-inks) A tubular airway that leads to the lungs. It contains vocal cords, where sound waves used in speech are produced.

latency Of viruses, a period of time during which viral genes remain inactive inside the host cell.

lens Of the eye, a saucer-shaped region behind the iris containing multiple layers of transparent proteins. Ligaments can move the lens, which functions to focus incoming light onto photoreceptors in the retina.

LH Luteinizing hormone, secreted by the anterior lobe of the pituitary gland. In males it acts on interstitial cells of the testes and prompts them to secrete testosterone. In females, LH stimulates follicle development in the ovaries.

ligament A strap of dense, regular connective tissue that connects two bones at a joint.

limbic system Brain regions that, along with the cerebral cortex, collectively govern emotions.

limiting factor Any essential resource that is in short supply and so limits population growth.

lineage (LIN-ee-age) A line of descent.

linkage The tendency of genes located on the same chromosome to end up in the same gamete. For any two of those genes, the probability that crossing over will disrupt the linkage is proportional to the distance separating them.

lipid A greasy or oily compound of mostly carbon and hydrogen that shows little tendency to dissolve in water, but that dissolves in nonpolar solvents (such as ether). Cells use lipids as energy stores and structural materials, especially in membranes.

lipid bilayer The structural basis of cell membranes, consisting of two layers of mostly phospholipid molecules. Hydrophilic heads force all fatty acid tails of the lipids to become sandwiched between the hydrophilic heads.

liver Glandular organ with roles in storing and interconverting carbohydrates, lipids, and proteins absorbed from the gut; maintaining blood; disposing of nitrogen-containing wastes; and other tasks.

local signaling molecules Secretions from cells in many different tissues that alter chemical conditions in the immediate vicinity where they are secreted, then are swiftly degraded.

locus (LOW-cuss) The specific location of a particular gene on a chromosome.

logistic population growth (low-JIS-tik) Pattern of population growth in which a low-density population slowly increases in size, goes through a rapid growth phase, then levels off once the carrying capacity is reached.

loop of Henle The hairpin-shaped, tubular region of a nephron that functions in reabsorption of water and solutes.

lung Saclike organ that serves as an internal respiratory surface.

lymph (limf) [L. *lympha*, water] Tissue fluid that has moved into the vessels of the lymphatic system.

lymph capillary A small-diameter vessel of the lymph vascular system that has no obvious entrance; tissue fluid moves inward by passing between overlapping endothelial cells at the vessel's tip.

lymph node A lymphoid organ that serves as a battleground of the immune system; each lymph node is packed with organized arrays of macrophages and lymphocytes that cleanse lymph of pathogens before it reaches the blood.

lymph vascular system [L. *lympha*, water, and *vasculum*, a small vessel] The vessels of the lymphatic system, which take up and transport excess tissue fluid and reclaimable solutes as well as fats absorbed from the digestive tract.

lymphatic system An organ system that supplements the circulatory system. Its vessels take up fluid and solutes from interstitial fluid and deliver them to the bloodstream; its lymphoid organs have roles in immunity.

lymphocyte Any of various white blood cells that take part in nonspecific and specific (immune) defense responses.

lymphoid organs The lymph nodes, spleen, thymus, tonsils, adenoids, and other organs with roles in immunity.

lysis [Gk. *lysis*, a loosening] Gross structural disruption of a plasma membrane that leads to cell death.

lysosome (LYE-so-sohm) The main organelle of digestion, with enzymes that can break down polysaccharides, proteins, nucleic acids, and some lipids.

lysozyme Infection-fighting enzyme that attacks and destroys various types of bacteria by digesting the bacterial cell wall. Present in mucous membranes that line the body's surfaces.

macroevolution The large-scale patterns, trends, and rates of change among groups of species.

macrophage One of the phagocytic white blood cells. It engulfs anything detected as foreign. Some also become antigen-presenting cells that serve as the trigger for immune responses by T and B lymphocytes. Compare antigen-presenting cell.

mammal A type of vertebrate; the only animal having offspring that are nourished by milk produced by mammary glands of females.

mandible The lower jaw, the largest facial bone.

mass extinction An abrupt rise in extinction rates above the background level; a catastrophic, global event in which major taxa are wiped out simultaneously.

mass number The total number of protons and neutrons in an atom's nucleus. The relative masses of atoms are also called atomic weights.

maternal chromosome One of the chromosomes bearing the alleles that are inherited from a female parent.

mechanoreceptor Sensory cell or cell part that detects mechanical energy associated with changes in pressure, position, or acceleration.

medulla oblongata Part of the brainstem with reflex centers for respiration, blood circulation, and other vital functions.

meiosis (my-OH-sis) [Gk. *meioun*, to diminish] Two-stage nuclear division process in which the chromosome number of a germ cell is reduced by half, to the haploid number. (Each daughter nucleus ends up with one of each type of chromosome.) Meiosis is the basis of gamete formation.

memory The storage and retrieval of information about previous experiences; underlies the capacity for learning.

memory cell Any of the various B or T lymphocytes of the immune system that are formed in response to invasion by a foreign agent and that circulate for some period, available to mount a rapid attack if the same type of invader reappears.

meninges Membranes of connective tissue that are layered between the skull bones and the brain and cover and protect the neurons and blood vessels that service the brain tissue.

menopause (MEN-uh-pozz) [L. *mensis*, month, and *pausa*, stop] End of the period of a human female's reproductive potential.

menstrual cycle The cyclic release of oocytes and priming of the endometrium (lining of the uterus) to receive a fertilized egg; the complete cycle averages about 28 days in female humans.

menstruation Periodic sloughing of the blood-enriched lining of the uterus when pregnancy does not occur.

mesoderm (MEH-so-derm) [Gk. *mesos*, middle, and *derm*, skin] In an embryo, a primary tissue layer (germ layer) between ectoderm and endoderm. Gives rise to muscle; organs of circulation, reproduction, and excretion; most of the internal skeleton (when present); and connective tissue layers of the gastrointestinal tract and body covering.

messenger RNA A linear sequence of ribonucleotides transcribed from DNA and translated into a polypeptide chain; the only type of RNA that carries protein-building instructions.

metabolic pathway One of many orderly sequences of enzyme-mediated reactions by which cells normally maintain, increase, or decrease the concentrations of substances. Different pathways are linear or circular, and often they interconnect.

metabolism (meh-TAB-oh-lizm) [Gk. *meta*, change] All controlled, enzyme-mediated chemical reactions by which cells acquire and use energy. Through these reactions, cells synthesize, store, break apart, and eliminate substances in ways that contribute to growth, survival, and reproduction.

metaphase Of mitosis or meiosis II, the stage when each duplicated chromosome has become positioned at the midpoint of the microtubular spindle, with its two sister chromatids attached to microtubules from opposite spindle poles. Of meiosis I, the stage when all pairs of homologous chromosomes are positioned at the spindle's midpoint, with the two members of each pair attached to opposite spindle poles.

metastasis The process in which cancer cells break away from a primary tumor and migrate (via blood or lymphatic tissues) to other locations, where they establish new cancer sites.

MHC marker Any of a variety of proteins that are self-markers. Some occur on all body cells of an individual; others are unique to the macrophages and lymphocytes.

microevolution Changes in allele frequencies brought about by mutation, genetic drift, gene flow, and natural selection.

microfilament [Gk. *mikros*, small, and L. *filum*, thread] One of a variety of cytoskeletal components. Actin and myosin filaments are examples.

microtubular spindle A bipolar structure composed of organized arrays of microtubules that forms during nuclear division and that moves the chromosomes.

microtubule Hollow cylinder of mainly tubulin subunits; a cytoskeletal element with roles in cell shape, motion, and growth and in the structure of cilia and flagella.

microvillus (my-crow-VILL-us) [L. *villus*, shaggy hair] A slender, cylindrical extension of the animal cell surface that functions in absorption or secretion.

midbrain A brain region that evolved as a coordination center for reflex responses to visual and auditory input; together with the pons and medulla oblongata, part of the brainstem, which includes the reticular formation.

mineral An inorganic substance required for the normal functioning of body cells.

mineralocorticoid Type of hormone, secreted by the adrenal cortex, that mainly regulates the concentrations of mineral salts in extracellular fluid.

mitochondrion (my-toe-KON-dree-on), plural mitochondria Organelle that specializes in ATP formation; it is the site of the second and third stages of aerobic respiration, an oxygen-requiring pathway.

mitosis (my-TOE-sis) [Gk. *mitos*, thread] Type of nuclear division that maintains the parental chromosome number for daughter cells. It is the basis of bodily growth and, in many eukaryotic species, asexual reproduction.

molecule A unit of matter in which chemical bonding holds together two or more atoms of the same or different elements.

monoclonal antibody Antibody produced in the laboratory by a population of genetically identical cells that are clones of a single "parent" antibody-producing cell.

monohybrid cross [Gk. *monos*, alone] An experimental cross in which offspring inherit a pair of nonidentical alleles for a single trait being studied, so that they are heterozygous.

monosaccharide (mon-oh-SAK-ah-ride) [Gk. *monos*, alone, single, and *sakharon*, sugar] The simplest carbohydrate, with only one sugar unit. Glucose is an example.

monosomy Abnormal condition in which one chromosome of diploid cells has no homologue.

morphogenesis (more-foe-JEN-ih-sis) [Gk. *morphe*, form, and *genesis*, origin] Processes by which differentiated cells in an embryo become organized into tissues and organs, under genetic controls and environmental influences.

motor neuron A type of neuron; it delivers signals from the brain and spinal cord that can stimulate or inhibit the body's effectors (muscles, glands, or both).

motor unit A motor neuron and the muscle cells under its control.

mouth The oral cavity; in digestion, the site where polysaccharide breakdown begins.

multiple allele system Three or more different molecular forms of the same gene (alleles) that exist in a population.

muscle fatigue A decline in tension of a muscle that has been kept in a state of tetanic contraction as a result of continuous, high-frequency stimulation.

muscle tension A mechanical force, exerted by a contracting muscle, that resists opposing forces such as gravity and the weight of objects being lifted.

muscle tissue Tissue having cells able to contract in response to stimulation, then passively lengthen and so return to their resting stage.

muscle twitch Muscle response in which the muscle contracts briefly, then relaxes, when a brief stimulus activates a motor unit.

mutagen (MEW-tuh-jen) An environmental agent that can permanently modify the structure of a DNA molecule. Certain viruses and ultraviolet radiation are examples.

mutation [L. *mutatus*, a change] A heritable change in DNA due to the deletion, addition, or substitution of one to several bases in the nucleotide sequence.

myelin sheath Of many sensory and motor neurons, an axonal sheath that affects how fast action potentials travel; formed from the plasma membranes of Schwann cells that are wrapped repeatedly around the axon and are separated from each other by a small node.

myocardium The cardiac muscle tissue.

myofibril (MY-oh-fy-brill) One of many threadlike structures inside a muscle cell; each is functionally divided into sarcomeres, the basic units of contraction.

myosin (MY-uh-sin) A type of protein with a head and long tail. In muscle cells, it interacts with actin, another protein, to bring about contraction.

NAD Nicotinamide adenine dinucleotide; a nucleotide coenzyme. When carrying electrons and unbound protons (H^+) between reaction sites, it is abbreviated NADH.

NADP Nicotinamide adenine dinucleotide phosphate; a phosphorylated nucleotide coenzyme. When carrying electrons and unbound protons (H^+) between reaction sites, it is abbreviated $NADPH_2$.

nasal cavity The region of the respiratory system where air is warmed, moistened, and filtered of airborne particles and dust.

natural killer cell Cell of the immune system, possibly a type of lymphocyte, that kills tumor cells (by lysis) or infected cells identified as abnormal.

natural selection A microevolutionary process; a difference in survival and reproduction among members of a population that vary in one or more traits.

negative feedback mechanism A homeostatic feedback mechanism in which an activity changes some condition in the internal environment and so triggers a response that reverses the changed condition.

nephron (NEFF-ron) [Gk. *nephros*, kidney] Of the vertebrate kidney, a slender tubule in which water and solutes filtered from blood are selectively reabsorbed and in which urine forms.

nerve Cordlike communication line of the peripheral nervous system, composed of axons of sensory neurons, motor neurons, or both packed within connective tissue. In the brain and spinal cord, similar cordlike bundles are called nerve pathways or tracts.

nerve tract A bundle of myelinated axons of interneurons inside the spinal cord and brain.

nervous system System of neurons oriented relative to one another in precise message-conducting and information-processing pathways.

nervous tissue A type of tissue composed of neurons.

net energy Of energy resources available to the human population, the amount of energy that is left over after subtracting the energy used to locate, extract, transport, store, and deliver energy to consumers.

neuroendocrine control center The portions of the hypothalamus and pituitary gland that interact to control many body functions.

neuroglial cell (nur-OH-glee-uhl) One of the cells that provide structural and metabolic support for neurons and that collectively represent about half the volume of the nervous system.

neuromodulator Type of signaling molecule that influences the effects of transmitter substances by enhancing or reducing membrane responses in target neurons.

neuromuscular junction Chemical synapses between axon terminals of a motor neuron and a muscle cell.

neuron A nerve cell; the basic unit of communication in nervous systems. Neurons collectively sense environmental change, integrate sensory inputs, then activate muscles or glands that initiate or carry out responses.

neurotransmitter Any of the class of signaling molecules that are secreted from neurons, act on immediately adjacent cells, and are then rapidly degraded or recycled.

neural mutation A gene mutation that has neither harmful nor helpful effects on the individual's ability to survive and reproduce.

neutron Unit of matter, one or more of which occupies the atomic nucleus, that has mass but no electric charge.

neutrophil Phagocytic white blood cell that takes part in inflammatory responses against bacteria.

niche (nitch) [L. *nidas*, nest] Of a species, the full range of physical and biological conditions under which its members can live and reproduce.

nitrification (nye-trih-fih-KAY-shun) A chemosynthetic process in which certain bacteria strip electrons from ammonia or ammonium present in soil. The end product, nitrite (NO_2^-), is broken down to nitrate (NO_3^-) by different bacteria.

nitrogen cycle Biogeochemical cycle in which the atmosphere is the largest reservoir of nitrogen.

nitrogen fixation Process by which a few kinds of bacteria convert gaseous nitrogen (N_2) to ammonia. This dissolves rapidly in their cytoplasm to form ammonium, which can be used in biosynthetic pathways.

NK cell Natural killer cell, possibly of the lymphocyte lineage, that reconnoiters and kills tumor cells and infected body cells.

nociceptor A receptor, such as a free nerve ending, that detects any stimulus causing tissue damage.

nondisjunction Failure of one or more chromosomes to separate properly during mitosis or meiosis.

nongonococcal urethritis (NGU) An inflammation of the urethra; often caused by infection by the bacterium that causes chlamydia and considered a sexually transmitted disease.

nonsteroid hormone A type of water-soluble hormone, such as a protein hormone, that cannot cross the lipid bilayer of a target cell. These hormones enter the cell by receptor-mediated endocytosis, or they bind to receptors that activate membrane proteins or second messengers within the cell.

nuclear envelope A double membrane (two lipid bilayers and associated proteins) that is the outermost portion of a cell nucleus.

nucleic acid (new-CLAY-ik) A long, single- or double-stranded chain of four different kinds of nucleotides joined one after the other at their phosphate groups. They differ in which nucleotide base follows the next in sequence. DNA and RNA are examples.

nucleic acid hybridization The base-pairing of nucleotide sequences from different sources, as used in genetics, genetic engineering, and studies of evolutionary relationship based on similarities and differences in the DNA or RNA of different species.

nucleoid Of bacteria, a region in which DNA is physically organized apart from other cytoplasmic components.

nucleolus (new-KLEE-oh-lus) [L. *nucleolus*, a little kernel] Within the nucleus of a nondividing cell, a site where the protein and RNA subunits of ribosomes are assembled.

nucleosome (new-KLEE-oh-sohm) Of chromosomes, one of many organizational units, each consisting of a small stretch of DNA looped twice around a "spool" of histone molecules, which another histone molecule stabilizes.

nucleotide (new-KLEE-oh-tide) A small organic compound having a five-carbon sugar (deoxyribose), nitrogen-containing base, and phosphate group. Nucleotides are the structural units of adenosine phosphates, nucleotide coenzymes, and nucleic acids.

nucleotide coenzyme A protein that transports hydrogen atoms (free protons) and electrons from one reaction site to another in cells.

nucleus (NEW-klee-us) [L. *nucleus*, a kernel] Of atoms, the central core of one or more positively charged protons and (in all but hydrogen) electrically neutral neutrons. In cells, a membranous organelle that physically isolates and organizes the DNA, out of the way of cytoplasmic machinery.

nutrition All those processes by which food is selectively ingested, digested, absorbed, and later converted to the body's own organic compounds.

obesity An excess of fat in the body's adipose tissues, caused by imbalances between caloric intake and energy output.

oligosaccharide A carbohydrate consisting of a short chain of two or more covalently bonded sugar units. One subclass, disaccharides, has two sugar units. Compare monosaccharide; polysaccharide.

omnivore [L. *omnis*, all, and *vovare*, to devour] An organism able to obtain energy from more than one source rather than being limited to one trophic level.

oncogene (ON-coe-jeen) Any gene having the potential to induce cancerous transformations in a cell.

oocyte An immature egg.

oogenesis (oo-oh-JEN-uh-sis) Formation of a female gamete, from a germ cell to a mature haploid ovum (egg).

orbital Volume of space around the nucleus of an atom in which electrons are likely to be at any instant.

organ A structure of definite form and function that is composed of more than one tissue.

organ of Corti Membrane region of the inner ear that contains the sensory hair cells involved in hearing.

organ system Two or more organs that interact chemically, physically, or both in performing a common task.

organelle Of cells, an internal, membrane-bounded sac or compartment that has a specific, specialized metabolic function.

organic compound In biology, a compound assembled in cells and having a carbon backbone, often with carbon atoms arranged as a chain or ring structure.

organogenesis Stage of development in which primary tissue layers (germ layers) split into subpopulations of cells, and different lines of cells become unique in structure and function; foundation for growth and tissue specialization, when organs acquire specialized chemical and physical properties.

origin The end of a muscle that is attached to the bone that remains relatively stationary when the muscle contracts.

osmosis (oss-MOE-sis) [Gk. *osmos*, act of pushing] Of cells, the tendency of water to move through channel proteins that span a membrane in response to a concentration gradient, fluid pressure, or both. Hydrogen bonds among water molecules prevent water itself from becoming more or less concentrated; but a gradient may exist when the water on either side of the membrane has more substances dissolved in it.

osteocyte A living bone cell.

osteon A set of thin, concentric layers of compact bone tissue surrounding a narrow canal carrying blood vessels and nerves; arrays of osteons make up compact bone.

ovary (OH-vuh-ree) In females, the primary reproductive organ in which eggs form.

oviduct (OH-vih-dukt) Duct through which eggs travel from the ovary to the uterus. Formerly called Fallopian tube.

ovulation (ahv-you-LAY-shun) During each turn of the menstrual cycle, the release of a secondary oocyte (immature egg) from an ovary.

ovum (OH-vum) A mature female gamete (egg).

oxidation-reduction reaction An electron transfer from one atom or molecule to another. Often hydrogen is transferred along with the electron or electrons.

oxidative phosphorylation (foss-for-ih-LAY-shun) Final stage of aerobic respiration, in which ATP forms after hydrogen ions and electrons (from the Krebs cycle) are sent through a transport system that gives up the electrons to oxygen.

oxyhemoglobin A hemoglobin molecule that has oxygen bound to it; HbO_2.

ozone hole A pronounced seasonal thinning of the ozone layer in the lower stratosphere above Antarctica.

palate Structure that separates the nasal cavity from the oral cavity. The bone-reinforced hard palate serves as a hard surface against which the tongue can press food as it mixes it with saliva.

pancreas (PAN-cree-us) Gland that secretes enzymes and bicarbonate into the small intestine during digestion, and that also secretes the hormones insulin and glucagon.

pancreatic islets Any of the two million clusters of endocrine cells in the pancreas, including alpha cells, beta cells, and delta cells.

parasite [Gk. *para*, alongside, and *sitos*, food] An organism that obtains nutrients directly from the tissues of a living host, which it lives on or in and may or may not kill.

parasympathetic nerve Of the autonomic nervous system, any of the nerves carrying signals that tend to slow the body down overall and divert energy to basic tasks; also work continually in opposition with sympathetic nerves to bring about minor adjustments in internal organs.

parathyroid glands (pare-uh-THY-royd) Endocrine glands embedded in the thyroid gland that secrete parathyroid hormone, which helps restore blood calcium levels.

passive immunity Temporary immunity conferred by deliberately introducing antibodies into the body.

passive transport Diffusion of a solute through a channel or carrier protein that spans the lipid bilayer of a cell membrane. Its passage does not require an energy input; the protein passively allows the solute to follow its concentration gradient.

paternal chromosome One of the chromosomes bearing alleles that are inherited from a male parent.

pathogen (PATH-oh-jen) [Gk. *pathos*, suffering, and *-gene–s*, origin]. An infectious, disease-causing agent, such as a virus or bacterium.

pattern formation Mechanisms responsible for specialization and positioning of tissues during embryonic development.

pectoral girdle Set of bones, including the scapula (shoulder blade) and clavicle (collarbone), to which the long bone of each arm attaches. The pectoral girdles form the upper part of the appendicular skeleton and are only loosely attached to the rest of the body by muscles.

pedigree A chart of genetic connections among individuals, as constructed according to standardized methods.

pelvic girdle Set of bones including coxal bones that form an open basin, the pelvis; the lower part of the appendicular skeleton. The upper portions of the two coxal bones are the "hipbones"; the thighbones (femurs) join the coxal bones at hip joints. The pelvic girdle bears the body's weight when a person stands and is much more massive than the pectoral girdle.

pelvic inflammatory disease (PID) Generally, a bacterially caused inflammation of the uterus, oviducts, and ovaries. Often a complication of gonorrhea, chlamydia, or some other sexually transmitted disease.

penetrance In a given population, the percentage of individuals in which a particular genotype is expressed (that is, the percentage of individuals who have the genotype and also exhibit the corresponding phenotype).

penis Male organ that deposits sperm into the female reproductive tract.

pepsin Any of several digestive enzymes that are part of gastric fluid in the stomach.

perforin A type of protein secreted by a natural killer cell of the immune system, and which creates holes (pores) in the plasma membrane of a target cell.

peripheral nervous system (per-IF-ur-uhl) [Gk. *peripherein*, to carry around] The nerves leading into and out from the spinal cord and brain and the ganglia along those communication lines.

peristalsis (pare-ih-STAL-sis) A rhythmic contraction of muscles that moves food forward through the gastrointestinal tract.

peritoneum A lining of the coelom that also covers and helps maintain the position of internal organs.

peritubular capillaries The set of blood capillaries that threads around the tubular parts of a nephron; they function in reabsorption of water and solutes back into the body and in secretion of hydrogen ions and some other substances in the forming urine.

peroxisome Enzyme-filled vesicle in which fatty acids and amino acids are digested first into hydrogen peroxide (which is toxic), then to harmless products.

PGA Phosphoglycerate (foss-foe-GLISS-er-ate). A key intermediate in glycolysis.

PGAL Phosphoglyceraldehyde. A key intermediate in glycolysis.

pH scale A scale used to measure the concentration of free hydrogen ions in blood, water, and other solutions; pH 0 is the most acidic, 14 the most basic, and 7, neutral.

phagocyte (FAYG-uh-sight) [Gk. *phagein*, to eat, and *-kytos*, hollow vessel] A macrophage or certain other white blood cells that engulf and destroy foreign agents.

phagocytosis (fayg-uh-sigh-TOE-sis) [Gk. *phagein*, to eat, and *-kytos*, hollow vessel] Engulfment of foreign cells or substances by amoebas and some white blood cells by means of endocytosis.

pharynx (FARE-inks) A muscular tube by which food enters the gastrointestinal tract; the dual entrance for the tubular part of the digestive tract and windpipe (trachea).

phenotype (FEE-no-type) [Gk. *phainein*, to show, and *-typos*, image] Observable trait or traits of an individual; arises from interactions between genes, and between genes and the environment.

pheromone (FARE-oh-moan) [Gk. *phero*, to carry, and *-mone*, as in hormone] A type of signaling molecule secreted by exocrine glands that serves as a communication signal between individuals of the same species.

phospholipid A type of lipid that is the main structural component of cell membranes. Each has a hydrophobic tail (of two fatty acids) and a hydrophilic head that incorporates glycerol and a phosphate group.

phosphorus cycle Movement of phosphorus from rock or soil through organisms, then back to soil.

phosphorylation (foss-for-ih-LAY-shun) The attachment of unbound (inorganic) phosphate to a molecule; also the transfer of a phosphate group from one molecule to another, as when ATP phosphorylates glucose.

photochemical smog A brown-air smog that develops over large cities when the surrounding land forms a natural basin.

photoreceptor A light-sensitive sensory cell.

phylogeny Evolutionary relationships among species, starting with most ancestral forms and including the branches leading to their descendants.

pigment A light-absorbing molecule.

pilomotor response Contraction of smooth muscle controlling the erection of body hair when outside temperature drops. This creates a layer of still air that reduces heat losses from the body. (It is most effective in mammals which have more body hair than humans do.)

pineal gland (py NEEL) A light-sensitive endocrine gland that secretes melatonin, a hormone that influences reproductive cycles and the development of reproductive organs.

pioneer species Typically small plants with short life cycles that are adapted to growing in exposed, often windy areas with intense sunlight, wide swings in air temperature, and soils deficient in nitrogen and other nutrients. By improving conditions in areas they colonize, pioneers invite their own replacement by other species.

pituitary gland An endocrine gland that interacts with the hypothalamus to coordinate and control many physiological functions, including the activity of many other endocrine glands. Its posterior lobe stores and secretes hypothalamic hormones; the anterior lobe produces and secretes its own hormones.

placenta (pluh-SEN-tuh) Of the uterus, an organ composed of maternal tissues and extraembryonic membranes (the chorion especially); it delivers nutrients to the fetus and accepts wastes from it, yet allows the fetal circulatory system to develop separately from the mother's.

plasma (PLAZ-muh) Liquid component of blood; consists of water, various proteins, ions, sugars, dissolved gases, and other substances.

plasma cell Of the immune system, any of the antibody-secreting daughter cells of a rapidly dividing population of B cells.

plasma membrane Of cells, the outermost membrane. Its lipid bilayer structure and proteins carry out most functions, including transport across the membrane and reception of extracellular signals.

plasmid Of many bacteria, a small, circular molecule of extra DNA that carries only a few genes and replicates independently of the bacterial chromosome.

plasticity Of the human species, the ability to remain flexible and adapt to a wide range of environments.

plate tectonics Arrangement of the earth's outer layer (lithosphere) in slab-like plates, all in motion and floating on a hot, plastic layer of the underlying mantle.

platelet (PLAYT-let) Any of the cell fragments in blood that release substances necessary for clot formation.

pleiotropy (pleye-AH-troe-pee) [Gk. *pleon*, more, and *trope*, direction] A type of gene interaction in which a single gene exerts multiple effects on seemingly unrelated aspects of an individual's phenotype.

pleura Thin, double membrane surrounding each lung.

polar body Any of three cells that form during the meiotic cell division of an oocyte; the division also forms the mature egg, or ovum.

pollutant Any substance with which an ecosystem has had no prior evolutionary experience in terms of kinds or amounts, and that can accumulate to disruptive or harmful levels. Can be naturally occurring or synthetic.

polymer (PAH-lih-mur) [Gk. *polus*, many, and *meris*, part] A molecule composed of three to millions of small subunits that may or may not be identical.

polymerase chain reaction or PCR DNA amplification method; DNA containing a gene of interest is split into single strands, which enzymes (polymerases) copy; the enzymes also act on the accumulating copies, multiplying the gene sequence by the millions.

polymorphism (poly-MORE-fizz-um) [Gk. *polus*, many, and *morphe*, form] Of a population, the persistence through the generations of two or more forms of a trait, at a frequency greater than can be maintained by new mutations alone.

polypeptide chain Three or more amino acids joined by peptide bonds.

polyploidy (PAHL-ee-ployd-ee) A change in the chromosome number following inheritance of three or more of each type of chromosome.

polysaccharide [Gk. *polus*, many, and *sakharon*, sugar] A straight or branched chain of hundreds of thousands of covalently linked sugar units of the same or different kinds. The most common polysaccharides are cellulose, starch, and glycogen.

polysome Of protein synthesis, several ribosomes all translating the same messenger RNA molecule, one after the other.

population A group of individuals of the same species occupying a given area.

population density The number of individuals of a population that are living in a specified area or volume.

population distribution The general pattern of dispersion of individuals of a population throughout their habitat.

population size The number of individuals that make up the gene pool of a population.

positive feedback mechanism Homeostatic mechanism by which a chain of events is set in motion that intensifies a change from an original condition; after a limited time, the intensification reverses the change.

post-translational controls Of eukaryotes, controls that govern modification of newly formed polypeptide chains into functional enzymes and other proteins.

prediction A claim about what you can expect to observe in nature if a theory or hypothesis is correct.

primary immune response Actions by white blood cells and their products elicited by a first-time encounter with an antigen; includes both antibody-mediated and cell-mediated responses.

primary productivity Of ecosystems, *gross* primary productivity is the rate at which the producer organisms capture and store a given amount of energy during a specified interval. *Net* primary productivity is the rate of energy storage in the tissues of producers in excess of their rate of aerobic respiration.

primate The mammalian lineage that includes prosimians, tarsioids, and anthropoids (monkeys, apes, and humans).

probability With respect to any chance event, the most likely number of times it will turn out a certain way, divided by the total number of all possible outcomes.

producers, primary Of ecosystems, the organisms that secure energy from the physical environment, as by photosynthesis or chemosynthesis.

progesterone (pro-JESS-tuh-rown) Female sex hormone secreted by the ovaries.

prokaryotic cell (pro-carry-OH-tic) [L. *pro,* before, and Gk. *karyon,* kernel] A bacterium; a single-celled organism that has no nucleus or any of the other membrane-bound organelles characteristic of eukaryotic cells.

promoter Of transcription, a base sequence that signals the start of a gene; the site where RNA polymerase initially binds.

prophase Of mitosis, the stage when each duplicated chromosome starts to condense, microtubules form a spindle apparatus, and the nuclear envelope starts to break up.

prophase I Of meiosis, the stage at which the microtubular spindle starts to form, the nuclear envelope starts to break up, and each duplicated chromosome also condenses and pairs with its homologous partner. At this time, their sister chromatids typically undergo crossing over and geneticz recombination.

prophase II Of meiosis, a brief stage after interkinesis during which each chromosome still consists of two chromatids.

prostaglandin Any of various lipids present in tissues throughout the body and that can act as local signaling molecules. Prostaglandins typically cause smooth muscle to contract or relax, as in blood vessels, the uterus, and respiratory airways.

prostate gland Gland in males that wraps around the urethra and ejaculatory ducts; its secretions become part of semen.

protein Large organic compound composed of one or more chains of amino acids held together by peptide bonds. Proteins have unique sequences of different kinds of amino acids in their polypeptide chains; such sequences are the basis of a protein's three-dimensional structure and chemical behavior.

proto-oncogene A gene sequence similar to an oncogene but that codes for a protein required in normal cell function; may trigger cancer, generally when specific mutations alter its structure or function.

proton Positively charged particle, one or more of which is present in the atomic nucleus.

proximal tubule Of a nephron, the tubular region that receives water and solutes filtered from the blood.

pulmonary circuit Blood circulation route leading to and from the lungs.

pulse Rhythmic pressure surge of blood flowing in an artery, created during each cardiac cycle when a ventricle contracts.

Punnett square A method to predict the possible outcome of a mating or an experimental cross in simple diagrammatic form.

purine Nucleotide base having a double ring structure. Adenine and guanine are examples.

pyrimidine (pie-RIM-ih-deen) Nucleotide base having a single ring structure. Cytosine and thymine are examples.

pyruvate (pie-ROO-vate) A compound with a backbone of three carbon atoms. Two pyruvate molecules are the end products of glycolysis.

r Designates net population growth rate; the birth and death rates are assumed to remain constant and so are combined into this one variable for population growth equations.

radioisotope An unstable atom that has dissimilar numbers of protons and neutrons and that spontaneously decays (emits electrons and energy) to a new, stable atom that is not radioactive.

radius One of two long bones of the forearm that extend from the humerus (at the elbow joint) to the wrist. The radius runs along the "thumb side" of the forearm, parallel to the ulna.

reabsorption In the kidney, the diffusion or active transport of water and usable solutes out of a nephron and into capillaries leading back to the general circulation; regulated by ADH and aldosterone.

receptor Of cells, a molecule at the surface of the plasma membrane or in the cytoplasm that binds molecules present in the extracellular environment. The binding triggers changes in cellular activities. Of nervous systems, a sensory cell or cell part that may be activated by a specific stimulus.

receptor protein Protein that binds a signaling molecule such as a hormone, then triggers alterations in cell behavior or metabolism.

recessive allele [L. *recedere,* to recede] In heterozygotes, an allele whose expression is fully or partially masked by expression of its partner; fully expressed only in the homozygous recessive condition.

recognition protein Protein at cell surface recognized by cells of like type; helps guide the ordering of cells into tissues during development and functions in cell-to-cell interactions.

recombinant DNA technology Procedures by which DNA (genes) from different species may be isolated, cut, spliced together, and the new recombinant molecules multiplied in quantity in a population of rapidly dividing cells such as bacteria.

rectum Final region of the gastrointestinal tract, which receives and temporarily stores undigested food residues (feces).

red blood cell Erythrocyte; an oxygen-transporting cell in blood.

red marrow A substance in the spongy tissue of many bones that serves as a major site of blood cell formation.

reflex [L. *reflectere,* to bend back] A simple, stereotyped movement elicited directly by sensory stimulation.

reflex arc [L. *reflectere,* to bend back] Type of neural pathway in which signals from sensory neurons directly stimulate or inhibit motor neurons without intervention by interneurons.

refractory period Of neurons, the period following an action potential at a given patch of membrane when sodium gates are shut and potassium gates are open, so that the patch is insensitive to stimulation.

regulatory protein A protein that enhances or suppresses the rate at which a gene is transcribed.

releasing hormone A hypothalamic signaling molecule that stimulates or slows down secretion by target cells in the anterior lobe of the pituitary gland.

renal corpuscle In a kidney, the site where the nephron wall balloons around the glomerulus (a cluster of capillaries). Compare glomerular capsule.

repressor protein Regulatory protein that provides negative control of gene activity by preventing RNA polymerase from binding to DNA.

reproduction In biology, processes by which a new generation of cells or multicelled individuals is produced. Sexual reproduction requires meiosis, formation of gametes, and fertilization. Asexual reproduction refers to the production of new individuals by any mode that does not involve gametes.

reproduction, sexual Mode of reproduction that begins with meiosis, proceeds through gamete formation, and ends at fertilization.

reproductive isolating mechanism Any aspect of structure, functioning, or behavior that restricts gene flow between two populations.

reproductive isolaion An absence of gene flow between populations.

reproductive success The survival and production of the offspring of an individual.

respiration [L. *respirare*, to breathe] The overall exchange of oxygen from the environment for carbon dioxide wastes from cells by way of circulating blood. Compare aerobic respiration.

respiratory bronchiole Smallest airway in the respiratory system; opens onto alveoli.

respiratory surface In alveoli of the lungs, the thin, moist membrane across which gases diffuse.

respiratory system An organ system that functions in respiration.

resting membrane potential Of neurons and other excitable cells that are not being stimulated, the steady voltage difference across the plasma membrane.

restriction enzymes Class of bacterial enzymes that cut apart foreign DNA injected into them, as by viruses; also used in recombinant DNA technology.

restriction fragment A piece of DNA that has been spliced out of a chromosome by restriction enzymes.

reticular activating system A branch of the brain's reticular formation that controls the changing levels of consciousness by sending signals to the spinal cord, cerebellum, and cerebrum, as well as back to itself.

reticular formation Of the brainstem, a major network of interneurons that helps govern activity of the whole nervous system.

reverse transcriptase Viral enzyme required for reverse transcription of mRNA into DNA; used in recombinant DNA technology.

reverse transcription Assembly of DNA on a single-stranded mRNA molecule by viral enzymes.

RFLPs Restriction fragment length polymorphisms. Of DNA samples from different individuals, slight but unique differences in the banding pattern of fragments of the DNA that have been cut with restriction enzymes.

Rh blood typing A method of characterizing red blood cells on the basis of a protein that serves as a self-marker at their surface; Rh^+ signifies its presence and Rh^-, its absence.

rhodopsin Substance in rod cells of the eye consisting of the protein opsin and a side group, cis-retinal. When the side group absorbs incoming light energy, a series of chemical events follow that result in action potentials in associated neurons.

ribosomal RNA (rRNA) Type of RNA molecule that combines with proteins to form ribosomes, on which the polypeptide chains of proteins are assembled.

ribosome In all cells, the structure at which amino acids are strung together in specified sequence to form the polypeptide chains of proteins. An intact ribosome consists of two subunits, each composed of ribosomal RNA and protein molecules.

RNA Ribonucleic acid. A category of single-stranded nucleic acids that function in processes by which genetic instructions are used to build proteins.

rod cell Of the retina, a photoreceptor sensitive to very dim light and that contributes to coarse perception of movement.

S-shaped curve A curve, obtained when population size is plotted against time, that is characteristic of logistic growth.

salinization A salt buildup in soil as a result of evaporation, poor drainage, and often the importation of mineral salts in irrigation water.

salivary amylase Starch-degrading enzyme in saliva.

salivary gland Any of the glands that secrete saliva, a fluid that initially mixes with food in the mouth and starts the breakdown of starch.

salt An ionic compound formed when an acid reacts with a base.

saltatory conduction In myelinated neurons, rapid, node-to-node hopping of action potentials.

sarcomere (SAR-koe-meer) The basic unit of muscle contraction; a region of myosin and actin filaments organized in parallel between two Z lines of a myofibril inside a muscle cell.

sarcoplasmic reticulum (sar-koe-PLAZ-mik reh-TIK-you-lum) In muscle cells, a membrane system that takes up, stores, and releases the calcium ions required for cross-bridge formation in sarcomeres, hence for contraction.

scapula Flat, triangular bone on either side of the pectoral girdle; the scapulae form the shoulder blades.

Schwann cells Specialized neuroglial cells that grow around neuron axons, forming a myelin sheath.

second messenger A molecule inside a cell that mediates and generally triggers amplified response to a hormone.

secondary immune response Rapid, prolonged response by white blood cells, memory cells especially, to a previously encountered antigen.

secondary oocyte An oocyte (unfertilized egg cell) that has completed meiosis I; it is this haploid cell that is released at ovulation.

secondary sexual trait Trait associated with maleness or femaleness, but not directly involved with reproduction. Beard growth in males and breast development in females are examples.

secretion Generally, the release of a substance for use by the organism producing it. (Not the same as excretion, the expulsion of excess or waste material.) Of kidneys, a regulated stage in urine formation in which ions and other substances move from capillaries into nephrons.

segregation, Mendelian principle of [L. *se-*, apart, and *grex*, herd] The principle that diploid organisms inherit a pair of genes for each trait (on a pair of homologous chromosomes) and that the two genes segregate during meiosis and end up in separate gametes.

selective gene expression Activation or suppression of a fraction of the genes in unique ways in different cells, leading to pronounced differences in structure and function among different cell lineages.

semen [L. *serere*, to sow] Sperm-bearing fluid expelled from the penis during male orgasm.

semicircular canals Fluid-filled canals positioned at different angles within the vestibular apparatus of the inner ear and that contain sensory receptors that detect head movements, deceleration, and acceleration.

semiconservative replication [Gk. *he–mi*, half, and L. *conservare*, to keep] Reproduction of a DNA molecule when a complementary strand forms on each of the unzipping strands of an existing DNA double helix, the outcome being two "half-old, half-new" molecules.

semilunar valve In each half of the heart; during each heartbeat, it opens and closes in ways that keep blood flowing in one direction from the ventricle to the arteries leading away from it.

seminiferous tubules Coiled tubes inside the testes where sperm develop.

senescence (sen-ESS-cents) [L. *senescere*, to grow old] Sum total of processes leading to the natural death of an organism or some of its parts.

sensation The conscious awareness of a stimulus.

sensory neuron Any of the nerve cells that act as sensory receptors, detecting specific stimuli (such as light energy) and relaying signals to the brain and spinal cord.

sensory system Element of the nervous system consisting of sensory receptors (such as photoreceptors), nerve pathways from the receptors to the brain, and brain regions that process sensory information.

septum Of the heart, a thick wall that divides the heart into right and left halves.

Sertoli cells Cells in seminiferous tubules that nourish and otherwise aid the development of sperm.

sex chromosome A chromosome whose presence determines a new individual's gender. Compare autosomes.

shifting cultivation The cutting and burning of trees, followed by tilling of ashes into the soil; once called "slash-and-burn agriculture."

sinoatrial node Region of conducting cells in the upper wall of the right atrium that generate periodic waves of excitation that stimulate the atria to contract.

sinus In the skull, an air-filled space lined with mucous membrane and that functions to lighten the skull.

sister chromatids Of a duplicated chromosome, two DNA molecules (and associated proteins) that remain attached at their centromere only during nuclear division. Each ends up in a separate daughter nucleus.

skeletal muscle Type of muscle that interacts with the skeleton to bring about body movements. A skeletal muscle typically consists of bundles of many long cylindrical cells encapsulated by connective tissue.

skull Bony structure that includes more than two dozen bones, including bones of the brain case and facial bones.

sliding filament mechanism Model of muscle contraction in which myosin filaments physically slide along and pull two sets of actin filaments toward the center of the sarcomere, which shortens. The sliding requires ATP energy and cross-bridge formation between the actin and myosin.

small intestine The portion of the digestive system where digestion is completed and most nutrients absorbed.

smog, industrial Gray-colored air pollution that predominates in industrialized cities with cold, wet winters.

smog, photochemical Form of brown, smelly air pollution occurring in large cities with warm climates.

sodium-potassium pump A transport protein spanning the lipid bilayer of the plasma membrane. When activated by ATP, its shape changes and it selectively transports sodium ions out of the cell and potassium ions in.

solute (SOL-yoot) [L. *solvere*, to loosen] Any substance dissolved in a solution. In water, this means spheres of hydration surround the charged parts of individual ions or molecules and keep them dispersed.

solvent Fluid in which one or more substances is dissolved.

somatic cell (so-MAT-ik) [Gk. *soma–*, body] Any body cell that is not a germ cell (which gives rise to gametes).

somatic nervous system Those nerves leading from the central nervous system to skeletal muscles.

somatic sensation Awareness of touch, pressure, heat, cold, pain, and limb movement.

speciation (spee-cee-AY-shun) The evolutionary process by which species originate. One speciation route starts with divergence of two reproductively isolated populations of a species. They become separate species when accumulated differences in allele frequencies prevent them from interbreeding successfully under natural conditions.

species (SPEE-ceez) [L. *species*, a kind] Of sexually reproducing organisms, a unit consisting of one or more populations of individuals that can interbreed under natural conditions to produce fertile offspring that are reproductively isolated from other such units.

sperm [Gk. *sperma*, seed] A type of mature male gamete.

spermatogenesis (sperm-at-oh-JEN-ih-sis) Formation of a mature sperm from a germ cell.

sphere of hydration Through positive or negative interactions, a clustering of water molecules around the individual molecules of a substance placed in water. Compare solute.

sphincter (SFINK-tur) Ring of muscle between regions of a tubelike system (as between the stomach and small intestine).

spinal cord Of the central nervous system, the portion threading through a canal inside the vertebral column and providing direct reflex connections between sensory and motor neurons as well as communication lines to and from the brain.

spindle apparatus A type of bipolar structure that forms during mitosis or meiosis and that moves the chromosomes. It consists of two sets of microtubules that extend from the opposite poles and that overlap at the spindle's equator.

spleen One of the lymphoid organs; it is a filtering station for blood, a reservoir of red blood cells, and a reservoir of macrophages.

spongy bone Type of bone tissue in which hard, needlelike struts separate large spaces filled with marrow. Spongy bone occurs at the ends of long bones and within the breastbone, pelvis, and bones of the skull.

stabilizing selection Of a population, a persistence over time of the alleles responsible for the most common phenotypes.

start codon Of protein synthesis, a base triplet in a strand of mRNA that serves as the start signal for mRNA translation.

stem cell Unspecialized cell that can give rise to descendants that differentiate into specialized cells.

sternum Elongated flat bone (also called the breastbone) to which the upper ribs attach and so form the rib cage.

steroid (STAIR-oid) A lipid with a backbone of four carbon rings and with no fatty acid tails. Steroids differ in their functional groups. Different types have roles in metabolism, intercellular communication, and cell membranes.

steroid hormone A type of lipid-soluble hormone synthesized from cholesterol. Many steroid hormones move into the nucleus and bind to a receptor for it there; others bind to a receptor in the cytoplasm, and the entire complex moves into the nucleus.

stimulus [L. *stimulus*, goad] A specific change in the environment, such as a variation in light, heat, or mechanical pressure, that the body can detect through sensory receptors.

stomach A muscular, stretchable sac that receives ingested food; the organ between the esophagus and intestine in which considerable protein digestion occurs.

stop codon Of protein synthesis, a base triplet in a strand of mRNA that serves as the stop signal for translation, so that no more amino acids are added to the polypeptide chain.

stromatolite Of shallow seas, layered structures formed from sediments and large mats of the slowly accumulated remains of photosynthetic populations.

substrate A reactant or precursor molecule for a metabolic reaction; a specific molecule or molecules that an enzyme can chemically recognize, bind briefly to itself, and modify in a specific way.

substrate-level phosphorylation The direct, enzyme-mediated transfer of a phosphate group from the substrate of a reaction to another molecule. An example is the transfer of phosphate from an intermediate of glycolysis to ADP, forming ATP.

succession, primary (suk-SESH-un) [L. *succedere*, to follow after] Orderly changes from the time pioneer species colonize a barren habitat through replacements by various species until the climax community, when the composition of species remains steady under prevailing conditions.

succession, secondary Orderly changes in a community or patch of habitat toward the climax state after having been disturbed, as by fire.

surface-to-volume ratio A mathematical relationship in which volume increases with the cube of the diameter, but surface area increases only with the square. Of growing cells, the volume of cytoplasm increases more rapidly than the surface area of the plasma membrane that must service the cytoplasm. Because of this constraint, cells generally remain small or elongated, or have elaborate membrane foldings.

survivorship curve A plot of the age-specific survival of a group of individuals in a given environment, from the time of their birth until the last one dies.

sympathetic nerve Of the autonomic nervous system, any of the nerves generally concerned with increasing overall body activities during times of heightened awareness, excitement, or danger; also work continually in opposition with parasympathetic nerves to bring about minor adjustments in internal organs.

synaptic integration (sin-AP-tik) The moment-by-moment combining of excitatory and inhibitory signals arriving at a trigger zone of a neuron.

synovial joint Freely movable joint in which adjoining bones are separated by a fluid-filled cavity and stabilized by straplike ligaments. An example is the ball-and-socket joint at the hip.

syphilis Clinical term for the sexually transmitted disease caused by infection by the spirochete bacterium *Treponema pallidum*. Untreated syphilis can lead to lesions in mucous membranes, the eyes, bones, skin, liver, and central nervous system.

systemic circuit (sis-TEM-ik) Circulation route in which oxygenated blood flows from the lungs to the left half of the heart, through the rest of the body (where it gives up oxygen and takes on carbon dioxide), then back to the right side of the heart.

systole Contraction phase of the cardiac cycle.

T lymphocyte A white blood cell with roles in immune responses.

target cell Any cell that has receptors for a specific signaling molecule and that may alter its behavior in response to the molecule.

tectorial membrane Inner ear structure against which sensory hair cells are bent, producing action potentials that travel to the brain via the auditory nerve.

telophase (TEE-low-faze) Of mitosis, the final stage when chromosomes decondense into threadlike structures and two daughter nuclei form. Of meiosis I, the stage when one of each pair of homologous chromosomes has arrived at one or the other end of the spindle pole. At telophase II, chromosomes decondense and four daughter nuclei form.

telophase II Of meiosis, final stage when four daughter nuclei form.

temperate pathway A viral infection that enters a latent period; the host is not killed outright.

temporal summation The adding together (summing) of several muscle contractions, resulting in a single, stronger contraction, when stimulatory signals arrive in rapid succession.

tendon A cord or strap of dense, regular connective tissue that attaches a muscle to bone or to another muscle.

test An attempt to produce actual observations that match predicted or expected observations.

testcross Experimental cross to reveal whether an organism is homozygous dominant or heterozygous for a trait. The organism showing dominance is crossed to an individual known to be homozygous recessive for the same trait.

testis, plural testes Male gonad; primary reproductive organ in which male gametes and sex hormones are produced.

testosterone (tess-TOSS-tuh-rown) In males, a major sex hormone that helps control reproductive functions.

tetany Condition in which a muscle motor unit is maintained in a state of contraction for an extended period.

theory A testable explanation of a broad range of related phenomena. In modern science, only explanations that have been extensively tested and can be relied on with a very high degree of confidence are accorded the status of theory.

thermal inversion Situation in which a layer of dense, cool air becomes trapped beneath a layer of warm air; can cause air pollutants to accumulate to dangerous levels close to the ground.

thermoreceptor Sensory cell that can detect radiant energy associated with temperature.

thirst center Cluster of nerve cells in the hypothalamus that can inhibit saliva production, resulting in mouth dryness that the brain interprets as thirst and leading a person to seek out drinking fluids.

threshold Of neurons and other excitable cells, a certain minimum amount by which the voltage difference across the plasma membrane must change to produce an action potential.

thymine Nitrogen-containing base in some nucleotides.

thymus A lymphoid organ with endocrine functions; lymphocytes of the immune system multiply, differentiate, and mature in its tissues, and its hormone secretions affect their functions.

thyroid gland Of the endocrine system, a gland that produces hormones that affect overall metabolic rates, growth, and development.

tidal volume Volume of air, about 500 milliliters, that enters or leaves the lungs in a normal breath.

tissue A group of cells and intercellular substances that function together in one or more specialized tasks.

tonicity The relative concentrations of solutes in two fluids, such as inside and outside a cell. When solute concentrations are isotonic (equal in both fluids), water shows no net osmotic movement in either direction. When one fluid is hypotonic (has less solutes than the other), the other is hypertonic (has more solutes) and is the direction in which water tends to move.

tracer A radioisotope used to label a substance so that its pathway or destination in a cell, organism, ecosystem, or some other system can be tracked, as by scintillation counters that detect its emissions.

trachea (TRAY-kee-uh) The windpipe, which carries air between the larynx and bronchi.

transcript-processing controls Controls that govern modification of new mRNA molecules into mature transcripts before shipment from the nucleus.

transcription [L. *trans*, across, and *scribere*, to write] Of protein synthesis, the assembly of an RNA strand on one of the two strands of a DNA double helix; the base sequence of the resulting transcript is complementary to the DNA region on which it was assembled.

transcriptional controls Controls influencing when and to what degree a particular gene will be transcribed.

transfer RNA (tRNA) Of protein synthesis, any of the type of RNA molecules that bind and deliver specific amino acids to ribosomes and pair with mRNA code words for those amino acids.

translation In protein synthesis, the conversion of the coded sequence of information in mRNA into a particular sequence of amino acids to form a polypeptide chain; depends on interactions of rRNA, tRNA, and mRNA.

translational controls Controls governing the rates at which mRNA transcripts that reach the cytoplasm will be translated into polypeptide chains at ribosomes.

translocation Of cells, a change in a chromosome's structure following the insertion of part of a nonhomologous chromosome into it.

transposable element DNA element that can spontaneously "jump" to new locations in the same DNA molecule or a different one. Such elements often inactivate the genes into which they become inserted and give rise to observable changes in phenotype.

trisomy (TRY-so-mee) Of diploid cells, the abnormal presence of three of one type of chromosome.

trophic level (TROE-fik) [Gk. *trophos*, feeder] All the organisms in an ecosystem that are the same number of transfer steps away from the energy input into the system.

trophoblast Surface layer of cells of the blastocyst that secrete enzymes that break down the uterine lining where the forthcoming embryo will implant.

tumor A tissue mass composed of cells that are dividing at an abnormally high rate.

tumor marker A substance that is produced by a specific type of cancer cell or by normal cells in response to cancer.

tumor suppressor gene A gene whose protein product operates to keep cell growth and division within normal bounds, or whose product has a role in keeping cells anchored in place within a tissue.

tympanic membrane The eardrum, which vibrates when struck by sound waves.

ulna One of two long bones of the forearm; the ulna extends along the "little finger" side of the forearm, parallel to the radius on the "thumb" side.

umbilical cord Structure containing blood vessels that connect a fetus to its mother's circulatory system.

uracil (YUR-uh-sill) Nitrogen-containing base found in RNA molecules; can base-pair with adenine.

ureter Tubular channel that carries urine from each kidney to the urinary bladder.

urethra Tube that carries urine from the bladder to the body surface.

urinary bladder Storage organ for urine.

urinary excretion A mechanism by which excess water and solutes are removed by way of the urinary system.

urinary system An organ system that adjusts the volume and composition of blood and so helps maintain extracellular fluid.

urine Fluid formed by filtration, reabsorption, and secretion in kidneys; consists of wastes, excess water, and solutes.

uterus (YOU-tur-us) [L. *uterus*, womb] Chamber in which the developing embryo is contained and nurtured during pregnancy.

vaccine Antigen-containing preparation injected into the body or taken orally; it elicits an immune response leading to the proliferation of memory cells that offer long-lasting protection against a particular pathogen.

vagina Part of a female reproductive system that receives sperm, forms part of the birth canal, and channels menstrual flow to the exterior.

variable Of a scientific experiment, the only factor that is not exactly the same in the experimental group as it is in the control group.

vasoconstriction Decrease in the diameter of an arteriole, so that blood pressure rises; may be triggered by the hormones epinephrine and angiotensin.

vasodilation Enlargement of arteriole diameter, so that blood pressure falls; may be triggered by hormones including epinephrine and angiotensin.

vein Of the circulatory system, any of the large-diameter vessels that lead back to the heart.

ventricle (VEN-tri-kul) Of the heart, one of two chambers from which blood is pumped out. Compare atrium.

venule Small blood vessel that receives blood from tissue capillaries and merges into larger-diameter veins; a limited amount of diffusion occurs across venule walls.

vertebra, plural vertebrae One of a series of hard bones arranged with intervertebral disks into a backbone.

vertebrate Animal having a backbone of bony segments, the vertebrae.

vesicle (VESS-ih-kul) [L. *vesicula,* little bladder] Within the cytoplasm of cells, one of a variety of small membrane-bound sacs that function in the transport, storage, or digestion of substances or in some other activity.

vestibular apparatus A closed system of fluid-filled canals and sacs in the inner ear that functions in the sense of balance. Compare semicircular canals.

villus (VIL-us), plural villi Any of several types of absorptive structures projecting from the free surface of an epithelium.

virus A noncellular infectious agent consisting of DNA or RNA and a protein coat; can replicate only after its genetic material enters a host cell and subverts its metabolic machinery.

vision Precise light focusing onto a layer of photoreceptive cells that is dense enough to sample details concerning a given light stimulus, followed by image formation in the brain.

vital capacity Maximum volume of air that can move out of the lungs after a person inhales as deeply as possible.

vitamin Any of more than a dozen organic substances that the body requires in small amounts for normal cell metabolism but generally cannot synthesize for itself.

vocal cords A pair of elastic ligaments on either side of the larynx wall. Air forced between them causes the cords to vibrate and produce sounds.

water table The upper limit at which the ground in a specified region is fully saturated with water.

watershed Any specified region in which all precipitation drains into a single stream or river.

wax A type of lipid with long-chain fatty acid tails that help form protective, lubricating, or water-repellent coatings.

white blood cell Leukocyte; any of the macrophages, eosinophils, neutrophils, and other cells that, together with their products, make up the immune system.

white matter Of the spinal cord, major nerve tracts so named because of the glistening myelin sheaths of their axons.

X chromosome A sex chromosome with genes that cause an embryo to develop into a female, provided that it inherits a pair of these.

X-linked gene Any gene on an X chromosome.

X-linked recessive inheritance Recessive condition in which the responsible, mutated gene occurs on the X chromosome.

Y chromosome A sex chromosome with genes that cause the embryo that inherited it to develop into a male.

Y-linked gene Any gene on a Y chromosome.

yellow marrow Bone marrow that consists mainly of fat and hence appears yellow. It can convert to red marrow and produce red blood cells if the need arises.

yolk sac Of land vertebrates, one of four extraembryonic membranes. In humans, part becomes a site of blood cell formation and some of its cells give rise to the forerunners of gametes.

zero population growth A population for which the number of births is balanced by the number of deaths over a specified period, assuming immigration and emigration also are balanced.

zygote (ZYE-goat) The first cell of a new individual, formed by the fusion of a sperm nucleus with the nucleus of an egg (fertilization).

CREDITS AND ACKNOWLEDGMENTS

Page 1 Eric Gravé/Science Source/Photo Researchers

Introduction

I.1 Frank Kaczmarek / **I.2** Art by Carlyn Iverson / **I.3** Rich Buzzelli/Tom Stack & Associates / **I.4** Art by American Composition and Graphics / **I.6** Daniel McDonald/Stock Shop / **I.8** Jon Feingersh/Tom Stack & Associates / **Page 12** Jan Halaska/Photo Researchers

Chapter 1

1.1 (a) Martin Rogers/FPG; (b) © Alan Craft/Photophile / **1.2** Jack Carey / **Page 18** (b) (left) Hank Morgan/Rainbow; (right) Dr. Harry T. Chugani, M.D., UCLA School of Medicine / **1.8** Photograph Richard Riley/FPG / **1.9** Art by Raychel Ciemma / **1.11** Michael Grecco/Picture Group / **1.12** Art by Precision Graphics / **1.17** (b) Micrograph Biophoto Associates/SPL/Photo Researchers / **1.18-1.21** Art by Precision Graphics / **1.23** Art by Palais/Beaubois / **1.24** (a) CNRI/SPL/Photo Researchers; (b), (c) art by Robert Demarest / **1.26** (b) A. Lesk/SPL/Photo Researchers

Chapter 2

2.1 Frank L. Lambrecht/Visuals Unlimited / **2.2** (a) Photograph Ed Reschke; (b) Photograph Carolina Biological Supply Company / **Page 39** (a) (left) National Library of Medicine; (right) Armed Forces Institute of Pathology; (b) George Musil/Visuals Unlimited; (c) Lennart Nilsson from *Behold Man*, © 1974 by Albert Bonniers Forlag and Little, Brown and Company, Boston; (d) CNRI/SPL/Photo Researchers; (e) David M. Phillips/Visuals Unlimited / **2.3** Art by Leonard Morgan; micrograph H. C. Aldrich / **2.4** Art by Raychel Ciemma / **2.5** Micrographs M. Sheetz, R. Painter, and S. Singer, *Journal of Cell Biology*, 70:193 (1976), by copyright permission of The Rockefeller University Press / **2.6** Art by Leonard Morgan after B. Alberts et al., *Molecular Biology of the Cell*, Second edition, Garland Publishing Co., 1989 / **2.7** Art by Raychel Ciemma and American Composition and Graphics / **2.8** Micrograph G. L. Decker; art by Raychel Ciemma and American Composition and Graphics / **2.9** Micrograph D. Fawcett, *The Cell*, Philadelphia: W. B. Saunders Co.; art by D. & V. Hennings / **2.10** Art by Raychel Ciemma and American Composition and Graphics / **2.11** (a), (b) Micrographs Don W. Fawcett/Visuals Unlimited; (below right) art by Robert Demarest / **2.12** (left) Art by Robert Demarest after a model by J. Kephart; micrograph Gary W. Grimes / **2.13** Micrograph Keith R. Porter / **2.14** (a) Art by Precision Graphics; (b) J. Victor Small and Gottfried Rinnerthaler / **2.15** (a) Art by D. & V. Hennings after B. Alberts et al., *Molecular Biology of the Cell*, Garland Publishing Co.; (b) Dianne Woodrum and Richard W. Linck / **2.16** Art by D. & V. Hennings / **2.18** (a), (b) Thomas A. Steitz; art by Palay/Beaubois / **2.19** Janeart/Image Bank; art by Precision Graphics

Page 62–63 CNRI/SPL/Photo Researchers

Chapter 3

3.1 © Richard B. Levine / **3.2** Photographs (a) Lennart Nilsson from *Behold Man*, © 1974 by Albert Bonniers Forlag and Little, Brown and Company, Boston; (b) Manfred Kage/Bruce Coleman Ltd.; (c) Ed Reschke/Peter Arnold, Inc. / **3.3** Focus on Sports; (inset) Manfred Kage/Bruce Coleman Ltd.; art by Palay/Beaubois / **Page 68** NASA/Johnson Space Center / **3.4** Art by Palay/Beaubois / **3.5** Photographs by Ed Reschke / **3.6** Fred Hossler/Visuals Unlimited / **3.7** (left) Art by L. Calver; (right) art by Joel Ito / **3.9** Photographs Ed Reschke / **3.10** Lennart Nilsson from *Behold Man*, © 1974 Albert Bonniers Forlag and Little, Brown and Company, Boston / **3.11** Art by L. Calver / **3.12** Art by Palay/Beaubois / **3.13** Art by Robert Demarest / **3.14** Ed Reschke / **3.15** CNRI/SPL/Photo Researchers / **Page 80** Michael Keller/FPG / **Page 83** (a) Manfred Kage/Bruce Coleman Ltd.; (b–d) Ed Reschke / **Page 85** (above) Cath Ellis/University of Hull/SPL/Photo Researchers; (below) Kent Wood/Photo Researchers

Chapter 4

4.1 Jeff Schultz/AlaskaStock Images / **4.2** Art by L. Calver / **4.3** (left) Art by L. Calver; (right) Ed Reschke; (below) art by Joel Ito / **4.4** Art by K. Kasnot / **4.5** National Osteoporosis Foundation / **4.6** Art by Kevin Somerville / **4.7** Art by John W. Karapelou / **4.8** Art by Ron Ervin / **4.9, 4.10** Art by John W. Karapelou / **4.11** Art by Precision Graphics; photographs Gary Head / **Page 98** (a) Photograph C. Yokochi and J. Rohen, *Photographic Anatomy of the Human Body*, Second edition, Igaku-Shoin Ltd., 1979 / **4.12** Mike Devlin/SPL/Photo Researchers / **4.13** Art by Robert Demarest / **4.14** Art by Kevin Somerville / **4.15** (a) Art by Robert Demarest; (b) John D. Cunningham; (c) micrograph D. W. Fawcett, *The Cell*, Philadelphia: W. B. Saunders Co., 1966 / **4.16, 4.17** Art by Nadine Sokol / **4.18** Art by Robert Demarest / **4.19** Adapted from R. Eckert and D. Randall, *Animal Physiology: Mechanisms and Adaptations*, Second edition, W. H. Freeman and Co., 1983 / **4.20** (a) Art by Kevin Somerville; (b) Ed Reschke / **Page 106** Photograph Michael Neveux

Chapter 5

5.1 Nancy J. Pierce/Photo Researchers / **5.2** Art by Kevin Somerville / **5.5** (a) (right), (b) after A. Vander et al., *Human Physiology: Mechanisms of Body Function*, Fifth edition, McGraw-Hill, 1990 / **5.6** (a),(b) Art by Carlyn Iverson; (c) art by Precision Graphics / **5.7, 5.8** Art by Carlyn Iverson / **5.9** (a) Art by Victor Royer; (b), (d) Lennart Nilsson, © by Boehringer Ingleheim International GmbH; (c) Biophoto Associates/SPL/Photo Researchers / **5.10** Art by Robert Demarest / **5.11** Art by Raychel Ciemma / **5.12** Art by Carlyn Iverson / **Page 123** Tom Tracy/FPG / **5.14** Photograph Steven Jones/FPG / **Page 127** Photograph CNRI/Phototake / **5.16** Modified after A. Vander et al., *Human Physiology*, Fourth edition, McGraw-Hill, 1985; photograph Ralph Pleasant/FPG / **Page 134** Joe Munroe/Photo Researchers / **Page 135** Wide World Photos

Chapter 6

6.1 (a) From A. D. Waller, *Physiology, The Servant of Medicine*, Hitchcock Lectures, University of London Press, 1910; (b) photograph courtesy of The New York Academy of Medicine Library / **6.3** Art by Palay/Beaubois / **6.4** (a) (above) CNRI/SPL/Photo Researchers; (below) Lennart Nilsson from *Behold Man*, copyright 1974 by Albert Bonniers Forlag and Little, Brown and Company, Boston; (b) Art by Palay/Beaubois / **6.5** (a) By Raychel Ciemma / **6.6** (below) Art by John W. Karapelou / **6.7** Art by Kevin Somerville / **6.8** (a) C. Yokochi and J. Rohen, *Photographic Anatomy of the Human Body*, Second edition, Igaku-Shoin Ltd., 1979; (b), (c) art by Kevin Somerville / **6.9, 6.10** Art by John Karapelou / **6.11** Art by Kevin Somerville / **6.12** Art by John Karapelou / **6.13** Micrograph Ed Reschke / **6.15** Sheila Terry/SPL/Photo Researchers / **6.16** Art by Robert Demarest based on A. Spence, *Basic Human Anatomy*, Benjamin-Cummings, 1982 / **6.18** Art by John W. Karapelou / **6.19** Art by Raychel Ciemma / **6.20** Art by Kevin Somerville / **6.21** Photograph Lennart Nilsson, © Boehringer Ingelheim International GmbH / **6.22** (a) After F. Ayala and J. Kiger, *Modern Genetics*, © 1980 Benjamin-Cummings; (b) Lester V. Bergman & Associates, Inc. / **6.23** Art by Nadine Sokol after Gerard J. Tortora and Nicholas P. Anagnostakos, *Principles of Anatomy and Physiology*, Sixth edition, 1990. © 1990 by Biological Sciences Textbooks, Inc., A & P Textbooks, Inc. and Elia-Sparta, Inc. Reprinted by permission of HarperCollins Publishers. / **6.24** (a) (above) Ed Reschke; (below) F. Sloop and W. Ober/Visuals Unlimited / **Page 161** (a) Lewis L. Lainey / **6.26** Art by Raychel Ciemma

Chapter 7

7.1 (a) The Granger Collection, New York; (b) Lennart Nilsson © Boehringer Ingelheim International GmbH / **7.2** Lennart Nilsson © Boehringer Ingelheim International GmbH / **7.3** (a) Art by Nadine Sokol; (b) micrograph Robert R. Dourmashkin, courtesy of Clinical Research Centre, Harrow, England / **7.4, 7.5, 7.6** Art by Raychel Ciemma / **7.8** Art by Hans & Cassady / **7.9** Art by Raychel Ciemma; micrograph Morton H. Nielsen and Ole Werdlin, University of Copenhagen / **7.11** Art by Palay/Beaubois after B. Alberts et al., *Molecular Biology of the Cell*, Garland Publishing Company, 1983 / **7.13** (left) Lowell Georgia/Science Source/Photo Researchers; (right) Matt Meadows/Peter Arnold, Inc. / **7.14** (b) © 1992 B.S.I.P./Custom Medical Stock Photo

Chapter 8

8.1 Galen Rowell/Peter Arnold, Inc. / **8.4, 8.5** Art by Kevin Somerville / **Page 193** (above) Larry Mulvehill/Photo Researchers / **8.7** CNRI/SPL/Photo Researchers / **Page 194** Lennart Nilsson from *Behold Man*, © 1974 by Albert Bonniers Forlag and Little, Brown and Company, Boston / **8.8** After A. Vander et al., *Human Physiology*, Fifth edition, © 1990. Used by permission of McGraw-Hill, Inc. / **8.9** Art by K. Kasnot / **8.10** From L. G. Mitchell, J. A. Mutchmor, and W. D. Dolphin, *Zoology*, © 1988, Benjamin-Cummings Publishing Company / **8.11** Art by Carlyn Iverson / **8.12** Art by Leonard Morgan / **Page 199** (a), (b) O. Auerbach/Visuals Unlimited / **8.13** (b) Art by Carlyn Iverson

Chapter 9

9.1 National Park Service; (inset) © Greg Mancuso/Jeroboam / **9.3-9.5** Art by Robert Demarest / **9.6** Art by Precision Graphics / **9.9** Art by Joel Ito / **9.10** Art by Precision Graphics / **Page 219** VU/SIU/Visuals Unlimited / **9.11** Photograph Colin Monteath, Hedgehog House, New Zealand; art by Kevin Somerville / **9.12** The Bettmann Archive / **9.13** Evan Cerasoli

Chapter 10

10.1 Comstock, Inc. / **10.2** Manfred Kage/Peter Arnold, Inc. / **10.3** (left) Art by Kevin Somerville; (right) art by L. Calver / **10.4** Art by Leonard Morgan / **10.8** Art by John W. Karapelou / **10.9** Photograph Carolina Biological Supply Company; art by Leonard Morgan / **10.10** Art by D. & V. Hennings / **10.12** Micrograph from *Tissues and Organs: A Text-Atlas of Scanning Electron Microscopy*, by R. G. Kessel and R. H. Kardon. Copyright © 1979 by W. H. Freeman and Company. Reprinted with permission. Art by Robert Demarest / **10.13** Art by Robert Demarest / **10.14-10.15** Art by Kevin Somerville / **10.17** Art by Robert Demarest; micrograph Manfred Kage/Peter Arnold, Inc. / **10.18** Art by Kevin Somerville / **Page 242** © Wiley/Wales/ProFiles West / **10.19** Photograph C. Yokochi and J. Rohen, *Photographic Anatomy of the Human Body*, Second edition, Igaku-Shoin Ltd., 1979 / **10.20** Photographs Marcus Raichle, Washington University School of Medicine / **10.21** Art by Palais/Beaubois after Wilder Penfield and Theodore Rasmussen, *The Cerebral Cortex of Man*, © 1950 Macmillan Publishing Company; copyright renewed © 1978 by Theodore Rasmussen. Reprinted with permission of Macmillan Publishing. / **10.22** After H. Jasper, 1941 / **10.23** Art by Robert Demarest / **Page 248** (above) (left) P. Gadomski/Photo Researchers; (right) James Prince/Photo Researchers; (below) (left) Oscar Burriel/Latin Stock/Science Photo Library/Photo Researchers; (right) Andy Levine/Photo Researchers

Chapter 11

11.1 Superstock/FourByFive / **11.2** Art by Kevin Somerville / **11.3** Art by Palais/Beaubois after Wilder Penfield and Theodore Rasmussen, *The Cerebral Cortex of Man*, © 1950 Macmillan Publishing Company; copyright renewed © 1978 by Theodore Rasmussen. Reprinted with permission of Macmillan Publishing. / **11.4** From Hensel and Bowman, *Journal of Physiology*, 23:564–568, 1960 / **11.5** Art by Ron Ervin; photograph Ed Reschke / **11.6** Art by D. & V. Hennings / **11.7** Art by Robert Demarest; micrograph Omikron/SPL/Photo Researchers / **11.8** Art by Kevin Somerville / **11.9** (a) Art by Robert Demarest / **11.10** Art by Robert Demarest / **Page 263** (a), (b) Robert E. Preston, courtesy Joseph E. Hawkins, Kresge Hearing Research Institute, University of Michigan Medical School / **11.11** (left) Art by Kevin Somerville; (right) art by Betsy Palay/Artemis / **11.12** Art by Robert Demarest / **11.13–11.14** Art by Kevin Somerville / **11.15** Micrograph Lennart Nilsson © Boehringer Ingelheim International GmbH / **Page 269** Photographs Gerry Ellis/The Wildlife Collection; art by Kevin Somerville / **11.16** (a) Art by Nadine Sokol; (b) art by Palais/Beaubois / **11.17** Art by Palais/Beaubois after S. Kuffler and J. Nicholls, *From Neuron to Brain*, Sinauer, 1977

Chapter 12

12.1 Evan Cerasoli / **12.2** Art by Kevin Somerville / **12.3-12.6** Art by Robert Demarest / **12.7** (a) Mitchell Layton; (b) Syndication International (1986) Ltd. / **12.8** Photographs courtesy of Dr. William H. Daughaday, Washington University School of Medicine. From A. I. Mendelhoff and D. E. Smith, eds., *American Journal of Medicine*, 20.133 (1956) / **12.10** The Mark Shaw Collection/Photo Researchers / **12.11** Art by Joel Ito / **12.12** The Bettmann Archive / **12.13** Biophoto Associates/SPL/Photo Researchers / **12.14** Art by Leonard Morgan / **Page 291** Thomas Zimmermann/FPG

Chapter 13

13.1 (left) Evan Cerasoli; (right) Art by Robert Demarest after Patten, Carlson, and others / **13.2** (left) Art by Kevin Somerville; (right) art by L. Calver / **13.3** (a) Micrograph Ed Reschke; (b) art by Kevin Somerville / **13.4** Art by Ron Ervin / **13.6** (a) Art by Kevin Somerville / **13.7** Photograph Lennart Nilsson from *A Child Is Born*, © 1966, 1977 Dell Publishing Company. Inc.; art by Robert Demarest / **13.9** Art by Robert Demarest / **Page 305** David Frazier/Photo Researchers

Chapter 14

14.1 Lennart Nilsson from *A Child Is Born*, © 1966, 1977 Dell Publishing Company. Inc / **14.3** Art by Palay/Beaubois adapted from R. G. Ham and M. J. Veomett, *Mechanisms of Development*, St. Louis: C. V. Mosby Co., 1980 / **14.4** Art by Robert Demarest; (left) micrograph from Lennart Nilsson from *A Child Is Born*, © 1966, 1977 Dell Publishing Company. Inc.; (right) photograph from Lennart Nilsson, *Behold Man*, © 1974 by Albert Bonniers Forlag and Little, Brown and Co., Boston / **14.5** Betsy Palais/Artemis / **14.6** Art by L. Calver; (c) after A. S. Romer and T. S. Parsons, *The Vertebrate Body*, Sixth edition, Saunders College Publishing, © CBS College Publishing / **14.7** Art by Raychel Ciemma after Carlson, Moore, and others / **14.8** Art by Robert Demarest; photographs from Lennart Nilsson, *A Child Is Born*, © 1966, 1977 Dell Publishing Company, Inc. / **14.9** Art by Robert Demarest / **14.10** Sketches after B. Burnside, *Developmental Biology*, 26:416–441, 1971; micrograph K. W. Tosney / **Page 321** (a) Art by Palay/Beaubois; (b) from Fran Heyl Associates, © Jaques Cohen, Computer enhanced by © Pix*Elation / **14.11** © Lutheran Hospital/Peter Arnold, Inc. / **14.12** Art by Carlyn Iverson / **14.13** Art by Robert Demarest / **14.14** Art by Ron Ervin / **14.15** (a) Art by Raychel Ciemma modified after Keith L. Moore, *The Developing Human: Clinically Oriented Embryology*, Fourth edition, W. B. Saunders Co., 1988; (b) James W. Hanson, M.D. / **14.16** Art by Raychel Ciemma adapted from L. B. Arey, *Developmental Anatomy*, Philadelphia, W. B. Saunders Co., 1965 / **Page 331** Gary Head / **Page 332** Jose Carillo/Photophile / **Page 333** (left) G. Musil/Visuals Unlimited; (right) Cecil Fox/Science Source/Photo Researchers

Chapter 15

15.1 Allen Russell/ProFiles West / **15.2** Art by Nadine Sokol / **15.3** Art by Palais/Beaubois / **15.4** Art by L. Calver / **15.5** Micrograph J. J. Cardamone, Jr./BPS; art by D. & V. Hennings / **15.6** Lennart Nilsson © Boehringer Ingelheim International GmbH / **15.7** Michael H. Merson. "Slowing the Speed of HIV: Agenda for the 1990s," *Science*, 260:1266. Copyright 1993 by the AAAS. / **15.9** Art by Nadine Sokol / **15.11** Z. Salahuddin, National Institutes of Health / **15.12** Chuen-mo To and C. C. Brinton / **15.13** (a) Biophoto Associates/Photo Researchers; (b) Ken Greer/Visuals Unlimited; (c) Science VU/Visuals Unlimited; (d) Joel B. Baseman / **15.14** (a) David M.

Phillips/Visuals Unlimited; (b) © M. Richards/PhotoEdit / **15.15** (a) George Musil/Visuals Unlimited; (b) Biophoto Associates/Photo Researchers / **15.16** E. Grau/SPL/Photo Researchers / **15.17** David M. Phillips/Visuals Unlimited

Page 354-355 David M. Phillips/Science Source/Photo Researchers

Chapter 16

16.1 (a-c),(e) Lennart Nilsson from *A Child Is Born* © 1966, 1967 Dell Publishing Company, Inc.; (d) Lennart Nilsson from *Behold Man*, © 1974 by Albert Bonniers Forlag and Little, Brown and Company, Boston / **16.2** C. J. Harrison et al., *Cytogenetics and Cell Genetics* 35:21-27, copyright 1983 S. Karger A.G., Basel / **16.3** CNRI/Photo Researchers / **16.5** Andrew S. Bajer, University of Oregon / **16.6** Micrographs Ed Reschke; art by Raychel Ciemma / **16.7** Micrographs H. Beams and R. G. Kessel, *American Scientist*, 64:279–290, 1976 / **16.9** Art by Raychel Ciemma / **16.10** Art by Raychel Ciemma and American Composition and Graphics / **16.14** Micrograph David M. Phillips/Visuals Unlimited

Chapter 17

17.1 (a) Frank Trapper/Sygma; (b) Focus on Sports; (c) Fabian/Sygma; (d) Moravian Museum, Brno / **17.2** Photograph Jean M. Labat/Ardea, London; art by Jennifer Wardrip / **17.4** Courtesy of Kirk Douglas/The Bryna Company / **17.10** Art by Raychel Ciemma / **17.13** (above) David M. Phillips/Visuals Unlimited; (below) Bill Longcore/Photo Researchers / **17.14** Photograph Bill Longcore/Photo Researchers / **17.15** Jim Stevenson/SPL/Photo Researchers / **17.16** (top to bottom) Frank Cezus; Frank Cezus; Michael Keller; Ted Beaudin; Stan Sholik/all FPG / **17.18** Robert Burroughs Photography / **Page 389** Evan Cerasoli

Chapter 18

18.1 Cystic Fibrosis Foundation / **18.3** Art by Palay/Beaubois; photograph Omikron/Photo Researchers / **18.4** (b) Reprinted by permission from page 109 of Michael Cummings, *Human Heredity: Principles and Issues*, Second edition. Copyright © 1991 by West Publishing Company. All rights reserved. Art by Robert Demarest / **18.5** (a), (b) Stuart Kenter Associates / **18.7** Photograph Dr. Victor A. McCusick / **18.10** Eddie Adams/AP Photo / **18.12** After Victor A. McKusick, *Human Genetics*, Second edition, copyright 1969. Reprinted by permission of Prentice-Hall, Inc., Englewood Cliffs, NJ; photograph The Bettmann Archive / **18.13** © Superstock, Inc. / **18.14** Courtesy of David D. Weaver, M.D. / **18.16** Science VU-NIL/Visuals Unlimited / **18.17** Photograph courtesy of B. R. Brinkley from D. E. Merry et al., *American Journal of Human Genetics*, 37.425–430, 1985. © 1985 by The American Society of Human Genetics / **18.18** Art by Raychel Ciemma / **18.19** (a) Cytogenetics Laboratory, University of California, San Francisco; (b) after Collman and Stoller, *American Journal of Public Health*, 52, 1962 / **18.20** (left and right) Courtesy of Peninsula Association for Retarded Children and Adults, San Mateo Special Olympics, Burlingame, CA; (center) used by permission of Carole Iafrate

Chapter 19

19.1 Photograph A. C. Barrington Brown © 1968 J. D. Watson; model A. Lesk/SPL/Photo Researchers / **19.2** Micrograph Biophoto Associates/SPL/Photo Researchers / **Page 413** Photograph Ken Greer/Visuals Unlimited / **19.5** (a) (left) C. J. Harrison et al., *Cytogenetics and Cell Genetics* 35:21–27, copyright 1983 S. Karger A.G., Basel; (right) U. K. Laemmli from *Cell*, 12:817–828, copyright 1977 by MIT/Cell Press; (b) B. Hamkalo; (c) O. L. Miller, Jr. and Steve L. McKnight; art by Nadine Sokol / **19.6** Art by Hans & Cassady, Inc. / **19.9** Art by Raychel Ciemma and American Composition and Graphics / **19.10** Art by Raychel Ciemma

Chapter 20

20.1 Billy E. Barnes/Jeroboam / **20.2** Betsy Palais/Artemis / **20.3** Lennart Nilsson © Boehringer Ingelheim GmbH / **20.4** Betsy Palais/Artemis / **20.6** (a) Sam Kittner/Greenpeace; (b) © Tony Freeman/PhotoEdit / **20.7** (a) Paul Shambroom/Photo Researchers; (b) Simon Fraser/SPL/Photo Researchers / **Page 437** Photograph Ralph Pleasant/ FPG / **20.9** Photograph Mills-Peninsula Hospitals / **20.10** (a) John D. Cunningham/Visuals Unlimited; (b) Alfred Pasieka/Peter Arnold, Inc. / **20.11** Art by Betsy Palais/Artemis; (a) Biophoto Associates/Science Source/Photo Researchers; (b) James Stevenson/ SPL/Photo Researchers; (c) Biophoto Associates/Science Source/Photo Researchers

Chapter 21

21.1 (a) Mark Clarke/SPL/Photo Researchers; (b) NCI/Science Source/Photo Researchers / **21.2** Dr. Huntington Potter and Dr. David Dressler / **21.4** Art by Jeanne Schreiber / **21.7** (a), (b) Monsanto Company; (c), (d) Calgene, Inc. / **Page 452** Michael Maloney/San Francisco Chronicle / **21.8** (a) R. Brinster and R. E. Hammer, School of Veterinary Medicine, University of Pennsylvania; (b) Courtesy of Tufts University School of Veterinary Medicine/Genzyme Corporation / **21.9** Courtesy of Victor A. McKusick and Joanna Strayer Amberger, Johns Hopkins University / **21.10** Damon Biotech, Inc. / **Page 457** (above) Secchi-Lecague/Roussel-UCLAF/CNRI/ SPL/Photo Researchers; (below) © David Young-Wolff/PhotoEdit

Page 460-461 NASA

Chapter 22

22.1 (left) Vatican Museums; (right) Martin Dohrn/SPL/Photo Researchers / **22.2** Alan Solem / **22.3** After F. Ayala and J. Valentine, *Evolving*, Benjamin-Cummings, 1979 / **22.4** (above) Art by Leonard Morgan; (below) art by Lloyd K. Townsend / **22.5** (a) Patricia G. Gensel; (b) Donald Baird, Princeton Museum of Natural History / **22.6** (below) From T. Storer et al., *General Zoology*, Sixth edition, McGraw-Hill, 1979. Reproduced by permission of McGraw-Hill, Inc. / **22.7** Art by Victor Royer / **22.9** (a) Art by Joel Ito; (b) art by Betsy Palais/Artemis / **22.10** (top) Douglas P. Wilson/Eric & David Hosking; (center) Superstock, Inc.; (bottom) E. R. Degginger / **22.12** Chesley Bonestell / **Page 473** (below) Bruce Coleman Ltd. / **22.13** Stanley W. Awramik / **22.15** (a) Sidney W. Fox; (b) W. Hargreaves and D. Deamer

Chapter 23

23.1 Douglas Mazonowicz/Gallery of Prehistoric Art / **23.3** (a) Bruce Coleman Ltd.; (b) Tom McHugh/ Photo Researchers; (c) Larry Burrows/Aspect Picture Library; (d) Stephen Rapley / **23.4, 23.6** Art by D. & V. Hennings / **23.7** (a) Louise M. Robbins; (b) Dr. Donald Johanson, Institute of Human Origins / **23.8, 23.9** Art by D. & V. Hennings

Chapter 24

24.1 Wolfgang Kaehler / **24.2** (a) Harlo H. Hadow; (b) Jack Carey / **24.3** Roger K. Burnard / **24.4** (a), (b) Roger K. Burnard; (c), (d) E. R. Degginger / **24.6** Photograph Sharon R. Chester / **24.10** Photograph © Gary Braasch / **24.12** (a) Art by Raychel Ciemma; photograph NASA / **Page 504** (a) After W. Dansgaard et al., *Nature*, 364:218–220, 15 July 1993; D. Raymond et al., *Science*, 259:926–933, February 1993; W. Post, *American Scientist*, 78:310–326, July–August 1990 / **24.13** Photograph William J. Weber/Visuals Unlimited

Chapter 25

25.1 Antoinette Jongen/FPG / **25.2** © 1983 Billy Grimes / **25.3, 25.4** Data from Population Reference Bureau after G. T. Miller, Jr., *Living in the Environment*, Eighth edition, Wadsworth, Inc. 1993 / **25.5** (c) Stanley Flegler/Visuals Unlimited / **25.7** Photograph NASA / **25.9** Data from G. T. Miller, Jr., *Living in the Environment*, Eighth edition, Wadsworth, Inc., 1993 / **25.10** United Nations / **25.11** (a) USDA Forest Service; (b) Heather Angel / **25.12** NASA / **25.13** Dr. Charles Heggeghien/Bruce Coleman Ltd. / **25.14** From Water Resources Council / **25.15** (a) Adolf Schmidecker/FPG; (b) Gerry Ellis/The Wildlife Collection; (c) (above) R. Bieregaard/Photo Researchers; (below) after G. T. Miller, Jr., *Living in the Environment*, Eighth edition, Wadsworth, Inc., 1993 / **Page 522** Richard Parker/Photo Researchers / **25.16** Data from G. T. Miller, Jr. / **25.17** Warren Gretz/NREL / **25.18** Art by Precision Graphics after Marvin H. Dickerson, "ARAC: Modeling an Ill Wind" in *Energy and Technology Review*, August 1987. Used by permission of University of California, Lawrence Livermore National Laboratory and U.S. Department of Energy / **25.19** World Bank

INDEX

Italic numerals refer to illustrations.

E